C0-ATX-380

ANNUAL REVIEW OF PLANT PHYSIOLOGY AND PLANT MOLECULAR BIOLOGY

Production Editor: HILDA E. GOULD
Subject Indexer: KYRA KITTS

Annual Review of Plant Physiology and Plant Molecular Biology

VOLUME 51, 2000

RUSSELL L. JONES, *Editor*
University of California, Berkeley

HANS J. BOHNERT, *Associate Editor*
University of Arizona, Tucson

DEBORAH P. DELMER, *Associate Editor*
University of California, Davis

www.AnnualReviews.org science@AnnualReviews.org 650-493-4400

ANNUAL REVIEWS
4139 El Camino Way • P.O. Box 10139 • Palo Alto, California 94303-0139

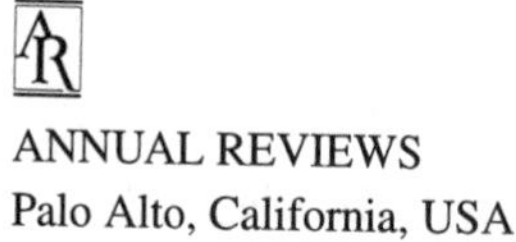

ANNUAL REVIEWS
Palo Alto, California, USA

International Standard Serial Number: 1040-2519
International Standard Book Number: 0-8243-0651-1
Library of Congress Catalog Card Number: 50-13143

∞ The paper used in this publication meets the minimum requirements of American National Standards for Information Sciences—Permanence of Paper for Printed Library Materials, ANSI Z39.48-1992.

Typeset by Techbooks, Fairfax, VA
Printed and Bound in the United States of America

CONTENTS

RELATED ARTICLES

From the ***Annual Review of Biochemistry***, Volume 69 (2000)

Cryptochrome: The Second Photoactive Pigment in the Eye and Its Role in Circadian Photoreception, Aziz Sancar

Protein Glucosylation and Its Role in Protein Folding, Armando J. Parodi

Spindle Assembly in Animal Cells, Duane A. Compton

Chromosome Cohension, Condensation, and Separation, Tatsuya Hirano

Two-Component Signal Transduction, Ann M. Stock, Victoria L. Robinson, and Paul N. Goudreau

Apoptosis Signaling, Andreas Strasser, Liam O'Connor, and Vishva M. Dixit

Yeast Homotypic Vacuole Fusion: A Window on Organelle Trafficking Mechanisms, William Wickner and Albert Haas

Structural Insights into Microtuble Function, Eva Nogales

Autophagy, Cytoplasm-to-Vacuole Targeting Pathway, and Pexophagy in Yeast and Mammalian Cells, John Kim and Daniel J. Klionsky

Coupling of Open Reading Frames by Translational Bypassing, Alan J. Herr, John F. Atkins, and Raymond F. Gesteland

Protein Tyrosine Kinase Structure and Function, Stevan R. Hubbard and Jeffrey H. Till

Import of Peroxisomal Matrix and Membrane Proteins, S. Subramani, Antonius Koller, and William B. Snyder

Protein Splicing and Related Forms of Protein Autoprocessing, Henry Paulus

DNA Replication Fidelity, Thomas A. Kunkel and Katarzyna Bebenek

Mechanisms and Control of mRNA Decapping in Saccharomyces cerevisiae, Roy Parker and Morgan Tucker

Ribozyme Structures and Mechanisms, Elizabeth A. Doherty and Jennifer A. Doudna

Structure and Function of Cytochrome bc Complexes, Edward A. Berry, Mariana Guergova-Kuras, Li-shar Huang, and Anthony R. Crofts

Synthesis of Native Proteins by Chemical Ligation, Philip E. Dawson and Stephen B. H. Kent

Critical Analysis of Antibody Catalysis, Donald Hilvert

Regulation of Chromosome Replication, Thomas J. Kelly and Grant W. Brown

Clathrin, Tomas Kirchhausen

Mediator of Transcriptional Regulation, Lawrence C. Myers and Roger D. Kornberg

Structure and Function of Hexameric Helicases, S. S. Patel and K. M. Picha

Swinging Arms and Swinging Domains in Multifunctional Enzymes: Catalytic Machines for Multistep Reactions, Richard N. Perham

Helical Membrane Protein Folding, Stability, and Evolution, Jean-Luc Popot and Donald M. Engelman

Aminoacyl-tRNA Synthesis, Dieter Soll and Michael Ibba

From the ***Annual Review of Biophysics and Biomolecular Structure***, Volume 28 (1999)

RNA Folds: Insights From Recent Crystal Structures, Adrian R. Ferré-D'Amaré and Jennifer Doudna

Structure and Conformation of Complex Carbohydrates of Glycoproteins, Glycolipids, and Bacterial Polysaccharides, C. Allen Buch, Manuel Martin-Pastor, and Anne Imberty

Membrane Protein Folding and Stability: Physical Principles, Stephen H. White and William C. Wimley

Multiprotein-DNA Complexes in Transcriptional Regulation, Cynthia Wolberger

DNA Repair Mechanisms for the Recognition and Removal of Damaged DNA Bases, Clifford D. Mol, Sudip S. Parikh, Christopher D. Putnam, Terence P. Lo, and John A. Tainer

The Lysosomal Cysteine Proteases, Mary E. McGrath

Rotational Coupling in the F_0F_1 ATP Synthase, Robert K. Nakamoto, Christian J. Ketchum, and Marwan K. Al-Shawi

The Proteasome, Matthias Bochtler, Lars Ditzel, Michael Groll, Claudia Hartmann, and Robert Huber

Closing in on Bacteriorhodopsin: Progress in Understanding the Molecule, Ulrich Haupts, Jörg Tittor, and Dieter Oesterhelt

Molecular Dynamics Simulations of Biomolecules: Long-Range Electrostatic Effects, Celeste Sagui and Thomas A. Darden

Raman Spectroscopy of Proteins and Nucleic Acid Assemblies, George J. Thomas, Jr.

Modern Applications of Analytical Ultracentrifugation, T. M. Laue and W. F. Stafford, III

From the ***Annual Review of Cell and Developmental Biology***, Volume 15 (1999)

Vacuolar Import of Proteins and Organelles from the Cytoplasm, Daniel J. Klionsky and Yoshinori Ohsumi

Blue-Light Photoreceptors in Higher Plants, Winslow R. Briggs and Eva Huala

Cooperation Between Microtubule- and Actin-Based Motor Proteins, Susan S. Brown

The Road Less Traveled: Emerging Principles of Kinesin Motor Utilization, Lawrence S. B. Goldstein and Alastair Valentine Philp

Proteins of the ADF/Cofilin Family: Essential Regulators of Actin Dynamics, James R. Bamburg

Biochemical Pathways of Caspases Activation During Apoptosis, Imawati Budihardjo, Holt Oliver, Michael Lutter, Xu Luo, and Xiaodong Wang

Regulation of Nuclear Localization: A Key to a Door, Arie Kaffman and Erin K. O'Shea

Actin-Related Proteins, D. A. Schafer and T. A. Schroer

Cell Polarity in Yeast, John Chant

SCF and Cullin/Ring H2-Based Ubiquitin Ligases, R. J. Deshaies

Integration of Signaling Networks That Regulate Dictyostelium *Differentiation*, Laurence Aubry and Richard Firtel

When to Switch to Flowering, Gordon G. Simpson, Anthony R. Gendall, and Caroline Dean

Regulation of Mammalian O_2 Homeostasis by Hypoxia-Inducible Factor 1, Gregg L. Semenza

Transport Between the Cell Nucleus and the Cytoplasm, Dirk Görlich and Ulrike Kutay

Adaptors for Clathrin-Mediated Traffic, Tomas Kirchhausen

The Translocon: A Dynamic Gateway at the ER Membrane, Arthur E. Johnson and Michael A. van Waes

From the ***Annual Review of Genetics***, Volume 33 (1999)

Shufflons: Multiple Inversion Systems and Integrons, Teruya Komano

Mechanisms of mRNA Surveillance in Eukaryotes, P. Hilleren and R. Parker

Ribosome Synthesis in Saccharomyces cerevisiae, Jaap Venema and David Tollervey

Mitochondrial Genome Evolution and the Origin of Eukaryotes, B. Franz Lang, Michael W. Gray, and Gertraud Burger

Plant Retrotransposons, Amar Kumar and Jeffrey Bennetzen

Meiotic Chromosomes: Integrating Structure and Function, Denise Zickler and Nancy Kleckner

From the ***Annual Review of Phytopathology***, Volume 38 (2000)

Control of Virulence and Pathogenicity Genes of Ralstonia solanacearum *by an Elaborate Sensory Array*, Mark Schell

The Induction and Modulation of Plant Defense Responses by Bacterial Lipopolysaccharides, Max Dow, Mari-Anne Newman, and Edda von Roepenack

Nematode Parasitism Genes, Eric L. Davis, Richard S. Hussey, Thomas J. Baum, Jaap Bakker, Arjen Schots, Marie-Noëlle Rosso, and Pierre Abad

Role of Mitochondrial DNA in the Senescence and Hypovirulence of Fungi and the Potential for Plant Disease Control, Helmut Bertrand

Advances in Imaging the Cell Biology of Plant-Microbe Interactions, Michèle C. Heath

The Photoactivated Cercospora Toxin Cersoporin: Contributions to Plant Disease and Fundamental Biology, Margaret E. Daub and Marilyn Ehrenshaft

Epidemiology of Wheat Leaf and Stem Rust in the Central Great Plains of the USA, M. G. Eversmeyer and C. L. Kramer

The Impact of the Food Quality Plant Protection Act on the Future of Plant Disease Management, Nancy N. Ragsdale

Adhesion of Fungal Pathogens: Relationship of Adhesins to Virulence, N. Talbot

ANNUAL REVIEWS is a nonprofit scientific publisher established to promote the advancement of the sciences. Beginning in 1932 with the *Annual Review of Biochemistry*, the Company has pursued as its principal function the publication of high-quality, reasonably priced *Annual Review* volumes. The volumes are organized by Editors and Editorial Committees who invite qualified authors to contribute critical articles reviewing significant developments within each major discipline. The Editor-in-Chief invites those interested in serving as future Editorial Committee members to communicate directly with him. Annual Reviews is administered by a Board of Directors, whose members serve without compensation.

Annual Reviews of

Anthropology
Astronomy and Astrophysics
Biochemistry
Biomedical Engineering
Biophysics and Biomolecular Structure
Cell and Developmental Biology
Earth and Planetary Sciences
Ecology and Systematics
Energy and the Environment
Entomology
Fluid Mechanics
Genetics
Genomics and Human Genetics
Immunology
Materials Science
Medicine
Microbiology
Neuroscience
Nuclear and Particle Science
Nutrition
Pharmacology and Toxicology
Physical Chemistry
Physiology
Phytopathology
Plant Physiology and Plant Molecular Biology
Political Science
Psychology
Public Health
Sociology

SPECIAL PUBLICATIONS
Excitement and Fascination of Science, Vols. 1, 2, 3, and 4

Ann Oaks

Annu. Rev. Plant Physiol. Plant Mol. Biol. 2000. 51:1–16

FIFTY YEARS OF PLANT SCIENCE: Was There Really No Place for A Woman?

Ann Oaks
Department of Botany, University of Guelph, Guelph, Ontario N1G 2W1, Canada; e-mail: aoaks@uoguelph.ca

Key Words nitrate assimilation, nitrate reductase, endopeptidases

CONTENTS

THE EARLY YEARS

I was born in June 1929; born, as my father HA "Doc" Oaks used to say, with a golden spoon in my mouth. I do not know whether he meant that I was endowed with riches or with luck. I have had a little of each.

My father was one of those young Canadians who learned to fly with the Royal Flying Corps in Britain during the First World War. After the war, he, like so many others who had learned the joys of flying, was bored with standard civilian life. He turned to the north where he flew with the Ontario Department of Lands and Forests. Later he founded an independent company that operated out of Sioux Lookout, a sectional stop on the railroad in northwestern Ontario. He (they; those daring young men) flew the uncharted wilderness of northern Canada without radio and without weather reports. They charted that wilderness and in addition looked for gold, flew in supplies to isolated communities, and searched for lost airplanes. Six months after my birth, the glow of that early success was gone with the stock market crash and the onset of the Great Depression. All that bustle of the late

1040-2519/00/0601-0001$14.00

1920s was gone, but the planes still flew. Short of cash, my father flew his plane without insurance, went on routine errands for the government, and helped fight forest fires.

Our house, a big green frame house with a veranda, was between the lake and the railroad tracks. My maternal grandfather George Farlinger lived next door with my grandmother and three of their five children. We had a big vegetable garden behind the two houses, which kept us supplied with produce both in summer and for the following winter. At that time there was no running water and no indoor plumbing. How the times have changed! We managed. Then came the Red Lake gold rush. Business picked up.

In 1935, my parents moved to Port Arthur (now Thunder Bay), in part because it had become a major center for the mining industry and in part because the school system was superior. However, we maintained strong connections with Sioux Lookout by returning every summer for years after our move. There was a lake with a nice sandy beach. I learned to swim and to handle boats; I watched the fall migration of shorebirds and other wonders of nature. I watched the small bush planes as they taxied up and down the lake, usually overloaded with supplies and gear for people working in the bush. I still hear that special noise and feel the wind and water on my face when I see a small plane overhead.

My maternal grandfather had been a contractor in eastern Canada, but had moved west with the railroad and settled in Sioux Lookout, where he built a mill to supply ties to the railroad. His mill and the railroad were the major industries in that small town. Now both are gone, replaced by a major center for the First Nations' People and by tourism. I remember with fondness the giant work-horses at the mill, their snorting and their odor, and the smell of wet sawdust.

My grandmother was also an important influence in that pioneer community. She was a major mover in establishing a Carnegie library. I can remember going with her to the library, helping her sort the books, and experiencing some of the wonder hidden between all those covers. In those days we did not speak of role models, but if we had, she would have been one for me.

After the move to Port Arthur I attended school, where I was always a good student. I was also active in sports (basketball, volleyball, swimming, tennis, and skiing). We were lucky in having good and dedicated teachers. The most influential for me was Claude Garton, my grade 8 teacher. He was a "professional" naturalist. He had us out watching the spring migration at 6 a.m., had us collecting wild flowers and identifying them. He even had us plant a kitchen garden. He also taught regular subjects like reading, writing, and arithmetic, but it is those outdoor activities that I remember and that I pursued, with encouragement from my parents, all through my high school years.

My father's family was located in Preston, a small town in southern Ontario. This branch of the family had migrated to Canada from Pennsylvania after the American Civil War. My paternal grandfather had been a schoolteacher and then a medical doctor. While at university I often spent weekends in Preston and during those visits got to know my aunt Marjorie. She had been a schoolteacher, and had joined the navy in the Second World War to teach the sailors how to cook. I am not quite sure how she subsisted after the war, but she never did go back to teaching. She was an independent, often abrasive woman, and stubborn, in character much like me. I liked her independence. She encouraged me to go on with university, on to graduate school, on to great things. Again, if there had been role models or mentors identified in those days, she would have been one.

THE UNIVERSITY YEARS

In September 1947, I found myself at the University of Toronto, in residence at Victoria College, 18, and away from home for the first time. In those days home was 24 hours away by train. There was no road along the north shore of Lake Superior and if there were airplane connections, they were too expensive. I had chosen Honours Science to develop hobbies that I had engaged in all through high school rather than English and history, other favorite subjects, which might have led to journalism or teaching.

Classes were overflowing with ex-soldiers intent on getting on with their lives. They were organized, knew why they were at the university, and they knew how to both work and play. I remember our chemistry professor saying to us, "Look to the right and now look to the left—of the three of you, only one will be back next year." I made the grade, but just by the skin of my teeth. I remember another time when my lab partner [Tove (Jensen) Finlay] and I were trying to piece together apparati for an experiment in electricity, the physics professor said, "What are you two girls doing in a science lab? You should be in home economics." We had no ready response as a student in the year 2000 might have. We did not even ask ourselves whether there was really no place for a woman in Science. We did not alter our course of studies.

In the second year, in Honours Biology now, we studied a very traditional biology: taxonomy, structure—memory work! Biochemistry and metabolism were in their infancy and were not even touched upon during the first two years of study. In that second year, we had about 1000 scientific names rattling around in our heads.

That year, the teaching assistant in Botany was Norman Good, developer of the "Good buffers"; these were the buffers that made research with organelles possible. He convinced me that Botany was a real science, a good alternative to Zoology, and even a good alternative to Medicine. My father said "No" to Medicine. I think it was Norman Good who convinced me that Physiology was an interesting field of study, which required more logic and less memory work.

In our third year, we took physiology, both plant and animal, and biochemistry. The Krebs Cycle was still an active area of research and controversy. Acetyl-CoA was still unknown and hence not identified as an integral part of the cycle. Physiology and biochemistry were more to my liking than more traditional offerings in biology. In fourth year, I took another course in biochemistry, much to the chagrin of the chairman of the Zoology Department. He thought evolution would be much more appropriate for a biologist. At that time, four bases were known to hold the code of life, but we did not understand how only four units could determine the structure of all the proteins found in living cells, or how changes in those bases might contribute to changes in form and function. Today we can modify genes in the laboratory. Such progress!

The 14 of us who made it through to third year were exposed to field courses—the last two weeks in September before our third and fourth years—hiking in Algonquin Park, learning the associations of living creatures in land and water environments, exposed to the cold autumn rains, evenings writing reports on what we had done during the day, and parties. We had a good time and we developed friendships that have lasted a lifetime. Students of today with their more flexible programs do not bond so effectively. At the graduation party, we broke a plate and each of us took a piece. We glued that plate together at a reunion ten years later. We still have those reunions, more often now that we are all retired.

In the last summer of my undergraduate program, I worked on a fire tower in the wilderness west of Lake Nipigon. I had convinced the district forester to hire me. He had suggested a tower that I could bike to, but I insisted on one in the wilderness. Two people were required to man (woman?) those towers. In June, my cousin Helen (St. John) Squires and I arrived by plane at Sauerbrei Lake to begin three glorious months in the wilderness. I think my brother, Stephen, never really forgave me for robbing two of his high school friends of good paying summer jobs. For the first week or so the silence of the wilderness was eerie; but we adapted to it. This silence of the wilderness has a different texture from the big city buzz. There were loons on our lake, and beavers. The bay adjacent to the one where our cabin was had an echo. The loons liked it and so did we: both to hear the loon-echoes and to hear our own. I collected plants where no botanist had been before and identified those that I could. I collected everything in duplicate, one set for Claude Garton and one set for me. At a later time I donated my set to the herbarium at the University of Toronto. For years after that donation, people familiar with the herbarium at the U of T asked me if I was the same person who had collected all those plants. They were always surprised when I said "Yes"; they saw me as an indoor biologist.

I also trapped small mammals that summer as part of a project required for graduation. It was fun buying 50 mouse traps at the local hardware store—they could not understand why I needed so many! One memorable morning, I caught 24 red-backed mice on a small island in the middle of the lake. The next morning I caught ten more mice on the same trap line, the next morning only three, and then no more. Either I had trapped all the mice, or the mice had learned not to trust my peanut butter and oatmeal mix. In the course of the summer, I also caught a flying squirrel and six baby partridges with that peanut butter and oatmeal mix. Ah, science in the hands of an amateur; an amateur in the hands of science. If there had been an opportunity, I would have become an ecologist but they would not take women on field trips—too expensive they said, and perhaps too immoral. In those days there really was no place for a woman in Ecology.

Of the 14 of us in the Class of '51, three became teachers, five entomologists, one a paleontologist, and the rest found work as biologists after their BA or MA programs. I was the only woman to go on for a PhD, and of the 14, the only one to become a botanist. There were four women in our class. Two became lab technicians, one a teacher, and then there was me. The other women interrupted working careers to raise children and, in fact, only the teacher went back to full-time paid employment. Seven of the total earned PhDs and were hired into a stable working career by the time of our ten-year reunion. It took me 14 years after my BA at the University of Toronto to land a "real" job at McMaster University, and even that job was very uncertain until I had earned tenure. Was there really no place for a woman in science?

In the intervening years, I spent one year as a technician at the Defense Research Board labs at Fort Churchill, Manitoba. I studied cold hardiness in Chironomids. What amazing creatures they were. One could bring a block of frozen muskeg into the lab and watch the appearance of red wriggly worms as the block thawed. Churchill, located between tundra and boreal forest, is a classical place for birders. I spent many enjoyable hours pursuing that long-time hobby of mine.

Following that, I spent two years on a graduate program at the University of Saskatchewan in Saskatoon, and in 1954 graduated with an MA degree. I studied the genetics and physiology of chlorophyll-deficient mutants in barley under the supervision of Tom Arnason and Michael Shaw. Then came one year as a technician in Roy Waygood's lab at the University of Manitoba studying the role of indoleacetic acid oxidase in controlling levels of indoleacetic acid in plant tissues. I could not see myself as a technician for the rest of my life, and I thought that a PhD degree in Botany was out of the question. As a result, there seemed to be no future for me in science. I turned to teaching, but after three months as a student at the College of Education in Toronto, I found that that was also really not for me. I remember one class I had to give on bread mold and the excited questions and discussion at the end of the class. This was one of the best lessons I have ever given, challenged only by the first-year lectures I gave at McMaster, which dealt with the path of carbon in photosynthesis. The evaluation by the supervising teacher was "too many questions." But is that not what teaching is all about? To get students to ask questions and then to try to answer them? That teacher was old, perhaps 40. I pictured myself at that age and did not like the image. So, what to do? I decided that teaching was not for me and that perhaps I should try research once again.

I returned to the University of Saskatchewan for my PhD—host-parasite relations with stem rust of wheat; the role of indoleacetic acid. Michael Shaw was my supervisor. The years in Saskatoon were good ones. Work time in the lab, and play time with badminton in the winter and tennis in the summer. There was also good birdwatching around Saskatoon and trips to the northern wilderness. Hunting was a major activity in the fall—grain-fed wild ducks. Stewart Whitney, also a student in Michael Shaw's lab, was an avid hunter. He often supplied his catch for a Saturday evening feast that was enjoyed by graduate students from the biology and chemistry labs. We, Sheila Hird and I, worked our way through the *Joy of Cooking* (first edition) recipes for preparing wild fowl.

While at the University of Saskatchewan I studied chemistry, strengthening the rather inferior offering of chemistries given to biology students at the University of Toronto, and I took a course in plant biochemistry given by Arthur Neish, who at that time was Director of the Prairie Regional Lab in Saskatoon. His course stimulated in me a life-long interest in biochemistry and metabolism. After graduating with my PhD, I spent two years in Otto Kandler's lab in Germany, supported by the Alexander von Humboldt Stiftung. I studied the path of carbon in photosynthesis in *Chlorella vulgaris*—using tracer methodology and inhibitors, worrying about the asymmetric labeling of glucose carbon identified by Gibbs and Kandler in the 1950s. Kandler had just returned from two years in the United States working with Calvin and Gibbs. His enthusiasm for the project and his understanding of science were both exciting and contagious. The lab was in Freising, with a cathedral that was built 1000 years ago—history, that was the first time it really struck me; for 1000 years people had walked up that narrow pathway to go to church; a brewery, and a beer garden right behind the lab; Munich, a city with plays, concerts, opera, art galleries, and museums, half an hour away by train; skiing or hiking in the mountains about two hours away by train. What an experience for someone who had grown up on the fringes of civilization. I had many good teachers of the German language and culture: Martin Busse, Lotte von Gavel, Annamarie Hund, Heinrich Kauss, and both Otto and Truadel Kandler. They were patient with my German and often amused by my strange sentence structures. Annamarie was a special friend. She even invited me, as a family member, to her wedding.

At the end of my time in Germany, September 1961, there was no job on the horizon. In those days there were no advertisements for jobs and although I had worked in two of the best labs in Canada, I was not even on the grapevine for hearing about possible openings. Undaunted, I pursued my career in science. I went to Harry Beevers' lab at Purdue University, to another excellent work experience. There I met maize seedlings, research objects that would occupy the rest of my research career, and end-product regulation of amino acid biosynthesis, my initiation to nitrogen metabolism, aspects of which would also occupy my attention for the greater part of my research career. During my stay at Purdue, I absorbed aspects of microbiology practiced by Frederich Neidhardt and Arthur Aronson, the first wave of the molecular revolution. I was impressed by the exactness with which one could trace the path of metabolites in unicellular organisms. Despite the power of doing experiments with such a system, I stayed with maize seedlings, which also provided the challenge of deciphering aspects of the metabolic interactions between different plant parts. After three years at Purdue and another year at the Oak Ridge National Lab, I was asked to apply for a position as Assistant Professor in the Biology Department at McMaster University. It was Dennis McCalla, a graduate student in Saskatoon for part of the time that I was there, who encouraged me to apply for the position. In the interim, he had finished a PhD at Caltech and was an Assistant Professor in the Research Unit in Biochemistry, Biophysics and Molecular Biology at McMaster University. I applied and was invited to give a seminar. Success: I was appointed Assistant Professor in the Research Unit and the Biology Department. Over the next 24 years I worked my way through the ranks.

THE MCMASTER YEARS

Science Aspects

Most of my time at McMaster I had a small research group: one or two senior undergraduate students, one or two MSc students, and perhaps a PhD student and a visitor from abroad. YP Abrol from India, Pierre Gadal from France, and Hiroki Nakagawa and Tom Yamaya from Japan were visitors and major contributors to my research endeavors. I also had a chance to work in their labs. Since I had a rather young research group, I required technical help primarily to tidy up experiments left undone by departing BSc and MSc students. I was lucky in this regard. Over the years Ingrid Boesel, Valerie Goodfellow, Fred Johnston, and Mike Winspear kept the operation afloat. I also had good sabbaticals with Oluf Gamborg and Klaus Hahlbrock—my initiation into the world of tissue culture, with Andy Kleinhofs and Bob Warner—learning how to purify nitrate reductase. I had a successful research career and for these efforts was inducted into the Royal Society of Canada in 1986 and in 1988 was awarded the Gold Medal of the Canadian Society of Plant Physiologists.

Teaching

While at McMaster I managed to attract good students to my fourth-year class in Plant Physiology. Peter Bzonec, Karen Jones, and Michael Dennis were awarded gold medals in their graduating years. Most of the undergraduates at McMaster wanted a professional career—medicine, dentistry, optometry, vet school. So I had a problem: what to teach these people that would really help them. I decided that I could teach them to read critically. Each year I chose six scientific papers from top labs, usually papers with a hidden flaw, of which they had to choose four. Each of the papers was central to one of the topics I covered in the lecture part of the course. They wrote a short critique of the paper, handed it in to me, and then we had a one-to-one discussion related to its contribution to research. Once they recovered from their original fear, we had many good discussions. I also learned a lot about the pressures and frustrations of the students.

One could only pursue such a course offering with small classes. The experiment was a success and each year I managed to attract at least one student to pursue a program in the plant sciences.

Political Aspects

The years at McMaster were not really comfortable years. At first I was never really sure whether it was because I was a woman, or because I was not good enough for the job. My self-confidence was boosted by Joe Varner when he nominated me to be a reviewer on the editorial board for *Plant Physiology*, a position I held for the next 20 years. At one of the editorial board meetings, Martin Gibbs was explaining how each associate editor was assigned so many review editors. This struck me as odd because I had been receiving manuscripts from all of them, but in particular from Clanton Black and Jack Preiss. When I queried Gibbs about this, there was some laughter. They must have thought I was a pretty good reviewer! Since I was rather isolated at McMaster, I found it a really beneficial experience to be involved with current research that was being submitted for publication in the top plant journal.

McMaster had a medical school, with a pioneering program that introduced potential doctors to the hospital wards early in their studies. This program exposed them to small study groups rather than large class lectures. In these study groups they learned co-operative problem solving. Everybody was enthusiastic about this program, administration, faculty, and students.

In this atmosphere, the teaching of botany and research in that area were considered to be a necessary evil and not an area of endeavor to be actively supported. For 24 years I talked about primary producers, food for the starving millions and all that stuff, but to no avail. I probably would not have survived the lack of respect from the administration had it not been for my good friend and colleague, Esther McCandless. At one point a chairman, who shall remain nameless, suggested that I was not pulling my weight in the department, that perhaps I should consider teaching a second course, perhaps in economic botany. I tried to explain to him that another plant course would not help, and that perhaps a course that combined plant and animal physiology would. He did not agree. I went swimming for an hour to relieve the frustration of that conversation and to help to clear the mind. However, I discussed the possibility of a joint course with Esther. She thought it was a great idea. We designed a course in nutrition. I was to give the first half in plant nutrition and to discuss how plant breeders modified agriculturally important crops to satisfy animal nutrition. I included aspects of the potentially very important biotech revolution. In the second term, Esther took over and dealt with various aspects of animal nutrition. With some difficulty we managed to get the course approved by the Biology Faculty. In the first year the course was offered, we had 100 students. Students from Phys Ed to Honours Biology—wow!! It was a success.

In 1989, when the opportunity arose, I took early retirement from McMaster University and moved with my research group to the University of Guelph. Derek Bewley was the newly appointed Chair of Botany and in this position was able to facilitate my move. I worked at the University of Guelph as an Adjunct Professor for the last 10 years of my career. It was fun doing collaborative work with various members of the department, something I had not experienced at McMaster. My collaborators in various research endeavors were Derek Bewley, John Greenwood, Wilf Rauser, and Steven Rothstein.

When I started my science career, metabolism was the "hot" area of research. However, during most of the time that I spent at McMaster, it was molecular genetics that was at the cutting edge of biological research—isolating genes, mapping them, modifying them, putting them back into organisms. My colleagues there considered me a little out of date because I did not jump on this new bandwagon. Today metabolism is back in vogue benefitting from all those new molecular techniques but also benefitting from those of us who remained "biochemists." Today we can, for example, ask very explicit questions about the function of isoenzymes, as Turlough Finan is doing from the bacterial side of the legume/*Rhizobium* complex, and as Gloria Coruzzi and Tom Yamaya are doing with enzymes active in nitrogen metabolism in higher plants. Good luck to them. It looks like an exciting new era. Good luck too to my former students (Xiu-Zhen Li, Debbie Long, Santosh Misra, Vipin Rastogi, and Shoba Sivasankar) who have taken the leap into this new area that is effectively binding physiology, genetics, and structure. I sort of chuckle with my replacement at McMaster, Elizabeth Weretilnyk. She was hired as a plant molecular geneticist, but is doing definitive experiments in plant biochemistry and metabolism, applying molecular techniques in a very sensible way. So I was not so out of date after all! I guess there was a place in Science for this woman after all.

THE RESEARCH YEARS: Nitrogen Assimilation

Regulation of Amino Acid Biosynthesis in Maize Seedling Roots

When I arrived at Purdue University, Harry Beevers had two major lab projects: one dealing with the newly discovered glyoxylate cycle in the castor bean endosperm and the other with general metabolism in various tissues. David MacLennon was examining the in vivo functioning of the Krebs Cycle in various plant tissues and was establishing the compartmentation of the organic acid pools. Widmar Tanner was trying to find out why the glyoxylate cycle was so efficient even though the Krebs Cycle enzymes were all present. This was the pre-glyoxysome era. Reasons for compartmentation of metabolites were a major focus in Harry's lab in those days.

My first project had to do with the fate of ^{14}C-acetate and ^{14}C-glucose in maize seedling roots. I saw labeling in only a few amino acids: glutamate, aspartate and their amides, and serine and glycine, and this was more intense with acetate than with glucose radioactive tracers. In addition, there was no conversion of acetate to glucose or other sugar derivatives. For these reasons, ^{14}C-acetate was the preferred precursor for examining drainage of carbon from the Krebs Cycle or, more specifically, for studying amino acid biosynthesis and utilization. Ion exchange chromatography was used to separate sugars, organic acids, and amino acids, and paper chromatography to separate various amino acids. These were techniques developed in Calvin's lab, but which were, by 1962, used in every respectable lab project dealing with metabolism.

During my stay at Purdue, the Biology Department bought a "Moore and Stein" amino acid analyzer. What an advance! In 24 h one could separate the soluble amino acids with precision. It took another day to calculate the results, counting the dots, using a slide rule. Today with various liquid chromatography techniques one can do up to 10 samples per day and with computers can quantitate the results almost simultaneously. At Purdue we had time to think about our results or to count the dots. Today we are always in a rush. So much data. So many papers to write.

I wondered what the labeling would look like in the "protein" fraction. Extraction of proteins was difficult and not usually done in a quantitative fashion in plant labs. The method involved first an extraction of the tissue with 80% ethanol to remove the soluble amino acids, organic acids, and sugars, followed by the removal of starch and cellulose and finally the hydrolysis of the remaining protein and nucleic acids. When radioactive tracers were involved, paper chromatographs were placed on X-ray film and exposed for the required time—possibly a month. All the protein amino acids potentially derived from the Krebs Cycle were labeled in my experiments. Interestingly, labeling was in the protein product but often not in the intermediate soluble amino acid fraction.

This was the start of an investigation on the biosynthesis of amino acids in the maize root tip and my studies on the amino acid pool. With acetate labeling and by adding a mixture of amino acids to maize root tips, I was able to establish that many of the amino acids were end product inhibited (arginine, proline, leucine, isoleucine, threonine, valine, and lysine). When I fed the unlabeled amino acid, the soluble pool of that amino acid became labeled with [^{14}C] from acetate. This was evidence, later backed up by time-course experiments which indicated that the amino acids were synthesized in the soluble fraction before incorporation into protein, and that the metabolic pool was only a small fraction of the total soluble pool.

The next question was, "Were the amino acids that were transported to the root tip responsible for end product inhibition in intact root tip sections?" I used ^{14}C-leucine to answer this question because it was present in high concentrations in the maize endosperm and because it was usually not extensively metabolized. Leucine was either synthesized in the root tips or transported from more mature regions of the seedling—mature root sections, or the endosperm via the scutellum. I tried feeding leucine (or other radioactive tracers) to specific parts of the seedling. This was done by incorporating the radioactive metabolite into agar blocks that I was then able to apply directly to the root tip or more mature regions of the root or to the scutellum. Oops! Did I say leucine was stable? When I treated the scutellum, the preferred location for establishing the fate of transport leucine, I found that the leucine carbon was metabolized—in fact, sucrose was a major product. This was good evidence that the glyoxylate cycle was active in the maize scutellum. With a little more work it was easy to establish that the two enzymes required for a functioning glyoxylate cycle were present in this tissue. Since I was in Harry's lab, surrounded by all that expertise, I tidied up these observations with a few well-designed experiments and had a publication (7).

But how to follow the transport of leucine when it was so effectively metabolized in the source tissue? In future experiments, I had to be satisfied by feeding the tracer to the base of the root where leucine was not metabolized. With this technique I found that labeled leucine was located in the protein fraction in the root tip. With direct feeding of the tip considerable leucine was trapped in the soluble fraction, and since it did not chase in pulse-chase experiments, it was probably sequestered in the vacuole. It appeared that leucine supplied by mature regions of the root was, in fact, in a separate pool that was destined for protein. This was similar to results from acetate feeding experiments where leucine derived from acetate was also detected principally in a protein precursor pool and in the protein fraction. This rather circumstantial evidence indicated that transport leucine and leucine synthesized in the root tip passed through the same pool, and hence the synthesis in the root tip could be inhibited by the transport amino acid. Later I established that (*a*) the effect was specific: that is, that leucine additions inhibited leucine biosynthesis and that proline additions inhibited proline biosynthesis; (*b*) micromolar levels of a particular amino acid were sufficient to cause the effect; and (*c*) the effect was more pronounced in meristematic than in mature tissue. This work is summarized in a review by Oaks & Bidwell (9).

The Maize Seedling System

Although not traditionally used as such, the seedling represents an ideal source/sink model system. One can analyze the products released by the source, in this case the maize endosperm, and the effect of these products on the sink tissues, the developing seedling. In the early 1960s, there was much research directed to understanding the growth of embryos in vitro and plant cell cultures. FC Steward, for example, was doing experiments that would demonstrate plant cells are totipotent. I examined the effect of endosperm removal on subsequent seedling growth in a sterile hydroponic system. The dry weight increase of the seedlings commenced immediately but the extent of that increase depended on the concentration of sucrose in the medium. In contrast, there was a three-day lag in the increase in alcohol-insoluble nitrogen—the protein fraction. A wounding effect? An endosperm factor? When I incubated excised endosperm in liquid culture and then grew excised seedlings in that medium, I found an immediate increase in protein nitrogen. The factor that caused this turned out to be amino acids, but not just any amino acid mix. I made a synthetic mixture of amino acids similar in concentration to the amino acids released by the endosperm and found that this mixture, but not a casein hydrolysate, enhanced the accumulation of alcohol-insoluble nitrogen in the excised embryos (8). I knew from those early experiments that small peptides were also released and I thought that they were probably important in the overall utilization of endosperm nitrogen, protecting leucine from degradation in the scutellum, for example (see Reference 4 for details).

At that time, research from Ed Umbarger's lab in Cold Spring Harbor was directed at elucidating the path of leucine biosynthesis and its regulation in

Salmonella typhimurium. Umbarger's research team had to make their own substrates and they generously gave me some for the second enzyme specific for the leucine pathway so that I could try my luck at enzymology. Dick Burns and Ed Umbarger were very helpful in offering me tips in these early experiments. The second enzyme in the pathway was stable and with that substrate I established that the leucine pathway was not repressed by leucine additions. I took heroic measures to assure myself that leucine was indeed the rate-limiting component, which it would have to be if I were to see a derepression. Very difficult without mutants! I looked at the first enzyme in the pathway without success, and with that the leucine project came to an end.

Two early experiments indicated to me that the initial hydrolysis of the endosperm proteins was regulated by the amino acid requirements of the seedling, and that the rate of loss of reduced nitrogen from the endosperm was not influenced by additions of nitrate to the medium even though there was an induction of nitrate reductase (NR) and an accumulation of nitrate in the seedling tissues. This early work and the rationale for subsequent experiments is summarized in two reviews (4, 9). From that time on, there were always two "experiments" in my lab: (*a*) the hydrolysis of endosperm proteins or, put another way, the supply of reduced nitrogen to the seedling; and (*b*) the regulation of the induction of NR, which I believed (still believe) is a good indicator of the capacity of the embryo to assimilate externally supplied nitrate. This broad approach was something that successive NSERC granting panels never really appreciated. Both the granting agency and many subsequent students preferred the nitrate reductase project because it was more amenable to molecular techniques.

Hydrolysis of Endosperm Proteins The induction of proteases in cereal endosperm tissue, in particular in the barley endosperm, is induced by gibberellic acid (GA) and repressed by abscisic acid (ABA) (13). In maize the response to hormones was more difficult to establish. However, we (Harvey & Oaks) did establish that with ABA there was an inhibition of the induction of α-amylase and a major acid endopeptidase. The inhibition was reversed by appropriate additions of GA. In addition, the levels of α-amylase and endopeptidase in GA-deficient mutants of maize increased substantially with additions of GA (reviewed in Reference 4).

Subsequently, our working hypothesis was that a specific endopeptidase was responsible for the initial cleavage of the water-insoluble zeins (prolamines) and that the products of this cleavage would be water soluble and sensitive to attack by less specialized endopeptidases and the various carboxy- and amino-peptidases. A series of students and visiting scientists worked on various aspects of this project (Barbara Harvey, Wataru Mitsuhashi, Vipin Rastogi, Bill Wallace, Michel Poulle, and Michael Winspear).

Initially, we (Rastogi & Oaks) thought this problem of multiple endopeptidases might be resolved more easily by using barley half-seeds that were responsive to GA. We found that there was an initial cleavage of the water-insoluble hordein polypeptide in the absence of GA, but that subsequent cleavages required GA.

Sometime later, Wataru Mitsuhashi identified 17 activity bands using acrylamide gel electrophoresis and an activity staining technique that he had developed (3). This lead was followed by Bill Wallace who established that an early endopeptidase activity, a cysteine endopeptidase, was converted to a second form both in vivo and in vitro and that zein was more sensitive to attack by the second modified endopeptidase (unpublished results). It should be quite easy to purify this initial endopeptidase since it is the only one present two days after the addition of water to the system (summarized in Reference 6). However, Bill ran out of time and I ran out of money so that is where the project stands. Valerie Goodfellow, a technician of mine for some 15 years, a chemist who liked fancy equipment, was to identify the amino acids at the cleavage site of Bill's endosperm endopeptidase. She was also to identify the small peptides released by the endosperm and found in the root exudates. But she (we) seemed to be kept busy tidying up critical experiments left by graduate students, and the endosperm experiments have not yet been performed.

Nitrate Reductase To analyze the role of reduced nitrogen on the acquisition of exogenous nitrogen, I needed a good marker enzyme in the embryo. Harold Evans had established in 1953 that there was such an enzyme and that pyridine nucleotides were required to reduce nitrate to nitrite. He was working with soybean where there is a significant amount of NR even in the absence of nitrate. In 1957, Tang and Wu established that in some species NR activity was induced by nitrate. While I was at Purdue, Dick Hageman and associates were establishing the environmental factors required for the induction and turnover of NR. Because of that work, I decided that NR should be an appropriate marker enzyme for defining the importance of reduced nitrogen in the capacity of the seedling to take up and reduce nitrate-nitrogen. [For appropriate reviews of these early events see Oaks & Hirel (10), Oaks (5), and Sivasankar & Oaks (12).] Bill Wallace was the first person in my lab to examine the properties of NR in maize seedling roots. We (Wallace & Oaks) examined the time course of the induction of NR and the effect of cordycepin, an inhibitor of mRNA production, and cycloheximide, an inhibitor of protein synthesis. The important points established by this research were:

1. NR was very unstable and this instability was the result of a major root endopeptidase. Later this endopeptidase was examined in some detail in my lab by Tom Yamaya and in collaboration with Larry Solomonson at the University of South Florida. We had what appeared to be a specific endopeptidase and Larry had a pure NR so that we could examine in detail the structural characteristics of NR (Larry's project) and the properties of the endopeptidase (my project). The properties of the endopeptidase are summarized by Goodfellow et al (1). NR activity was higher in root tips and the enzyme was more stable. The endopeptidase was higher in mature regions of the root. Because of the stability problem we worked largely with root tips. In view of the phosphorylation of NR that was identified

recently by Werner Kaiser (summarized in Reference 2), I wonder if the enzyme in the root tip NR was truly active in vivo.

2. The induction was fast, minutes instead of the hours which we thought at that time (circa 1970) were required for responses in plant tissues.
3. The induction process involved the synthesis of both mRNA and protein. These points have been confirmed at the molecular level and, in fact, people in Steven Rothstein's lab at the University of Guelph have been able to identify the promoter region of nitrite reductase (NiR) and to establish that it responds to nitrate additions in transgenic tobacco plants using constructs of GUS with the NiR promoter from spinach (see Reference 12 for a review of current knowledge related to the regulation of both NR and NiR).

Sometime later Debbie Long, in my lab, examined the stability problem of NR once again and demonstrated that the root endopeptidase was effectively inhibited by chymostatin. With this inhibitor, activity of NR was higher in mature root sections than in the root tips, and the turnover was much lower (reviewed in 12). In a sense this observation solved an old riddle—if amino acids play a role in the regulation of NR, this enzyme should be lower in the root tip because amino acids seem to be transported to the meristems, and certainly this was where amino acid biosynthesis was most sensitive to the effects of exogenous amino acids. In addition, early work from Beevers' lab had shown that the hexose monophosphate shunt was more active in mature regions of the root, and that this pathway was probably required for the production of NAD(P)H. The shunt could then supply the reductant required by the root NR, which is primarily an NAD(P)H-requiring NR.

I had two major collaborations in research related to NR, with Larry Solomonson and later Andy Kleinhofs and Bob Warner at Washington State University. Telephones are as good as e-mail in facilitating long-distance collaborations.

The actual role of the amino acids was more difficult to establish, in part because of a "mistake" with the corn mixture of amino acids. I had hydrolyzed the amino acid fraction of the endosperm leachates, thereby converting the amides present to their respective amino acids. Mohammad Aslam looked valiantly for an amino acid effect and found that there was a minimal inhibition even when millimolar amounts of amino acid were used. This was true whether we looked directly at NR or followed the assimilation of ^{15}N-nitrate. He did establish, however, that glucose or sucrose additions led to elevated levels of NR in both root tip and mature root segments, but not in green leaves. More recently, Shoba Sivasankar tackled the amino acid problem once again. This time she/we started by examining the amino acid content in root exudates. She found that asparagine and glutamine were present at high concentrations in exudates prepared from seedlings three days after germination, and in seedling root and leaf tissues. After another three days the asparagine had almost disappeared from the root exudates, and there was a significant increase in NR. In her hands, the induction of NR was repressed by micromolar levels of either glutamine or asparagine (reviewed in References 5 and 12). With this work we had established that amides were significant regulators in maize seedling roots. In view of the fact that sucrose inhibits the production of asparagine in seedling tissues (8) and enhances the induction of NR, it may be that the sucrose released from the endosperm is responsible for an inhibition of asparagine production and thus, indirectly leads to the induction of NR by five days post germination. However, there was still another problem. During seedling development levels of NR were low in both root and shoot tissue at three days. When Xiu-Zhen Li added glutamine back, she saw a clear inhibition of the induction of NR in maize roots, but not in the shoots, this despite the fact that the actual levels of glutamine increased in shoot tissues of treated plants (5).

The next problem was that when we added one amide, there was an increase in both amides in the soluble fraction. We should have anticipated this event from research done in Ken Joy's lab. Which one was effective? What to do with no mutants? Well, we could always resort to inhibitors. We tried methionine sulfoximine (MSX), which should result in lower levels of glutamine and hence an overexpression of NR. We also tried azaserine (AZA), which should have led to higher levels of glutamine, but lower levels of glutamate and asparagine. Thus, with AZA, if glutamine were the effector molecule an inhibition in the induction of NR; and if asparagine were the effective molecule an increase in the induction of NR. Surprise!! With MSX we found only marginally lower levels of glutamine and in root tissues minor increases in NR. In the leaves, however, there was a dramatic increase in the NAD(P)H isoform of NR (11). When we tried AZA there was a significant increase in asparagine in both root and leaf tissue, and a dramatic reduction in levels of NR. Was asparagine really the effective regulator of NR? Was there another asparagine synthetase? Another can of worms!! The asparagine synthetase that used NH_4^+; the one described by Stulen & Oaks (1977) and by Oaks & Ross (1984) and condemned to obscurity because of experiments performed by Rognes (1975) on an asparagine synthetase purified from lupin cotyledons (reviewed in References 5 and 10). Different species, different enzyme?—so often the explanation of controversial results! No time; no money. There should be no controversial results at the end of one's scientific career!

CHASING WINDMILLS

During my time at McMaster, I often felt a little like Don Quixote and his windmills. I was a thorn in the side of several department chairmen and the occasional dean. I tackled granting agencies, tackled scientists from afar over the roles of glutamate dehydrogenase and the NH_4^+-dependent asparagine synthetase. I won some and lost some. I enjoyed that role. And now? I am taking on the biotech industry. We should not be introducing herbicide and pesticide resistance into crop plants. That is only an extension of the chemical agriculture introduced after the second World War. At that time there was not enough testing of the chemicals as health hazards. We know now that the wide usage of DDT was wrong. We do not know of potential side effects of Round-up and other potential herbicides. We should be looking for modifications of plants that will enhance metabolic events to increase yield, for example, improving the efficiency of photosynthesis, or of nitrogen use. We should be looking for mechanisms of increasing resistance to disease, drought, or salt stress. These projects are more difficult than herbicide/pesticide resistance, but are the ones that should bring real rewards for a world threatened with overpopulation and the poisoning of our limited agricultural lands. This task should keep me busy for awhile. My colleagues in this endeavor include the Council of Canadians, and, in Guelph, The Rabble Rousers.

ACKNOWLEDGMENTS

I thank Derek Bewley for many stimulating discussions and for editing an early version of this manuscript.

LITERATURE CITED

1. Goodfellow VJ, Solomonson LP, Oaks A. 1993. Characterization of a maize root protease. *Plant Physiol.* 101:415–19
2. Kaiser WM, Weiner H, Huber SC. 1999. Nitrate reductase in higher plants: A case study for transduction of environmental stimuli into control of catalytic activity. *Physiol. Plant.* 105:385–90
3. Mitsuhashi W, Oaks A. 1994. Development of endopeptidase activities in maize (*Zea mays*) endosperms. *Plant Physiol.* 104: 401–7
4. Oaks A. 1983. Regulation of nitrogen assimilation during early seedling growth. In *Mobilization of Reserves in Germination*, ed. C Nozzolillo, PJ Lea, FA Loewus, pp. 53–75. New York/London: Plenum
5. Oaks A. 1994. Primary nitrogen assimilation in higher plants and its regulation. *Can. J. Bot.* 72:739–75
6. Oaks A. 1997. Strategies for nitrogen assimilation in *Zea mays*: early seedling growth. *Maydica.* 42:203–10
7. Oaks A, Beevers H. 1964. The glyoxylate cycle in maize scutella. *Plant Physiol.* 39:431–34
8. Oaks A, Beevers H. 1964. The requirement for organic nitrogen in *Zea mays* embryos. *Plant Physiol.* 39:37–43
9. Oaks A, Bidwell RGS. 1970. Compartmentation of intermediary metabolites. *Annu. Rev. Plant Physiol.* 21:43–66
10. Oaks A, Hirel B. 1985. Nitrogen metabolism in roots. *Annu. Rev. Plant Physiol.* 36:345–65
11. Oaks A, Sivasankar S, Goodfellow VJ. 1998. The specificity of methionine sulfoximine and azaserine inhibition in plant tissues. *Phytochemistry* 49:355–57
12. Sivasankar S, Oaks A. 1996. Nitrate assimilation in higher plants: the effect of metabolites and light. A review. *Plant Physiol. Biochem.* 74:609–20
13. Varner JF, Ho DTH. 1976. The role of hormones in the integration of seedling growth. In *The Molecular Biology of Hormone Action*, ed. J. Papaconstantinow, pp. 173–94. New York: Academic

Annu. Rev. Plant Physiol. Plant Mol. Biol. 2000. 51:17–47

BIOTIN METABOLISM IN PLANTS

Claude Alban, Dominique Job, and Roland Douce
Laboratoire Mixte CNRS/Aventis (UMR 1932), Aventis CropScience, Lyon, France; e-mail: rdouce@cea.fr

Key Words biotin, biotin-containing proteins, biotin biosynthesis, protein biotinylation, plants

■ **Abstract** Biotin is an essential cofactor for a small number of enzymes involved mainly in the transfer of CO_2 during HCO_3^--dependent carboxylation reactions. This review highlights progress in plant biotin research by focusing on the four major areas of recent investigation: the structure, enzymology, and localization of two important biotinylated proteins (methylcrotonoyl-CoA carboxylase involved in the catabolism of leucine and noncyclic isoprenoids; acetyl-CoA carboxylase isoforms involved in a number of biosynthetic pathways); the biosynthesis of biotin; the biotinylation of biotin-dependent carboxylases, including the characterization of biotin holocarboxylase synthetase isoforms; and the detailed characterization of a novel, seed-specific biotinylated protein. A central challenge for plant biotin research is to determine in molecular terms how plant cells regulate the flow of biotin to sustain the biotinylation of biotin-dependent carboxylases during biosynthetic reactions.

CONTENTS

INTRODUCTION

Biotin is a water-soluble vitamin called vitamin H or B_8 that is essential for life in all organisms. Its biosynthesis is limited to most bacteria, some fungi, and plants. Thus, many fungi and bacteria and all animals must obtain biotin from their environment. In all organisms, biotin serves as an essential cofactor for a small number of enzymes involved in the transfer of CO_2 during carboxylation, decarboxylation,

1040-2519/00/0601-0017$14.00

and transcarboxylation reactions (41, 93). Biotinylated proteins are extremely rare in nature. For example, the only biotin-dependent carboxylase in *Escherichia coli* is acetyl-CoA carboxylase (ACCase; EC 6.4.1.2), a multisubunit enzyme in which one polypeptide is biotinylated and corresponds to the biotin carboxyl carrier protein (BCCP). Other bacteria contain one to three biotinylated proteins (60). Eukaryotic cells appear to contain a slightly greater number of biotinylated proteins. For example, *Saccharomyces cerevisiae* contains four or five biotinylated proteins depending on growth conditions (107), whereas mammals (30, 134) are reported to contain four biotinylated proteins. [All these enzymes have crucial cellular housekeeping functions. More specifically, ACCase that catalyzes the ATP- and HCO_3^--dependent carboxylation of acetyl-CoA is recognized as the regulatory enzyme of lipogenesis; methylcrotonoyl-CoA carboxylase (MCCase; EC 6.4.1.4) catalyzes the conversion of methylcrotonoyl-CoA to methylglutaconyl-CoA, a key reaction in the Leu degradation pathway; propionyl-CoA carboxylase (PCCase; EC 6.4.1.3) is a key enzyme in the catabolism of odd-numbered fatty acids and the amino acids Ile, Thr, Met, and Val; and pyruvate carboxylase (PyrCase; EC 6.4.1.1) has an anaplerotic role in the formation of oxaloacetate.] The common feature of these reactions is the transfer of a carboxyl group from bicarbonate to an acceptor substrate, utilizing biotin as a carboxyl carrier. The reactions catalyzed by these enzymes take place in two discrete steps [1] and [2], resulting in overall catalysis [3]:

$$HCO_3^- + \text{enzyme-biotin} + \text{ATP-Mg} \rightarrow \text{enzyme-biotin-}CO_2^- + \text{ADP-Mg} + \text{Pi} \quad [1]$$

$$\text{Enzyme-biotin-}CO_2^- + \text{acceptor} \rightarrow \text{acceptor-}CO_2^- + \text{enzyme-biotin} \quad [2]$$

$$HCO_3^- + \text{acceptor} + \text{ATP-Mg} \rightarrow \text{acceptor-}CO_2^- + \text{ADP-Mg} + \text{Pi} \quad [3]$$

The features distinguishing the reactions of each of these enzymes are the acceptor substrates (115). The family of biotin enzymes also includes oxaloacetate, methylmalonyl-CoA, and glutaconyl-CoA decarboxylases, which are involved in sodium transport in anaerobic prokaryotes (47), as well as transcarboxylase (EC 2.13.1), which participates in propionic acid fermentation in *Propionibacterium shermanii* (178). The latter two classes of enzymes do not require ATP as a substrate. In all biotin enzymes described to date, the biotin is covalently linked to the ε-amino group of a specific Lys residue located within a highly conserved (Ala/Val)-Met-Lys-(Met/Leu) tetrapeptide motif.

The presence of ACCase activity in plants was first documented in 1961 (79). The first direct evidence for the occurrence of biotin enzymes other than ACCase in plants was provided in 1990 by Wurtele & Nikolau (179). These researchers detected MCCase, PCCase, and PyrCase activities, in addition to ACCase activity, in cell-free extracts of various monocot and dicot plant species. Thus, the four biotin-dependent activities found in mammals are also present in plants. Information on the structure, regulation, and function of plant biotin-dependent carboxylases, biotin biosynthesis, and protein biotinylation processes in higher plants has burgeoned worldwide since 1990 (see below and Figure 1).

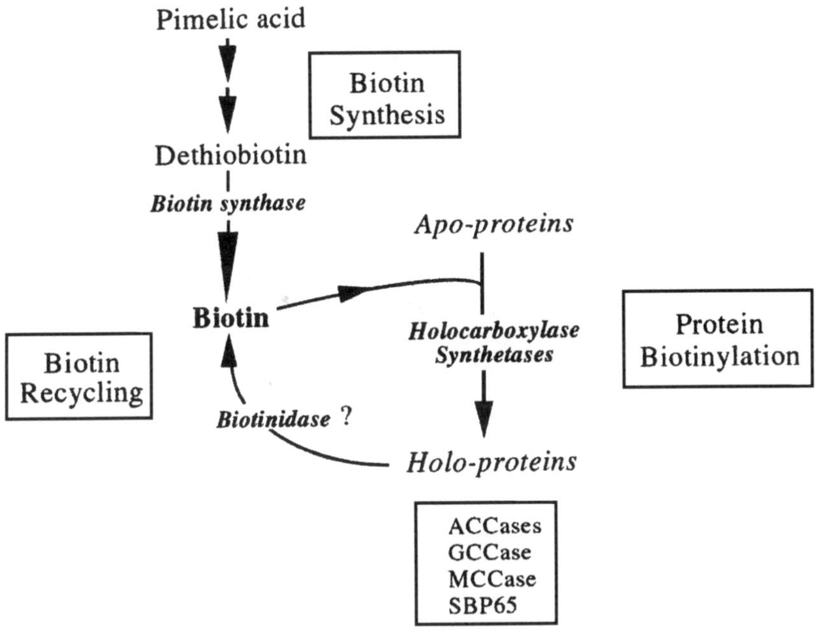

Figure 1 Biotin metabolism in plants: the biotin cycle. See Figure 2 for the details of biotin synthesis.

Methylcrotonoyl-CoA Carboxylase

MCCase catalyzes the ATP-Me^{2+}-dependent carboxylation of 3-methylcrotonoyl-CoA to form 3-methylglutaconoyl-CoA, a reaction essential for Leu catabolism in animals and microorganisms (reviewed in 177). In addition, MCCase has been implicated in the recycling of carbon from mevalonic acid, a process called the "mevalonate shunt" (127). In humans, the fatal inherited disorder 3-methylcrotonylglycinuria has been shown to result from a deficiency in MCCase (62, 69, 172). In plants, the first detailed characterization of MCCase was obtained by Baldet et al (14); they demonstrated that in pea leaves all the biotin present in the matrix space of the mitochondrion (at a concentration of approx. 13 μM) was protein-bound and associated with MCCase. These authors also observed that plant MCCase was mostly, if not exclusively, located in mitochondria (14). These data were consistent with the previously determined mitochondrial location of MCCase activity in mammals (80). MCCase activity was also purified to homogeneity from pea leaf and potato tuber mitochondria by use of avidin affinity columns (2). Avidin binds biotin specifically and extremely tightly, with a dissociation constant (K_d) of 10^{-15} M. Because the avidin-biotin interaction is so strong, it was necessary, prior to chromatography, to denature the native tetrameric, avidin-bound protein to its monomeric form, which has reduced affinity for biotin (K_d of 10^{-7} M) (94).

The use of purified mitochondria and affinity chromatography appeared to be the most convenient way to purify plant MCCase to homogeneity (2). Because plant MCCase is very unstable, attempts to purify the enzyme from crude extracts of carrot somatic embryos and maize leaves only yielded less pure and less active preparations. Presumably, this was because other biotin proteins were present in these extracts and the enzyme was not protected (by the mitochondrial membranes) from proteases released from the vacuolar space during tissue homogenization (32, 46). Structural and biochemical properties of MCCase from pea leaves and potato tubers were determined (2) and found to be similar to those reported for bacterial and mammal MCCases (8, 61, 81, 93, 100, 138, 145, 146). Pea leaf and potato tuber MCCases are composed of two nonidentical subunits. The larger α-subunit is biotinylated and has a molecular mass of 74–76 kDa, whereas the smaller β-subunit is biotin-free and has a molecular mass of 53–54 kDa. This heteromeric structure was subsequently confirmed with the corresponding enzymes from carrot and maize (32, 46). The native enzyme behaves as a heterooctamer composed of four large subunits and four small subunits ($\alpha_4\beta_4$). Gel filtration experiments determined a M_r in the range of 500,000 to 530,000 for both purified pea and potato MCCases. In addition, the biotin content of plant MCCase was consistent with a stoichiometry of four molecules per $\alpha_4\beta_4$ octamer, presumably with each α-subunit containing one biotin prosthetic group (2). In this respect, pea and potato MCCases resemble bacterial MCCase (61), but differ from mammalian, carrot, and maize MCCases, which seem to adopt a dodecameric structure ($\alpha_6\beta_6$) (32, 46, 100). Steady-state kinetic analyses demonstrated that the plant MCCase-catalyzed reaction proceeds by a double-displacement mechanism (Bi Bi Uni Uni ping-pong), where ATP and bicarbonate bind to the enzyme in the first half-reaction and react to form the carboxybiotinyl enzyme derivative (see Reaction [1]). ADP and Pi produced by hydrolysis of ATP are then released. In the second partial reaction, 3-methylcrotonoyl-CoA binds to the carboxybiotinyl enzyme and becomes carboxylated to form 3-methylglutaconoyl-CoA (2). The purified enzyme is substantially inhibited by the reaction end products, ADP and Pi (2), and by acetoacetyl-CoA, an end product of Leu degradation (32). Plant MCCase is also inactivated by *N*-ethylmaleimide, a sulfhydryl-modifying reagent, and phenylglyoxal, a reagent that modifies Arg residues. Preincubation of MCCase with ATP and 3-methylcrotonoyl-CoA prevented inactivation, which suggests that Cys and Arg residues are involved in catalysis or regulation of enzyme activity (2). In tomato, the MCCase activity found in leaves is lower than that in roots (169). However, the steady-state levels of the biotinyl subunit of MCCase (measured using specific antibodies) and its mRNA are approximately equal in both tissues. Labeling experiments with ^{125}I-streptavidin suggested that the lower activity of MCCase in leaves could be attributed to the reduced biotinylation of its biotin-containing subunit. Consistent with this hypothesis, a pool of non-biotinylated enzyme was found in leaves (169). These data suggest, therefore, that the relative biotinylation of a biotin-containing enzyme can be a potential mechanism for regulating its activity. However, direct measurement of the biotin holo-carboxylase synthetase (biotin ligase) activity showed no detectable biotin incorporation into protein from pea leaf extracts at any stage of the plant's development, unless an exogenous apo-carboxylase substrate was added to the reaction medium (165), which indicates that the apo-carboxylases in pea leaves are at very low concentrations. On the other hand, in seedling organs and developing cotyledons of soybean, differences in the abundance of the biotin prosthetic group during plant development closely parallel differences in the abundance of the biotinyl subunit of MCCase (7). Thus, MC-Case does not appear to be regulated by biotinylation in soybean. The isolation of cDNAs and genes coding for the biotinylated subunit of MCCase revealed that, in addition to the biotinylation domain, this subunit contains the functional domains for the first half-reaction catalyzed by all biotin-dependent carboxylases, namely the carboxylation of biotin (158, 168, 170). These domains are arranged serially on the polypeptide, with the biotin carboxylase domain at the amino terminus and the biotin-carboxyl carrier domain at the carboxyl terminus. Cloning of the cDNA and gene coding for the biotin-free subunit of this enzyme will be very useful in studying how the coordinate expression of the two MCCase subunits is regulated.

The metabolic functions of MCCase in plants are still not fully understood, notwithstanding these biochemical and molecular characterizations. As in mammals, MCCase may be involved in the mitochondrial catabolism of Leu and/or in the so-called "mevalonate shunt" by which mevalonate can be metabolized to nonisoprenoid compounds. Previous radiotracer studies of plants grown in the presence of either [^{14}C]Leu (159) or [^{14}C]mevalonate (117) indicated that Leu catabolism and the mevalonate shunt operate in vivo. Whether these two pathways involve MCCase remains to be demonstrated. The degradation of amino acids in plants in general and in particular the degradation pathway(s) of the branched-chain amino acids (Leu, Ile, and Val) are poorly understood. The first step in degradation leads, by transamination, to branched-chain 2-oxo acids (2-oxoisocaproate from Leu), in which oxidative decarboxylation (to isovaleryl-CoA with Leu) and further oxidation (to 3-methylcrotonoyl-CoA in from Leu) take place in peroxisomes (66). Gerbling & Gerhardt (66) suggested that the subsequent steps could also involve peroxisomal enzymes. However, the fate of 3-methylcrotonoyl-CoA was not clearly identified. These authors (66) proposed that an extraperoxisomal pathway is involved in Leu catabolism since peroxisomes were unable to carboxylate 3-methylcrotonoyl-CoA. Recent metabolic radiotracer studies in which extracts from isolated soybean seedling mitochondria were used have shown that all the enzymes necessary for Leu degradation to acetoacetate, and presumably acetyl-CoA, are present in plant mitochondria and that this pathway likely involves MCCase (7). Despite the lack of direct evidence implicating plant MCCase in Leu catabolism, some circumstantial evidence indicates that the pathway may be similar to that in animals and in bacteria. Plant MCCase is constitutively expressed, but its activity increases markedly during leaf senescence, when intense degradation of proteins occurs (2). This accumulation of MCCase activity is correlated with a decline in Leu content (7). Leu is the second-most abundant amino acid in plants after Asn, and it transiently accumulates in plant cells during prolonged carbon starvation when the induction of proteolytic activities leads to a massive breakdown of proteins (27, 65). These amino acids, as well as the fatty acids derived from lipid breakdown, are used in place of sugar to fuel the respiration of mitochondria spared by autophagy (11). In addition, MCCase activity is induced in pea cotyledons during the mobilization of storage proteins for seedling growth (51). It is also noteworthy that MCCase activity is strongly induced in starved sycamore cells (up to a sixfold increase in specific activity after four days of sucrose starvation), which correlates with an accumulation of the enzyme in mitochondria (10). Taken together, these observations support the findings that plant MCCase, like its animal counterpart, is involved in the degradation of Leu derived from protein mobilization.

Plant MCCase could also be implicated in noncyclic isoprenoid catabolism. In some bacterial species, noncyclic isoprenoids such as geranoyl-CoA are catabolized to acetyl-CoA via a set of reactions analogous to ß-oxidation; one reaction in this pathway is the carboxylation of methylcrotonoyl-CoA (149). Little is known about the biochemistry of isoprenoid degradation in plants, although recent findings indicate that plants contain geranoyl-CoA carboxylase (GCCase; EC 6.4.1.5), a biotin-containing enzyme previously characterized in *Pseudomonas* species (71). Thus, the catabolism of noncyclic isoprenoids in plants may be analogous to that described for bacterial species, i.e. after the carboxylation of the γ-methyl group of geranoyl-CoA, the carboxymethyl branch group of the product carboxygeranoyl-CoA is then eliminated. The resulting ß-keto thioester would then be amenable to ß-oxidation. Ultimately, such a process would generate methylcrotonoyl-CoA, which would require MCCase for its subsequent catabolism (71). Since GCCase was reported to occur in chloroplasts (71), this catabolic pathway would necessitate coordination of at least three intracellular compartments: plastids (the location of phytol, carotenoids, and GCCase), microbodies (the major location of ß-oxidation), and mitochondria (the location of MCCase).

Acetyl-CoA Carboxylase

ACCase [acetyl-CoA: carbon dioxide ligase (ADP-forming)] catalyzes the ATP-Mg-dependent carboxylation of acetyl-CoA to form malonyl-CoA, with bicarbonate as the source of inorganic carbon. The reaction is the first committed step in the synthesis of fatty acids. In plant cells, large amounts of malonyl-CoA are needed in the plastids to sustain de novo fatty acid synthesis (C16:0, C18:0, C18:1), but malonyl-CoA is also needed in the cytosol for a variety of reactions including the elongation of very long-chain fatty acids, the synthesis of secondary metabolites such as flavonoids, anthocyanins, and stilbenoids, and the malonylation of some amino acids and secondary metabolites (reviewed in 75). The occurrence of cytosolic reactions requiring malonyl-CoA and the fact that the plastid envelope is not permeable to malonyl-CoA led to the hypothesis that at least two isoenzymes of ACCase, a plastidic isoenzyme and a cytosolic isoenzyme, are present in plants (reviewed in 77).

Interest in plant ACCase has increased with the discovery that this enzyme is the primary target site of several major classes of grass-specific herbicides that affect fatty acid synthesis, principally the cyclohexanediones (CHD) and the aryloxyphenoxypropionates (APP) (29, 76). These herbicides are reversible inhibitors of grass ACCase, and there is kinetic evidence that these compounds inhibit the carboxyltransferase partial reaction by acting as transition-state analogues of the complex formed at this site (175). This latter observation is interesting because it concerns the partial reaction unique to individual biotin-dependent carboxylases (see Reaction [2]), and inhibition here is in keeping with the specificity of these compounds for plant ACCase. The other partial reaction, the carboxylation of biotin (see Reaction [1]), is common to all biotin enzymes, and inhibition at this step would render these other carboxylases sensitive, which is, in fact, not the case (2). Kinetic analyses also revealed that the CHD and APP are mutually exclusive inhibitors of ACCase, i.e. the binding of a herbicide from one class prevents binding of a herbicide from the other class (28, 131), either because the two binding sites overlap, or through allosteric effects. Most grass species are susceptible to CHD and APP herbicides, whereas broad-leaved plants and monocotyledonous species other than grasses are resistant (77). With few exceptions, the selectivity of these compounds is expressed at the level of the target site. In contrast, herbicide selectivity is usually determined by the ability of plants to metabolize these pesticides, or by differences in uptake and/or translocation of the active compound. There are several examples where herbicide resistance in grasses correlates with the presence of a tolerant form of ACCase, such as in natural populations of red fescue (*Festuca rubra*) and annual meadow grass (*Poa annua*) (83, 106). Moreover, widespread herbicide use has led to the selection of resistant biotypes of weeds that were previously susceptible (reviewed in 45). Again, the main mechanism of resistance appears to involve target site modification. Such modifications of the target ACCase resulting in a herbicide-resistant enzyme have been documented in a number of weed biotypes following selection in the field (for example, see 109), and in maize tissue culture following in vitro selection (111, 123). Resistance of a biotype of *Lolium rigidum* (resulting from 10 consecutive years of selection with diclofop-methyl) to ACCase-inhibiting herbicides correlates with the possession of a modified, resistant form of ACCase that is controlled by a single major gene (162). However, the structural and molecular basis for the difference in herbicide sensitivity of the enzymes from grasses and other plants, particularly dicotyledons, was unknown until the recent elucidation of the two different types of structural organization of plant ACCase, together with the discovery of ACCase isoforms in different cell and tissue compartments. These findings have improved our understanding of the selectivity of ACCase-inhibiting herbicides.

The molecular organization of ACCase differs depending on the source of the enzyme. In *E. coli*, ACCase is composed of four distinct subunit types that readily dissociate into three components: a homodimer of 49-kDa subunits (biotin carboxylase module), a homodimer of 17-kDa subunits (BCCP), and a carboxyltransferase heterotetramer containing two 33-kDa and two 35-kDa ($\alpha_2\beta_2$) subunits.

These are encoded by four separate genes, which in *E. coli* are named *acc*C, *acc*B, *acc*A, and *acc*D, respectively (6, 72, 95, 103, 104). In contrast, in animals, fungi and yeasts, these entities are located on a single, multifunctional polypeptide that has a molecular mass exceeding 200 kDa (5, 12, 108). The functional ACCase enzyme in these organisms is composed of multimers of this large polypeptide. The structure of plant ACCase has been the subject of considerable confusion in the past. Early experiments indicated that ACCase in spinach and barley chloroplasts and avocado plastids had a multisubunit structure similar to that found in prokaryotes (90, 91, 113). Plants may also contain distinct isozymes of ACCase in separate spatial and temporal compartments to generate the malonyl-CoA needed for a variety of phytochemicals (160). However, the subsequent inclusion of proteinase inhibitors in purification media and the development of avidin-affinity matrices, which allowed rapid purification of the enzyme, led to the isolation of a homomeric ACCase of high molecular mass comparable to that found in other eukaryotes (31, 54, 55). Based on this finding, it was claimed and generally agreed that the occurrence of low-molecular-mass biotinyl polypeptides in purified ACCase preparations was largely, if not exclusively, due to severe degradation of the high-molecular-mass polypeptide form during enzyme isolation (reviewed in 75, 77). Consequently, the concept of a prokaryotic form of ACCase in plants fell into disfavor. However, because there was no evidence for compartmentation of ACCase, the second hypothesis was rejected too. However, it is now evident that both hypotheses are valid. The confusion about the exact nature of plant ACCase arose in large part because (*a*) plants contain structurally different forms of the enzyme, one of which rapidly loses activity during purification, not because of inactivation by proteases but rather because it readily dissociates into its component subunits, and (*b*) the enzymes from Gramineae were mainly studied. Most plants other than Gramineae are now known to have a multisubunit plastidial ACCase (frequently referred to as the prokaryotic form) and a multifunctional, extraplastidial ACCase (also called the eukaryotic form), whereas members of the Gramineae have only the multifunctional-type ACCase in both plastids and cytosol (3, 56, 99, 141). To date, only two exceptions to this generalization have been reported: Chloroplasts from *Brassica napus* appear to contain both multisubunit and multifunctional ACCases (59, 148); and chloroplasts from the dicot *Erodium moschatum* lack the multisubunit ACCase but instead contain the multifunctional-type enzyme (35). The first major finding that revived the idea that plant ACCase can adopt a multisubunit structure was the discovery of a putative *acc*D homologue in the pea chloroplast genome (105). It was subsequently shown that the product of this gene in pea binds at least two other polypeptides, one of which was biotinylated (140). Shortly thereafter, direct biochemical evidence established the presence of a prokaryotic form of ACCase in pea mesophyll chloroplasts (3, 98) composed of different-sized subunits. One of these, apparently 38 kDa in size, is biotinylated, which gives a native 600-kDa size. This form of the enzyme represents about 80% of the total ACCase activity in the whole leaf (3) and is totally insensitive to herbicides of the CHD and APP classes (3, 43, 98). These enzyme subunits are organized into two functional domains that interact through ionic interactions and are readily dissociable and reassociable according to the law of mass action (3, 157). A similar multisubunit organization was recently reported for soybean chloroplastic ACCase (132). This dissociation/reassociation property of the multisubunit form of ACCase is thought to play a role in regulating ACCase activity in chloroplasts (132). The chloroplastic ACCase is able to carboxylate free D-biotin as an alternate substrate in lieu of BCCP and in the absence of acetyl-CoA. This specific property was used to purify the biotin carboxylase component of the enzyme (4). The minimal structure of the active biotin carboxylase domain corresponds to a complex of two, tightly interacting polypeptides (the 38-kDa biotinylated subunit and a 32-kDa biotin-free polypeptide) (4). This contrasts with the organization of bacterial ACCase in which all subunit components are freely separable (72). Steady-state kinetic analyses of pea biotin carboxylase were compatible with an ordered mechanism in which MgATP binds first, followed by free biotin and then bicarbonate. Consistent with this mechanism, bicarbonate-dependent ATP hydrolysis by the enzyme could be observed only in the presence of added D-biotin (4). These data suggest the existence of functional and structural differences between the multisubunit isoform of ACCase from plants and bacteria (163). cDNAs and/or genes encoding BCCP and biotin carboxylase subunits from various plants have been cloned (19, 34, 59, 132, 156). These proteins are nuclear encoded and share substantial sequence similarity with the corresponding bacterial ACCase subunits. Furthermore, their expression is coordinated during seed and leaf development (135). The second functional domain of ACCase found in pea or in soybean chloroplasts is composed of the α- and β-subunits of carboxyltransferase (132, 157). The α-subunit (73 to 91 kDa) is nuclear encoded and possesses a structural motif similar to a prokaryotic membrane lipoprotein lipid attachment site. As this polypeptide was previously identified as an inner-envelope membrane constituent of pea chloroplasts, this observation suggests that the prokaryotic form of ACCase in pea may be at least partly associated with chloroplast membranes (157). However, the insoluble nature of the carboxyltransferase subunits of ACCase could be artifactual, resulting from a redistribution of the ACCase components among organellar subfractions following disruption of the complex during enzyme purification (132). In fact, the multisubunit ACCase complex remains soluble after chloroplast lysis if organelle fractionation is done rapidly (3). The gene encoding the β-subunit (49 to 80 kDa) has been identified in the plastid genome by its similarity to one of the carboxyltransferase subunits of *E. coli* ACCase (59, 105, 132, 140). This is the only component of plant lipid metabolism known to be encoded by the plastid genome.

In addition to this major prokaryotic form, pea leaves also contain a minor eukaryotic form of ACCase that is mainly, if not exclusively, concentrated in the epidermis (3). By contrast, in mature pea seeds this ACCase comprises the major portion of the enzyme activity (23, 43). This enzyme is a homodimer of a single biotinylated polypeptide of about 220 kDa that is sensitive to APP herbicides (3, 43). Several genes and cDNA clones have been isolated for this type of ACCase from other dicotyledonous plants. None of them seems to have typical organelle-targeting presequences (reviewed in 78, 118). All of these clones encode proteins with the biotin carboxylase domain at the N-terminus, the biotin carboxyl carrier domain in the middle, and the carboxyltransferase domain at the C-terminus. A substantial portion of the polypeptide that is located between the two latter domains has no correspondence with *E. coli* ACCase, and may have a structural role. Grasses such as maize and wheat also contain two ACCase isoforms, the major one in the chloroplasts of mesophyll cells, the other in an extraplastidial compartment (probably the cytosol) of another cell type(s) (56, 70, 84). However, in contrast to dicotyledonous plants, both isoforms are composed of a single type of high-molecular-mass, biotin-containing polypeptide of 227 (ACCase 1, mesophyll chloroplast enzyme) or 219 kDa (ACCase 2, non-mesophyll-chloroplast enzyme) (9, 56). Antibodies raised against the 227-kDa form poorly recognized the minor 219-kDa isoform. These ACCase isoforms are encoded by distinct nuclear genes (9). Finally, both enzymes are inhibited by CHD and APP herbicides, although ACCase 1 is much more sensitive than ACCase 2 (56, 82, 84). Thus, it now appears that all plants contain two ACCase isoforms in two different subcellular and tissue locations. However, the Gramineae family of plants is different in that both the plastid and cytosolic ACCase isozymes are of the eukaryotic type. Coincident with this evolutionary difference, a recent finding noted that probably all plants except the Gramineae contain the plastid-encoded *acc*D gene (99). Consistent with this finding, both western and Southern blot analyses failed to detect subunits of the prokaryotic form of ACCase in all grass species studied to date, whereas prokaryotic and eukaryotic forms of ACCase were detected in all the other plants analyzed, including monocotyledons and dicotyledons (99, 135). Thus, this difference in ACCase molecular organization provides an explanation for the action of grass-specific herbicides, which inhibit only the eukaryotic and not the prokaryotic form of the enzyme (3, 98). The chloroplastic ACCase, thought to play a key role in de novo fatty acid biosynthesis, is strongly inhibited by these compounds in the Gramineae (eukaryotic form) but not in other plants (prokaryotic form). In contrast, in all plant species the cytosolic ACCase (eukaryotic form) is only a minor form in leaves, probably involved in secondary metabolic pathways, and is much less affected by CHD and APP herbicides. Interestingly, *Erodium moschatum*, the only dicot known to be sensitive to APP herbicides, lacks the herbicide-insensitive prokaryotic form of ACCase in its plastids (35).

In addition to these differences in sensitivity to herbicides, the two structurally distinct ACCase isozymes exhibit important differences in their biochemical properties. The multifunctional enzyme has a much lower K_m for acetyl-CoA than the multisubunit complex (3, 56). The multifunctional enzyme is able to carboxylate propionyl-CoA at substantial rates, whereas the multisubunit complex is not active with this substrate (43). Furthermore, no other PCCase activity, different from that catalyzed by the eukaryotic forms of ACCase, can be detected in either reproductive or vegetative organs of pea plants or from maize leaves at any stage of development (43, 84). Steady-state kinetic analysis of the multifunctional form of ACCase, purified from mature pea seeds, with respect to substrate specificity and inhibition by quizalofop, a member of the APP class of herbicides, demonstrated that both reactions catalyzed by the enzyme (acetyl-CoA and propionyl-CoA carboxylations) proceed at separate sites on the enzyme and are inhibited by the herbicide. The two sites, however, show different catalytic properties: One site binds either acetyl-CoA or propionyl-CoA and is inhibited by quizalofop, whereas the other is specific for acetyl-CoA and is much less affected by quizalofop. Owing to these two catalytically distinct sites, the enzyme obeyed Michaelis-Menten kinetics with respect to propionyl-CoA, but exhibited kinetic cooperativity in the presence of acetyl-CoA. Also, the kinetics of PCCase activity exhibited hyperbolic inhibition in the presence of quizalofop, but cooperative inhibition when measuring the ACCase activity of the enzyme. These results indicate that the higher the substrate specificity, the lower the quizalofop sensitivity of the active site, which suggests that the nature of the interactions between the two identical enzyme subunits may play a role in governing the degree of herbicide sensitivity. Presumably, such interactions introduce some distortions in the structure of the individual active sites, for example, generating kinetic cooperativity with substrates and inhibitors (43). Similar conclusions were drawn from kinetic studies with maize ACCases (84). Also, this apparent correlation between substrate specificity and sensitivity of ACCase toward quizalofop was confirmed by kinetic analysis of the prokaryotic form of ACCase from pea leaf chloroplasts. This enzyme, which is insensitive to quizalofop inhibition, is unable to carboxylate propionyl-CoA (43).

The discovery of the occurrence of ACCase isoforms in separate plant cell compartments raises the question of their physiological significance. The main role of ACCase in plastids is to provide malonyl-CoA, the precursor for fatty acid biosynthesis. Indeed, plastids represent the major site for de novo fatty acid biosynthesis in plants (118, 119). In support of this conclusion, we note, for example, that the activity level of the prokaryotic form of ACCase in pea shows considerable variation during development; it is maximal in young pea leaves, presumably reflecting a high demand for fatty acids required for the biosynthesis of thylakoid membranes (43). Also, recent investigations on fatty acid biosynthesis in pea suggest that the enzymes of fatty acid synthesis are organized within the chloroplast into a multienzyme assembly that channels acetate, through acetyl-CoA synthetase and ACCase, into long-chain fatty acids, glycerides, and CoA esters (137). Considerable in vivo and in vitro evidence suggests that chloroplastic ACCase is involved in the regulation of plant fatty acid synthesis (ARPP-1997). For example, analysis of substrate and product pool sizes implicated ACCase in the light/dark regulation of fatty acid synthesis in spinach leaves and chloroplasts (128, 129). The biochemical environment of the chloroplast stroma undergoes numerous reversible changes upon illumination, and there are increases particularly in pH, ATP, NADPH, and Mg^{2+} concentrations and in the ATP/ADP ratio (75). These differences can largely account for the stimulation of ACCase in the light and therefore fatty acid synthesis (87). Also, it was recently shown that reduced thioredoxin activates multisubunit chloroplastic ACCase from pea and spinach plants, which suggests that this mechanism is a link between light activation and fatty acid synthesis (87, 142). Herbicide inhibition of ACCase was used to determine flux control coefficients, which led to the conclusion that chloroplastic ACCase exerts a major control over fatty acid synthesis rates in barley and maize leaves (121). ACCase was also the apparent site of feedback inhibition of fatty acid synthesis in tobacco suspension cells supplemented with exogenous fatty acids (153). Finally, recent in vivo evidence for a regulatory role of ACCase in oilseeds was obtained by genetic engineering approaches (136, 154). Now that clones are available for various chloroplastic ACCases, plant transformant experiments will provide tools to assess the role of this enzyme in controlling flux through the fatty acid biosynthetic pathway. At present, there is no evidence that de novo fatty acid synthesis occurs in the cytosol of plant cells. However, a recent finding indicated that plant mitochondria contain not only the acyl carrier protein but also all the enzymes required for de novo fatty acid synthesis from malonate, but not from acetate (167). Thus, in this case, ACCase is probably not required. Indeed, plant mitochondria do not have detectable ACCase activity (2).

Although malonyl-CoA is required, in addition to fatty acid synthesis, in a number of biosynthetic pathways, the physiological role(s) of the multifunctional, cytosolic ACCase is not yet clearly established. There are indications that this enzyme might be involved in the biosynthesis of very long-chain fatty acids required for cuticular waxes, or the biosynthesis of flavonoids via chalcone synthase. Interestingly, in pea leaves, both processes occur in the cytosol of epidermal cells, thus matching the tissue localization of the eukaryotic extraplastidial form of pea leaf ACCase (3). Cuticular wax and flavonoids are important in the interaction of plants with their environment, for example, for protection against UV radiation and pathogens. Two recent findings indicate that this isozyme of ACCase probably helps to control the synthesis of such protective compounds: First, the transcript for the eukaryotic form of alfalfa ACCase is induced by yeast or fungal elicitors of isoflavonoid phytoalexin synthesis (155); and second, the cytosolic eukaryotic form (but not the chloroplastic prokaryotic form) of ACCase is induced by UV-B irradiation of fully expanded pea leaves, in parallel to the induction of some of the enzymes involved in flavonoid synthesis, such as chalcone synthase (97). The products of two different ACCase genes have been identified in the cytosol of human and rat cells (1, 176). The presence of multiple cytosolic ACCase genes in mammals also reflects the need for differential expression of the enzyme in response to varying environmental or developmental cues. The mammalian and yeast ACCases are regulated by reversible phosphorylation through the action of an AMP-activated protein kinase (92). Recent investigations demonstrated that the carboxyltransferase β-subunit of pea chloroplast ACCase is phosphorylated, which suggests that such a regulatory mechanism may also operate in plants (143).

BIOTIN BIOSYNTHETIC PATHWAY

Plants, like micro organisms, can synthesize biotin, whereas other multicellular eukaryotic organisms are biotin auxotrophs. Biotin biosynthesis has been well characterized in *E. coli* and, more recently, in other bacteria such as *Bacillus sphearicus*, through combined biochemical and genetic studies. In these bacteria, the *bio* (*ABFCD*) locus contains the genes required for biotin biosynthesis (57, 67, 89). These genes encode the enzymes catalyzing the synthesis of biotin from pimeloyl-CoA and Ala: 7-keto-8-aminopelargonic acid synthase (KAPA synthase);

7,8-diaminopelargonic acid aminotransferase (DAPA aminotransferase); dethiobiotin (DTB) synthase; and biotin synthase, coded by *bio*F, *bio*A, *bio*D, and *bioB*, respectively (Figure 2). Until recently, little was known about biotin metabolism in plants. Initial information on biotin synthesis and transport in plants came from analysis of the *bio1* biotin auxotroph of *A. thaliana*, which was discovered by

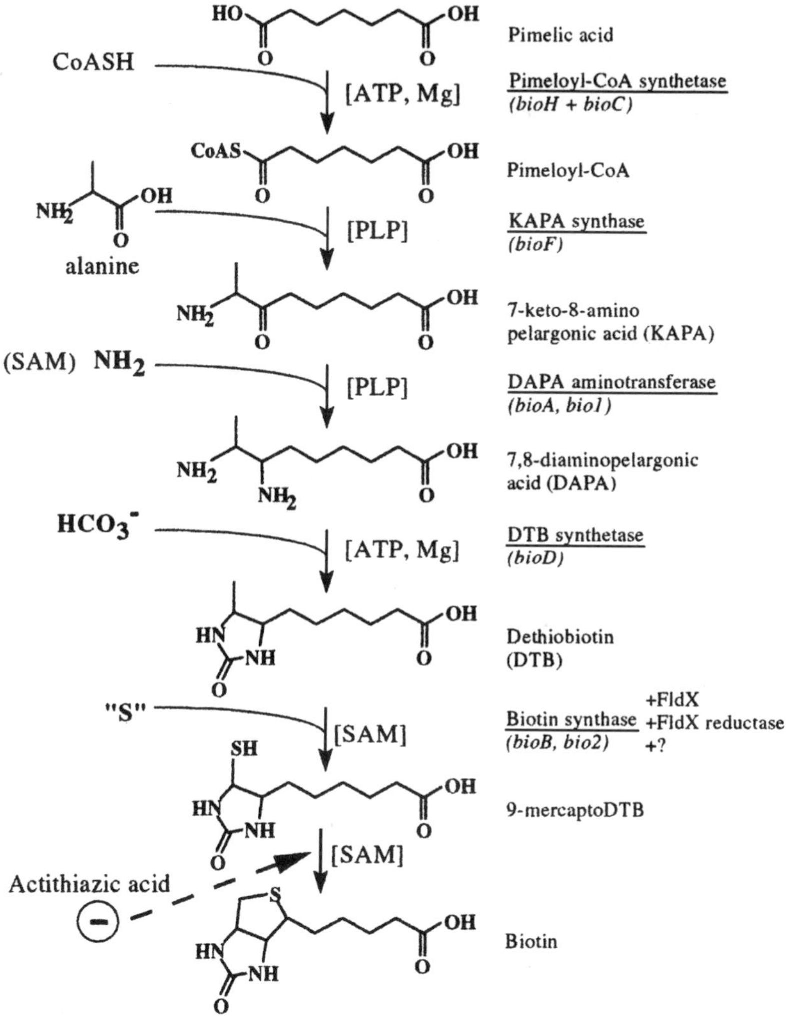

Figure 2 Biotin biosynthetic pathway in plants and microorganisms. Positions where mutations block biotin synthesis in *E. coli* (*bioH, bioC, bioF, bioA, bioD,* and *bioB*) and *Arabidopsis* (*bio1* and *bio2*) are shown, as is inhibition by actithiazic acid (−). The immediate precursor of pimeloyl-CoA appears to be pimelic acid in *B. subtilis*, *B. sphaericus*, and *Arabidopsis* but not in *E. coli*. SAM: S-adenosyl methionine; PLP: pyridoxal phosphate; FldX: flavodoxin; FldX reductase: flavodoxin reductase.

Schneider et al (147). Seeds homozygous for the mutation failed to develop unless exogenous biotin, dethiobiotin, or DAPA, but not KAPA, was supplied to the plant (151). Recently, Patton et al (126) have shown that the *E. coli bioA* gene, which codes for DAPA aminotransferase, can genetically complement the *bio1* mutation, demonstrating that *bio1/bio1* mutant plants are defective in this enzyme. A second biotin auxotroph of *A. thaliana* has subsequently been identified. Arrested embryos from this *bio2* mutant are defective in the final step of biotin synthesis, i.e. the conversion of dethiobiotin to biotin (125). Treatment of a biotin-overexpressing strain of lavender cells with [^{3}H]pimelic acid showed that all the intermediates of biotin synthesis established in bacteria, plus the novel metabolite 9-mercaptodethiobiotin (9-mDTB), accumulate in plants (17), which demonstrates that the pathway of biotin synthesis in bacteria is conserved in plants (Figure 2). Also, this study indicated that the reaction catalyzed by the plant biotin synthase proceeds in two distinct steps involving 9-mDTB as an intermediate and that actithiazic acid, an inhibitor of biotin biosynthesis, specifically blocks the conversion of 9-mDTB to biotin, i.e. the formation of the thiophene ring of biotin (17) (Figure 2). Subsequent experimental evidence suggested that 9-mDTB may be an intermediate in the bacterial biotin biosynthetic pathway as well (110). Inhibition of biotin synthesis by actithiazic acid is lethal to plants and can be prevented by supplementation with nanomolar concentrations of biotin (13), findings that demonstrate both the specificity of action of this inhibitor and the fact that plants, like all organisms, need only trace amounts of biotin for growth. The catalytic mechanism of the last step of biotin biosynthesis is still unclear. It has been demonstrated in *E. coli* that the conversion of dethiobiotin to biotin is catalyzed by a complex involving 2 (or 3) proteins in addition to the *bioB* gene product, which by itself is totally inactive (24, 89, 139). The term biotin synthase, previously used to designate the *bioB* gene product, should therefore be reserved for this multisubunit complex. A cDNA encoding an *A. thaliana* homologue of the *bioB* gene product has been isolated by functional complementation of a *bioB* biotin auxotroph mutant of *E.coli* (18). This cDNA shows specific regions of similarity with the corresponding genes from bacteria and yeast. In particular, the predicted amino acid sequence of the plant protein contains the consensus region (GXCXEDCXYCXQ) involved in binding a [2Fe-2S] cluster (16). Interestingly, the plant sequence contains an N-terminal extension of about 40 amino acids that is not found in its bacterial counterparts, which suggests an organellar location for this enzyme (18). Computer analyses of the primary structure of the *A. thaliana bioB* gene product predict that this protein might be targeted to mitochondria (171). Indeed, western blot analyses with antibodies raised against the purified plant recombinant protein demonstrated such a mitochondrial location (16). This subcellular location is intriguing since most of the free biotin pool in plant mesophyll cells accumulates in the cytosol to a concentration of about 11 μM (15) (note that in bacteria free biotin never accumulates above a nM concentration range, see later). Thus, if biotin is synthesized within mitochondria, it must be exported to accumulate in the cytosol. The precise role of this pool of free biotin in the cytosol is not known. One hypothesis is that, as in bacteria, the level of free biotin controls the expression of genes encoding the biotin-containing enzymes and/or the enzymes involved in biotin synthesis. In support of this suggestion, it was demonstrated that the expression of the gene encoding the *A. thaliana bioB* gene product is strongly induced under biotin-limiting conditions (124). Finally, the recombinant plant biotin synthase containing an iron-sulfur cluster was shown to catalyze the direct conversion of dethiobiotin into biotin (C Alban & R Douce, unpublished results). This reconstituted in vitro system requires an electron donor [NAD(P)H], S-adenosylmethionine (SAM) acting as a radical-forming molecule (74), and a soluble bacterial extract containing flavodoxin and flavodoxin reductase (89).

BIOTIN HOLOCARBOXYLASE SYNTHETASE

Biotinylation of biotin-dependent carboxylases permits the transformation of inactive apo-carboxylases into active holo-forms (138). In *E. coli*, the biotinylation of apo-ACCase (the unique biotin enzyme found in this organism) is carried out by biotin ligase. The enzyme catalyzes the posttranslational attachment of D-biotin to a specific Lys residue of the apo-enzyme, via an amide linkage between the biotin carboxyl group and the ε-amino group of Lys (138). This covalent attachment occurs in two discrete steps ([4] and [5]):

$$\text{D-biotin} + \text{ATP} \rightarrow \text{D-biotinyl 5}'\text{-AMP} + \text{PPi} \qquad [4]$$

$$\text{D-biotinyl 5}'\text{-AMP} + \text{apo-BCCP} \rightarrow \text{holo-BCCP} + \text{AMP} + \text{H}_2\text{O} \qquad [5]$$

The first step is the activation of D-biotin by ATP, which yields D-biotinyl 5′-AMP, followed by the covalent attachment of the biotinyl group to the ε-amino-group of a specific Lys residue of the apo-BCCP, with release of AMP. Biotin ligase has been purified from *E. coli* and its gene cloned (58, 85). This enzyme, also called BirA, is a 33.5-kDa protein (85) that also acts as a repressor of the biotin operon in the presence of D-biotinyl 5′-AMP as the co-repressor (39). Its three-dimensional structure has been determined at 2.3 Å resolution (174). The corresponding enzymes from various mammalian species have been purified and are referred to as biotin holo-carboxylase synthetase (HCS) (33, 181). Recently, clones encoding the *Saccharomyces cerevisiae* HCS gene (40) and human HCS cDNAs have been obtained (102, 161). These latter enzymes show some sequence similarity to the biotin ligases from bacteria, but are more than twice their size. The first direct evidence for the existence of HCS activity in plants came from the recent purification and characterization of HCS activity from pea leaves (165). The enzyme was able to use bacterial apo-BCCP as substrate, which demonstrates that plant HCS acts across species barriers. This cross-species activity revealed a molecular mechanism common to these enzymes. In contrast, plant HCS showed a very high specificity for its biotin substrate, exhibiting an apparent K_m value of 28 nM (165). It was subsequently shown that in plants, HCS activity is associated with several subcellular compartments (164). Thus, three enzyme forms can be separated by anion-exchange chromatography of a pea leaf protein extract. The major form was found to be specific for the cytosolic compartment, whereas the two minor forms were present in mitochondria and chloroplasts, respectively. The high purity and latency values of HCS activity measured in Percoll-purified chloroplasts and mitochondria, together with the observed protection of enzyme activity in these organelles during thermolysin treatment, demonstrated that HCS is a genuine constituent of chloroplasts and mitochondria (164). The existence of HCS isoforms in various plant cell compartments suggests that the different biotin-dependent carboxylases localized in chloroplasts, mitochondria, and cytosol are biotinylated in the cell compartment within which they are localized. One possible explanation for this feature is that the active site of HCS recognizes a particular three-dimensional folded structure of apo-carboxylases rather than primary amino acid structure alone (see below). By functional complementation of the *birA 215 E. coli* mutant with an *A. thaliana* cDNA expression library, a full-length HCS cDNA encoding a 41-kDa polypeptide was isolated (164). Although this plant HCS shows specific regions of similarity with other known biotin ligases of bacterial, yeast, and human origin, the similarities are restricted to the ATP- and biotin-binding domains. These motifs, which are located in the central part of the protein, are interconnected and thus reflect the requirement for ATP and biotin to be spatially close to allow the formation of the biotinyl-5′-AMP intermediate. Interestingly, the eight amino acid residues that have been shown to be in direct contact with biotin in *E. coli* BirA by X-ray crystallography (174) are strictly conserved in *A. thaliana* HCS. In contrast, plant HCS does not contain the helix-turn-helix DNA binding domain of BirA from *E. coli* (174) or *B. subtilis* (25). In bacteria, this DNA-binding domain mediates the repressor function of the BirA protein. Consistent with this observation, the cloned plant HCS proved unable to substitute for BirA as a repressor of the biotin operon in a *birA* derepressed mutant of *E. coli* (C Alban & R Douce, in preparation). The N-terminal region of *A. thaliana* HCS exhibits the characteristic features of an organelle transit peptide. Also, the occurrence of two Met residues close together in this region suggests the possible existence of cytosolic and "organelle-targeted" forms of HCS, synthesized from a single species of mRNA by alternative translational initiation. Such a possibility, together with an alternative splicing mechanism, has been suggested to account for the synthesis of human mitochondrial and cytosolic HCS isoforms (102). Two translation products of the expected sizes were obtained by in vitro transcription-translation experiments with the plant HCS cDNA, which is consistent with this hypothesis (164). However, further studies were needed to definitively assign the cellular localization of this clone to a specific compartment. To address this question, the cloned *A. thaliana* HCS cDNA was used to create an overproducing *E. coli* strain (166). Polyclonal antibodies raised against pure recombinant HCS were then produced to elucidate the subcellular localization of this protein. Both immunodetection by western blotting of proteins from isolated pea leaf subcellular compartments, and immunocytology on tissue sections of tobacco leaves expressing the complete coding sequence of *A. thaliana* HCS demonstrated that the enzyme encoded by this cDNA corresponds to the chloroplastic isoform (166). Thus, this enzyme form is probably responsible for the biotinylation of chloroplastic ACCase. Some physicochemical, biochemical, and kinetic properties of the pure recombinant HCS were also determined. The native recombinant protein is a 37-kDa monomer. The enzyme is able to efficiently biotinylate apo-BCCP from *E. coli* and also apo-MCCase from *A. thaliana*, but with less efficiency (166). Such broad substrate specificity was previously noted for HCS from various origins. However, neither apo-SBP65, the unbiotinylated form of the 65-kDa pea seed-specific biotinyl protein [see below and (51)], nor a synthetic peptide produced as a maltose binding fusion protein in *E. coli*, which was previously shown to be efficiently biotinylated by *E. coli* biotin ligase in vivo (144), can serve as substrate for the recombinant chloroplastic *A. thaliana* HCS (166). Steady-state kinetic analyses of the HCS-catalyzed reaction were conducted to determine the reaction mechanism and inhibition by reaction products. These experiments indicated that the reaction proceeds by a double-displacement "Bi Uni Uni Bi ping-pong Ter Ter" mechanism whereby ATP and D-biotin bind to the enzyme in an ordered fashion during the first half-reaction and react to form biotinyl-AMP. Then, the PPi produced by hydrolysis of ATP is released. In the second half-reaction, apo-BCCP binds to the biotinyl-AMP-enzyme complex and is biotinylated to form holo-BCCP. Finally, the holo-BCCP and AMP produced are released (166).

Owing to the presence of HCS activity in plant cell mitochondrial and cytosolic fractions as well as in chloroplasts, possible structurally distinct HCS isoforms need investigation in these subcellular compartments if the mechanisms governing the requirements and regulation of biotinylation in plants are to be understood. This information will also help elucidate why HCS activity has to be compartmentalized in plant cells. Interestingly, a recent discovery revealed that human serum biotinidase, an enzyme involved in biotin recycling in mammals, has biotinyl-transferase activity in addition to biotinidase hydrolase activity (88). Although biotinidase activity is undetectable in plant extracts (13), the possible existence of such biotinyl-transferase activity in plants merits exploration in the future.

STRUCTURAL REQUIREMENTS FOR PROTEIN LIPOYLATION AND COMPARISON WITH PROTEIN BIOTINYLATION

Lipoic acid (1,2-dithiolane-3-valeric acid) is the prosthetic group of the E2-protein component (dihydrolipoylacyltransferase) of 2-oxoacid dehydrogenases, a family of multienzyme complexes that catalyze the irreversible oxidative decarboxylation of 2-oxoacids (pyruvate, 2-oxoglutarate, and the three short-branched chain 2-oxoacids produced by transamination of Leu, Val, and Ile) (22, 112). The E2-protein component forms the structural core of the complexes on the surface of which oxoacid decarboxylase (E1-protein) and dihydrolipoamide dehydrogenase (E-3 protein) are bound in a noncovalent manner (22, 112). Lipoic acid is also the prosthetic group of the H-protein component of the glycine decarboxylase complex which catalyzes the oxidative decarboxylation and deamination of Gly with the formation of CO_2, NH_3, and methylene-tetrahydrofolate (48, 120). This complex consists of four protein components (P-, H-, T-, and L-proteins). The H-protein component forms the mechanistic "heart" of the Gly decarboxylase complex with the P-, T-, and L-proteins (dihydrolipoamide dehydrogenase) interacting on its surface. The H-protein from pea leaf mitochondria has been crystallized and the X-ray crystal structure has been determined at 2 Å resolution (37, 122). The core of this structure consists of two antiparallel β-sheets forming a hybrid barrel-sandwich structure. One sheet has six β-strands and the other has three strands and two adjacent antiparallel strands joined by a loop (hairpin β-motif) in which the lipoate cofactor is attached to a specific Lys residue. Despite minimal amino acid sequence similarity, the structure of the lipoyl domain in the E2 protein component of 2-oxo acid dehydrogenase multienzyme complexes and in H-protein closely resembles that of the biotinyl domain of BCCP (26, 36, 133). How, therefore, do the lipoate protein ligase and biotin holocarboxylase synthetase distinguish lipoyl and biotinyl apo-domains as substrates for lipoylation and biotinylation, respectively, avoiding aberrant modifications? The correct positioning of the target Lys in the protruding β-turn is in fact essential. In the lipoyl-containing protein the configuration of the hairpin loop has the type I conformation, as defined by Wilmot & Thornton (173). In contrast, in BCCP, the configuration of the hairpin loop has a type I′ conformation. Consequently, the superimposition of H-protein with the biotin domain of BCCP revealed a mirror inversion between the hairpin loops in the two structures (36). We may assume, therefore, that this structural difference is important and plays a crucial role in the recognition by the lipoate and biotin ligases. In support of this observation, mutagenesis studies of the hairpin region around the lipoylated or biotinylated Lys indicate that primary structure is not sufficient for the selective recognition process by the ligases (73, 101, 133, 152). Rather, these proximal amino acids, particularly define Met in BCCP, facilitate the carboxylation reaction and have a critical role in the carboxyl transfer reaction (96). Consistent with this finding, Schatz (144) reported the isolation from a peptide library of synthetic peptides that did not contain the consensus sequence found in natural biotinylated proteins but which were efficiently biotinylated in *E. coli*. This suggests that these peptides somehow mimic the folded structure formed by the natural substrates. The specificity constant (k_{cat}/K_m) governing BirA-catalyzed biotinylation of these peptides, including a 14-residue minimal substrate, is nearly identical to that measured for the natural substrate apo-BCCP (21). Likewise, Val and Ala residues surrounding the lipoyl-lysine (H-protein) play an important role in the molecular events that govern the reaction between the P- and H-proteins but do not intervene in the recognition of the binding site of lipoic acid by lipoyl ligase (73).

SEED-SPECIFIC BIOTINYLATED PROTEINS

A salient trait typical of the plant kingdom is the existence of a unique, seed-specific, biotinylated protein that was first documented in pea and called SBP65 (for Seed Biotinylated Protein of 65 kDa) (51). SBP65, which is the major biotinylated protein in mature pea seeds, is localized to the cytosol of embryonic cells. This protein behaves as a "sink" for free biotin during late stages of embryo development and is rapidly degraded during germination (51). In support of a possible function for SBP65 is the fact that this protein is devoid of any biotin-dependent carboxylase activity, presumably because, in breaking the rule for protein biotinylation, covalent binding of biotin to the apoprotein does not occur within the consensus tetrapeptide sequence of Val/Ala-Met-Lys-Met/Leu (50, 53). In particular, the two Met residues flanking the modified Lys in most biotin enzymes are absent in the pea seed protein. The existence of SBP65 is not restricted to pea. Homologues of pea SBP65 have been identified in developing castor seeds (135), carrot somatic embryos (180), and developing soybean seeds (63, 86, 150). A cDNA encoding the soybean protein has also been cloned (86). Analysis of its sequence revealed that the similarity between the soybean and pea proteins is about 72% and the identity 54%. As in pea embryos (51), little or no seed-specific biotinylated protein can be detected in young soybean seeds, and the protein accumulates in the later stages of seed development. When immature soybean seeds are artificially dried, the seed-specific biotinylated protein increases rapidly to a high level, similar to that found in mature seeds. Thus, in seeds but not in leaves, desiccation can elicit the accumulation of the seed-specific biotinylated proteins, together with the expression of the biotinylation system responsible for biotinylation of these proteins. Seed-specific biotinylated proteins have also recently been purified from mature seeds of soybean, lentil, peanut, cabbage, rape, tomato, and carrot and their biotinylation site biochemically characterized (C Job & D Job, unpublished results). For all of these proteins, the biotinylated Lys residue is within a conserved pentapeptide sequence of (M/V)GKF(E/Q/V).

SBP65 shares many properties in common with the Lea (Late embryogenesis abundant) proteins, a group of ubiquitous seed proteins that are very abundant in late embryogenesis and disappear rapidly during early germination. Both types of proteins are highly hydrophilic and their amino acid sequences have many repeated motifs (49, 52, 53). The Lea proteins and their mRNAs can be induced to accumulate in young excised embryos by pretreatment with the plant hormone abscisic acid (ABA), and in non-seed tissues upon ABA treatment and/or water stress (49, 64, 68, 116, 130). Thus, some Lea proteins are assumed to play a protectant role during desiccation, which occurs in the later stages of embryo development and in vegetative tissues in response to water stress (49). However, SBP65 is not an abundant protein, representing only about 10^{-4} of the total proteins in the mature pea seeds (51), which raises questions about its role in a protectant mechanism. Since SBP65 covalently binds biotin, which is an essential cofactor for basic metabolic activity, its role, presumably, is different. Furthermore, owing to its highly atypical biotinylation site (53), SBP65 defines a new class of biotinylated protein, with a role likely to be different from that of the well-characterized biotin enzymes from bacteria, animals, and plants. Taken together, these findings implicate a possible crucial role for biotin and protein biotinylation in plant embryo development that might have no counterpart in other systems. As outlined above, the importance of biotin in seed development was first pointed out by Meinke and coworkers, after they identified and characterized the *bio*1 biotin auxotroph mutant of *Arabidopsis* (151). They showed that virtually no biotin was detectable in the seeds of this embryo-lethal mutant, but that these embryos could be rescued to produce phenotypically normal plants when cultured in the presence of either biotin, dethiobiotin, or diaminopelargonic acid (151). Interestingly, Shellhammer & Meinke (151) also observed that maternal sources of biotin are insufficient to rescue mutant embryos produced by heterozygous plants grown in the absence of supplemental biotin. This means that embryo development cannot be achieved without some biotin synthesis occurring within the embryo itself.

The spatial and temporal expression of *sbp*, the gene encoding SBP65, was characterized by northern blot analysis of mRNA from both in vivo and in vitro developing pea embryos (44). As for *lea* genes, the expression of the *sbp* gene is induced by ABA in immature embryos in culture. Accordingly, the *sbp* promoter region contains several potential *cis*-acting elements, some of which are analogous to those found in promoters of ABA-regulated plant genes. Note, in this context, that developing embryos from the viviparous *vip-1* pea mutant, which contain reduced ABA levels during mid-development compared with wild-type embryos, failed to accumulate *sbp* mRNA and the corresponding biotinylated polypeptide (44). Furthermore, mature seeds from the ABA-deficient "wilty" pea mutant, which accumulate five times less ABA than wild-type seeds during development (42), have a reduced SBP65 content, corresponding to only half the amount present in the wild-type seeds (M Duval & D Job, unpublished results). These results strongly support the hypothesis that ABA is involved in *sbp* gene expression during in vivo embryo development. However, in contrast to other ABA-regulated genes [e.g. some *lea* genes (49) and the *Em* gene (114)], ABA was unable to induce *sbp* mRNA accumulation at stages other than embryo maturation, e.g. during germination of mature seeds and seedling growth. Furthermore, during severe desiccation of pea leaves when ABA content increases 30-fold (20), the *sbp* mRNA level remained undetectable in either the control or stressed leaves. Thus, in addition to ABA, expression of the *sbp* gene is dependent on some tissue-specific factors.

There are indications that a specific biotin ligase is involved in biotinylation of the apo-SBP65. Attempts to express the holo form of recombinant SBP65 in bacteria transformed with the *sbp* cDNA have been unsuccessful; evidence of the apoprotein form was found only in the transformed bacteria. This means that the bacterial biotin ligase, despite showing wide reactivity toward naturally occurring apoprotein substrates from eukaryotic cells (101) and synthetic peptides (144), does not accept apo-SBP65 as a substrate (44, 50). Similar results have been obtained for the seed-specific biotinylated protein from soybean (86).

SBP65 might function in at least two different roles. A possible role for SBP65 in the well-characterized storage systems (e.g. proteins, starch, triglycerides) that are crucial for the long-term survival of plant species would be for the biotin bound to this protein to be used by the embryo during germination and initial seedling growth. Although the resources in a seed and the requirements for germination may differ considerably in a natural environment from those in plants grown in the laboratory, some data do not fully agree with this hypothesis. For example, immature pea embryos cultured in a medium lacking biotin can germinate precociously and regenerate seedlings even though they contained undetectable amounts of *sbp* mRNA and holoprotein. Also, *vip-1* mutant pea seeds, which are capable of germinating precociously on the mother plant, contain normal levels of free biotin and are viable if removed from the pods and planted (44). Finally, for a protein-bound biotin storage system to be efficient implies the participation of biotinidase, the only enzyme known to cleave the covalent bond between Lys and biotin in biocytin (38). However, the existence of this enzyme in the plant kingdom has not yet been reported. Thus, these results implicate other mechanisms in which this protein somehow enables the developing embryonic cells to shift from a metabolically active state to a resting state. One possibility would be that during late seed development this protein traps the essential cofactor biotin in an inactive form. Such a role would be detrimental to cells requiring biotin for basic metabolism, especially for young tissues where the demand for biotin seems to be higher because of the increased anabolism associated with cell division and expansion (51).

CONCLUSIONS

Over the past several years, our understanding of the enzymes that impact upon biotin metabolism in plants has advanced dramatically, including the characterization of biotin-containing carboxylases, biotin synthesizing enzymes, biotin ligases, and novel biotinylated proteins distinct from the well-characterized carboxylases. Most of these proteins have been, or soon will be, purified and/or their genes cloned, which should further our understanding of the regulation and interconnection between these different pathways. This knowledge will allow for more rational and directed efforts at their manipulation by genetic engineering. Indeed, some of the processes involving biotin generate biochemicals that serve a broad range of nutritional and industrial purposes. For example, plant storage oils, the biosynthesis of which requires ACCase, are a major resource for both human and animal nutrition, as well as nonfood uses including pharmaceuticals, cosmetics, detergents, and even fuels. We anticipate that the future elucidation of the structure of these enzymes will allow new inhibitor families to be designed with herbicidal activities that affect plants in a specific manner and therefore with reduced environmental impact.

ACKNOWLEDGMENT

This article is dedicated to Professor Paul Stumpf, tireless champion of lipid metabolism in plants.

Visit the Annual Reviews home page at www.AnnualReviews.org

LITERATURE CITED

1. Abu-Elheiga L, Jayakumar A, Baldini A, Chirala SS, Wakil SJ. 1995. Human acetyl-CoA carboxylase: characterization, molecular cloning, and evidence for two isoforms. *Proc. Natl. Acad. Sci. USA* 92:4011–15
2. Alban C, Baldet P, Axiotis S, Douce R. 1993. Purification and characterization of 3-methylcrotonoyl-coenzyme A carboxylase from higher plant mitochondria. *Plant Physiol.* 102:957–65
3. Alban C, Baldet P, Douce R. 1994. Localization and characterization of two structurally different forms of acetyl-CoA carboxylase in young pea leaves, of which one is sensitive to aryloxyphenoxypropionate herbicides. *Biochem. J.* 300:557–65
4. Alban C, Jullien J, Job D, Douce R. 1995. Isolation and characterization of biotin carboxylase from pea chloroplasts. *Plant Physiol.* 109:927–35
5. Al-Feel W, Chirala SS, Wakil SJ. 1992. Cloning of the yeast *FAS3* gene and primary structure of yeast acetyl-CoA carboxylase. *Proc. Natl. Acad Sci. USA* 89:4534–38
6. Alix JH. 1989. A rapid procedure for cloning genes from lambda libraries by complementation of *E. coli* defective mutants. Application to the FabE region of the *E. coli* chromosome. *DNA* 8:779–89
7. Anderson MD, Che P, Song J, Nikolau BJ, Wurtele ES. 1998. 3-Methylcrotonyl-coenzyme A carboxylase is a component of the mitochondrial leucine catabolic pathway in plants. *Plant Physiol.* 118:1127–38
8. Apitz-Castro R, Rehn K, Lynen F. 1970. β-Methylcrotonyl-CoA carboxylase. Kristallisation und einige physikalische Eigenschaften. *Eur. J. Biochem.* 16:71–79
9. Ashton AR, Jenkins CLD, Whitfeld PR. 1994. Molecular cloning of two different cDNAs for maize acetyl-CoA carboxylase. *Plant Mol. Biol.* 24:35–49
10. Aubert S, Alban C, Bligny R, Douce R. 1996. Induction of β-methylcrotonyl-coenzyme A carboxylase in higher plant cells during carbohydrate starvation: evidence for the role of MCCase in leucine catabolism. *FEBS Lett.* 383:175–80
11. Aubert S, Gout E, Bligny R, Marty-Mazars D, Barrieu F, et al. 1996. Ultrastructural and biochemical characterization of autophagy in higher plant cells subjected to carbon deprivation: control by the supply of mitochondria with respiratory substrates. *J. Cell Biol.* 133:1251–63
12. Bailey A, Keon J, Owen J, Hargreaves J. 1995. The *ACC1* gene, encoding acetyl-CoA carboxylase, is essential for growth in *Ustilago maydis*. *Mol. Gen. Genet.* 249:191–201
13. Baldet P. 1993. *Quelques observations sur le métabolisme de la biotine chez les plantes supérieures.* PhD thesis, Univ. Grenoble, France
14. Baldet P, Alban C, Axiotis S, Douce R. 1992. Characterization of biotin and 3-methylcrotonyl-Coenzyme A carboxylase in higher plant mitochondria. *Plant Physiol.* 99:450–55
15. Baldet P, Alban C, Axiotis S, Douce R. 1993. Localization of free and bound biotin in cells from green pea leaves. *Arch. Biochem. Biophys.* 303:67–73
16. Baldet P, Alban C, Douce R. 1997. Biotin synthesis in higher plants: purification of *bio* B gene product equivalent from

Arabidopsis thaliana overexpressed in *Escherichia coli* and its subcellular localization in pea leaf cells. *FEBS Lett.* 419: 206–10

17. Baldet P, Gerbling H, Axiotis S, Douce R. 1993. Biotin biosynthesis in higher plant cells. Identification of intermediates. *Eur. J. Biochem.* 217:479–85
18. Baldet P, Ruffet ML. 1996. Biotin synthesis in higher plants: isolation of a cDNA encoding *Arabidopsis thaliana bioB*-gene product equivalent by functional complementation of a biotin auxotroph mutant *bioB105* of *Escherichia coli* K12. *C. R. Acad. Sci. Paris* 309:99–106
19. Bao X, Shorrosh BS, Ohlrogge JB. 1997. Isolation and characterization of an *Arabidopsis* biotin carboxylase gene and its promoter. *Plant Mol. Biol.* 35:539–50
20. Barratt DHP, Clark JA. 1991. Proteins arising during the late stages of embryogenesis in *Pisum sativum* L. *Planta* 184:14–23
21. Beckett D, Kovaleva E, Schatz PJ. 1999. A minimal peptide substrate in biotin holoenzyme synthetase-catalyzed biotinylation. *Protein Sci.* 8:921–29
22. Berg A, de Kok A. 1997. 2-Oxo acid dehydrogenase multienzyme complexes. The central role of the lipoyl domain. *Biol. Chem.* 378:617–34
23. Bettey M, Ireland RJ, Smith AM. 1992. Purification and characterization of acetyl-CoA carboxylase from developing pea embryos. *J. Plant Physiol.* 140:513–20
24. Birch OM, Fuhrmann M, Shaw NM. 1995. Biotin synthase from *Escherichia coli*, an investigation of the low molecular weight and protein components required for activity *in vitro*. *J. Biol. Chem.* 270:19158–65
25. Bower S, Perkins J, Yocum RR, Serror P, Sorokin A, et al. 1995. Cloning and characterization of the *Bacillus subtilis birA* gene encoding a repressor of the biotin operon. *J. Bacteriol.* 177:2572–75
26. Brocklehurst SM, Perham RN. 1993. Prediction of the three-dimensional structures of the biotinylated domain from yeast pyruvate decarboxylase and of the lipoylated H-protein from the pea leaf glycine cleavage system: a new automated method for the protection of protein tertiary structure. *Protein Sci.* 2:626–39
27. Brouquisse R, James F, Pradet A, Raymond P. 1992. Asparagine metabolism and nitrogen distribution during protein degradation in sugar-starved maize root tips. *Planta* 188:384–95
28. Burton JD, Gronwald JW, Keith RA, Somers DA, Gengenbach BG, Wyse BL. 1991. Kinetics of inhibition of acetyl-coenzyme A carboxylase by Sethoxydim and Haloxyfop. *Pestic. Biochem. Physiol.* 39:100–9
29. Burton JD, Gronwald JW, Somers DA, Connelly JA, Gengenbach BG, Wyse DL. 1987. Inhibition of plant acetyl-CoA carboxylase by the herbicides Sethoxydim and Haloxyfop. *Biochem. Biophys. Res. Commun.* 148:1039–44
30. Chandler CS, Ballard FJ. 1988. Regulation of the breakdown rates of biotin-containing proteins in Swiss 3T3-L1 cells. *Biochem. J.* 251:749–55
31. Charles DJ, Cherry JH. 1986. Purification and characterization of acetyl-CoA carboxylase from developing soybean seeds. *Phytochemistry* 25:1067–71
32. Chen Y, Wurtele ES, Wang X, Nikolau BJ. 1993. Purification and characterization of 3-methylcrotonyl-CoA carboxylase from somatic embryos of *Daucus carota*. *Arch. Biochem. Biophys.* 305:103–9
33. Chiba Y, Suzuki Y, Aoki Y, Ishida Y, Narisawa K. 1994. Purification and properties of bovine liver holocarboxylase synthetase. *Arch. Biochem. Biophys.* 313:8–14
34. Choi JK, Yu F, Wurtele ES, Nikolau BJ. 1995. Molecular cloning and characterization of the cDNA coding for the biotin-containing subunit of the chloroplastic acetyl-CoA carboxylase. *Plant Physiol.* 109:619–25
35. Christopher JT, Holtum JAM. 1998. The

dicotyledonous species *Erodium moschatum* (L) L'Hér. ex Aiton is sensitive to haloxyfop herbicide due to herbicide sensitive acetyl-coenzyme A carboxylase. *Planta* 207:275–79

36. Cohen-Addad C, Faure M, Neuburger M, Ober R, Sieker L, et al. 1997. Structural studies of the glycine decarboxylase complex from pea leaf mitochondria. *Biochimie* 79:637–44
37. Cohen-Addad C, Pares S, Sieker L, Neuburger M, Douce R. 1995. The lipoamide arm in the glycine decarboxylase complex is not freely swinging. *Nature Struct. Biol.* 2:63–68
38. Craft DV, Goss NH, Chandramouli N, Wood HG. 1985. Purification of biotinidase from human plasma and its activity on biotinyl peptides. *Biochemistry* 24:2471–76
39. Cronan JE Jr. 1989. The *E. coli bio* operon: transcriptional repression by an essential protein modification enzyme. *Cell* 58:427–29
40. Cronan JE Jr, Wallace JC. 1995. The gene encoding the biotin-apoprotein ligase of *Saccharomyces cerevisiae. FEMS Microbiol. Lett.* 130:221–30
41. Dakshinamurti K, Cauhan J. 1989. Biotin. *Vitam. Horm.* 45:337–84
42. de Bruijn SM, Buddendorf CJJ, Vreugdenhil D. 1993. Characterization of the ABA-deficient *Pisum sativum* 'wilty' mutant. *Acta Bot. Neerl.* 42:491–103
43. Dehaye L, Alban C, Job C, Douce R, Job D. 1994. Kinetics of the two forms of acetyl-CoA carboxylase from *Pisum sativum.* Correlation of the substrate specificity of the enzymes and sensitivity towards aryloxyphenoxypropionate herbicides. *Eur. J. Biochem.* 225:1113–23
44. Dehaye L, Duval M, Viguier D, Yaxley J, Job D. 1997. Cloning and expression of the pea gene encoding SBP65, a seed-specific biotinylated protein. *Plant Mol. Biol.* 35:605–21
45. Devine MD, Shimabukuro RH. 1994. Resistance to acetyl coenzyme A carboxylase inhibiting herbicides. In *Herbicide Resistance in Plants: Biology and Biochemistry,* ed. SB Powles, JAM Holtum, pp. 141–69. Boca Raton, FL: Lewis
46. Diez TA, Wurtele ES, Nikolau BJ. 1994. Purification and characterization of 3-methylcrotonyl-coenzyme A carboxylase from leaves of *Zea mays. Arch. Biochem. Biophys.* 310:64–75
47. Dimroth P. 1985. Biotin-dependent decarboxylases as energy transducing systems. *Ann. NY Acad. Sci.* 447:72–85
48. Douce R, Neuburger M. 1989. The uniqueness of plant mitochondria. *Annu. Rev. Plant Physiol. Plant Mol. Biol.* 40:371–414
49. Dure L III. 1993. The LEA proteins of higher plants. See Ref. 166a, pp. 325–35
50. Duval M, DeRose RT, Job C, Faucher D, Douce R, Job D. 1994. The major biotinyl protein from *Pisum sativum* seeds covalently binds biotin at a novel site. *Plant Mol. Biol.* 26:265–73
51. Duval M, Job C, Alban C, Douce R, Job D. 1994. Developmental patterns of free and protein-bound biotin during maturation and germination of seeds of *Pisum sativum.* Characterization of a novel seed-specific biotinylated protein. *Biochem. J.* 299:141–50
52. Duval M, Loiseau J, Dehaye L, Pépin R, LeDeunff Y, et al. 1996. SBP65, a seed-specific biotinylated protein behaves as a LEA protein in developing pea embryos. *C. R. Acad. Sci. Paris* 319:585–94
53. Duval M, Pépin R, Job C, Derpierre C, Douce R, Job D. 1995. Ultrastructural localization of the major biotinylated protein from *Pisum sativum* seeds. *J. Exp. Bot.* 46:1783–86
54. Egin-Bühler B, Ebel J. 1983. Improved purification and further characterization of acetyl-CoA carboxylase from cultured cells of parsley (*Petroselinum hortense*). *Eur. J. Biochem.* 133:335–39
55. Egin-Bühler B, Loyal R, Ebel J. 1980. Comparison of acetyl-CoA carboxylases

from parsley cell cultures and wheat germ. *Arch. Biochem. Biophys.* 203:90–100

56. Egli MA, Gengenbach BG, Gronwald JW, Somers DA, Wyse DL. 1993. Characterization of maize acetyl-CoA carboxylase. *Plant Physiol.* 101:499–506
57. Eisenberg MA. 1987. Biosynthesis of biotin and lipoic acid. In *Escherichia coli and Salmonella typhimurium*. In *Cellular and Molecular Biology*, ed. FC Neidhardt, JL Ingraham, KB Low, B Magasanik, M Schaechter, ME Umbarger, pp. 544–50. New York: Am. Soc. Microbiol.
58. Eisenberg MA, Prakash O, Hsiung S-C. 1982. Purification and properties of the biotin repressor. A bifunctional protein. *J. Biol. Chem.* 257:15167–73
59. Elborough KM, Winz R, Deka RK, Markham JE, White AJ, et al. 1996. Biotin carboxyl carrier protein and carboxyltransferase subunit form of acetyl-CoA carboxylase from *Brassica napus*: cloning and analysis of expression during oilseed rape embryogenesis. *Biochem. J.* 315:103–12
60. Fall RR 1979. Analysis of microbial biotin proteins. *Methods Enzymol.* 62:390–98
61. Fall RR, Hector ML. 1977. Acyl-coenzyme A carboxylases. Homologous 3-methylcrotonyl-CoA and geranyl-CoA carboxylases from *Pseudomonas citronellolis*. *Biochemistry* 16:4000–5
62. Finnie MDA, Cottrall K, Seakins JWT, Snedden W. 1976. Massive excretion of 2 oxoglutaric acid and 3 hydroxy iso valeric-acid in a patient with a deficiency of 3-methylcrotonyl coenzyme A carboxylase. *Clin. Chim. Acta* 73:513–19
63. França Neto JB, Shatters RG, West SH. 1998. Developmental pattern of biotinylated proteins during embryogenesis and maturation of soybean seed. *Seed Sci. Res.* 7:377–84
64. Galau GA, Hughes DW, Dure L III. 1986. Abscisic acid induction of cloned late embryogenesis-abundant (*Lea*) mRNAs. *Plant Mol. Biol.* 7:155–70
65. Genix P, Bligny R, Martin JB, Douce R. 1990. Transient accumulation of asparagine in sycamore cells after a long period of sucrose starvation. *Plant Physiol.* 94:717–22
66. Gerbling H, Gerhardt B. 1989. Peroxisomal degradation of branched-chain 2–oxo acids. *Plant Physiol.* 91:1387–92
67. Gloecker R, Ohsawa I, Speck D, Ledoux C, Bernard S, et al. 1990. Cloning and characterization of the *Bacillus sphaericus* genes controlling the bioconversion of pimelate into dethiobiotin. *Gene* 87:63–70
68. Gomez J, Sanchez-Martinez D, Stiefel V, Rigau J, Pages PM. 1988. A gene induced by the plant hormone abscisic acid in response to water stress encodes a glycine-rich protein. *Nature* 334:262–64
69. Gompertz D, Goodey PA, Barlett K. 1973. Evidence for the enzymatic defect in beta-methylcrotonylglycinuria. *FEBS Lett.* 32:13–14
70. Gornicki P, Haselkorn R. 1993. Wheat acetyl-CoA carboxylase. *Plant Mol. Biol.* 22:547–52
71. Guan X, Diez T, Prasad TK, Nikolau BJ, Wurtele ES. 1999. Geranoyl-CoA carboxylase: a novel biotin-containing enzyme in plants. *Arch. Biochem. Biophys.* 362:12–21
72. Guchhait RB, Polakis SE, Dimroth P, Stoll E, Moss J, Lane MD. 1974. Acetyl coenzyme A carboxylase system of *Escherichia coli*. Purification and properties of the biotin carboxylase, carboxyltransferase, and carboxyl carrier protein components. *J. Biol. Chem.* 249:6633–45
73. Gueguen V, Macherel D, Neuburger M, Saint Pierre C, Jaquinod M, et al. 1999. Structural and functional characterization of H protein of the glycine decarboxylase complex. *J. Biol. Chem.* 274:26344–52
74. Guianvarc'h D, Florentin D, Tse Sum Bui B, Nunzi F, Marquet A. 1997. Biotin synthase, a new member of the family of enzymes which use *S*-adenosylmethionine as a source of deoxyadenosyl radical. *Biochem. Biophys. Res. Commun.* 236:402–6

75. Harwood JL. 1988. Fatty acid metabolism. *Annu. Rev. Plant Physiol. Plant Mol. Biol.* 39:101–38
76. Harwood JL. 1989. The properties and importance of acetyl-coenzyme A carboxylase in plants. In *Weeds, Brighton Crop Prot. Conf.*, pp. 155–62. Br. Crop Prot. Counc., Farnham, UK
77. Harwood JL. 1991. Lipid synthesis. In *Target Sites for Herbicide Action*, ed. RC Kirkwood, pp. 57–94. New York: Plenum
78. Harwood JL. 1996. Recent advances in the biosynthesis of plant fatty acids. *Biochim. Biophys. Acta* 1301:7–56
79. Hatch MD, Stumpf PK. 1961. Fat metabolism in higher plants. XVI. Acetyl coenzyme A carboxylase and acyl coenzyme A-malonyl coenzyme A transcarboxylase from wheat germ. *J. Biol. Chem.* 236:2879–85
80. Hector ML, Cochran BC, Logue EA, Fall RR. 1980. Subcellular localization of 3-methylcrotonyl-coenzyme A carboxylase in bovine kidney. *Arch. Biochem. Biophys.* 199:28–36
81. Hector ML, Fall RR. 1976. Evidence for distinct 3-methylcrotonyl-CoA and geranyl-CoA carboxylases in *Pseudomonas citronellolis*. *Biochem. Biophys. Res. Commun.* 71:746–53
82. Herbert D, Alban C, Cole DJ, Pallett KE, Harwood JL. 1994. Characteristics of two forms of acetyl-CoA carboxylase from maize leaves. *Biochem. Soc. Trans.* 22:261
83. Herbert D, Cole DJ, Pallett KE, Harwood JL. 1996. Susceptibilities of different test systems from maize (*Zea mays*), *Poa annua*, and *Festuca rubra* to herbicides that inhibit the enzyme acetyl-coenzyme A carboxylase. *Pestic. Biochem. Physiol.* 55:129–39
84. Herbert D, Price LJ, Alban C, Dehaye L, Job D, et al. 1996. Kinetic studies on two isoforms of acetyl-CoA carboxylase from maize leaves. *Biochem. J.* 318:997–1006
85. Howard PK, Shaw J, Otsuka AJ. 1985. Nucleotide sequence of the *birA* gene encoding the biotin operon repressor and biotin holoenzyme synthetase of *Escherichia coli*. *Gene* 35:321–31
86. Hsing YC, Tsou CH, Hsu TF, Chen ZY, Hsieh KL, et al. 1998. Tissue- and stage-specific expression of a soybean (*Glycine max* L.) seed-maturation, biotinylated protein. *Plant Mol. Biol.* 38:481–90
87. Hunter SC, Ohlrogge JB. 1998. Regulation of spinach chloroplast acetyl-CoA carboxylase. *Arch. Biochem. Biophys.* 359:170–78
88. Hymes J, Fleischhauer K, Wolf B. 1995. Biotinylation of histones by human serum biotinidase: assessment of biotinyl-transferase activity in sera from normal individuals and children with biotinidase deficiency. *Biochem. Mol. Med.* 56:76–83
89. Ifuku O, Koga N, Haze S, Kishimoto J, Wachi Y. 1994. Flavodoxin is required for conversion of dethiobiotin to biotin in *Escherichia coli*. *Eur. J. Biochem.* 224:173–78
90. Kannangara CG, Jensen CJ. 1975. Biotin carboxyl carrier protein in barley chloroplast membranes. *Eur. J. Biochem.* 54:25–30
91. Kannangara CG, Stumpf PK 1972. Fat metabolism in higher plants. A prokaryotic type acetyl-CoA carboxylase in spinach chloroplasts. *Arch. Biochem. Biophys.* 152:83–91
92. Kim KH, Lopez-Casillas F, Bai DH, Luo X, Pape ME. 1989. Role of reversible phosphorylation of acetyl-CoA carboxylase in long-chain fatty acid synthesis. *FASEB J.* 89:2250–56
93. Knowles JR. 1989. The mechanism of biotin-dependent enzymes. *Annu. Rev. Biochem.* 58:195–21
94. Kohansky RA, Lane DL. 1985. Homogeneous functional insulin receptor from 3T3–L1 adipocytes. Purification using NαB1-(biotinyl-ε-aminocaproyl) insulin and avidin-Sepharose. *J. Biol. Chem.* 260:5014–25

95. Kondo H, Shiratsuchi K, Yoshimoto T, Masuda T, Kitazono A, et al. 1991. Acetyl-CoA carboxylase from *Escherichia coli*: gene organization and nucleotide sequence of the biotin carboxylase subunit. *Proc. Natl. Acad. Sci. USA* 88: 9730–33
96. Kondo H, Uno S, Komizo Y, Sunamoto J. 1984. Importance of methionine residues in the enzymatic carboxylation of biotin-containing peptides representing the local biotinyl site of *E. coli* acetyl-CoA carboxylase. *Int. Pept. Protein Res.* 23:559–64
97. Konishi T, Kamoi T, Matsuno R, Sasaki Y. 1996. Induction of cytosolic acetyl-coenzyme A carboxylase in pea leaves by ultraviolet-B irradiation. *Plant Cell Physiol.* 37:1197–200
98. Konishi T, Sasaki Y. 1994. Compartmentalization of two forms of acetyl-CoA carboxylase in plants and the origin of their tolerance toward herbicides. *Proc. Natl. Acad. Sci. USA* 91:3598–601
99. Konishi T, Shinohara K, Yamada K, Sasaki Y. 1996. Acetyl-CoA carboxylase in higher plants: most plants other than Gramineae have both the prokaryotic and the eukaryotic forms of this enzyme. *Plant Cell Physiol.* 37:117–22
100. Lau EP, Cochran BC, Fall RR. 1980. Isolation of 3-methylcrotonyl-coenzyme A carboxylase from bovine kidney. *Arch. Biochem. Biophys.* 205:352–59
101. León-Del-Rio A, Gravel RA. 1994. Sequence requirements for the biotinylation of carboxyl-terminal fragments of human propionyl-CoA carboxylase α subunit expressed in *Escherichia coli*. *J. Biol. Chem.* 269:22964–68
102. León-Del-Rio A, Leclerc D, Akerman B, Wakamatsu N, Gravel RA. 1995. Isolation of a cDNA encoding human holocarboxylase synthetase by functional complementation of a biotin auxotroph of *Escherichia coli*. *Proc. Natl. Acad. Sci. USA* 92:4626–30
103. Li S, Cronan JE. 1992. The gene encoding the biotin carboxylase subunit of *Escherichia coli* acetyl-CoA carboxylase. *J. Biol. Chem.* 267:855–63
104. Li S, Cronan JE. 1992. The genes encoding the two carboxyltransferase subunits of *E. coli* acetyl-CoA carboxylase. *J. Biol. Chem.* 267:16841–47
105. Li S, Cronan JE. 1992. Putative zinc finger protein encoded by a conserved chloroplast gene is very likely a subunit of a biotin-dependent carboxylase. *Plant Mol. Biol.* 20:759–61
106. Lichtenthaler HK, Focke M, Golz A, Hoffman S, Kobek K, Motel A. 1992. Investigation of the starting enzymes of plant fatty acid biosynthesis and particular inhibitors. In *Metabolism Structure and Utilization of Plant Lipids. Proc. Int. Symp. Plant Lipids, 10th*, ed. A Sherif, DB Miled-Daoul, B Marzouk, A Smaoui, M Zarrouk, pp. 103–12. Tunisia: Cent. Natl. Pédagog.
107. Lim F, Rohde M, Morris CP, Wallace JC. 1987. Pyruvate carboxylase in the yeast *pyc* mutant. *Arch. Biochem. Biophys.* 258:259–64
108. Lopez-Casillas F, Bai DH, Luo X, Kong IS, Hermodson MA, Kim KH. 1988. Structure of the coding sequence and primary amino acid sequence of acetyl-CoA carboxylase. *Proc. Natl. Acad. Sci. USA* 85:5784–88
109. Maneechote C, Holtum JAM, Preston C, Powles SB. 1994. Resistant acetyl-CoA carboxylase is a mechanism of herbicide resistance in a biotype of *Avena sterilis* ssp. *ludoviciana*. *Plant Cell Physiol.* 35:627–35
110. Marquet A, Frappier F, Guillerm G, Azoulay M, Florentin D, Tabet JC. 1993. Biotin biosynthesis: synthesis and biological evaluation of the putative intermediate thiols. *J. Am. Chem. Soc.* 115:2139–45
111. Marshall LC, Somers DA, Dotray PD, Gengenbach BG, Wyse DL, Gronwald

JW. 1992. Allelic mutations in acetyl-coenzyme A carboxylase confer herbicide tolerance in maize. *Theor. Appl. Genet.* 83:435–42

112. Mattevi A, de Kok A, Perham RN. 1992. The pyruvate dehydrogenase complex. *Curr. Opin. Struct. Biol.* 2:277–87
113. Mohan SB, Kekwick RGO. 1980. Acetyl-coenzyme A carboxylase from avocado (*Persea americana*) plastids and spinach (*Spinacia oleracea*) chloroplasts. *Biochem. J.* 187:667–76
114. Morris PC, Kumar A, Bowles DJ, Cuming AC. 1990. Osmotic stress and abscisic acid induce expression of the wheat *Em* genes. *Eur. J. Biochem.* 190: 625–30
115. Moss J, Lane MD. 1971. The biotin-dependent enzymes. *Adv. Enzymol.* 35: 321–42
116. Mundy J, Chua N-H. 1988. Abscisic acid and water stress induce the expression of a novel rice gene. *EMBO J.* 7:2279–86
117. Nes WD, Bach TJ. 1985. Evidence for a mevalonate shunt in a tracheophyte. *Proc. R. Soc. London Ser. B* 225:425–44
118. Ohlrogge JB, Browse J. 1995. Lipid biosynthesis. *Plant Cell* 7:957–70
119. Ohlrogge JB, Kuhn DN, Stumpf PK. 1979. Subcellular localization of acyl carrier protein in leaf protoplasts of *Spinacia oleracea*. *Proc. Natl. Acad. Sci. USA* 76:1194–98
120. Oliver DJ. 1994. The glycine decarboxylase complex from plant mitochondria. *Annu. Rev. Plant Physiol. Plant Mol. Biol.* 45:323–37
121. Page RA, Okada S, Harwood JL. 1994. Acetyl-CoA carboxylase exerts strong flux control over lipid synthesis in plants. *Biochim. Biophys. Acta* 1210:369–72
122. Pares S, Cohen-Addad C, Sieker L, Neuburger M, Douce R. 1994. X-ray structure determination at 2.6Å resolution of a lipoate-containing protein: the H-protein of the glycine decarboxylase complex from pea leaves. *Proc. Natl. Acad. Sci. USA* 91:4850–53
123. Parker WB, Marshall LC, Burton JD, Somers DA, Wyse DL, et al. 1990. Dominant mutations causing alterations in acetyl-coenzyme A carboxylase confer tolerance to cyclohexanedione and aryloxyphenoxypropionate herbicides in maize. *Proc. Natl. Acad. Sci. USA* 87:7175–79
124. Patton DA, Johnson M, Ward ER. 1996. Biotin synthase from *Arabidopsis thaliana*. cDNA isolation and characterization of gene expression. *Plant Physiol.* 112:371–78
125. Patton DA, Schetter AL, Franzmann LH, Nelson K, Ward ER, Meinke DW. 1998. An embryo-defective mutant of *Arabidopsis* disrupted in the final step of biotin synthesis. *Plant Physiol.* 116:935–46
126. Patton DA, Volrath S, Ward ER. 1996. Complementation of an *Arabidopsis thaliana* biotin auxotroph with an *Escherichia coli* biotin biosynthetic gene. *Mol. Gen. Genet.* 251:261–66
127. Popjak G. 1971. Specificity of enzymes of sterol biosynthesis. *Harvey Lect.* 65:127–56
128. Post-Beittenmiller D, Jaworski JG, Ohlrogge JB. 1991. *In vivo* pools of free and acylated acyl carrier proteins in spinach: evidence for sites of regulation of fatty acid biosynthesis. *J. Biol. Chem.* 266:1858–65
129. Post-Beittenmiller D, Roughan G, Ohlrogge JB. 1992. Regulation of plant fatty acid biosynthesis: analysis of acyl-CoA and acyl-ACP substrate pools in spinach and pea chloroplasts. *Plant Physiol.* 100:923–30
130. Quatrano RS, Guiltinan MJ, Marcotte WR. 1993. Regulation of gene expression by abscisic acid. See Ref. 166a, pp. 69–90
131. Rendina AR, Craig-Kennard AC, Beaudoin JD, Breen MK. 1990. Inhibition of acetyl-coenzyme A carboxylase by two classes of grass-selective herbicides. *J. Agric. Food Chem.* 38:1282–87
132. Reverdatto S, Beilinson V, Nielsen N.

1999. A multisubunit acetyl coenzyme A carboxylase from soybean. *Plant Physiol.* 119:961–78

133. Roberts EL, Shu N, Howard MJ, Broadhurst RW, Chapman-Smith A, et al. 1999. Solution structures of apo and holo biotinyl domains from acetyl coenzyme A carboxylase of *Escherichia coli* determined by triple-resonance nuclear magnetic resonance spectroscopy. *Biochemistry* 38:5045–53
134. Robinson BH, Oei J, Saunders M, Gravel R. 1983. ^{3}H-biotin-labeled proteins in cultured human skin fibroblasts from patients with pyruvate carboxylase deficiency. *J. Biol. Chem.* 258:6660–64
135. Roesler KR, Savage LJ, Shintani DK, Shorrosh BS, Ohlrogge JB. 1996. Co-purification, co-immuno-precipitation, and coordinate expression of acetyl-coenzyme A carboxylase activity, biotin carboxylase, and biotin carboxyl carrier protein of higher plants. *Planta* 198: 517–25
136. Roesler K, Shintani D, Savage L, Boddupalli S, Ohlrogge JB. 1997. Targeting of the *Arabidopsis* homomeric acetyl-CoA carboxylase to plastid rapeseeds. *Plant Physiol.* 113:75–81
137. Roughan PG, Ohlrogge JB. 1996. Evidence that isolated chloroplasts contain an integrated lipid-synthesizing assembly that channels acetate into long-chain fatty acids. *Plant Physiol.* 110:1239–47
138. Samols D, Thornton CG, Murtif VL, Kumar GK, Hasse FC, Wood HG. 1988. Evolutionary conservation among biotin enzymes. *J. Biol. Chem.* 263:6461–64
139. Sanyal I, Cohen G, Flint DH. 1994. Biotin synthase: purification, characterization as a [2Fe-2S] cluster protein, and *in vitro* activity of the *Escherichia coli bioB* gene product. *Biochemistry* 33: 3625–31
140. Sasaki Y, Hakamada K, Suama Y, Nagano Y, Furusawa I, Matsuno R. 1993. Chloroplast-encoded protein as a subunit of acetyl-CoA carboxylase in pea plant. *J. Biol. Chem.* 268:25118–23
141. Sasaki Y, Konishi T, Nagano Y. 1995. The compartmentation of acetyl-coenzyme A carboxylase in plants. *Plant Physiol.* 108:445–49
142. Sasaki Y, Kozaki A, Hatano M. 1997. Link between light and fatty acid synthesis: thioredoxin-linked reductive activation of plastidic acetyl-CoA carboxylase. *Proc. Natl. Acad. Sci. USA* 94:11096–101
143. Savage LJ, Ohlrogge JB. 1999. Phosphorylation of pea chloroplast acetyl-CoA carboxylase. *Plant J.* 18:521–27
144. Schatz PJ. 1993. Use of peptide libraries to map the substrate specificity of a peptide-modifying enzyme: a 13 residue consensus peptide specifies biotinylation in *Escherichia coli. BioTechnol.* 11:1138–43
145. Schiele U, Lynen F. 1981 3-Methylcrotonyl-CoA carboxylase from *Achromobacter. Methods Enzymol.* 71:781–91
146. Schiele U, Niedermeier R, Stürzer M, Lynen F. 1975. Investigations of structure of 3-methylcrotonyl-CoA carboxylase from *Achromobacter. Eur. J. Biochem.* 60:259–66
147. Schneider T, Dinkins R, Robinson K, Shellhammer J, Meinke DW. 1989. An embryo-lethal mutant of *Arabidopsis thaliana* is a biotin auxotroph. *Dev. Biol.* 131:161–67
148. Schulte W, Töpfer R, Stracke R, Schell J, Martini N. 1997. Multi-functional acetyl-CoA carboxylase from *Brassica napus* is encoded by a multi-gene family: indication for plastidic localization of at least one isoform. *Proc. Natl. Acad. Sci. USA* 94:3465–70
149. Seubert W, Remberger U. 1963. Untersuchungen uber den bakteriellen Abbau von Isoprenoiden. II. Die Rolle der Kohlensaure. *Biochem. Z.* 338:245–63
150. Shatters RG, Boo SP, França Neto JB, West SH. 1997. Identification of biotinylated proteins in soybean [*Glycine max*

(L.) Merrill] seeds and their characterization during germination and seedling growth. *Seed Sci. Res.* 7:373–76

151. Shellhammer J, Meinke D. 1990. Arrested embryos from the *bio1* auxotroph of *Arabidopsis thaliana* contain reduced levels of biotin. *Plant Physiol.* 93:1162–67
152. Shenoy BC, Paranjape S, Murtif VL, Kumar GK, Samols D, Wood HG. 1988. Effect of mutation at Met-88 and Met-90 on the biotinylation of Lys-89 of the apo 1.3S subunit of transcarboxylase. *FASEB J.* 2:2505–11
153. Shintani DK, Ohlrogge JB. 1995. Feedback inhibition of fatty acid synthesis in tobacco suspension cells. *Plant J.* 7:577–87
154. Shintani D, Roesler K, Shorrosh B, Savage L, Ohlrogge J. 1997. Antisense expression and overexpression of biotin carboxylase in tobacco leaves. *Plant Physiol.* 114:881–86
155. Shorrosh BS, Dixon RA, Ohlrogge JB. 1994. Molecular cloning, characterization, and elicitation of acetyl-CoA carboxylase from alfalfa. *Proc. Natl. Acad. Sci. USA* 91:4323–27
156. Shorrosh BS, Roesler KR, Shintani D, van de Loo FJ, Ohlrogge JB. 1995. Structural analysis, plastid localization, and expression of the biotin carboxylase subunit of acetyl-coenzyme A carboxylase from tobacco. *Plant Physiol.* 108:805–12
157. Shorrosh BS, Savage LJ, Soll J, Ohlrogge JB. 1996. The pea chloroplast membrane-associated protein, IEP96, is a subunit of acetyl-CoA carboxylase. *Plant J.* 10:261–68
158. Song J, Wurtele ES, Nikolau BJ. 1994. Molecular cloning and characterization of the cDNA coding for the biotin-containing subunit of 3-methylcrotonyl-CoA carboxylase: identification of the biotin carboxylase and biotin carrier domains. *Proc. Natl. Acad. Sci. USA* 91:5779–83
159. Stewart CR, Beevers H. 1967. Gluconeogenesis from amino acids in germinating castor bean endosperm and its role in transport to the embryo. *Plant Physiol.* 42:1587–95
160. Stumpf PK. 1980. Biosynthesis of saturated and unsaturated fatty acids, In *The Biochemistry of Plants*, ed. PK Stumpf, 10:177–204. New York: Academic Press
161. Suzuki Y, Aoki Y, Ishida Y, Chiba Y, Iwamatsu A, et al. 1994. Isolation and characterization of mutations in the human holocarboxylase synthetase cDNA. *Nature Genet.* 8:122–28
162. Tardif FJ, Preston C, Holtum JAM, Powles SB. 1996. Resistance to acetyl-Coenzyme A carboxylase-inhibiting herbicides endowed by a single major gene encoding a resistant target site in a biotype of *Lolium rigidum*. *Aust. J. Plant Physiol.* 23:15–23
163. Tipton PA, Cleland WW. 1988. Catalytic mechanism of biotin carboxylase: steady-state kinetic investigations. *Biochemistry* 27:4317–25
164. Tissot G, Douce R, Alban C. 1997. Evidence for multiple forms of biotin holocarboxylase synthetase in pea (*Pisum sativum*) and in *Arabidopsis thaliana*: subcellular fractionation studies and isolation of a cDNA clone. *Biochem. J.* 323:179–88
165. Tissot G, Job D, Douce R, Alban C. 1996. Protein biotinylation in higher plants: characterization of biotin holocarboxylase synthetase activity from pea (*Pisum sativum*) leaves. *Biochem. J.* 314:391–95
166. Tissot G, Pépin R, Job D, Douce R, Alban C. 1998. Purification and properties of the chloroplastic form of biotin holocarboxylase synthetase from *Arabidopsis thaliana* overexpressed in *Escherichia coli*. *Eur. J. Biochem.* 258:586–96

166a. Verma DPS, ed. 1993. *Control of Plant Gene Expression*. Boca Raton, FL: CRC Press

167. Wada H, Shintani D, Ohlrogge JB. 1997. Why do mitochondria synthesize fatty acids? Evidence for involvement in lipoic acid production. *Proc. Natl. Acad. Sci. USA* 94:1591–96
168. Wang X, Wurtele ES, Keller G, McKean AL, Nikolau BJ. 1994. Molecular cloning of cDNAs and genes coding of ß-methylcrotonyl-CoA carboxylase of tomato. *J. Biol. Chem.* 269:11760–69
169. Wang X, Wurtele ES, Nikolau BJ. 1995. Regulation of β-methylcrotonyl-coenzyme A carboxylase activity by biotinylation of the apoenzyme. *Plant Physiol.* 108:1133–39
170. Weaver LM, Lebrun L, Franklin A, Huang L, Hoffman N, et al. 1994. Molecular cloning of the biotinylated subunit of 3-methylcrotonyl-CoA carboxylase of *Arabidopsis thaliana*. *Plant Physiol.* 107:113–14
171. Weaver LM, Yu F, Wurtele ES, Nikolau BJ. 1996. Characterization of the cDNA and gene coding for the biotin synthase of *Arabidopsis thaliana*. *Plant Physiol.* 110:1021–28
172. Weyler W, Sweetman L, Maggio DC, Nyhan WL. 1977. Deficiency of propionyl coenzyme A carboxylase and methylcrotonyl coenzyme A carboxylase in a patient with methylcrotonylglycinuria. *Clin. Chim. Acta* 76:321–28
173. Wilmot CM, Thornton 1990. β-Turns and their distorsions: a proposed new nomenclature. *Protein Eng.* 3:479–93
174. Wilson KP, Shewchuk LM, Brennan RG, Otsuka AJ, Matthews BW. 1992. *Escherichia coli* biotin holoenzyme synthetase-*bio* repressor crystal structure delineates the biotin and DNA-binding domains. *Proc. Natl. Acad. Sci. USA* 89:9257–61
175. Winkler DA, Liepa AJ, Anderson-McKay JE, Hart NK. 1989. A molecular graphics study of factors influencing herbicidal activity of oximes of 3-acyl-tetrahydro-2*H*-pyran-2,4,-diones. *Pestic. Sci.* 27, 45–63
176. Winz R, Hess D, Aebersold R, Brownsey RW. 1994. Unique structural features and differential phosphorylation of the 280-kDa component (isozyme) of rat liver acetyl-CoA carboxylase. *J. Biol. Chem.*, 269:14438–45
177. Wood HG, Barden RE. 1977. Biotin enzymes. *Annu. Rev. Biochem.* 46:385–13
178. Wood HG, Kumar GK. 1985. Transcarboxylase: its quaternary structure and the role of biotinyl subunit in the assembly of the enzyme and in catalysis. *Ann. NY Acad. Sci.* 447:1–22
179. Wurtele ES, Nikolau BJ. 1990. Plants contains multiple biotin enzymes: discovery of 3-methylcrotonyl-CoA carboxylase, propionyl-CoA carboxylase and pyruvate carboxylase in the plant kingdom. *Arch. Biochem. Biophys.* 278:179–86
180. Wurtele ES, Nikolau BJ. 1992. Differential accumulation of biotin enzymes during carrot somatic embryogenesis. *Plant Physiol.* 99:1699–703
181. Xia W-L, Zhang J, Ahmad F. 1994. Biotin holocarboxylase synthetase: purification from rat liver cytosol and some properties. *Biochem. Mol. Biol. Int.* 34:225–32

Annu. Rev. Plant Physiol. Plant Mol. Biol. 2000. 51:49–81

SUGAR-INDUCED SIGNAL TRANSDUCTION IN PLANTS

Sjef Smeekens
Department of Molecular Plant Physiology, University of Utrecht, Padualaan 8, 3584 CH Utrecht, The Netherlands; e-mail: j.c.m.smeekens@bio.uu.nl

Key Words hexose sensing, sucrose sensing, hexokinase, phytohormones, Arabidopsis, signal transduction

■ **Abstract** Sugars have important signaling functions throughout all stages of the plant's life cycle. This review presents our current understanding of the different mechanisms of sugar sensing and sugar-induced signal transduction, including the experimental approaches used. In plants separate sensing systems are present for hexose and sucrose. Hexokinase-dependent and -independent hexose sensing systems can further be distinguished. There has been progress in understanding the signal transduction cascade by analyzing the function of the SNF1 kinase complex and the regulatory PRL1 protein. The role of sugar signaling in seed development and in seed germination is discussed, especially with respect to the various mechanisms by which sugar signaling controls gene expression. Finally, recent literature on interacting signal transduction cascades is discussed, with particular emphasis on the ethylene and ABA signal transduction pathways.

CONTENTS

1040-2519/00/0601-0049$14.00

INTRODUCTION

Sugars such as sucrose, glucose, and fructose have an essential function in plant metabolism. These sugars are important for intermediary and respiratory metabolism and are the substrate for synthesizing complex carbohydrates such as starch and cellulose. Moreover, sugars provide the building blocks for amino acid and fatty acid biosynthesis and essentially all other compounds present in plants. These metabolic processes have long been studied in depth but another aspect of plant sugar biology has recently become the focus of intense research efforts: the signaling function of sugars. Sugars as such can signal alterations in gene expression similar to the concepts developed for hormones. Whereas hormones are purpose-built molecules that are functional in the nano- to micromolar range, sugars take part in intermediary metabolism and are present in the millimolar range. Sugar sensing can be defined as the interaction between a sugar molecule and a sensor protein in such a way that a signal is generated. This signal then initiates signal transduction cascades that result in cellular responses such as altered gene expression and enzymatic activities. The metabolic and signaling functions of sugars are not always easy to separate but in many cases, convincing evidence for a signaling function has been obtained. Questions addressed in sugar sensing and signaling are similar to those of other signal transduction cascades. These relate to which sugars are being sensed, their interaction with sensor molecules, the molecular nature and cellular location of these sensors, the transduction of the signal and, finally, the way in which gene expression, enzymatic activities, or other cellular processes are altered.

Sugars as signaling compounds have profound effects in all stages of the plant's life cycle from germination and vegetative growth to reproductive development and seed formation. There has been extensive research effort in bacteria, yeast, and animal systems to understand sugar sensing in molecular detail; the yeast glucose repression system in particular has provided a wealth of information (18, 40, 79). This information is important for other eukaryotes as well. Sugar sensing in yeast is by definition a cell-autonomous process, a characterization most likely true for plant cells as well. However, in multicellular plants with specialized metabolic organs and metabolite transport systems, integrative responses to plant sugar status are needed.

Several excellent reviews on sugar sensing have recently appeared (47, 74, 88, 132, 149, 150). The review by Koch (88) presented and discussed an extensive list of sugar-regulated genes. This review aims to provide an update on recent literature and an integrated view of sugar signaling in plants. First, the experimental approaches to studying sugar signaling are introduced, followed by a discussion on the different aspects of known sugar sensing mechanisms and sugar mediated signal transduction. In the final part, sugar signaling and its integration with other plant signaling and developmental pathways are discussed.

APPROACHES TO STUDYING SUGAR SENSING AND SIGNALING

The various strategies for studying sugar signaling can be divided into genetic, molecular, biochemical, and physiological approaches. The creative combination of these methodologies has been productive in many fields and is also used successfully to study sugar-signaling pathways in plants.

Mutants are important tools to analyze the physiological function of complex sensing and signaling systems. Moreover, mutants allow the study of functional interactions between genes. Importantly, in several plant systems, the technology is now available to clone the relevant genes and study their function. Arabidopsis is being used extensively in sugar sensing research for mutant identification, and several laboratories have established different mutant identification protocols (Table 1). Several groups are using reporter-based screening protocols in which promoters of sugar-induced or sugar-repressed genes are linked to reporters like β-glucuronidase (*GUS*, *iudA*) or luciferase (*LUC*) genes. These constructs are introduced into plants and used as tools to select sugar-unresponsive or sugar hyperresponsive mutants. The plastocyanin (*PC*) gene of Arabidopsis can be repressed by sugars (31) and in seedlings carrying a *PC*-promoter luciferase reporter

TABLE 1 Strategies used by different groups to select for sugar sensing mutants in Arabidopsis

		Screen	Reference
Reduced sensitivity			
cai	*carbohydrate insensitive*	Low nitrogen, 100 mM sucrose	(15)
gin	*glucose insensitive*	Growth on 330 mM glucose	(188)
lba	*low-level beta amylase*	Amylase act., 175 mM sucrose	(104)
mig	*mannose insens. germin.*	Growth on 7.5 mM mannose	(119)
ram	*reduced beta amylase*	*pgm* mutant on sucrose	(32)
rsr	*reduced sugar response*	*Pat(B33)-GUS*, 90 mM sucrose	(101)
sis	*sugar insensitive*	300 mM sucrose or glucose	(43)
sun	*sucrose-uncoupled*	*PC-LUC*, 88 mM sucrose	(30)
sig	*sucrose-insens. growth*	Growth on 350 mM sucrose	(119)
Enhanced sensitivity			
gss	*glucose super sensitive*	Growth on 56 mM glucose	(119)
sss	*sucrose super sensitive*	Growth on 350 mM sucrose	(119)
hba	*high-level beta amylase*	Amylase act., 175 mM sucrose	(105)
prl1	*pleiotopic regul. locus*	Growth on 175 mM sucrose	(111)

gene construct, luciferase activity is similarly repressed by sugars. Mutants defective in sucrose repression were identified on the basis of normal luminescence when grown on plates with 3% sucrose (30). Such *sucrose uncoupled* (*sun*) mutants show no or reduced sucrose repression of luminescence. In these mutants, endogenous *PC*, *CAB*, and *RBCS* mRNA levels were similarly insensitive to sugar repression.

A similar strategy was used to select for mutants in sugar induction. The patatin class I (B33) promoter is induced by sugars, and signaling mutants were selected by using transgenic Arabidopsis plants harboring the *Pat* (*B33*)-*iudA* construct and a nondestructive GUS activity assay (101). In this way, *reduced sugar response* (*rsr*) mutants were identified in which sucrose-induced expression of patatin is perturbed. Genetic analysis suggests that one of these mutants, *rsr4*, is codominant (101) and likely encodes an activator, whereas most other sugar sensing mutants that were isolated are recessive and probably encode repressing functions.

The Arabidopsis β-amylase gene is induced by sugars, and mutants that display either an increased or a reduced sugar sensitivity have been isolated in amylase activity screens (32, 104, 105). A mutant was identified showing elevated β-amylase expression (*hba1*, *high level β-amylase*) independent of the presence of sugars in the medium. Conversely, the *low-level beta amylase* (*lba*) mutants show reduced induction of β-amylase gene expression in response to sugars. Remarkably, the Arabidopsis Landsberg *erecta* (L*er*) ecotype represents a natural *lba* mutant (105). A single recessive L*er* locus, named *lba2*, reduced the sugar responsiveness of β-amylase gene expression. Sucrose-induced accumulation of anthocyanins is reduced in both *lba1* and *lba2* mutants. Moreover, the expression of only a subset of sugar-regulated genes is affected in these mutants.

An equally effective mutant isolation strategy relies on the observation that Arabidopsis seedling development is arrested at high (6%) glucose concentrations. Mutant seedlings that develop more or less normally in the presence of 6% glucose have been isolated and are named *glucose insensitive* (*gin*) (188). Other mutants with reduced sugar sensitivity are the *carbohydrate insensitive* (*cai*) mutants. Arabidopsis seedlings grown on a high-sucrose/low-nitrogen medium show enhanced sugar signaling due to increased intracellular sugar concentrations. Wild-type seedlings grown under these conditions accumulate high levels of anthocyanin and are low in chlorophyll. A number of *cai* mutants have been isolated that do not accumulate anthocyanin and that show higher levels of chlorophyll (15).

The glucose epimer mannose is phosphorylated by hexokinase (HXK) to mannose-6-phosphate (Figure 1), which only slowly enters glycolytic metabolism (87). In Arabidopsis mannose and also the glucose analog 2-deoxy glucose (2-dGlc) inhibit seed germination in a process that involves HXK signaling (121). Non-HXK substrates such as 3-O-methyl glucose (3-O-mGlc) and 6-deoxyglucose (6-dGlc) have no effect on germination, whereas the HXK inhibitor mannoheptulose relieves the mannose-induced block of germination. This mannose inhibition of germination can be reversed by the addition of metabolizable sugars.

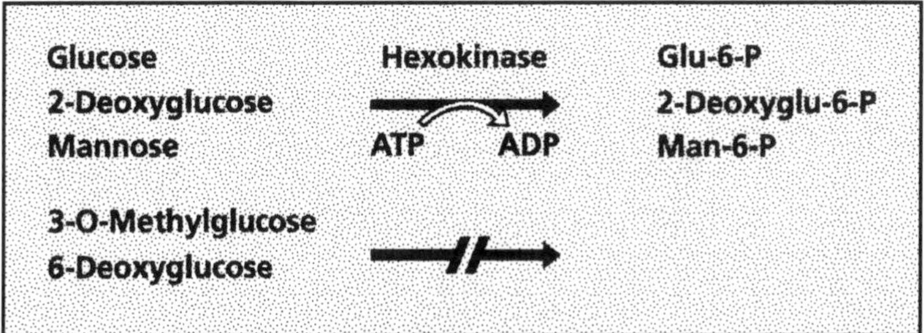

Figure 1 Mannose and 2-deoxy glucose are substrates for hexokinase. The resulting hexose phosphates can to some extent be further metabolized (87). 3-O-methyl glucose and 6-deoxy glucose are not metabolized by hexokinase.

Arabidopsis is an oilseed and during germination the lipids are converted to sucrose in a process that involves the glyoxylate cycle. Thus HXK signaling may inhibit mobilization of lipids and, possibly, other storage compounds as well, and in this way prevent germination (121). This observation was used to isolate *mannose-insensitive-germination* (*mig*) mutants potentially defective in HXK activity or in HXK-induced signaling (119). Not surprisingly, several of the mutants isolated in a particular screen turned out to be mutants under other selective conditions as well, e.g. selected *sun* and *cai* mutants are *mig* and/or *gin* as well (15, 119). Other screening methods have been used (summarized in Table 1). Several of the mutant isolation protocols described allow for screening of T-DNA and transposon-tagged seed collections. This greatly accelerates the isolation of genes affected using PCR techniques, and a number of genes have been identified in this way. Notwithstanding their usefulness, these mutant selection approaches are rather crude and do not address the intricate complexities associated with sugar transport, sugar inter- and intracellular compartmentation, and plant development and differentiation. The challenge is to devise selective mutant identification procedures addressing these points.

In addition to the mutant isolation approach, genes have been cloned from several different plant species that are homologous to genes encoding known components of sugar sensing pathways in microorganisms, especially yeast (*Saccharomyces cerevisiae*). For example, yeast HXK is thought to play a central role as a sugar sensing molecule and plant HXK genes have been cloned and analyzed for their function in sugar sensing as well (25, 73). Also, plant genes have been isolated that encode homologues of the yeast heterotrimeric SNF1 kinase complex involved in derepression of glucose-repressed genes (14, 53, 93). In addition, many pharmacological and other compounds are known that interfere with signal transduction steps, often with reasonable specificity. The effect of such compounds in inhibiting or promoting specific sugar responses is indicative of the involvement of specific intermediary steps in signaling. Thus, protein kinases, protein phosphatases, Ca^{2+}, and calmodulin have been implicated in sugar signaling.

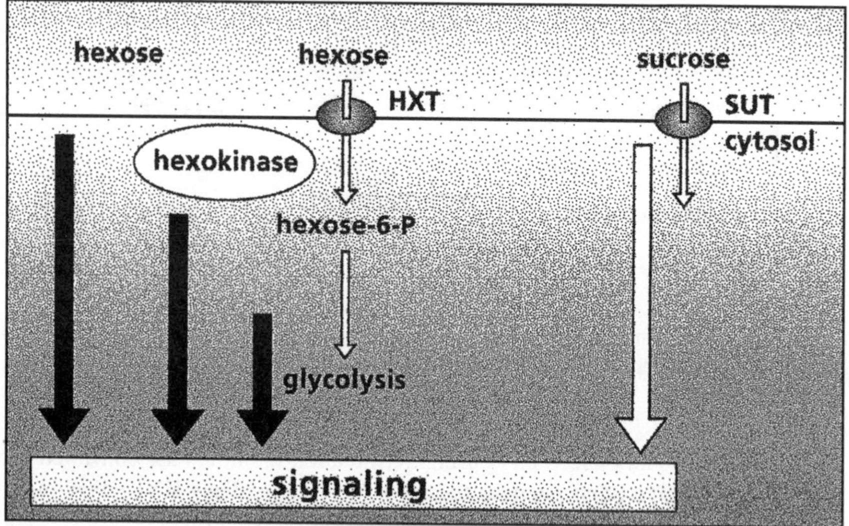

Figure 2 Sugar sensing mechanisms in plants. Hexose sensing can occur via separate hexokinase-independent or hexokinase-dependent systems. Sucrose is sensed via a separate system.

WHICH SUGARS ARE BEING SENSED?

In principle, any neutral sugar or glycolytic intermediate could have a signaling function but so far this has only been shown for hexoses and sucrose (Figure 2). Next to sugars, the cellular energy status must have an important signaling function for the control of metabolism.

Sucrose Sensing

A signaling function for molecular sucrose was proposed in the past but only recently has experimental evidence for this role of sucrose become available. The problem is that sucrose can readily be hydrolyzed in glucose and fructose, and in the absence of specific non-hydrolyzable sucrose analogs, it is difficult to establish a direct function for the sucrose molecule. Sucrose-specific induction of gene expression has been reported for the patatin promoter and the phloem-specific rolC promoter, among others (76, 177, 184). Glucose and fructose were less effective inducers. For these promoters, however, the function of sucrose as inducing agent has not been exhaustively investigated; sucrose is efficiently transported and hydrolyzed in plants and the resulting monosaccharides may be the signal. For example, the glucose analog 3-O-mGlc is an effective inducer of the patatin promoter (101).

Recently, an Arabidopsis basic leucine zipper gene, *ATB2*, was described that is controlled by light and sucrose via transcriptional and translational control, respectively (135, 136). *ATB2* expression is associated with newly established sink tissues. It is rapidly induced in funiculi upon fertilization of the ovules and is expressed in association with vascular tissue in developing leaves. Transcription of the gene is controlled by light through the *DET1* and *COP1* gene products. Remarkably, translation of the *ATB2* mRNA is repressed specifically by sucrose at physiological concentrations. Other sugars and also combinations of glucose and fructose were ineffective in this repression. The *ATB2* mRNA has a complex leader containing small open reading frames. Deletion of this leader abolishes sucrose repression, which shows that a sucrose-specific signal controls translation. Possibly, in this system it is the influx of sucrose into the cell that is being sensed rather than the actual cytosolic sucrose concentration, since Arabidopsis seedlings synthesize sucrose efficiently when glucose or fructose is added (31), although this sucrose apparently is not sensed. A signaling function for molecular sucrose was also suggested by experiments with excised sugar beet leaves (23). In this system, sucrose repressed mRNA levels and transport activity of the proton-sucrose symporter. Glucose and fructose had no effect on activity. In conclusion, these findings point to the presence of sucrose-specific sensing and signaling pathways in plants.

Hexose Sensing

Experimental evidence suggests the presence of at least two different systems for hexose sensing (Figure 2). One system senses hexose as such while the other requires substrate phosphorylation by a hexose kinase for signaling.

Hexose Sensor Proteins In yeast, membrane proteins with homology to hexose transporters are present that function as glucose sensors (116, 117). These SNF3 and RGT2 proteins sense low and high levels of glucose, respectively. Dominant mutations in the SNF3 and RGT2 proteins have been identified that initiate signaling in the absence of glucose. Also in yeast, a membrane-bound glucose sensor GPR1 has recently been identified (89). GPR1 is a G protein–coupled receptor specifically required for glucose activation of the cAMP pathway.

Such hexose sensing proteins are also present in plants, although their molecular nature and cellular location are still obscure. Glucose analogs like 3-O-mGlc and 6-dGlc can initiate signaling but are not phosphorylated by hexokinase (Figure 1). In a cell suspension–culture of *Chenopodium rubrum*, the addition of either glucose or 6-dGlc induces the expression of genes for extracellular invertase and sucrose synthase (46, 133). In the unicellular green alga *Chlorella kessleri*, glucose and 6-dGlc induce several genes, including a glucose transporter gene (60). The sugar- and amino acid–induced patatin class I pat (B33) promoter is also induced by the glucose analogs 6-dGlc and 3-O-mGlc in transgenic Arabidopsis plants harboring the Pat (B33)-iudA construct (101). These results suggest that plant cells sense the presence of hexoses as such, independently of hexose phosphorylation.

Hexokinase An extensive body of literature suggests that hexose phosphorylation by hexokinase (HXK) is an important sugar sensing mechanism in yeast and animal systems (18, 36, 40, 51, 79, 102). Somehow the active hexokinase initiates a signaling cascade that leads to altered gene expression. It was proposed that the yeast Hxk2 protein is itself located in the nucleus as part of a DNA-protein complex that binds to glucose-repressed genes (59).

In plants, a similar HXK-dependent sugar sensing mechanism controls many processes and metabolic pathways. Sugar-induced feedback inhibition of photosynthesis has been described for many species and this overrides regulation by light, tissue type, and developmental stage (75, 91, 144). Increased carbohydrate levels lead to inhibition of photosynthesis and a decrease in ribulose-1,5-bisphosphate carboxylase (Rubisco) protein, other Calvin-cycle enzymes, and chlorophyll. Moreover, this inhibition of photosynthesis is sustained by repression of many genes encoding proteins involved in photosynthesis (PS-related genes). For example, decreased Rubisco small subunit (RBCS) transcript levels were observed in a *C. rubrum* photoautotrophic cell suspension when cultured in the presence of glucose (91). Glucose phosphorylation is essential for repression since non-phosphorylatable analogs such as 6-dGlc and 3-O-mGlc have no effect. Jang & Sheen (75) used a maize protoplast transient expression system to monitor the effects of various sugars on promoter activity of photosynthesis genes. They observed that HXK substrates such as glucose and 2-dGlc induce repression, whereas various metabolic intermediates were ineffective. The involvement of HXK was further suggested by the observation that the HXK-inhibitor mannoheptulose (MNH) blocks the 2-dGlc-mediated repression (75).

A transgenic approach was recently taken to provide more direct evidence for the involvement of HXK in repression of photosynthesis genes (73). Two HXK genes from Arabidopsis, AtHXK1 and AtHXK2, have been cloned and used in overexpression and antisense experiments to investigate the in vivo function of HXK in sugar sensing. Germination of wild-type Arabidopsis seeds on a medium containing 6% glucose inhibits hypocotyl elongation and greening of the seedlings, and represses expression of photosynthesis genes. Antisense plants with reduced expression of AtHXK1 and AtHXK2 are less sensitive to these effects of glucose than wild-type Arabidopsis, whereas enhanced glucose sensitivity was observed in HXK-overexpressing plants. In a separate experiment, the yeast *HXK2* gene was introduced in Arabidopsis and glucose sensitivity was tested. Such transgenic *HXK2*-overexpressing lines showed reduced glucose sensitivity similar to the HXK antisense lines. The explanation given for this observation is that the yeast enzyme phosphorylates cellular glucose, thereby reducing enzymatic and signaling activity of the endogenous HXKs, which results in reduced glucose sensitivity. Moreover, yeast HXK apparently has no signaling effect in plants. Overexpression of the Arabidopsis *AtHXK1* in transgenic tomato plants leads to a phenotype that includes reduced photosynthetic activity; a regulatory role for HXK was also proposed (25).

HXK-mediated sugar sensing is present in non-green tissues as well. In a cucumber cell culture system, the glyoxylate cycle genes malate synthase (*MS*) and isocitrate lyase (*ICL*) were shown to be repressed by the addition of glucose to the growth medium. 2-dGlc and mannose could mimic this effect but 3-O-mGlc, which is not a HXK substrate, could not (48, 49). Moreover, mannose or 2-dGlc inhibit germination of Arabidopsis seedlings (121). Adding the HXK inhibitor MNH to the growth medium could relieve this inhibition. In this system 3-O-mGlc and 6-dGlc had no effect, which shows that hexose uptake per se is not involved. It was concluded that mannose inhibits Arabidopsis germination via a hexokinase-mediated step. In celery, the activity of the mannitol-catabolizing enzyme mannitol dehydrogenase (MTD) is repressed by sugars that are substrate for HXK but not by 3-O-mGlc (124). This inhibition of MTD activity could be relieved by MNH. These studies are in agreement with the notion that HXK is of major importance for hexose sensing in the plant's life cycle.

The function of HXK as a hexose sensor in plants has not been generally accepted (35, 54, 58). Herbers et al (58) suggested that hexose sensing occurs in association with the secretory (Golgi-ER) system. Experiments in which a yeast invertase was expressed in the plant apoplast or vacuole result in monosaccharide release, which leads to repression of PS-related genes such as *CAB*, encoding chlorophyll-a/b binding protein. Since the cytosolic expression of yeast invertase did not induce these changes in gene expression, it was concluded that sensing occurs in association with the endomembrane system independent from HXK. Moreover, Halford et al (54) have questioned the sensing function of HXK. These authors argue that the reduced energy status of the cell due to HXK activity may feed into signaling systems, as was found for the AMP-activated protein kinase (AMPK), the animal homologue of the yeast SNF1 kinase.

Clearly, the molecular details on the sugar sensing function of HXKs and its signaling to downstream components must be resolved (107), e.g. by functionally separating the enzymatic and signaling function of HXK via mutation analysis. Dominant signaling mutations such as those present in mutant SNF3 and RGT2 hexose binding proteins have not yet been identified for plant HXKs, but for yeast there are reports on the separation of enzymatic and signaling functions (61, 90). Such a separation of functions is essential to understand the way in which HXK operates as a sensor. Also unclear is how the activated HXK interacts with downstream components of the signaling pathway. More plant-specific questions relate to the diversity of hexokinase genes, their regulation, tissue-specific expression patterns, intracellular localization, and the possible control systems that operate on HXKs, such as the trehalose system (45, 157). Mutants should help in elucidating the role of HXK in sugar sensing; recently, the first Arabidopsis hexokinase mutants were isolated (107).

Several other sugar kinases in addition to HXK are present in plants; fructokinase, galactokinase, and arabinose kinase. The Arabidopsis galactokinase (*GAL1*) and arabinose kinase (*ARA1*) genes have been cloned and an arabinose kinase deficient mutant (*ara1*) has also been identified (81, 145). Whether these enzymes have signaling function similar to the proposed HXK function is unclear. The recent identification of an Arabidopsis mutant with a disrupted fructokinase 2 gene should help in answering this question (120).

Galactokinase (Gal1p) has been implicated in monosaccharide sensing in yeast. Gal1p can bind to Gal80p, which is the inhibitor protein of the transcriptional regulator Gal4p. Gal1p binds to Gal80p in the presence of galactose and ATP, allowing Gal4p to activate expression of the *GAL* genes (123, 186). Galactokinase enzymatic activity was dispensable for the Gal1p-Gal80p interaction, as shown by a kinase-negative Gal1p mutant that retains its regulatory function.

Other products of intermediary metabolism are likely being sensed. For example, in the maize protoplast system it was found that in addition to glucose, acetate also inhibits expression of photosynthesis genes (75). However, it has been argued that intracellular acidification may cause the acetate-induced repression of photosynthesis genes (35).

Trehalose Trehalose biosynthetic enzymes have a regulatory function in yeast that somehow controls HXK activity and signaling. However, the molecular details of this regulatory mechanism are unclear (157). It has now become apparent that all flowering plants are capable of trehalose biosynthesis and degradation (11, 44, 45, 167). Genes encoding TPS- and TPP-like proteins have been cloned from several plant species and these are present as multigene families (11, 45, 167). These plant *TPS* and *TPP* genes can functionally complement yeast *tps* and *tpp* mutants, respectively. Overexpression of bacterial and yeast *TPS* and *TPP* genes in plants leads to opposite phenotypes (45, 62, 134). These phenotypes, and the detailed analysis of the transgenic plants, suggest that plant trehalose metabolism also has a function in sugar sensing in plants. Interestingly, it was reported that trehalose addition to soybean induced sucrose synthase and alkaline invertase activity (110).

THE SIGNAL TRANSDUCTION CASCADE

The sugar sensors feed information into signal transduction cascades that lead to different plant responses. Our knowledge of this process is limited but progress is being made through different approaches. The involvement of protein kinases, protein phosphatases, and other signal transduction mediators such as Ca^{2+} and calmodulin have been proposed (Figure 3). Furthermore, the importance in plants of SNF1-like protein kinase complexes and interacting proteins in sugar sensing is now being established. In addition, other kinases and phosphatases have been identified that control activity of enzymes in intermediary metabolism such as sucrose phosphate synthase (SPS), sucrose synthase (SS), and nitrate reductase (NR). These protein kinases and phosphatases are most likely connected to sugar signaling pathways.

Protein Kinases, Protein Phosphatases, Ca^{2+}, and Calmodulin

Protein Kinases, Protein Phosphatases, Ca^{2+}, and Calmodulin Protein phosphorylation and dephosphorylation, Ca^{2+}, and calmodulin have been implicated in sugar-mediated signaling of the sweet potato and Arabidopsis genes encoding β-amylase, sporamin, and the small subunit of AGPase (106, 113, 154). Specific inhibitors of protein-Ser/Thr phosphatases 1 (PP1) and 2A (PP2A) such as okadaic acid, microcystin-LR, and calyculin A blocked the sugar induction of these genes

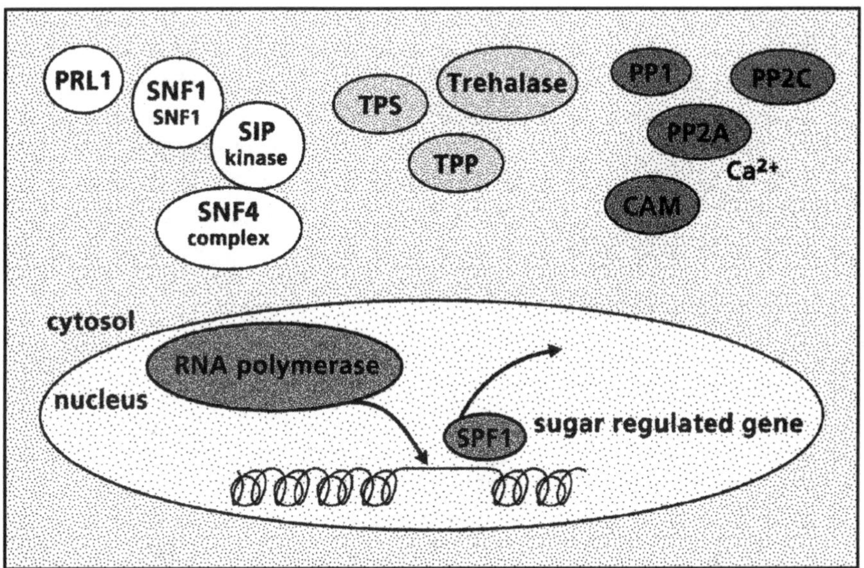

Figure 3 Possible intermediates in sugar-induced signaling in plants. The individual components are discussed in the text.

in sweet potato as well as reporter gene expression in β-amylase promoter-*iudA* (*AMY-GUS*) fusion genes in tobacco (154). In addition, inhibitors of Ser/Thr protein kinases, staurosporine and K-252a, inhibited the sugar induction of the *AMY-GUS* gene in tobacco (115). These authors reported on a sugar-inducible calcium-dependent (calmodulin-domain) Ser/Thr protein kinase (CDPK) associated with the plasma membrane in leaf tissue of tobacco.

The involvement of Ca^{2+} and calmodulin in sugar-induced β-amylase and sporamin expression was suggested from experiments with calmodulin inhibitors La^{3+} and EGTA and the Ca^{2+}-channel blockers diltiazem and nicardipine (113). Moreover, cytoplasmic Ca^{2+} concentrations increase upon incubation with sugars, as was demonstrated in experiments with transgenic tobacco plants that expressed a Ca^{2+}-sensitive photoprotein of jellyfish, aequorin (113).

The *C. rubrum* cell culture system has been used to study the sugar-regulated expression of the *RBCS*, *RBCS*, *CIN1* (encoding cell wall invertase), and *PAL* (encoding phenylalanine ammonium lyase). In this system, the source-specific *RBCS* gene is repressed by sugars, whereas the sink-specific *CIN1* gene and the pathogen-induced PAL gene are sugar induced (35). These three genes were found to be coordinately regulated by glucose in an HXK-independent way. Four different protein phosphatase inhibitors were able to mimic the glucose-mediated regulation of these three genes. Thus protein dephosphorylation is involved in transducing the sugar signal. Moreover, it appears that sugar signaling requires both de novo protein synthesis and the activation of MAP kinases. The glucose-regulated expression of these three genes is mimicked by stress-related stimuli such as addition of the fungal elicitor chitosan. Interestingly, the glucose- and elicitor-induced regulation of these three genes involves different perception and signal transduction systems since the protein kinase inhibitor staurosporine inhibits elicitor-induced but not glucose-induced gene expression response. Thus protein kinase activity is essential for transmission of the elicitor signal, whereas protein phosphatase activity is essential for transmission of the glucose signal. These results do not support the model that elevated glucose concentrations as such are the primary signal for induction of stress-related genes (58, 75).

The SNF1 Kinase Complex

The glucose-repressed state in yeast is relieved by the action of the SNF1 kinase complex, a protein-serine/threonine kinase. A shift to low glucose concentrations somehow activates the SNF1 kinase and this results in the phosphorylation of the DNA binding protein MIG1. When glucose is available, MIG1 interacts with the repressor complex SSN6/TUP1 to maintain the glucose-repressed state (163). At low glucose concentrations, MIG1 is phosphorylated and translocated to the cytosol (18, 29, 164). The dissociation of the repressive complex enables the activation of glucose-repressed genes through the involvement of the SNF/SWI chromatin remodeling machine. These glucose-repressed genes encode functions that allow growth of the cells on alternative fermentable carbon sources such as sucrose and galactose. The SNF1 kinase complex is a heterotrimeric protein that consists of the SNF1, SNF4, and a member of the SNF-interacting (SIP) protein family (SIP1-4, GAL83). The SNF1 protein harbors the catalytic function, whereas SNF4 is the activating subunit (77). The SIP proteins function as adapters between the SNF1 and SNF4 subunits. In yeast the function of the different subunits in the SNF1 complex in relation to glucose sensing has been thoroughly investigated and a detailed model presented (18).

The SNF1 kinase complex is evolutionarily conserved and has been found in animals and plants (Table 2). The animal complex is the AMP-activated protein kinase (AMPK) (55, 56, 83). AMPK is also a heterotrimeric complex and the subunits share considerable homology at the amino acid level with the yeast SNF1 kinase complex subunits. SNF1 kinase homologues have been cloned from many plant species [for a compilation see (53)]. The classification of known plant

TABLE 2 Subunit composition and nomenclature of yeast, mammalian, and plant SNF1 kinase complexes

Yeast	Mammals	Plant	MW plant enzyme[a]	Function
SNF1	α	SnRK[b]	~58 kD	Catalytic
SNF4	γ	SNF4 homologue	~40 kD	Activator
SIPs	β	SIP homologue	~30 kD	SNF1 + 4 bridging

[a]Approximate molecular weights (MW) are provided for the plant subunits.

[b]SnRK, SNF1-related kinase.

SNF1-related sequences and their expression patterns has been reviewed recently (53). The other components of the plant SNF1 kinase complex have also been identified. For example, the potato SNF1 kinase was used in a yeast two-hybrid system and interacting proteins were found homologous to the yeast GAL83 protein (93). Arabidopsis *SNF4* and *SIP* genes have also been cloned (14). Several of these plant *SNF1*, *SNF4*, and *SIP* genes complement the corresponding yeast mutations, showing the high degree of functional conservation of the complex. In both plants and animals the constituents of the kinase complex are encoded by multigene families (14, 53, 56, 93).

The function of the AMPK complex in animals was proposed to be that of a fuel gauge (55, 56, 83). Activation of the complex leads to energy preservation by inactivating ATP-consuming anabolic enzymes via phosphorylation. Target proteins of the AMPK complex include acetyl-CoA carboxylase (fatty acid synthesis) and 3-hydroxy-3-methylglutaryl-coenzyme A (HMG-CoA) reductase (isoprenoid and sterol synthesis). Phosphorylation of these enzymes leads to their inactivation. AMPK activity can be assayed in vitro by determining the phosphorylation of the SAMS peptide. This peptide is derived from the phosphorylation site of the rat acetyl-CoA carboxylase (His Met Arg Ser Ala Met Ser Gly Leu His Leu Val Lys Arg Arg). Plant SNF1-related protein kinases also phosphorylate this peptide (4, 100).

Molecular and physiological analysis of the SNF kinase complex in plants is still limited. However, antisense suppression of a potato *SNF1* homologous gene was found to result in loss of sucrose-inducibility of sucrose synthase (125). Moreover, the SAMS peptide kinase activity was reduced in these plants. Two different kinase complexes have been biochemically identified in spinach (153). One such complex may be involved in diverse phosphorylation functions. Moreover, the exchangeable SIP/GAL83 proteins may direct the kinase to different cellular substrates. This would explain the variety of processes in yeast in which the SNF1 kinase complex is active: thermotolerance, sporulation, cell cycle progression, and peroxisome biogenesis (18).

In addition to this proposed function in coupling sugar perception to altered gene expression, the SNF kinase complex likely controls the activity of several enzymes in plant metabolism and in this way allows for rapid changes in metabolism. The activity of plant nitrate reductase (NR) is controlled by phosphorylation at a specific serine residue (in spinach, serine-543). Phosphorylation per se does not alter NR activity but it allows for the binding of 14-3-3 proteins (Figure 4). Once complexed with a 14-3-3 protein, the phosphoserine-NR is inactivated (3). It appears that many enzymes in intermediary metabolism are controlled through protein phosphorylation and 14-3-3 protein binding. These proteins include NR, SPS, SS, and HMG CoA reductase (4, 34, 67, 153, 161). The phosphorylation of these proteins by SNF1-like and other protein kinases, followed by binding of 14-3-3 proteins, results in rapid adaptation of enzymatic activities and metabolic pathways to changing conditions. A family consisting of at least ten 14-3-3 genes has been cloned from Arabidopsis (181), and the encoded proteins are clearly involved in controlling diverse cellular processes in plants. A number of 14-3-3-binding

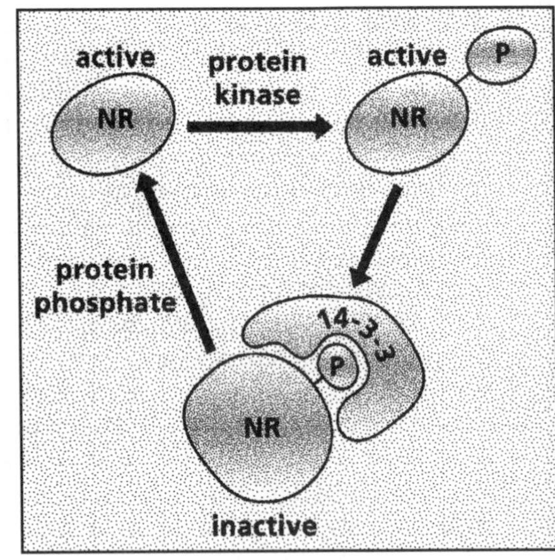

Figure 4 Control of enzymatic activity by protein kinase activity and 14-3-3 protein binding to the phosphorylated enzyme. As an example nitrate reductase (NR) is shown but this model has been proposed for several enzymes in plant carbohydrate metabolism.

proteins have been isolated from cauliflower by using affinity chromatography on immobilized 14-3-3 proteins and specific elution with a 14-3-3-binding phosphopeptide (108). These proteins were identified as being active in diverse cellular processes. One of these proteins is a calcium-dependent (calmodulin domain) protein kinase (CDPK) that phosphorylates NR and thereby makes it a target for inhibition by 14-3-3 proteins. Thus it appears that this CDPK is itself a target for phosphorylation and 14-3-3 protein binding.

Interestingly, the nitrate reductase phosphoprotein-14-3-3 protein complex is AMP sensitive. Addition of 5′-AMP or homologous compounds leads to the dissociation of the complex, apparently through the binding of the 5′-AMP to a domain on the 14-3-3 protein (2, 80). Although the physiological relevance of this observation is unclear, enzyme activities may be directly linked in this way to energy charge of the cell.

Other protein kinases in addition to the SNF kinase complex have been identified that also phosphorylate enzymes in intermediary metabolism. A single enzyme can be a target for different protein kinases that regulate its activity in opposite ways. For example, SPS can be both activated and repressed by site-specific protein phosphorylation (161). Diurnal regulation of SPS is controlled by serine-158 phosphorylation through the activity of a SNF1-like kinase that leads to inactivation. This control is overridden by stress-induced activation via a CDPK-mediated serine-424 phosphorylation (161). Moreover, enzymes such as NR, SPS, HMG CoA reductase, and SS can also be phosphorylated by CDPK (33, 34, 67, 162). Phosphorylation of SS selectively activates the sucrose cleavage reaction, thereby releasing UDP-glucose and fructose for intermediary metabolism (67). A spinach NR serine-543 phosphorylating enzyme has recently been purified and its identity established by partial amino acid sequencing of the protein (33). The peptide sequences of the spinach protein were very similar to the Arabidopsis *CDPK6/CPK3* gene (63, 64).

These kinases and phosphatases are themselves targets for regulation. What emerges is a highly complex interactive web that is functional in fine-tuning the assimilatory and respiratory processes, and adapting to continuously changing conditions. Biochemical and molecular knowledge about these kinases is rapidly increasing but the links with sugar signaling cascades need strengthening. The identification and analysis of mutants such as *prl1* (see below) should fill this gap.

The biochemical and molecular details of the dephosphorylating enzymes remain to be established. Mammalian protein phosphatases PP2A and PP2C can inactivate plant SNF1 kinase activity (34). Interestingly, the plant PP2C-type protein phosphatase activity encoded by the *ABI1* locus can also inactivate SNF1 kinase. This led MacKintosh (100) to suggest a function of other plant PP2Cs, like the serine/threonine receptor–associated kinase-associated protein phosphatases (KAPP) (16, 179) in regulating plant SNF1-like kinases. SNF1-like protein kinases probably are involved in other metabolite signaling pathways as well. A SNF1-like kinase was implicated in *Chlamydomonas* in the responses to sulfur limitation (26).

PRL1, A SNF1 Kinase Complex Interacting Protein

A T-DNA tagged Arabidopsis mutant that shows growth defects on media containing 175 mM sucrose or glucose was discovered by Koncz and colleagues (111, 138). The T-DNA tag allowed identification of this *PLEIOTROPIC REGULATORY LOCUS 1* (*PRL1*). *PRL1* encodes a WD protein that is localized in the nucleus. The *prl1* mutant shows a rather pleiotropic phenotype that includes developmental alterations such as a short root. Moreover, *prl1* is hypersensitive to ethylene, auxin, cytokinin, ABA, and cold. The presence of 0.1 micromolar ABA already results in *prl1* bleaching and growth reduction, whereas this concentration has no effect on wild type. Many genes that are up-regulated by sugar or cytokinin are overexpressed in the mutant, leading to overproduction of anthocyanin and starch.

Remarkably, in a yeast two-hybrid screen for interacting partners, the PRL1 protein was found to interact with AKIN10 and AKIN11, the Arabidopsis homologues of yeast SNF1 protein kinase (9, 42). In yeast this PRL1-AKIN interaction was dependent on the presence of glucose in the medium. Low glucose increased the strength of the interaction, which suggests that these plant proteins are responsive to yeast endogenous glucose-derived signals. PRL1 probably inhibits the phosphorylating activity of AKIN10 and 11, as shown in in vitro experiments with a peptide substrate. In yeast the SNF1 kinase is activated by glucose starvation but, in contrast, sugar feeding to light-grown plants stimulated peptide-substrate kinase activity. This sugar-induced kinase activity is independent of the presence of PRL1 since it is observed both in wild-type and in the *prl1* mutant. These observations make it difficult to construct a model on PRL1 function. In addition to the AKIN10 and 11 proteins, PRL1 was found to interact with other proteins in yeast two-hybrid screens. This and the pleiotropic nature of the mutation suggest that PRL1 is a central regulator in several processes (9).

Nuclear Processes

Sugar signaling can result in altered transcriptional activity of target genes, and for most genes documented this seems to be the mode of control. Transcriptional regulation is not the only response to sugars, and cytosolic targets for control have been identified. Enzyme activity can be directly regulated, as proposed for the SNF1 kinase complex. Moreover, the above-mentioned sucrose-regulated *ATB2* gene is controlled at translation (135). In addition, modulating mRNA stability is a major control element for cereal α-amylase gene expression (20).

Several sugar-inducible promoters have been analyzed in some detail to locate sugar-responsive *cis*-elements. In mutagenesis experiments, promoter elements were located that confer sucrose-inducible expression when fused to a heterologous core promoter sequence. Such sucrose-responsive elements (SURE) in the patatin class I promoter interact in gelshift assays with Sucrose Response Factors (50, 97). Moreover, SURE elements may show similarity to the SP8 motifs in the promoter region of the sucrose-induced β-amylase and sporamin genes from sweet potato (71). These SP8 motifs are recognized by the SP8BF nuclear factor (70, 71). A nuclear factor with similar binding specificity to SP8BF also binds to the SURE elements in the patatin promoter. The promoter of the sugar-inducible potato SUS4 gene encoding sucrose synthase contains SURE elements (39). Moreover, SURE homologous sequences were also observed in other sucrose-inducible sucrose synthase genes such as Arabidopsis (*ASUS1*), maize (*SUS1*), and rice (*SUS1*) [for references, see (39)]. Direct experimental evidence for a function of SURE homologous sequences in these promoters is lacking. Moreover, the SP8/SURE elements are not present in the 5′-upstream regions of all sugar-inducible genes. Nuclear factors that are potato tuber-specific or induced by sucrose in leaves were identified that bind to the class-I patatin promoter (85). The binding of these factors was localized to four different regions of this promoter including a SURE-like region.

Plant genes encoding isocitrate lyase and malate synthase genes are developmentally regulated and are induced by starvation and germination. These genes are also responsive to sugars, and the promoter elements involved in repression have been localized. Distinct *cis*-acting elements have been identified for developmental control and sugar repression in these genes (27, 48, 49, 126, 140). The sugar repression elements seem not to resemble the SURE elements.

A cDNA clone encoding a new type of DNA-binding protein, SPF1, that binds the SP8 motif was isolated from a sweet potato petiole cDNA library (71). Interestingly, SPF1 transcript levels decreased when leaf-petiole cuttings were treated with sucrose concentrations that induce accumulation of sporamin and β-amylase mRNAs. This observation suggests that SPF1 is a negative regulator. A putative SPF1 homologue has recently also been isolated from cucumber (84).

CEREAL SEED GERMINATION

The cereal seed germination system provides an interesting system for the analysis of sugar-regulated gene expression. During seed germination starch is mobilized by the action of α-amylases. Cereal α-amylases are encoded by multigene families. In rice nine members comprise the α-amylase gene family (159). During cereal seed germination α-amylases are produced by the scutellar layer of the embryo and by the aleurone layer of the endosperm. The different members of the gene family are expressed in a tissue-specific and developmentally specific way. *RAmy3D* [nomenclature according to (159)] is the major gene expressed in the rice scutellum during germination. At the rice seedling elongation stage *RAmy1A*, *RAmy3B*, *RAmy3C*, and *RAmy3E* are expressed (159).

Glucose generated by the α-amylases is transported to the embryo for growth of the seedling. Interestingly, when glucose levels exceed demand α-amylase gene expression is down-regulated in a process that involves sugar sensing. This regulatory feedback system has been studied in detail in intact tissues and in suspension cultured cells (69, 82, 185). For the *RAmy3D* gene it was found that HXK is most likely involved in transmitting the glucose signal. The hexokinase substrate 2-dGlc can induce signaling and this signaling is inhibited by the HXK inhibitor glucosamine. Moreover, 3-O-mGlc and 6-dGlc are not effective in signaling (165). Also, it was found that α-amylase expression is inhibited in the barley seed germination system by hexoses that are substrates for HXK but not by other hexoses (122).

Pharmacological studies have also been performed with sugar-repressed genes. In cultivated rice cells, the expression of α-amylase genes like *RAmy3D* is repressed by sugars (99). In this system, protein phosphatase inhibitors strongly induce *amy3D* expression and an AMP-activated protein kinase may be involved in induction.

Several, but not all, of the α-amylase genes are GA responsive. GA induces these α-amylase genes but sugars override the GA signal and repress gene expression. The sugar- and GA-responsive elements in the promoter of the *RAmy1A* gene appear to overlap, which indicates that the two signal transduction pathways communicate at a point upstream of the promoter elements (109). The promoters of the α-amylase genes contain important *cis* elements for developmentally specific expression and for sugar regulation, as was found in promoter-reporter constructs and nuclear run-on experiments (146). In addition, regulation of mRNA stability appears to be important as well since the *RAmy3E* mRNA half-life was reduced from 12 h in sugar-starved cells to less than 1 h when sugar was added to the rice suspension cultured cells (146). The major mRNA stability determinants were mapped to specific regions in the 3′-UTR region of the mRNA. These regions did not affect transcription of the gene (19, 20).

GA induces α-amylase activity in the scutellum and aleurone of germinating barley seeds. This GA-mediated α-amylase induction is repressed by sugars in the scutellum only (122). As noted above, sugars that are substrate for HXK are effective in this repression and sugar repression overrides the GA-inductive effect.

Sugars also repress α-amylase expression in the barley GA-constitutive response *slender* mutant, and it was concluded that sugars negatively interfere with GA signal transduction (122). The sugar repression of GA-induced α-amylase activity and of constitutive α-amylase activity in the *slender* mutant is mimicked by ABA application, which prompted the suggestion that ABA mediates the glucose effect (see Interacting Signaling Pathways, below). However, glucose decreased ABA concentrations in barley embryos, and ABA sensitivity in this system also seemed to be unaltered, as indicated by expression of the ABA-sensitive *Rab16A* gene (122).

SEED DEVELOPMENT

Sugar import and utilization during leguminous seed development have been particularly well studied. Hexoses and sucrose serve specialized functions in different phases of seed development. This conclusion was based on analysis of the spatial and temporal expression of genes encoding sucrose metabolizing enzymes and hexose and sucrose transporters (156, 172–174, 176, 180). Hexose metabolism is associated with meristematic activity (cell division) in the developing embryo, whereas sucrose metabolism is associated with starch and protein storage functions. Meristematic versus storage functions could be manipulated by incubating embryos in media with different hexose-to-sucrose ratios and by seed-specific expression of the yeast invertase gene in *Vicia narbonensis* (175). This model was further supported by high-resolution histographical mapping of glucose concentrations in developing cotyledons of *Vicia faba*. Glucose co-mapped with regions of meristematic activity, and it was suggested that glucose functions as a developmental trigger molecule or morphogen (13). The importance of monosaccharides for seed development was also shown in maize, where the small seed *miniature1* mutant has a defect in the gene encoding extracellular invertase (22). Moreover, carbohydrate metabolism is disturbed in the Arabidopsis *wrinkled1* mutant during seed development. This mutant shows reduced HXK activity (38).

Sugars as signals for developmental switches in seed development must act in concert with other factors and phytohormones. In Arabidopsis such factors may include the more general regulators of seed development ABI3, LEC1, and FUS3 (28, 129, 180). The action of these proteins may not be restricted to seed development as originally proposed. For example, the expression pattern of the Arabidopsis *ABI3* gene suggests a function in vegetative quiescence processes (131).

INTERACTING SIGNALING PATHWAYS

Sugar-signaling pathways do not operate in isolation but are part of cellular regulatory networks. Recent results clearly show cross talk between different signaling systems, especially those of sugars, phytohormones, and light.

Ethylene

A close interaction between HXK- and ethylene-mediated signaling pathways was revealed by the analysis of the Arabidopsis *gin1* mutant (188). This mutant is resistant to elevated (6%) glucose amount. Moreover, elevated sugar levels do not repress PS-related genes as they do in wild-type plants. The *gin1* mutation is epistatic to HXK in the glucose signaling pathways since combining *gin1-1* with *AtHXK1* overexpressing lines produced plants that showed the same glucose insensitivity phenotype and had the same appearance as *gin1-1*. The *gin1* mutants germinate faster, are smaller, and have darker green rosettes than wild-type, and this phenotype is reminiscent of wild-type plants treated with ethylene. Interestingly, the *gin1-1* phenotype could be copied in wild-type plants by treatment with the ethylene precursor ACC. In addition, the ethylene-overproducing mutant *eto1-1* and the ethylene constitutive response mutant *ctr1-1* are also glucose insensitive. Conversely, the ethylene-insensitive mutant *etr1-1* shows a glucose hypersensitive phenotype. In the *etr1-1* mutant an ethylene receptor is mutated, which results in a dominant ethylene-unresponsive phenotype (66). Further investigations showed that *GIN1* acts downstream of *ETR1* in the ethylene-signaling pathway. These findings reveal a close interaction between the glucose- and ethylene-signaling pathways. Glucose signaling through HXK and GIN1 down-regulates a branch of the ethylene-signaling pathways that stimulates germination and cotyledon and leaf development (188).

Abscisic Acid and Gibberellic Acid

The cloning and analysis of the *sun6* mutation (30, 68) led to the discovery that an intact ABA signal transduction chain is important for hexokinase-dependent glucose signaling (68). As well as being sucrose-insensitive, the *sun6* mutant is also insensitive to glucose and mannose. Moreover, elevated sugar levels in *sun6* do not repress photosynthesis genes. In accordance with these results, whole plant photosynthesis in mature *sun6* rosettes was found to be more resistant to the glucose analog 2-dGlc than in wild-type plants (166). The identification of a *sun6* allele in a transposon-tagged seed collection allowed cloning of the gene. *SUN6* is identical to the previously cloned *ABI4* gene (37). The SUN6/ABI4 protein falls in the group of AP2-domain transcription factor genes and the *sun6* mutant bears a stop codon in the AP2 domain. The *sun6* mutation is allelic to *abi4* and, like *abi4*, germinates on ABA-containing medium. Remarkably, all Arabidopsis *aba* and *abi* mutants are, to varying degrees, sugar sensing mutants (68). These results suggest that hexokinase-mediated sugar signaling requires an intact ABA signal transduction chain. It is unclear whether sugars enhance cellular ABA sensitivity or increase ABA levels. The ABI4/SUN6 protein is important during germination and photosynthetic growth. Down-regulation of photosynthesis genes is mediated through ABI4/SUN6, which allows sugar supply to be matched with demand. At germination, ABI4/SUN6 is involved in controlling the mobilization of seed reserves in an ABA- and sugar-dependent way. That ABA inhibits germination by restricting reserve mobilization was also shown by Garciarrubbio et al (41). These findings may explain the sensitivity of Arabidopsis seeds to mannose (121). Mannose activates HXK signaling, which activates the *ABI4/SUN6* gene or gene product through the ABA pathway, thereby restricting reserve mobilization.

The close relation between HXK-mediated sugar sensing and ABA signal transduction may explain many earlier observations on the similar effects of ABA and sugar application on gene expression. For example, in addition to sugars, ABA also inhibits expression of light- (phytochrome) dependent PS-related genes (6, 21, 92, 103). This relationship with light and sugars is also apparent for other phytochrome-regulated genes. Interestingly, phytochrome signaling seems able to control plant ABA amounts, and one can speculate that phytochrome and sugars regulate photosynthesis-related and other genes in an integrated way by modulating ABA levels (170; 171).

Many plants respond to elevated CO_2 by repression of photosynthesis, possibly due to increasing sugar concentrations. Remarkably, this is not a cell-autonomous response but is controlled at the whole plant level (148). Phytohormones, like ABA, may mediate such systemic responses. Studies on the mechanism by which viral movement proteins affect assimilate allocation led to similar conclusions on systemic regulatory systems (98).

Sugars and ABA activate "sink-related" genes such as the sporamin and β-amylase genes of sweet potato (114). Moreover, sugars and ABA promote tuber development (182). In both cases, gibberellic acid (GA) has the opposite effect. Interestingly, modulation of phytochrome B levels in potato greatly affects tuber induction and development (72, 158), and one can speculate whether ABA is a mediator of this response.

ABA inhibits phloem loading of sucrose, whereas GA promotes export of assimilates (1, 168). Thus the ABA/GA balance may regulate cellular sugar levels. The opposite biological effect observed for ABA and GA may be due to interacting signal transduction pathways, as suggested by recent findings on SPINDLY (SPY) function. SPY is a negative regulator of GA action that shows homology to an O-linked N-acetylglucosamine transferase (128). In a barley aleurone transient assay system, SPY repressed the GA-induced activity of α-amylase. Surprisingly, SPY also increased promoter activity of the ABA-inducible dehydrin promoter in the absence of ABA. Thus GA- and ABA-responsive signal transduction chains may interact at a point upstream of SPY. For α-amylase the ABA/GA response has been shown to act through a single *cis* element (52, 130). In Arabidopsis ABA sensitivity depends on the activity of a farnesyltransferase, the product of the *ERA1* gene (24). Interestingly, sucrose and glucose inhibit the expression of a protein farnesyltransferase in pea and thus might increase ABA sensitivity (187).

Sugars and ABA can also act in opposite ways, as was observed in a transgenic tobacco line harboring the *Phaseolus* phaseolin promoter-GUS gene. Applied ABA could induce this transgene and the endogenous 12S globulin gene in prematuration zygotic embryos. Sucrose repressed this induction but addition of Ca^{2+} to the medium overcame the repression (17).

Cytokinins and Plant Development

Cyclins are central regulators of the cell cycle and are targets for hormonal and metabolic control. In Arabidopsis cytokinin can activate the cell cycle through the D-type cyclin CycD3, which operates at the G1-S phase transition. Constitutive expression of CycD3 in transgenic plants alleviated the cytokinin requirement for cell division (127). Addition of sucrose to quiescent, sugar-starved Arabidopsis cells triggers expression of CycD3 and G1-S phase transition. The *CycD2* gene is also induced by sugars (151). Whether sugar has a signaling function in this system or whether its metabolism induces the phase transition has not yet been closely investigated, but it is safe to predict such a signaling function of sugars in the control of the cell cycle.

Plants grown at elevated CO_2 show an increase in apical meristem cell number and size, and a more rapid progression through the cell cycle has been observed in different plant species (78, 86). Growth at elevated CO_2 leads to increased sugar levels that will activate sugar-sensing systems. In the apical meristems this may stimulate increased cell division rates through control of cyclin gene expression.

Many developmental processes are likely tightly linked to sugar signaling mechanisms, e.g. floral induction is dependent on a multifactorial signal that is transported to the vegetative apical meristem. Sucrose is an important component of this inductive signal (8, 96). Interesting in this respect is that the flowering-promoting gene *LEAFY* is induced by sucrose (10). Results obtained by Tang et al (155) also suggested the importance of sucrose and hexose as signal molecules in plant development. Interference with sucrose metabolism through antisense repression of invertase activity leads to developmental aberrations. These could be relieved by the addition of glucose and fructose to the medium. This effect of hexoses is reminiscent of the situation in developing bean seeds (180).

Many sugar-induced genes are also responsive to jasmonate, e.g. both jasmonate and sugars induce the expression of soybean vegetative storage protein (*VSP*) genes (7). Wounding, light, and phosphate also control expression of these *VSP* genes.

Carbon and nitrogen metabolism are tightly linked and it seems obvious that nitrogen-signaling pathways interact with sugar-signaling pathways (95, 118, 141, 152). Similar links can be expected for other metabolites as well (26, 112, 137). Such interactions in turn may be controlled or mediated by phytohormones.

Light

Metabolizable sugars alter the responsiveness of plants to light; this has been described for the far-red light, PHYA-specific, signaling pathway. Sugars can block the so-called far-red light-induced block of greening that is caused by the PHYA-specific repression of protochlorophyllide oxidoreductase (5). Sugars negatively interfere with PHYA signaling, thereby protecting the plant against far-red–induced damage. Interestingly, in the *sun6* mutant, sugars are not sensed and PHYA signaling is no longer inhibited by sugars (30). Arabidopsis lines that overexpress PHYB show reduced far-red light–induced PHYA signaling only if metabolizable sugars are present in the growth medium (147). These and other findings (65, 178) suggest a close interaction between sugar- and light-signaling pathways. For example, the sugar-induced expression of the Arabidopsis β-amylase is greatly enhanced by light, most likely through phytochrome signaling (106, 143).

Phytochrome affects activities of many enzymes in intermediary carbohydrate metabolism and anthocyanin biosynthesis. Overexpression of phytochrome in potato and tobacco leads to anthocyanin biosynthesis, activation of the Calvin cycle assimilatory genes, and the sucrose phosphate synthase gene (142, 183).

Stress

Elevated sugar concentrations can induce resistance to pathogen attack by inducing stress-related genes such as the PR proteins (132). Expression of yeast invertase in the plant apoplast or direct sugar feeding to leaves lead to increased levels of monosaccharides and PR proteins (58). Similarly, it was found that elevated sugar levels during fruit ripening in grape induce antifungal protein accumulation (139). It has been argued that monosaccharides directly induce stress-related genes, resulting in a systemic protective response (58). However, an alternative model suggests that a separate signaling pathway induces these stress-related genes (35). The sugar-responsive sporamin and β-amylase genes are also induced by elicitor and this induction is GA repressible (114). Since ABA induction of these genes is also GA repressible, sugars and elicitor could well have the ABA pathway in common.

In micorrhizal plant-fungus symbiosis, fungal hyphae penetrate the cortical cells of the host plant root to form arbuscules. This specialized structure allows for carbon transfer from the plant to the fungus, and gibberellic acid may be involved in this mobilization process (12). In the arbusculated cells assimilate unloading is probably maintained by the localized expression of sucrose synthase and soluble invertase. It is tempting to speculate on a signaling function for sugars in this specialized developmental process.

Many environmental stresses, most notably drought and cold, lead to major alteration in carbohydrate metabolism (57, 160, 169), and most likely sugar signaling pathways interact with stress pathways to modulate metabolism. In a screen of selected sugar sensing mutants, one mutant was found to be impaired in the cold acclimation response (L Wanner & S Smeekens, unpublished).

CONCLUSION

Significant progress has been made in sugar sensing research in recent years. Many Arabidopsis mutants have been identified with various sugar sensing or signaling phenotypes. Efficient cloning techniques will allow for the identification of the genes involved. Sugar-sensing pathways are clearly closely linked to other signaling pathways, most notably those of hormones. As these links are strengthened and worked out in molecular detail, analysis of mutants and the corresponding genes should provide new insights and identify connections to other pathways. More sophisticated mutant identification strategies are needed to address specific aspects of plant sugar sensing and signaling. For example, important and well-characterized mutants can be used as starting material to identify suppressor and enhancer mutations that are pathway specific.

Furthermore, the role of sugar transporters in sugar uptake and distribution over cellular compartments and its relation with sugar sensing and signaling must be better understood (94). The compartments in the cell where sugars are sensed must be identified. Clearly, sugar-signaling pathways are intimately woven into cellular signaling webs and what has been discovered so far gives only a glimpse of the complexity of the system overall.

ACKNOWLEDGMENTS

I am most grateful to Henriette Schlüpmann and Anne Kortstee for comments on the manuscript. I would also like to thank the other members of the Molecular Plant Physiology laboratory for valuable discussion.

Visit the Annual Reviews home page at www.AnnualReviews.org

LITERATURE CITED

1. Aloni B, Daie J, Wyse RE. 1986. Enhancement of [^{14}C] sucrose export from source leaves of *Vicia faba* by gibberellic acid. *Plant Physiol.* 82:962–67
2. Athwal GS, Huber JL, Huber SC. 1998. Phosphorylated nitrate reductase and 14-3-3 proteins. Site of interaction, effects of ions, and evidence for an AMP-binding site on 14-3-3 proteins. *Plant Physiol.* 118:1041–48
3. Bachmann M, Huber JL, Liao PC, Gage DA, Huber SC. 1996. The inhibitor protein of phosphorylated nitrate reductase from spinach (*Spinacia oleracea*) leaves is a 14-3-3 protein. *FEBS Lett.* 387:127–31
4. Barker JHA, Slocombe SP, Ball KL, Hardie DG, Shewry PR, Halford NG. 1996. Evidence that barley 3-hydroxy-3-methylglutaryl-coenzyme A reductase kinase is a member of the sucrose nonfermenting-1-related protein kinase family. *Plant Physiol.* 112:1141–49
5. Barnes SA, Nishizawa NK, Quaggio RB, Whitelam GC, Chua NH. 1996. Far-red light blocks greening of Arabidopsis seedlings via a phytochrome A-mediated change in plastid development. *Plant Cell* 8:601–15
6. Bartholomew DM, Bartley GE, Scolnik PA. 1991. Abscisic acid control of *rbcS* and *cab* transcription in tomato leaves. *Plant Physiol.* 96:291–96
7. Berger S, Bell E, Sadka A, Mullet JE. 1995. *Arabidopsis thaliana Atvsp* is homologous to soybean *VspA* and *VspB*, genes encoding vegetative storage protein acid phosphatases, and is regulated similarly by methyl jasmonate, wounding, sugars, light and phosphate. *Plant Mol. Biol.* 27:933–42
8. Bernier G, Havelange A, Houssa C, Petitjean A, Lejeune P. 1993. Physiological signals that induce flowering. *Plant Cell* 5:1147–55
9. Bhalerao RP, Salchert K, Bako L, Okresz L, Szabados L, et al. 1999. Regulatory interaction of PRL1 WD protein with Arabidopsis SNF1-like protein kinases. *Proc. Natl. Acad. Sci. USA* 96:5322–27

10. Blazquez MA, Green R, Nilsson O, Sussman MR, Weigel D. 1998. Gibberellins promote flowering of Arabidopsis by activating the *LEAFY* promoter. *Plant Cell* 10:791–800
11. Blazquez MA, Santos E, Flores CL, Martinez-Zapater JM, Salinas J, Gancedo C. 1998. Isolation and molecular characterization of the Arabidopsis *TPS1* gene, encoding trehalose-6-phosphate synthase. *Plant J.* 13:685–89
12. Blee KA, Anderson AJ. 1999. Regulation of arbuscule formation by carbon in the plant. *Plant J.* 16:523–30
13. Borisjuk L, Walenta S, Weber H, Mueller-Klieser W, Wobus U. 1998. High-resolution histographical mapping of glucose concentrations in developing cotyledons of *Vicia faba* in relation to mitotic activity and storage processes: glucose as a possible developmental trigger. *Plant J.* 15:583–91
14. Bouly J-P, Gissot L, Lessard P, Kreis M, Thomas M. 1999. *Arabidopsis thaliana* proteins related to the yeast SIP and SNF4 interact with AKINα1: an SNF1-like protein kinase. *Plant J.* 18:541–50
15. Boxall S, Martin T, Graham IA. 1996. A new class of Arabidopsis mutants that is carbohydrate insensitive. *Int. Conf. Arabidopsis Res., 7th, Norwich, UK.* (Abstr.)
16. Braun DM, Stone JM, Walker JC. 1997. Interaction of the maize and Arabidopsis kinase interacting domains with a subset of receptor-like protein kinases: implications for transmembrane signaling in plants. *Plant J.* 12:83-95
17. Bustos MM, Iyer M, Gagliardi SJ. 1998. Induction of a β-phaseolin promoter by exogenous abscisic acid in tobacco: developmental regulation and modulation by external sucrose and Ca^{2+} ions. *Plant Mol. Biol.* 37:265–74
18. Carlson M. 1998. Regulation of glucose utilization in yeast. *Curr. Opin. Genet. Dev.* 8:560–64
19. Chan MT, Yu SM. 1998. The 3′ untranslated region of a rice α-amylase gene functions as a sugar-dependent mRNA stability determinant. *Proc. Natl. Acad. Sci. USA* 95:6543–47
20. Chan MT, Yu SM. 1998. The 3′ untranslated region of a rice α-amylase gene mediates sugar-dependent abundance of mRNA. *Plant J.* 15:685–95
21. Chang YC, Walling LL. 1991. Abscisic acid negatively regulates expression of chlorophyll *a/b* binding protein genes during soybean embryogeny. *Plant Physiol.* 97:1260–64
22. Cheng W, Tallercio EW, Chourey PS. 1996. The *miniature1* seed locus of maize encodes a cell wall invertase required for normal development of endosperm and maternal cells in the pedicel. *Plant Cell* 8:971–83
23. Chiou TJ, Bush DR. 1998. Sucrose is a signal molecule in assimilate partitioning. *Proc. Natl. Acad. Sci. USA* 95:4784–88
24. Cutler S, Ghassemian M, Bonetta D, Cooney S, McCourt P. 1996. A protein farnesyl transferase involved in abscisic acid signal transduction in Arabidopsis. *Science* 273:1239–41
25. Dai N, Schaffer A, Petreikov M, Shahak Y, Giller Y, et al. 1999. Overexpression of Arabidopsis hexokinase in tomato plants inhibits growth, reduces photosynthesis, and induces rapid senescence. *Plant Cell* 11:1253–66
26. Davies JP, Yildiz FH, Grossman AR. 1999. Sac3: an Snf1-like serine/threonine kinase that positively and negatively regulates the responses of *Chlamydomonas* to sulfur limitation. *Plant Cell* 11:1179–90
27. De Bellis L, Ismail I, Reynolds SJ, Barrett MD, Smith SM. 1997. Distinct cis-acting sequences are required for the germination and sugar responses of the cucumber isocitrate lyase gene. *Gene* 197:375–78
28. de Bruijn SM, Ooms JJJ, Karssen CM, Vreugdenhil D. 1997. Effects of abscisic acid on reserve deposition in developing *Arabidopsis* seeds. *Acta Bot. Neerl.* 46:263–77

29. De Vit MJ, Waddle JA, Johnston M. 1997. Regulated nuclear translocation of the Mig1 glucose repressor. *Mol. Biol. Cell* 8:1603–18
30. Dijkwel PP, Huijser, C, Weisbeek PJ, Chua N-H, Smeekens SCM. 1997. Sucrose control of phytochrome A signalling in Arabidopsis. *Plant Cell* 9:583–95
31. Dijkwel PP, Kock P, Bezemer R, Weisbeek P, Smeekens SCM. 1996. Sucrose represses the developmentally controlled transient activation of the plastocyanin gene in *Arabidopsis thaliana* seedlings. *Plant Physiol.* 110:455–63
32. Donggiun K, Laby RJ, Gibson SI. 1998. Regulation of sugar responses and characterization of *Arabidopsis thaliana* mutants with reduced beta-amylase activity. *Int. Conf. Arabidopsis Res., Madison, WI, 9th*. (Abstr.)
33. Douglas P, Moorhead G, Hong Y, Morrice N, MacKintosh C. 1998. Purification of a nitrate reductase kinase from *Spinacea oleracea* leaves, and its identification as a calmodulin-domain protein kinase. *Planta* 206:435–42
34. Douglas P, Pigaglio E, Ferrer A, Halfords NG, MacKintosh C. 1997. Three spinach leaf nitrate reductase-3-hydroxy-3-methylglutaryl-CoA reductase kinases that are required by reversible phosphorylation and/or Ca^{2+} ions. *Biochem. J.* 325:101–9
35. Ehness R, Ecker M, Godt DE, Roitsch TH. 1997. Glucose and stress independently regulate source and sink metabolism and defence mechanisms via signal transduction pathways involving protein phosphorylation. *Plant Cell* 9:1825–41
36. Epstein PN, Boschero AC, Atwater I, Cai X, Overbeek PA. 1992. Expression of yeast hexokinase in pancreatic β cells of transgenic mice reduces blood glucose, enhances insulin secretion, and decreases diabetes. *Proc. Natl. Acad. Sci. USA* 89:12038–42
37. Finkelstein RR, Wang ML, Lynch TJ, Rao S, Goodman HM. 1998. The Arabidopsis abscisic acid response locus *ABI4* encodes an APETALA 2 domain protein. *Plant Cell* 10:1043–54
38. Focks N, Benning C. 1998. *wrinkled1*: A novel, low-seed-oil mutant of Arabidopsis with a deficiency in the seed-specific regulation of carbohydrate metabolism. *Plant Physiol.* 118:91–101
39. Fu H, Kim SY, Park WD. 1995. High-level tuber expression and sucrose inducibility of a potato *Sus4* sucrose synthase gene require 5′ and 3′ flanking sequences and the leader intron. *Plant Cell* 7:1387–94
40. Gancedo JM. 1998. Yeast carbon catabolite repression. *Microbiol. Mol. Biol. Rev.* 62:334–61
41. Garciarrubio A, Legaria JP, Covarrubias AA. 1997. Abscisic acid inhibits germination of mature Arabidopsis seeds by limiting the availability of energy and nutrients. *Planta* 203:182–87
42. Gibson SI, Graham IA. 1999. Another player joins the complex field of sugar-regulated gene expression in plants. *Proc. Natl. Acad. Sci. USA* 96:4746–48
43. Gibson SI, Laby RJ, Donggiun K. 1999. Sugar-insensitive mutants of Arabidopsis with defects in phytohormone metabolism and/or response. *Int. Conf. Arabidopsis Res., 10th, Melbourne, Aust.*, (Abstr.)
44. Goddijn O, Smeekens SCM. 1998. Sensing trehalose biosynthesis in plants. *Plant J.* 14:143–46
45. Goddijn OJ, Van Dun K. 1999. Trehalose metabolism in plants. *Trends Plant Sci.* 4:315–19
46. Godt DE, Riegel A, Roitsch T. 1995. Regulation of sucrose synthase expression in *Chenopodium rubrum*: characterization of sugar induced expression in photoautotrophic suspension cultures and sink tissue specific expression in plants. *Plant Physiol.* 146:231–38
47. Graham IA. 1996. Carbohydrate control of gene expression in higher plants. *Res. Microbiol.* 147:572–80

48. Graham IA, Baker CJ, Leaver CJ. 1994. Analysis of the cucumber malate synthase gene promoter by transient expression and gel retardation assays. *Plant J.* 6:893–902
49. Graham IA, Denby KJ, Leaver CJ. 1994. Carbon catabolite repression regulates glyoxylate cycle gene-expression in cucumber. *Plant Cell* 6:761–72
50. Grierson C, Du J-S, de Torres Zabala M, Beggs K, Smith C, et al. 1994. Separate cis sequences and trans factors direct metabolic and developmental regulation of a potato tuber storage protein gene. *Plant J.* 5:815–26
51. Grupe A, Hultgren B, Ryan A, Ma YH, Bauer M, Stewart TA. 1995. Transgenic knockouts reveal a critical requirement for pancreatic β cell glucokinase in maintaining glucose homeostasis. *Cell* 83:69–78
52. Gubler F, Jacobsen JV. 1992. Gibberellin-responsive elements in the promotor of a barley high-pI α-amylase gene. *Plant Cell* 4:1435–41
53. Halford NG, Hardie DG. 1998. SNF1-related protein kinases: global regulators of carbon metabolism in plants? *Plant Mol. Biol.* 37:735–48
54. Halford NG, Purcell P, Hardie DG. 1999. Is hexokinase really a sugar sensor in plants? *Trends Plant Sci.* 4:117–20
55. Hardie DG, Carling D. 1997. The AMP-activated protein kinase-fuel gauge of the mammalian cell? *Eur. J. Biochem.* 246:259–73
56. Hardie DG, Carling D, Carlson M. 1998. The AMP-activated/SNF1 protein kinase subfamily: metabolic sensors of the eukaryotic cell? *Annu. Rev. Biochem.* 67:821–55
57. Hare PD, Cress WA, van Staden J. 1998. Dissecting the roles of osmolyte accumulation during stress. *Plant Cell Environ.* 21:535–54
58. Herbers K, Meuwly P, Frommer W, Métraux J-P, Sonnewald U. 1996. Systemic acquired resistance mediated by the ectopic expression of invertase: possible hexose sensing in the secretory pathway. *Plant Cell* 8:793–803
59. Herrero P, Martinez-Campa C, Moreno F. 1998. The hexokinase 2 protein participates in regulatory DNA-protein complexes necessary for glucose repression of the *SUC2* gene in *Saccharomyces cerevisiae*. *FEBS Lett.* 434:71–76
60. Hilgarth C, Sauer N, Tanner W. 1991. Glucose increases the expression of the ATP/ADP translocator and the glyceraldehyde-3-phosphate dehydrogenase genes in Chlorella. *J. Biol. Chem.* 266:24044–47
61. Hohmann S, Winderickx J, de Winde WH, Valckx D, Cobbaert P, et al. 1999. Novel alleles of yeast hexokinase PII with distinct effects on catalytic activity and catabolite repression of *SUC2*. *Microbiology* 145:703–14
62. Holmstrom KO, Mantyla E, Welin B, Mandal A, Palva ET. 1996. Drought tolerance in tobacco. *Nature* 379:683–84
63. Hong Y, Takano M, Liu CM, Gasch A, Chye ML, Chua NH. 1996. Expression of three members of the calcium-dependent protein kinase gene family in *Arabidopsis thaliana*. *Plant Mol. Biol.* 30:1259–75
64. Hrabak EM, Dickmann LJ, Satterlee JS, Sussman MR. 1996. Characterization of eight new members of the calmodulin-like domain protein kinase gene family from *Arabidopsis thaliana*. *Plant Mol. Biol.* 31:405–12
65. Hsiao AI, Quick WA. 1997. Roles of soluble sugars in protecting phytochrome- and gibberellin A3-mediated germination control in skotodormant lettuce seeds. *J. Plant Growth Regul.* 16:141–46
66. Hua J, Meyerowitz EM. 1998. Ethylene responses are negatively regulated by a receptor gene family in *Arabidopsis thaliana*. *Cell* 94:261–71
67. Huber SC, Huber JL, Liao PC, Gage DA, McMichael RW Jr, et al. 1996. Phosphorylation of serine-15 of maize leaf sucrose synthase. Occurrence in vivo and

possible regulatory significance. *Plant Physiol.* 112:793–802

68. Huijser C, Kortstee A, Pego JV, Wisman E, Weisbeek P, Smeekens SCM. 1999. The Arabidopsis *SUN6* gene is identical to *ABI4*. Submitted
69. Hwang YS, Karrer EE, Thomas BR, Chen L, Rodriguez RL. 1998. Three cis-elements required for rice α-amylase Amy3D expression during sugar starvation. *Plant Mol. Biol.* 36:331–41
70. Ishiguro S, Nakamura, K. 1992. The nuclear factor SP8BF binds to the 5′-upstream regions of three different genes coding for major proteins of sweet potato tuberous roots. *Plant Mol. Biol.* 18:97–108
71. Ishiguro S, Nakamura K. 1994. Characterization of a cDNA encoding a novel DNA-binding protein, SPF1, that recognizes SP8 sequences in the 5′ upstream regions of genes coding for sporamin and α-amylase from sweet potato. *Mol. Gen. Genet.* 244:563–71
72. Jackson SD, Heyer A, Dietze J, Prat S. 1996. Phytochrome B mediates the photoperiodic control of tuber formation in potato. *Plant J.* 9:159–66
73. Jang J-C, Leon P, Zhou L, Sheen J. 1997. Hexokinase as a sugar sensor in higher plants. *Plant Cell* 9:5–19
74. Jang J-C, Sheen J. 1997. Sugar sensing in higher plants. *Trends Plant Sci.* 2:208–14
75. Jang JC, Sheen J. 1994. Sugar sensing in higher plants. *Plant Cell* 6:1665–79
76. Jefferson R, Goldsbrough A, Bevan M. 1990. Transcriptional regulation of a patatin-1 gene in potato. *Plant Mol. Biol.* 14:995–1006
77. Jiang R, Carlson M. 1996. Glucose regulates protein interactions within the yeast SNF1 protein kinase complex. *Genes Dev.* 10:3105–15
78. Jitla D, Rogers G, Seneweera S, Basra A, Oldfield R, Conroy J. 1997. Accelerated early growth of rice at elevated CO_2. *Plant Physiol.* 115:15–22
79. Johnston M. 1999. Feasting, fasting and fermenting. Glucose sensing in yeast and other cells. *Trends Genet.* 15:29–33
80. Kaiser WM, Huber SC. 1994. Modulation of nitrate reductase in vivo and in vitro: effects of phosphoprotein phosphatase inhibitors, free Mg^{2+} and 5′ AMP. *Planta* 193:358–64
81. Kaplan CP, Tugal HB, Baker A. 1997. Isolation of a cDNA encoding an Arabidopsis galactokinase by functional expression in yeast. *Plant Mol. Biol.* 34:497–506
82. Karrer EE, Rodriguez RL. 1992. Metabolic regulation of rice α-amylase and sucrose synthase genes in planta. *Plant J.* 2:517–23
83. Kemp BE, Mitchelhill KI, Stapleton D, Michell BJ, Chen ZP, Witters LA. 1999. Dealing with energy demand: the AMP-activated protein kinase. *Trends Biochem. Sci.* 24:22–25
84. Kim D-J, Smith SM, Leaver CJ. 1997. A cDNA encoding a putative SPF1-type DNA-binding protein from cucumber. *Gene* 185:265–69
85. Kim SY, May GD, Park WD. 1994. Nuclear protein factors binding to a class I patatin promoter region are tuber-specific and sucrose-inducible. *Plant Mol. Biol.* 26:603–15
86. Kinsman E, Lewis C, Davies M, Young J, Francis D, et al. 1997. Elevated CO_2 stimulates cells to divide in grass meristems: a differential effect in two natural populations of *Dactylis glomerata*. *Plant Cell Environ.* 20:1309–16
87. Klein D, Stitt M. 1998. Effects of 2-deoxyglucose on the expression of RBCS and the metabolism of *Chenopodium rubrum* cell suspension cultures. *Planta* 205:223–34
88. Koch KE. 1996. Carbohydrate-modulated gene expression in plants. *Annu. Rev. Plant Physiol.Plant Mol. Biol.* 47:509–40
89. Kraakman L, Lemaire K, Ma P, Teunissen AWRH, Donaton M, et al. 1999. A *Saccharomyces cerevisiae* G-coupled receptor, Gpr1, is specifically required for glucose activation of the cAMP pathway during the

transition to growth on glucose. *Mol. Microbiol.* 32:1002–12

90. Kraakman LS, Winderickx J, Thevelein JM, de Winde JH. 1999. Structure-function analysis of yeast hexokinase: structural requirements for triggering cAMP signalling and catabolite repression. *Biochem. J.* 343:159–68
91. Krapp A, Hofmann B, Schafer C, Stitt M. 1993. Regulation of expression of *rbcS* and other photosynthetic genes by carbohydrates: a mechanism for the 'sink regulation' of photosynthesis? *Plant J.* 3:817–28
92. Kusnetsov V, Herrmann RG, Kulaeva ON, Oelmuller R. 1998. Cytokinin stimulates and abscisic acid inhibits greening of etiolated *Lupinus luteus* cotyledons by affecting the expression of the light-sensitive protochlorophyllide oxidoreductase. *Mol. Gen. Genet.* 259:21–28
93. Lakatos L, Klein M, Hofgen R, Banfalvi Z. 1999. Potato StubSNF1 interacts with StubGAL83: a plant protein kinase complex with yeast and mammalian counterparts. *Plant J.* 17:569–74
94. Lalonde S, Boles E, Hellmann H, Barker L, Patrick JW, et al. 1999. The dual function of sugar carriers. Transport and sugar sensing. *Plant Cell* 11:707–26
95. Lam H-M, Coschigano KT, Oliveira IC, Melo-Oliveira R, Coruzzi GM. 1996. The molecular-genetics of nitrogen assimilation into amino acids in higher plants. *Annu. Rev. Plant Physiol. Plant Mol. Biol.* 47:569–93
96. Levy YY, Dean C. 1998. The transition to flowering. *Plant Cell* 10:1973–90
97. Liu XJ, Prat S, Willmitzer L, Frommer WB. 1990. Cis regulatory elements directing tuber-specific and sucrose-inducible expression of a chimeric class I patatin promoter/GUS-gene fusion. *Mol. Gen. Genet.* 23:401–6
98. Lucas WJ, Wolf S. 1999. Connections between virus movement, macromolecular signaling and assimilate allocation. *Curr. Opin. Plant Biol.* 2:192–97
99. Lue M-Y, Lee H. 1994. Protein phosphatase inhibitors enhance the expression of an α-amylase gene, αAmy3, in cultured rice cells. *Biochem. Biophys. Res. Commun.* 205:807–16
100. MacKintosh C. 1998. Regulation of cytosolic enzymes in primary metabolism by reversible protein phosphorylation. *Curr. Opin. Plant Biol.* 1:224–29
101. Martin T, Hellmann H, Schmidt R, Willmitzer L, Frommer WB. 1997. Identification of mutants in metabolically regulated gene expression. *Plant J.* 11:53–62
102. Matschinsky F, Liang Y, Kesavan P, Wang L. 1993. Glucokinase as pancreatic β cell glucose sensor and diabetes gene. *J. Clin. Invest.* 92:2092–98
103. Medford JI, Sussex IM. 1989. Regulation of chlorophyll and Rubisco levels in embryonic cotyledons of *Phaseolus vulgaris*. *Planta* 179:309–15
104. Mita S, Hirano H, Nakamura K. 1997. Negative regulation in the expression of a sugar-inducible gene in *Arabidopsis thaliana*; a recessive mutation causing enhanced expression of a gene for β-amylase. *Plant Physiol.* 114:575–82
105. Mita S, Murano N, Akaike M, Nakamura K. 1997. Mutants of *Arabidopsis thaliana* with pleiotropic effects on the expression of the gene for β-amylase and on the accumulation of anthocyanin that are inducible by sugars. *Plant J.* 11:841–51
106. Mita S, Suzuki-Fujii K, Nakamura K. 1995. Sugar-inducible expression of a gene for β-amylase in *Arabidopsis thaliana*. *Plant Physiol.* 107:895–904
107. Moore BD, Sheen J. 1999. Plant sugar sensing and signaling—a complex reality. *Trends Plant Sci.* 4:250
108. Moorhead G, Douglas P, Cotelle V, Harthill J, Morrice N, et al. 1999. Phosphorylation-dependent interactions between enzymes of plant metabolism and 14-3-3 proteins. *Plant J.* 18:1–12
109. Morita A, Umemura T, Kuroyanagi M, Futsuhara Y, Perata P, Yamaguchi J. 1998.

Functional dissection of a sugar-repressed α-amylase gene (RAmy1 A) promoter in rice embryos. *FEBS Lett.* 423:81–85
110. Müller J, Boller T, Wiemken A. 1998. Trehalose affects sucrose synthase and invertase activities in soybean (*Glycine max* L. Merr.) roots. *J. Plant Physiol.* 153:255–57
111. Nemeth K, Salchert K, Putnoky P, Bhalerao R, Koncz-Kalman Z, et al. 1998. Pleiotropic control of glucose and hormone responses by PRL1: a nuclear WD protein, in Arabidopsis. *Genes Dev.* 12:3059–73
112. Nielsen TH, Krapp A, Roeper-Schwarz U, Stitt M. 1998. The sugar-mediated regulation of genes encoding the small subunit of Rubisco and the regulatory subunit of ADP glucose pyrophosphorylase is modified by phosphate and nitrogen. *Plant Cell Environ.* 21:443–54
113. Ohto M, Hayashi K, Isobe M, Nakamura K. 1995. Involvement of Ca^{2+}-signalling in the sugar-inducible expression of genes coding for sporamin and β-amylase of sweet potato. *Plant J.* 7:297–307
114. Ohto M, Nakamura-Kito K, Nakamura K. 1992. Induction of expression of genes coding for sporamin and β-amylase by polygalacturonic acid in leaf-petiole cuttings of sweet potato. *Plant Physiol.* 99:422–27
115. Ohto M, Nakamura K. 1995. Sugar-induced increase of calcium-dependent protein kinases associated with the plasma membrane in leaf tissues of tobacco. *Plant Physiol.* 109:973–81
116. Özcan S, Dover J, Johnston M. 1998. Glucose sensing and signaling by two glucose receptors in the yeast *Saccharomyces cerevisiae*. *EMBO J.* 17:2566–73
117. Özcan S, Dover J, Rosenwald AG, Wolfl S, Johnston M. 1996. Two glucose transporters in *Saccharomyces cerevisiae* are glucose sensors that generate a signal for induction of gene expression. *Proc. Natl. Acad. Sci. USA* 93:1–5
118. Paul MJ, Driscoll SP. 1997. Sugar repression of photosynthesis: the role of carbohydrates in signalling nitrogen deficiency through source:sink imbalance. *Plant Cell Environ.* 20:110–16
119. Pego JV, Kortstee A, Huijser C, Smeekens SCM. 2000. Photosynthesis, sugars and the regulation of gene expression. *J. Exp. Bot.* 51:407–16
120. Pego JV, Krapp A, Wobbes B, Weisbeek P, Stitt M, Smeekens SCM. 2000. Arabidopsis fructokinase 2 and its involvement in metabolite-mediated regulation of gene expression. Submitted
121. Pego JV, Weisbeek PJ, Smeekens SCM. 1999. Mannose inhibits Arabidopsis germination via a hexokinase-mediated step. *Plant Physiol.* 119:1017–23
122. Perata P, Matsukura C, Vernieri P, Yamaguchi J. 1997. Sugar repression of a gibberellin-dependent signaling pathway in barley embryos. *Plant Cell* 9:2197–208
123. Platt A, Reece RJ. 1998. The yeast galactose genetic switch is mediated by the formation of a Gal4p-Gal80p-Gal3p complex. *EMBO J,* 17:4086–91
124. Prata RTN, Williamson JD, Conkling MA, Pharr DM. 1997. Sugar repression of mannitol dehydrogenase activity in celery cells. *Plant Physiol.* 114:307–14
125. Purcell PC, Smith AM, Halford NG. 1998. Antisense expression of a sucrose non-fermenting-1-related protein kinase sequence in potato results in decreased expression of sucrose synthase in tubers and loss of sucrose-inducibility of sucrose synthase transcripts in leaves. *Plant J.* 14:195–202
126. Reynolds SJ, Smith SM. 1995. Regulation of expression of the cucumber isocitrate lyase gene in cotyledons upon seed germination and by sucrose. *Plant Mol. Biol.* 29:885–96
127. Riou-Khamlichi C, Huntley R, Jacqmard A, Murray JA. 1999. Cytokinin activation of Arabidopsis cell division through a D-type cyclin. *Science* 283:1541–44
128. Robertson M, Swain SM, Chandler PM,

Olszewski NE. 1998. Identification of a negative regulator of gibberellin action, HvSPY, in barley. *Plant Cell* 10:995–1007

129. Robinson CK, Hill SA. 1999. Altered resource allocation during seed development in Arabidopsis by the *abi3* mutation. *Plant Cell Environ.* 22:117–23
130. Rogers JC, Rogers SW. 1992. Definition and functional implications of gibberellin and abscisic acid *cis*-acting hormone response complexes. *Plant Cell* 4:1443–51
131. Rohde A, van Montagu M, Boerjan W. 1999. The *ABSCISIC ACID-INSENSITIVE 3* (*ABI3*) gene is expressed during vegetative quiescence processes in Arabidopsis. *Plant Cell Environ.* 22:261–70
132. Roitsch T. 1999. Source-sink regulation by sugar and stress. *Curr. Opin. Plant Biol.* 2:198–206
133. Roitsch T, Bittner M, Godt DE. 1995. Induction of apoplastic invertase of *Chenopodium rubrum* by D-glucose and a glucose analog and tissue-specific expression suggest a role in sink-source regulation. *Plant Physiol.* 108:285–94
134. Romero C, Bellés JM, Vayaz JL, Serrano R, Culianez-Macià FA. 1997. Expression of the yeast *trehalose-6-phosphate synthase* gene in transgenic tobacco plants: pleiotropic phenotypes include drought tolerance. *Planta* 201:293–97
135. Rook F, Gerrits N, Kortstee A, van Kampe M, Borrias M, et al. 1998. Sucrose-specific signalling represses translation of the Arabidopsis *ATB2* bZIP transcription factor gene. *Plant J.* 15:253–63
136. Rook F, Weisbeek PJ, Smeekens SCM. 1998. The light-controlled *Arabidopsis* bZIP transcription factor gene *ATB2* encodes a protein with an unusually long leucine zipper domain. *Plant Mol. Biol.* 37:171–78
137. Sadka A, DeWald DB, May GD, Park WD, Mullet JE. 1994. Phosphate modulates transcription of soybean *VspB* and other sugar inducible genes. *Plant Cell* 6:737–49
138. Salchert K, Bhalerao R, Koncz-Kalman Z, Koncz C. 1998. Control of cell elongation and stress responses by steroid hormones and carbon catabolic repression in plants. *Philos. Trans. R. Soc. London Ser. B* 353:1517–20
139. Salzman RA, Tikhonova I, Bordelon BP, Hasegawa PM, Bressan RA. 1998. Coordinate accumulation of antifungal proteins and hexoses constitutes a developmentally controlled defense response during fruit ripening in grape. *Plant Physiol.* 117:465–72
140. Sarah CJ, Graham IA, Reynolds SJ, Leaver CJ, Smith SM. 1996. Distinct cis-acting elements direct the germination and sugar responses of the cucumber malate synthase gene. *Mol. Gen. Genet.* 250:153–61
141. Scheible W-R, Gonzalez-Fontez A, Lauerer M, Müller-Röber B, Caboche M, Stitt M. 1997. Nitrate acts as a signal to induce organic acid metabolism and repress starch metabolism in tobacco. *Plant Cell* 9:809–24
142. Sharkey TD, Vassey TL, Vanderveer PJ, Vierstra RD. 1991. Carbon metabolism and photosynthesis in transgenic tobacco (*Nicotiana tabacum* L.) having excess phytochrome. *Planta* 185:287–96
143. Sharma R, Schopfer P. 1987. Phytochrome-mediated regulation of β-amylase mRNA level in mustard (*Sinapsis alba* L.) cotyledons. *Planta* 171:313–20
144. Sheen J. 1990. Metabolic repression of transcription in higher plants. *Plant Cell* 2:1027–38
145. Sherson S, Gy I, Medd J, Schmidt R, Dean C, et al. 1999. The arabinose kinase, *ARA1*, gene of Arabidopsis is a novel member of the galactose kinase gene family. *Plant Mol. Biol.* 39:1003–12
146. Sheu J, Jan S, Lee H, Yu S. 1994. Control of transcription and mRNA turnover as

mechanisms of metabolic repression of α-amylase gene expression. *Plant J.* 5:655–64
147. Short TW. 1999. Overexpression of Arabidopsis phytochrome B inhibits phytochrome A function in the presence of sucrose. *Plant Physiol.* 119:1497–506
148. Sims DA, Luo Y, Seemann JR. 1998. Importance of leaf versus whole plant CO_2 environment for photosynthetic acclimation. *Plant Cell Environ.* 21:1189–96
149. Smeekens SCM. 1998. Sugar regulation of gene expression in plants. *Curr. Opin. Plant Biol.* 1:230–34
150. Smeekens SCM, Rook F. 1997. Sugar sensing and sugar-mediated signal transduction in plants. *Plant Physiol.* 115:7–13
151. Soni R, Carmichael JP, Shah ZH, Murray JA. 1995. A family of cyclin D homologs from plants differentially controlled by growth regulators and containing the conserved retinoblastoma protein interaction motif. *Plant Cell* 7:85–103
152. Stitt M. 1999. Nitrate regulation of metabolism and growth. *Curr. Opin. Plant Biol.* 2:178–86
153. Sugden C, Donaghy PG, Halford NG, Hardie DG. 1999. Two SNF1-related protein kinases from spinach leaf phosphorylate and inactivate 3-hydroxy-3-methylglutaryl-coenzyme A reductase, nitrate reductase, and sucrose phosphate synthase in vitro. *Plant Physiol.* 120:257–74
154. Takeda S, Mano S, Ohto M, Nakamura K. 1994. Inhibitors of protein phosphatases I and 2A block the sugar-inducible gene expression in plants. *Plant Physiol.* 106:567–74
155. Tang GQ, Luscher M, Sturm A. 1999. Antisense repression of vacuolar and cell wall invertase in transgenic carrot alters early plant development and sucrose partitioning. *Plant Cell* 11:177–89
156. Tegeder M, Wang XD, Frommer WB, Offler CE, Patrick JW. 1999. Sucrose transport into developing seeds of *Pisum sativum* L. *Plant J.* 18:151–61
157. Thevelein JM, Hohmann S. 1995. Trehalose synthase: guard to the gate of glycolysis in yeast? *Trends Biochem Sci.* 20:3–10
158. Thiele A, Herold M, Lenk I, Quail PH, Gatz C. 1999. Heterologous expression of arabidopsis phytochrome B in transgenic potato influences photosynthetic performance and tuber development. *Plant Physiol.* 120:73–82
159. Thomas BR, Rodriguez RL. 1994. Metabolite signals regulate gene expression and source/sink relations in cereal seedlings. *Plant Physiol.* 106:1235–39
160. Thomashow MF. 1999. Plant cold acclimation: freezing tolerance genes and regulatory mechanisms. *Annu. Rev. Plant Physiol. Plant Mol. Biol.* 50:571–99
161. Toroser D, Huber SC. 1997. Protein phosphorylation as a mechanism for osmotic-stress activation of sucrose-phosphate synthase in spinach leaves. *Plant Physiol.* 114:947–55
162. Toroser D, Huber SC. 1998. 3-Hydroxy-3-methylglutaryl-coenzyme A reductase kinase and sucrose-phosphate synthase kinase activities in cauliflower florets: Ca^{2+} dependence and substrate specificities. *Arch. Biochem. Biophys.* 355:291–300
163. Treitel MA, Carlson M. 1995. Repression by SSN6-TUP1 is directed by MIG1: a repressor/activator protein. *Proc. Natl. Acad. Sci. USA* 92:3132–36
164. Treitel MA, Kuchin S, Carlson M. 1998. Snf1 protein kinase regulates phosphorylation of the Mig1 repressor in *Saccharomyces cerevisiae*. *Mol. Cell. Biol.* 18:6273–80
165. Umemura T, Perata P, Futsuhara Y, Yamaguchi J. 1998. Sugar sensing and α-amylase gene repression in rice embryos. *Planta* 204:420–28
166. van Oosten JJ, Gerbaud A, Huijser C, Dijkwel PP, Chua N-H, Smeekens SCM. 1997. An Arabidopsis mutant showing

reduced feedback inhibition of photosynthesis. *Plant J.* 12:1011–20

167. Vogel G, Aeschbacher RA, Müller J, Boller T, Wiemken A. 1998. Trehalose-6-phosphate phosphatases from *Arabidopsis thaliana*: identification by functional complementation of the yeast *tps2* mutant. *Plant J.* 13:673–83
168. Vreugdenhil D. 1983. Abscisic acid inhibits phloem loading of sucrose. *Physiol. Plant.* 57:463–67
169. Wanner LA, Junttila O. 1999. Cold-induced freezing tolerance in Arabidopsis. *Plant Physiol.* 120:391–99
170. Weatherwax SC, Ong MS, Degenhardt J, Bray EA, Tobin EM. 1996. The interaction of light and abscisic acid in the regulation of plant gene expression. *Plant Physiol.* 111:363–70
171. Weatherwax SC, Williams SA, Tingay S, Tobin EM. 1998. The phytochrome response of the *Lemna gibba NPR1* gene is mediated primarily through changes in abscisic acid levels. *Plant Physiol.* 116:1299–305
172. Weber H, Borisjuk L, Heim U, Sauer N, Wobus U. 1997. A role for sugar transporters during seed development: molecular characterization of a hexose and a sucrose carrier in fava bean seeds. *Plant Cell* 9:895–908
173. Weber H, Borisjuk L, Wobus U. 1996. Controlling seed development and seed size in *Vicia faba*: a role for seed coat-associated invertases and carbohydrate state. *Plant J.* 10:823–34
174. Weber H, Buchner P, Borisjuk L, Wobus U. 1996. Sucrose metabolism during cotyledon development of *Vicia faba* L. is controlled by the concerted action of both sucrose-phosphate synthase and sucrose synthase: expression patterns, metabolic regulation and implications for seed development. *Plant J.* 9:841–50
175. Weber H, Heim U, Golombek S, Borisjuk L, Manteuffel R, Wobus U. 1998. Expression of a yeast-derived invertase in developing cotyledons of *Vicia narbonensis* alters the carbohydrate state and affects storage functions. *Plant J.* 16:163–72
176. Weber H, Wobus U. 1997. Sugar import and metabolism during seed development. *Trends Plant Sci.* 2:169–74
177. Wenzler HC, Mignery G, Fisher L, Park W. 1989. Sucrose-regulated expression of a chimeric potato tuber gene in leaves of transgenic tobacco plants. *Plant Mol. Biol.* 13:347–54
178. Whitelam GC, Johnson E, Peng J, Carol P, Anderson ML, et al. 1993. Phytochrome A null mutants of Arabidopsis display a wild-type phenotype in white light. *Plant Cell* 5:757–68
179. Williams RW, Wilson JM, Meyerowitz EM. 1997. A possible role for kinase-associated protein phosphatase in the Arabidopsis CLAVATA1 signaling pathway. *Proc. Natl. Acad. Sci. USA* 94:10467–72
180. Wobus U, Weber H. 1999. Seed maturation: genetic programmes and control signals. *Curr. Opin. Plant Biol.* 2:33–38
181. Wu K, Rooney MF, Ferl RJ. 1997. The Arabidopsis 14-3-3 multigene family. *Plant Physiol.* 114:1421–31
182. Xu X, van Lammeren AA, Vermeer E, Vreugdenhil D. 1998. The role of gibberellin, abscisic acid, and sucrose in the regulation of potato tuber formation in vitro. *Plant Physiol.* 117:575–84
183. Yanovsky MJ, Alconada-Magliano TM, Mazzella MA, Gatz C, Thomas B, Casal JJ. 1998. Phytochrome A affects stem growth, anthocyanin synthesis, sucrose-phosphate-synthase activity and neighbour detection in sunlight-grown potato. *Planta* 205:235–41
184. Yokoyama R, Hirose T, Fujii N, Aspuria ET, Kato A, Uchimiya H. 1994. The rolC promoter of Agrobacterium rhizogenes Ri plasmid is activated by sucrose in transgenic tobacco plants. *Mol. Gen. Genet.* 244:15–22
185. Yu SM, Kuo YH, Sheu G, Sheu YJ, Liu LF. 1991. Metabolic derepression of

α-amylase gene expression in suspension-cultured cells of rice. *J. Biol. Chem.* 266:21131–37

186. Zenke FT, Engles R, Vollenbroich V, Meyer J, Hollenberg CP, Breunig KD. 1996. Activation of Gal4p by galactose-dependent interaction of galactokinase and Gal80p. *Science* 272:1662–65 [erratum in *Science* 273:417]

187. Zhou D, Qian D, Cramer CL, Yang Z. 1997. Developmental and environmental regulation of tissue- and cell-specific expression for a pea protein farnesyltransferase gene in transgenic plants. *Plant J.* 12:921–30

188. Zhou L, Jang JC, Jones TL, Sheen J. 1998. Glucose and ethylene signal transduction crosstalk revealed by an Arabidopsis glucose-insensitive mutant. *Proc. Natl. Acad. Sci. USA* 95:10294–99

Annu. Rev. Plant Physiol. Plant Mol. Biol. 2000. 51:83–109

THE CHLOROPLAST ATP SYNTHASE: A Rotary Enzyme?

R. E. McCarty, Y. Evron, and E. A. Johnson
Department of Biology, Johns Hopkins University, Baltimore, Maryland 21218; e-mail: REM1@jhu.edu

Key Words ATPase, ATP synthesis, mitochondria, F1Fo, CF1CFo

■ **Abstract** The chloroplast adenosine triphosphate (ATP) synthase is located in the thylakoid membrane and synthesizes ATP from adenosine diphosphate and inorganic phosphate at the expense of the electrochemical proton gradient formed by light-dependent electron flow. The structure, activities, and mechanism of the chloroplast ATP synthase are discussed. Emphasis is given to the inherent structural asymmetry of the ATP synthase and to the implication of this asymmetry to the mechanism of ATP synthesis and hydrolysis. A critical evaluation of the evidence in support of and against the notion that one part of the enzyme rotates with respect to other parts during catalytic turnover is presented. It is concluded that although rotation can occur, whether it is required for activity of the ATP synthase has not been established unequivocally.

CONTENTS

INTRODUCTION

The adenosine triphosphate (ATP) synthases couple the transmembrane movement of protons to the synthesis and hydrolysis of ATP. These enzymes are also called F-ATPases and are present in the energy transducing membranes of chloroplasts,

1040-2519/00/0601-0083$14.00

mitochondria, and some bacteria. The "F" of F-ATPases is derived from F1, the first factor isolated from mitochondria shown to be required for oxidative phosphorylation (65). A similar factor, designated CF1 for chloroplast F1, was extracted from spinach chloroplast thylakoids (7, 85). F1 from the inner membrane of *Escherichia coli* (31) has properties similar to those of F1 and CF1 and is often called ECF1.

The chloroplast ATP synthase is located in the energy-converting thylakoid membrane and makes ATP at the expense of the electrochemical gradient established by photoelectron transfer. In this article we review the structure and activities of the chloroplast ATP synthase. A prominent feature of this discussion is consideration of symmetric and asymmetric aspects of the structure of the enzyme. Special emphasis is given to the concept that rotation of one segment of the enzyme with respect to another is part of the catalytic mechanism.

STRUCTURE

Subunit Composition and Asymmetry

ATP synthases are composed of two parts; F1 and Fo. Thus, the ATP synthase is sometimes referred to as F1Fo or FoF1. F1 is a large, hydrophilic protein that is associated with the membrane by virtue of its interactions with Fo. After its release from the membrane, F1 is very soluble in water. In contrast, Fo is hydrophobic and detergents must be used to effect its solubilization. F1 is also called coupling factor, since its removal from the membrane uncouples ATP synthesis from electron transport and its reconstitution restores, or recouples, ATP synthesis. Fo is not a coupling factor and was first described as a factor that restored sensitivity of the ATPase activity of mitochondrial F1 (MF1) to oligomycin, from which the letter "o" was derived (42). The part of the chloroplast ATP synthase that is an integral to the thylakoid membrane is called CFo for chloroplast Fo. Thus, the chloroplast ATP synthase may be denoted CF1CFo and that from *E. coli*, ECF1ECFo.

CF1 The nucleotide binding sites of the chloroplast ATP synthase are present in CF1, which consists of nine polypeptide chains and five different proteins, labeled α-ε in order of decreasing molecular mass. The α and β polypeptides are present in three copies each, whereas the γ, δ and ε subunits are present in single copies. The α and β subunits alternate to form a heterohexameric ring that in itself has threefold symmetry (82). This symmetry is, however, broken by the presence of the single copy subunits that lack threefold symmetry. The interactions of a particular α/β pair with the γ, δ, and ε subunits cannot be equivalent to those of the other two α/β pairs. Thus, CF1 is inherently structurally asymmetric.

The subunit structures of CF1, MF1, and ECF1 are compared in Table 1 (40, 43, 86, 87). The primary structure of the β subunit is strongly conserved,

TABLE 1 Comparison of subunits from various ATP synthases. Sequences aligned using CLUSTALW and similarity of aligned residues compared to CFlCFo. Molecular mass (kDa) of subunit listed in parentheses

	Sequence similarities				
	α	β	γ	δ	ε
CF1	(55.5)	(53.7)	(35.8)	(20.4)	(14.7)
ECF1	80% (55.2)	82% (50.2)	65% (31.6)	57% (19.3)	70% (14.9)
F1	82% (55.3)	86% (51.6)	59% (30.3)	45% (15.1)	17% (5.7)
	IV(a)	**I(b)***	**II(b)***	**III(c)**	
CFo	(25.1)	(21.0)	(16.5)	(8.0)	
ECFo	58% (30.3)	49% (17.3)	52% (17.3)	68% (8.3)	
Fo	54% (24.8)	49% (24.7)	46% (24.7)	68% (7.6)	

*Both subunits I and II are similar to subunit b in ECFo and Fo.

whereas that of the α subunits is less well conserved. The amino acid sequence identity among the γ, δ and ε subunits is much less than that for the larger subunits. The ε subunit of MF1 shows no amino acid sequence similarity to the ε subunit of either CF1 or ECF1.

The six nucleotide binding sites in CF1 are located at the interfaces between the α and β subunits (9). If CF1 were symmetric, there should be just two types of binding sites that could differ with respect to rates of nucleotide association or dissociation as well as dissociation constants. Instead, CF1 has three distinct classes of nucleotide binding sites. Also, the properties of two sites within a class may be quite different. These differences among the nucleotide binding sites are likely to be induced by the asymmetric interactions between the α/β subunits and the γ subunit.

The α subunits of CF1 are chemically equivalent, but structurally asymmetric. Lys-378 of just one of the three α subunits of CF1 reacts rapidly with Lucifer yellow vinyl sulfone (49). The extent of labeling of αLys378 by Lucifer yellow vinyl sulfone in α/β preparations was close to three times higher than that in CF1 preparations. Since reconstitution of the α/β preparation with the γ subunit restored the selective labeling of one α subunit (49), interactions with the γ subunit are responsible for the asymmetry of the α subunits.

Although all of the CF1 subunits are required for ATP synthesis, neither the δ nor the ε subunit is required for ATPase activity (61). In solution intact CF1 can form substoichiometric amounts of bound ATP from bound ADP and medium P_i (25). Although soluble CF1 cannot catalyze the net synthesis of ATP, it will hydrolyze ATP at a low rate. The ε subunit is an inhibitor of the ATPase activity and fulfills a regulatory role as well as being a part of the mechanism that couples proton flow to ATP synthesis (69). The δ subunit has no effect on ATPase activity, but is required for the functional binding of CF1 to CFo (24, 91).

CFo CF1CFo (26, 64) and CFo (27) have been purified, but are difficult to work on because of their insolubility in the absence of detergents. CFo binds CF1 and transports protons across the thylakoid membrane. CFo contains four different polypeptides that are labeled I–IV (Table 1). Subunits I, II, and III are in the order of decreasing molecular mass. Subunit IV was found to be a part of the complex after subunits I, II, and III had been detected (30). ECFo contains three different polypeptides (a–c), whereas MFo has as many as nine different proteins. The stoichiometry of the CFo subunits has not been established firmly. It is , however, clear that subunit III of CFo, which is equivalent to subunit c of ECFo, is present in multiple copies (from 6 to 12) per complex. Subunits I, II, and IV may be present in single copies. On the basis of deduced amino acid sequences, subunit IV of CFo is related to subunit a of ECFo and CFo subunit I, to ECFo subunit b. The b subunit of ECFo is a dimer (54) and subunits I and II of CF1 may interact in a manner that is functionally equivalent to the b subunit dimer.

Analysis of mutants of *Chlamydomonas reinhardtii* revealed that subunits I, II, and III were required for CF1 binding (48). Although subunit IV is required for activity, it is not required for the binding of CF1 to CFo (28). A subunit III–CF1 complex has been isolated, demonstrating directly that subunit III binds to CF1 (89). Chemical cross-linking has revealed the physical proximity of some of the CF1 subunits to those of CFo. The following cross-links have been detected: subunit III-ε, subunit I-β, subunit I-γ, subunit II-α, subunit II-β, subunit II-γ (81), and subunit I-δ (10). The cross-linking of CFo subunit I with the larger CF1 subunits is interesting in that subunit I is predicted to have a large polar domain and is anchored to the membrane by a single transmembrane α-helix. The polar domain of subunit I likely extends from the membrane to the α/β heterohexamer. With the exception of the cross-links between subunits I and II with the γ subunits, the cross-linking data are consistent with the model of CF1 CFo shown in Figure 6.

Subunit III has a molecular weight of just 8×10^3 and is predicted to form two transmembrane helices connected by a short polar loop that projects slightly from the stromal side of the thylakoid membrane. Subunit c of ECF1 assumes a hairpin structure as determined by NMR (29).

Low-Resolution Structures

Electron Microscopy F1 and F1Fo have been examined by electron microscopy. Cryomicroscopy and single molecule averaging of ECF1ECFo revealed a 2.5 to 3 nm wide mass (a stalk) that connected the bulk of ECF1 to ECFo (34). The length of the stalk in these reconstructed images was 4 to 4.5 nm. Negatively stained CF1CFo images were subjected to single molecule averaging. The appearance of CF1CFo was more compact than that of ECF1ECFo, with the length of the stalk estimated to be 3 nm (13). A splitting of density in the stalk region of CF1CFo was taken as evidence for two separate stalks (14). The shorter stalk(s) observed in the chloroplast ATP synthase seems to be more consistent with structural information gleaned from biochemical and distance mapping by fluorescence resonance energy transfer (FRET). The ε subunit of ECF1 cross-links with subunit c of ECFo and the α and β subunits (4, 93). Therefore, the stalk(s) that connects CF1 to CFo with a connecting stalk of 4 to 4.5 nm is likely to be shorter than that estimated by cryomicroscopy or ECF1-ECFo.

Electron microscopy of the F1 part of the ATP synthase revealed the hexagonal ring of α/β subunits and a central location within this ring of at least part of the γ subunit. Immunocryoelectron microscopy has been used to localize the smaller subunits of ECF1. Cryomicroscopy revealed that 1.4-nm gold particle attached to the ε subunit could assume two positions relative to the α subunits tagged with Fab fragments of monoclonal antibodies against the α subunit (3). Analysis of negatively stained preparations of CF1 suggested that the δ or ε subunits of CF1 could be in structurally different positions (13). These microscopic observations suggest that the smaller ECF1 and CF1 subunits are not locked into one asymmetric position relative to the α/β heterohexamer.

Structural Mapping by Fluorescence Resonance Energy Transfer The nonradiative transfer of energy from a donor molecule to an acceptor molecule depends on a number of factors, including the quantum yield of the donor, the extinction coefficient of the acceptor, the overlap of the fluorescence emission spectrum of the donor and the absorption spectrum of the acceptor, and the orientation of the two molecules. The efficiency of energy transfer between two probes decreases with distance, R, as R^{-6}. By measuring energy transfer, distances between two probes may be determined. A number of specific residues or sites within CF1 have been labeled with reagents that are good donors or acceptors (51). A structural map based on more than thirty individual measurements is shown in Figure 1. Several structural features of CF1 are apparent. The asymmetry of the enzyme as well as the central location of part of the γ subunit are seen. That Lucifer yellow vinyl sulfone reacts with just one of the three αLys378 residues in CF1 is confirmed by the fact that seven different FRET measurements to bound Lucifer yellow converge on a single location.

Although electron micrographic studies suggest that the ε subunit may exist in more than one position (13), the FRET map indicates that this subunit has a preferred orientation in CF1 in solution. Six independent measurements from various sites on CF1 subunits to reagents attached to εCys6 converge to a single location. Recently, a distance of about 3.5 nm from εTrp57 to pyrenylmaleimide attached to γCys322 was determined (EA Johnson, unpublished observations). This distance is in accord with those determined previously.

The distance between γCys89 to γCys322 has not been measured directly, but a best-fit distance was computed to be within 8 Å. This distance is in accord with the evidence that these two residues may be cross-linked, either directly or by bisfunctional maleimides (52). However, the amino acids in the same positions

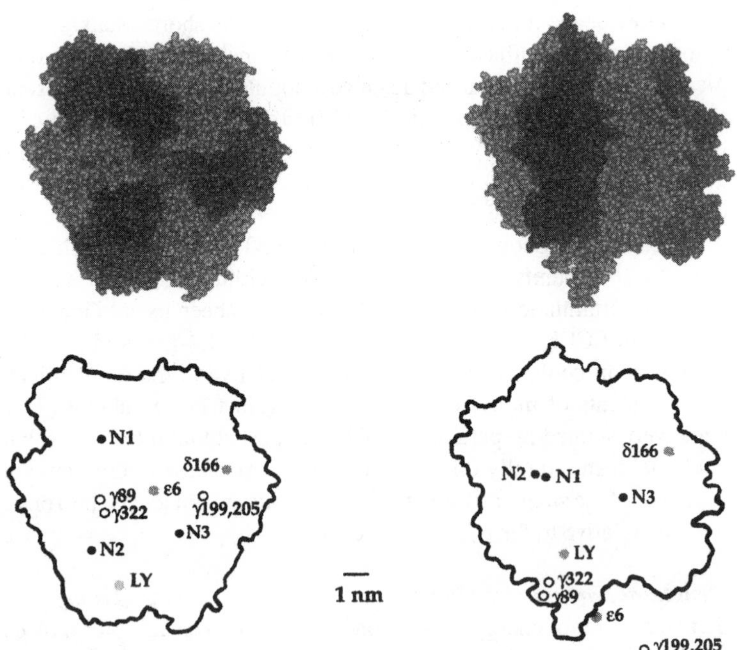

Figure 1 Structural map of CF1. Distances between specific sites on CF1 were determined by fluorescence resonance energy transfer. The coordinates of the δ subunit were from ref. (23) and the remainder, from ref. (72). LY stands for Lucifer yellow attached to αLys378. N1, N2 and N3 are nucleotide binding sites. γ322, γ89 and the γ199,205 are Cys residues of the γ subunit. ε 6 is εCys 6. For clarity, only one of the four residues (166) of the δ subunit mapped is shown. Heart MF1 is shown at the same scale. The α subunits are shown in darker gray than the β subunits. Left: (upper) MF1 viewed from the top; (lower) top view of the FRET map superimposed on the outline of MF1. Right: (upper) side view of MF1; (lower) side view of the FRET map superimposed on the outline of the MF1. Molecules were rendered using Insight II software.

in MF1 are more than 70 Å from each other in the crystal structure. The 3.5-nm distance between εTrp57 and γCys322 is also consonant with γCys322 being close to γCys89. The distance between Glu65 of subunit III of CFo and γCys89 was estimated to be 41 Å, whereas that from Glu65 of subunit III to γCys199,205 was 50 Å (57). Glu65 of subunit III was determined to be 6 to 10 Å from the bilayer surface and, if it assumed that γCys89 is directly above this residue, γCys89 would be from 30 to 35 Å from the surface of the membrane. This distance is more consistent with a stalk of 30 Å than of 40 to 45 Å.

Based on its requirement for the functional binding to CFo, the δ subunit's position in CF1CFo was assumed to be beneath the α/β subunits in contact with CFo. The δ subunit contains neither Trp nor Cys residues and was difficult to label specifically with fluorescent probes. This problem was neatly solved by using recombinant δ subunit, expressed in *E. coli*, into which Ser residues were individually changed to Cys residues. FRET distances were determined between the labeled δ and several sites on CF1 (23). The FRET results are consistent with the δ subunit being in association with one $\alpha\beta$ pair, well removed from the membrane surface (Figure 1). The close association of the δ subunit with α/β explains why the δ subunit stabilizes the ATPase activity of CF1-ε from inactivation in the cold. α/β-δ interactions would stabilize the α/β heterohexamer. The δ subunit forms a zero-length cross-link with subunit I of CFo (10), suggesting that the δ subunit is in close proximity to subunit I. The subunit I-δ subunit complex, possibly together with subunit II of CFo, could be a second stalk.

High-Resolution Structures

A high resolution structure is not available for any part of the chloroplast ATP synthase. CF1 has proven difficult to crystallize. Two structures of MF1 mitochondria, one from rat liver (12) and the other from beef heart (1), have been solved to a resolution of 2.85 Å. Although the two MF1 structures differ in some respects, especially in the occupancy of the nucleotide binding sites and the symmetry of the crystals, the structures resemble each other. Representations of the beef heart MF1 structure are shown in Figures 1 and 2. Major structural features of MF1 are the alternating α/β subunits (80 × 100 Å) surrounding two long α-helices that are the N- and C-terminal regions of the γ subunit. These helices are arranged as an antiparallel coiled coil, with the C-terminal α-helix, reaching to near the top of the α/β heterohexamer. A third short α-helix is attributed to the part of the γ subunit that contains the Cys residue (Cys 89 in CF1 γ) that is present in all eukaryotic F1s. Although they were present in the crystals, the δ and ε subunits as well as 145 residues of the γ subunit were not resolved.

The crystal structures confirmed that the nucleotide binding sites are at the α/β subunit interfaces and revealed residues in the α and β subunits that are involved in the binding of nucleotides. The P-loop motif (GXXXGKT/S), predicted to be involved in the binding of the phosphates of the nucleotide, binds the di- and triphosphates of nucleotides in both the α and β sites. An explanation for why the α subunits are noncatalytic was based on the lack in the α binding sites of a Glu residue which in the β subunits (βGlu188) is in close contact to the terminal phosphate of AMP-PNP (1). The P-loop motifs and the Glu residue in the β subunits involved in catalysis are also found in CF1.

Four of the six sites of the beef heart F1 in the crystals contained AMP-PMP. A fifth contained ADP, and the last contained no nucleotide (1). Rat liver MF1, which was crystallized under very different conditions from those used for heart MF1, contained ATP in all three α sites and ADP in all three β sites. Two of the β sites also contain Pi, which was included in the crystallization buffer (12). Also evident from the structure of the heart MF1 is the asymmetry of the α and β subunits. Although the structure of the β subunits is very similar overall, they differ significantly in the region of nucleotide binding regions. These differences are induced, at least in part, by the γ subunit. The α subunits also show asymmetry in the nucleotide binding areas. Interestingly, the α/β pairs in liver F1 exhibit threefold symmetry.

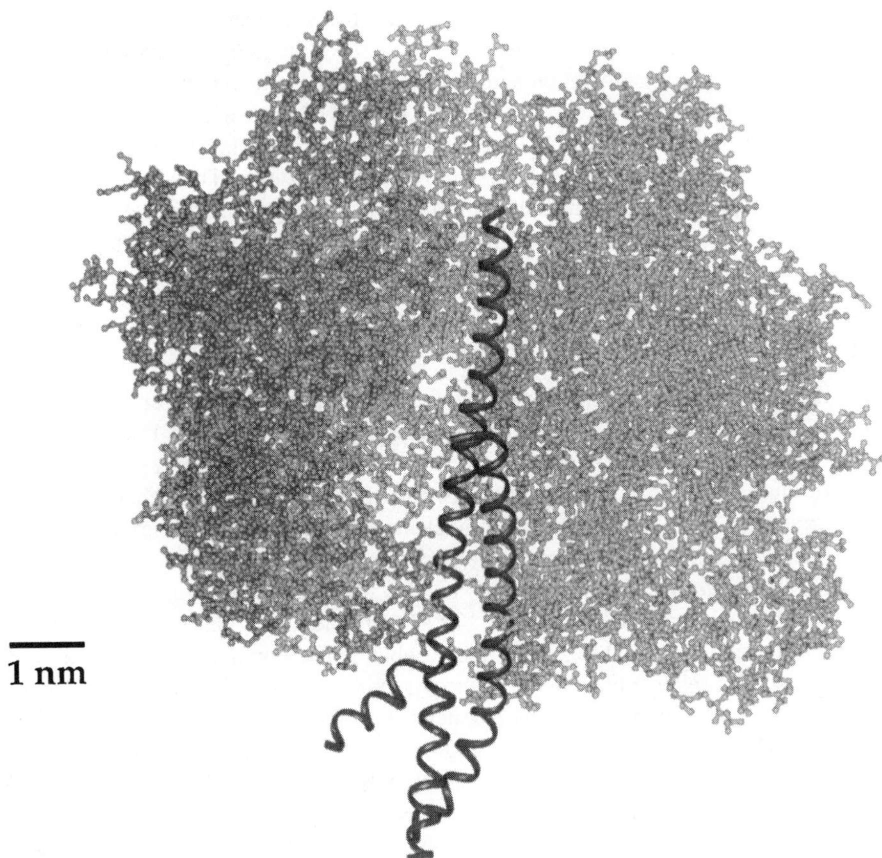

Figure 2 A representation of heart MF1 structure emphasizing the central α-helices of the γ subunit. The C-terminal helix extends almost to the top of the α/β heterohexamer which has been partially cut away to reveal the γ subunit. Molecule was rendered using Insight II software.

A crystal structure (resolution, 2.3 Å) (84) and a solution structure (90) of the ε subunit of ECF1 have been reported. The ε subunit of CF1 has 70% sequence similarity to the ε subunit of ECF1 (Table 1). The protein is folded into two domains; an N-terminal β-sandwich and a C-terminal domain composed of two α-helices. It has been difficult to "dock" the ε subunit to MF1. The ε subunit strongly interacts with the γ subunit (22) and has been shown to cross-link to the α and β subunits as well as to subunit c (81). The structure of the ε subunit may be altered when bound to ECF1. There may also be differences between the structures of MF1 and ECF1.

These high resolution structures are exciting, remarkable advances in the study of ATP synthases. There is still much to be done. A glance at Figure 3 shows that atomic resolution structure of significant parts of F1 and all of Fo is lacking.

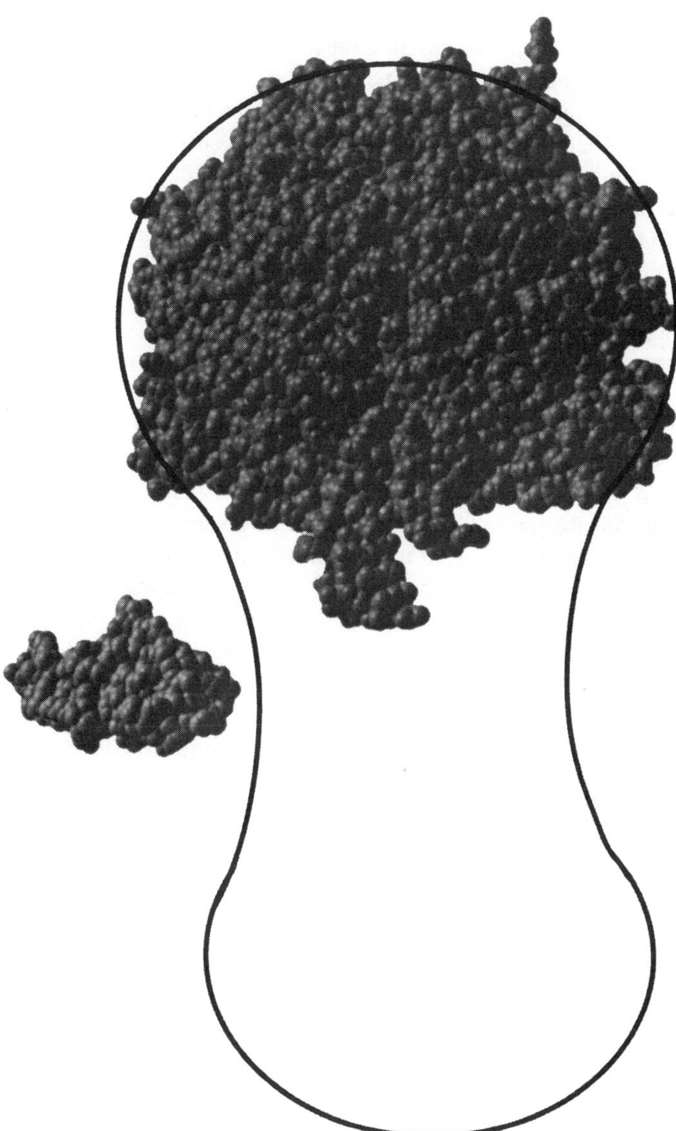

Figure 3 The crystal structures of MF1 and ECF1 ε subunit are shown superimposed on an outline of an image at the same scale of CF1CFo obtained by electron microscopy.

NUCLEOTIDE BINDING AND ACTIVITIES

Catalytic and Noncatalytic Sites

There are six nucleotide binding sites in CF1 (33, 73). From cross-linking studies it was established that at least some of these sites are at the interfaces of the α and β subunits (9). The sequence similarities between the α and β subunits of MF1 and CF1 suggest that the MF1 crystal structure may apply to the α and β subunits of CF1. These structures also show that the binding sites are at α-β interfaces. Since a Glu residue is present in the β subunits of CF1 that is equivalent to βGlu188 in MF1, it is likely that the β subunits of CF1 contain the catalytic sites. It is often assumed that three of the nucleotide binding sites are catalytically active, whereas the remaining three are noncatalytic. For CF1, clear evidence in support of this supposition has not been obtained. The study of nucleotide binding to CF1 is complicated by the presence of tightly bound nucleotides on the isolated enzyme and by the phenomenon of nucleotide exchange. Nonetheless, a somewhat coherent view of the property of the nucleotide binding sites of CF1 has emerged.

As isolated, CF1 contains from 1.2 to 1.6 mol of ADP per mol. ADP is released very slowly from the enzyme, with a half-time on the order of many weeks. Incubation of CF1 with ADP in the presence of EDTA increases the content of tightly bound ADP to 2 mol/mol CF1. Part of the bound ADP (close to 1 mol ADP/mol of CF1) undergoes an apparent exchange with ATP or ADP in the medium. The remaining bound ADP also exchanges, but very slowly (18). Thus, there are two nucleotide binding sites, N1 and N4, that bind ADP tightly in the absence of divalent cations. Two sites on CF1 bind ATP or AMP-PNP tightly in the presence of Mg^{2+}. A third ATP binding site also binds Mg-ATP, but unlike the two tight Mg-ATP sites, ATP bound to this site, possibly N1, is hydrolyzed to ADP (73). The tight Mg-ATP sites (N2 and N5) are noncatalytic, since the Mg-ATP remains bound to CF1 after extensive hydrolysis of medium ATP. Filling the tight Mg-ATP sites may activate the Mg-ATPase of F1 (19, 55) and stabilizes the enzyme from heat inactivation (88) and dissociation in the cold (36). Mg-ADP also binds to N2 and N5, but it dissociates readily from these sites. Mg-ATP remains bound to CF1 indefinitely unless the enzyme is incubated with EDTA in the presence of high concentrations of ammonium sulfate. Tightly bound ATP is absent in CF1 stored as an ammonium sulfate precipitate in the presence of EDTA. This treatment is, however, much less effective in the removal of bound ADP. Extended incubation of CF1-ε with alkaline phosphatase in the presence of sulfite is required for the removal of the ADP (19). Sulfite promotes the release of bound ADP and alkaline phosphatase hydrolyzes the released ADP to AMP and P_i. AMP binds very weakly, if at all, to CF1.

The remaining two sites, N3 and N6, bind nucleotides in a freely dissociable manner. A variety of nucleoside di- and triphosphates bind to N3 in the presence or absence of divalent cations. The Kd value for the binding of nucleotide to N3 is in the low micromolar range (16). N6 is less well characterized, but is likely to

TABLE 2 Characteristics of different nucleotide binding sites within the chloroplast ATP synthase. See Figure 4 for possible alignment of sites within structure

Bound nucleotide	Site	Activity	Binding property
Tightly bound	N2	Regulatory	
Mg-ATP	N5	Regulatory	
Tightly bound ADP	N1	Catalytic, & regulatory (?)	Exchangeable with free nucleotide, will bind ATP and hydrolyze to ADP
	N4	Regulatory	Slow exchange with free nucleotide
Freely dissociable	N3	Catalytic	Binding affinity ~10 μM
nucleotide	N6	Catalytic	Binding affinity >200 μM

have a Kd value in excess of 200 μM. The properties of the nucleotide binding sites of CF1 are summarized in Table 2.

There are four sites that could be catalytic: the two tight ADP sites (N1 and N4) and the two dissociable sites (N3 and N6). The most likely candidates are N1, N3, and N6. The rate of release of ADP bound to N4 is very slow, even under conditions that promote rapid turnover. The exchange of nucleotide bound to N1 with nucleotide in the medium is slow, but treatments that increase its ATPase activity accelerate the rate of exchange (18, 67). The rate of release of tightly bound ADP was found by Larson et al (46) to be slower than that of ATP hydrolysis, arguing against a role of the site to which ADP was bound in catalysis. However, Du & Boyer (20) found a lag in the onset of Mg^{2+}-ATP hydrolysis, during which most of the ADP at N1 had exchanged, and suggested that the site to which ADP was bound could be catalytic. These experiments were carried out with CF1 in thylakoid membranes that had been incubated with radiolabeled ADP, to load a tight binding site. Different labeling conditions were used by the two groups and different sites, not necessarily N1, may have been labeled. Recent results with CF1 in solution suggest that the exchange is fast enough for N1 to be involved in catalysis (19).

The binding of Mg-ADP to CF1, either in solution (19, 58) or membrane bound (8), inhibits the activity of the enzyme. ADP is released from CF1 in thylakoid membranes very shortly after turning on the light (35). Release is a consequence of the generation of the electrochemical proton gradient by light-dependent electron flow. It is not known whether the ADP is released from N1 or N4. If the ADP is released from N1, N1 could have the interesting property of being both a regulatory and a catalytic site.

It has been difficult to assign the CF1 nucleotide binding sites defined by their properties to specific sites occupied in the crystal structures of MF1. Only three of the six nucleotide binding sites of CF1 have been mapped reliably by the FRET method. A fourth site (putatively N4) was also mapped (72), but the calculated distances are not correct, because of the difficulty in labeling this site specifically with

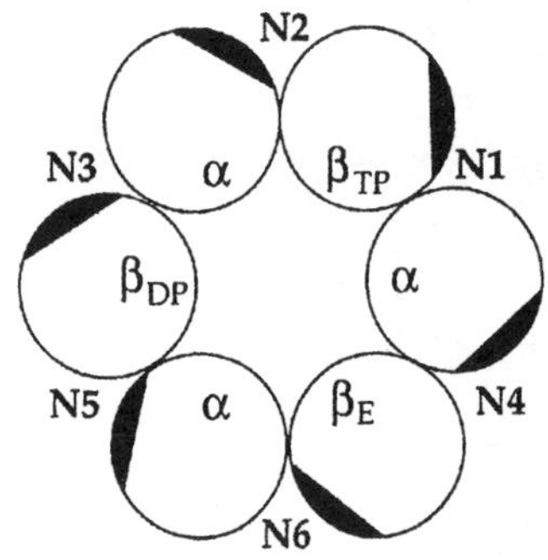

Figure 4 Speculative alignment of the nucleotide binding sites of CF1 with those of MF1. βTP corresponds to the heart MF1 β subunit that contained AMP-PNP, βDP to the β subunit that contained ADP and βE to the β subunit that did not contain bound nucleotide.

trinitrophenyl-ADP. A model (Figure 4) gives our attempt to assign the nucleotide binding sites of CF1 to those in heart MF1. The assignment of sites N1, N2, and N3 to adjacent α/β pairs is consistent with N1 and N3 being catalytic sites. The coordinates of the FRET map for these three sites fit reasonably well with those of the nucleotide binding site in MF1 (WD Frasch, personal communication; EA Johnson, unpublished observations). The arrangement of the three remaining sites can only be surmised. Assuming N6 is a catalytic site, it would occupy the last of the three sites in which β contacts with the nucleotide would predominate. If N4 is located between N6 and N1, N5 would have to be on the α site between N3 and N6. N4 binds ADP tightly and has properties quite different from those of the other noncatalytic sites, the two tight Mg-ATP sites (N2 and N5). N4 may be involved in regulation of activity. Thus, if these site assignments are correct, the α subunits of CF1 would be asymmetric with respect to the properties of their nucleotide binding sites. This asymmetry is likely to be induced by the interactions of the α/β heterohexamer with the γ subunit.

ATPase and ATP Synthesis

The chloroplast ATP synthase couples the synthesis of ATP from ADP and Pi to the efflux of protons from the thylakoid lumen down the electrochemical proton gradient established by photoelectron transfer. Under some conditions, the ATP synthase can act as an H^+-ATPase, hydrolyzing ATP and pumping protons into the lumen. Several mechanisms combine to assure that CF1CFo in situ does not hydrolyze ATP to a significant extent (56). Although CF1 in solution cannot catalyze the net synthesis of ATP from medium ADP and Pi, ADP tightly bound to CF1 may be phosphorylated to ATP that remains tightly bound to the enzyme (25).

Purified CF1CFo contains *p*-nitrophenylphosphatase activity that has been suggested (17) to be an intrinsic activity of the complex. However, the *p*-nitrophenylphosphatase activity of CF1 deficient in the ε subunit (CF1-ε) was undetectable (less than 1 nmol of *p*-nitrophenylphosphate hydrolyzed/min/mg of CF1), making it likely that the phosphatase activity of CF1CFo results from a contaminant in that preparation.

CF1CFo in thylakoid membranes in the dark has very low ATPase activity, less than 1% of the rate of ATP synthesis in the light. Two treatments that stimulate the MgATPase activity of CF1CFo in thylakoids alter the γ subunit of CF1. Reduction of the disulfide bond in the γ subunit (between γCys199 and γCys205) by either dithiothreitol or reduced thioredoxin in the light permits the hydrolysis of MgATP in a manner that is coupled to inward proton translocation (53). The treatment of thylakoid membranes with trypsin in the light specifically cleaves the γ subunit and activates the MgATPase activity (37). Reduction of the disulfide and proteolytic cleavage of the γ subunit reduce the affinity of CF1 for the inhibitory ε subunit (79). The incubation of thylakoid membranes with certain detergents (63) and alcohols (6) also stimulates MgATPase activity, probably by weakening the interactions between the γ and ε subunits.

Purified CF1 resembles the ATP synthase in situ in that it also has low ATPase activity. CF1 isolated from thylakoid membranes treated with dithiothreitol (53) or with trypsin (37) in the light has much higher ATPase activity. Alcohols, some detergents, and heat activate the ATPase of CF1 by either removing or selectively denaturing the ε subunit, an ATPase inhibitor (61). The γ and ε subunits cooperate to regulate the expression of wasteful ATP hydrolysis by the ATP synthase (79).

Divalent cations are absolutely required for ATP hydrolysis and synthesis by CF1 and CF1CFo. Although CF1 in solution hydrolyzes MgATP and CaATP, CF1CFo shows specificity only for MgATP or MgADP. The steady-state rate of ATP hydrolysis by soluble CF1 deficient in the ε subunit (CF1-ε) in the presence of Ca^{2+} is about 50 to 100 times that in the presence of Mg^{2+} when the cation to ATP concentration ratio was 1:1. The rate of MgATP hydrolysis by CF1-ε is strongly enhanced by oxyanions such as sulfite (46), alcohols (71), or detergents (63). CF1-ε in which the γ subunit disulfide had been reduced can hydrolyze MgATP at 37°C at rates exceeding 60 μmol ATP cleaved/min/mg protein when 100 mM sulfite is present.

Mg^{2+} is a potent inhibitor of the ATPase activity of CF1 (38). The inhibition by Mg^{2+} is very complex and depends on the occupancy of the nucleotide binding sites as well as on the free Mg^{2+} concentration in the assay mixture (19). The most significant inhibition is by tightly bound MgADP. The MgATPase activity of CF1-ε that contains bound ADP, but no Mg^{2+}, is initially rapid and decays within a minute to a slow, steady-state rate. Incubation of CF1-ε containing bound ADP with Mg^{2+} prior to assay in the absence of sulfite abolishes the fast phase of hydrolysis. The fast phase was eliminated when the enzyme contained 2 mol of MgADP/mol, probably in N1 and N4. The apparent exchange of ADP in N1 with nucleotide in the medium is also inhibited by the incubation of CF1-ε with Mg^{2+} prior to initiation of exchange by addition of nucleotide to the medium. The inhibition of exchange by Mg^{2+} was reversed when sulfite was added together with the ATP. These results indicate that the release of MgADP is the rate-limiting step in ATP hydrolysis in reaction mixtures that contain MgATP. Rather than inhibiting exchange, Ca^{2+} stimulates exchange at the concentrations of ATP and Ca^{2+} used in ATPase assays (19).

ATP synthesis can occur at high, sustained rates in the presence of free Mg^{2+} at concentrations that would very rapidly inhibit the ATPase activity of CF1-ε. The difference in the sensitivity of the activity of soluble and membrane bound CF1 to Mg^{2+} is puzzling. However, membrane-associated CF1 is subject to the effects of the electrochemical proton gradient when it is active. The proton gradient likely prevents the tight binding of MgADP to one or more sites on CF1, thereby overcoming Mg^{2+} inhibition. Why Ca^{2+} is a poor substitute for Mg^{2+} in supporting the activity of the intact ATP synthase is unknown. CF1-ε bound to CFo in thylakoid membranes is an active CaATPase, suggesting that the interaction of CF1 with the CFo subunits is not the reason for the strong preference of the bound enzyme for Mg^{2+}.

PROPOSED MECHANISMS OF CATALYSIS

There are aspects of the mechanism of the ATP synthase that are accepted widely (15). Among them are:

- The phosphorylation of ADP bound to a catalytic site to bound ATP has an equilibrium constant close to 1. In contrast, this reaction in solution has an equilibrium constant of 10^{-6} M.
- The energy of the electrochemical proton gradient is required not for anhydride bond formation, but mostly for the release of product ATP.
- Protons are not likely to be involved directly in the chemical mechanism of ATP synthesis and hydrolysis. For example, ATP synthesis by the ATP synthase of *Propionigenium modestum* (47) is driven by the electrochemical Na^+ gradient.
- The proton gradient induces conformational changes in the synthase that promote product release and substrate binding.
- Nucleotide binding sites interact during catalysis. The binding of nucleotide to one site promotes catalysis and product release from another site.

The prevailing mechanism, the alternating sites or binding change mechanism, was developed by Paul Boyer and coworkers. The mechanism evolved from Boyer's ideas on conformational coupling and was refined by analysis of ^{18}O exchanges and the kinetics of MF1 and CF1. In this mechanism (15), illustrated in Figure 5, three catalytic sites are thought to interact cooperatively, alternating their properties among tight binding, loose binding and unbound states during one catalytic cycle. These sites are labeled T (tight), L (loose) and O (open). In MF1, the Ka for binding of ATP to a tight site was estimated to about 10^9 that of the Ka for the open site (62). In ATP synthesis, the energy of the electrochemical proton gradient is proposed to be used to elicit structural changes that cause the tight site to release product ATP and become an open site. Coupled to this event, a second site that contains bound ADP and Pi (a loose site) becomes a tight site

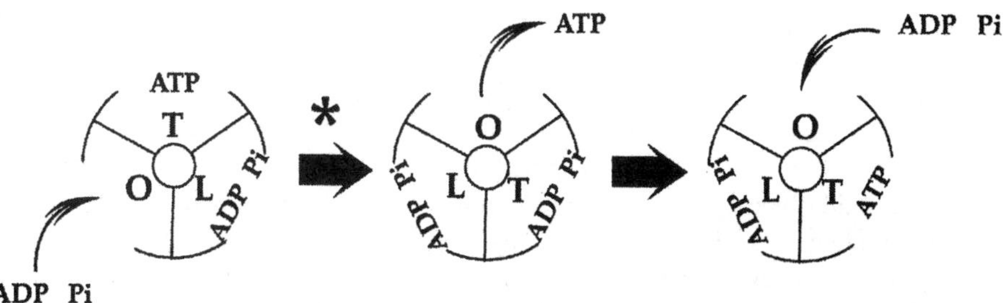

Figure 5 The binding change or alternating sites mechanism of the ATP synthase. Three sites interact cooperatively to alternate between tight (T), loose (L), and open (O) states. The energy (indicated by an asterisk) required for the synthesis of ATP is used for substrate binding and product release.

and the site that was unoccupied binds ADP and Pi to become a loose site. A model that involves three catalytic sites, but does not involve an open site has been proposed (12).

The heart MF1 structure shows that the β subunit can assume three different states with respect to the nucleotide binding sites. The β to which AMP-PNP is bound has been equated to the tight site, the β that contains ADP, the loose site and the β that does not contain nucleotide, the open site (1). Bianchet et al (12), however, point out that the heart MF1 was crystallized in the absence of Pi, a substrate, and in the presence of AMP-PNP and azide, both of which are inhibitors. Although the site occupancy of the heart MF1 β subunits seems nicely consistent with the alternating site model, it may be that the enzyme crystallized in an inhibited form that is not on the main catalytic pathway.

The properties of the three potential catalytic sites of CF1 are also in line with the alternating site proposal. Site N1 binds nucleotide (ADP or ATP) tightly, whereas N3 is characterized by a Kd of 2 to 5 μM and N6, has a much lower affinity for nucleotide, and would be empty under the conditions used to crystallize the heart MF1. There is also evidence for site-site interaction in CF1. The presence of nucleotides in the medium dramatically increases the rate of release of ADP from N1. Medium nucleotide is tightly bound to the enzyme in the process. These findings are consonant with the idea that medium nucleotide binds to dissociable sites, promoting the release of ADP bound to N1 and the conversion of the dissociable site to a tight site. It cannot, however, be excluded that medium ADP binds to the same site from which the ADP was released. The fungal toxin, tentoxin, is a potent inhibitor of ATP hydrolysis, and ATP synthesis by spinach CF1 also inhibits nucleotide exchange, which suggests that exchange may be part of the catalytic cycle (39).

The structure of CF1 is altered by the binding of MgATP or MgAMP-PNP. The properties of two nucleotide sites (N1 and N3) in CF1 switch as a result of the binding of MgATP or MgAMP-PNP (75). Site switching was demonstrated using CF1 that had been labeled specifically by Lucifer yellow vinyl sulfone at a single α subunit at Lys378. After exposure of the enzyme to either MgATP or MgAMP-PNP, the efficiency of energy transfer from Lucifer yellow to TNP-ADP in N1 increased, whereas that for transfer to TNP-ADP in N3 decreased. Thus, for the distance between the nucleotide binding sites and Lucifer yellow to change, nucleotide binding sites N1 and N3 must switch their properties as the result of binding MgATP. Site switching likely results from structural changes in CF1 caused by MgATP binding. ATP hydrolysis is not required to effect site switching. Changes in the interactions among the α and β subunits with the γ subunit are probably involved.

The number of sites involved in catalysis is not established with certainty. Evidence for two (11) or three (66) sites operating in catalysis has been obtained. The fact that there are three α/β pairs in CF1 suggests, simply on the basis of symmetry, that there are three catalytic sites and three noncatalytic sites in the enzyme. The structural asymmetry of the enzyme, however, clouds this simple appraisal.

For all three β subunit sites to participate in catalysis some mechanism would have to exist that allows each of the β subunits to become equivalent during a catalytic cycle. If two sites are involved, only two of the three β subunits would need to switch their properties during turnover. As shown by the crystal structure of heart MF1 and by biochemical studies, interactions of the α/β heterohexamer with the γ subunit are responsible for the asymmetry of the α and β subunits. A way to distribute the asymmetry evenly among the γ subunit and the α/β pairs is for the γ subunit to rotate relative to the α/β heterohexamer during catalysis. A model for rotational catalysis is shown in Figure 5.

ANALYSIS OF THE EVIDENCE FOR ROTATION

The approaches used to detect rotation vary from direct observation of individual molecules to biochemical and biophysical studies. The most visually appealing approach, developed by Noji et al (60), used epiflourescence microscopy to monitor the motion of fluorescent actin filaments attached to the γ subunit of modified $\alpha_3\beta_3\gamma$ from the thermophilic bacterium, PS 3. The subunits were expressed in *E. coli*. The β subunit contained a "His tag" of ten His residues on the N terminus. The His tag allowed the attachment of the $\alpha_3\beta_3\gamma$ to a slide coated with a Ni^{2+}-nitrilotriacetic acid complex. The γ subunit was biotinylated at a Cys residue introduced at position 107 and was attached to biotinylated, fluorescent actin filaments through strepavidin.

The addition of ATP caused movement of the attached actin filaments. On the average, 1 molecule of 70 rotated in a counter clockwise direction and 15 of 70 showed "... irregular to- and fro- fluctuation around one fixed point." Other filaments were immobile. Azide prevented the rotation induced by ATP addition. The fastest rotation speed was four revolutions/s, whereas the enzyme in solution

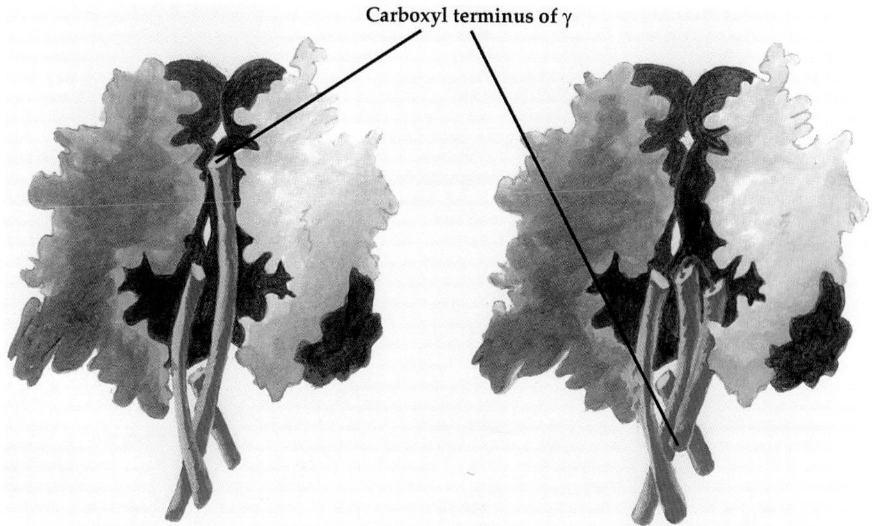

Figure 7 The γ subunit of heart MF1 (left) and of CF1 are shown in orange, surrounded by the α and β polypeptides in the same orientation as in Figure 2. The C-terminal α-helix of the γ subunit of CF1 is shown to be oriented such that γCys322, the penultimate amino acid, is close to γCys89, located on the short piece of orange helix. FRET and cross-linking data support this orientation.

hydrolyzed 52 ATP/s. For three catalytic sites, there should be $52/3 =$ about 17 rotations/s. The discrepancy in rates of hydrolysis and rotation was ascribed to the load created by frictional drag of the relatively large actin filament.

The rate of rotation was compared to the rate of ATP hydrolysis (45). Again, only a few percent of the actin filaments rotated. Since the authors were unable to measure reliably the ATPase activity of the enzyme attached to a cover slip, rotation rates were compared to rates of ATPase activity of the complex, without the actin filament attached, in solution. At concentrations of ATP below 2 μM, rotation and hydrolysis showed similar dependence on ATP concentration. At saturating ATP levels, however, the two curves deviated, with rotation saturating at about 2 μM ATP and ATPase activity at about 2 mM. At saturating ATP levels, the rate of ATP hydrolysis was 15 times that of rotation. A comparison of the ATPase activity of enzyme that does not contain actin to the rotation of actin filaments linked to F1 attached to a glass cover slip is incongruous. A minority of the molecules are seen to rotate, but most, if not all, of the enzyme molecules in solution are active. It is likely that the ATPase activity of the actin-modified enzyme would be lower than that of the unmodified enzyme, if rotation is required for ATPase activity. The rate of ATP hydrolysis at low ATP concentrations, therefore, should be slower than rotation.

That a small percentage of the actin filaments rotates may be rationalized in part by misattachment of either the complex to the cover slip or of the actin filament to the F1. The fact that the number of molecules showing to- and fro- movement of the actin filament was 15 times that of those that rotate is disturbing. Also, the relation of the rotation of the few molecules that do rotate to catalysis needs to be examined further. For example, azide was reported to prevent ATP-induced actin rotation but whether nonhydrolyzable ATP analogs such as AMP-PNP are effective has not been determined.

Actin was also attached to the ε subunit of TF1 and less than 1% of the actin filaments were observed to rotate, but at a slower rate than that reported before for the γ subunit (44). It was concluded that the γ and ε subunits rotate together and the difference between the rotation rates of the two subunits was ascribed to an unspecified experimental artifact.

Sabbert et al (70) used polarized absorption relaxation after photobleaching to investigate rotation in CF1. γCys322 was labeled with eosin maleimide and the labeled enzyme bound to DEAE-Sephadex A-50. A polarized laser beam bleached irreversibly a small percentage of the bound eosin. If there is rotation of the eosin probe, a decay over time between the absorption of light that is polarized in parallel versus that polarized perpendicularly should be evident. No change was observed in the presence of AMP-PNP, but when 5 mM ATP was present, a slow decay (half-time about 100 ms) was seen.

It is remarkable that an effect of ATP was seen. The immobilized enzyme (0.5 mg/ml) was incubated for 15 min in the dark, presumably at room temperature, prior to repetitive laser flashes. In addition to the CF1, 5 mM ATP, 2 mM Mg^{2+}, and 30 mM octylglucoside were present. The samples were then warmed to 37°C and laser excitation and measurement started. The rate of ATP hydrolysis at

37°C by CF1 in solution under these conditions is about 25 μmol/min/mg protein. At room temperature (23°C), the rate is 6 μmol/min/mg protein. CF1-ε bound to DEAE-cellulose hydrolyzed ATP at 23°C at a rate of about 5 μmol/min/mg protein (RE McCarty, unpublished data). Under the conditions used by Sabbert et al (70), 5mM ATP would have been completely hydrolyzed in about 2 min. Thus, there would be very little ATP left in the sample by the time data collection begins. A rapid inactivation of the ATPase activity of CF1 in solution when the enzyme is incubated in the presence of MgATP and octylglucoside was seen (RE McCarty, unpublished data). The increased mobility of the eosin could reflect partial dissociation of the CF1. These considerations complicate the interpretation of the experiments of Sabbert et al (70). The results show that bound eosin rotates, but they do not, however, prove that the γ subunit rotates together with the eosin during catalysis.

A third approach uses cross-linking between specific Cys residues in the γ and β subunits of ECF1 as a test of the movement of the γ subunit in relation to the β subunits. The region of the β subunit that contains the sequence DELSEED (residues 380 to 386 in ECF1 β subunit) interacts with the part of γ that contains the conserved Cys residue (Cys 87 in ECF1 γ) (1). A disulfide bond may be formed between γCys87 and a β subunit in which D380 was mutated to a Cys residue (21). The cross-linked ECF1 was dissociated in the cold and mixed with an equimolar ratio of radiolabeled enzyme (βD380C) that had not been cross-linked. The assembled enzyme can contain the γ subunit cross-linked to a particular β subunit that is unlabeled as well as one or more labeled β subunits. The disulfide bond between the γ and unlabeled β subunit was reduced and the enzyme exposed to various conditions prior to the addition of an oxidant to reform the disulfide bond. Exposure of the hybrid enzyme to MgATP prior to reformation of the disulfide bond resulted in increased disulfide bond formation between the γ subunit and the radiolabeled β polypeptides. Thus, the γ subunit had switched its β subunit partner as a result of exposure to MgATP. Whether this switching is part of a full rotation and the kinetic competence of this movement cannot be established. Since exposure of the enzyme to MgAMP-PNP gives 75% of the effect seen with MgATP (RL Cross, personal communication), the movement seen may be related more to substrate binding than to catalytic turnover. Therefore, although this experiment is consistent with rotation, movements of γ other than rotation induced by the binding of MgATP could also explain the results.

Nucleotide-dependent changes in the structure of ECF1 and CF1 have been reported. For example, the binding of MgATP has been reported to cause a shift in the position of the ε subunit of ECF1 (2). Different cross-links are formed between the ε and β subunits when ADP is bound than when ATP is present (4). Immunogold cryomicroscopy revealed the ε subunit to be in more than one position in ECF1(5). Also, ATP addition causes changes in the γ subunit of ECF1 that were detected by ATP-induced alterations in the fluorescence of a bound probe (83) and by cross-linking experiments. The ATP-dependent effects on the γ subunit were lost when the ε subunit was removed, even though the ε-deficient enzyme is a very highly active ATPase. The evidence for nucleotide-induced changes in CF1 includes bound nucleotide release induced by the addition of nucleotides to the medium (18), site switching (74), and the influence of nucleotides on the reaction of αLys378 with Lucifer yellow vinyl sulfone (KM Lowe & RE McCarty, unpublished observations). It is likely that the changes in the α and β subunits elicited by nucleotides are transmitted by the γ subunit. Rotation of the γ subunit could effect these changes, but other kinds of changes in the structure of the γ subunit cannot be excluded.

It is clear that rotation of the α/β subunit relative to the γ is not required for ATP hydrolysis. For example, $\alpha_3\beta_3\delta$, formed from purified subunits of F1 from the thermophilic bacterium, PS3, hydrolyzed MgATP at a rate of 1.1 μmol/min/mg protein at 23°C, a rate about half that of the $\alpha_3\beta_3\gamma$ (92). Subcomplexes of CF1 that contain only the α and β subunits in a stoichiometry of 1:1 have a low, but significant ATPase activity (32, 76). Tentoxin, a highly specific inhibitor of the chloroplast ATP synthase of some species, inhibits the ATPase activity of spinach CF1. At concentrations of tentoxin above 100 μM, ATPase activity is stimulated (80). The ATPase activity of the α/β complex was not inhibited by low concentrations of tentoxin, but was stimulated as much as 50-fold by high tentoxin concentrations (76). Tentoxin also stabilized an $\alpha_3\beta_3$ structure. Rates of ATP hydrolysis as high as 5 μmole/min/mg protein (27 ATP/s) were detected, even though the γ subunit was absent. Therefore, the γ subunit of CF1 is not required for ATP hydrolysis.

In work that seems to have been mostly forgotten, Musier & Hammes (59) used cross-linkers with a maleimide on one end and a photoactivatable group, benzoylphenyl, on the other. A spacer of either 6 or 12 carbon atoms linked the two groups. CF1 specifically labeled by the maleimide part of the cross-linkers at γCys322, γCys199, and γCys205 was exposed to long-wave UV light to promote cross-linking. Greater than 50% of the γ subunit was cross-linked to the α or β (or both) subunits with about 7% loss in ATPase activity for the cross-linker with a 12-carbon spacer and about 12% loss for the 6-carbon one. The rates of ATP hydrolysis at 37°C were in excess of 150 ATP/s. The authors conclude that "... large movements of the α-β polypeptides with respect to the smaller γ-, δ, and ε-polypeptides are not required during the catalytic cycle of CF1" (59).

The C-terminal α-helix of the γ subunit of MF1 was proposed to be a major part of the hydrophobic shaft that rotates during catalysis within a bearing made up of the hydrophobic interior surfaces of the α and β subunits (1). Recombinant γ subunit of CF1 combines with the α/β subunits to form a fully active $\alpha_3\beta_3\gamma$ complex. As many as 20 amino acids could be deleted from the C terminus of the γ subunit without a strong effect on the assembly or the ATPase activity of the complex (77). Thus, even after removal of a major portion of the purported shaft, ATP is still hydrolyzed at high rates.

Monoclonal antibodies with epitopes against the polar loop region of the c subunit of ECFo bind to subunit c in ECF1ECFo in inside out vesicles of the coupling membrane of *E. coli*. Both ECF1 and the antibody were bound to the same ECFo, showing that not all of the c subunits are involved in the binding of

ECF1. The antibody had no effect on the ATPase activity of the enzyme and, more importantly, on ATP-dependent proton pumping, making it rather unlikely that the oligomer of c subunits undergoes complete rotation during turnover (12a).

The b subunit of ECFo contains an N-terminal hydrophobic domain that anchors the protein in the membrane. Deletion of this domain allows the expression of the b subunit in soluble form. Soluble b subunit has a dimerization domain as well as a C-terminal domain that cross-links to the δ subunit. The C-terminal region of subunit I of CFo was also shown to cross-link to the δ subunit. The second stalk is proposed to be made up at least in part of the b subunit dimer and attached δ subunit. Further, this structure has been suggested to be a stator, stabilizing the α/β heterohexamer as the γ, ε and c subunits rotate during catalysis. (See Figure 6). Mutants of *E. coli* that contain b subunits from which as many as 11 amino acids were deleted from the middle of the molecule retain their ability to grow in media containing succinate as the sole energy source (78). When 12 or more amino acids were deleted from the b subunits, ATP synthase activity was lost. These results indicate that the b subunit may be flexible. Rather than play a static role in preventing the rotation of the α/β heterohexamer as part of a stator, the b subunits could be part of the mechanism that transmits conformational changes induced by proton flux to ECF1.

The extent of incorporation of Lucifer yellow vinyl sulfone into αLys378 of CF1 in thylakoid membranes was unaffected by illumination of the membranes in the presence of substrate levels of ADP and the Pi analog, arsenate (KM Lowe & RE McCarty, unpublished observations). Arsenate was used in place of Pi because the

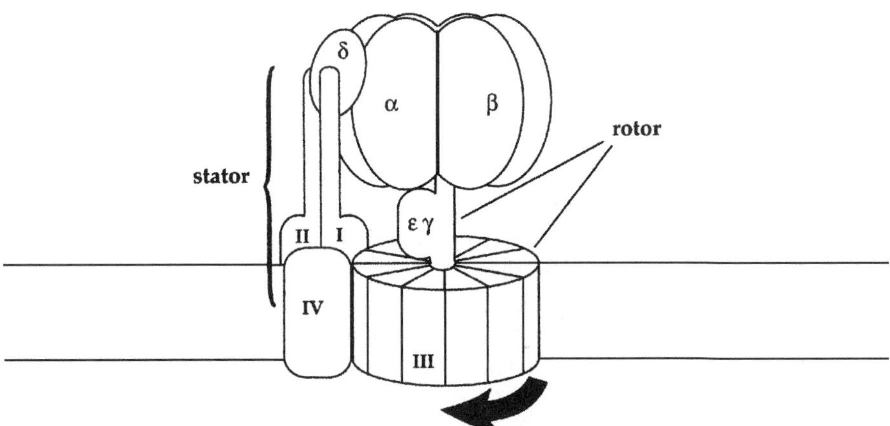

Figure 6 A model of rotational catalysis. The subunit III oligomer of the ATP synthase is proposed to rotate together with the γ and ε subunits as protons flow through Fo. This rotation distributes the asymmetrical interactions among the α/β subunits and the γ subunit, allowing the catalytic sites to switch their properties. The central α-helices of γ subunit are thought to be a rotor and subunits I and II, together with the δ subunit, a stator that holds the α/β heterohexamer.

ADP-arsenate product is very unstable and hydrolyzes rapidly after its release from the enzyme, regenerating ADP and arsenate. The synthase underwent about 10^5 turnovers during the period of illumination. If the γ subunit were to rotate 360° during turnover, the γ subunits should become equivalent during turnover. The extent of Lucifer yellow incorporation into the α subunits should have increased during turnover, as is the case when the γ subunit is removed (49).

CONCLUSIONS AND SOME UNRESOLVED QUESTIONS

Much progress has been made in the past two decades in the delineation of the structure and mechanism of the chloroplast ATP synthase. This fact is amply demonstrated by a glance at a review that appeared in this series 21 years ago (50). The subunit composition of CF1 was thought to be $\alpha_2\beta_2\gamma\delta\varepsilon_2$. The accepted molecular weight of CF1 in 1979 was nearly 20% lower than the actual value. Very little was known about CFo, but CF1CFo had been reconstituted into membrane vesicles and shown to be active.

The renditions of the ATP synthase in action [see, for example, (41)] may convey the impression that a great deal more is known about the way this enzyme works than actually is. These depictions should properly be construed as speculative models. There is, for example, no direct evidence for the rotation of the oligomer of subunit c (or III); nor has the stoichiometry of subunit III been established beyond the range of 6 to 12/CF1.

The highest resolution structure of any part of the chloroplast ATP synthase is the FRET distance map of CF1. Although it is reasonable to attempt to model CF1 on the basis of the MF1 partial structures, it would be far preferable to have a structure of CF1 at atomic resolution. There are significant differences between the mitochondrial and chloroplast enzymes. The structure of the γ subunit in CF1 likely differs from that in MF1. Based on FRET data and cross-linking studies, the C terminus of the γ subunit is not located at the top of the α/β heterohexamer, but is instead close to γCys89. This difference is illustrated in Figure 7 (see color plate). The two enzymes should also have quite different structures with respect to the δ and ε subunits.

The title of this article is a question: The Chloroplast ATP Synthase: a Rotary Enzyme? Regrettably, the answer to this question is not clear. Is rotation of γ relative to the α and β subunits involved in the mechanism of ATP hydrolysis? There is some evidence that the γ subunit can rotate, but the small fraction of molecules seen to rotate as well as questionable analyses cast some doubt on the relation of the rotation to catalysis. The interpretation of data from other experiments that are thought to support rotation of the γ subunit is also questioned in this article. Kinetic competence has not been shown in any of the experiments in support of rotation.

Rotation is not required for ATP hydrolysis by subcomplexes of F1 that lack the γ subunit. There is also evidence against the rotary mechanism as currently conceived. Other mechanisms for the enzyme are probably worth considering.

For example, a model in which two catalytic sites alternate between tight and loose configurations is feasible (68). The γ subunit could then move back and forth between the two catalytic β subunits. This would amount to rotation over 120°.

Most of the experiments relating to rotation have been done with F1, either in solution or immobilized in some way. It is preferable to carry out experiments with the entire ATP synthase in its natural environment when the enzyme is synthesizing ATP. It will be difficult to detect changes in the structure of parts of the enzyme elicited by the electrochemical proton gradient and nucleotide binding and release. However, experiments of this kind will be needed as further tests of the mechanism of ATP synthesis. There is much to do.

ACKNOWLEDGMENTS

We gratefully acknowledge the support of the National Science Foundation for our work on the chloroplast ATP synthase (MCB 9723945). We thank Lynn Scott for the rendition of mitochondrial and chloroplast ATP synthases.

Visit the Annual Reviews home page at www.AnnualReviews.org

LITERATURE CITED

1. Abrahams JP, Leslie AG, Lutter R, Walker JE. 1994. Structure at 2.8 Å resolution of F1-ATPase from bovine heart mitochondria. *Nature* 370:621–28
2. Aggeler R, Capaldi RA 1996. Nucleotide-dependent movement of the epsilon subunit between alpha and beta subunits in the *Escherichia coli* F1Fo-type ATPase. *J. Biol. Chem.* 271:13888–91
3. Aggeler R, Capaldi RA, Dunn S, Gogol EP 1992. Epitope mapping of monoclonal antibodies to the *Escherichia coli* F1 ATPase alpha subunit in relation to activity effects and location in the enzyme complex based on cryoelectron microscopy *Arch. Biochem. Biophys* 296:685–90
4. Aggeler R, Haughton MA, Capaldi RA. 1995. Disulfide bond formation between the COOH-terminal domain of the beta subunits and the gamma and epsilon subunits of the *Escherichia coli* F1-ATPase. Structural implications and functional consequences. *J. Biol. Chem.* 270:9185–91
5. Aggeler R, Weinreich F, Capaldi RA. 1995. Arrangement of the epsilon subunit in the *Escherichia coli* ATP synthase from the reactivity of cysteine residues introduced at different positions in this subunit. *Biochim. Biophys. Acta* 1230:62–68
6. Anthon GE, Jagendorf AT 1983. Effect of methanol on spinach thylakoid ATPase. *Biochim. Biophys. Acta* 723:358–65
7. Avron M. 1963. A coupling factor in photophosphorylation. *Biochim. Biophys. Acta* 77:699–702
8. Bar-Zvi D, Shavit N. 1982. Modulation of the chloroplast ATPase by tight ADP binding. Effect of uncouplers and ATP. *J. Bioenerg. Biomembr.* 14:467–78
9. Bar-Zvi D, Tiefert MA, Shavit N. 1983. Interaction of the chloroplast ATP-synthetase with the photoreactive nucleotide 3′-O-(4-benzoyl)benzoyl adenosine 5′-diphosphate. *FEBS Lett.* 160:233–38
10. Beckers G, Berzborn RJ, Strotmann H. 1992. Zero-length crosslinking between

subunits delta and I of the H^+- translocating ATPase of chloroplasts. *Biochim. Biophys. Acta* 1101:97–104
11. Berden JA, Hartog AF, Edel CM. 1991. Hydrolysis of ATP by F1 can be described only on the basis of a dual-site mechanism. *Biochim. Biophys. Acta* 1057:151–56
12. Bianchet MA, Hullihen J, Pedersen PL, Amzel LM. 1998. The 2.8 Å structure of rat liver F1-ATPase: configuration of a critical intermediate in ATP synthesis/hydrolysis. *Proc. Natl. Acad. Sci. USA* 95:11065—G70
12a. Birkenhäger R, Greie J-C, Altendorf K, Deckers-Hebestreit G. 1999. Fo complex of the *Escherichia coli* ATP synthase. Not all monomers of the subunit c oligomer are involved in F1 interaction. *Eur. J. Biochem.* 264:385–96
13. BoekemaEJ, Schmidt G, Gräber P, Berden JA. 1988. Structure of the ATP-synthase from chloroplasts and mitochondria studied by electron microscopy. *Z. Naturforsch. Teil C* 43:219–25
14. Böttcher B, Schwarz L, Gräber P. 1998. Direct indication for the existence of a double stalk in CF0F1. *J. Mol. Biol.* 281:757–62
15. Boyer PD. 1997. The ATP synthase—a splendid molecular machine. *Annu. Rev. Biochem.* 66:717–49
16. Bruist MF, Hammes GG. 1981. Further characterization of nucleotide binding sites on chloroplast coupling factor one. *Biochemistry* 20:6298–305
17. Creczynskipasa TB, Gräber P, Alves EW, Ferreira AT, Scofano HM. 1997. Phosphatase activity of H^+-ATPase from chloroplasts. *Biochim. Biophys. Acta* 1320:58–64
18. Digel JG, McCarty RE. 1995. Two tight binding sites for ADP and their interactions during nucleotide exchange in chloroplast coupling factor 1. *Biochemistry* 34:14482–89
19. Digel JG, Moore ND, McCarty RE. 1998. Influence of divalent cations on nucleotide exchange and ATPase activity of chloroplast coupling factor 1. *Biochemistry* 37:17209–15
20. Du ZY, Boyer PD. 1990. On the mechanism of sulfite activation of chloroplast thylakoid ATPase and the relation of ADP tightly bound at a catalytic site to the binding change mechanism. *Biochemistry* 29:402–7
21. Duncan TM, Bulygin VV, Zhou Y, Hutcheon ML, Cross RL. 1995. Rotation of subunits during catalysis by *Escherichia coli* F1-ATPase. *Proc. Natl. Acad. Sci. USA* 92:10964–68
22. Dunn SD. 1982. The isolated gamma subunit of *Escherichia coli* F1 ATPase binds the epsilon subunit. *J. Biol. Chem.* 257:7354–59
23. Engelbrecht S, Giakas E, Marx O, Lill H. 1998. Fluorescence resonance energy transfer mapping of subunit delta in spinach chloroplast F1 ATPase. *Eur. J. Biochem.* 252:277–83
24. Engelbrecht S, Junge W. 1990. Subunit delta of H^+-ATPases: at the interface between proton flow and ATP synthesis. *Biochim. Biophys. Acta* 1015:379–90
25. Feldman RI, Sigman DS. 1982. The synthesis of enzyme-bound ATP by soluble chloroplast coupling factor 1. *J. Biol. Chem.* 257:1676–83
26. Feng Y, McCarty RE. 1990. Chromatographic purification of the chloroplast ATP synthase (CF0-CF1) and the role of CF0 subunit IV in proton conduction. *J. Biol. Chem.* 265:12474–80
27. Feng Y, McCarty RE. 1990. Purification and reconstitution of active chloroplast Fo. *J. Biol. Chem.* 265:5104–9
28. Feng Y, McCarty RE. 1990. Subunit interactions within the chloroplast ATP synthase (CF0-CF1) as deduced by specific depletion of CF0 polypeptides. *J. Biol. Chem.* 265:12481–85
29. Fillingame RH. 1992. Subunit c of F1Fo ATP synthase: structure and role in

transmembrane energy transduction. *Biochim. Biophys. Acta* 1101:240–43
30. Fromme PP, Gräber P, Salnikow J. 1987. Isolation and identification of a fourth subunit in the membrane part of the chloroplast ATP synthase. *FEBS Lett.* 210:27–30
31. Futai M, Kanazawa H. 1983. Structure and function of proton-translocating adenosine triphosphatase (FoF1): Biochemical and molecular biological approaches. *Microbiol. Rev.* 47:285–312
32. Gao F, Lipscomb B, Wu I, Richter ML. 1995. *In vitro* assembly of the core catalytic complex of the chloroplast ATP synthase. *J. Biol. Chem.* 270:9763–69
33. Girault G, Berger G, Galmiche JM, Andre F. 1988. Characterization of six nucleotide-binding sites on chloroplast coupling factor 1 and one site on its purified beta subunit. *J. Biol. Chem.* 263:14690–95
34. Gogol EP, Lucken U, Burk T, Capaldi RA. 1989. Molecular architecture of *Escherichia coli* F1 adenosine triphosphatase. *Biochemistry* 28:7709–16
35. Gräber P, Schlodder E, Witt HT. 1977. Conformational change of chloroplast ATPase induced by a transmembrane electric-field and its correlation to phosphorylation. *Biochim. Biophys. Acta* 461:426–40
36. Hightower KE, McCarty RE. 1996. Influence of nucleotides on the cold stability of chloroplast coupling factor 1. *Biochemistry* 35:10051–57
37. Hightower KE, McCarty RE. 1996. Proteolytic cleavage within a regulatory region of the gamma subunit of chloroplast coupling factor 1. *Biochemistry* 35:4846–51
38. Hochman Y, Lanir A, Carmeli C. 1976. Relations between divalent-cation binding and ATPase activity in coupling factor from chloroplasts. *FEBS Lett.* 61: 255–59
39. Hu N, Mills DA, Huchzermeyer B, Richter ML. 1993. Inhibition by tentoxin of cooperativity among nucleotide binding sites on chloroplast coupling factor 1. *J. Biol. Chem.* 268:8536–40
40. Hudson GS, Mason JG. 1988. The chloroplast genes encoding subunits of the H^+-ATP synthase. *Photosyn. Res.* 18:205–22
41. Junge W. 1999. ATP synthase and other motor proteins. *Proc. Natl. Acad. Sci. USA* 96:4735–37
42. Kagawa Y, Racker E. 1971. Reconstitution of vesicles catalysing ^{32}P-adenosine triphosphate exchange. *J. Biol. Chem.* 246: 5477–87
43. Kanazawa H, Kayano T, Kiyasu T, Futai M. 1982. Nucleotide sequence of the genes for beta and epsilon subunits of proton-translocating ATPase from *Escherichia coli*. *Biochem. Biophys. Res. Commun.* 105:1257–64
44. Kato-Yamada Y, Noji H, Yasuda R, Kinosita K Jr, Yoshida M. 1998. Direct observation of the rotation of epsilon subunit in F1-ATPase. *J. Biol. Chem.* 273:19375–77
45. Kinosita K Jr, Yasuda R, Noji H, Ishiwata S, Yoshida M. 1998. F1-ATPase: a rotary motor made of a single molecule. *Cell* 93:21–24
46. Larson EM, Umbach A, Jagendorf AT. 1989. Sulfite-stimulated release of [^{3}H] ADP bound to chloroplast thylakoid ATPase. *Biochim. Biophys. Acta* 973:78–85
47. Laubinger W, Dimroth P. 1987. Characterization of the Na^+-stimulated ATPase of *Propionigenium modestum* as an enzyme of the F1Fo type. *Eur. J. Biochem.* 168:475–80
48. Lemaire C, Wollman FA. 1989. The chloroplast ATP synthase in *Chlamydomonas reinhardtii*. II. Biochemical studies on its biogenesis using mutants defective in photophosphorylation *J. Biol. Chem.* 264:10235–42
49. Lowe KM, McCarty RE. 1998. Asymmetry of the alpha subunit of the chloroplast ATP synthase as probed by the binding of Lucifer yellow vinyl sulfone. *Biochemistry* 37:2507–14
50. McCarty RE. 1979. Roles of a coupling factor for photophosphorylation in

chloroplasts. *Annu. Rev. Plant Physiol.* 30:79–104
51. McCarty RE. 1997. Applications of fluorescence resonance energy transfer to structure and mechanism of chloroplast ATP synthase. *Methods Enzymol.* 278:528–38
52. McCarty RE, Moroney JV. 1985. Functions of the subunits and regulation of chloroplast coupling factor 1 In *The Enzymes of Biological Membranes,* ed. A Martonosi, pp. 383–413. New York: Plenum
53. McCarty RE, Racker E. 1968. Partial resolution of the enzymes catalyzing photophosphorylation. 3. Activation of adenosine triphosphatase and ^{32}P-labeled orthophosphate-adenosine triphosphate exchange in chloroplasts. *J. Biol. Chem.* 243:129–37
54. McLachlin DT, Dunn SD. 1997. Dimerization interactions of the b subunit of the *Escherichia coli* F1Fo–ATPase. *J. Biol. Chem.* 272:21233–39
55. Milgrom YM, Ehler LL, Boyer PD. 1990. ATP binding at noncatalytic sites of soluble chloroplast F1-ATPase is required for expression of the enzyme activity. *J. Biol. Chem.* 265:18725–28
56. Mills JD. 1996. The regulation of chloroplast ATP synthase, CFoCF1. See Ref. 60a, pp. 469–85
57. Mitra B, Hammes GG. 1989. Structural map of the dicyclohexylcarbodiimide site of chloroplast coupling factor determined by resonance energy transfer. *Biochemistry* 28:3063–69
58. Murataliev MB, Milgrom YM, Boyer PD. 1991. Characteristics of the combination of inhibitory Mg^{2+} and azide with the F1 ATPase from chloroplasts. *Biochemistry* 30:8305–10
59. Musier KM, Hammes GG. 1987. Rotation of nucleotide sites is not required for the enzymatic activity of chloroplast coupling factor 1. *Biochemistry* 26:5982–88
60. Noji H, Yasuda R, Yoshida M, Kinosita K Jr. 1997. Direct observation of the rotation of F1-ATPase. *Nature* 386:299-302
60a. Ort D, Yocum C, eds. 1996. *Oxygenic Photosynthesis: The Light Reactions.* Dordrecht/Boston/London: Kluwer
61. Patrie WJ, McCarty RE. 1984. Specific binding of coupling factor 1 lacking the delta and epsilon subunits to thylakoids. *J. Biol. Chem.* 259:11121–28
62. Penefsky HS, Cross RL. 1991. Structure and mechanism of FoF1-type ATP synthases and ATPases. *Adv. Enzymol. Relat. Areas Mol. Biol.* 64:173–214
63. Pick U, Bassilian S. 1982. Activation of magnesium ion specific adenosinetriphosphatase in chloroplast coupling factor 1 by octyl glucoside. *Biochemistry* 21:6144–52
64. Pick U, Racker E. 1979. Purification and reconstitution of the N,N′-dicyclohexylcarbodiimide-sensitive ATPase complex from spinach chloroplasts. *J. Biol. Chem.* 254:2793–99
65. Pullman ME, Penefsky HS, Datta A, Racker E. 1960. Purification and properties of soluble, dinitrophenol-stimulated adenosine triphosphatase. *J. Biol. Chem.* 235:3322–29
66. Rao R, Senior AE. 1987. The properties of hybrid F1-ATPase enzymes suggest that a cyclical catalytic mechanism involving three catalytic sites occurs. *J. Biol. Chem.* 262:17450–54
67. Richter ML, Gao F. 1996. The chloroplast ATP synthase: structural changes during catalysis. *J. Bioenerg. Biomembr.* 28:443–49
68. Richter ML, Mills DA. 1996. The relationship between the structure and catalytic mechanism of the chloroplast ATP synthase. See Ref. 60a, pp. 453–68
69. Richter ML, Patrie WJ, McCarty RE. 1984. Preparation of the epsilon subunit and epsilon subunit-deficient chloroplast coupling factor 1 in reconstitutively active forms. *J. Biol. Chem.* 259:7371–73
70. Sabbert D, Engelbrecht S, Junge W. 1996.

Intersubunit rotation in active F-ATPase. *Nature* 381:623–25
71. Sakurai H, Hisabori T, Shinohara K. 1981. Enhancement of adenosine triphosphatase activity of purified chloroplast coupling factor 1 in an aqueous organic solvent. *J. Biochem.* 90:95–99
72. Shapiro AB, Gibson KD, Scheraga HA, McCarty RE. 1991. Fluorescence resonance energy transfer mapping of the fourth of six nucleotide-binding sites of chloroplast coupling factor 1. *J. Biol. Chem.* 266:17276–85
73. Shapiro AB, Huber AH, McCarty RE. 1991. Four tight nucleotide binding sites of chloroplast coupling factor 1. *J. Biol. Chem.* 266:4194–200
74. Shapiro AB, McCarty RE. 1988. Alteration of the nucleotide-binding site asymmetry of chloroplast coupling factor 1 by catalysis. *J. Biol. Chem.* 263:14160–65
75. Shapiro AB, McCarty RE. 1990. Substrate binding-induced alteration of nucleotide binding site properties of chloroplast coupling factor 1. *J. Biol. Chem.* 265:4340–47
76. Sokolov M, Gromet-Elhanan Z. 1996. Spinach chloroplast coupling factor CF1-alpha 3 beta 3 core complex: structure, stability, and catalytic properties. *Biochemistry* 35:1242–48
77. Sokolov M, Lu L, Tucker W, Gao F, Gegenheimer PA, Richter ML. 1999. The 20 C-terminal amino acid residues of the chloroplast ATP synthase gamma subunit are not essential for activity. *J. Biol. Chem.* 274:13824–29
78. Sorgen PL, Caviston TL, Perry RC, Cain BD. 1998. Deletions in the second stalk of F1Fo-ATP synthase in *Escherichia coli*. *J. Biol. Chem.* 273:27873–78
79. Soteropoulos P, Süss KH, McCarty RE. 1992. Modifications of the gamma subunit of chloroplast coupling factor 1 alter interactions with the inhibitory epsilon subunit. *J. Biol. Chem.* 267:10348–54
80. Steele JA, Uchytil TF, Durbin RD. 1978. The stimulation of coupling factor 1 ATPase by tentoxin. *Biochim. Biophys. Acta* 504:136–41
81. Süss KH. 1986. Neighbouring subunits of CFo and between CF1 and CFo of the soluble chloroplast ATP synthase (CF1-CFo as revealed by chemical cross-linking. *FEBS Lett.* 201:63–68
82. Tiedge H, Schafer G, Mayer F. 1983. An electron-microscopic approach to the quaternary structure of mitochondrial-F1-ATPase. *Eur. J. Biochem.* 132:37–45
83. Turina P, Capaldi RA. 1994. ATP binding causes a conformational change in the gamma-subunit of the *Escherichia-coli* F1-ATPase which is reversed on bond-cleavage. *Biochemistry* 33:14275–80
84. Uhlin U, Cox GB, Guss JM. 1997. Crystal structure of the epsilon subunit of the proton-translocating ATP synthase from *Escherichia coli*. *Structure* 5:1219–30
85. Vambutas VK, Racker E. 1965. Stimulation of photophosphorylation by a preparation of a latent, Ca^{++}-dependent adenosine triphosphatase from chloroplasts. *J. Biol. Chem.* 240:2660–67
86. Walker JE, Fearnley IM, Gay NJ, Gibson BW, Northrop FD, et al. 1985. Primary structure and subunit stoichiometry of F1-ATPase from bovine mitochondria. *J. Biol. Chem.* 184:677–701
87. Walker JE, Lutter R, Dupuis A, Runswick MJ. 1991. Identification of the subunits of FIFo-ATPase from bovine heart-mitochondria. *Biochemistry.* 30:5369–78
88. Wang ZY, Freire E, McCarty RE. 1993. Influence of nucleotide binding site occupancy on the thermal stability of the F1 portion of the chloroplast ATP synthase. *J. Biol. Chem.* 268:20785–90
89. Wetzel CM, McCarty RE. 1993. Aspects of subunit interactions in the chloroplast ATP synthase 2. Characterization of a chloroplast coupling factor 1–subunit-III complex from spinach thylakoids. *Plant Physiol.* 102:251–59
90. Wilkens S, Dahlquist FW, McIntosh LP, Donaldson LW, Capaldi RA. 1995.

Structural features of the epsilon subunit of the *Escherichia coli* ATP synthase determined by NMR spectroscopy. *Nat. Struct. Biol.* 2:961–67

91. Xiao J, McCarty RE. 1989. Binding of chloroplast coupling factor 1 deficient in the delta subunit to thylakoid membranes. *Biochim. Biophys. Acta* 976:203–9
92. Yokoyama K, Hisabori T, Yoshida M. 1989. The reconstituted alpha-3-beta-3-delta complex of the thermostable F1-ATPase. *J. Biol. Chem.* 264:21837–41
93. Zhang Y, Fillingame RH. 1995. Subunits coupling H^+ transport and ATP synthesis in the *Escherichia coli* ATP synthase. Cys-Cys cross-linking of F1 subunit epsilon to the polar loop of Fo subunit c. *J. Biol. Chem.* 270:24609–14

Annu. Rev. Plant Physiol. Plant Mol. Biol. 2000. 51:111–40

NONPHOTOSYNTHETIC METABOLISM IN PLASTIDS

H. E. Neuhaus and M. J. Emes

Pflanzenphysiologie, University of Osnabrück, Barbarastrasse 11, D-49069 Osnabrück, Germany; School of Biological Sciences, University of Manchester, Oxford Road, M13 9PL, Manchester, United Kingdom; e-mail: mike.emes@man.ac.uk

Key Words starch, fatty acids, nitrogen, metabolite transport, gene expression

■ **Abstract** Nonphotosynthetic plastids are important sites for the biosynthesis of starch, fatty acids, and the assimilation of nitrogen into amino acids in a wide range of plant tissues. Unlike chloroplasts, all the metabolites for these processes have to be imported, or generated by oxidative metabolism within the organelle. The aim of this review is to summarize our present understanding of the anabolic pathways involved, the requirement for import of precursors from the cytosol, the provision of energy for biosynthesis, and the interaction between pathways that share common intermediates. We emphasize the temporal and developmental regulation of events, and the variation in mechanisms employed by different species that produce the same end products.

CONTENTS

INTRODUCTION

Plastids are found in all living cells of higher plants except pollen and their very name indicates their metabolic, structural, and functional plasticity. They are the biosynthetic and assimilatory powerhouses of plant cells, responsible for the fixation of CO_2, manufacture of starch, fatty acids and pigments, and the synthesis of amino acids from inorganic nitrogen. Broadly, plastids can be categorized into a number of types based on color, structure, and stage of development. All plastids are surrounded by a double envelope, the outer membrane representing a barrier to the movement of proteins and the inner a barrier to small metabolites. Their classification into different types arises from their internal structure and origin and has been reviewed elsewhere (58a). Proplastids or eoplasts are colorless, occur in the meristematic cells of shoots, roots, embryos, and endosperm, and are the progenitors of other plastids. They have no distinctive morphology, vary in shape, and sometimes contain lamellae and starch granules. Chloroplasts are green, lens-shaped organelles, 5–10μ in diameter, the site of the photochemical apparatus with a distinctive internal membrane organization of thylakoid discs. Although obviously associated with photosynthesis in leaves, they are also found in the outer layers of unripe fruits and in storage cotyledons and embryos. Chromoplasts contain relatively high levels of carotenoids and give rise to the red, orange, and yellow colors associated with petals, fruits, and senescing leaves. They are often derived from chloroplasts and contain osmophilic globuli, which represent deposits of carotenoids and lipids. Etioplasts are formed in leaf cells grown entirely in the dark and appear yellow due to the presence of protochlorophyll. They possess crystalline centers known as prolamellar bodies and rapidly differentiate into chloroplasts on exposure of shoots to light. Leucoplast is a general term applied to colorless plastids, but, unlike proplastids and etioplasts, they are not regarded as progenitors of other plastids. Within this group are the amyloplasts and elaioplasts. The former contain substantial starch granules and are found in roots and storage tissues such as cotyledons, endosperm, and tubers, whereas elaioplasts are largely filled with oil and found in epidermal cells of some monocotyledenous families such as the Orchidaceae.

Most of the research effort has been expended in understanding assimilatory and biosynthetic metabolism in chloroplasts, and especially CO_2 fixation, which utilize energy derived from photochemistry. On the other hand, in plastids of developing seeds, fruits, tubers, and roots, even where there is some capacity for photosynthesis, metabolism is essentially heterotrophic and the accumulation of one or more end-products is dependent on a supply of precursors and energy from the cytosol. This is illustrated in Figure 1, which also emphasizes the interaction between several metabolic pathways. Such tissues provide the bulk of dietary intake in humans. The advent of the genetic tools to modify pathways calls for a detailed understanding of the processes that occur, their diversity and ontogeny, as an essential prerequisite to improved productivity and variety. In this review we summarize recent developments in our understanding of metabolism in nonphotosynthetic plastids and chloroplasts of storage organs in relation to anabolic pathways of synthesis and assimilation (starch, fatty acids, and nitrogen); the provision of energy for those pathways; and the spatial and developmental interactions that occur within the plastid and with the cytosol.

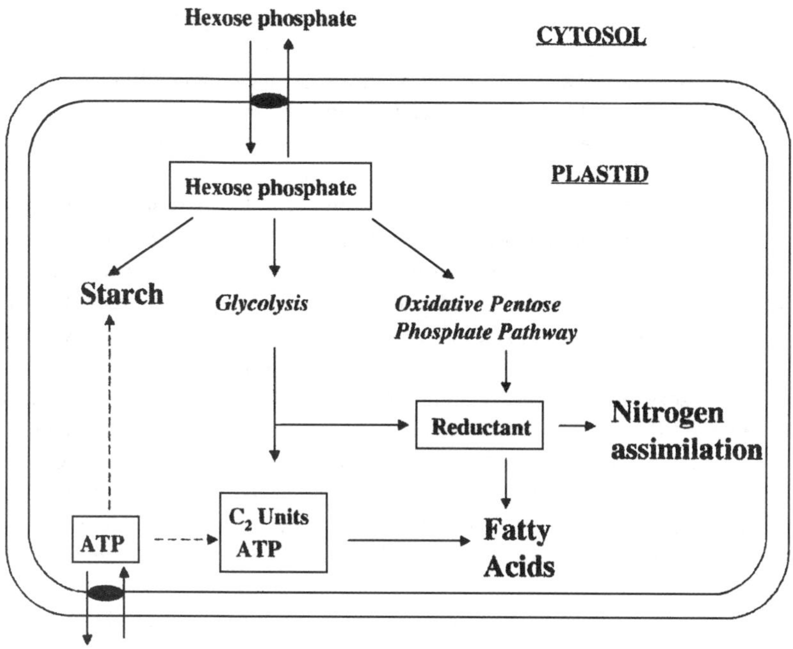

Figure 1 Interaction between pathways of anabolism and catabolism in nonphotosynthetic plastids. Biosynthesis and assimilation depend upon the import of precursors, such as hexose phosphate and ATP, from the cytosol and the generation of co-factors and intermediates by oxidative metabolism within the organelle. Inner envelope transporters are indicated by a solid oval here and in subsequent diagrams.

ANABOLIC PATHWAYS

Starch Synthesis

In nearly all types of plants, starch occurs in two forms, the transitory starch in photosynthetically active leaves, and the storage starch in specialized heterotrophic tissues such as roots, tubers, fruits, embryos, or endosperm (101, 152). In general, nucleotide sugars are required to elongate polysaccharides; for starch, this is ADPglucose (ADPGlc). ADPGlc is synthesized from Glc1P and ATP, and catalyzed by the enzyme ADPglucose pyrophosphorylase (AGPase). This metabolite is used as a substrate by the isoforms of starch synthase, either granule-bound starch synthases (GBSS), that are most likely involved in the elongation of amylose (142), or soluble starch synthases (SSS), the activities of which are coordinated with branching enzyme to elongate amylopectin. The biochemistry of the starch synthases and branching enzymes, and the formation of the starch granule have been extensively reviewed (99, 100, 120), hence other aspects relevant to the general regulation of metabolism within nonphotosynthetic plastids are emphasized here.

Starch Synthesis in Amyloplasts Considerable progress has been made in the molecular characterization of a number of transporters of phosphorylated intermediates in chloroplasts and this subject was reviewed recently (39). In the past ten years, considerable effort has been expended on attempts to determine the nature of the metabolites that are imported into amyloplasts; the picture that has emerged is complex and suggests no single paradigm.

The triose phosphate transporter from chloroplasts represents by far the best-analyzed transporter in the plastid envelope (39). Although amyloplasts are capable of catalyzing a similar counterexchange (36), there is uncertainty as to whether the same gene is expressed in both chlorophyllous and non-green tissue (63, 114). Nonetheless, various studies speculated that this protein might catalyze triose phosphate (TrioseP) import into storage plastids during starch synthesis (31). However, the observation that nearly all heterotrophic plastids from storage tissues lack the enzyme fructose1,6-bisphosphatase (37) argues against the likelihood that triose phosphates can be converted to hexose phosphates, and hence ADPGlc, within the organelle. Elegant labeling studies with intact tissues demonstrated that redistribution of carbon atoms from imported sugars into starch was not consistent with an involvement of triose phosphates in the direct pathway of synthesis (58, 141).

Consequently, other metabolites have been considered as cytosolic precursors for starch synthesis in the amyloplast. Using highly enriched amyloplasts from developing pea embryos, Hill & Smith (50) demonstrated glucose 6-phosphate (Glc6P)-dependent starch biosynthesis in the presence of exogenous ATP. Based on measurements of starch synthesis at saturating substrate concentrations, other types of plastid, including amyloplasts from cauliflower buds and developing rapeseed embryos, have been shown to use Glc6P as the most effective precursor for starch synthesis (56, 83). There are also examples of plastids from dicots that appear to convert exogenously supplied glucose 1-phosphate (GlclP) to starch (65, 73). In potato amyloplasts, it has been demonstrated that exogenously supplied GlclP, but not Glc6P, is able to support starch synthesis (82), although ample activity of phosphoglucomutase is present in these organelles to catalyze the interconversion of the two. The results cannot be explained by contamination of preparations with cytosolic phosphoglucomutase, since Glc6P did not support starch synthesis, nor by a reversal of starch phosphorylase reaction, since the process was ATP-dependent. A more recent study of starch synthesis in potato amyloplasts (148a) showed a strict dependence upon Glc6P but not GlclP. However, the recovery of organelles in these preparations was only 0.3% of the starting material, and may therefore represent a different population of amyloplasts compared with those studied by Naeem et al (82) where the recovery was 10%. In a different approach, studies of proteoliposomes reconstituted from potato amyloplast preparations showed that Glc6P could be transported in exchange for inorganic phosphate, at rates several-fold higher than GlclP (113). This does not resolve the question, but adds weight to the view that Glc6P is imported into potato amyloplasts and would therefore be expected to be a substrate for starch synthesis when imported from the cytosol.

The results in the discussion above are consistent with the view that the crucial reaction during conversion from hexose phosphates to starch, inside amyloplasts, is catalyzed by ADPGlc pyrophosphorylase (AGPase), which utilizes Glc1P and ATP (101). The presence of a hexose phosphate/Pi exchanging carrier has been demonstrated for many types of storage plastids (14, 73, 134). Flügge et al (54) identified a plastidic Glc6P/Pi transporter at the molecular level. This transporter exhibits only 36% identity to the triose phosphate transporter of leaves and belongs to the large group of solute carriers exhibiting 2×6 transmembrane helices in the functional state (39). Interestingly, it does not display substantial similarity to the inducible Glc6P/Pi exchanger from *Escherichia coli* (52). The plastidic Glc6P/Pi exchanger cloned from pea roots (54) is unable to transport Glc1P. Nonetheless, proteoliposomes harboring envelope proteins from wheat endosperm amyloplasts catalyze the transport of Glc1P in a 1:1 stoichiometric exchange with inorganic orthophosphate (134), indicating that there is still another type of hexose phosphate transporter to be identified.

Starch-accumulating plastids have to prevent accumulation of inorganic phosphate. The enzyme AGPase is highly regulated by the allosteric activator 3-phosphogylceric acid (3-PGA) and is inhibited by inorganic phosphate (99). Interestingly, an imbalance in the stromal phosphate status occurs during hexose phosphate–driven starch biosynthesis. Hexose phosphate is imported in a 1:1 exchange with stromal Pi (14, 134), and ATP required for the AGPase reaction enters the stroma in counterexchange with ADP (115). This leads to the import of a total of four phosphate moieties (A**T**P and hexose-**P**) and the release of only three phosphate moieties (A**D**P and **Pi**). This is distinct from the situation in chloroplasts, where the turnover of adenylates and phosphorylated intermediates is exactly balanced. The accumulation of Pi within the amyloplast stroma would lead to the inhibition of AGPase and therefore ultimately limit starch synthesis. This suggests that amyloplasts possess a mechanism for removing excess inorganic phosphate. An analysis of the rate of unidirectional Pi release from cauliflower bud amyloplasts revealed that the ratio between the rate of unidirectional Pi release to bi-directional Pi/Pi exchange is significantly higher in heterotrophic plastids than previously reported for chloroplasts (85). In addition, the rate of unidirectional Pi release was sufficient to account for the export of all the inorganic phosphate liberated during starch biosynthesis. Furthermore, phosphate did not accumulate in wheat amyloplasts actively synthesizing starch from exogenous Glc1P and ATP (133), which reinforces the view that amyloplasts possess an as-yet unidentified mechanism to remove inorganic orthophosphate. As discussed later, this problem is avoided in endosperm of monocotyledenous species where ADPGlc is synthesized in the cytosol.

Regulation of Starch Accumulation When studying starch synthesis in isolated amyloplasts, several metabolites have to be present in the incubation medium to obtain optimal rates of synthesis. In addition to the obvious requirement for hexose phosphate and ATP, amyloplasts from cauliflower buds require exogenously supplied 3-PGA as a positive activator (7). This stimulatory effect is due to the allosteric activation of AGPase by 3-PGA (99) and indicates that imported Glc6P is not readily converted to 3-PGA within the stroma. A similar stimulatory effect of 3-PGA on starch biosynthesis has also been made for amyloplasts enriched from sycamore cell-suspension cultures (97), and for plastids from sweet pepper and tomato fruits (8, 21). A role for 3-PGA in controlling the rate of starch synthesis in heterotrophic tissues is further underlined by the analysis of transgenic potato plants with reduced activity of the mitochondrial NAD malic enzyme. Tubers harvested from these plants exhibit higher starch contents than their corresponding wild types; the only substantial metabolic change that could be correlated with this observation was an increase in the concentration of 3-PGA (51).

It is highly likely therefore that changes in the cytosolic concentrations of hexose phosphates, ATP, and 3-PGA ultimately influence the rate of starch synthesis in amyloplasts, but information about such events in storage tissues is limited. Using a cell-suspension culture from *Chenopodium rubrum* as a model of a heterotrophic system, Stitt and colleagues demonstrated that glucose feeding induces a very rapid accumulation of starch (46). This accumulation of starch is accompanied by increased respiratory activity, with higher levels of Glc6P, ATP, and 3-PGA. The situation is thus comparable to the control of starch biosynthesis in green tissues, where transitory starch synthesis is also governed by metabolic events in the cytosol (87).

The content of starch in storage tissues is a function of the rate of synthesis and degradation, and a brief consideration of the latter is therefore pertinent. Amyloplasts from sycamore cell-suspension cultures, carrying out [^{14}C]-Glc6P-driven starch biosynthesis, are able to export radioactively labeled glucose previously incorporated into starch (97). Similarly, high starch–containing amyloplasts from cauliflower buds show ATP-dependent starch degradation (84). However, this degradation is nearly completely abolished in the presence of 3-PGA in the incubation medium, further highlighting the function of 3-PGA in the control of starch metabolism.

A study by ap Rees and colleagues revealed that rapid starch turnover is possible in potato tubers. Following heterologous expression of a permanently active AGPase in tubers (*glg*C16, AGPase from *Escherichia coli*), the rate of sucrose to starch conversion was increased (129). However, no net gain in overall starch content of the tubers was observed since the rate of degradation, revealed by pulse-chase experiments, also increased in the transformed state. The factors regulating this turnover in amyloplasts are still unknown.

Starch Synthesis in Storage Tissues from Monocotyledenous Plants In endosperms of monocotyledenous species, AGPase is found in both the cytosol and amyloplasts. Smith and co-workers demonstrated that the bulk of this enzyme's activity in barley and maize is associated with the soluble and not the particulate fraction of tissue extracts (27, 136); a similar association has been found in wheat (32). AGPase is a heterotetramer made of large and small subunits. Immunoblot analysis of subcellular fractions demonstrated that both subunits were detected in the amyloplast fraction and had a molecular mass similar to that of their chloroplast counterparts. However, the AGPase polypeptides associated with the soluble fraction were larger than those detected in amyloplasts. Further, the sequences cloned from maize appeared not to encode a transit peptide for targeting the subunits to plastids (43). The amyloplast form of the large subunit is present in wheat throughout endosperm development, whereas the cytosolic form appeared only after 10 days post anthesis; this implies developmental regulation of two spatially distinct forms (MM Burrell, unpublished information).

Thus the enzyme apparently has a dual location, which raises some intriguing issues, not the least of which is why the starch-storing tissues of endosperm, to which this phenomenon seems to be confined, operate differently from those of dicots such as potato and pea. In particular, these observations underpin the notion that ADPGlc can be transported into amyloplasts, a suggestion first made by Akazawa and colleagues (98). Studies of isolated wheat amyloplasts have shown that this substrate, supplied to intact organelles, can support 20-fold higher rates of starch synthesis in vitro than those obtained with hexose phosphates, which suggests that there is a distinct ADPGlc translocator in the amyloplast envelope (132). Amyloplasts from maize are also able to sustain starch synthesis with imported ADPGlc (75). It has been argued that the maize mutant *bt1* (brittle 1), in which the content of starch is reduced, arises as a result of a lesion in an amyloplastidic ADPGlc transporter. These mutants show an elevated concentration of ADPGlc (118), as a result of the loss of one or more major amyloplast envelope polypeptides between 39–44 kDa (24), and amyloplasts from the mutant lines are unable to synthesize starch from exogenous ADPGlc (117). These results are consistent with the view that BT1 is associated with the process of ADPGlc transport, but as yet there is no direct proof that BT1 is the transporter. Surprisingly, antibodies to the carboxy terminus of the maize protein were unable to detect a similar protein in other species, including various monocots. BT1 homologues from other species may be too dissimilar at their carboxy termini for epitopic recognition by the antibodies, or BT1 may be distinct to maize and, although associated with ADPGlc transport, is not the transporter itself. The latter possibility has some support in that cross-linking studies using azido-labeled adenylates show that amyloplasts from *bt1* still retain the ability to bind ADPGlc (IJ Tetlow, MJ Emes, unpublished). The expression of BT1 in maize is under developmental control, with transcripts not being detected until about 10 days post pollination (23). This situation is similar to that in wheat endosperm (Figure 2), from which an ADPGlc transporter has been purified (32) and which is expressed in development at approximately the same time as the cytosolic subunits for AGPase (IJ Tetlow, MJ Emes unpublished).

The existence of an alternative route of starch synthesis in cereal endosperm, different from (and possibly in addition to) that found in starch-storing tissues of dicotyledonous plants, also raises a number of novel possibilities for the regulation of this pathway. The hydrolysis of inorganic pyrophosphate by alkaline pyrophosphatase, in the amyloplast, pulls the equilibrium reaction catalyzed by

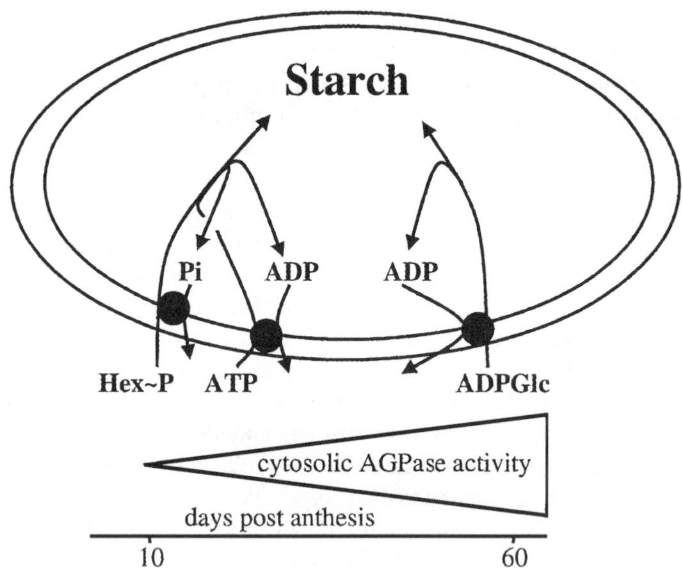

Figure 2 Starch synthesis in amyloplasts from endosperm of monocotyledenous plants. Species such as wheat, on which the model is based, are able to synthesise starch either from hexose phosphate or ADPglucose imported from the cytosol. The capacity for the synthesis of ADPGlc in the cytoplasm and its import into the amyloplast increases during development and is probably the main route of starch synthesis during grain filling. AGPase, ADPglucose pyrophosphorylase; ADPGlc, ADPglucose.

plastidic AGPase in the direction of ADPGlc synthesis. Localization of AGPase in the cytosol requires the removal of one or more products of the reaction (ADPGlc or pyrophosphate) to sustain starch synthesis in the amyloplast. Thus, the activity of an ADPGlc transporter will have an important bearing in two respects: Not only may it represent the major portal of entry of carbon into the organelle, but the rate at which it operates in supplying carbon to the starch-synthesizing machinery may also feed back onto the rate of synthesis of the soluble substrate in the cytosol, since an accumulation of ADPGlc would decrease its own synthesis as a consequence of the equilibrium constant of the reaction.

The breakdown of a mole of sucrose imported from the phloem leads to the formation of a mole of UDPglucose (UDPglc) and a mole of hexose phosphate. Although the turnover of UDPglc can be linked, via pyrophosphate, to the synthesis of ADPGlc in the cytosol, the synthesis of a second mole of ADPGlc from hexose phosphate will lead to net synthesis of pyrophosphate (Figure 3). Thus turnover of pyrophosphate is also needed in the cytosol, with the most obvious route via the pyrophosphate-dependent phosphofructokinase (PFP) (126), which is regulated by fructose 2,6-bisphosphate. Since the flux of carbon into starch represents the predominant activity of endosperm tissue, this may provide PFP with

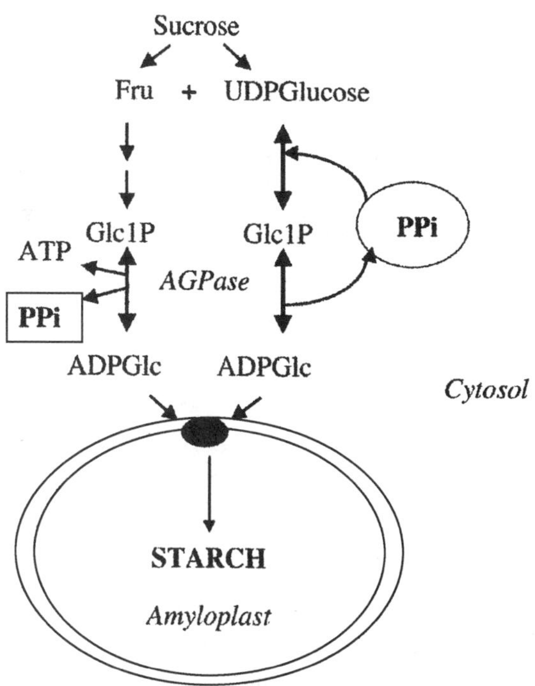

Figure 3 Turnover of inorganic pyrophosphate and ADPGlc in the cytoplasm during starch synthesis in developing endosperm of monocotyledenous plants. Note that the biosynthesis of two moles of ADPGlc from a mole of sucrose leads to the net generation of a mole of inorganic pyrophosphate (boxed). Fru, fructose; Glc1P, glucose 1-phosphate; ADPGlc, ADPglucose; PPi, inorganic pyrophosphate.

a distinct role, linking events in the cytosol to the synthesis of storage product in the amyloplast and emphasizing the requirement to integrate events between the two compartments. Whether the two alternative routes of starch synthesis operate in (*a*) the same cells or (*b*) the same amyloplast awaits resolution.

Fatty Acid Synthesis in Storage Tissues

Lipids occur in two major forms, as a metabolic reserve in the form of triacylglycerides or as membrane lipids involved in structuring the cell, carrying a hydrophilic headgroup at the sn 3 position of the glycerol backbone (128). The synthesis of fatty acids is strictly limited to the plastidic compartment (90, 91). In several heterotrophic tissues, notably oilseeds, lipids accumulate to comparably high levels, constituting up to 40% of dry weight (119). Since such storage tissues are essentially nonphotosynthetic (however, see section on plastid development), substantial carbon fluxes must enter the plastidic compartment during lipid biosynthesis. These are summarized in Figure 4.

Studies of purified plastids have demonstrated a wide range of intermediates capable of supporting fatty acid synthesis. As with most plastidic reactions, chloroplasts were the first type of plastids to be analyzed in detail. Various studies showed that acetate, previously liberated by mitochondrial acetylCoA hydrolase activity

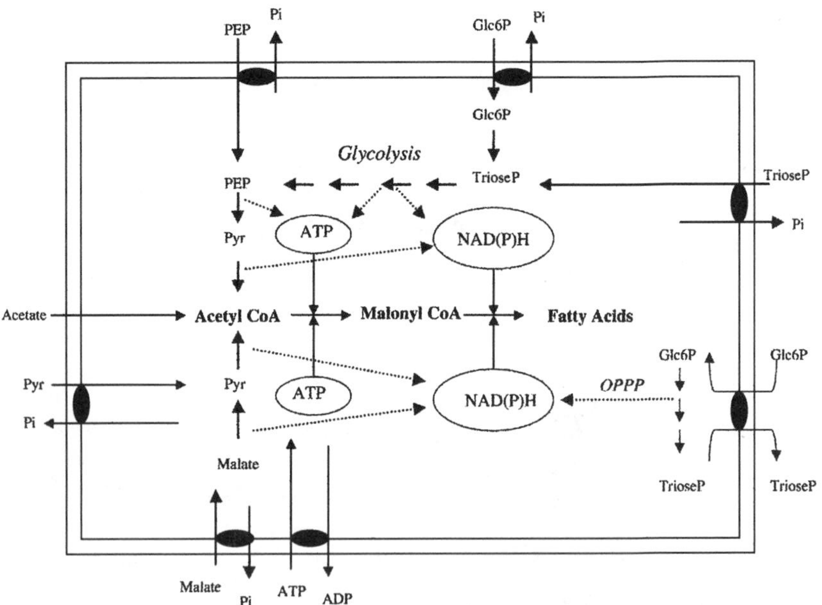

Figure 4 Fatty acid synthesis in nonphotosynthetic plastids. A wide array of different precursors have been shown to be capable of supporting fatty acid synthesis depending upon species and stage of tissue development. PEP, phospho*enol*pyruvate; Glc6P, glucose 6-phosphate; Pyr, pyruvate; TrioseP, triosephosphate; Pi, inorganic orthophosphate; OPPP, oxidative pentose phosphate pathway.

(44), enters the stroma by an unknown mechanism, perhaps by diffusion, and serves as a primary carbon source for de novo fatty acid synthesis (81, 102, 109). Similarly, heterotrophic plastids from cauliflower buds, maize endosperm, and, to a lesser extent, from developing wheat endosperm, have been shown to use acetate as the major carbon source (53, 77).

However, although all plastids seem able to use acetate as a source of carbon for fatty acid synthesis, other cytosolic precursors can also be imported for this purpose, and in some cases support a higher rate of synthesis than acetate, depending on species, availability of reductant, and stage of development. Some oil-rich plastids, such as those from *Cuphea wrightii* seeds, use acetate as the most effective precursor (47), whereas others do not. The castor bean (*Ricinus communis*) endosperm is characterized by high oil concentrations and, in the corresponding plastids, malate is the substrate that is incorporated at the highest rate into lipids at any given concentration (122) and is taken up by a plastidic malate-phosphate antiporter (28). Plastids from pea roots and from developing embryo tissue of rapeseed (*Brassica napus*) use a wide range of exogenous precursors to support fatty acid synthesis, including Glc6P, triose phosphates, malate, pyruvate, and acetate (57, 103). Rapeseed plastids are particularly interesting because the utilization of substrates appears to be developmentally regulated. In plastids from early stage embryos, Glc6P is the most effective substrate for fatty acid synthesis (and for starch synthesis) and pyruvate is a very poor precursor. This situation changes during development when Glc6P utilization and import decrease (associated with a decline in the capacity for plastidial glycolysis), and pyruvate utilization increases, coincident with increased capacity of a pyruvate transporter (105; P Eastmond & S Rawsthorne, unpublished results). The basis and function of these developmental changes remain to be resolved.

Co-Factors for Fatty Acid Synthesis Apart from a supply of carbon, fatty acid synthesis depends upon the stromal provision of ATP and reducing equivalents (128). Fatty acid synthesis in chloroplasts is substantially promoted by continuous illumination, as the supply of ATP and NADPH, which is mediated by photosynthesis, is limiting. In contrast, storage plastids must either take up these compounds from the cytosol or generate them internally. ATP is required for the carboxylation of acetylCoA to malonylCoA and can be maintained by two processes: A complete or partial glycolytic pathway within the organelle would allow ATP synthesis from either Glc6P, triose phosphates (GAP, DHAP), or phosphoenol pyruvate (PEP); alternatively, ATP may be imported direct from the cytosol.

Acetate-dependent fatty acid synthesis in developing castor bean leucoplasts is stimulated fourfold by the addition of PEP when compared with exogenous provision of ATP. This implies that internally generated ATP (resulting from plastidic pyruvate kinase activity) is more effective in supporting fatty acid biosynthesis (20). Stromal synthesis of ATP via a triose phosphate shuttle has been reported for acetate-dependent fatty acid synthesis in chromoplasts from daffodil flowers (59). The same phenomenon has been demonstrated in pea root plastids, where Glc6P was also able to substitute for ATP (60, 103), emphasizing the multiplicity of possibilities and interactions that take place. Nonetheless, in many cases, optimal rates of fatty acid synthesis in purified plastids are maintained by the provision of exogenous ATP, and the uptake of adenylates from the cytosol has a critical role (42, 56, 74, 103).

In addition to ATP, large amounts of reducing equivalents are required to synthesize fatty acids. Fatty acid synthesis is catalyzed by a multienzyme complex, fatty acid synthase (FAS), and during conversion of acetyl CoA to an 18-carbon fatty acid, 14 NAD(P)H are required (25, 90). The stimulatory effect of exogenous reducing equivalents on acetate-dependent fatty acid synthesis in isolated plastids has been observed in many different species (41, 74, 103). However, plastid envelopes are generally regarded as impermeable to pyridine nucleotides in vivo, and this stimulation is almost certainly an artefact due to a high artificial concentration gradient between the incubation medium and the stroma.

Storage plastids possess several ways to synthesize NADPH endogenously from organic acids and phosphorylated intermediates. Malate-driven fatty acid synthesis also brings about the indirect import of reducing equivalents (122). During conversion of malate to acetyl CoA, two moles of reduced pyridine nucleotide are liberated: one during the reaction catalyzed by NADP-dependent malic enzyme, and the other at the pyruvate dehydrogenase reaction. Bearing in mind that the elongation of a fatty acid by a C2 unit requires 2 moles of reducing equivalents, the stoichiometric balance between carbon and reducing equivalent supply is maintained. This does not hold true when compounds like pyruvate or acetate are utilized and therefore the oxidative metabolism of sugar phosphates seems to be required.

In general, sugar phosphates can be oxidized via glycolytic enzymes or via the oxidative pentose phosphate pathway (34). Heise and colleagues showed that oxidation of Glc6P could stimulate fatty acid synthesis in plastids isolated from seeds of *Cuphea wrightii* (41, 47). That this was due to the generation of reductant by the oxidative pentose phosphate pathway (OPPP) was confirmed by studies of plastids from the embryos of rapeseed, which followed the release of CO_2 from carbon atom 1 of Glc6P (57). The organization of the OPPP is discussed in a subsequent section.

Nitrogen Assimilation and Amino Acid Synthesis

While the initial step of inorganic nitrogen assimilation, catalyzed by nitrate reductase (NR), occurs in the cytosol of plant cells, all subsequent reactions of primary assimilation and amino acid biosynthesis occur within plastids, whether in green or non-green tissue (Figure 5). The extent to which nitrate assimilation occurs in roots is a function of the species and growth conditions. For instance, in legumes such as lupin, virtually 100% of the nitrate taken up by roots enters into organic combination before transport to the upper parts of the plant, whereas in species such as maize, roots may contribute only 30% to this process (93). Nitrogen assimilation places a high demand for energy and reducing power on the tissue, and in nonphotosynthetic plastids, this has to be met by carbohydrate oxidation. The requirement for ATP could be met either by import from the cytosol (115) or by glycolytic activity within the organelle (103). However, since plastid envelopes are impermeable to pyridine nucleotides, reducing power has to be generated oxidatively within the same compartment where it will be utilized.

There is substantial evidence that the reductant required for the conversion of nitrite to glutamate is generated by the oxidative pentose phosphate pathway (OPPP). Studies with isolated root amyloplasts have shown that the oxidation of glucose 6-phosphate by the OPPP is tightly coupled to the reactions catalyzed by nitrite reductase (NiR) and glutamate synthase (GOGAT) (16, 17). Glc6P is imported from the cytosol by a hexose phosphate translocator (54) that is distinct from the triosephosphate translocator of chloroplasts. Within the plastid, the OPPP operates cyclically, allowing the oxidation of each mole of glucose 6-phosphate to three moles of CO_2 and a mole of triose phosphate. The latter is exported in exchange for an incoming molecule of glucose 6-phosphate and so the stoichiometric balance is maintained (15, 45). The first enzyme of the OPPP, glucose 6-phosphate dehydrogenase (G6PDH), is strongly inhibited by the ratio of NADPH/NADP. There

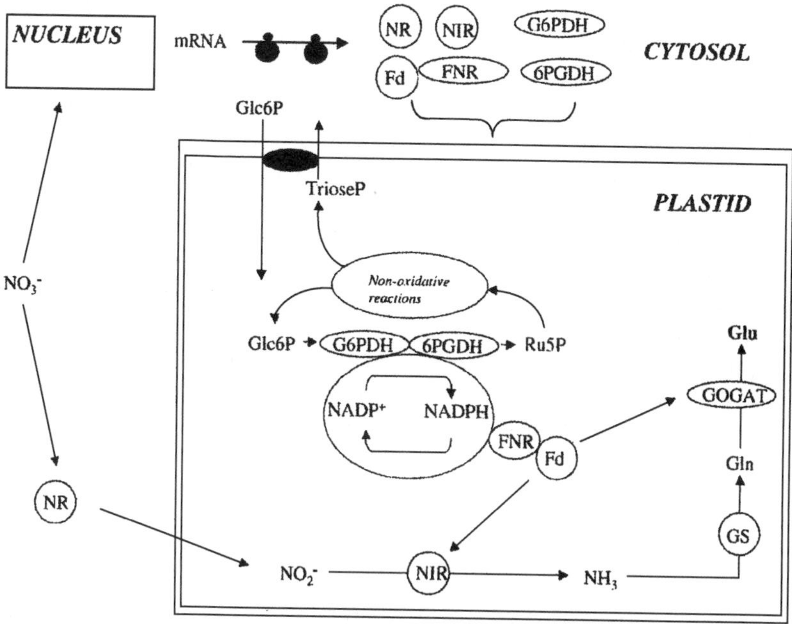

Figure 5 Interaction between nitrogen assimilation and the OPPP in root amyloplasts. The availability of nitrate brings about an induction of the nuclear-encoded enzymes of nitrate assimilation and also components involved in the generation of reductant which sustain assimilation inside the organelle. The plastidic OPPP utilizes Glc6P imported from the cytosol to generate NADPH. This is used to reduce a root-specific, nitrate-induced form of Fd, the immediate electron donor for NiR and GOGAT. NR, nitrate reductase; NIR, nitrite reductase; Fd, ferredoxin; FNR, ferredoxin-$NADP^+$ reductase; G6PDH, glucose 6-phosphate dehydrogenase; 6PGDH, 6-phosphogluconate dehydrogenase; GS, glutamine synthetase; GOGAT, glutamate synthase; Glc6P, glucose 6-phosphate; TrioseP, triosephosphate; Ru5P, ribulose 5-phosphate; Gln, glutamine; Glu, glutamate.

is therefore a balance to be struck, in that NADPH is required for assimilation and biosynthesis, but too high a concentration will inhibit its own production. Studies with barley have shown that the NADPH/NADP ratio increased from 0.9 to 2.0 when Glc6P was supplied to root plastids (149). The subsequent addition of nitrite led to a decline in this ratio to 1.5, and it was calculated that the G6PDH remained 50% active under these conditions, enabling it to sustain nitrogen assimilation.

The immediate source of reducing power for both NiR and GOGAT is ferredoxin (Fd), and electrons are transferred from NADPH via a ferredoxin-NADP reductase (FNR). This is, effectively, in the opposite direction of electron flow to that which occurs in photosystem I of the chloroplast and at first glance seems energetically unfavorable. However, the ferredoxin found in root plastids has a substantially lower negative midpoint potential than its leaf counterpart (131), at about −320 mV instead of −387 mV, facilitating transfer of electrons from NADPH. The primary sequence of the root FNR is also substantially different from the leaf enzyme (4), and this almost certainly reflects their different roles.

This close coupling between the pathways that generate and utilize reductant is also reflected through changes in transcription and protein levels. Both G6PDH and 6-phosphogluconate dehydrogenase (6PGDH) actvity increase in root plastids of pea (33) and maize (107) during nitrate assimilation. A cDNA to a root 6-phosphogluconate dehydrogenase hybridized with a transcript that accumulated rapidly and transiently in response to low concentrations of external nitrate (107). Ferredoxin in roots proved difficult to detect until it was realized that the protein is also induced by nitrate assimilation, as is FNR (18). Two forms of ferredoxin have subsequently been cloned from maize roots: one constitutive, the other transcribed within 2 h of the application of nitrate to plants (71). In rice roots, this is coincident with the appearance of the mRNAs for NiR and FNR (4). The promotor sequences for NiR (130) and the inducible forms of FNR (3), ferredoxin (71) and G6PDH (JS Knight, PD Debnam, MJ Emes, unpublished data) possess the NIT-2 motif, which is a global, regulatory factor of nitrogen metabolism in fungi (68). This observation of conservation of *cis* elements is consistent with experimental observations, all of which point to coordinated expression of several metabolically linked root plastid enzymes. However, until the promotor has been analyzed in detail and the common transcription factors have been identified, the role of this or any other putative regulatory sequence in these higher plant genes remains speculative.

ENERGY SUPPLY

Carbohydrate Oxidation

To different degrees, depending on species and stage of development, both glycolysis and the OPPP have a dual location in non-green plastids and the cytosol (35). The regulation of glycolysis in plastids was reviewed recently (96) and our understanding of this pathway has not changed substantively since then. However, there has been a significant development for the OPPP. Schnarrenberger and colleagues (112) demonstrated that in spinach leaves, while the first two reactions (G6PDH and 6PGDH) are located in both cytosol and chloroplast, the nonoxidative reactions of the pathway (catalyzed by transketolase, transaldolase, pentose phosphate isomerase, and epimerase) are confined to chloroplasts. A subsequent study showed that this was true for roots as well as leaves, and holds true for spinach, pea, and maize, although all the enzymes were found in both compartments of tobacco (26). The plastid thus becomes the sole source for de novo production of ribose 5-phosphate, needed for nucleic acid synthesis, which implies that large amounts will be exported, particularly during periods of cell growth and division. The termination of the OPPP in the cytosol means that ribulose 5-phosphate will have to be imported into plastids if it is to be further metabolized. There is indirect evidence that plastids are capable of transporting pentose phosphates (26, 45) and that this is likely to occur via a phosphate translocator-type protein (39). The absence of the nonoxidative reactions in the cytosol also affects the regulation of glycolysis since it would then become impossible for the oxidation of fructose 6-phosphate to bypass the pyrophosphate- and ATP-dependent phosphofructokinases (32), both of which are highly regulated enzymes (96).

At the molecular level, our knowledge of the OPPP enzymes lags considerably behind that of glycolysis. cDNA clones have now been described for several plastidic forms of OPPP enzymes (9, 62, 72, 89, 107, 143). G6PDH in chloroplasts is regulated by thiol modulation of two cysteine residues (145), which does not seem to be the case for the enzyme from nonphotosynthetic plastids (R Hauschild & A von Schaewen, personal communication). Given the changes in expression of the root plastid enzymes in response to nitrate (see above), and the fact that some of the enzymes catalyzing the nonoxidative reactions are found in chloroplasts where they also function in the Calvin cycle, the regulation of any of these steps is not necessarily the same in different tissues.

The Plastidic ATP/ADP Transporter

As outlined previously, heterotrophic plastids require uptake of ATP to energize various anabolic reactions. Because of its charge and size, ATP cannot penetrate biomembranes, and all higher plant plastids (chloroplasts and all types of heterotrophic plastids) analyzed so far possess an inner envelope protein that mediates ATP/ADP exchange (147). The biochemical properties of a plastidic adenylate transporter were characterized in detail about 30 years ago. By analyzing the uptake of adenylates into isolated chloroplasts, it became evident that this type of transporter preferentially imports ATP—rather than ADP—in exchange for ADP, although the maximal import rate does not suffice to support photosynthetic CO_2 fixation (48, 108). The equivalent ATP/ADP-transporter in heterotrophic plastids exhibits essentially similar biochemical properties to its chloroplastic counterpart (115). One notable difference is that, at least in the case of pea roots plastids, the maximal rates for both ATP and ADP uptake are similar, indicating that under in vivo conditions the preferred substrate depends upon the cytosolic ATP/ADP ratio.

Several other cellular compartments require the provision of adenylates from the cytosol. Mitochondria have to import ADP for the synthesis of ATP via oxidative phosphorylation, and this is mediated by a functional equivalent, the ADP/ATP carrier (AAC) (61). Other compartments such as the Golgi apparatus and the endoplasmatic reticulum also have to import ATP (1). Of these, only the mitochondrial adenylate transporter has been characterized at the molecular level (61), and, indeed, represents one of the best characterized eukaryotic membrane proteins.

Prior to its being cloned, several features of the plastidic ATP/ADP transporter indicated that the molecular nature of this protein must differ substantially from the functional equivalent in mitochondria. For example, the mitochondrial transporter specifically imports ADP in strict counterexchange to ATP, whereas, although the plastidic carrier can catalyze import of both adenylates at the same rate (115), the physiological imperative will be to import ATP into heterotrophic plastids. Further, the mitochondrial transporter is very strongly inhibited by compounds like carboxyatractyloside and bongkrekic acid (127), whereas the plastidic ATP/ADP transporter is not affected by these compounds (88, 115). Finally, polyclonal antisera raised against the mitochondrial AAC crossreact with all mitochondrial homologues from different species, but do not recognize proteins located in the inner envelope of plastids of either autotrophic or heterotrophic origin (88, 115).

In 1995, Kampfenkel and co-workers discovered a cDNA in an *A. thaliana* cDNA library encoding a highly hydrophobic membrane protein, with 12 predicted transmembrane domains, exhibiting 66% similarity to the ATP/ADP transporter from the gram-negative bacterium *Rickettsia prowazekii* (55). *R. prowazekii* is the causative agent of the epidemic typhus; it lives freely in the cytosol of vertebrate cells and exploits the host cell cytosol by uptake of various compounds such as NAD, AMP, lysine, and ATP (147, 148). The uptake of ATP into the bacterial pathogen is mediated by an ATP/ADP transporter protein (RpTlc) (146, 148). As both *R. prowazekii* and higher plant plastids reside in the same cellular niche, the cytosol, and import ATP in exchange for ADP (95, 115), it was postulated that the *A. thaliana* cDNA also encoded an ATP/ADP transporter and the plastid protein was named AATP1(At) (**ATP**/**ADP** **t**rans**p**orter **1**, **A**. **t**haliana).

AATP1 is a nuclear-encoded protein with an N-terminal transit domain (55) that allows integration into the inner plastid envelope membrane and processing of the precursor protein to the mature form (88). The high degree of similarity between the plastidic and bacterial homologues (more than 66%) suggested that it might be possible to express the AATP1 cDNA in *Escherichia coli*. This was the first plant solute transporter to be properly integrated into the bacterial cytosol membrane as a functionally active protein (138). As such, this heterologous expression system proved ideal as the biochemical features of the recombinant protein were unaltered compared to the authentic protein in isolated plastids (76, 137, 138). The apparent affinities of various plastidic ATP/ADP transporters (the isoforms AATP1 and AATP2 in *A. thaliana*, and a first potato homologue AATP1,St) for ATP and ADP are all in the micromolar range (ranging from 12 to 35 μM) and the specificity for transport of ATP and ADP is absolute (138).

The evolutionary relationship between the plastidic and the bacterial ATP/ADP transporters is complex. *R. prowazekii* belongs to the α-class of proteobacteria, which represent the phylogenetic origin of mitochondria (2). However, it is not clear whether ancient rickettsial bacteria, which lived outside host cells, already possessed an ATP/ADP transporter, although ATP would be expected to be virtually absent in such environments (147). Recently, two further homologues to the rickettsial transporter have been identified in the intracellular bacterium *Chlamydia trachomatis* (139), and this species does not cluster into the group of α-proteobacteria (140).

Physiological Significance of ATP Import The importance of plastidic adenylate transport has been investigated in transgenic plants. The cDNA coding for AATP1(At) was introduced into the genome of potato to increase the abundance of the transporter in developing tubers, while cloning of the endogenous potato cDNA in antisense orientation led to transgenic plants with reduced plastidic ATP/ADP transporter activity (137).

Remarkably, the overexpressing plants showed a substantial increase in the yield of starch, and an increase in the ratio of amylose to amylopectin. This observation clearly indicates that the activity of the plastidic ATP/ADP transporter limits the rate of end-product synthesis in potato amyloplasts, and concurs with previous experiments on isolated amyloplasts from other species that indicated the limitations of ATP supply in storage plastids (74).

The cytosol from plants (as well as from other eukaryotic cells) is characterized by a high ATP/ADP ratio. For example, in photosynthetic mesophyll cells the cytosolic ATP/ADP ratio is about 10, and therefore fourfold higher than in the stroma, despite the obvious impact of photophosphorylation on chloroplast ATP content (124, 125). Assuming a similar relationship in cells of storage tissues, and that the cytosolic concentration of ATP is saturating, increasing AATP1 activity will translate a high cytosolic ATP/ADP ratio more efficiently into the stroma of storage plastids. Elevation of the stromal ATP concentration presumably leads to a higher content of ADPglucose and enhanced starch synthesis. The heterologous expression of a bacterial AGPase also enhances starch synthesis (123) and emphasizes the importance of the ADPglucose pool in controlling starch synthesis within the amyloplast.

By contrast, AATP1 antisense potatoes were characterized by less starch and a decrease in the amylose to amylopectin ratio, with less than 60% amylose when compared to the wild type (137). This observation further illustrates that changes in the plastidic ATP/ADP transporter activity have a profound effect on not only the carbohydrate yield of a storage tissue, but also the composition of the metabolic end-product starch. That the amylose to amylopectin ratio is affected by the supply of energy is most likely a reflection of the biochemical properties of the enzymes involved in starch biosynthesis and the importance of the ADPglucose pool. Storage tissues such as potato possess at least two isoforms of starch synthase. SSS, responsible for amylopectin elongation (66), has a higher affinity for ADPglucose than GBSS, which is responsible for amylose synthesis (120). When the ATP supply is altered, the change in ADPglucose content will affect GBSS disproportionately and thus effect a greater change in amylose content compared to amylopectin. Still to be determined is whether altering the expression of this transporter in an oil-rich storage tissue also affects lipid level and composition.

SPATIAL AND TEMPORAL INTEGRATION

Metabolic Cross-Talk in Storage Tissues and Plastids

Given the capacity of different storage plastids to carry out several different anabolic and catabolic processes simultaneously, and the complexity of metabolites that can permeate the envelope of these organelles, it is important to consider how these reactions influence each other and how they are integrated. That such communication takes place can be illustrated by observations made on intact tissues and on isolated plastids.

During the course of rapeseed embryo growth, for example, the relative fluxes of carbon to starch, lipids, and protein are dependent upon the developmental stage and, clearly, are programmed genetically. For example, activation of lipid synthesis in the middle phase of development correlates with reduced rates of starch accumulation (80). This predisposition can be influenced by altering the expression of key enzymes. Expression of an unregulated AGPase from *E. coli* in transgenic rapeseed (canola) led to a diversion of carbon from lipid into starch (13). Conversely, pea embryos, recessive at the rugosus locus (*rr*) that codes for a starch-branching enzyme (11), contain less starch and double the content of seed lipid (10). However, decreasing one end product does not always lead to the accumulation of an alternative. Transgenic potato plants, with a reduced capacity for tuber starch synthesis, also show a decrease in storage protein (79).

Alternatively, perturbation of lipid metabolism can influence starch accumulation, at least transiently. Mature *A. thaliana* seeds do not contain substantial amounts of starch. A recently identified wrinkled seed mutant (*wri1*) from this model plant exhibits 80% less lipids in the mature seed than the wild type, but elevated starch levels during development (40). Nonetheless, the mature seeds from mutants and wild-type plants contain the same amount of starch, presumably because of enhanced turnover, though this is not able to compensate for the loss of precursors responsible for the reduction in triacylglycerol synthesis.

Interestingly, it is also possible to create *A. thaliana* plants with high starch levels in the mature seed. A novel shrunken seed mutant (*sse1*), identified from stocks of T-DNA mutated lines, lacks the ability to form peroxisomes (67). The fully developed seeds contain very high starch levels, whereas lipids are drastically reduced. This suggests that starch synthesis in plastids represents a default pathway in the case of flux limitation into the major sink.

Fatty acid and starch synthesis can and do occur at the same time within the same organelle during the early mid-stage of rapeseed embryo development (57). Although embryo plastids are photosynthetic, they are not thought to be capable of net CO_2 fixation in vivo because the pod wall shields the tissue from photosynthetically active radiation (29). Experiments with purified plastids from this tissue demonstrated that supplementing Glc6P with pyruvate, in the presence of ATP, led to a threefold increase in fatty acid synthesis without affecting starch synthesis. This implies that these biosynthetic pathways do not compete for ATP. When the same organelles were utilizing Glc6P alone, the flux through the OPPP was sufficient to account for the requirements for reducing power for fatty acid synthesis. However, during the simultaneous oxidation of both Glc6P and pyruvate, NADPH generated by the OPPP was insufficient to meet the demands of fatty acid synthesis, suggesting that other intraplastidial sources were being utilized.

By contrast, amyloplasts from cauliflower buds show marked interdependency on the same pool of ATP for fatty acid and starch synthesis, such that activation of starch synthesis by PGA, the allosteric regulator of AGPase, inhibits fatty acid synthesis (74). Similar negative interactions between pathways can be seen in amyloplasts of developing wheat endosperm. When hexose phosphate (as opposed to ADPGlc, see earlier) is imported, these amyloplasts utilize Glc1P rather than Glc6P for the biosynthesis of starch (132). Within these organelles, the OPPP is stimulated when there is a demand for reducing power, as is required, for example, for the glutamate synthase reaction during amino acid production, a prerequisite for the deposition of storage protein. Under these conditions, Glc1P is diverted toward the OPPP with a concomitant 75% reduction in starch synthesis (19). Despite obvious limitations on how far such in vitro experiments can be interpreted, they illustrate the integrated nature of metabolism within nonphotosynthetic plastids and the need for a more detailed understanding of such cross-talk if attempts to manipulate major end products are to succeed.

Physiological Changes During Plastid Development

As indicated previously, plastids are often broadly categorized as either chlorophyllous, autotrophic chloroplasts or as heterotrophic plastids covering various types like elaioplasts, leucoplasts, amyloplasts, and chromoplasts (49, 58a). The description of only a few examples should accentuate the inadequacy of too simple a classification, and in this review emphasis has been placed on the specific physiological properties of plastids at various developmental stages of a tissue rather than on classification according to the location of the organelle in a specific type of tissue or its pigmentation.

Characterization of chloroplasts has mainly been carried out using plastids isolated from leaf tissues. Chloroplasts are also present in a wide range of fruits and in developing embryos, and due to the sink character of these tissues, they are thought to behave heterotrophically, i.e. they import metabolites to sustain anabolic reactions. However, various examples demonstrate that photosynthesis of fruit plastids is important for crop yield, but is species dependent. The ability to use light energy clearly depends upon the presence of light-absorbing pigments and upon the capacity for photosynthetic electron flow, leading to synthesis of both ATP and NADPH (12). The relationship between heterotrophic and autotrophic metabolism in green plastids of storage tissues has been examined in detail in plastids from rapeseed and pea embryos (57, 121), sweet pepper (8), tomato fruits (21, 116), and olive fruits (111). It is likely that all types of green-fruit or seed plastids are involved in CO_2-fixation, but that they operate a mixed economy, with the capacity to simultaneously utilize autotrophic and heterotrophic pathways of energy production to sustain anabolic reactions.

During development of fruits that pass through a green stage, extensive changes occur in the physiological properties of the corresponding plastids (5). In early stages of fruiting, green chloroplasts are present and function in CO_2-fixation (12). Later, chlorophyll degradation takes place and carotenoid biosynthesis becomes one of the dominant anabolic pathways (22, 78). The concomitant alteration of the levels of proteins involved in the Calvin cycle is complex. The activity of sweet pepper fruit Rubisco decreases substantially during conversion from green to red sweet pepper fruits (151), whereas other enzymes that are also involved in CO_2-fixation (e.g. fructose bisphosphatase, aldolase) increase in activity (135). These differences show that the regulation of individual enzymes that ostensibly belong to the same pathway is specifically regulated during tissue development.

Plastids within the developing embryos of oilseed rape contain thylakoid-like membranes and have rates of CO_2-dependent O_2 evolution per unit chlorophyll that are comparable to those of leaf chloroplasts, and which correlate with chlorophyll content during development (29). Illuminated embryo plastids (using a photon flux density estimated to represent the maximum that would be received by plastids at the outer face of the embryo in vivo) synthesized starch and fatty acids equally well from either CO_2 alone, or from Glc6P plus ATP (30). However, in the absence of an exogenous supply of ATP, the rates of Glc6P-dependent storage product synthesis fell by some 80%, even though the plastids were illuminated. This suggests that while light energy can contribute toward the accumulation of storage products, its contribution is likely to be small, and that embryo plastids, although photosynthetic in character, are more reliant on provision of energy from the cytosol through the oxidation of carbohydrate. Interpretation of such data is complicated by the uncertainty as to whether such preparations consist of more than one type of plastid; in general, we need to know a great deal more about the heterogeneity of plastids within storage tissues.

In leaves grown in the absence of light (e.g. from germination), etioplasts develop that lack both internal membrane structures and chlorophyll. However, etioplasts contain large amounts of starch (144), which raises the question of how the required carbon enters the stroma. Etioplasts from leaves of dark-grown barley plants, although heterotrophic, do not import hexose phosphates, but rather use cytosolic triose phosphates for starch biosynthesis (6). Similar observations have been made on etioplasts enriched from developing pinyon seedlings (150), clearly demonstrating that heterotrophy per se is not responsible for the presence of hexose phosphate transport activity across the plastid envelope. In etioplasts, an active FBPase allows conversion of triose to hexose phosphates (6, 150) and thus supports heterotrophic starch accumulation. The presence of a highly active triose phosphate/phosphate transporter in etioplasts probably represents a genetically programmed commitment toward photosynthetic activity at an early stage of development, where its role will be in carbon export. However, there is ample evidence that a number of fully developed source chloroplasts are also competent for hexose phosphate transport or are able to develop this capacity. These include chloroplasts from leaves of CAM plants such as *Sedum praealtum* (94) and *Mesembryanthemum crystallinum* (64, 86), the alga *Codium fragile* (110), sweet pepper and tomato fruits (8, 116), and from guard cells isolated from pea leaf epidermis (92).

The molecular basis for the regulation of expression of the hexose phosphate transporter is unknown. Kammerer et al (54) demonstrated that mRNA coding for the Glc6P/Pi antiporter accumulates mainly in heterotrophic plant tissues.

Nonetheless, chloroplasts from spinach leaves, which do not possess the capacity for Glc6P transport when plants are maintained under normal growth conditions (38), exhibit Glc6P/Pi exchange after feeding the leaves with glucose (104). Such glucose feeding induces a rapid conversion from autotrophy to heterotrophy and may indicate a role for sugars in signaling this switch. Although most autotrophic leaf chloroplasts lack a Glc6P/Pi antiporter, the fact that it is induced during the transition from C3 to CAM in *Mesembryanthemum* (64) suggests that it is likely to be linked to this novel mechanism of CO_2 fixation and is not confined to cells carrying out heterotrophic metabolism. Its presence in guard cell chloroplasts is probably a reflection of the fact that starch turnover occurs during opening and closing of stomata. Although guard cell chloroplasts possess a functional light-harvesting apparatus, since it is unlikely that they have anything but the most limited capacity for CO_2 fixation, and lack fructose 1,6-bisphosphatase (106), the starch that is formed within these organelles must arise from hexose phosphates imported from the cytosol. In this case, it would appear that the presence of the hexose phosphate transporter is a reflection of carbohydrate metabolism and not a function of the capacity for energy provision within the organelle.

This contrasts with the situation during senescence when leaf plastids degrade chlorophyll (69). This is part of a strictly controlled process that depends upon the provision of the stroma with energy in the form of ATP. In chloroplasts from *Hordeum vulgare,* gerontoplasts or "old chloroplasts" that still possess an intact envelope membrane exhibit high rates of chlorophyll degradation when provided with physiological concentrations of hexose phosphates, leading to stromal ATP generation (70). The observation that a hexose phosphate transporter is expressed at later stages of leaf development, when there is a need to generate energy within the plastid by oxidation (contrast this with guard cell chloroplasts) emphasizes the complexity of controls and plasticity of metabolism in these organelles.

SUMMARY AND PERSPECTIVE

Plastids not involved in the primary harvesting of light energy nonetheless play a central role in the synthesis of important end products and storage compounds. The metabolic pathways that contribute to this diversity of activities interact at the level of control of gene expression and through the uptake and utilization of common intermediates, and these relationships change throughout development of an organ. Much of the progress to date has necessarily involved biochemical studies of purified plastids from different species and tissues. This approach has provided important but, inevitably, limited information, since it presumes that the preparations are homogeneous, whereas in most cases they are likely to represent a heterogeneous mixture of organelles derived from different cell types. A more detailed picture of the molecular components and mechanisms that contribute to their breadth of function is now emerging. One of the most significant challenges in the future will be to refine our understanding of metabolism in plastids down to the level of the single cell. This will require a bedrock of biochemical information, coupled with the tools of cell biology and molecular biology, to produce a clear picture of the function of an individual plastid in the context of its cellular environment in a particular species.

ACKNOWLEDGMENTS

We are very grateful to N Amrhein, MM Burrell, KP Heise, J Preiss, S Rawsthorne, A von Schaewen, and IJ Tetlow for providing us with unpublished information and preprints used in the preparation of this article.

Visit the Annual Reviews home page at www.AnnualReviews.org

LITERATURE CITED

1. Abeijon C, Mandon EC, Hirschberg CB. 1997. Transport of nucleotide sugars, nucleotide sulfate and ATP in the Golgi apparatus. *Trends Biochem. Sci.* 22:203–7
2. Andersson SGE, Zomorodipour A, Andersson JO, Sicheritz-Ponten T, Alsmark UCM, et al. 1998. The genome sequence of *Rickettsia prowazekii* and the origin of mitochondria. *Nature* 396:133–40
3. Aoki H, Tanaka K, Ida S. 1995. The genomic organisation of the gene encoding a nitrate-inducible ferredoxin-NADP$^+$ oxidoreductase from rice roots. *Biochim. Biophys. Acta* 1229:389–92
4. Aoki H, Ida S. 1994. Nucleotide sequence of a rice root ferredoxin-NADP$^+$ reductase and its induction by nitrate. *Biochim. Biophys. Acta* 1183:553–56
5. Bathgate B, Purton ME, Grierson D, Goodenough PW 1985. Plastid changes during the conversion of chloroplasts to chromoplasts in ripening tomatoes. *Planta* 165: 197–204
6. Batz O, Scheibe R, Neuhaus HE. 1992. Transport processes and corresponding changes in metabolite levels in relation to starch synthesis in barley (*Hordeum vulgare* L.) etioplasts. *Plant Physiol.* 100:184–90
7. Batz O, Scheibe R, Neuhaus HE. 1994. Glucose- and ADPGlc-dependent starch synthesis in isolated cauliflower-bud amyloplasts. Analysis of the interaction of various potential precursors. *Biochim. Biophys. Acta* 1200:148–54
8. Batz O, Scheibe R, Neuhaus HE. 1995. Purification of chloroplasts from fruits of green-pepper (*Capsicum annuum* L.) and characterization of starch synthesis. Evidence for a functional hexose-phosphate translocator. *Planta* 196:50–57
9. Bernacchia G, Schwall G, Lottspeich F, Salamini F, Bartels D. 1995. The transketolase gene family of the resurrection plant *Craterostigma plantagineum:* differential experession during the rehydration phase. *EMBO J.* 14:610–18
10. Bettey M, Smith AM. 1990. Nature of the effect of the *r* locus on the lipid content of embryos of peas (*Pisum sativum* L.). *Planta* 180:420–28
11. Bhattacharyya M, Martin C, Smith AM. 1993. The importance of starch biosynthesis in the wrinkled seed shape character of peas studied by Mendel. *Plant Mol. Biol.* 22:525–31
12. Blanke MM, Lenz F. 1989. Fruit photosynthesis. *Plant Cell Environ.* 12:31–46
13. Boddupalli S, Stark DM, Barry GF, Kishore GM. 1995. Effect of overexpressing ADPGlc pyrophosphorylase on oil biosynthesis in canola. In *Biochemistry Molecular Biology Plant Fatty Acids and*

Glycerolipids, ed. JB Ohlrogge, JG Jaworski, P-102 (Abstr.) South Lake Tahoe, CA: Natl. Plant Lipid Coop.
14. Borchert S, Grosse H, Heldt HW. 1989. Specific transport of inorganic phosphate, glucose 6-phosphate, dihydroxyacetone phosphate and 3-phosphoglycerate into amyloplasts. *FEBS Lett.* 253:183–86
15. Borchert S, Harborth J, Schünemann D, Hoferichter P, Heldt HW. 1993. Studies of the enzymic capacities and transport properties of pea root plastids. *Plant Physiol.* 101:303–12
16. Bowsher CG, Boulton EL, Rose J, Nayagam S, Emes MJ. 1992. Reductant for glutamate synthase is generated by the oxidative pentose phosphate pathway in non-photosynthetic root plastids. *Plant J.* 2:893–98
17. Bowsher CG, Hucklesby DP, Emes MJ. 1989. Nitrite reduction and carbohydrate metabolism in plastids purified from roots of *Pisum sativum* L. *Planta* 177:359–66
18. Bowsher CG, Hucklesby DP, Emes MJ. 1993. Induction of ferredoxin-NADP$^+$ oxidoreductase and ferredoxin synthesis in pea root plastids during nitrate assimilation. *Plant J.* 3:463–7
19. Bowsher CG, Tetlow IJ, Lacey AE, Hanke GT, Emes MJ. 1996. Integration of metabolism in non-photosynthetic plastids of higher plants. *C. R. Acad. Sci.* 319:853–60
20. Boyle SA, Hemmingsen SM, Dennis DT. 1990. Energy requirement for the import of proteins from developing endosperm of *Ricinus communis* L. *Plant Physiol.* 92:151–54
21. Büker M, Schünemann D, Borchert S. 1998. Enzymic properties and capacities of developing tomato (*Lycopersicon esculentum*) fruit plastids. *J. Exp. Bot.* 49:681–91
22. Camara B, Hugueney P, Bouvier F, Kuntz M, Monéger R. 1995. Biochemistry and molecular biology of chromoplast development. *Int. Rev. Cytol.* 163:175–247
23. Cao H, Shannon JC. 1997. BT1, a possible adenylate translocator, is developmentally expressed in maize endosperm but not detected in starchy tissues from several other species. *Physiol. Plant.* 100:400–6
24. Cao H, Sullivan TD, Boyer CD, Shannon JC. 1995. *Bt1*, a structural gene for the major 39–44 kDa amyloplast membrane polypeptides. *Physiol. Plant.* 95:176–86
25. Caughey I, Kekwick RGO. 1982. The characteristics of some components of the fatty acid synthetase system in the plastids from the mesocarp of avocado (*Persea americana*) fruit. *Eur. J. Biochem.* 132:553–61
26. Debnam PM, Emes MJ. 1999. Subcellular distribution of enzymes of the oxidative pentose phosphate pathway in root and leaf tissues. *J. Exp. Bot.* 50:1653–61
27. Denyer K, Dunlap F, Thorbjornsen T, Keeling P, Smith AM. 1996. The major form of ADP-glucose pyrophosphorylase in maize endosperm is extra-plastidial. *Plant Physiol.* 112:779–85
28. Eastmond PJ, Dennis DT, Rawsthorne S. 1997. Evidence that a malate-inorganic phosphate exchange translocator imports carbon across the leucoplast envelope for fatty acid synthesis in developing castor seed endosperm. *Plant Physiol.* 114:851–56
29. Eastmond PJ, Kolâcnâ L, Rawsthorne S. 1996. Photosynthesis by developing embryos of oilseed rape (*Brassica napus* L.). *J. Exp. Bot.* 47:1763–69
30. Eastmond PJ, Rawsthorne S. 1998. Comparison of the metabolic properties of plastids isolated from developing leaves or embryos of *Brassica napus* L. *J. Exp. Bot.* 49:1105–11
31. Echeverria E, Boyer CD, Thomas PA, Liu K-C, Shannon JC. 1988. Enzyme activities associated with maize kernel amyloplasts. *Plant Physiol.* 86:786–92
32. Emes MJ, Bowsher CG, Debnam PM, Dennis DT, Hanke G, et al. 1999. Implications of inter- and intracellular compartmentation for the movement of metabolites in plant cells. In *Plant Carbohydrate*

Biochemistry, eds. JA Bryant, MM Burrell, NJ Kruger, 16:231–44. Oxford: Bios

33. Emes MJ, Fowler MW. 1983. The supply of reducing power for nitrite reduction in plastids of seedling pea roots (*Pisum sativum* L.). *Planta* 158:97–102
34. Emes MJ, Neuhaus HE. 1997. Metabolism and transport in non-photosynthetic plastids. *J. Exp. Bot.* 48:1995–2005
35. Emes MJ, Tobin AK. 1993. Control of metabolism and development in higher plant plastids. *Int. Rev. Cytol.* 145:149–216
36. Emes MJ, Traska M. 1987. Uptake of inorganic phosphate by plastids purified from the roots of *Pisum sativum* L. *J. Exp. Bot.* 38:1781–88
37. Entwistle G, ap Rees T. 1990. Lack of fructose-1,6-bisphosphatase in a range of higher plants that store starch. *Biochem. J.* 271:467–72
38. Fliege R, Flügge UI, Werdan K, Heldt HW. 1978. Specific transport of inorganic phosphate, 3-phosphoglycerate and triose phosphates across the inner membrane of the envelope in spinach chloroplasts. *Biochim. Biophys. Acta* 502:232–47
39. Flügge UI. 1999. Phosphate translocators in plastids. *Annu. Rev. Plant Physiol. Plant Mol. Biol.* 50:27–45
40. Focks N, Benning C. 1998. *Wrinkled 1*: a novel, low-seed-oil mutant of Arabidopsis with a deficiency in the seed-specific regulation of carbohydrate metabolism. *Plant Physiol.* 118:91–101
41. Fuhrmann J, Heise KP. 1993. Factors controlling medium-chain fatty acid synthesis in plastids from maturing *Cuphea embryos. Z. Naturforsch.* 48:616–22
42. Fuhrmann J, Johnen T, Heise K-P. 1994. Compartmentation of fatty acid metabolism in zygotic rape embryos. *J. Plant Physiol.* 143:565–69
43. Giroux MJ, Hannah LC. 1994. ADP-glucose pyrophosphorylase in *shrunken-2* and *brittle-2* mutants of maize. *Mol. Gen. Genet.* 243:400–8
44. Givan CV, Hodgson JM. 1983. The source of acetyl-CoA in chloroplasts of higher plants. *Plant Physiol.* 57:311–16
45. Hartwell J, Bowsher CG, Emes MJ. 1996. Recycling of carbon in the oxidative pentose phosphate pathway in non-photosynthetic plastids. *Planta* 200:107–12
46. Hatzfeld W-D, Dancer J, Stitt M. 1990. Fructose-2,6-bisphosphate, metabolites and 'coarse' control of pyrophosphate: fructose-6-phosphate phosphotransferase during triose-phosphate cycling in heterotrophic cell-suspension cultures of *Chenopodium rubrum*. *Planta* 180:205–11
47. Heise KP, Fuhrmann J. 1994. Factors controlling medium-chain fatty acid synthesis in plastids from *Cuphea* embryos. *Prog. Lipid Res.* 33:87–95
48. Heldt HW. 1969. Adenine nucleotide translocation in spinach chloroplasts. *FEBS Lett.* 5:11–14
49. Hermann RG, Westhoff P, Link G. 1992. Biogenesis of plastids in higher plants. In *Cell Organelles*, ed. RG. Hermann, pp.112–87. Heidelberg: Springer-Verlag
50. Hill LM, Smith AM. 1991. Evidence that glucose 6-phosphate is imported as the substrate for starch synthesis by the plastids of developing pea embryos. *Planta* 185:91–96
51. Hill SA, Jenner H, Winning BM, Leaver CJ. 1997. NAD-malic enzyme and the regulation of starch synthesis in potato. *Plant Physiol.* 114:285(S)
52. Island MD, Wei BY, Kadner JJ. 1992. Structure and function of the *uhp* genes for the sugar phosphate transport system in *E. coli* and *Salmonella typhinurium*. *J. Bacteriol.* 174:2754–62
53. Journet E-P, Douce R. 1985. Enzymic capacities of purified cauliflower bud plastids for lipid synthesis and carbohydrate metabolism. *Plant Physiol.* 79:458–67
54. Kammerer B, Fischer K, Hilpert B, Schubert S, Gutensohn M et al. 1998. Molecular characterisation of a carbon

transporter in plastids from heterotrophic tissues: the glucose 6-phosphate antiporter. *Plant Cell* 10:105–17

55. Kampfenkel K, Möhlmann T, Batz O, van Montagu M, Inzé D, et al. 1995. Molecular characterisation of an *Arabidopsis thaliana* cDNA encoding a novel putative adenylate translocator of higher plants. *FEBS Lett.* 374:351–55
56. Kang F, Rawsthorne S. 1994. Starch and fatty acid synthesis in plastids from developing embryos of oilseed rape (*Brassica napus* L.). *Plant J.* 6:795–805
57. Kang F, Rawsthorne S. 1996. Metabolism of glucose-6-phosphate and utilization of multiple metabolites for fatty acid synthesis by plastids from developing oilseed rape embryos. *Planta* 199:321–27
58. Keeling PL, Wood JR, Tyson HW, Bridges IG. 1988. Starch biosynthesis in developing wheat grain. Evidence against the direct involvement of triose phosphates in the metabolic pathway. *Plant Physiol.* 87:311–19

58a. Kirk JTO, Tilney-Bassett RAE. 1978. *The Plastids: Their Chemistry, Structure, Growth and Inheritance.* Amsterdam/Oxford: Elsevier. 2nd ed.

59. Kleinig H, Liedvogel B. 1980. Fatty acid synthesis by isolated chromoplasts from the daffodil. Energy source and distribution patterns of the acids. *Planta* 150:166–69
60. Kleppinger-Sparace KF, Stahl RJ, Sparace SA. 1992. Energy requirements for fatty acid synthesis and glycerolipid biosynthesis from acetate by isolated pea root plastids. *Plant Physiol.* 98:723–27
61. Klingenberg M, 1989. Molecular aspects of the adenine nucleotide carrier from mitochondria. *Arch. Biochem. Biophys.* 270:1–14
62. Knight J, Emes MJ. 1996. Isolation of a tobacco (*Nicotiana tabacum*) chloroplast glucose 6-phosphate dehydrogenase cDNA (Accession No. X99405). *Plant Physiol.* 112:861
63. Knight JS, Gray JC. 1994. Expression of genes encoding the tobacco chloroplast phosphate translocator is not light regulated and is repressed by sucrose. *Mol. Gen. Genet.* 242:586–94
64. Kore-eda S, Kanai R. 1997. Induction of glucose 6-phosphate transport activity in chloroplasts of *Mesembryanthemum crystallinum* by the C3-CAM transition. *Plant Cell Physiol.* 38:895–901
65. Kosegarten H, Mengel K. 1994. Evidence for a glucose 1-phosphate translocator in storage tissue amyloplasts of potato (*Solanum tuberosum*) suspension-cultured cells. *Physiol. Plant.* 91:111–20
66. Kuipers AGJ, Jacobson E, Visser RGF. 1994. Formation and deposition of amylose in the potato tuber starch granule are affected by the reduction of granule-bound starch synthase gene expression. *Plant Cell* 6:43–52
67. Lin Y, Sun L, Nguyen LV, Rachubinski RA, Goodman HM. 1999. The pex16b homolog SSE1 and storage organelle formation in Arabidopsis seeds. *Science* 284:328–30
68. Marzluf GA. 1993. Regulation of sulfur and nitrogen and metabolism in filamentous fungi. *Annu. Rev. Microbiol.* 47:31–55
69. Matile P, Hörtensteiner S, Thomas H. 1999. Chlorophyll degradation. *Annu. Rev. Plant Physiol. Plant Mol. Biol.* 50:67–95
70. Matile P, Schellenberg M, Peisker C. 1992. Production and release of chlorophyll catabolites in isolated senescent chloroplasts. *Planta* 187:230–35
71. Matsumara T, Sakakibara H, Nakano R, Kimata Y, Sugiyama T et al. 1997. A nitrate-inducible ferredoxin in maize roots. Genomic organisation and differential expression of two nonphotosynthetic ferredoxin isoproteins. *Plant Physiol.* 114:653–60
72. Moehs CP, Allen PV, Friedman M, Belknap WR. 1996. Cloning and expression of transaldolase from potato. *Plant Mol. Biol.* 32:447–52

73. Möhlmann T, Batz O, Maass U, Neuhaus HE. 1995. Analysis of carbohydrate transport across the envelope of isolated cauliflower-bud amyloplasts. *Biochem J.* 307:521–26
74. Möhlmann T, Scheibe R, Neuhaus HE. 1994. Interaction between starch synthesis and fatty-acid synthesis in isolated cauliflower-bud amyloplasts. *Planta* 194:492–97
75. Möhlmann T, Tjaden J, Henrichs G, Quick WP, Hausler R, et al. 1997. ADPglucose drives starch synthesis in isolated maize-endosperm amyloplasts. Characterisation of starch synthesis and transport properties across the amyloplastidic envelope. *Biochem. J.* 324:503–9
76. Möhlmann T, Tjaden J, Schwöppe C, Winkler HH, Kampfenkel K, et al. 1998. Occurence of two plastidic ATP/ADP transporters in *Arabidopsis thaliana*: molecular characterisation and comparative structural analysis of homologous ATP/ADP translocators from plastids and *Rickettsia prowazekii*. *Eur. J. Biochem.* 252:353–59
77. Möhlmann T, Neuhaus HE. 1997. Analysis of the precursor and effector dependency of lipid synthesis in amyloplasts isolated from developing wheat- or maize-endosperm tissue. *J. Cereal Sci.* 26:161–67
78. Morano MR, Serra EC, Orellano EG, Carrillo N. 1993. The path of chromoplast development in fruits and flowers. *Plant Sci.* 94:1–17
79. Müller-Röber B, Sonnewald U, Willmitzer L. 1992. Inhibition of the ADP-glucose pyrophosphorylase in transgenic potatoes leads to sugar-storing tubers and influences tuber formation and expression of tuber storage protein genes. *EMBO J.* 11:1229–38
80. Murphy DJ, Cummins I. 1989. Biosynthesis of seed storage products during embryogenesis in rape seed, *Brassica napus*. *J. Plant Physiol.* 135:63–69
81. Murphy DJ, Leech RM. 1981. Photosynthesis of lipids from $^{14}CO_2$ in *Spinacia oleracea*. *Plant Physiol.* 68:762–65
82. Naeem M, Tetlow IJ, Emes MJ. 1997. Starch synthesis in amyloplasts purified from developing potato tubers. *Plant J.* 11:101–9
83. Neuhaus HE, Henrichs G, Scheibe R. 1993. Characterization of glucose-6-phosphate incorporation into starch by isolated intact cauliflower-bud plastids. *Plant Physiol.* 101:573–78
84. Neuhaus HE, Henrichs G, Scheibe R. 1995. Starch degradation in intact amyloplasts from cauliflower buds (*Brassica oleracea* L.). *Planta* 195:496–504
85. Neuhaus HE, Maass U. 1996. Unidirectional transport of orthophosphate across the envelope of isolated cauliflower-bud amyloplasts. *Planta* 198:542–48
86. Neuhaus HE, Schulte N. 1996. Starch degradation in chloroplasts isolated from C3 or CAM induced *Mesembryanthemum crystallinum* L. *Biochem. J.* 318:945–53
87. Neuhaus HE, Stitt M. 1990. Control analysis of photosynthate partitioning. Impact of reduced activity of ADP-glucose pyrophosphorylase or plastid phosphoglucomutase on the fluxes to starch and sucrose in *Arabidopsis thaliana* (L.) Henyh. *Planta* 182:445–54
88. Neuhaus HE, Thom E, Möhlmann T, Steup M, Kampfenkel K. 1997. Characterization of a novel ATP/ADP transporter from *Arabidopsis thaliana* L. *Plant J.* 11:73–82
89. Nowitzki U, Wyrich R, Westhoff P, Henze K, Schnarrenberger C, et al. 1995. Cloning of the amphibolic Calvin cycle OPPP enzyme D-ribulose-5-phosphate 3-epimerase (EC 5.1.3.1) from spinach chloroplasts: functional and evolutionary aspects. *Plant Mol. Biol.* 29:1279–91
90. Ohlrogge JB, Jaworski JG. 1997. Regulation of fatty acid synthesis. *Annu. Rev. Plant Physiol. Plant Mol. Biol.* 48:109–36
91. Ohlrogge JB, Kuhn DN, Stumpf PK. 1979. Subcellular localization of acyl carrier protein in leaf protoplasts of *Spinach*

oleraceae. Proc. Natl. Acad. Sci. USA 76:1194–98

92. Overlach S, Dickmann W, Raschke K. 1993. Phosphate translocator of isolated guard-cell chloroplasts from *Pisum sativum* transports glucose-6-phosphate. *Plant Physiol.* 101:1201–7
93. Pate JS. 1980. Transport and partitioning of nitrogenous solutes. *Annu. Rev. Plant Physiol.* 31:313–40
94. Piazza GJ, Smith MG, Gibbs M. 1982. Characterization of the formation and distribution of photosynthetic products by *Sedum praealtum* chloroplasts. *Plant Physiol.* 70:1748–58
95. Plano GV, Winkler HH. 1989. Solubilization and reconstitution of the *Rickettsia prowazekii* ATP/ADP translocase. *J. Membr. Biol.* 110:227–33
96. Plaxton WC. 1996. The organisation and regulation of plant glycolysis. *Annu. Rev. Plant Physiol. Plant Mol. Biol.* 47:185–214
97. Pozueta-Romero J, Akazawa T. 1993. Biochemical mechanism of starch biosynthesis in amyloplasts from cultured cells of sycamore (*Acer pseudoplatanus*). *J. Exp. Bot.* 44(S):297–306
98. Pozueta-Romero J, Frehner M, Viale AM, Akazawa T. 1991. Direct transport of ADPglucose by an adenylate translocator is linked to starch biosynthesis in amyloplasts. *Proc. Natl. Acad. Sci. USA* 88:5769–73
99. Preiss J, Sivak M. 1998. Biochemistry, molecular biology and regulation of starch synthesis. In *Genetic Engineering*, ed. JK Setlow, 20:177–223. New York: Plenum
100. Preiss J, Sivak M. 1999. Starch and glycogen biosynthesis. In *Comprehensive Natural Products Chemistry, Vol 3. Carbohydrates and Their Derivatives Including Tannins, Cellulose and Related Lignins*, ed. BM Pinto, pp. 441–95. Amsterdam, the Netherlands: Elsevier/Pergamon
101. Preiss J. 1991. Biology and molecular biology of starch synthesis and its regulation. *Oxford Surv. Plant Mol. Cell Biol.* 7:59–114
102. Preiss M, Rosidi B, Hoppe P, Schultz G. 1993. Competition of CO_2 and acetate as substrates for fatty acid synthesis in immature chloroplasts of barley seedlings. *J. Plant Physiol.* 142:525–30
103. Qi Q, Kleppinger-Sparace KF, Sparace SA. 1994. The role of the triose-phosphate shuttle and glycolytic intermediates in fatty-acid and glycerolipid biosynthesis in pea root plastids. *Planta.* 194:193–99
104. Quick PW, Scheibe R, Neuhaus HE. 1995. Induction of a hexose-phosphate translocator activity in spinach chloroplasts. *Plant Physiol.* 109:113–21
105. Rawsthorne S, Eastmond PJ, Kang F, Da Silva PMR, Smith AM, et al. 1997. Carbon partitioning in plastids during development of *B. napus* embryos. In *Physiology, Biochemistry and Molecular Biology of Plant Lipids*, ed. JP Williams, MU Khan, NW Len, pp. 307–9. Dordrecht: Kluwer
106. Reckman U, Scheibe R, Raschke K. 1990. Rubisco activity in guard cells compared with solute requirement for stomatal opening. *Plant Physiol.* 92:246–53
107. Redinbaugh MG, Campbell WH. 1998. Nitrate regulation of the oxidative pentose phosphate pathway in maize (*Zea mays* L.) root plastids: induction of 6-phosphogluconate dehydrogenase activity, protein and transcript levels. *Plant Sci.* 134:129–40
108. Robinson SP, Wiskich JT. 1977. Uptake of ATP analogs by isolated pea chloroplasts and their effect on CO_2 fixation and electron transport. *Biochim. Biophys. Acta* 461:131–40
109. Roughan G, Post-Beitenmiller D, Ohlrogge J, Browse J. 1993. Is acetylcarnitine a substrate for fatty acid synthesis in plants? *Plant Physiol.* 101:1157–62

110. Rutter CJ, Cobb AH. 1983. Translocation of orthophosphate and glucose-6-phosphate in *Codium fragile* chloroplasts. *New Phytol.* 95:559–68
111. Sanchez J. 1994. Lipid photosynthesis in olive fruit. *Prog. Lipid Res.* 33:97–104
112. Schnarrenberger C, Flechner A, Martin W. 1995. Enzymatic evidence for a complete oxidative pentose phosphate pathway in chloroplasts and an incomplete pathway in the cytosol of spinach leaves. *Plant Physiol.* 108:609–14
113. Schott K, Borchert S, Müller-Röber B, Heldt HW. 1995. Transport of inorganic phosphate, and C_3^- and C_6^- sugar phosphates across the envelope membranes of potato tuber amyloplasts. *Planta* 196:647–52
114. Schulz B, Frommer WB, Flügge U-I, Fischer K, Willmitzer L. 1993. Expression of the triose phosphate translocator gene from potato is light dependent and restricted to green tissues. *Mol. Gen. Genet.* 238:357–61
115. Schünemann D, Borchert S, Flügge UI, Heldt HW. 1993. ATP/ADP translocator from pea root plastids. Comparison with translocators from spinach chloroplasts and pea leaf mitochondria. *Plant Physiol.* 103:131–37
116. Schünemann D, Borchert S. 1994. Specific transport of inorganic phosphate and C3– and C6 sugar-phosphates across the envelope membranes of tomato (*Lycopersicon esculentum*) leaf-chloroplasts, tomato fruit-chloroplasts and fruit-chromoplasts. *Bot. Acta* 107:461–67
117. Shannon JC, Pien F-M, Cao H, Liu K-C. 1998. Brittle-1, an adenylate translocator, facilitates transfer of extra plastidial synthesised ADP-glucose into amyloplasts of maize endosperms. *Plant Physiol.* 117:1235–52
118. Shannon JC, Pien F-M, Liu K-C. 1996. Nucleotides and nucleotide sugars in developing maize endosperms. Synthesis of ADP-glucose in *brittle-1*. *Plant Physiol.* 110:835–43
119. Slabas AR, Fawcett T. 1992. The biochemistry and molecular biology of plant lipid biosynthesis. *Plant Mol Biol.* 19:169–91
120. Smith AM, Denyer K, Martin C. 1997. The synthesis of the starch granule. *Annu. Rev. Plant Physiol. Plant Mol. Biol.* 48: 67–87
121. Smith AM, Quinton-Tulloch J, Denyer K. 1990. Characteristics of plastids responsible for starch synthesis in developing pea embryos. *Planta* 180:517–23
122. Smith RG, Gauthier DA, Dennis DT, Turpin DH. 1992. Malate- and pyruvate-dependent fatty acid synthesis in leucoplasts from developing castor endosperm. *Plant Physiol.* 98:1233–38
123. Stark DM, Timmerman KP, Barry GF, Preiss J, Kishore GM. 1992. Role of ADPglucose pyrophosphorylase in regulating starch levels in plant tissues. *Science* 258:287–92
124. Stitt M, McC Lilley R, Gerhardt R, Heldt HW. 1989. Metabolite levels in specific cells and subcellular compartments of plant leaves. *Methods Enzymol.* 174:518–52
125. Stitt M, McC Lilley R, Heldt HW. 1982. Adenine nucleotide levels in the cytosol, chloroplasts, and mitochondria of wheat leaf protoplasts. *Plant Physiol.* 70:971–77
126. Stitt, M. 1990. Fructose 2, 6-bisphosphate as a regulatory molecule in plants. *Annu. Rev. Plant Physiol. Plant Mol. Biol.* 41: 153–85
127. Stubbs M. 1981. Inhibitors of the adenine nucleotide translocase. *Int. Encycl. Pharm. Ther.* 107:283–304
128. Stumpf PK. 1980. Biosynthesis of saturated and unsaturated fatty acids. *Biochem. Plants* 4:177–99
129. Sweetlove LJ, Burrell MM, ap Rees T. 1996. Starch metabolism in tubers of transgenic potato (*Solanum tuberosum*)

with increased ADPglucose pyrophosphorylase. *Biochem. J.* 320:493–98

130. Tanaka T, Ida A, Irifune K, Oeda K, Morikawa H. 1994. Nucleotide sequence of a gene for nitrite reductase from *Arabidopsis thaliana. J. DNA Seq. Mapp.* 5:57–61
131. Taniguchi I, Miyahara A, Iwakiri, K, Hirakawa Y, Hayashi K, et al. 1997. Electrochemical study of biological functions of particular evolutionary conserved amino acid residues using mutated molecules of maize ferredoxin. *Chem. Lett.* 9:929–30
132. Tetlow IJ, Blissett KJ, Emes MJ. 1994. Starch synthesis and carbohydrate oxidation in amyloplasts from developing wheat endosperm. *Planta* 194:454–60
133. Tetlow IJ, Blissett KJ, Emes MJ. 1998. Metabolite pools during starch synthesis and carbohydrate oxidation in amyloplasts isolated from wheat endosperm. *Planta* 204:100–8
134. Tetlow IJ, Bowsher CG, Emes MJ. 1996. Reconstitution of the hexose phosphate translocator from the envelope membranes of wheat endoperm amyloplasts. *Biochem. J.* 319:717–23
135. Thom E, Möhlmann T, Quick WP, Camara B, Neuhaus HE. 1998. Sweet pepper plastids: enzymic equipment, characterisation of the oxidative pentose phosphate pathway, and transport of phosphorylated intermediates across the envelope membrane. *Planta* 204:226–33
136. Thorbjørnsen T, Villand P, Denyer K, Olsen OA, Smith AM. 1996. Distinct isoforms of ADPglucose pyrophosphorylase occur inside and outside the amyloplasts in barley endoeperm. *Plant J.* 10:243–50
137. Tjaden J, Möhlmann T, Kampfenkel K, Henrichs G, Neuhaus HE. 1998. Altered plastidic ATP/ADP-transporter activity influences potato (*Solanum tuberosum*) morphology, amount and composition of tuber starch, and tuber morphology. *Plant J.* 16:531–40
138. Tjaden J, Schwöppe C, Möhlmann T, Neuhaus HE. 1998. Expression of the plastidic ATP/ADP transporter gene in *Escherichia coli* leads to the presence of a functional adenine nucleotide transport system in the bacterial cytosolic membrane. *J. Biol. Chem.* 273:9630–36
139. Tjaden J, van der Laan M, Schwöppe C, Möhlmann T, Winkler HH, et al. 1999. Two nucleotide transport proteins in *Chlamydia trachomatis*. One for net nucleoside triphosphate uptake and the other for the transport of energy. *J. Bacteriol.* 181:1196–1202
140. Viale AM, Arakaki AA, Soncini FC, Ferreyra RG. 1994. Evolutionary relationship among eubacterial groups as inferred from GroEL (chaperonin) sequence comparisons. *Int. J. Syst. Bacteriol.* 44:527–33
141. Viola R, Davies HV, Chudeck AR. 1991. Pathways of starch and sucrose biosynthesis in developing tubers of potato (*Solanum tuberosum* L) and seeds of faba bean (*Vicia faba* L.). Elucidation by ^{13}C-nuclear-magnetic-resonance spectroscopy. *Planta* 183:202–8
142. Visser RGF, Somhorst I, Kuipers GJ, Ruys NJ, Feenstra WJ, et al. 1991. Inhibition of the expression of the gene for granule-bound starch synthase in potato by antisense constructs. *Mol. Gen. Genet.* 225:289–96
143. von Schaewen A, Langenkämper G, Graeve K, Wenderoth I, Scheibe R. 1995. Molecular characterisation of the plastidic glucose-6-phosphate dehydrogenase from potato in comparison to its cytosolic counterpart. *Plant Physiol.* 109:1327–35
144. Wellburn AR. 1982. Bioenergetic and ultrastructural changes associated with chloroplast development. *Int. Rev. Cytol.* 80:133–89
145. Wenderoth I, Scheibe R, von Schaewen A. 1997. Identification of the cysteine residues involved in redox modification of plant plastidic glucose-6-phosphate dehydrogenase. *J. Biol. Chem.* 272:26985–90

146. Williamson LR, Plano GV, Winkler HH, Krause DC, Wood DO. 1989. Nucleotide sequence of the *Rickettsia prowazekii* ATP/ADP translocase-encoding gene. *Gene* 80:269–78
147. Winkler HH. 1991. Molecular biology of Rickettsia. *Eur. J. Epidemiol.* 7:207–12
148. Winkler HH. 1976. Rickettsial permeability:an ADP-ATP transport system. *J. Biol. Chem.* 251:389–96
148a. Wischmann B, Nielsen TH, Moller BL. 1999. In vitro biosynthesis of phosphorylated starch in intact potato amyloplasts. *Plant Physiol.* 119:455–62
149. Wright DP, Huppe HC, Turpin DH. 1997. *In vivo* and *in vitro* studies of glucose 6-phosphate dehydrogenase from barley root plastids in relation to reductant supply for NO_2^- assimilation. *Plant Physiol.* 114:1413–19
150. Zang F, Murphy JB. 1996. A phosphate translocator is present in photosynthetically inactive plastids of dark-germinated pinyon seedlings. *Plant Physiol.* 111(Suppl.):108
151. Ziegler H, Schäfer E, Schneider MM. 1983. Some metabolic changes during chloroplast-chromoplast transition in *Capsicum annuum*. *Physiol. Vég.* 21:485–94
152. Ziegler P, Beck E. 1989. Biosynthesis and degradation of starch in higher plants. *Annu. Rev. Plant Physiol. Plant Mol. Biol.* 40:95–117

Annu. Rev. Plant Physiol. Plant Mol. Biol. 2000. 51:141–65

PATHWAYS AND REGULATION OF SULFUR METABOLISM REVEALED THROUGH MOLECULAR AND GENETIC STUDIES

Thomas Leustek[1], Melinda N. Martin[1], Julie-Ann Bick[1], and John P. Davies[2]

[1]*Biotechnology Center for Agriculture and the Environment, Rutgers University, New Brunswick, New Jersey 08901-8520; e-mail: leustek@aesop.rutgers.edu, mnmartin@aesop.rutgers.edu, bick@aesop.rutgers.edu*

[2]*Department of Botany, Iowa State University, 459 Bessey Hall, Ames, Iowa 50011-1020; e-mail: jdavies@iastate.edu*

Key Words sulfate assimilation, cysteine metabolism, sulfation, sulfate reduction, sulfate uptake and transport, glutathione metabolism

■ **Abstract** Sulfur is essential for life. Its oxidation state is in constant flux as it circulates through the global sulfur cycle. Plants play a key role in the cycle since they are primary producers of organic sulfur compounds. They are able to couple photosynthesis to the reduction of sulfate, assimilation into cysteine, and further metabolism into methionine, glutathione, and many other compounds. The activity of the sulfur assimilation pathway responds dynamically to changes in sulfur supply and to environmental conditions that alter the need for reduced sulfur. Molecular genetic analysis has allowed many of the enzymes and regulatory mechanisms involved in the process to be defined. This review focuses on recent advances in the field of plant sulfur metabolism. It also emphasizes areas about which little is known, including transport and recycling/degradation of sulfur compounds.

CONTENTS

1040-2519/00/0601-0141$14.00

INTRODUCTION

Historical Perspective and Context of the Review

In the previous review of sulfur metabolism appearing in this series, Schmidt & Jäger (117) described the many questions outstanding about the plant sulfur assimilation pathway. The review was published precisely at the time when the methods of molecular biology and genetics were first applied to plant sulfur metabolism. In the intervening years many of the open questions were addressed with new methods (7, 49, 75). Thus,it is now appropriate to review the advances and the present understanding of plant sulfur metabolism, and to underscore the remaining questions.

The Biological Functions of Sulfur

Sulfur is the least abundant of the macroelements found in plants. It is present at approximately 0.1% of dry matter compared with approximately 1.5% for nitrogen and 45% for carbon. Unlike carbon and nitrogen, sulfur is generally not a structural component of biomolecules. Rather, it is nearly always directly involved in the catalytic or electrochemical functions of the molecules of which it is a component. Since the biological functions of sulfur are numerous, only the most significant are mentioned here. The reduced sulfur found in cysteine, termed a thiol group, is strongly nucleophilic. It is this property that defines the function of cysteine in all its forms. Within the context of proteins, two cysteine thiol groups can become oxidized to form a stable, covalent disulfide bond. But the thiol groups can be restored by reduction. The reversibility of the reaction serves well the regulatory function that cysteines often play in proteins (2). Thiols readily react with electrophilic compounds. In some cases, the reaction is necessary for catalytic activity, as is true for metalloenzymes in which cysteines act as the metal ligand. However, when metal ions are present in excess, are of the incorrect type, or when cells are exposed to any of a wide range of poisons, the reaction with protein thiols results in enzyme inactivation. Thus, protein thiols can be sites of vulnerability to toxins.

In most aerobic organisms, intracellular proteins exist in a milieu that is buffered in the reduced state by another thiol compound known as glutathione. Glutathione is an enzymatically synthesized tripeptide composed of glutamate, cysteine, and glycine. Upon oxidation, one reduced glutathione (GSH) can react with another to produce the disulfide form (GSSG); GSH is restored by the activity of NADPH-glutathione reductase. Intracellular GSH is normally greater than 90% of the total glutathione content, but the ratio of reduced to oxidized glutathione decreases under conditions that consume reducing equivalents, such as oxidative stress (90, 96). Glutathione has critical functions aside from its role as a redox buffer. Recent information shows that it may serve as a signal of stress and as a trigger for development (90). It also serves for detoxification of xenobiotics. For example, herbicide detoxification is mediated by glutathione-*S*-transferases (105). Glutathione is the precursor of phytochelatins, polymers with the general structure (γ-glutamylcysteine)$_{n(2\rightarrow 11)}$glycine. These play an important role in detoxification of heavy metals (103). Glutathione also serves as a storage and transported form of reduced sulfur and as a signal for regulation of sulfur assimilation (96).

Sulfur is central to the function of thioether and thioester compounds. The sulfur moiety of coenzyme A and *S*-adenosyl-L-methionine (SAM) imparts reactivity onto the acyl or methyl group of these molecules, which allows them to serve in group transfer reactions (116). The sulfur of methionine is vulnerable to oxidation and can readily form methionine sulfoxide. Oxidation of protein methionines can be inactivating (92); but because the reaction is reversible, it can serve as a mechanism for enzyme regulation (125).

Overview of Plant Sulfur Metabolism

Plants, bacteria, and fungi occupy a prominent position in the biological sulfur cycle. Unlike animals that have a dietary requirement for sulfur amino acids, plants are able to assimilate inorganic sulfate, which is reduced to sulfide and is then incorporated into cysteine. The reductive assimilation pathway used by plants is depicted in Figure 1. Cysteine is the central intermediate from which most sulfur compounds are synthesized. The level of free cysteine in plants is very low ($\leq$10 μM) but the flux is quite high (41), owing to its rapid utilization for methionine, protein, and glutathione synthesis. Although most research has focused on assimilation and synthesis of sulfur-containing compounds, this is only part of the story. Degradation and recycling of reduced sulfur compounds is potentially very significant, but less well understood. The specific properties and regulation of sulfate transport, assimilation, and degradation/recycling of sulfur compounds are described in the subsequent sections. Methionine biosynthesis and its regulation are complex topics that are beyond the scope of the current article. However, the topic has recently been reviewed (104).

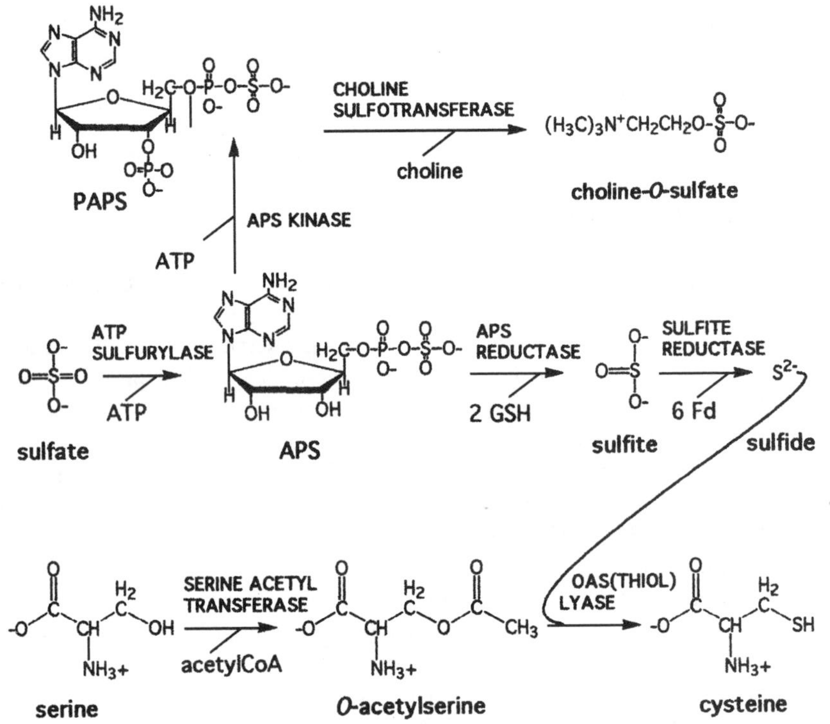

Figure 1 After adenylation of sulfate APS is used for sulfation reactions or for reduction and assimilation into cysteine. PAPS-dependent sulfotransferases catalyze the *O*-sulfation of various metabolites. The sulfation of choline by choline sulfotransferase is illustrated as an example (top row). PAPS is formed by ATP-dependent phosphorylation of APS. Cysteine synthesis requires coordination of the sulfate reduction pathway (middle row) and the pathway leading to OAS synthesis (bottom row). Enzymes are indicated in upper case and metabolites in lower case letters.

SULFATE TRANSPORT

Plasma Membrane Transporters

Anionic sulfate (SO_4^{2-}) is relatively abundant in the environment and serves as the primary sulfur source for plants. It is actively transported into roots where it can remain or be distributed to other sites. Transport into cells is mediated by plasma membrane–localized H^+/SO_4^{2-} co-transporters that are driven by the electrochemical gradient established by the plasma membrane proton ATPase (21). Seven genes encoding sulfate transporters have been identified from *Arabidopsis thaliana* (129, 130, 137). Individual transporters show differing affinities for sulfate. They are expressed in specific tissues (121, 130) and some are strongly regulated at the mRNA level by changes in the sulfur nutritional status of the plant

(10, 121, 130). The diversity of transporters in plants likely reflects the need to obtain sulfate from a varying supply in the soil and the necessity of transporting it throughout the plant. However, the precise function of each type of sulfate transporter is not yet fully understood.

Transport Into Plastids

Sulfate must be transported into plastids where reduction and most of assimilation takes place. Sulfide or thiosulfide are probably exported from plastids since isoenzymes for cysteine synthesis, but not for sulfate reduction, are localized outside of plastids. The nature of the plastid sulfate transporter has been the subject of much speculation but it has not been conclusively identified (21). Recently, two candidates have been found. One is an Arabidopsis cDNA product with homology to plasma membrane sulfate transporters (GenBank accession number AB008782), but containing an amino terminal sequence that resembles a plastid transit peptide. The putative transit peptide is able to direct the localization of a reporter protein to chloroplasts (128). The high degree of similarity to plasma membrane transporters suggests that the activity of the putative plastid sulfate transporter requires an electrochemical proton gradient. However, this idea is at odds with previous results showing that sulfate import into isolated chloroplasts is not stimulated by an acidic bathing solution and is accompanied by efflux of phosphate at a 1:1 ratio with sulfate (93).

A second possible plastid sulfate transporter was identified in the nonvascular plant *Marchantia polymorpha* (97), and in the eukaryotic alga *Chlorella vulgaris* (134). The chloroplast genomes of these organisms contain homologs of the bacterial *cysA* and *cysT* genes (GenBank accession numbers P26246, P10091, P56343, and P56344) encoding integral, plasma membrane proteins that function as a complex for ATP-dependent sulfate and thiosulfate transport (58, 67). The CysT protein likely forms the ion channel whereas CysA provides the energy for ion transport by binding and hydrolyzing ATP. Both belong to an ATPase transporter superfamily that includes the ABC-type transporters of eukaryotes (32). ABC-type solute transporters are known to exist in plant organelles (42), although sulfate-specific transporters of this type have not yet been described. Transport activity of the CysA/CysT complex in *Escherichia coli* is enhanced by soluble, extracellular proteins that bind sulfate or thiosulfate and are encoded by *sbp* and *cysP*. Sequences with homology to CysA/CysT have not been found in the chloroplast genomes of vascular plants, nor have CysA sequences yet been identified in nuclear genomes. However, there are many coding sequences with homology to CysT in the Arabidopsis nuclear genome, owing to the presence of motifs for ABC transporters. The enzyme rhodanese (thiosulfide sulfurtransferase) is thought to be an extracellular protein involved in sulfate transport in *Synechoccus* sp. PCC 7942 (66). It catalyzes the cleavage of the S-S bond in thiosulfide, transferring the thiol to any one of a number of thiophilic acceptor molecules. Two Arabidopsis genes with homology to rhodanese have been identified (99; GenBank accession numbers W43228 and T03983).

SULFATE ASSIMILATION

General Aspects of Assimilation

Once within cells, sulfate can be stored in the vacuole, can serve as a key water-structuring component of the cytoplasm, or can enter the metabolic stream. Metabolism is initiated by an adenylation reaction catalyzed by ATP sulfurylase (Figure 1, Reaction 1).

$$SO_4^{2-} + MgATP \leftrightarrow MgPPi + 5'\text{-adenylylsulfate (APS)} \qquad \text{Reaction 1}$$

This enzyme is localized primarily in plastids but there is also a minor cytosolic form (81, 107, 110). In the vascular plants that have been studied a gene family encodes ATP sulfurylase isoenzymes, but only a single gene exists in the green alga *Chlamydomonas reinhardtii* (57, 94, 139). The reaction product 5′-adenylylsulfate (APS) is a branch point intermediate, which can be channeled toward reduction or sulfation.

Sulfate reduction is the dominant route for assimilation and is carried out exclusively in plastids (14, 110) (Figure 1). It is carried out in two steps. In the first step, APS reductase transfers two electrons to APS to produce sulfite (Reaction 2). The best evidence so far is that the electrons are derived from GSH (5).

$$APS + 2\ GSH \rightarrow SO_3^{2-} + 2\ H^+ + GSSG + AMP \qquad \text{Reaction 2}$$

In the second reaction, sulfite reductase transfers 6 electrons from ferredoxin to produce sulfide (Reaction 3).

$$SO_3^{2-} + 6\ \text{ferredoxin}_{red} \rightarrow S^{2-} + 6\ \text{ferredoxin}_{ox} \qquad \text{Reaction 3}$$

Sulfide then reacts with *O*-acetylserine (OAS) to form cysteine (Figure 1) catalyzed by serine acetyltransferase and OAS thiol-lyase (Reactions 4 and 5).

$$\text{Serine} + \text{acetylCoA} \rightarrow \text{OAS} + \text{CoA} \qquad \text{Reaction 4}$$

$$O\text{-acetylserine (OAS)} + S^{2-} \rightarrow \text{cysteine} + \text{acetate} \qquad \text{Reaction 5}$$

It is now evident that the interaction between serine acetyltransferase and OAS thiol-lyase plays a central role in regulation of cysteine synthesis (9, 33, 56).

Sulfation

An alternate assimilation pathway is the incorporation of sulfate into organic compounds. Sulfation is catalyzed by sulfotransferases that are located in the cytosol and use the phosphorylated derivative of APS, 3′-phospho-5′-adenylylsulfate (PAPS) as the sulfuryl donor (132). In Arabidopsis at least three genes encode APS kinase, the enzyme that catalyzes PAPS synthesis. At least one APS kinase is localized in the chloroplast (69, 115). The sulfation pathway leading to choline-*O*-sulfate is depicted in Figure 1 as an example. Other compounds sulfated by

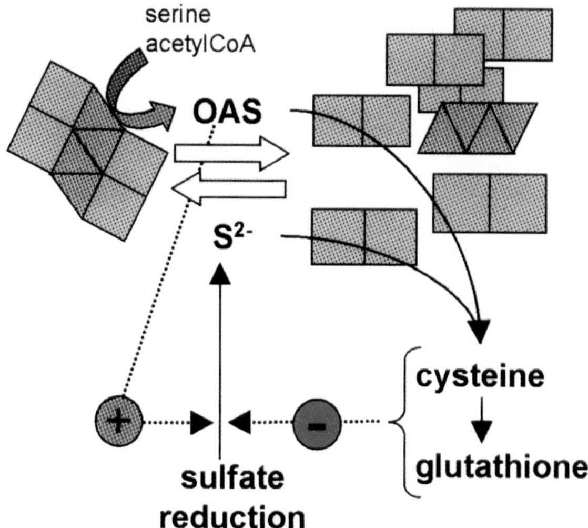

Figure 2 Regulation of cysteine synthesis by OAS and thiol compounds. Yellow boxes indicate OAS thiol-lyase dimers that are present in excess over serine acetyltransferase tetramers, indicated as blue triangles. The enzymes associate through specific interaction domains. The cartoon shows that sulfide promotes formation of the complex, thereby stimulating OAS formation. OAS positively regulates expression of proteins for sulfate assimilation. If OAS accumulates owing to an insufficiency of sulfide, the complex is destabilized, thereby reducing OAS synthesis. OAS also reacts with sulfide to form cysteine catalyzed by free OAS thiol-lyase dimers. The resulting increased level of cysteine and glutathione represses expression of sulfate assimilation proteins. The system serves to coordinate the synthesis of OAS with the activity of the sulfate reduction pathway.

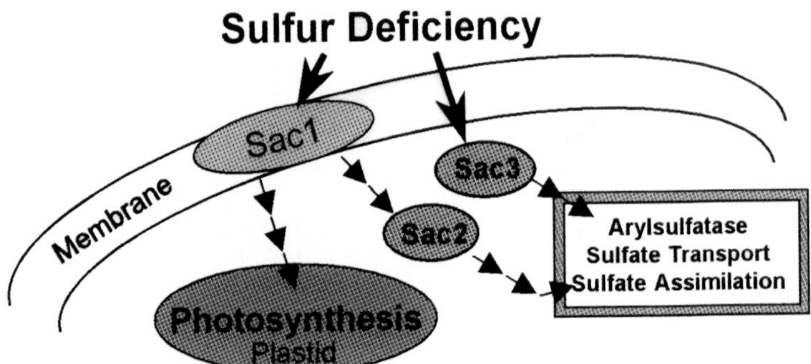

Figure 3 Model for the regulation of responses to sulfur deficiency in Chlamydomonas. Sac1 is an integral membrane protein that may sense the sulfur status of the cell and initiate a signal transduction system that includes Sac2 as an intermediate component. This system controls induced Ars expression, and increases in sulfate uptake and assimilation in response to sulfur deprivation. The Sac3 protein is a serine/threonine kinase that is required for repression of Ars expression during sulfate-sufficient growth and induction of sulfate uptake during sulfur deprivation. Sac3 appears to act through a signaling pathway independent of Sac1 and Sac2. Sac1, but not Sac2 or Sac3, controls the down-regulation of photosynthesis during sulfur deprivation.

this pathway include brassinosteroids, flavanol, gallic acid glucoside, glucosinolates, peptides, and extracellular polysaccharides. Sulfation has long been known in animal systems as a mechanism for regulating cytokine and hormone activity (51). Recent findings indicate that sulfation is similarly important in regulating plant growth and development. The activities of brassinosteroids (112) and gallate glucoside, which controls seismonastic leaf movement in *Mimosa pudica*, are controlled by *O*-sulfation (131). Sulfation of the Nod factor from *Rhizobium meliloti* limits host-range of nodule development in legumes (74). Phytosulfokine, a pentapeptide identified as a cell proliferation factor in asparagus and rice cell cultures, is *O*-sulfated on tyrosine (89, 137a). The sulfation-state of oligosaccharides derived from the extracellular matrix of a marine alga serves as a signal in host/pathogen interactions (12, 101).

Sulfate Reduction

Historical context is important for understanding the genesis of the current ideas on sulfate reduction. Before 1996, the leading hypothesis centered on the existence of a pathway in which the sulfur of APS is transferred to a thiol compound by APS sulfotransferase (114, 117). Then ferredoxin-dependent thiosulfonate reductase completes the reduction to thiosulfide. Neither enzyme had been purified despite having been identified 25 years earlier. Researchers therefore hypothesized that sulfur is reduced in plants via free intermediates, as it is in *E. coli* and *Saccharomyces cerevisiae* (119). In these organisms PAPS is reduced to free sulfite by thioredoxin-dependent PAPS reductase, and then NADPH-sulfite reductase completes the reduction to free sulfide. Plants contain two of the three activities needed for the free intermediate pathway, APS kinase and a ferredoxin-sulfite reductase.

In an attempt to identify the plant PAPS reductase, two groups independently isolated a family of cDNAs from Arabidopsis that functionally complemented an *E. coli* PAPS reductase mutant. However, detailed analysis revealed that the proteins encoded by these cDNAs catalyze APS, not PAPS reduction (5, 43, 120). APS reductase genes have subsequently been cloned from a number of plant and bacterial species (6, 38, 48, 102, 127). Kinetic similarities between recombinant Arabidopsis APS reductase and an APS sulfotransferase purified from the red marine alga, *Porphyra yezoensis* (55), suggested that they are the same enzyme (7). More recently, an amino acid sequence from the APS sulfotransferase of *Lemna minor* unambiguously confirmed that the two are indeed the same enzyme (127).

The amino acid sequence of plant APS reductase revealed a multidomain structure. It is synthesized as a precursor with an amino terminal plastid transit peptide. The amino terminal domain of the mature protein is homologous with PAPS reductase and the carboxyl terminal domain is homologous with thioredoxin (Trx), a redox enzyme. APS reductase is able to use GSH or dithiothreitol as an electron source. By contrast, PAPS reductase, which lacks the C-terminal domain of APS reductase, requires Trx as a cofactor. The implication is that the C-terminal domain functions in catalysis as an exclusive redox cofactor. This idea is supported by the finding that the C-domain functions as glutaredoxin (Grx), rather than Trx (5). Although they are functionally related, Trx and Grx differ in the manner by which they are reduced. Whereas Grx is reduced by GSH, Trx is reduced by NADPH-thioredoxin reductase or ferredoxin-thioredoxin reductase. An *E. coli* complementation assay provided evidence that GSH may be the natural electron donor for plant APS reductase. The APS reductase cDNAs are able to complement the *E. coli* PAPS reductase mutant only if the strain produces glutathione, but they are fully able to complement a strain lacking thioredoxin reductase (5). Another line of evidence indicating that the C-domain functions in catalysis is that neither the reductase domain nor the C-domain, expressed as separate proteins, has APS reductase activity. However, when they are mixed, activity is partly restored. Moreover, Trx can substitute for the C-domain, although it functions less efficiently (102).

The exact catalytic mechanism of APS reductase is uncertain. As a reductase the enzyme would be expected to form sulfite through the transfer of two electrons to APS. By contrast, as a sulfotransferase the enzyme would transfer sulfate from APS to the thiol group of GSH to form *S*-sulfoglutathione. The sulfotransferase model amounts to a single electron transfer. There is evidence that the bacterial assimilatory-type APS and PAPS reductases function as reductases (3, 5). However, neither the sulfotransferase nor reductase mechanisms can be ruled out based on the kinetic behavior of plant APS reductase (JA Bick & T Leustek, unpublished results). The question of catalytic mechanism is an important one because it would indicate whether sulfite reductase is sufficient to complete reduction to sulfide, or whether there is a requirement for thiosulfonate reductase. Recall that thiosulfonate reductase is a central component of the hypothetical bound sulfur reduction pathway that has not been conclusively demonstrated.

APS reductase is thought to be a key regulator of the sulfate reduction pathway. Its activity and steady-state mRNA level increase markedly and coordinately in response to sulfate starvation (43, 68, 130), oxidative stress (JA Bick & T Leustek, unpublished results), or heavy metal exposure (48, 68). The later two stresses increase the demand for glutathione, and hence, the cysteine necessary for glutathione synthesis. Other sulfate assimilation enzymes are regulated to a lesser degree (ATP sulfurylase) or are constitutively expressed (sulfite reductase) (11).

Cysteine Synthesis—The Serine Acetyltransferase/OAS Thiol-Lyase Bi-Enzyme Complex

Serine acetyltransferase and OAS thiol-lyase can associate through protein-protein interactions into a bi-enzyme complex (9). The complex does not synthesize cysteine efficiently although free OAS thiol lyase does (33). Rather, complex formation alters the kinetic behavior of serine acetyltransferase from Michaelis-Menten-type in the free form to one showing positive cooperativity with respect to its substrates when associated in the complex (33). Positive cooperativity is likely very significant considering the limiting concentration of acetylCoA in plastids (111). In contrast, OAS thiol-lyase, which is active in the free form, is nearly inactive in the complex. The findings suggest that (*a*) free OAS thiol-lyase is responsible for cysteine synthesis and (*b*) it also functions as a regulatory subunit of serine acetyltransferase. Although the complex has previously been termed cysteine synthase, this name is inappropriate given these findings.

In chloroplasts the ratio of OAS thiol-lyase to serine acetyltransferase is 300:1 (33), so the majority of the OAS thiol-lyase is in the free form. Complex formation is controlled by sulfide, which promotes complex formation, and OAS, which disrupts it. The effect is to regulate OAS formation in response to the activity of the sulfate reduction pathway. As depicted in the top half of Figure 2 (see color plate), the hypothesis suggests that if the concentration of OAS increases due to insufficient sulfide production, as might occur during sulfur starvation (56), dissolution of the complex would slow further production of OAS. By contrast, if the production of OAS does not keep pace with sulfate reduction, then accumulated sulfide would stimulate complex formation, thereby stimulating OAS production. However, such a system requires some mechanism for controlling sulfate reduction, since runaway cysteine synthesis could result from uncontrolled overproduction of sulfide. As is described in the next section, OAS and reduced sulfur compounds have reciprocal effects on expression of sulfate transporters and assimilation enzymes.

It is important to note that the model described above is based primarily on in vitro enzymological data. However, a key prediction of the model, that OAS limits cysteine synthesis, is supported by a transgenic plant experiment. Overexpression of chloroplast OAS thiol-lyase in tobacco had no effect on cysteine content, but when OAS was fed to chloroplasts isolated from the plants, cysteine was overproduced (113). It is also important to note that the enzymes of the complex also are localized in the cytosol and mitochondria, where their ratios differ markedly from that in chloroplasts. What role the extra-chloroplastic isoforms play in overall cysteine synthesis is not known.

Multiple Mechanisms for Regulation of Sulfate Assimilation Enzymes

Sulfur assimilation is regulated to maintain the rate of cysteine synthesis in the face of conditions that perturb sulfur homeostasis. For example, stresses such as exposure to reactive oxygen or heavy metals, which increase the demand for cysteine, glutathione, and phytochelatin, cause an increase in the activity of the sulfur assimilation pathway. Similarly, withholding sulfur from the nutrient medium activates the pathway, while feeding reduced sulfur compounds to plants represses it (8, 64, 65).

Withholding sulfur induces a variety of responses. Sulfate transport and assimilation activities are increased (75). The expression of seed storage proteins is modulated (25, 36, 39, 56). Changes in the rate of photosynthesis (136) and protein turnover (34, 40) occur. In some organisms sulfur-scavenging mechanisms are induced (28). The responses to sulfur limitation can be reversed by the feeding of sulfate or reduced sulfur. For example, the rate of sulfate transport and assimilation increases coordinately with the transcripts encoding specific sulfate transporters, ATP sulfurylase and APS reductase, during sulfur limitation (50, 68, 70, 121, 129). But they decline after plants are fed glutathione or cysteine. Similarly, expression of the gene encoding the β subunit of β-conglycinin, a seed storage protein of soybean with low content of sulfur amino acids, is increased by sulfur limitation (36, 39) and is repressed by methionine (36). These results suggest that the need for reduced sulfur regulates expression of certain genes at the level of transcript abundance.

An experimental design known as the split root method provided evidence that glutathione functions as a transported signal of plant sulfur status in *Brassica napus*. When one portion of the root mass was deprived of sulfur while the other portion was supplied with adequate sulfate, the levels of glutathione and cysteine declined in the sulfate-fed portion of the root system and sulfate uptake and ATP sulfurylase activity increased (64, 65). In contrast, when one portion of the roots was fed with glutathione or cysteine instead of sulfate, the increase in sulfate uptake and ATP sulfurylase activity was prevented. The effect of cysteine could be blocked with buthionine sulfoximine (BSO), an inhibitor of glutathione synthesis, suggesting that the cellular level of glutathione, not cysteine, acts as the repressor signal of plant sulfur status. However, similar experiments with *Zea mays* led to the conclusion that cysteine rather than glutathione acts as the repressor signal in this species (10). In support of this conclusion, cysteine rather than glutathione appears to be the major sulfur metabolite transported between bundle-sheath and mesophyll cells in *Z. mays* (15).

The response to sulfur starvation can be attenuated by limiting nitrogen, which suggests that some nitrogen-containing compound is necessary for de-repression of sulfur transport and assimilation. The first report of this phenomenon was with tobacco cell cultures (108). But more recently, it was found that nitrogen limitation blocks the accumulation of transcripts for ATP sulfurylase and APS reductase normally induced by limiting sulfur to whole Arabidopsis and *Brassica juncea* plants (68, 137). During sulfur limitation OAS accumulates, but the increase is blocked when nitrogen is limiting (56). Thus, accumulation of OAS may act as a signal for insufficient sulfide production. However, it can work as a positive signal even if sulfur is not limiting. Addition of OAS to *L. minor* caused APS reductase activity to increase (95). More recently it was found that application of OAS to barley caused a rapid increase in sulfate transport activity and coordinate accumulation of transcript for the high-affinity sulfate transporter (121). Both responses occurred even though the plants were grown with high levels of sulfate. In addition, OAS caused cysteine and glutathione to accumulate. A similar effect of OAS was observed on expression of the β subunit of β-conglycinin (56). In total, the results indicate that OAS acts as a positive signal for de-repression of sulfate assimilation and can, to some extent, override repression by thiol compounds. Application of sucrose to Arabidopsis coordinately induces both nitrate reductase and APS reductase (58a), further indicating the regulatory link between these assimilatory pathways.

An integrated model for regulation of cysteine synthesis in plants is shown in Figure 2. In it OAS, sulfide, cysteine, and glutathione are depicted as substrates and as regulatory molecules that modulate the activity of serine acetyltransferase and the expression of proteins for sulfate transport and assimilation. OAS represses its own synthesis by disrupting the serine acetyltransferase/OAS thiol-lyase complex and it stimulates sulfide production by de-repressing sulfate reduction. Sulfide acts in a reciprocal manner. It promotes complex formation, thereby stimulating OAS synthesis, but as cysteine and glutathione accumulate they repress sulfate reduction. The simultaneous operation of both positive and negative signals maintains the rate of cysteine synthesis in proportion to the plant's demand for cysteine and glutathione.

The genes responsible for coordinating expression of sulfur-regulated genes in flowering plants are unknown. However, genetic analysis of mutants has identified components of a signal transduction pathway that controls sulfur-regulated gene expression in Chlamydomonas. Similar to vascular plants, sulfur-deprived Chlamydomonas increases sulfate uptake (138) and assimilation (72, 73, 139) and down-regulates photosynthesis (29, 136). An extracellular arylsulfatase (Ars) that cleaves sulfate from aromatic sulfate esters and releases inorganic sulfate for import into the cell is also induced. Ars is expressed only during sulfur limitation (27, 31, 79) and this property provided the means to isolate mutants altered in regulation of Ars expression (29). The mutants, designated *sac* for sulfur acclimation, are also deficient in other aspects of the response to sulfur limitation (29, 139), which indicates that many of the responses to sulfur deficiency are regulated by the same processes. Sulfur-deprived *sac1* and *sac2* mutants do not synthesize Ars or accumulate the high-affinity sulfate transport system to the same extent as wild-type cells (29). In addition, the *sac1* mutant is unable to increase ATP sulfurylase transcript accumulation (139), or to decrease photosynthetic electron transport during sulfur-limited growth (29, 136). Continued photosynthetic electron transport during sulfur deprivation leads to cell death. Unlike *sac1* and *sac2*, the *sac3* mutant exhibits constitutive, low-level synthesis of Ars. During sulfur deprivation, the *sac3* mutant is unable to produce elevated levels of the high-affinity sulfate transport system, but it accumulates Ars (28) and is able to down-regulate photosynthetic electron transport. Analysis of Ars activity in double mutants revealed that *sac1* is epistatic to *sac2*, but that no clear epistatic relationship exists between *sac3* and either *sac1* or *sac2* (29). This suggests that *Sac1* and *Sac2* may function in the same regulatory pathway, whereas *Sac3* may control Ars activity through a separate regulatory mechanism.

The genes disrupted in the *sac1* and *sac3* mutants have been cloned and sequenced (29, 30). The deduced amino acid sequence of the Sac1 protein suggests that it is an integral membrane protein with at least eight membrane-spanning domains. The sequence is similar to hypothetical protein sequences of unknown function in *Synechocystis* PCC6803, and *E. coli*. Sac1 and the bacterial proteins share a low degree of similarity with the sodium dicarboxylate transporter family from vertebrates (20, 98). Mutations in *Sac1* cause profound effects on the cell's responses to sulfur limitation, which suggests that it regulates the responses. But how might this transporter-like protein serve in regulation? There are several possibilities. It may function as a transporter of a compound that signals the responses to sulfate deprivation or it may act as a sensor of a cellular metabolite that signals sulfur levels within the cell. It is also possible that it functions both for transport and in sulfur sensing. The level of the *Sac1* transcript is similar in cells grown on sulfur-replete and sulfur-deficient media.

The transcript encoding Sac3 is also present at similar levels in cells grown in sulfur-replete and sulfur-deficient media. The deduced amino acid sequence of the Sac3 protein is similar to a large family of Snf1 (sucrose non-fermenting mutant of yeast)-related kinases (SnRKs) (29). It is not known how the activity of the Sac3 kinase is regulated or what the substrate for this kinase is. The prototypical SnRK is the Snf1 kinase of *S. cerevisiae*. Both Sac3 and Snf1 function in regulating the response to nutrient stress; Sac3 regulates Chlamydomonas' response to sulfur deprivation, and *SNF1* regulates *S. cerevisiae*'s response to glucose starvation (18, 19). Within the plant kingdom, the SnRKs fall into three groups. All are similar within their N-terminal catalytic domain and are distinguished by their C-terminal domains (47). The Chlamydomonas Sac3 protein is a member of the SnRK2 family, as are at least 15 other proteins that have been identified in flowering plants (29). These proteins are all about 40 kDa and have short C-terminal domains that contain a highly acidic region. Although the biological functions of the flowering plant kinases have not been demonstrated, transcripts of several of these genes accumulate when plants are exposed to hyper-osmotic conditions (1, 100, 140). The high degree of identity shared between the flowering plant kinases and Sac3, along with the observation that some of the flowering plant kinases are induced during osmotic and salt stress, suggest that they may function in signal transduction pathways controlling the responses to various stresses. Some of these kinases may be involved in mediating responses to sulfur limitation. Based on the genetic properties of *Sac1*, *Sac2* and *Sac3* a model has been developed for the role of these genes in regulation of responses to sulfur deficiency (Figure 3, see color plate).

METABOLISM OF GLUTATHIONE AND GLUTATHIONE *S*-CONJUGATES

The glutathione pool has been estimated to be as high as 10 mM in plants. Even so, large and transient changes in the total glutathione pool and in the GSH/GSSG ratio occur during development and in response to biotic and abiotic stimuli. Glutathione homeostasis is a complex interplay of synthesis, transport, storage, oxidation/reduction, further metabolism, and catabolism (Figure 4). Many aspects of

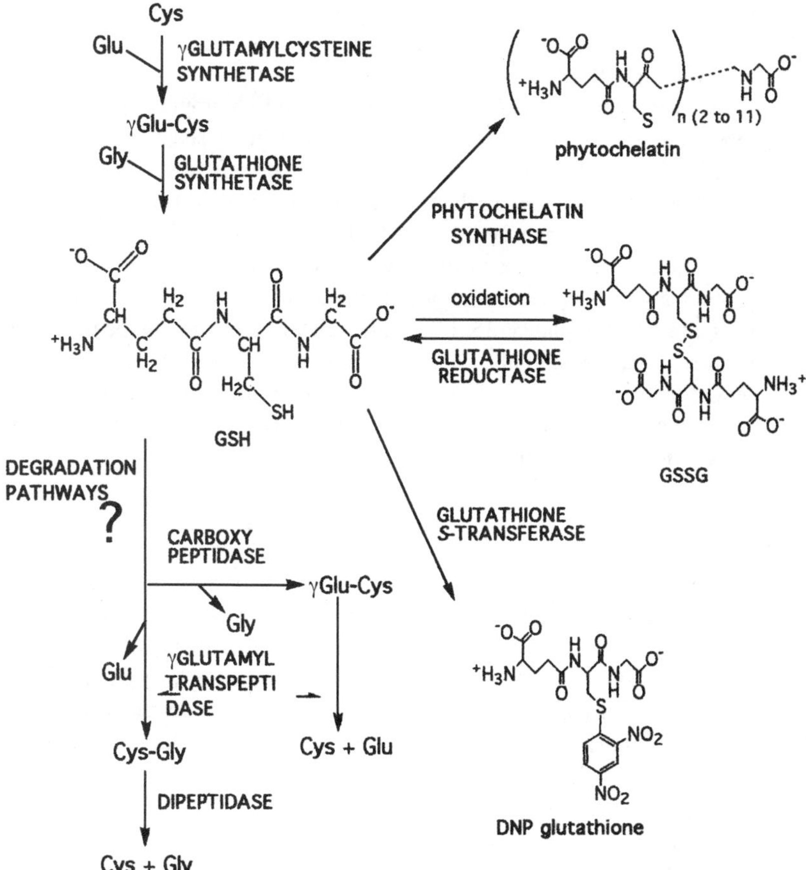

Figure 4 Glutathione synthesis and metabolism. Glutathione is synthesized in two steps and undergoes oxidation/reduction (equilibrium between GSH and GSSG). The pathway shows how glutathione is synthesized, as well as the various metabolic fates of glutathione including as a redox buffer, detoxicant for xenobiotics (conjugation to dinitrophenol is shown as an example), phytochelatin synthesis, and degradation. Enzymes are indicated in upper-case letters and metabolites in lower-case letters.

these processes are poorly understood. Their potential importance is underscored in this review, based on analogies to animal systems.

Synthesis and Transport of Glutathione

Recent advances in the molecular and biochemical characterization of the enzymes and genes responsible for the synthesis of GSH and reduction of GSSG in plants (γ-glutamylcysteine synthetase, glutathione synthetase, and glutathione reductase) have enabled the examination of their roles in sulfur metabolism and response to environmental stresses. This work has been the focus of several recent articles and reviews (26, 90, 96).

GSH is synthesized in both the cytosol and chloroplast. It is exported from leaves and redistributed through the phloem to fruits, seeds, and roots (4, 8, 126). To elucidate the role of transport in GSH homeostasis new approaches are needed to identify the transporters that mediate GSH export from the chloroplast, efflux across the plasma membrane, and GSH import into cells and organelles. Glutathione transport has been directly measured in tobacco cells (118) and bean protoplasts (54). Several high-affinity, low-specificity peptide transporters have been identified in plants (35, 122), although a role in glutathione transport has not been reported. In animals, one glutathione transporter has been identified in the mitochondrial membrane (84). However, glutathione is not the only form in which reduced sulfur is transported in plants. Recently, *S*-methlymethionine was found to be a major transported form also (13).

Degradation of Glutathione

Although the role of degradation in the regulation of the glutathione pool in plants is poorly understood, it likely plays a role in transport and delivery of cysteine to specific sites within cells and whole plants. The pathway of reactions has been characterized in animals (91). Hydrolysis is initiated by a γ-glutamyl transpeptidase, which cleaves the N-terminal γ-linked glutamic acid and transfers it to an acceptor amino acid. The new peptide is cleaved by a γ-glutamylcyclotransferase, which converts it to 5-oxo-proline that is hydrolyzed to glutamate by 5-oxo-prolinase. The Cys-Gly peptide is hydrolyzed by Cys-Gly dipeptidases. Both the γ-glutamyl transpeptidases and the Cys-Gly dipeptidases occur primarily on the outer surface of cell membranes and provide a means of delivering free cysteine for transport into the cell (45, 46, 91). γ-Glutamyl transpeptidases are the only enzymes known to hydrolyze the unique γ-Glu-Cys bond. γ-Glutamyl transpeptidase knockout mice require a cysteine-supplemented diet in order to survive (78).

There is uncertainty as to the sequence of reactions for glutathione degradation in plants. γ-Glu-Cys has been identified as an intermediate in the degradation of [^{35}S]GSH in some species, consistent with C-terminal hydrolysis by a carboxypeptidase (123). In other species, a Cys-Gly intermediate has been identified, consistent with N-terminal hydrolysis by a γ-glutamyl transpeptidase (4, 123). Carboxypeptidase and Cys-Gly dipeptidase activities have been measured in a few plant species (4). γ-Glutamylcyclotransferase and oxo-prolinase have been measured only in tobacco (106, 124). In contrast, γ-glutamyl transpeptidase activity is widely distributed in species ranging from *L. gibba* to *Z. mays*, Arabidopsis, and *Lycopersicon esculentum*. In *L. esculentum* a constitutive level of activity has been detected in all tissues examined (M Martin, unpublished results). In tomato fruit, γ-glutamyl transpeptidase activity increases through the course of development and ripening (85). Four γ-glutamyl transpeptidases differing in localization, substrate specificity, or degree of glycosylation were resolved in ripe tomato fruit.

Two were purified to homogeneity and shown to be 43-kD monomeric proteins exhibiting K_m values of approximately 100 μM for glutathione and several GS-conjugates (87). By contrast, a 56.7-kD γ-glutamyl transpeptidase partially purified from onion exhibited a K_m value for GSH of 5 mM (63). A gene with homology to animal γ-glutamyl transpeptidase has been isolated by screening an Arabidopsis cDNA library for clones that confer tolerance of yeast to the thiol-oxidizing drug diamide (59). Whether the gene product has γ-glutamyl transpeptidase activity has not been investigated.

Formation and Degradation of Glutathione *S*-Conjugates

The glutathione *S*-transferases (GST) catalyze the reaction of the thiol group on GSH with a wide range of hydrophobic and electrophilic toxins. The GSTs are integral to the mechanisms by which plants and animals defend themselves against toxins. Many GSTs are induced in plants by chemical treatments and environmental stresses. The increase in demand for glutathione results in an increase in the glutathione pool (24, 82). In animals, GSTs are part of a four-phase detoxification pathway. First, functional groups are introduced on the target substrate by the action of cytochrome P450-monooxygenases. The substrate is then conjugated to GSH by GSTs, followed by transport of the GS-conjugate by ABC transporters. Finally, the GS-conjugate is degraded. Conjugation to GSH occurs in the cytosol or on microsomal membranes. Transport occurs to the external surface of the plasma membrane (53). GS-conjugates are hydrolyzed by the sequential action of γ-glutamyl transpeptidases and Cys-Gly dipeptidases. The resulting Cys-conjugates are either metabolized to a mercapturate by an N-acetyl transferase or cleaved to a mercaptan by a C-S-β-lyase. The mercaptan may undergo methylation to form methylthio-metabolites or glucuronydation to form a thioglucuronide (24).

Far less is known about the synthesis and fate of GS-conjugates in plants. GSTs were initially identified in plants owing to their role in herbicide detoxification. Numerous genes with homology to GSTs have been identified in plants. The structure, organization, and regulation of this superfamily was recently reviewed in this series (82). The catalytic properties of these GSTs are generally unknown and for some no catalytic function has been demonstrated.

GS-conjugate transporters of the ABC type have been identified in the vacuolar membrane of plants (80, 105). The transcript levels are up-regulated along with GSTs upon exposure to some xenobiotics. They are also able to transport several xenobiotics with high affinity (37, 77, 88). ABC transporters have not yet been found in other plant membranes although there is evidence that some GS-conjugates are exported to the apoplast.

GS-conjugates of herbicides and pesticides are rapidly metabolized in plants to Cys-conjugates (60–62). The fate of the Cys-conjugate is likely species- and xenobiotic-dependent and may include cleavage to a mercaptan by a C-S-β-lyase or formation of N-malonylcysteine conjugates. Two enzymes with a possible role in GS-conjugate hydrolysis have been purified. γ-Glutamyl transpeptidases that are able to hydrolyze GS-conjugates from the N terminus have been purified from soybean and tomato (86, 87), and a vacuolar carboxypeptidase able to hydrolyze a GS-conjugate from the C terminus has been partially purified from barley (135). Fully elucidating this pathway is important because pesticide and herbicide metabolites are stored in plants and ultimately enter the food supply. In animals, GS-conjugate degradation sometimes generates highly toxic intermediates.

Glutathione as a Carrier

GSH conjugation, transport, and hydrolysis serve to modulate the activity of hormones and other biologically active compounds in animals. Prostaglandins are an example of hormones that are inactivated, transported from the cell, and excreted by the same pathway used to clear xenobiotics (53). By contrast, the activity of the cysteinyl leukotrienes, potent mediators of inflammatory response, is modulated by this reaction sequence. Conjugation with GSH targets the leukotriene for export to the external surface of the membrane, where it is activated through hydrolysis by γ-glutamyl transpeptidase. Subsequent hydrolysis by a Cys-Gly dipeptidase leads to inactivation. Both the GST and the γ-glutamyl transpeptidase are highly specific for leukotrienes (16, 17, 71).

Not all plant GSTs are induced by environmental stress and xenobiotics. In fact, constitutively expressed GSTs comprise up to 1% of the soluble proteins in some tissues of plants (109). Many members of the GST superfamily are spatially, temporally, and hormonally regulated. All phytohormones induce subpopulations of GSTs (82). Although they are primarily soluble, cytosolic proteins, GSTs localized to the apoplast nucleus, and plasma membrane have also been identified. Taken together, this information suggests that GSTs play a major role in plant growth. To date, in vivo substrates for plant GSTs have not been identified with certainty. GS-conjugation of the anthocyanin, cyanidin-3-glucoside, may occur. The maize *bronze-2* mutant, which accumulates anthocyanins in the cytosol, was found to have a defect in a gene with homology to GSTs. In wild-type maize, the GS-anthocyanin conjugate is transported into the vacuole, where it is metabolized (83). GS-conjugation of medicarpin, an isoflavone, is formed in vitro and is transported with high affinity by the GS-conjugate vacuolar pump (76).

Phytochelatins Are Synthesized from Glutathione

Phytochelatins are a class of peptides involved in the detoxification of heavy metals, particularly cadmium, in plants and yeast. They are synthesized from GSH by a γ-glutamylcysteine dipeptidyl transpeptidase (phytochelatin synthase). This enzyme transfers γ-Glu-Cys units to form a polymer $(\gamma\text{-Glu-Cys})_n$-Gly (141). Recently, three laboratories have independently cloned genes encoding phytochelatin synthase. In one case, the isolation of two cadmium-sensitive Arabidopsis mutants (*cad1* and *cad2-1*) led to the cloning of genes involved in phytochelatin synthesis. The CAD2 gene was shown to encode γ-glutamylcysteine synthetase, one of the enzymes required for glutathione synthesis (23). The *cad2-1* mutants have less glutathione than wild type. In contrast, the *cad1* mutant has normal levels of glutathione but is unable to synthesize phytochelatins (52). A positional cloning method was used to isolate the CAD1 gene (44). The recombinant enzyme was used to show that the gene product catalyzes phytochelatin synthesis, confirming that CAD1 encodes phytochelatin synthase. The other two groups cloned Arabidopsis and wheat phytochelatin synthase by selecting for cDNAs that confer tolerance of yeast to cadmium (22, 133).

FUTURE DIRECTIONS

This review has highlighted the recent advances in understanding plant sulfur metabolism. The state of knowledge has been significantly influenced by the isolation of genes for each of the metabolic steps, but not all areas have benefited equally from molecular methods. There is still a clear opportunity for applying gene-cloning methods to learn more about how glutathione and glutathione S-conjugates are transported and degraded. Although significant progress has been made toward elucidating the structure, organization, and regulation of GSTs, the in vivo catalytic function of most GSTs is unknown. It has become increasingly apparent that sulfation reactions play a critical role in controlling developmental signals, but this process is still poorly understood; only a few systems have been described. Important questions remain concerning the transport of most sulfur compounds. To date, no specific membrane transport systems for glutathione have been described, and the transport of cysteine and *S*-methylmethionine, two recently identified transported forms of sulfur, remain uncharacterized. Moreover, even with the multitude of sulfate transporters identified, there is still no conclusive identification of plastid and vacuolar-localized transporters.

Despite attempts to dissect sulfur metabolism pathways with antisense mutants, no severe phenotypes have been reported. This possibly reflects the strong interaction and co-regulation of these enzyme pathways and the influence of other pathways such as N metabolism. The metabolic signals used to communicate the nutritional status of vascular plants are beginning to be elucidated. However, the details of the response mechanisms and the isolation of null mutants in these systems are completely lacking despite the use of Arabidopsis as a genetic model. In contrast, there has been significant progress in defining through genetics the signaling pathway that regulates sulfur response in Chlamydomonas. Nevertheless, there is little understanding of the role that OAS and reduced sulfur compounds play in the signaling pathways of this organism. Whether Arabidopsis and Chlamydomonas share common pathways for regulating sulfur assimilation awaits determination.

ACKNOWLEDGMENTS

We thank T Fujiwara, B Kloareg, and K Saito for sharing results prior to publication and M Droux and R Hell for helpful comments on the manuscript. The authors' work has been supported by grants from the National Science Foundation, the United States Department of Agriculture, and Pioneer Hi-Bred International, Inc.

Visit the Annual Reviews home page at www.AnnualReviews.org

LITERATURE CITED

1. Anderberg RJ, Walker-Simmons MK. 1992. Isolation of a wheat cDNA clone for an abscisic acid-inducible transcript with homology to protein kinases. *Proc. Natl. Acad. Sci. USA* 89:10183–87
2. Åslund F, Beckwith J. 1999. Bridge over troubled waters: sensing stress by disulfide bond formation. *Cell* 96:751–53
3. Berendt U, Haverkamp T, Prior A, Schwenn JD. 1995. Reaction mechanism of thioredoxin: 3′-phosphoadenylylsulfate reductase investigated by site-directed mutagenesis. *Eur. J. Biochem.* 233:347–56
4. Bergmann L, Rennenberg H. 1993. Glutathione metabolism in plants. See Ref. 31a, pp. 109–23
5. Bick JA, Åslund F, Chen Y, Leustek T. 1998. Glutaredoxin function for the carboxyl terminal domain of the plant-type 5′-adenylylsulfate (APS) reductase. *Proc. Natl. Acad. Sci. USA* 95:8404–9
6. Bick JA, Dennis JJ, Zylstra GJ, Nowack J, Leustek T. 2000. Identification of a new class of 5′-adenylylsulfate (APS) reductases from sulfate assimilating bacteria. *J. Bacteriol.* 182:135–42
7. Bick JA, Leustek T. 1998. Plant sulfur metabolism—the reduction of sulfate to sulfite. *Curr. Opin. Plant Biol.* 1:240–44
8. Blake-Kalff MMA, Harrison KR, Hawkesford MJ, Zhao FJ, McGrath SP. 1998. Distribution of sulfur within oilseed rape leaves in response to sulfur deficiency during vegetative growth. *Plant Physiol.* 118: 1337–44
9. Bogdanova N, Hell R. 1997. Cysteine synthesis in plants. Protein-protein interactions of serine acetyltransferase from *Arabidopsis thaliana*. *Plant J.* 11:251–62
10. Bolchi A, Petrucco S, Tenca PL, Foroni C, Ottonello S. 1999. Coordinate modulation of maize sulfate permease and ATP sulfurylase mRNAs in response to variations in sulfur nutritional status: stereospecific down-regulation by L-cysteine. *Plant Mol. Biol.* 39:527–37
11. Bork C, Schwenn JD, Hell R. 1998. Isolation and characterization of a gene for assimilatory sulfite reductase from *Arabidopsis thaliana*. *Gene* 212:147–53
12. Bouarab K, Potin P, Correa J, Kloareg B. 1999. Sulfated oligosaccharides mediate cross-talk between a marine red alga and its green algal pathogenic endophyte. *Plant Cell* 11:1635–50
13. Bourgis F, Roje S, Nuccio ML, Fisher DB, Tarczynski MC, et al. 1999. *S*-Methylmethionine plays a major role in phloem sulfur transport and is synthesized by a novel type of methyltransferase. *Plant Cell* 11:1485–98
14. Brunold C, Suter M. 1989. Localization of enzymes of assimilatory sulfate reduction in pea roots. *Planta* 179:228–34
15. Burgener M, Suter M, Jones S, Brunold C. 1998. Cyst(e)ine is the transport metabolite of assimilated sulfur from bundle-sheath to mesophyll cells in maize leaves. *Plant Physiol.* 116:1315–22
16. Carter BZ, Shi Z-Z, Barrios R, Liberman MW. 1998. γ-Glutamyl leukorienase, a γ-glutamyl transpeptidase gene family member, is expressed primarily in spleen. *J. Biol. Chem.* 273:28277–85
17. Carter BZ, Wiseman AL, Orkiszewski R, Ballard KD, Ou C-N, Liberman MW. 1997. Metabolism of leukotriene C_4 in γ-glutamyl transpeptidase-deficient mice. *J. Biol. Chem.* 272:12305–10
18. Celenza JL, Carlson M. 1984. Cloning and genetic mapping of SNF1, a gene required for expression of glucose-repressible genes in *Saccharomyces cerevisiae*. *Mol. Cell Biol.* 4:49–53
19. Celenza JL, Carlson M. 1986. A yeast gene

that is essential for release from glucose repression encodes a protein kinase. *Science* 233:1175–80

20. Chen XZ, Shayakul C, Berger UV, Tian W, Hediger MA. 1998. Characterization of a rat Na^+-dicarboxylate cotransporter. *J. Biol. Chem.* 273:20972–81
21. Clarkson DT, Hawkesford MJ, Davidian J-C. 1993. Membrane and long-distance transport of sulfate. See Ref. 31a, pp. 3–20
22. Clemens S, Kim EJ, Neumann D, Schroeder JI. 1999. Tolerance to toxic metals by a gene family of phytochelatin synthases from plants and yeast. *EMBO J.* 18:3325–33
23. Cobbett CS, May MJ, Howden R, Rolls B. 1998. The glutathione-deficient, cadmium-sensitive mutant, *cad2-1*, of *Arabidopsis thaliana* is deficient in gamma-glutamylcysteine synthetase. *Plant J.* 16:73–78
24. Commandeur JNM, Stijntjes GJ, Vermeulen NPE. 1995. Enzymes and transport systems involved in the formation and disposition of glutathione *S*-conjugates: roles in bioactivation and detoxification mechanisms of xenobiotics. *Pharmacol. Rev.* 47:271–330
25. Creason G, Holowach L, Thompson J, Madison J. 1983. Exogenous methionine depresses the level of mRNA for a soybean storage protein. *Biochem. Biophys. Res. Commun.* 117:658–62
26. Creissen G, Firmin J, Fryer M, Kular B, Leyland N, et al. 1999. Elevated glutathione biosynthetic capacity in the chloroplasts of transgenic plants paradoxically causes increased oxidative stress. *Plant Cell* 11:1277–91
27. Davies JP, Grossman AR. 1998. Survival during macronutrient limitation. In *The Molecular Biology of Chloroplasts and Mitochondria in Chlamydomonas*, ed. J-D Rochaiz, M Goldschmidt-Clermont, S Merchant, pp. 613–35. Dordrecht: Kluwer
28. Davies JP, Yildiz FH, Grossman AR. 1994. Mutants of *Chlamydomonas* with aberrant responses to sulfur deprivation. *Plant Cell* 6:53–63
29. Davies JP, Yildiz FH, Grossman AR. 1996. SAC1, a putative regulator that is critical for survival of *Chlamydomonas reinhardtii* during sulfur deprivation. *EMBO J.* 15:2150–59
30. Davies JP, Yildiz FH, Grossman AR. 1999. Sac3, an Snf1-like Serine/Threonine kinase that positively and negatively regulates the responses of *Chlamydomonas* to sulfur limitation. *Plant Cell* 11:1179–90
31. de Hostos EL, Togasaki RK, Grossman AR. 1988. Purification and biosynthesis of a derepressible periplasmic arylsulfatase from *Chlamydomonas reinhardtii*. *J. Cell. Biol.* 106:29–37

31a. de Kok LJ, Stulen I, Rennenberg H, Brunold C, Rauser WE, eds. 1993. *Sulfur Nutrition and Sulfur Assimilation in Higher Plants*. The Hague: SPB Academic Publ.

32. Dioge CA, Ames GF-L. 1993. ATP-dependent transport systems in bacteria and humans: relevance to cystic fibrosis and multidrug resistance. *Annu. Rev. Microbiol.* 47:291–319
33. Droux M, Ruffet ML, Douce R, Job D. 1998. Interactions between serine acetyltransferase and *O*-acetylserine(thiol)lyase in higher plants—structural and kinetic properties of the free and bound enzymes. *Eur. J. Biochem.* 255:235–45
34. Ferreira RM, Teixeira AR. 1992. Sulfur starvation in *Lemna* leads to degradation of ribulose-bisphosphatecarboxylase without plant death. *J. Biol. Chem.* 267:7253–57
35. Frommer WB, Hummel S, Rentsch D. 1994. Cloning of an Arabidopsis histidine transporting protein related to nitrate and peptide transporters. *FEBS Lett.* 347:185–89
36. Fujiwara T, Hirai M, Chino M, Komeda Y, Naito S. 1992. Effects of sulfur nutrition

on expression of the soybean seed storage protein genes in transgenic Petunia. *Plant Physiol.* 99:263–68

37. Gaillard C, Dufaud A, Tommasini R, Kreuz K, Amrhein N, Martinoia E. 1994. A herbicide antidote (safener) induces the activity of both the herbicide detoxifying enzyme and of a vacuolar transporter for the detoxified herbicide. *FEBS Lett.* 352:219–21
38. Gao Y, Leustek T. 1999. Cloning and characterization of 5′-adenylylsulfate (APS) reductase from marine algae. *Am. Soc. Plant Physiol. Plant Biol.* 99:641 (Abstr.)
39. Gayler K, Sykes G. 1985. Effects of nutritional stress on the storage proteins of soybeans. *Plant Physiol.* 78:582–85
40. Gilbert SM, Clarkson DT, Cambridge M, Lambers H, Hawkesford M. 1997. SO_4^{2-} deprivation has an early effect on the content of ribulose-1,5-bisphosphate carboxylase/oxygenase and photosynthesis in young leaves of wheat. *Plant Physiol.* 115:1231–39
41. Giovanelli J, Mudd SH, Datko AH. 1980. Sulfur amino acids in plants. In *The Biochemistry of Plants*, ed. BJ Miflin, 5:454–500. New York: Academic
42. Goldman BS, Beckman DL, Bali A, Monika EM, Gabbert KK, Kranz RG. 1997. Molecular and immunological analysis of an ABC transporter complex required for cytochrome c biogenesis. *J. Mol. Biol.* 268:724–38
43. Gutierrez-Marcos JF, Roberts MA, Campbell EI, Wray JL. 1996. Three members of a novel small gene-family from *Arabidopsis thaliana* able to complement functionally an *Escherichia coli* mutant defective in PAPS reductase activity encode proteins with a thioredoxin-like domain and "APS reductase" activity. *Proc. Natl. Acad. Sci. USA* 93:13377–82
44. Ha SB, Smith AP, Howden R, Dietrich WM, Bugg S, et al. 1999. Phytochelatin synthase genes from *Arabidopsis* and the yeast *Schizosaccharomyces pombe*. *Plant Cell* 11:153–63
45. Habib GM, Barrios R, Shi Z-Z, Lieberman MW. 1996. Four distinct membrane-bound dipeptidase RNAs are differentially expressed and show discordant regulation with γ-glutamyl transpeptidase. *J. Biol. Chem.* 271:16273–80
46. Habib GM, Shi Z-Z, Cuevas AA, Guo Q, Matzuk MM, Lieberman MW. 1998. Leukotriene D_4 and cystinyl-bisglycine metabolism in membrane-bound dipeptidase-deficient mice. *Proc. Natl. Acad. Sci. USA* 95:4859–63
47. Halford NG, Hardie DG. 1998. SNF1-related protein kinases: global regulators of carbon metabolism in plants? *Plant Mol. Biol.* 37:735–48
48. Heiss S, Schäfer H, Haag-Kerwer A, Rausch T. 1999. Cloning sulfur assimilation genes of *Brassica juncea* L. Cadmium differentially affects the expression of a putative low affinity sulfate transporter and isoforms of ATP sulfurylase and APS reductase. *Plant Mol. Biol.* 39:847–57
49. Hell R. 1997. Molecular physiology of plant sulfur metabolism. *Planta* 202:138–48
50. Herschbach C, Rennenberg H. 1994. Influence of glutathione (GSH) on net uptake of sulfate and sulfate transport in tobacco plants. *J. Exp. Bot.* 45:1069–76
51. Hooper LV, Manzella SM, Baenziger JU. 1996. From legumes to leukocytes: biological roles for sulfated carbohydrates. *FASEB J.* 10:1137–46
52. Howden R, Goldsbrough PB, Anderson CR, Cobbett CS. 1995. Cadmiun-sensitive *cad1* mutants of *Arabidopsis thaliana* are phytochelatin deficient. *Plant Physiol.* 107:1059–66
53. Ishikawa Y. 1992. The ATP-dependent glutathione *S*-conjugate export pump. *Trends Biol. Sci.* 17:463–68

54. Jamai A, Tommasini R, Martinoia E, Delrot S. 1996. Characterization of glutathione uptake in broad bean leaf protoplasts. *Plant Physiol.* 111:1145–52 leftskip-4.5pt
55. Kanno N, Nagahisa E, Sato M, Sato Y. 1996. Adenosine 5′-phosphosulfate sulfotransferase from the marine macroalga *Porphyra yezoensis* Ueda (Rhodophyta): stabilization, purification, and properties. *Planta* 198:440–6
56. Kim H, Hirai MY, Hayashi H, Chino M, Naito S, Fujiwara T. 1999. Role of *O*-acetyl-L-serine in the coordinated regulation of the expression of a soybean seed storage-protein gene by sulfur and nitrogen nutrition. *Planta* 209:282–89
57. Klonus D, Hofgen R, Willmitzer L, Riesmeier JW. 1994. Isolation and characterization of two cDNA clones encoding ATP-sulfurylases from potato by complementation of a yeast mutant. *Plant J.* 105–12
58. Kohn C, Schumann J. 1993. Nucleotide sequence and homology comparison of two genes of the sulfate transport operon from the cyanobacterium *Synechocystis* sp. PCC 6803. *Plant Mol. Biol.* 21:409–12
58a. Kopriva S, Muheim R, Koprivova A, Trachsel N, Catalano C, et al. 1999. Light regulation of assimilatory sulphate reduction in *Arabidopsis thaliana. Plant J.* 20:37–44
59. Kushnir S, Babiychuk E, Kampfenkel K, Bellesboix E, Van Montagu M, Inzé D. 1995. Characterization of *Arabidopsis thaliana* cDNAs that render yeasts tolerant toward the thiol-oxidizing drug diamide. *Proc. Natl. Acad. Sci. USA* 92:10580–84
60. Lamoureux GL, Rusness DG. 1983. Malonyl cysteine conjugates as end-products of glutathione conjugate metabolism in plants. In *Pesticide Chemistry: Human Welfare and the Environment,* ed. J Miyamoto, PC Kearnery, pp. 295–300. New York: Pergamon
61. Lamoureux GL, Rusness DG. 1993. Glutathione in the metabolism and detoxification of xenobiotics in plants. See Ref. 31a, pp. 221–37
62. Lamoureux GL, Shimabukuro RH, Frear DS. 1991. Glutathione and glucoside conjugation in herbicide selectivity. In *Herbicide Resistance in Weeds and Crops* ed. JC Caseley, GW Cussans, RK Atkin, pp. 227–61. Oxford: Butterworth-Heinemann
63. Lancaster JE, Shaw ML 1994. Characterization of purified γ-glutamyl transpeptidase in onions: evidence for in vivo role as a peptidase. *Phytochemistry* 36:1351–58
64. Lappartient A, Touraine B. 1996. Demand-driven control of root ATP sulfurylase activity and sulfate uptake in intact canola. *Plant Physiol.* 111:147–57
65. Lappartient AG, Vidmar JJ, Leustek T, Glass ADM, Touraine B. 1999. Inter-organ signaling in plants: regulation of ATP sulfurylase and sulfate transporter genes expression in roots mediated by phloem-translocated compounds. *Plant J.* 18:89–95
66. Laudenbach DE, Ehrhardt D, Green L, Grossman A. 1991. Isolation and characterization of a sulfur-regulated gene encoding a periplasmically localized protein with sequence similarity to rhodanese. *J. Bacteriol.* 173:2751–60
67. Laudenbach DE, Grossman AR. 1991. Characterization and mutagenesis of sulfur-regulated genes in a cyanobacterium: evidence for function in sulfate transport. *J. Bacteriol.* 173:2739–50
68. Lee S. 1999. *Molecular analysis of sulfate assimilation in higher plants: effect of cysteine, sulfur and nitrogen nutrients, heavy metal stress, and genomic DNA cloning.* PhD thesis. Rutgers State Univ. NJ, New Brunswick. 155 pp.
69. Lee S, Leustek T. 1998. APS kinase from *Arabidopsis thaliana*: genomic organization, expression, and kinetic analysis of the

recombinant enzyme. *Biochem. Biophys. Res. Commun.* 247:171–75

70. Lee S, Leustek T. 1999. The affect of cadmium on sulfate assimilation enzymes in *Brassica juncea. Plant Sci.* 141:201–7
71. Leier I, Jedlitschky G, Buchholz U, Cole SPC, Deeley RG, Keppler D. 1994. The MRP gene encodes an ATP-dependent export pump for leukotriene C_4 and structurally related conjugates. *J. Biol Chem.* 269:27807–10
72. León J, Romero L, Galván F. 1988. Intracellular levels and regulation of *O*-acetyl-L-serine sulfhydrylase activity in *Chlamydomonas reinhardtii. J. Plant Physiol.* 132:618-22
73. León J, Vega J. 1991. Separation and regulatory properties of *O*-acetyl-L-serine sulfhydrylase isoenzymes from *Chlamydomonas reinhardtii. Plant Physiol. Biochem.* 29:595–99
74. Lerouge P, Roche P, Faucher C, Maillet F, Truchet G, et al. 1990. Symbiotic host specificity of *Rhizobium meliloti* is determined by a sulphated and acylated glucosamine oligosaccharide signal. *Nature* 344:781–84
75. Leustek T, Saito K. 1999. Sulfate transport and assimilation in plants. *Plant Physiol.* 120:637–44
76. Li Z-S, Alfenito M, Rea PA, Walbot V, Dixon RA. 1997. Vacuolar uptake of the phytoalexin medicarpin by the glutathione conjugate pump. *Phytochemistry* 45:689–93
77. Li Z-S, Zhen RG, Rea PA. 1995. 1-Chlorodinitrobenzene-elicited increase in vacuolar glutathione *S*-conjugate transport activity. *Plant Physiol.* 109:177–85
78. Lieberman MW, Weisman AL, Shi Z-Z, Carter BZ, Barrios R, et al. 1996. Growth retardation and cysteine deficiency in γ-glutamyl transpeptidase-deficient mice. *Proc. Natl. Acad. Sci. USA* 93:7923–26
79. Lien T, Schreiner O. 1975. Purification of a derepressible arlysulfatase from *Chlamydomonas reinhardtii. Biochem. Biophys. Acta* 384:168–79
80. Lu Y-P, Li Z-S, Rea PA. 1997. AtMRP1 gene of Arabidopsis encodes a glutathione *S*-conjugate pump: isolation and functional definition of a plant ATP-binding cassette transporter gene. *Proc. Natl. Acad. Sci. USA* 94:8243–48
81. Lunn JE, Droux M, Martin J, Douce R. 1990. Localization of ATP sulfurylase and *O*-acetylserine(thiol)lyasein spinach leaves. *Plant Physiol.* 94:1345–52
82. Marrs KA. 1996. The functions and regulation of glutathione *S*-transferases in plants. *Annu. Rev. Plant Physiol. Plant Mol. Biol.* 47:127–58
83. Marrs KA, Alfenito MR, Lloyd AM, Walbot V. 1995. A glutathione *S*-transferase involved in vacuolar transfer encoded by the maize gene *Bronze*-2. *Nature* 375:397–400
84. Martensson J, Lai JCK, Meister A. 1990. High affinity transport of glutathione is part of a multicomponent system essential for mitochondrial function. *Proc. Natl. Acad. Sci. USA* 87:7185–89
85. Martin MN, Cohen JD, Saftner RA. 1995. A new 1-aminocyclopropane-1-carboxylic acid-conjugating activity in tomato fruit. *Plant Physiol.* 109:917–26
86. Martin MN, Slovin JP. 1996. Purification of a γ-glutamyl transpeptidase from soybean seeds. *Am. Soc. Plant Physiol. Plant Biol.* 111:52(Abstr.)
87. Martin MN, Slovin JP. 2000. Purified γ-glutamyl transpeptidases from tomato exhibit high affinity for glutathione *S*-conjugates. *Plant Physiol.* In press
88. Martinoia E, Grill E, Tommasini R, Kreuz K, Amrhein N. 1993. An ATP-dependent glutathione S-conjugate "export" pump in the vacuolar membrane of plants. *Nature* 364:247–49
89. Matsubayashi Y, Takagi L, Sakagami Y. 1997. Phytosulfokine-α, a sulfated pentapeptide, stimulates the proliferation of rice cells by means of specific high- and

low-affinity binding sites. *Proc. Natl. Acad. Sci. USA* 94:13357–62

90. May MJ, Vernoux T, Leaver C, Van Montagu M, Inzé D. 1998. Glutathione homeostasis in plants: implications for environmental sensing and plant development. *J. Exp. Bot.* 49:649–67
91. Meister A. 1988. Glutathione metabolism and its selective modification. *J. Biol. Chem.* 263:17205–8
92. Moskovitz J, Berlett BS, Poston JM, Stadtman ER. 1999. Methionine sulfoxide reductase in antioxidant defense. *Methods Enzymol.* 300:239–44
93. Mourioux G, Douce R. 1979. Sulfate transport across the limiting double membrane or envelope of spinach chloroplasts. *Biochimie* 61:1283–92
94. Murillo M, Leustek T. 1995. ATP sulfurylase from *Arabidopsis thaliana* and *Escherichia coli* are functionally equivalent but structurally and kinetically divergent. Nucleotide sequence of two ATP sulfurylase cDNAs from *Arabidopsis thaliana* and analysis of a recombinant enzyme. *Arch. Biochem. Biophys.* 323:195–204
95. Neuenschwander U, Suter M, Brunold C. 1991. Regulation of sulfate assimilation by light and *O*-acetyl-L-serine in *Lemna minor L. Plant Physiol.* 97:253–58
96. Noctor G, Arisi ACM, Jouanin L, Kunert KJ, Rennenberg H, Foyer CH. 1998. Glutathione: biosynthesis, metabolism and relationship to stress tolerance explored in transformed plants. *J. Exp. Bot.* 49:623–47
97. Ohyama K, Fukuzawa H, Kohchi T, Shirai H, Sano T, et al. 1986. Chloroplast gene organization deduced from complete sequence of liverwort *Marchantia polymorpha* chloroplast DNA. *Nature* 322:572–74
98. Pajor AM. 1995. Sequence and functional characterization of a renal sodium/dicarboxylate cotransporter. *J. Biol. Chem.* 270:5779–85
99. Papenbrock J, Schmidt A. 2000. Characterization of a sulfur transferase from *Arabidopsis thaliana. Eur. J. Biochem.* In press
100. Park YS, Hong SW, Oh SA, Kwak JM, Lee HH, Nam HG. 1993. Two putative protein kinases from *Arabidopsis thaliana* contain highly acidic domains. *Plant Mol. Biol.* 22:615–24
101. Potin P, Bouarab K, Kupper F, Kloareg B. 1999. Oligosaccharide recognition signals and defence reactions in marine plant-microbe interactions. *Curr. Opin. Microbiol.* 2:276–83
102. Prior A, Uhrig JF, Heins L, Wiesmann A, Lillig CH, et al. 1999. Structural and kinetic properties of adenylylsulfate reductase from *Catharanthus roseus* cell cultures. *Biochim. Biophys. Acta* 1430:25–38
103. Rauser WE. 1995. Phytochelatins and related peptides. Structure, biosynthesis, and function. *Plant Physiol* 109:1141–49
104. Ravanel S, Gakiere B, Job D, Douce R. 1998. The specific features of methionine biosynthesis and metabolism in plants. *Proc. Natl. Acad. Sci. USA* 95:7805–12
105. Rea PA, Li ZS, Lu YP, Drozdowicz YM, Martinoia E. 1998. From vacuolar GS-X pumps to multispecific ABC transporters. *Annu. Rev. Plant Physiol. Plant Mol. Biol.* 49:727–60
106. Rennenberg H, Steinkamp R, Kesselmeier J. 1981. 5-Oxo-prolinase in *Nicotiana tabacum*: catalytic properties and subcellular localization. *Physiol. Plant.* 62:211–16
107. Renosto F, Patel HC, Martin RL, Thomassian C, Zimmerman G, Segel IH. 1993. ATP sulfurylase from higher plants: kinetic and structural characterization of the chloroplast and cytosol enzymes from spinach leaf. *Arch. Biochem. Biophys.* 307:272–85
108. Reuveny Z, Filner P. 1977. Regulation of adenosine triphosphate sulfurylase in

cultured tobacco cells. *J. Biol. Chem.* 252:1858–64

109. Rossini L, Pe ME, Frova C, Hein K, Sari-Gorla M. 1995. Molecular analysis and mapping of two genes encoding maize glutathione *S*-transferase (GSTI) and (GSTII). *Mol. Gen. Genet.* 248:535–39
110. Rotte C. 1998. *Subcellular localization of sulfur assimilation enzymes in Arabidopsis thaliana (L.) HEYNH.* Diplomarbeit thesis. Carl von Ossietzky Univ. Oldenburg, Germany. 87 pp.
111. Roughan PG. 1997. Stromal concentrations of coenzyme A and its esters are insufficient to account for rates of chloroplast fatty acid synthesis: evidence for substrate channeling within the chloroplast fatty acid synthase. *Biochem. J.* 327:267–73
112. Rouleau M, Marsolais F, Richard M, Nicolle L, Voigt B, et al. 1999. Inactivation of brassinosteroid biological activity by a salicylate-inducible steroid sulfotransferase from *Brassica napus*. *J. Biol. Chem.* 274:20925–30
113. Saito K, Kurosawa M, Tatsuguchi K, Takagi Y, Murakoshi I. 1994. Modulation of cysteine biosynthesis in chloroplasts of transgenic tobacco overexpressing cysteine synthase [*O*-acetylserine(thiol)lyase]. *Plant Physiol.* 106:887–95
114. Schiff JA, Hodson RC. 1973. The metabolism of sulfate. *Annu. Rev. Plant Physiol.* 24:381–414
115. Schiffmann S, Schwenn JD. 1998. Isolation of cDNA clones encoding adenosine-5′-phosphosulfate kinase (EC2.7.1.25) from *Catharanthus roseus* (Accession No. AF044285) and an isoform (akn2) from *Arabidopsis*(Accession No. AF043351) (PGR98-116). *Plant Physiol.* 117:1125
116. Schlenk F. 1965. The chemistry of biological sulfonium compounds. *Fortschr. Chemie Org. Naturst.* 23:61–112
117. Schmidt A, Jäger K. 1992. Open questions about sulfur metabolism in plants. *Annu. Rev. Plant. Physiol. Plant Mol. Biol.* 43:325–49
118. Schneider A, Martini N, Rennenberg H. 1992. Reduced glutathione (GSH) transport in cultured tobacco cells. *Plant Physiol. Biochem.* 30:29–38
119. Schwenn J. 1997. Assimilatory reduction of inorganic sulphate. In *In Sulfur Metabolism in Higher Plants*, ed. WJ Cram, LJ De Kok, I Stulen, C Brunold, H Rennenberg, pp. 39–58. Leiden: Backhuys Publ.
120. Setya A, Murillo M, Leustek T. 1996. Sulfate reduction in higher plants: molecular evidence for a novel 5′-adenylylphosphosulfate (APS) reductase. *Proc. Natl. Acad. Sci. USA* 93:13383–88
121. Smith FW, Hawkesford MJ, Ealing PM, Clarkson DT, Vanden Berg PJ, et al. 1997. Regulation of expression of a cDNA from barley roots encoding a high affinity sulphate transporter. *Plant J.* 12:875–84
122. Song W, Steiner H-Y, Zhang L, Naider F, Stacey G, Becker M. 1996. Cloning of a second Arabidopsis peptide transporter gene. *Plant Physiol.* 110:171–78
123. Steinkamp R, Rennenberg H. 1985. Degradation of glutathione in plant cells: evidence against the participation of a γ-glutamyltranspeptidase. *Z. Naturforsch.* 40c:29–33
124. Steinkamp R, Schweihofen B, Rennenberg H. 1987. γ-Glutamyl-cyclotransferase in tobacco suspension cultures: catalytic properties and subcellular localization. *Physiol. Plant.* 69:499–503
125. Sun H, Gao J, Ferrington DA, Biesiada H, Williams TD, Squier TC. 1999. Repair of oxidized calmodulin by methionine sulfoxide reductase restores ability to activate the plasma membrane Ca-ATPase. *Biochemistry* 38:105–12
126. Sunarpi, Anderson JW. 1997. Allocation of S in generative growth of soybean. *Plant Physiol.* 114:687–93

127. Suter M, von Ballmoos P, Kopriva S, den Camp RO, Schaller J, et al. 2000. Adenosine 5′-phosphosulfate sulfotransferase and adenosine 5′-phosphosulfate reductase are identical enzymes. *J. Biol. Chem.* 275:930–36

128. Takahashi H, Asanuma W, Saito K. 1999. Cloning of an Arabidopsis cDNA encoding a chloroplast localizing sulfate transporter isoform. *J. Exp. Bot.* 50:1713–14

129. Takahashi H, Sasakura N, Noji M, Saito K. 1996. Isolation and characterization of a cDNA encoding a sulfate transporter from *Arabidopsis thaliana*. *FEBS Lett.* 392:95–99

130. Takahashi H, Yamazaki M, Sasakura N, Watanabe A, Leustek T, et al. 1997. Regulation of cysteine biosynthesis in higher plants: a sulfate transporter induced in sulfate-starved roots plays a central role in *Arabidopsis thaliana*. *Proc. Natl. Acad. Sci. USA* 94:11102–7

131. Varin L, Chamberland H, Lafontaine JG, Richard M. 1997. The enzyme involved in sulfation of the turgorin, gallic acid 4-O-(beta-D-glucopyranosyl-6′-sulfate) is pulvini-localized in *Mimosa pudica*. *Plant J.* 12:831–37

132. Varin L, Marsolais F, Richard M, Rouleau M. 1997. Biochemistry and molecular biology of plant sulfotransferases. *FASEB J.* 11:517–25

133. Vatamaniuk OK, Stéphane M, Lu Y-P, Rea PA. 1999. AtPCS1, a phytochelatin synthase from *Arabidopsis*: isolation and in vitro reconstitution. *Proc. Natl. Acad. Sci. USA* 96:7110–15

134. Wakasugi T, Nagai T, Kapoor M, Sugita M, Ito M, et al. 1997. Complete nucleotide sequence of the chloroplast genome from the green alga *Chlorella vulgaris*: the existence of genes possibly involved in chloroplast division. *Proc. Natl. Acad. Sci. USA* 94:5967–72

135. Wolf AE, Dietz KJ, Schroder P. 1996. Degradation of glutathione *S*-conjugates by a carboxypeptidase in the plant vacuole. *FEBS Lett.* 384:31–34

136. Wykoff DD, Davies JP, Melis A, Grossman AR. 1998. The regulation of photosynthetic electron transport during nutrient deprivation in *Chlamydomonas reinhardtii*. *Plant Physiol.* 117:129–39

137. Yamaguchi Y, Nakamura T, Harada E, Koizumi N, Sano H. 1999. Differential accumulation of transcripts encoding sulfur assimilation enzymes upon sulfur and/or nitrogen deprivation in *Arabidopsis thaliana*. *Biosci. Biotechnol. Biochem.* 63:762–66

137a. Yang H, Matsubayashi Y, Nakamura K, Sakagami Y. 1999. *Oryza sativa* PSK gene encodes a precursor of phytosulfokine-alpha, a sulfated peptide growth factor found in plants. *Proc. Natl. Acad. Sci. USA* 96:13560–65

138. Yildiz FH, Davies JP, Grossman AR. 1994. Characterization of sulfate transport in *Chlamydomonas reinhardtii* during sulfur-limited and sulfur-sufficient growth. *Plant Physiol.* 104:981–87

139. Yildiz FH, Davies JP, Grossman AR. 1996. Sulfur availability and the SAC1 gene control adenosine triphosphate sulfurylase gene expression in *Chlamydomonas reinhardtii*. *Plant Physiol.* 112:669–75

140. Yoon HW, Kim MC, Shin PG, Kim JS, Kim CY, et al. 1997. Differential expression of two functional serine/threonine protein kinases from soybean that have an unusual acidic domain at the carboxy terminus. *Mol. Gen. Genet.* 255:359–71

141. Zenk MH. 1996. Heavy metal detoxification in higher plants—a review. *Gene* 179:21–30

Annu. Rev. Plant Physiol. Plant Mol. Biol. 2000. 51:167–94

(TRANS)GENE SILENCING IN PLANTS: How Many Mechanisms?

M. Fagard and H. Vaucheret

Laboratoire de Biologie Cellulaire, INRA, 78026 Versailles Cedex, France; e-mail: vauchere@versailles.inra.fr

Key Words transgene, virus, methylation, transcription, RNA degradation

■ **Abstract** Epigenetic silencing of transgenes and endogenous genes can occur at the transcriptional level (TGS) or at the posttranscriptional level (PTGS). Because they can be induced by transgenes and viruses, TGS and PTGS probably reflect alternative (although not exclusive) responses to two important stress factors that the plant's genome has to face: the stable integration of additional DNA into chromosomes and the extrachromosomal replication of a viral genome. TGS, which results from the impairment of transcription initiation through methylation and/or chromatin condensation, could derive from the mechanisms by which transposed copies of mobile elements and T-DNA insertions are tamed. PTGS, which results from the degradation of mRNA when aberrant sense, antisense, or double-stranded forms of RNA are produced, could derive from the process of recovery by which cells eliminate pathogens (RNA viruses) or their undesirable products (RNA encoded by DNA viruses). Mechanisms involving DNA-DNA, DNA-RNA, or RNA-RNA interactions are discussed to explain the various pathways for triggering (trans)gene silencing in plants.

CONTENTS

INTRODUCTION

Plants are subject to various endogenous and environmental stimuli that may lead to changes in genome structure and/or genome expression. Because plants are not able to move and cannot escape from their environment, they have developed defenses to limit the potentially deleterious effects resulting from such stimuli.

The movement of transposable elements (TEs) is activated by many stresses (32). Plants with a small genome, like Arabidopsis, carry a limited number of copies of TEs, whereas plants with a large genome, like maize, consist to more than 83% of TEs (9, 80). In both cases, the great majority of these elements are silent, which indicates that plants have developed efficient defenses that limit the expression and mobility of TEs (53, 56).

Many pathogens infect plants by using the cellular machinery for their own purposes. Plants have developed race-specific defenses against particular pathogens, which lead to localized cell death and necrosis around the site of infection; these defenses prevent further spread of the pathogen in the plant (66). However, more than two thirds of the reported defenses against virus infection do not involve a hypersensitive response (HR), but rather are associated with other mechanisms (30). In one mechanism, which is observed with RNA viruses, plants trigger the sequence-specific degradation of the viral RNA. Alternatively, the virus persists in a noninfectious form which is observed with DNA viruses (1, 13, 77, 78).

The genome structure of plants can also be altered by genetic transformation. Organisms such as *Agrobacterium tumefaciens* integrate part of their genome into the genome of susceptible species. Recently, genetic transformation techniques have begun to modify significantly the organization of the genome. Indeed, introducing transgenes into plants can both modify the number of copies of a given sequence and affect gene expression. Because the expression of a transgene cannot always be predicted, interest in studying the consequences of genetic transformations at the genome level has increased considerably over the past ten years (reviewed in 17, 20, 27, 29, 55, 56, 61, 83, 92). Transgenes can become silent after a (more or less) long phase of expression, and can sometimes silence the expression (at least partially) of homologous elements located at ectopic positions in the genome. In some cases, the silencing of transgenes also triggers resistance against homologous viruses; in other cases, infection by viruses triggers silencing of homologous transgenes (5, 6).

The silencing of transgenes probably results from the activation of defense mechanisms, indicating that plants possess systems for controlling genome structure and gene expression (56). The transgene itself or its product(s) are probably perceived as endogenous stimuli that activate this machinery. The study of transgene silencing provides an appropriate way to understand the different mechanisms controlling plant genome structure and expression. This review summarizes current knowledge on silencing events mediated by stably integrated transgenes and DNA and RNA viruses.

METHODS AND RULES FOR THE CLASSIFICATION OF SILENCING EVENTS

One factor that makes it difficult to determine the precise number of mechanisms involved in silencing is the diversity of analytical methods used by different research groups. It is important to define the largest number of parameters and criteria that allow one mechanism to be discriminated from others and then to analyze each silencing event according to these parameters and criteria.

Sources and Targets of Silencing

In analyzing silencing events, it is important to distinguish the source leading to silencing from the target that is being silenced. Four scenarios are described in the literature.

1. An element can be exclusively the source of silencing without being subjected to the silencing process it triggers in *trans*. Examples of transgenes or viruses that silence homologous genes but are not affected themselves have been reported (58, 78, 79) and are described in detail below.
2. An element can be a source of silencing but affect only itself, in which case, it is said to occur in *cis*. Examples of transgenes that are silenced when inserted into a particular structure or into a particular location of the genome, and that do not affect the expression of any other element have been reported (4, 49, 51, 64, 100) and are detailed below.
3. An element can be a source of silencing for itself and for homologous ectopic elements, i.e. silencing occurs in both *cis* and *trans*. Examples of transgenes or viruses that are simultaneously sources and targets of silencing, and that trigger silencing of homologous ectopic elements have been reported (3, 14, 15, 25, 34, 38, 50, 60, 68, 71, 73, 82, 86, 88, 91) and are detailed below.
4. An element can be exclusively a target for silencing by trans-acting elements, i.e. it is not a source leading to silencing. Many examples exist of transgenes, endogenous genes, and viruses in which expression is silenced only when the element is brought into the presence of other homologous silenced elements.

Molecular Parameters

Many molecular criteria can help to classify silencing events. Ideally, analysis of silencing would include all the molecular characteristics listed below. Unfortunately, as shown in Tables 1 and 2, none of the silencing events reported in the literature has been analyzed with all these criteria. Consequently, it is a matter of speculation to determine if separate silencing events rely on the same type of mechanism.

TGS versus PTGS Silencing may result from a block of transcription (TGS; 60), or from the degradation of RNA (PTGS; 15). Northern blot assays performed on cytoplasmic RNA combined with run-on transcription or RNase protection assays performed with isolated nuclei enable TGS to be distinguished from PTGS. For TGS, the absence of transcription in the nucleus and the failure of RNA to accumulate in the cytosol provide the result, while for PTGS transcription occurs but RNA fails to accumulate. Not all silencing events have been analyzed at the nuclear level (by run-on or RNase protection assays); thus the question of which type of silencing occurs remains unresolved in many cases (for example, see 14, 35). Other types of analyses may help to solve this ambiguity; for example, TGS correlates with methylation in the promoter, whereas PTGS correlates with methylation in the coding sequence; TGS is both mitotically and meiotically heritable, whereas PTGS is meiotically reversible. However, such short-cuts may be dangerous because they could prevent the discovery of counter-examples, such as TGS events not associated with methylation as in yeast and Drosophila.

Copy Number In only a few cases have transgene loci been recloned and sequenced (37, 58, 67, 74). Two analyses indicated that a single transgene copy can be subjected to silencing in *cis* (37, 74), whereas two other cases demonstrated clearly the requirement of a particular inverted repeat structure to trigger silencing in *trans* (58, 67). It is thus unresolved whether a single transgene copy can trigger silencing in *trans*. The studies of *trans*-silencing loci that suggested the presence of a single transgene copy were all based on southern blot analysis (19, 22, 71, 76, 99). This is an imperfect method to score for small rearrangements of the transgene such as partial duplications of the inserted DNA that are thought to play a role in triggering silencing (4, 25, 34, 42, 51, 58, 81, 84).

Transcription Silencing loci may be transcribed at a high level, low level, or not at all (i.e. below detectable levels). Whether the level of transcription is important for triggering silencing is an important question, one that may be addressed by introducing transgenes driven either by promoters of different strengths or without a promoter, and then comparing their effect (75, 88). However, due to position effect, this approach does not provide definite proof. Transcription of transgenes driven by the 35S promoter may be blocked by the 35S-specific silencing locus of the tobacco line 271 (86, 91), which allows the requirement for transcription

TABLE 1 Characteristics of representative TGS events triggered by transgenes or viruses

Examples	Trans-silenced homologous targets		Molecular characteristics of silencing locus					Genetic modifiers			
								Viruses		Mutations	
	Endogenous genes or transgenes	Viruses	Minimal copy number	Transcription (run-on)	Transcription required	Methylation of promoter region	Systemic acquired silencing	CMV	TEV, PVY	*sgs*	*ddm, som, hog, sil*
Surrounding heterochromatin											
p35S-A1	—	—	1 (74)	—	—	—	—	—	—	—	—
Repetitive sequences (RPS)											
RPS/p35S-GUS	—	—	1 (85)	—	—	N (85)	—	—	—	—	—
Transgene-genomic junctions											
p35S-A1	Y (60)	—	1 (74)	N (60)	—	Y (60)	—	—	—	—	—
p35S-HPT	—	—	1 (37)	—	—	—	—	—	—	—	—
(Trans)gene repeats											
p35S-HPT	—	—	>1 (62)	—	—	Y (63)	—	—	—	N (21)	Y (63)
p35S-GUS	N (u)	—	>1 (u)	N (u)	—	Y (u)	—	—	—	N (u)	Y (u)
pNos-NPT	—	—	>1 (14)	—	—	Y (31)	—	—	—	—	Y (31)
pNos-OCS	Y (54)	—	>1 (54)	—	—	Y (54)	—	—	—	—	—
p35S-RiN	Y (91)	—	>1 (91)	N (93)	—	Y (91)	N (u)	N (u)	N (u)	—	—
PAI1-PAI4	Y (39)	—	>1 (8)	Y (8)	N (52)	Y (52)	—	—	—	—	Y (39)
Aberrant promoter transcripts											
p35S-pNos	Y (58)	—	>1 (58)	Y* (58)	Y (58)	—	—	—	—	—	—
DNA viruses											
CaMV	Y (1)	—	—	Y (1)	—	—	—	—	—	—	—

Y: yes; N: no; —: not determined; Y*: determined by northern; (number): reference; (u): unpublished data from our lab.

TABLE 2 Characteristics of representative PTGS events triggered by transgenes or viruses

	Trans-silenced homologous targets		Molecular characteristics of silencing locus					Genetic modifiers			
								Viruses		Mutations	
Examples	Endogenous genes or transgenes	Viruses	Minimal copy number	Transcription (run-on)	Transcription required	Methylation of transcribed region	Systemic acquired silencing	CMV	TEV, PVY	*sgs*	*ddm, som, hog, sil*
Sense transgenes											
p35S-NIA	Y (94)	—	1 (71)	Y (94)	Y^1 (94)	Y (u)	Y (70)	Y (7)	—	Y (21)	—
p35S-GUS	Y (22)	Y (24)	1 (22)	Y (22)	Y^1 (24)	Y (21)	Y (70)	Y (10)	Y (2)	Y (21)	—
p35S-TEV	—	Y (50)	1 (19)	Y (50)	—	—	N* (19)	—	—	—	—
Δpro-CHS	Y (88)	—	>1 (88)	N (84)	—	Y (84)	—	—	—	—	—
Antisense transgenes											
p35S-aMET1	Y (28)	—	—	—	—	—	—	—	—	N (u)	—
p35S-aGUS	Y (u)	—	—	—	—	—	N (u)	—	—	N (u)	—
p35S-aPVY	—	Y (99)	>1 (99)	—	—	—	—	—	—	—	—
p35S-aNII	Y (86)	—	>1 (91)	N (93)	—	Y (91)	N (u)	Y (u)	Y (u)	—	—
Sense/antisense transgenes											
p35S-ACC-aACC	Y (34)	—	—	—	—	—	—	—	—	—	—
p35S-GUS-aGUS	Y (99)	—	—	—	Y^2 (99)	—	—	—	—	N (u)	—
p35S-PVY-aPVY	—	Y (99)	1 (99)	—	—	—	—	—	—	—	—
Viruses inducing recovery											
CaMV	Y (1)	—					—	—	—	—	—
TBRV	—	Y (77)					—	—	—	—	—
TRV	Y (78)	Y (78)					—	—	—	—	—
Viruses not inducing recovery											
TGMV	Y (46)	—					Y* (46)	—	—	—	—
PVX	Y (79)	Y (78)					Y* (79)	—	—	—	—

Y: yes; N: no; —: not determined; Y^1: construct brought into the presence of the 35S-silencing locus 271; Y^2: transformation with a promoterless contruct; N*: primary determination could be to be evaluated independently of position effect (24, 94). However, such a strong and specific promoter silencer exists only in tobacco. Transcription may also be controlled by inducible promoters. Surprisingly, no examples of silencing events triggered by such promoters have been reported in the literature. Finally, the promoter of transgenic silencing loci may be eliminated by using the Cre-*lox* system. However, site-directed deletions such as these may also trigger structural changes in the transgene locus that could modify its silencing properties.

Production of a Systemic Silencing Signal Basic grafting experiments have demonstrated clearly that a sequence-specific systemic silencing signal is produced in some cases of PTGS, which allows PTGS to propagate and become amplified throughout the plant (70, 72, 95, 96). Whether this is a common aspect of all PTGS events is unknown. Also to be determined is whether TGS could rely on the production of a presumably diffusible or transported molecule (98).

Methylation Methylation has often been associated with silencing. Although methylation can sometimes affect a large part of a transgene locus, TGS correlates mainly with methylation of the promoter sequence (4, 18, 49, 51, 54, 57, 60, 63, 73, 74, 84, 91, 100), whereas PTGS correlates with methylation of coding sequences (21, 25, 38, 41, 84). However, whether methylation is a cause or a consequence of silencing is not known. Furthermore, methylation has usually been scored using methylation-sensitive enzymes, and rarely by genomic sequencing in which methylation of all sites is assayed. It is thus difficult to conclude that methylation is not involved in silencing when only a limited number of methylation-sensitive enzymes have been used.

Genetic Modifiers

Mutants affected in TGS or PTGS have been identified recently (16, 21, 31, 39, 63), allowing a genetic classification of silencing events. To date, few silenced loci have been transferred to these mutants to test whether release of silencing occurs, mainly because the mutants were obtained in Arabidopsis whereas a larger number of silencing events were identified in crop species.

The release of PTGS by non-homologous viruses has also been reported recently (2, 7, 10, 44), indicating that viruses can interfere with the plant silencing machinery. Therefore, an additional criterion to classify silencing relies on the analysis of their sensitivity to infection by such viruses.

TRANSCRIPTIONAL GENE SILENCING

TGS corresponds to a block of transcription. TGS has been shown to affect sequences that are integrated in the genome and not extrachromosomal DNA. However, it has been reported that artificially methylated sequences introduced transiently in plant cells are not expressed, even when using methylation-target-free promoters. This indicates that methylation of the coding sequence is sufficient to block expression (35). Since it has not been determined whether these methylated sequences are transcribed or not, it is not possible to classify this type of silencing event as TGS or PTGS.

As shown in Table 1, TGS of integrated sequences can be classified into six classes according to the nature of the source of silencing. Whether TGS occurs in *cis*, simultaneously in *cis* and *trans*, or in *trans* only is discussed individually.

TGS Mediated by Surrounding Heterochromatin

Transgenes insert randomly into the genome. Depending on the position of their insertion, they may be surrounded by euchromatin or heterochromatin. In the latter case, the transgene adopts the structure of the surrounding transcriptionally silent chromatin, thus leading to TGS. This phenomenon can affect transgenes present even as a single copy (74).

TGS Mediated by Endogenous Repetitive Sequences

Repetitive sequences (RPS) exist in the genome of most plant species and are often methylated. The association of a methylated RPS element from petunia with a 35S-GUS transgene destabilizes its expression in transgenic tobacco and petunia plants, leading to variegation (85). This RPS element probably attracts repressive chromatin complexes, which then spread into the neighboring 35S-GUS transgene. Although de novo methylation of the RPS element has been observed, there is no evidence for methylation of the 35S-GUS transgene (85).

To determine whether or not this TGS effect relies on a *trans*-effect of endogenous RPS elements on the RPS-associated transgenes, the RPS-p35S-GUS transgene was introduced into Arabidopsis, which lacks this RPS element. Methylation occurred at the RPS element, even when present as a single copy, which suggests that a stem loop region present in this RPS element is a target for de novo methylation by the cellular machinery (P Meyer, personal communication). A protein was characterized that binds to this RPS element. It shows similarities to proteins that form repressive chromatin complexes in yeast and Drosophila (two organisms that show TGS but lack methylation), suggesting that methylation per se is not necessary to repress transcription. Rather, methylation of the RPS element probably recruits chromatin components that induce TGS of neighboring transgenes (P Meyer, personal communication).

Of interest will be the resolution of whether the *cis*-TGS effect mediated by RPS can be modified in the Arabidopsis *ddm1* mutant, which is impaired in synthesizing a chromatin remodeling factor (39, 40) or in the *ddm2* mutant, which is affected in synthesizing the MET1 DNA-methyltransferase (26, 28; E Richards, personal communication). This analysis will allow a precise determination of the respective roles of methylation and chromatin structure on TGS.

TGS Mediated by Transgene-Genomic Junctions

Integration of a single transgene copy in a nonmethylated area of the genome generally allows transgene expression. However, expression may be unstable, leading to variegation when part of the plant genome is silenced, for example by environmental factors, or to non-Mendelian segregation when the DNA of part of the progeny is silenced. Transcriptionally silenced individuals show methylation and a condensed chromatin structure (60, 87). Molecular analysis of such unstable TGS events affecting single transgene copies indicated that either the GC content of the transgene differed significantly from that of the surrounding genomic sequences (23, 60), or the presence of backbone plasmid DNA unexpectedly transferred with the transgene (37). It was therefore hypothesized that such a local discrepancy may disorganize chromatin structure and contribute to destabilizing gene expression (48, 60, 74). Surprisingly, one of these TG-Silenced loci was able to silence the expression of an active allelic copy brought in by crossing; this copy then was able to silence another active allelic copy (60). This phenomenon is reminiscent of paramutation in plants, a phenomenon involving conversion of the epigenetic state of an endogenous allele (paramutator) which is silent and methylated to an active allele (paramutable) that suggests cross-talk between homologous chromosomes in somatic tissues.

TGS Mediated by (Trans)Gene Repeats

Integration of multiple copies of a transgene in a particular spatial arrangement may lead to methylation and TGS (4, 14, 49, 63). In one case, TGS was shown to correlate with chromatin condensation (4, 100). The implication of repeats in this process was elegantly demonstrated by analyzing internal deletions within this transgene locus that eliminate TGS (4), thus ensuring that TGS was not mediated by *cis*-surrounding sequences or by particular transgene-genomic junctions, as outlined above. In other cases, the contribution of repeats versus that of surrounding sequences remains unclear because either no internal deletions were identified (14, 49) or internal deletions that eliminate TGS could also have modified transgene-genomic junctions (62).

Two transgenic lines hypothesized to be TG-Silenced (14, 62) and one transgenic line in which run-on assays clearly identified TGS (P Mourrain & H Vaucheret, unpublished data) were used to identify mutants and/or to test the effect of previously identified genetic modifiers. Mutants impaired in the *SGS1* or *SGS2* genes, which control PTGS (21), failed to release TGS from the two tested loci (21; P Mourrain & H Vaucheret, unpublished data), suggesting that *SGS* genes play a role specific to PTGS. Conversely, mutants impaired in the *DDM1* gene encoding a chromatin remodeling factor (40) released TGS from the three loci (31, 63; P Mourrain & H Vaucheret, unpublished data). Mutants impaired in the *DDM2* gene encoding the major DNA-methyltransferase of Arabidopsis (also termed *MET1*; 26, 28: E Richards, personal communication) or transgenic plants expressing an antisense *MET1* RNA failed to release TGS from one line (63), had very little effect on another line (I Furner, personal communication), but released TGS from the third line (P Mourrain & H Vaucheret, unpublished data). These results suggest a range of efficiency in TGS that might be due to methylation alone or a combination of methylation and chromatin remodeling (see conclusions on TGS below). Analysis of the effect of *som* (63), *hog*, and *sil* (31) mutants on the different reporter loci, as well as characterization of the corresponding genes, should ensure a more complete analysis of the genetic determinism of TGS.

In two cases, integration of multiple copies of a transgene in a particular spatial arrangement led to methylation and TGS in both *cis* and *trans*, i.e. transgenic loci were able to silence ectopic target transgenes driven by homologous promoters (54, 57, 86, 91). The molecular mechanism of transmission of TGS from these two silencing loci to their targets remains unclear. It may involve transient DNA-DNA pairing between the silencing loci and their targets, followed by the imposition of a mitotically and meiotically heritable transcriptionally repressive state on the targets (54, 73, 94). Alternatively, it may result from the production of specific molecules by the silencing loci that impose such a mitotically and meiotically heritable transcriptionally repressive state on the targets (73, 98). The molecules required to trigger TGS may be below detectable amounts. In addition, the diffusion of putative silencing molecules would certainly be restricted to the cell, and these molecules would be unable to propagate from cell to cell, as there is no evidence for graft-transmission of *trans*-TGS from silenced rootstocks to target scions (H Vaucheret, unpublished data). *Trans*-TGS seems to require a specific arrangement of transgene copies and a specific degree of methylation of the silencing locus, because hypomethylated epigenetic variants as well as mutants with a rearranged hypomethylated locus are unable to trigger *trans*-TGS (P Mourrain & H Vaucheret, unpublished data). *Trans*-TGS does not require the presence of symmetrical methylation sites in the targeted promoters, whereas symmetrical sites are required to maintain silencing after meiotic elimination of the silencing locus (18). This latter experiment shows that methylation plays a role in maintaining *trans*-TGS rather than in its establishment.

Strong evidence for a DNA-DNA directed trans-methylation mechanism was suggested by the analysis of the effect of an endogenous inverted repeat of the *PAI1* and *PAI4* genes carried by the Ws strain of Arabidopsis on the unlinked *PAI2* and *PAI3* single copies (8). When introduced by crossing into the Col strain (carrying single nonmethylated *PAI1*, *PAI2*, and *PAI3* copies), this inverted repeat triggers methylation of unlinked endogenous *PAI2* and *PAI3* copies (52). Surprisingly, one of the *PAI* genes of the endogenous inverted repeat of the Ws strain is expressed at a high level despite being methylated (8; J Bender, personal communication), thus leaving open the possibility that it produces silencing RNA molecules. However, introduction of a transgene consisting in a promoterless *PAI1-PAI4* inverted repeat in the Col strain also triggers methylation of unlinked endogenous *PAI2* and *PAI3* copies. The absence of fortuitous expression of transgene RNA (checked by RT-PCR) led the authors to suggest a direct DNA-DNA pairing mechanism for the transmission of methylation (52). Methylation of the multigene *PAI* family requires *DDM1* and *DDM2* genes. Indeed, when brought into the Ws strain, the *ddm1* mutation strongly reduces methylation of *PAI2* and *PAI3* (80% reduction), but has little effect on the *PAI1-PAI4* inverted repeat (20% reduction). Conversely, when brought into the Ws strain, the *ddm2* mutation reduces methylation of *PAI2, PAI3*, and the *PAI1-PAI4* inverted repeat (70% reduction), which suggests that the *PAI1-PAI4* inverted repeat is in a more open chromatin configuration than the singlet *PAI2* and *PAI3* genes, and is thus less dependent on *DDM1* for access of the DNA to methylation (J Bender, personal communication). Methylation of *PAI2* is accompanied by silencing (39). Both *ddm1* and *ddm2* cause a loss of *PAI2* methylation and silencing when brought in the Ws *pai* mutant background (which carries a deletion of the *PAI1-PAI4* inverted repeat). This indicates that the maintenance of *PAI2* silencing in the absence of the *PAI1-PAI4* inverted repeat requires the integrity of both *DDM1* and *DDM2* genes (39; J Bender, personal communication). Here again, analysis of the effect of other genetic modifiers (*som*, *hog*, *sil*, and *sgs*) is needed.

TGS Mediated by Aberrant Promoter Transcripts

The production of diffusible silencing RNA molecules that trigger TGS in *trans* was shown when a transgene made of the Nos promoter sequences (pNos) under the control of the 35S promoter was constructed for this purpose (58). Plants expressing polyadenylated pNos RNA failed to silence pNos-driven transgenes, whereas one plant producing truncated non-polyadenylated pNos RNA triggered *trans*-TGS and methylation. This plant carries two incomplete copies of the transgene arranged as an inverted repeat (IR), with pNos sequences at the center. This transgene locus produces RNA that could potentially adopt a hairpin conformation. The production of this distinctive RNA is required for *trans*-TGS of pNos-driven target transgenes since *trans*-TGS does not occur when transcription from the 35S promoter is impeded by the tobacco line 271-locus (58). This is the first evidence for *trans*-TGS mediated by an RNA, and it is not known whether other previously described *trans*-TGS events involve the production of an aberrant RNA that triggers methylation of the promoter of target transgenes and TGS. Once again, introduction of this system into Arabidopsis and confrontation with the previously identified genetic modifiers *ddm*, *som*, *hog*, *sil*, and *sgs* should provide insight into the mechanisms involved.

TGS Mediated by DNA Viruses

One example of *trans*-TGS mediated by a nuclear DNA virus was reported recently (1). Wild-type *Brassica napus* plants recover naturally from CaMV-infection by a PTGS-like mechanism, i.e. 19S and 35S RNA encoded by CaMV are degraded while replication of CaMV DNA is occurring in the nucleus (see PTGS section). CaMV-infection of transgenic *B. napus* plants expressing a p35S-GUS transgene with a 35S or Nos terminator leads to recovery from CaMV infection and PTGS or TGS of the p35S-GUS transgene, respectively. These results led the authors to suggest that, in the presence of homology in both promoter and transcribed regions, PTGS preferentially occurs, whereas TGS occurs only if the homology is restricted to the promoter region (1). Such *trans*-TGS mediated by DNA viruses resembles *trans*-TGS mediated by the tobacco transgenic line that expresses an aberrant RNA homologous to the Nos promoter (58). In both cases, the source of *trans*-TGS (CaMV, p35S-pNos transgene) is not subjected to TGS, and TGS involves the production of RNA either of aberrant structure (p35S-pNos transgene) or targeted for degradation by the cellular machinery (CaMV).

Conclusions on TGS

TGS can be triggered in *cis* or in *trans*. *Cis*-acting elements may be endogenous heterochromatin surrounding the transgene locus (74), endogenous repeated and methylated elements located close to the transgene locus (85), transgene-genomic junctions that disturb chromatin organization (37, 60, 87), or particular arrangements of transgene repeats that create heterochromatin locally (4, 100). *trans*-acting elements may be allelic or ectopic homologous loci that potentially transfer their epigenetic state by direct DNA-DNA pairing or protein-mediated DNA-DNA interactions (52, 54, 60, 86), or ectopic transgenes (58) or nuclear DNA viruses (1) that produce a diffusible signal (aberrant RNA, PTGS-targeted viral RNA) that potentially imposes an epigenetic silent state by interaction with the homologous promoter of target transgenes.

In all cases, TG-Silenced transgenes show hypermethylation (4, 18, 49, 51, 54, 57, 60, 63, 73, 74, 84, 91, 100). In cases where it was tested, chromatin condensation was also observed (87, 100). Some, but not all, TG-Silenced (trans)genes are reactivated in the methylation-deficient mutant *ddm2* or in plants expressing a *MET1* antisense RNA (63; J Bender, personal communication; I Furner, personal communication; P Mourrain & H Vaucheret, unpublished data), suggesting that methylation plays a critical role in some but not all TGS events. At these loci, transgene methylation could constitute the primary determinant that allows the attraction of nuclear factors, such as MeCP2, which specifically bind to methylated cytosines and assemble local chromatin into a repressive complex (43). Since the *DDM2* gene encodes the major DNA methyltransferase activity (MET1; 26, 28: E Richards, personal communication), *ddm2* mutants could release TGS only from loci in which the formation of repressive chromatin complexes depends essentially on the presence of methylation. Conversely, at other loci, repressive complexes could be formed independently of methylation, and methylation could be an indirect consequence of this chromatin state. The *DDM1* gene encodes a protein of the SWI2/SNF2 family that plays a role in various functions including transcriptional co-activation, transcriptional co-repression, chromatin assembly, and DNA repair (40). Both the repressive chromatin state and hypermethylation associated with TGS are expected to be lost in *ddm1* mutants, allowing the release of TGS from any locus.

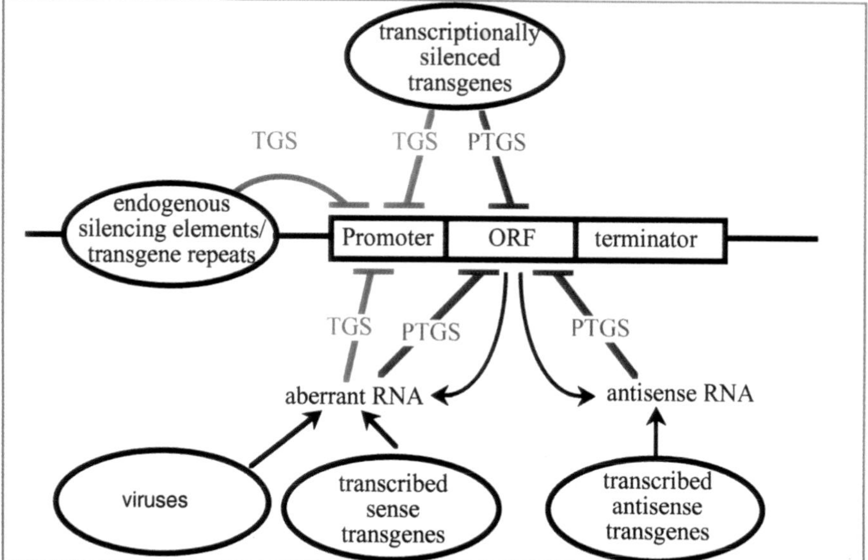

Figure 1 Putative mechanisms of (trans)gene silencing in plants. TGS (in red) could occur in *cis* because of the presence of neighboring endogenous silencing elements or because transgene repeats can interact to create a silencing structure. TGS could also occur in *trans* when active (trans)genes are brought into the presence of homologous TG-Silenced transgenes that can transfer their epigenetic silent state through DNA-DNA interactions, or when either viruses or transcribed transgenes produce an aberrant form of RNA that impedes transcription through DNA-RNA interactions. PTGS (green) could occur in *cis* owing to the production of aberrant sense, antisense, or double-stranded forms of RNA by the transgene itself, leading to the degradation of homologous mRNA. PTGS could also occur in *trans* when the RNA encoded by active (trans)genes share homology with viruses or transgenes that themselves produce aberrant forms of RNA that activate PTGS. Alternatively, PTGS could occur in *trans* when active (trans)genes are brought into the presence of homologous TG-Silenced transgenes that can transfer their epigenetic silent state through DNA-DNA interactions, thus impairing the regular production of mRNA.

POSTTRANSCRIPTIONAL GENE SILENCING

Three papers published in 1990 (68, 82, 89) demonstrated that introduction of transcribed sense transgenes could down-regulate the expression of homologous endogenous genes, a phenomenon called co-suppression (68). Co-suppression results in the degradation of endogenous gene and transgene RNA after transcription (15, 36, 38, 88, 90, 94). Because posttranscriptional RNA degradation can affect a wide range of transgenes expressing plant, bacterial, or viral sequences, it was more generally renamed PTGS. This section explores whether related silencing phenomena occurring with sense transgenes, antisense transgenes, and viruses rely on the same mechanism as the originally described co-suppression.

As in TGS (Table 1), PTGS may be classified according to the nature of the silencing source (Table 2), which can be a sense transgene, an antisense transgene, simultaneously expressed sense/antisense transgenes, or viruses. Many PTGS events have been reported in the literature, but only a few representative examples of PTGS events targeting endogenous sequences, foreign sequences, or viral sequences are presented for each class (when available). PTGS, like TGS, can occur in *cis* (only the RNA transcribed from the silencing source is degraded), simultaneously in *cis* and *trans* (RNA transcribed from the silencing source and all homologous RNA are degraded), or in *trans* (only RNA that is homologous to RNA transcribed from the silencing source is degraded, but not the RNA transcribed from the source).

PTGS Mediated by Sense Transgenes

Strongly Transcribed Sense Transgenes Comprehensive analysis of PTGS events with strongly transcribed sense transgenes allows the characteristics of this phenomenon to be defined precisely. Once initiated against the RNA of a given transgene, PTGS leads to the degradation of homologous RNA from either endogenous genes (co-suppression; 36, 68, 88), transgenes (*trans*-inactivation; 22, 25, 38), or RNA viruses (RNA-mediated virus resistance; 19, 25, 50, 81). In RNA-mediated virus resistance, plants can be either immune, i.e. virus resistance is established prior to the infection (19, 25, 81), or can recover from infection in newly emerging leaves (19, 50).

A single transgene copy appears to be sufficient to trigger this type of PTGS (19, 22, 71, 76). Transgene transcription seems to be required, since the frequency of silencing correlates with the strength of the promoter used to drive the transgene (75), and since transcriptional silencing of 35S-driven transgenes mediated by the tobacco locus 271 (86, 91) impedes co-suppression of homologous endogenous genes (94) as well as resistance against homologous RNA viruses (24).

The production of aberrant RNA by PTG-Silenced transgenes is evoked in many models that try to explain the mechanism of PTGS (6, 17, 20, 50, 56, 59, 92, 98). Because PTGS depends on active transcription of the transgene itself, it is unlikely that aberrant RNA is directly produced by readthrough transcription from neighboring transgenes beyond their terminators, or from transcription from neighboring endogenous promoters. However, such unintended transcription events could interfere with regular transcription of transgenes, leading to the production of aberrant RNA instead of regular mRNA, or could produce antisense RNA that could interact with regular mRNA to form aberrant (partially) double-stranded RNA. Alternatively, transgenes could produce directly single-stranded aberrant RNA because they are methylated. Indeed, in some cases, PTGS correlates with methylation of the transgene coding sequence (21, 25, 38, 41, 81, 84). In addition, de novo methylation of the transgene appeared to precede the onset of PTGS-mediated virus resistance (41). Since *de novo* methylation can be triggered in sequence-specific transgenes by introduction of homologous viroid RNA (97), an RNA signal is suggested to trigger transgene methylation and subsequently trigger PTGS (41, 98). Despite these data, it is still not clear whether methylation plays an active role in the triggering and/or the maintenance of PTGS, or whether it is an indirect consequence of PTGS. Analysis of the effect of methylation mutants like *ddm* on PTG-Silenced transgenes should help clarify this issue.

Grafting experiments revealed that PTG-Silenced plants produce a sequence-specific systemic silencing signal that propagates long distance from cell to cell and triggers PTGS in non-silenced graft-connected tissues of the plant (70, 72, 95). Because of its sequence-specificity and its mobility, this signal is assumed to be (part of) a transgene product, probably the putative aberrant RNA hypothesized above, that could migrate alone or within a ribonucleoprotein complex.

In one case of RNA-mediated virus resistance, PTGS was found not to be graft transmissible (19). However, transmission was scored by infection with a virus (TEV) that is itself a source and a target of silencing. In addition, the propagation and/or maintenance of PTGS is counteracted by viruses like TEV, PVY, or CMV, even when they do not exhibit any homology with the PTG-Silenced transgene (2, 7, 10, 44). The HC-Pro protein of potyviruses (TEV, PVY) and the 2b protein of cucumoviruses (CMV) are the genetic determinants of this PTGS-inhibitory effect (2, 10, 44). These proteins could either interact directly with proteins of the cellular machinery involved in PTGS, and/or they could impede the propagation of the systemic silencing signal. Viruses such as TEV, PVY, and CMV do not enter the meristems and are not transmitted through the seeds. Note that PTGS is also absent from meristems (7, 96), a result consistent with the absence of transmission of PTGS through meiosis (15, 16, 22, 36, 71). These observations therefore reinforce the similarities between the movement of the silencing signal and the movement of viruses.

The efficiency of PTGS is increased in Arabidopsis *egs* mutants that define two genetic loci (16). Conversely, PTGS is released in Arabidopsis *sgs* mutants that define three genetic loci (21; C Beclin & H Vaucheret, unpublished data). These *sgs* loci are not allelic to the *ddm1*, *ddm2*, *hog1*, and *sil1* loci (I Furner, E Richards & H Vaucheret, unpublished data). Methylation of the transgene coding sequence is lost in *sgs* mutants (21; F Feuerbach & H Vaucheret, unpublished data). Nevertheless, these mutants are unlikely to be methylation mutants since they do not show demethylation of repeated genomic sequences (21). In addition, sensitivity to RNA viruses is modified in *sgs* mutants (C Beclin & H Vaucheret, unpublished data), indicating that *SGS* genes are likely to act at the RNA level. Characterization of the functions encoded by the *EGS* and *SGS* genes should provide insight into the mechanism(s) involved in PTGS. Mutants that are defective in quelling (a mechanism related to PTGS in *Neurospora crassa*) also define three genetic loci called *qde* (12). The cloning of the *QDE-1* gene revealed that it encodes an RNA-dependent RNA-polymerase (RdRp) for which at least four homologous genes exist in Arabidopsis (11). This enzyme is presumed to play a key role in PTGS, either through the production of aberrant RNA using mRNA or unintended transcripts as a matrix, or by the amplification of aberrant RNA up to a threshold level that would activate the cellular RNA degradation machinery (6, 20, 50, 92, 98, 99). Whether one of the Arabidopsis genes encoding a RdRp plays a role in PTGS awaits the cloning of the *SGS* genes, as well as the identification of knockouts of each of the plant RdRp genes.

Very Weakly Transcribed or Untranscribed Sense Transgenes A deviation from classic PTGS came from the analysis of plants showing co-suppression of endogenous *CHS* genes by sense transgenes that are not transcribed at a high level despite the presence of a 35S promoter, or by promoterless transgenes (84, 88). All plants of this type showed complex transgene arrangements, which contain at least one inverted repeat and are methylated (84). These observations led the authors to propose that such structures could efficiently pair with homologous endogenous genes, thereby impairing the regular production of RNA (84). Alternatively, this type of structure could be as efficient as a strongly transcribed single transgene to produce the amount of aberrant RNA that is hypothesized to activate the RNA degradation machinery. In the absence of data on the actual requirement for transcription from these loci, on the production of a systemic silencing signal, and on the release of this type of PTGS by viruses or *sgs* mutations, it is not possible to determine if this type of co-suppression event relies on a different mechanism from that triggered by strongly transcribed sense transgenes.

PTGS Mediated by Antisense Transgenes

Transcribed Antisense Transgenes Before the discovery of co-suppression by sense transgenes, down-regulation of endogenous genes was usually achieved using antisense transgenes. It was therefore hypothesized that PTGS could result from the unintended production of antisense RNA by those sense transgene loci that trigger PTGS, leading to antisense-like inhibition (33). However, a precise comparison of sense and antisense inhibition reveals many differences, suggesting that few, if any, steps are common to these two processes . Although antisense inhibition is efficient against endogenous genes and foreign transgenes (28, 42, 76), patterns of silencing produced by antisense transgenes are usually different from those produced by sense transgenes (42, 68). This pattern was elegantly demonstrated by conversion of a sense transgene into an antisense one using the Cre-*lox* system (76), thereby avoiding interference of position effect. In addition, an antisense 35S-aGUS transgene that is able to silence a sense 35S-GUS transgene when it is present in the same cell fails to produce a graft-transmissible silencing signal that would silence a sense 35S-GUS transgene present in another cell, which suggests that the PTGS systemic signal is not made strictly of antisense RNA (M Fagard & H Vaucheret, unpublished data). Moreover, antisense inhibition of the endogenous *MET1* gene or of a p35S-GUS transgene occurs efficiently in *sgs* mutants, impaired in PTGS (C Beclin, F Feuerbach & H Vaucheret, unpublished data). Finally, antisense transgenes generally fail to inhibit virus infection (99). Although other characterizations are still required to determine if there are any common steps between sense and antisense inhibition, they clearly exhibit distinct steps. The identification of mutants impaired in antisense inhibition and the analysis of PTGS in such mutants will help to identify possible common steps.

Untranscribed Antisense Transgenes One instance of silencing by an antisense transgene that closely resembles PTGS mediated by sense transgenes was observed in the transgenic tobacco line 271 (86, 91, 93). Silencing of homologous endogenous genes in line 271 showed several characteristics of co-suppression mediated by transcribed sense transgenes: transcription of endogenous genes in the nucleus without accumulation of the corresponding RNA in the cytoplasm (73, 93), meiotic resetting, triggering of silencing during development, and release by viruses that counteract PTGS (C Beclin, M Fagard & H Vaucheret, unpublished data). However, run-on assays failed to detect transcription of the antisense transgene from the heavily methylated 271 locus (93). In this case, and perhaps also in promoterless sense transgenes (84, 88), silencing could result from an actual pairing of the transgene locus with the homologous endogenous genes and their subsequent modification, leading directly to the production of degradable endogenous RNA. Alternatively, aberrant sense RNA could be produced by the 271 locus, which cannot be distinguished by run-on assays from that produced by the endogenous genes.

PTGS Mediated by Sense/Antisense Transgenes

Although the data presented in the section above point to significant differences between antisense inhibition and sense inhibition, recent models explaining PTGS predict a key role for double-stranded RNA (59, 99). These models take into account data showing that injection of double-stranded RNA in worms, flies, and trypanosomes inhibits expression of the homologous endogenous genes (45, 65, 69). In addition, intermediates of RNA degradation were identified in co-suppressed petunia plants, corresponding to a region of the RNA that could potentially form a secondary structure due to internal complementarity (59). This result led the authors to propose a catalytic model that predicts the pairing of these degradation products with endogenous RNA, followed by cleavage and self-regeneration of these small RNA molecules, which therefore increase in number at each cycle and could eventually propagate from cell to cell (59). Furthermore, small antisense RNA complementary to the targeted RNA were detected in PTG-Silenced plants (33a). However, the role of these small antisense RNA in PTGS is still not known.

In particular, whether these small RNA could propagate from a PTG-Silenced stock to a non-silenced scion through a graft-union, and whether these small RNA are still present in plants in which PTGS is released by non-homologous viruses (2, 7, 10, 44) or in PTGS-deficient *sgs* mutants (21) has not been determined.

To test the hypothesis of a role of double-stranded RNA structures, a p35S-ACC sense transgene carrying a small inverted repeat in the 5′ UTR region was introduced in tomato. Co-suppression of endogenous *ACC* genes occurred at a higher frequency in these plants than in plants carrying the regular p35S-ACC sense transgene without the inverted repeat (34). In a similar approach, sense and antisense transgenes expressing part of a viral genome that, alone, failed to trigger resistance to the corresponding RNA virus (PVY) were simultaneously expressed in tobacco (99). Although sense and antisense RNA were still detectable, plants were immune to infection by PVY. In addition, plants carrying a single copy of a p35S-PVY-aPVY transgene expressing an RNA that potentially can form a secondary structure due to the presence of homologous sequences linked together in sense and antisense orientation were also immune to infection by PVY. Similarly, a p35S-GUS-aGUS transgene silenced an endogenous p35S-GUS sense transgene more efficiently than newly introduced sense or antisense transgenes could. The authors then proposed that the production of double-stranded RNA is required to trigger PTGS, and that RdRp could be involved in such production (99).

Whether these events of co-suppression (34), *trans*-inactivation (99), or virus resistance (99) mediated by sense or antisense transgenes rely on the same mechanism as PTGS mediated by sense transgenes alone awaits the analysis of methylation, graft-transmissibility, and release by viruses that counteract PTGS mediated by sense transgenes. Nevertheless, simultaneous expression of sense p35S-GUS and antisense p35S-aGUS transgenes triggers silencing in *sgs* mutants (C Beclin & H Vaucheret, unpublished data), which suggests that at least the three steps controlled by *SGS* genes are specific to PTGS mediated by sense transgenes, and are not involved in sense- or antisense-mediated silencing.

PTGS Mediated by DNA and RNA Viruses

As with transgenes, viruses can be either the source, the target, or both source and target of silencing. PTGS mediated by viruses can occur with DNA viruses, which replicate in the nucleus, and with RNA viruses, which replicate in the cytoplasm. These viruses can be inoculated into plants at a specific stage of their development, or can be expressed within plants throughout development by stably integrated virus-expressing transgenes.

Viruses That Trigger Recovery Infection of nontransgenic *Brassica napus* plants by CaMV (a DNA pararetrovirus) leads to recovery by a PTGS-like mechanism, i.e. 19S and 35S RNA encoded by CaMV are degraded while CaMV DNA is still replicating in the nucleus. Infection of *B. napus* plants expressing a p35S-GUS transgene with a 35S terminator by CaMV leads to recovery from CaMV infection and induction of PTGS of the p35S-GUS transgene (1). CaMV is primarily a target of the cellular silencing machinery since the 19S and 35S RNA are degraded. However, CaMV can also be considered as a source (or at least as an inducer) of PTGS for transgenes sharing homology with the virus within their transcribed regions because it activates the cellular RNA degradation machinery against them.

Infection of nontransgenic *Nicotiana clevelandii* plants by TBRV (an RNA nepovirus) also leads to recovery by a PTGS-like mechanism, i.e. TBRV RNA is degraded (77). Plants that have recovered are sensitive to infection by PVX (an unrelated RNA virus). However, they are immune to infection by a recombinant PVX virus in which TBRV sequences have been cloned. Similarly, nontransgenic *Nicotiana benthamiana* plants can recover from infection by TRV (an RNA tobravirus). Plants that have recovered from infection by a recombinant TRV-GFP virus are sensitive to infection by PVX but are immune to infection by a recombinant PVX virus in which GFP sequences have been cloned. In addition, plants that have recovered exhibit PTGS of a newly introduced 35S-GFP transgene. This indicates that viruses that induce recovery also induce PTGS against (at least partially) homologous viruses and transgenes (78).

Additional analyses are needed to determine whether PTGS mediated by viruses relies on the same mechanism as PTGS mediated by sense transgenes. Required will be analyses of transgene methylation, over-infection by viruses that counteract PTGS, introduction into *sgs* mutants, and the characterization of mutants impaired in recovery.

Viruses That Do Not Trigger Recovery Infection of *N. benthamiana* by TGMV (a DNA geminivirus) is followed by high-level replication in the nucleus and accumulation of viral RNA in the cytoplasm. Infection by a recombinant TGMV virus carrying the coding sequence of the sulfur (*SU*) gene in either sense or antisense orientation leads to PTGS of the endogenous *SU* gene, i.e. the endogenous *SU* RNA is degraded (46). However, TGMV-SU RNA is not degraded, suggesting that TGMV-SU behaves only as a source of PTGS. Infection of transgenic *N. benthamiana* expressing a p35S-LUC transgene by a recombinant TGMV virus carrying the coding sequence of the *LUC* gene in either sense or antisense orientation leads to PTGS of the *LUC* transgene. In this case, both *LUC* and TGMV-LUC RNA fail to accumulate. Although viral infections are nonuniform, silencing of the *LUC* transgene seems to be complete in infected leaves, whereas silencing of the endogenous *PDS* gene is incomplete, leading to variegation. These results suggest that, in nontransgenic plants, silencing of endogenous genes requires the permanent presence of the virus. Conversely, transgenes that behave initially as targets of PTGS induced by viruses may become maintainers of PTGS through the production of a systemic silencing signal; this allows degradation of transgene and viral RNA in infected cells, and degradation of transgene RNA in noninfected cells.

Infection of *N. benthamiana* by PVX (a single-stranded RNA potexvirus) or TMV (a single-stranded RNA tobamovirus) leads to virus replication and accumulation of viral RNA in the cytoplasm. Infection by recombinant PVX or TMV viruses carrying the coding sequence of the phytoene desaturase (*PDS*) gene in either sense or antisense orientation leads to PTGS of the endogenous *PDS* gene, i.e. the endogenous *PDS* RNA is degraded, a phenomenon called VIGS (virus-induced gene silencing) (47, 79). However, PVX-PDS RNA accumulates at a high level, suggesting that the virus is not targeted by VIGS. Infection of transgenic *N. benthamiana* expressing a p35S-GFP transgene by a recombinant PVX virus carrying the coding sequence of the *GFP* gene in either sense or antisense orientation leads to VIGS of the *GFP* transgene. In this case, both endogenous *GFP* and PVX-GFP RNA are efficiently and uniformly degraded, as in endogenous *LUC* and TGMV-LUC RNA (47). These results suggest that the continuous presence of the inducing virus is required to maintain VIGS of endogenous genes, whereas the presence of a transgene targeted by VIGS is sufficient to maintain VIGS, thus allowing the degradation of target viral RNA as well as systemic propagation of VIGS.

These results are reminiscent of data showing that RNA of endogenous genes can be degraded in nontransgenic plants grafted onto transgenic rootstocks exhibiting co-suppression of the homologous endogenous genes and sense transgenes. Here, silencing is not maintained when the source of silencing (the rootstock) is removed, which suggests that although transgenes are dispensable for the RNA degradation step of co-suppression, their presence is required to maintain silencing (72). In explanation, it was hypothesized that only some transgenes can undergo epigenetic changes that lead to re-amplification of this signal and maintenance of PTGS (72, 92), whereas endogenous genes cannot. Similarly, infection of transgenic plants by recombinant TGMV, TMV, or PVX viruses would trigger degradation of both transgene and viral RNA because transgenes would undergo epigenetic changes that allow production of the silencing signal to be maintained. Conversely, infection of nontransgenic plants by recombinant viruses would require the continuous presence of the inducing viruses to sustain silencing of endogenous genes. Therefore, the mechanism of VIGS is likely to be the same as PTGS mediated by sense transgenes, but additional molecular and genetic evidence is still required, using *sgs* mutants, for example.

Stably Integrated Viruses That Do Not Trigger Recovery Expression of a PVX-GUS recombinant virus from a stably integrated nuclear transgene, a construct referred to as an amplicon, allows 100% efficient triggering of PTGS of both PVX-GUS viruses and homologous GUS transgenes (3). Indeed, such amplicon has all the components required for efficient PTGS mediated by sense transgenes: The threshold level of transgene/viral RNA that triggers PTGS is obtained by a combination of high transcription from a p35S-driven transgene and replication of the viral RNA, whereas PTGS is maintained through transcription from a transgene, thus allowing a permanent production of the silencing signal (see above). This system therefore provides a powerful strategy for consistent silencing of endogenous genes in transgenic plants.

Conclusions on PTGS

PTGS can be triggered by transgenes and viruses, leading to the degradation of homologous RNA encoded by endogenous genes, transgenes, and, in some cases, by the virus itself. Because some plant species can recover from infection by some viruses (caulimo-, nepo-, and tobraviruses), by a PTGS-like mechanism (1, 13, 77, 78), PTGS is likely to be primarily a defense response of the plant against viruses. Once activated against such viruses, the RNA degradation machinery of PTGS becomes naturally efficient against endogenous gene or transgene RNA if it shares homology with the targeted virus (1, 78). Other viruses for which recovery is not observed (such as gemini-, potex-, and tobamoviruses) can also trigger silencing of endogenous genes and transgenes sharing homology at the RNA level (47, 78, 79), suggesting that although recovery does not occur, these viruses activate the plant's PTGS defense machinery. Finally, viruses of two other families (poty- and cucumoviruses) can counteract PTGS of nonhomologous transgenes (2, 7, 10, 44). The fact that these viruses have developed strategies to counteract PTGS suggests that they are also targets of PTGS, a hypothesis confirmed by the observation that *sgs* mutants, which are deficient for PTGS, are hypersensitive to CMV (C Beclin & H Vaucheret, unpublished data). These results suggest that plants use PTGS as a strategy to combat viruses, and that viruses have more or less succeeded in escaping this defense: Poty- and cucumoviruses are able to knock-out PTGS; gemini-, potex-, and tobamoviruses are able to infect plants although they activate PTGS; caulimo-, nepo-, and tobraviruses are still targeted by PTGS in some species.

Why do sense transgenes trigger PTGS in the absence of viruses? Many characteristics of virus-induced PTGS (VIGS) are shared with sense transgene-mediated PTGS. Sense transgene loci that trigger PTGS likely produce an aberrant form of RNA that resembles the type of viral RNA that activates recovery of the plant from infection. This RNA is subsequently targeted for degradation (1, 13, 77, 78). The mechanistic resemblance may be related to secondary structure, cellular compartmentalization, and/or affinity for cellular components (such as RdRp), and may lead to recognition by the cellular machinery that targets this type of RNA for degradation. The characterization of the whole process of recognition and degradation will require characterization of the function of proteins encoded by genes in which mutation confers either impairment of PTGS (*SGS* genes and others to be identified) or virus resistance (to be identified).

Inhibition of gene expression by antisense RNA or simultaneous expression of sense and antisense RNA seems not to rely on exactly the same mechanism as virus- or sense transgene-induced PTGS since antisense inhibition occurs efficiently in sgs mutants (C Beclin & H Vaucheret, unpublished data). However, some steps might be common to these processes and could be revealed by identifying and characterizing mutants impaired in antisense inhibition, as well as mutants impaired in both antisense inhibition and PTGS (if any).

GENERAL CONCLUSION: How Many Mechanisms of Gene Silencing?

As concluded in the TGS and PTGS sections, (trans)gene silencing cannot be explained by a single mechanism. Rather, multiple mechanisms involving DNA-DNA, DNA-RNA, or RNA-RNA interactions (55) may be evoked (Figure 1, see color plate). Nevertheless, there may well be common steps between these different mechanisms. Interestingly, a complex transgene locus that undergoes TGS triggers both TGS of promoter-homologous target transgenes and PTGS of coding sequence-homologous target (trans)genes (93). Similarly, a virus that undergoes RNA degradation during the PTGS-like process of recovery was shown to trigger either TGS or PTGS of homologous transgenes, depending on whether they share homology within their promoter or the coding sequence (1).

These two specific cases clearly demonstrate that both TGS and PTGS events affecting (trans)genes can be triggered as alternative (although not exclusive) responses to two important pathological conditions that plants have to face, i.e. the stable integration of additional pieces of DNA into chromosomes, and the extrachromosomal replication of a viral genome. Additional pieces of DNA can be added to chromosomes owing to the movement of transposable elements (TEs) or to the integration of (part of) the genome of pathogens like *A. tumefaciens*. These processes must be tightly regulated to avoid deleterious effects. Both TEs and T-DNA insertions can contribute to increasing the size of the genome, can deregulate the expression of neighboring endogenous genes, and could cause chromosomal rearrangements through recombination between homologous ectopic sequences. The extrachromosomal replication of a viral genome must also be regulated because viruses use the cellular machinery to their own advantage, thus limiting the availability of enzymes and subsequently of metabolites for growth.

Epigenetic silencing of plant transgenes may therefore reflect diverse cellular defense responses (56). TGS, which results from the impairment of transcription initiation by methylation and/or chromatin condensation, could derive from the mechanism by which additional pieces of DNA (TEs, T-DNA) are tamed by the genome. PTGS, which results from RNA degradation, could derive from the process of recovery by which cells eliminate undesirable pathogens (RNA viruses) or their undesirable products (RNA encoded by DNA viruses).

TGS is therefore expected to occur when transgenes insert near or within endogenous *cis*-acting silencing elements like heterochromatin, repeated and methylated elements (74, 85), or when they disorganize chromatin structure locally owing to a drastically different GC content (37, 60, 87) or the formation of secondary structures by *cis* DNA-DNA interactions between transgene repeats (4, 100). Transgenes would therefore undergo an epigenetic change (involving methylation and/or chromatin condensation) that impedes the initiation of transcription. Active transgenes could also be subjected to TGS when they are brought into the presence of promoter-homologous *trans*-acting silencing elements that may impose an epigenetic change and impedes the initiation of transcription. These *trans*-acting elements could be allelic or ectopic (trans)genes already subjected to TGS if their epigenetic silent state is transferred through direct or protein-mediated DNA-DNA interactions (52, 54, 60, 86). They could also be viral RNA (1) or (aberrant) RNA produced by transcribed transgenes that resemble viral RNA (58), and that are able to impose a transcriptionally repressive state on the homologous promoter sequences through DNA-RNA interactions.

On the other hand, PTGS is expected to occur when transgenes produce an aberrant form of RNA that mimics either viral RNA or viral RNA degradation products after infection. Transgene RNA would therefore be targeted for degradation, as is RNA from viruses inducing recovery (1, 13, 77, 78). Endogenous genes, transgenes, or viruses that are not themselves able to activate PTGS or recovery could also be subjected to PTGS when their RNA shares homology with the targeted RNA sequences of transgenes that induce PTGS (25, 34, 68, 99), or of viruses that induce a PTGS-like response (1, 77, 78). Surprisingly, endogenous genes can also be subjected to PTGS when brought into the presence of TG-Silenced transgenes that could transfer their epigenetic silent state through DNA-DNA interactions. The newly imposed epigenetic state would fail to inhibit transcription initiation because of the absence of homology within the promoter region, but would impair the regular transcription of mRNA and thus lead to degradation (84, 93).

Since PTGS is a mechanism leading to the degradation of viral RNA, it is not expected to involve any step at the DNA level (78). However, the fact that not all transgenes induce PTGS probably means that not all produce aberrant RNA, or at least not in sufficient quantities. PTGS mediated by sense transgenes most likely involves an additional step or steps at the DNA level compared to PTGS mediated by viruses. The production of the aberrant form of RNA could depend on the ability of a transgene locus to undergo readthrough transcription, transcription from a cryptic promoter, premature termination, and/or unintended production of antisense RNA. Alone, or in combination with regular mRNA, these types of molecules could therefore activate PTGS, as does viral RNA. In some cases, the plant RdRp enzyme (11) could be required to amplify these molecules in order to reach a threshold level of aberrant molecules capable of activating PTGS.

Whether the production of aberrant RNA relies only on the primary structure of DNA, i.e. the arrangement of transgene copies within the genome, or also depends on epigenetic changes is unclear. Changes in the methylation state of PTG-Silenced transgenes have been observed (21, 25, 38, 41, 84), but whether as a cause or a consequence of PTGS is not known. Introgression of PTG-Silenced transgenes into the Arabidopsis *ddm1* and *ddm2* mutants (or in plants expressing antisense *MET1* RNA will be critical in determining whether methylation (28) and/or chromatin remodeling (39, 40) play a role in PTGS. If an effect of *ddm1* and/or *ddm2* on PTGS were found, the hypothesis that epigenetic changes affecting transgenes play an active role in the triggering and/or the maintenance of PTGS would be confirmed. The role of such changes was already suggested by the requirement for the presence of a transgene to maintain grafting-induced PTGS in plants (72) and to degrade viral RNA in VIG-Silenced plants (46, 79). Only epigenetic changes (such as methylation) occurring through interactions between aberrant/viral RNA and the corresponding transgene DNA (92, 97, 98) could explain the maintenance of RNA degradation after the initial source of silencing (virus or PTG-Silenced rootstock) has been eliminated (72, 79). Similarly, only DNA-DNA interactions allowing transmission of an epigenetic silent state from TG-Silenced transgenes to homologous endogenous genes could explain the impairment of regular transcription and the subsequent degradation of endogenous RNA (93).

As mentioned throughout this review, we do not yet have enough information to understand the mechanisms of gene silencing in plants. The identification of viruses that are targets or sources of TGS and/or PTGS, and of Arabidopsis mutants impaired in TGS and/or PTGS will help to classify silencing events on a genetic basis, and determine how many mechanisms exist and the steps common to the different silencing pathways.

ACKNOWLEDGMENTS

We thank Judith Bender, Ian Furner, Rich Jorgensen, Jan Kooter, Peter Meyer and Eric Richards for communicating unpublished results. We also thank our colleagues from the lab and colleagues of the European Network on Gene Silencing for fruitful discussion.

LITERATURE CITED

1. Al-Kaff NS, Covey SN, Kreike MM, Page AM, Pinder R, Dale PJ. 1998. Transcriptional and posttranscriptional plant gene silencing in response to a pathogen. *Science* 279:2113–15
2. Anandalakshmi R, Pruss GJ, Ge X, Marathe R, Mallory AC, et al.1998. A viral suppressor of gene silencing in plants. *Proc. Natl. Acad. Sci. USA* 95:13079–84
3. Angell SM, Baulcombe DC. 1997. Consistent gene silencing in transgenic plants expressing a replicating potato virus X RNA. *EMBO J.* 16:3675–84
4. Assaad FF, Tucker KL, Signer ER. 1993. Epigenetic repeat-induced gene silencing (RIGS) in arabidopsis. *Plant Mol. Biol.* 22:1067–85
5. Baulcombe DC. 1996. Mechanisms of pathogen-derived resistance to viruses in transgenic plants. *Plant Cell* 8:1833–44
6. Baulcombe DC. 1999. Fast forward genetics based on virus-induced gene silencing. *Curr. Opin. Plant Biol.* 2:109–13
7. Beclin C, Berthome R, Palauqui JC, Tepfer M, Vaucheret H. 1998. Infection of tobacco or Arabidopsis plants by CMV counteracts systemic post-transcriptional silencing of nonviral (trans)genes. *Virology* 252:313–17
8. Bender J, Fink GR. 1995. Epigenetic control of an endogenous gene family is revealed by a novel blue fluorescent mutant of Arabidopsis. *Cell* 83:725–34
9. Bevan M, Bancroft I, Bent E, Love K,

Goodman H, et al. 1998. Analysis of 1.9 Mb of contiguous sequence from chromosome 4 of *Arabidopsis thaliana*. *Nature* 391:485–88

10. Brigneti G, Voinnet O, Li W-X, Ji L-H, Ding S-W, Baulcombe DC. 1998. Viral pathogenicity determinants are suppressors of transgene silencing in *Nicotiana benthamiana*. *EMBO J.* 17:6739–46
11. Cogoni C, Macino G. 1999. Gene silencing in *Neurospora crassa* requires a protein homologous to RNA-dependent RNA polymerase. *Nature* 399:166–69
12. Cogoni C, Macino G. 1997. Isolation of quelling-defective (*qde*) mutants impaired in posttranscriptional transgene-induced gene silencing in *Neurospora crassa*. *Proc. Natl. Acad. Sci. USA* 94:10233–38
13. Covey SN, Al-Kaff NS, Langara A, Turner DS. 1997. Plants combat infection by gene silencing. *Nature* 387:781–82
14. Davies GJ, Sheikh MA, Ratcliffe OJ, Coupland G, Furner IJ. 1997. Genetics of homology-dependent gene silencing in Arabidopsis; a role for methylation. *Plant J.* 12:791–804
15. de Carvalho F, Gheysen G, Kushnir S, Van Montagu M, Inzé D, Castresana C. 1992. Suppression of b-1,3-glucanase transgene expression in homozygous plants. *EMBO J.* 11:2595–602
16. Dehio C, Schell J. 1994. Identification of plant genetic loci involved in a posttranscriptionnal mechanism for meiotically reversible transgene silencing. *Proc. Natl. Acad. Sci. USA* 91:5538–42
17. Depicker A, Van Montagu M. 1997. Posttranscriptional gene silencing in plants. *Curr. Opin. Cell Biol.* 9:373–82
18. Diéguez MJ, Vaucheret H, Paszkowski J, Mittelsten Scheid O. 1998. Cytosine methylation at CG and CNG sites is not a prerequisite for the initiation of transcriptional silencing in plants, but is required for its maintenance. *Mol. Gen. Genet.* 259:207–15
19. Dougherty WG, Lindbo JA, Smith HA, Parks TD, Swaney S, Proebsting WM. 1994. RNA-mediated virus resistance in transgenic plants: exploitation of a cellular pathway possibly involved in RNA degradation. *Mol. Plant Microbe Interact.* 7:544–52
20. Dougherty WG, Parks TD. 1995. Transgenes and gene suppression: telling us something new? *Curr. Opin. Cell Biol.* 7:399–405
21. Elmayan T, Balzergue S, Béon F, Bourdon V, Daubremet J, et al. 1998. Arabidopsis mutants impaired in cosuppression. *Plant Cell* 10:1447–57
22. Elmayan T, Vaucheret H. 1996. Expression of single copies of a strongly expressed 35S transgene can be silenced posttranscriptionally. *Plant J.* 9:787–97
23. Elomaa P, Helariutta Y, Griesbach RJ, Kotilainen M, Seppänen P, Teeri TH. 1995. Transgene inactivation in *Petunia hybrida* is influenced by the properties of the foreign gene. *Mol. Gen. Genet.* 248:649–56
24. English JJ, Davenport GF, Elmayan T, Vaucheret H, Baulcombe DC. 1997. Requirement of sense transcription for homology-dependent virus resistance and transinactivation. *Plant J.* 12:597–603
25. English JJ, Mueller E, Baulcombe DC. 1996. Suppression of virus accumulation in transgenic plants exhibiting silencing of nuclear genes. *Plant Cell* 8:179–88
26. Finnegan EJ, Genger RK, Peacock WJ, Dennis ES. 1998. DNA methylation in plants. *Annu. Rev. Plant Physiol. Plant Mol. Biol.* 49:223–47
27. Finnegan EJ, McElroy D. 1994. Transgene inactivation: Plants fight back! *BioTechnology* 12:883–88
28. Finnegan EJ, Peacock WJ, Dennis ES. 1996. Reduced DNA methylation in *Arabidopsis thaliana* results in abnormal plant development. *Proc. Natl. Acad. Sci. USA* 93:8449–54
29. Flavell RB. 1994. Inactivation of gene expression in plants as a consequence of

specific sequence duplication. *Proc. Natl. Acad. Sci. USA* 1:3490–96

30. Fraser RSS. 1990. The genetics of plant-virus interaction: mechanisms controlling host range, resistance and virulence. In *Recognition and Response in Plant-Virus Interactions*, ed. RSS Fraser, pp. 71–93. Berlin: Springer-Verlag
31. Furner IJ, Sheikh MA, Collett CE. 1998. Gene silencing and homology-dependent gene silencing in Arabidopsis: genetic modifiers and DNA methylation. *Genetics* 149:651–62
32. Grandbastien M-A. 1998. Activation of plant retrotransposons under stress conditions. *Trends Plant Sci.* 3:181–87
33. Grierson D, Fray RG, Hamilton AJ, Smith CJS, Watson CF. 1991. Does co-suppression of sense genes in transgenic plants involve antisense RNA? *Trends Biotechnol.* 9:122–23

33a. Hamilton A, Baulcombe D. 1999. A species of small antisense RNA in posttranscriptional gene ssilencing in plants. *Science* 286:950–52

34. Hamilton AJ, Brown S, Yuanhai H, Ishizuka M, Lowe A, et al.1998. A transgene with repeated DNA causes high frequency, post-transcriptional suppression of ACC-oxidase gene expression in tomato. *Plant J.* 15:737–46
35. Hohn T, Corsten S, Rieke S, Muller M, Rothnie H. 1996. Methylation of coding region alone inhibits gene expression in plant protoplasts. *Proc. Natl. Acad. Sci. USA* 93:8834–39
36. Holtorf H, Schob H, Kunz C, Waldvogel R, Meins F Jr. 1999. Stochastic and nonstochastic post-transcriptional silencing of chitinase and beta-1,3-glucanase genes involves increased RNA turnover—possible role for ribosome-independent RNA degradation. *Plant Cell* 11:471–84
37. Iglesias VA, Moscone EA, Papp I, Neuhuber F, Michalowski S, et al. 1997. Molecular and cytogenetic analyses of stably and unstably expressed transgene loci in tobacco. *Plant Cell* 9:1251–64
38. Ingelbrecht I, Van Houdt H, Van Montagu M, Depicker A. 1994. Posttranscriptional silencing of reporter transgenes in tobacco correlates with DNA methylation. *Proc. Natl. Acad. Sci. USA* 91:10502–6
39. Jeddeloh JA, Bender J, Richards EJ. 1998. The DNA methylation locus DDM1 is required for maintenance of gene silencing in Arabidopsis. *Genes Dev.* 12:1714–25
40. Jeddeloh JA, Stokes TL, Richards EJ. 1999. Maintenance of genomic methylation requires a SWI2/SNF2-like protein. *Nat. Genet.* 22:94–97
41. Jones AL, Thomas CL, Maule AJ. 1998. De novo methylation and co-suppression induced by a cytoplasmically replication plant RNA virus. *EMBO J.* 17:6385–93
42. Jorgensen RA, Cluster P, Que Q, English J, Napoli C. 1996. Chalcone synthase co-suppression phenotypes in Petunia flowers: comparison of sense vs. antisense constructs and single copy vs. complex T-DNA sequences. *Plant Mol. Biol.* 31:957–73
43. Kass SU, Pruss D, Wolffe AP. 1997. How does DNA methylation repress transcription? *Trends Genet.* 13:444–49
44. Kasschau KD, Carrington JC. 1998. A counterdefensive strategy of plant viruses: suppression of posttranscriptional gene silencing. *Cell* 95:461–70
45. Kennerdell JR, Carthew RW. 1998. Use of dsRNA-mediated genetic interference to demonstrate that frizzled and frizzled 2 act in the wingless pathway. *Cell* 95:1017–26
46. Kjemtrup S, Sampson KS, Peele CG, Nguyen LV, Conkling MA, et al.1998. Gene silencing from plant DNA carried by a geminivirus. *Plant J.* 14:91–100
47. Kumagai MH, Donson J, Della-Cioppa G, Harvey D, Hanley K, Grille LK. 1995. Cytoplasmic inhibition of carotenoid biosynthesis with virus-derived RNA. *Proc. Natl. Acad. Sci. USA* 92:1679–83

48. Kumpatla SP, Chandrasekharan MB, Iyer LM, Li G, Hall TC. 1998. Genome intruder scanning and modulation systems and transgene silencing. *Trends Plant Sci.* 3:97–104
49. Kumpatla SP, Teng W, Buchholz WG, Hall TC. 1997. Epigenetic transcriptional silencing and 5-azacytidine-mediated reactivation of a complex transgene in rice. *Plant Physiol.* 115:361–73
50. Lindbo JA, Silva-Rosales L, Proebsting WM, Dougherty WG. 1993. Induction of highly specific antiviral state in transgenic plants: implications for regulation of gene expression and virus resistance. *Plant Cell* 5:1749–59
51. Linn F, Heidmann I, Saedler H, Meyer P. 1990. Epigenetic changes in the expression of the maize A1 gene in *Petunia hybrida*: role of numbers of integrated copies and state of methylation. *Mol. Gen. Genet.* 222:329–36
52. Luff B, Pawlowski L, Bender J. 1999. An inverted repeat triggers cytosine methylation of identical sequences in Arabidopsis. *Mol. Cell* 3:505–11
53. Martienssen R. 1998. Transposons, DNA methylation and gene control. *Trends Genet.* 14:263–64
54. Matzke AJM, Neuhuber F, Park YD, Ambros PF, Matzke MA. 1994. Homology-dependent gene silencing in transgenic plants: epistatic silencing loci contain multiple copies of methylated transgenes. *Mol. Gen. Genet.* 244:219–29
55. Matzke MA, Matzke AJM. 1995. How and why do plants inactivate homologous (trans)genes? *Plant Physiol.* 107:679–85
56. Matzke MA, Matzke AJM. 1998. Epigenetic silencing of plant transgenes as a consequence of diverse cellular defense responses. *Cell. Mol. Life Sci.* 54:94–103
57. Matzke MA, Primig M, Trnovsky J, Matzke AJM. 1989. Reversible methylation and inactivation of marker genes in sequentially transformed tobacco plants. *EMBO J.* 8:643–49
58. Mette MF, van der Winden J, Matzke MA, Matzke AJ. 1999. Production of aberrant promoter transcripts contributes to methylation and silencing of unlinked homologous promoters in trans. *EMBO J.* 18:241–48
59. Metzlaff M, O'Dell M, Cluster PD, Flavell RB. 1997. RNA-mediated RNA degradation and chalcone synthase A silencing in Petunia. *Cell* 88:845–54
60. Meyer P, Heidmann I, Niedenhof I. 1993. Differences in DNA-methylation are associated with a paramutation phenomenon in transgenic petunia. *Plant J.* 4:89–100
61. Meyer P, Saedler H. 1996. Homology-dependent gene silencing in plants. *Annu. Rev. Plant Physiol. Plant Mol. Biol.* 47:23–48
62. Mittelsten Scheid O, Afsar K, Paszkowski J. 1994. Gene inactivation in *Arabidopsis thaliana* is not accompanied by an accumulation of repeat-induced point mutations. *Mol. Gen. Genet.* 244:325–30
63. Mittelsten Scheid O, Afsar K, Paszkowski J. 1998. Release of epigenetic gene silencing by trans-acting mutations in Arabidopsis. *Proc. Natl. Acad. Sci. USA* 95:632–37
64. Mittelsten Scheid O, Paszkowski J, Potrikus I. 1991. Reversible inactivation of a transgene in *Arabidopsis thaliana*. *Mol. Gen. Genet.* 228:104–12
65. Montgomery MK, Xu S, Fire A. 1998. RNA as a target of double-stranded RNA-mediated genetic interference in Caenorhabditis elegans. *Proc. Natl. Acad. Sci. USA* 95:15502–7
66. Morel JB, Dangl JL. 1997. The hypersenssitive response and the induction of cell death in plants. *Cell Death Differ.* 4:671–83
67. Morino K, Olsen OA, Shimamoto K. 1999. Silencing of an aleurone-specific gene in transgenic rice is caused by a rearranged transgene. *Plant J.* 17:275–85
68. Napoli C, Lemieux C, Jorgensen R. 1990. Introduction of a chimeric chalcone synthase gene into petunia results in reversible

co-suppression of homologous gene in trans. *Plant Cell* 2:279–89

69. Ngo H, Tschudi C, Gull K, Ullu E. 1998. Double-stranded RNA induces mRNA degradation in *Trypanosoma brucei*. *Proc. Natl. Acad. Sci. USA* 95:14687–92
70. Palauqui J-C, Elmayan T, Pollien J-M, Vaucheret H. 1997. Systemic acquired silencing: Transgene specific post-transcriptional silencing is transmitted by grafting from silenced stocks to non-silenced scions. *EMBO J.* 16:4738–45
71. Palauqui J-C, Vaucheret H. 1995. Field trial analysis of nitrate reductase: a comparative study of 38 combinations of transgene loci. *Plant Mol. Biol.* 29:149–59
72. Palauqui J-C, Vaucheret H. 1998. Transgenes are dispensable for the RNA degradation step of cosuppression. *Proc. Natl. Acad. Sci. USA* 95:9675–80
73. Park YD, Papp I, Moscone EA, Iglesias VA, Vaucheret H, et al.1996. Gene silencing mediated by promoter homology occurs at the level of transcription and results in meiotically heritable alterations in methylation and gene activity. *Plant J.* 9:183–94
74. Pröls F, Meyer P. 1992. The methylation patterns of chromosomal integration regions influence gene activity of transferred DNA in *Petunia hybrida*. *Plant J.* 2:465–75
75. Que Q, Wang H-Y, English JJ, Jorgensen RA. 1997. The frequency and degree of co-suppression by sense chalcone synthase transgenes are dependent on transgene promoter strength and are reduced by premature nonsense codons in the transgene coding sequence. *Plant Cell* 9:1357–68
76. Que Q, Wang HY, Jorgensen RA. 1998. Distinct patterns of pigment suppression are produced by allelic sense and antisense chalcone synthase transgenes in petunia flowers. *Plant J.* 13:401–9
77. Ratcliff FG, Harrison BD, Baulcombe DC. 1997. A similarity between viral defense and gene silencing in plants. *Science* 276:1558–60
78. Ratcliff FG, MacFarlane SA, Baulcombe DC. 1999. Gene silencing without DNA: RNA-mediated cross-protection between viruses. *Plant Cell* 11:1207–15
79. Ruiz MT, Voinnet O, Baulcombe DC. 1998. Initiation and maintenance of virus-induced gene silencing. *Plant Cell* 10:937–46
80. SanMiguel P, Tikhonov A, Jin YK, Motchoulskaia N, Zakharov D, et al. 1996. Nested retrotransposons in the intergenic regions of the maize genome. *Science* 274:765–68
81. Sijen T, Wellink J, Hiriart JB, Van Kammen A. 1996. RNA-mediated virus resistance: role of repeated transgenes and delineation of targeted regions. *Plant Cell* 8:2277–94
82. Smith CJS, Watson CF, Bird CR, Ray J, Schuch W, Grierson D. 1990. Expression of a truncated tomato polygalacturonase gene inhibits expression of the endogenous gene in transgenic plants. *Mol. Gen. Genet.* 224:477–81
83. Stam M, Mol JNM, Kooter JM. 1997. The silence of genes in transgenic plants. *Ann. Bot.* 79:3–12
84. Stam M, Viterbo A, Mol JN, Kooter JM. 1998. Position-dependent methylation and transcriptional silencing of transgenes in inverted T-DNA repeats: implications for posttranscriptional silencing of homologous host genes in plants. *Mol. Cell. Biol.* 18:6165–77
85. ten Lohuis M, Muller A, Heidmann I, Niedenhof I, Meyer P. 1995. A repetitive DNA fragment carrying a hot spot for de novo DNA methylation enhances expression variegation in tobacco and petunia. *Plant J.* 8:919–32
86. Thierry D, Vaucheret H. 1996. Sequence homology requirements for transcriptional silencing of 35S transgenes and post-transcriptional silencing of nitrite reductase (trans)genes by the tobacco 271 locus. *Plant Mol. Biol.* 32:1075–83
87. van Blokland R, ten Lohuis M, Meyer P. 1997. Condensation of chromatin in

transcriptional regions of an inactivated plant transgene: evidence for an active role of transcription in gene silencing. *Mol. Gen. Genet.* 257:1–13

88. van Blokland R, van der Geest N, Mol JNM, Kooter JM. 1994. Transgene-mediated suppression of chalcone synthase expression in *Petunia hybrida* results from an increase in RNA turnover. *Plant J.* 6:861–77
89. van der Krol AR, Mur LA, Beld M, Mol JNM, Stuitje AR. 1990. Flavonoid genes in petunia: addition of a limited number of genes copies may lead to a suppression of gene expression. *Plant Cell* 2:291–99
90. van Eldik GJ, Litiere K, Jacobs JJ, Van Montagu M, Cornelissen M. 1998. Silencing of beta-1,3-glucanase genes in tobacco correlates with an increased abundance of RNA degradation intermediates. *Nucleic Acids Res.* 26:5176–81
91. Vaucheret H. 1993. Identification of a general silencer for 19S and 35S promoters in a transgenic tobacco plant: 90 pb of homology in the promoter sequences are sufficient for trans-inactivation. *C. R. Acad. Sci.* 316:1471–83
92. Vaucheret H, Beclin C, Elmayan T, Feuerbach F, Godon C, et al.1998. Transgene-induced gene silencing in plants. *Plant J.* 16:651–59
93. Vaucheret H, Elmayan T, Mourrain P, Palauqui J-C. 1996. Analysis of a tobacco transgene locus that triggers both transcriptional and post-transcriptional silencing. *Epigenetic Mechanisms of Gene Regulation*, ed. VE Russo, RA Martienssen, AD Riggs, pp. 403–14. Cold Spring Harbor, NY: Cold Spring Harbor Lab. Press
94. Vaucheret H, Nussaume L, Palauqui J-C, Quilleré I, Elmayan T. 1997. A transcriptionally active state is required for post-transcriptional silencing (co-suppression) of nitrate reductase host genes and transgenes. *Plant Cell* 9:1495–504
95. Voinnet O, Baulcombe DC. 1997. Systemic signalling in gene silencing. *Nature* 389:553
96. Voinnet O, Vain P, Angell S, Baulcombe DC. 1998. Systemic spread of sequence-specific transgene RNA degradation in plants is initiated by localized introduction of ectopic promoterless DNA. *Cell* 95:177–87
97. Wassenegger M, Heimes S, Riedel L, Sänger HL. 1994. RNA-directed de novo methylation of genomic sequences in plants. *Cell* 76:567–76
98. Wassenegger M, Pélissier T. 1998. A model for RNA-mediated gene silencing in higher plants. *Plant Mol. Biol.* 37:349–62
99. Waterhouse PM, Graham MW, Wang M-B. 1998. Virus resistance and gene silencing in plants can be induced by simultaneous expression of sense and antisense RNA. *Proc. Natl. Acad. Sci. USA* 95:13959–64
100. Ye F, Signer ER. 1996. RIGS (repeat-induced gene silencing) in *Arabidopsis* is transcriptional and alters chromatin configuration. *Proc. Natl. Sci. Acad. USA* 93:10881–86

Annu. Rev. Plant Physiol. Plant Mol. Biol. 2000. 51:195–222

CEREAL CHROMOSOME STRUCTURE, EVOLUTION, AND PAIRING

Graham Moore
John Innes Centre, Colney, Norwich, United Kingdom; e-mail: tracie.foote@bbsrc.ac.uk

Key Words polyploidy, centromeres, telomeres, *Ph1* locus, gene order

■ **Abstract** The determination of the order of genes along cereal chromosomes indicates that the cereals can be described as a single genetic system. Such a framework provides an opportunity to combine data generated from the studies on different cereals, enables chromosome evolution to be traced, and sheds light on key structures involved in cereal chromosome pairing. Centromeric and telomeric regions have been highlighted as important in these processes.

CONTENTS

INTRODUCTION

More than 50% of angiosperms are polyploids that occur either by multiplication of a basic set of chromosomes (autopolyploidy) or as a result of combining two parental genomes (allopolyploidy). The introduction of alien variation into

1040-2519/00/0601-0195$14.00

polyploids during plant breeding clearly would benefit from an understanding of the genome relationships of polyploid crop species and their wild diploid relatives. As a result, the genomic relationships within the *Triticeae* (wheat and its wild relatives) have been extensively studied (110). By 1952, the work of Kihara and colleagues had established many of the genomic relationships of the diploid and polyploid *Triticum* and *Aegilops* species on the basis of chromosome pairing in hybrids (110). These studies have continued to the present and have accurately classified chromosome translocations (92, 115, 135, 136). Chromosome pairing is assessed by squashing meiocytes at metaphase I and classifying the specific chromosome configurations at this stage. The association of chromosomes implies some conservation of chromosomal structure (including genes) and suggests a high degree of conservation of gene order on the chromosomes of wheat and its wild relatives. Clearly, such studies could not be extended beyond species from which hybrids could be generated and in which chromosomes paired. Thus for much of the past 50 years, major cereals and their wild relatives (maize, wheat, rice, sorghum, and the millets) have been studied largely in isolation. Detection of variation between individuals at the DNA sequence level in the form of restriction fragment length polymorphisms (RFLPs) (97, 107) and the concept of an RFLP linkage map (158) enabled these comparisons to be extended and the actual structure of the *Triticeae* chromosomes to be assessed in more detail. The analysis of these comparisons revealed a framework by which all data generated on the cereals can be collated (127, 130, 132). The cereal genomes can be described by a series of conserved units based on the linkage of genes found within their genomes (130). The structure of the rice genome is pivotal to the analysis of other cereal genomes (131, 132).

GENOME ORGANIZATION

Genome Size

Although mammals possess different chromosomes, their genome sizes are similar. In contrast, cereals have different chromosome numbers and vary greatly in genome size (21). It was the size of barley, wheat, and rye genomes that initially restricted their molecular analysis. The genomes of barley (*Hordeum vulgare*) and hexaploid bread wheat (*Triticum aestivum*) are relatively large (5×10^9 bp and 1.7×10^{10} bp per haploid nucleus, respectively). Maize (*Zea mays*) is intermediate in size 2.4×10^9 bp, and sorghum (*Sorghum bicolor*) is 8×10^8 bp (21). In contrast, other cereals and wild grass genomes are relatively small; for example, those of rice (*Oryza sativa*) and slender false-brome (*Brachypodium sylvaticum*) are 4×10^8 bp and 5×10^8 bp in size, respectively (21). Clearly, major changes have occurred in genome size since their speciation from a common ancestor. Most of this additional DNA consists of repetitive sequences that evolve rapidly and hence diverge substantially with speciation (65, 66).

Base Composition

Nuclear DNA can be heavily methylated at cytosine residues. The base composition of genomic regions can therefore influence the distribution of methylation within them. Large mammalian genomes have a CpG content of 1%. This is lower than the 4% expected for a genome that is 40% G + C rich. Thus there is a marked underrepresentation of the CpG content in these genomes (26, 71, 148, 163). However, there are small regions, several hundred bases long, that have a CpG/GpC dinucleotide ratio of approximately one. Such regions (termed CpG islands) coincide with the promoters of genes (29). They are also marked by clusters of unmethylated CpG dinucleotides, which therefore contain recognition sequences for methylation-sensitive restriction enzymes. In mammalian genomes, methylation is largely confined to the sequence m5CpG. In mammalian DNA, 70% to 80% of CpG dinucleotides are methylated. More than 80% of CpG dinucleotides are also methylated in wheat. In plants, 5-methylcytosine is not confined to CpG dinucleotides but is also present in more than 80% of the trinucleotides CpXpGs (83). Therefore, in the nuclear DNA of wheat (and other higher plants), a higher proportion of the cytosine residues is methylated: 30% compared with 1% to 8% in vertebrates. A single study has assessed the dinucleotide composition of a cereal nuclear genome, namely *Triticum aestivum* (bread wheat) (163). The G + C content of its genome is 45%. In contrast to mammalian genomes, the genome of wheat exhibits only a slight reduction in CpG content (83, 148). Therefore, the observed/expected ratio of CpG content is 0.77 for bulk wheat genomic DNA compared with 0.2 for human genomic DNA, whereas the CpG/GpC ratio for wheat is 0.78 and for human DNA is 0.23. Because there is little suppression of the CpG dinucleotide in the wheat genome, both repetitive and single-copy sequences would be expected to exhibit similar CpG contents. However, sequence analysis of 60 Kb of the barley genome indicates that the promoter regions of the genes, as with mammalian genes, were marked by a CpG/GpC ratio of more than one (141).

Gene Distribution

Long-range mapping within a 4-Mb length of the wheat genome indicated five clusters of unmethylated CG-rich recognition sites for methylation-sensitive restriction enzymes defining active genes (36). The distribution of unmethylated sites between repetitive sequences and single-copy sequences is therefore not random (128). Also the distribution of unmethylated restriction sites along the chromosomes of the *Triticeae* is evidence of a reduced density in the proximal regions and a higher percentage in the distal regions (128). This implies that the gene distribution along a cereal chromosome is also not random, with a higher density of genes concentrated in the distal regions. Extensive physical mapping of genes with deletion stocks has been undertaken on wheat chromosomes by Gill and colleagues (62, 74). The data indicate that on all the wheat chromosomes, very few genes are located in the proximal third of the chromosome arm. They also reported that the genes located in the distal regions occur in clusters.

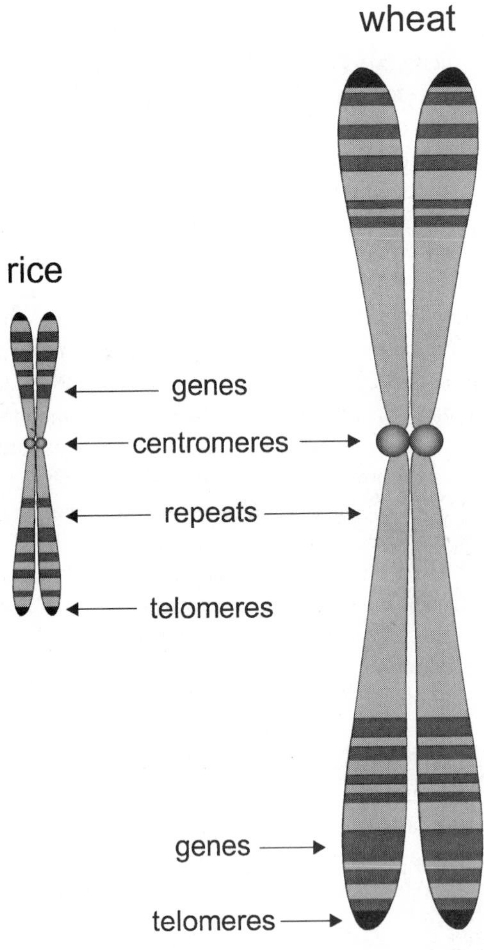

Figure 1 Schematic representation of metaphase chromosomes corresponding to rice chromosome 1 and a chromosome from wheat group 3. These chromosomes exhibit a similar gene content and extensive conservation in gene order. The rice chromosome has a more even distribution of clusters of gene-rich regions. The gene density in these gene-rich regions on the rice chromosome will be higher than in gene-rich regions in wheat. The proximal regions of wheat chromosomes are composed largely of repetitive blocks, with the distal regions composed of both clusters of gene-rich regions and repetitive blocks.

Although the sequence analysis of the 60-Kb region of the barley genome indicated a fivefold lower gene density than in *Arabidopsis*, the density was still six- to tenfold higher than expected from an equidistant gene distribution in the complex barley genome (141). This is consistent with the observations of Gill and colleagues. A schematic representation of the organization of a wheat chromosome compared to a rice chromosome is shown in Figure 1. If few genes are located in the proximal third of the chromosome arm, what is the nature of these regions? Analysis of repetitive sequence distribution indicated that they could be composed of blocks of amplified repetitive units that on restriction digestion reduce to discrete fragments of 200 Kb and upwards (24, 128). The rice genome, however, does not possess such large blocks of repetitive sequences. The proximal regions composed of large blocks of amplified repetitive sequences not only have reduced gene densities but exhibit a reduced level of recombination. Therefore, no direct correlation exists between physical and genetic distances in species such as wheat and barley (111, 157).

Dispersed Repetitive Sequences

The majority of the large genomes of cereals consist of dispersed repeats (65, 66). The most abundant elements are the retrotransposons (177). Retrotransposons can be divided into two classes, those flanked by long terminal repeats (LTRs) and those without LTRs, which terminate with poly (A) sequences. The non-LTR retrotransposons are also called LINES elements and possess an ORF that codes for proteins with homology to reverse transcriptase. The first such elements to be identified in plants were the *cin4* from maize (153). LINES elements can be found in great abundance, for example, *del2* identified in lily (108). Recent studies have suggested that they are ubiquitous in plant genomes (137). LTR retrotransposons have been classified according to Drosophila-type elements as either gypsy-like or copia-like, based on the order of the coding proteins between the LTRs. Gypsy elements, like retroviruses, possess LTR-gag-proteinase-rt-in-LTR and copia elements are characterized by a LTR-gag-proteinase-in-rt-LTR arrangement. Gypsy and copia elements are also ubiquitous in cereal genomes and can be found in great abundance (67, 162). BIS-1 constitutes in excess of 5% of the barley genome and probably a similar proportion of the genomes of wheat and rye (129). BARE1 is not only highly abundant in barley genome (160), it is also active (161). Retrotransposons constitute at least 50% of the maize genome (150). They are found in the spaces between genes (151, 159). Retrotransposons have increased the size of the maize genome two- to fivefold since the divergence of maize and sorghum from a common ancestor about 16 mya (150). Comparison of the *sh1–a1* regions in the genomes of maize and sorghum indicated that the retrotransposons were absent from the corresponding region in the sorghum genome (35). The gene density of the gene-rich regions in small genomes, such as rice, will be higher than in larger genomes, such as wheat (68). Virtually all of the retrotransposons had inserted in the past 6 million years and most in the past 3 million years in the maize genome (150). Sequencing of the *Arabidopsis* genome reveals little intergenic interspersion of retrotransposons; most are associated with the pericentromeric regions (121). The implication for the evolution of the cereal genome is that the common ancestral genome probably possessed few retrotransposons interspersed among their genes but that the retrotransposons were located in the pericentromeric regions. Genome expansion results from the accumulation of retrotransposons in such regions. Note, however, that not only can genomes expand, they can also contract, although it is unclear how this is achieved (23).

Other transposable elements, such as Ac in maize, possess inverted terminal repeats and transpose by excision and reintegration. They are present in only tens to hundreds of copies per genome and therefore do not make a major contribution to the total DNA content of the genome. They have, however, received special attention due to their value in gene tagging (15). No active endogenous elements have yet been found from the *Triticeae* genomes. Another type of mobile inverted repeat element has been recently identified, MITE, that is associated with the genes of many cereals (30, 31).

Long Tandem Arrays

Long tandem arrays of essentially identical sequences are found in the subtelomeric (142, 171) and pericentromeric heterochromatin regions (91) and the array of ribosomal RNA genes within the nucleolus organizer regions (73). There is a close correlation between the site of C and N bands and the presence of long tandem arrays of short, repeating sequences in chromosomal DNA (66). In rye, the subtelomeric heterochromatin constitutes from 15% to 18% of the genome, 60% of which comprises four repetitive sequences (18). These can be organized in a head-to-head arrangement or in head-to-tail organization (170). Birchler and colleagues demonstrated that sequences in maize knob regions possess strong homology to a repeat family specific to the centric region of maize B chromosomes (4). Under certain conditions, these knobs can be activated to function as "neocentromeres" that form kinetochores during meiosis. The presence of several classes of tandem and nontandem repeats has been identified in the centromere regions. Two classes (CCS1 and pSau3A9) of conserved sequences are located at all cereal centromeric sites studied (maize, wheat, rice, oats, sorghum, barley, and rye) (8, 95). Recent studies indicate that both sequences are part of the same DNA element; CCS1 is the LTR region and pSau3A9 is the integrase region of a retrotransposon (5, 51, 124, 138, 144). The insertion of these retrotransposons into the satellite sequences present at the centromere regions creates a unique pattern and hence specificity to this region (5). Retrotransposons have also inserted into the satellite sequences present in the knob regions (6). Given the rapid evolution of tandem arrays, it is likely that retrotransposons inserting into such sequences are steadily being deleted. These families of retrotransposons must be continually active in order to provide further elements for reinsertion.

Telomeres are the termini of chromosomes and in most species are made up of a short tandem array of similar DNA repeat sequences (182). In plants, they are mostly composed of many tandemly repeated copies of basic oligonucleotides of the sequence (TTTAGGG)n characterized by the G-rich strand running in the 5′ to 3′ direction toward the end of the chromosome and the C-rich strand running toward the centromere (37, 154, 176). The conservation of the basic sequence has been used as a basis for cloning the repeat families adjacent to the telomeres in rice, barley, and wheat (11, 37, 99, 101, 147). The repetitive families identified from these studies have been used to develop RFLP markers with which to mark the ends of the linkage maps. Many of these repetitive families adjacent to the telomere termini are specific to these regions. Shorter tandem arrays also occur throughout the cereal genomes composed of reiterating units of tens of basepairs (minisatellites) or even shorter simpler repeats of several basepairs (microsatellites) (174). Markers derived from these loci are highly polymorphic and prove useful for genetic mapping and fingerprinting varieties (120). Particular classes of microsatellite markers are associated with LTRs of retrotransposons (145).

Gene Order

The advent of RFLP markers in the early 1980s resulted in the development of RFLP linkage maps for many of the major cereals, including wheat (48, 69), rice (34, 104), barley (82, 103), sorghum (38), and maize (32, 33). By the mid-1990s, these genetic maps were relatively dense and were generated with markers that detected genic regions. Thus, they began to provide an indication of the gene order along particular chromosomes. The genetic map of hexaploid wheat that emerged was a comparative map of its three constituent ancestral genomes (A, B, and D genomes) (48, 69). Many markers map colinearily across these genomes and are separated by similar recombination distances. The comparative analysis revealed that the progenitor genomes had undergone some rearrangements, particularly chromosomes 4A, 5A, and 7B. In the diploid A genome progenitor, translocations involving 5AL and 4AL occurred. On polyploidization, a translocation involving 4AL and 7BS occurred, followed by an inversion of 4AL (106, 112). This was consistent with the predicted translocations proposed from analyzing chromosome pairing (136).

The availability of such genetic maps for the cereals permitted genome comparisons made over the past 70 years to be extended beyond species in which chromosome pairing studies could be undertaken. Chromosome pairing studies for mammals were inappropriate, so studies on their comparative genomics began later with the advent of somatic cell hybrids. The field of comparative mapping has spawned terms such as linkage or synteny, conserved linkage, conserved synteny, homology segment, colinearity, and microsynteny. Scientists studying plants or mammals have used these terms in different contexts. The basic observation is that groups of genes located together in the ancestral genome are still located together in the genomes of species that arise from speciation of this ancestor. Clearly, during evolution, rearrangements can occur that disrupt the order of genes along a chromosome but maintain their linkage, or disrupt the order so that the genes are no longer even linked. The important issue is how much disruption has occurred: Can linkage of genes still be observed between distantly related species and can this be exploited? If the gene order observed in the small genome of rice was sufficiently conserved with that in larger cereal genomes, the rice genome would provide a tool for gene isolation strategies in the other cereals. The rice genome would be, in effect, "the wheat genome without the repetitive sequences." Moreover, comparisons of genome organization across the different cereal genomes and the solution for these comparisons would mean that the cereals could be thought of as a single genetic system (22, 132).

The initial comparisons using RFLP maps involved wheat, barley, and rye (47, 49), and maize and sorghum (122, 143, 178). Although there had been some major translocations in rye after its speciation from the progenitors of hexaploid wheat and barley, the gene order is essentially similar in these species. The comparison of related species, maize, and sorghum, which have speciated within the past 16 to 20 million years, also revealed a degree of conservation of gene order. Even comparisons of species that have been isolated by more than 60 million years (180), such as rice and wheat and rice and maize, indicated that genes have been maintained in a similar order despite gross differences in the genome size and chromosome number of these species (2, 3, 105). From these initial comparisons of gene order in the rice, maize, and wheat genomes, genes on the rice genetic map could be grouped into sets and the genetic maps of wheat and maize could be described by the same sets of genes (rice linkage segments) (131). Furthermore, this analysis could also be extended to include the sorghum, foxtail millet, and sugarcane genomes (127, 130). Thus, a series of sets of genes could be used to describe most of the major cereal genomes as shown in Figure 2 (see color plate). The collation of all the information generated from studies on the different cereals requires a common framework. A limited number of combinations of linkage segments were noted in the genomes of the various cereals studied. It was therefore possible to create a generalized genome structure using the rice linkage segments that, when cleaved in the case of each cereal at a different number of junctions between the linkage segments, produced structures that describe the gross chromosome structures found in the two sets of maize chromosomes, the wheat and barley chromosomes and the sorghum and millet chromosomes (70, 127, 130). The breakage and fusion of rice linkage segments to create the different cereal chromosomes are reminiscent of the chromosome evolution of the holocentric chromosomes, which are the closest relatives of the cereals (116, 139). The rushes and sedges possess holocentric/polycentric chromosomes with many sites along their chromosome length for microtubule attachment. Such chromosomes naturally fragment to create smaller viable chromosomes. In hybrids between parental lines carrying the original larger chromosomes and these small, fragmented chromosomes, the large and small chromosomes align during meiosis, which suggests that these species are very adept at rearranging their chromosomes.

A large number of more detailed comparative mapping studies have now been undertaken (50, 56–59, 84, 125, 134, 149, 166–168, 179, 183). In essence, these studies confirm the basic framework outlined previously (127, 130). The gene order of chromosome regions covering several megabases have also been compared and indicate a high level of conservation (25, 60, 68, 85, 100, 102). However, it is also apparent that genes can be "transposed" to other regions, that they have become duplicated or deleted, or that they have diverged significantly. The comparative analysis based on linkage segments provides a framework for gene order; however, there are imperfections in this framework. The more detailed analyses indicated that in maize, for example, there have been inversions of regions and some of the linkage segments can be subdivided further. Analysis of these translocations and other gross rearrangements will be helpful in further classifying the relationships between the cereals (98).

The Maize Genome

Maize and sorghum are members of the Andropogoneae tribe consisting of over 900 species. Of more than 500 species analyzed from this tribe, 90% of the species possess a chromosome number that is a multiple of five. A basic number of nine occurs sporadically throughout the tribe. Consistent with this, the ten chromosomes of maize could be divided into two sets of five chromosomes, based on the linkage segment analysis (86, 131) (Figure 2, see color plate). The two sets of five chromosomes possess a different arrangement of the linkage segments. The maize chromosomes are divided into a set of the largest and a set of the smallest chromosomes. Moreover, the chromosome arms of one set are all larger than the corresponding homoeologous arms in the other set (Figure 3, see color plate). The two ancestral genomes of maize diverged some 20 mya, and one of the genomes is more closely related to sorghum, which diverged some 16 mya (72). The allotetraploidization took place some 11 million years ago (72). If the structure of the ancestral progenitor genomes of maize were similar to other small genomes (*Arabidopsis* and rice) studied to date, most of the repetitive sequences would be localized in the pericentromeric regions with the genic regions containing few repetitive elements. Bennetzen and colleagues have indicated that most of the major expansion of maize genome has taken place in the past 3 million years and has been intergenic (150). Since tetraploidization, it is unlikely that there has been preferential expansion of one of the sets of chromosomes. Thus, one set of chromosomes must have already possessed larger chromosomes than the other set.

The segmental relationship of the two chromosome sets of maize in which homoeology relationships are confined to chromosome arms generates a "circular organizational relationship" based around the centromeres (Figure 3). Such an arrangement is distinctive but not unique to plants. It bears a striking similarity to the chromosome relationships of translocation heterozygotes observed in diploid *Oenothera* and *Rhoeo* species (44). Their homologous chromosomes can be ordered into two sets, one structured as A○B, C○D, E○F (○ being the centromere) and the other set as F○A, B○C, D○E. At meiosis the homologous chromosomes will pair as a ring or a chain. Some *Oenothera* species possess two sets of, for example, A○B, C○D, E○F chromosomes and will pair as homologous chromosome pairs (bivalents). F1 progeny of two such ring-pairing species can occur and can undergo spontaneous amphidiploidy (45). Some 60 years ago, prior to any knowledge of its genome structure, cytogenetists considered the possibility of maize being a translocation heterozygote (28). Thus maize could be the amphidiploid of a F1 hybrid between two related ring-pairing (translocation heterozygotes) species. Our knowledge of the structure of the maize genome does not rule out this possibility. This event would have diploidized the two differently structured haploid genomes that were originally capable of recombining. McClintock and others reported nonhomologous chromosome pairing in maize (75, 76, 119). From our knowledge of the structure of the two sets of five chromosomes, this nonhomologous pairing can be reinterpreted as pairing between homoeologous segments derived from the two ancestral genomes. It is also consistent with the observations that there has been shuffling of genes between the homoeologous segments (72). Gaut & Doebley also observed from the sequence divergence of pairs of genes mapping on homoeologous segments that the genes fell into two classes, which is consistent with the hypothesis that they were derived originally from two diverged genomes (72). Translocation heterozygotes have been observed to have an unusual behavior of their centromeric heterochromatin regions during prophase. During early meiotic prophase in most species, the telomeric heterochromatin aggregates together into a bouquet that results in the intimate alignment (synapsis) of chromosomes. In translocation heterozygotes, the centromere heterochromatin also fuses as a single site or chromocenter during meiotic prophase (39). The aggregation of centromeres of paired bivalents and some knobs (neocentromeres) has also been observed in maize at meiotic prophase (75).

There has been some discussion as to whether there is homoeology between maize chromosome 3, maize chromosome 10, and rice chromosome 12, as some groups have failed to cross-map any markers between these regions in particular segregating populations (179). However, researchers using different segregating populations have identified markers mapping to both the centromere regions of maize 3 and maize 10 (57). This group of markers (including *UMC18*) mapping on both chromosomes is also located on rice chromosome 12. It is not the relationship of maize 3 and maize 10 that is unclear, but whether there is a duplicate region for maize 4S, which has homoeology with rice chromosome 11. This maize chromosome arm carries a large number of resistance genes and storage proteins. Disease-resistance genes are subject to rapid evolution. As a class of genes they tend not to reflect conservation in gene order (109). Many of these genes probably have diverged significantly from those of other species and probably from the duplicate region in maize. However, there are eight RFLP loci from maize 4S duplicated on 7S, which may indicate an ancient relationship (86) (Figure 3).

Centromeres, Telomeres, and Chromosome Evolution

Comparison of the location of the junctions of the linkage segments in the different cereal genomes, in particular rice (156), indicates that the borders fall in centromere and telomere regions (133). The region in one species is a centromere site and in another species a telomere site. Moreover, telomeric heterochromatin in maize, rye, wheat, and Bromus can, under certain conditions, function as neocentromeres (reviewed in 133). Centromere/telomere sites have been the major focus of rearrangements, probably indicating why the markers flanking these regions exhibit a loss of gene order. The structure of the maize genome indicates the importance of centromeres and telomeric regions in the evolution of its genomes (Figure 3). Because the centromeric sites have been subject to breakage and fusion events, the comparative relationship of specific centromere and telomere sites across species remains unclear. Among the cereals, the locations of centromere and telomere sites have been conserved at the gross level. Thus the potential location of these structures is not random but rather is limited to a number of sites. In mammals, in contrast, the locations of centromere sites have not been found to be conserved across species (126).

The rice genome has been a useful tool for determining the structural relationships of the cereal genomes. One explanation is that its structure is similar to that of the ancestral progenitor genome. An alternative explanation is that it is the diploid with the highest basic number of chromosomes that has been analyzed. The rice genome reveals more potential (centromere and telomere) sites involved in breakage and fusion events of chromosomes than other species. However, still unclear is how and why certain sites in the cereal genomes are activated as centromere sites in some species and not in other closely related genomes and how this activity is modified. This issue is important in the debate about whether sorghum with its 10 chromosomes is an ancient tetraploid (38). In situ hybridization shows that five of the sorghum centromeres are distinct from the other five sites. It has been argued that this observation supports the ancient tetraploid concept (81). However, comparative mapping with rice indicates that the gene order in the sorghum genome is similar to that of rice (57). There is no clear duplication of the sorghum genome with respect to the rice genome. Sorghum chromosomes are metacentric but share homoeology with whole or parts of chromosome arms of maize (57). Thus, active centromeric sites do not map comparatively between sorghum and the two maize chromosome sets. The lack of knowledge on how sites are activated or suppressed during chromosome evolution makes it difficult to interpret the different sorghum centromere structures.

CHROMOSOME PAIRING

Chromosome pairing is the process by which homologous chromosomes (termed homologues) start in a premeiotic somatic nucleus randomly organized with respect to each other but end up during the pachytene stage of meiotic prophase in close association. Homologue pairing is important for the correct segregation of the chromosomes to gametes. The process of bringing homologues together involves their reciprocal recognition, coalignment, and synapsis. The term chromosome pairing has often been applied to one or more of these individual stages. An S phase occurs between premeiotic interphase and meiotic prophase in which the chromosomes are replicated, generating two sister chromatids. To that effect the intimate association of the homologues, which are composed of sister chromatids, is facilitated by a protein structure, the synaptonemal complex (177a). A protein structure, the lateral element, is formed by the two sister chromatids of each homologue. The lateral elements of each homologue are then aligned and associated by a third protein structure forming between them.

Chromosome-pairing studies provided early indications as to the genome relationships between polyploid species and diploid relatives. In essence, these were the first comparative genomic studies. Conversely, have the recent comparative genomic studies contributed to our understanding of the chromosome-pairing process? As described, these studies reveal that two chromosome structures have been important in cereal evolution, the telomeres and the centromeres. Both chromosome structures in plants such as wheat, rye and barley are located on the nuclear membrane. In interphase cells the centromeres are at one pole of the nucleus and the two chromosome arms extend to the other pole where the telomeres are dispersed over the membrane (1, 9, 10). This chromosome organization has been described as a "Rabl" configuration (144a).

In Diploids

Early studies indicated an important role for telomeres/subtelomeric regions during the pairing process. Subtelomeric heterochromatin knobs are clearly visible in interphase nuclei in rye (27, 164). Light microscopy revealed that early during meiotic prophase, the telomere regions of diploid species form a single cluster or bouquet. Meiocytes visualized in premeiotic interphase exhibited no association of telomeric regions. This is consistent with the observations made on *Lilium longiflorum*. This species' chromosomes undergo a preleptotene contraction and become visible. There was no clear association of chromosomes at this stage, which suggests that there is no premeiotic alignment in this diploid species (172, 173). The intimate alignment of rye chromosomes during meiotic prophase was assessed by spreading and squashing meiocytes at the zygotene stage to reveal their synaptonemal complex structures (77). The initiation of synapsis occurred after the bouquet had formed. The telomere regions are among the first sites to undergo synapsis. However, many other sites along the chromosome are also involved in synapsis of the bivalents (pair of homologues). The intimate alignment process could not be explained simply by zipping up of these initiation sites from the telomeres (77). The sizes of heterochromatin knobs in rye can vary greatly. Synapsis between such homologous chromosomes differing in heterochromatin knob size is largely unaffected in rye (80). One lateral element was slightly longer than the other, which resulted in an unpaired telomeric end. However, there was only slight reduction in the level of recombination. Importantly, the length of the lateral elements did not correlate directly with the difference in size of the chromosomes. This is more marked in hybrids between two Lolium (diploid) species, which differ by 50% in their chromosome size (94). These different-sized chromosomes (homoeologues) intimately associate, which results in the alignment of lateral elements of similar size. The synapsed structures resemble perfectly paired bivalents. There is no indication of a substantial correction to the lateral element length during the process of synapsis. Thus the chromosomes, which are substantially different in size, produce only lateral elements of a length similar to the chromosome with which it will become associated. The implication is that the pair of chromosomes that are going to be synapsed as a bivalent are already associated at more than one site prior to synapsis. If the searching process for chromosomes took place after lateral elements had been formed, synapsed chromosomes with different lengths of lateral elements would be observed. Two chromosome structures, telomeres and centromeres, both located on the nuclear membrane, could be involved in the initial chromosome searching process. The clustering of the telomeres to form a bouquet brings the telomeres together. Data presented by Martinez-Perez at the recent 9th Botanical Congress indicates that in diploid *Triticum* and *Aegilops* species the centromeres also associate in pairs at this stage (118). Other sites along the chromosome arms would also have to be involved. It is unclear how these sites search and recognize each other without an elaborate system of motors and pulleys moving the chromosome sites around the nucleus. An attractive hypothesis proposed by Cook is that these sites are genes that are being associated at transcriptional sites (40). Homologous genes would be transcribed by being looped out to the same transcription factory, thereby enabling association to occur. Thus, chromosomes of different sizes but possessing similar gene orders would associate through their telomeres and centromeres on the nuclear membrane and the genic regions in the transcriptional factories.

In Autotetraploids

Synaptonemal complex spreading studies have also been undertaken on autotetraploids of *Triticum monococcum* (the A genome donor of hexaploid wheat) and autotetraploids of rye (*Secale cereale*) (79, 152). These plants contain seven basic chromosomes with four homologous chromosomes for each chromosome group (i.e. 28 chromosomes in all). The spreading data indicated that the chromosomes could synapse as seven cross-like structures. Thus, although the telomere regions had the potential to associate with three other homologous regions, they only formed pair-wise associations. This was the case for all the regions along the chromosome arm except for one site halfway along the chromosome. Most sites on the *Triticeae* chromosomes can only synapse in pairs, and these associations are with the same chromosome. Only a single site on the chromosome engages in multiple interactions. Martinez-Perez also observed that centromeres associate premeiotically in autotetraploids of the *Triticeae* species, reducing to approximately seven sites during meiotic prophase (118). The implication is that the centromeres of four homologous chromosomes associate, forming cross-like structures of synapsed chromosomes.

In Allopolyploids

Synaptonemal complex spreading studies on allotetraploids and allohexaploids (species possessing two or three sets of related but not identical genomes) including allotetraploid *Aegilops* species sharing D genomes (42), *Aegilops* species sharing U genomes (41), allotetraploid oats, and allohexaploid oats (96) all show that the vast majority of chromosomes are synapsed as bivalents. Multivalent structures are occasionally observed at low frequency. A study on hexaploid wheat also reported predominantly bivalent formation (165). The regions adjacent to the telomeres were among the first to synapse. However, other sites were also involved in initiating synapsis, producing bivalent structures resembling beads on a necklace. Thus, although chromosomes have the potential to associate with three other chromosomes (two homoeologues and an homologue), they associate with their homologue to form a bivalent structure. In this case, centromere regions are associated as pairs at meiotic prophase in hexaploid wheat (118). Thus the homoeologous and homologous centromere regions are distinguished in the allotetraploid and allohexaploid situations and resolve as pairs.

Bennett and colleagues showed that the length of meiotic prophase is shorter in polyploids compared to their diploid progenitors, which is the opposite from what would be expected (20). The more genomes or chromosomes are present, the shorter the meiotic prophase, and hence the time required to sort them. The implication is that the chromosome-sorting process in autotetraploids and allotetraploids has been extended outside meiotic prophase. In that regard, it is now apparent that centromeres associated in pairs in floral tissues of all the *Triticeae* allopolyploids studied including hexaploid wheat (118). Thus centromere association is occurring prior to telomere association in these species (9, 10, 117). However, the diploid progenitors do not associate their centromeres until meiotic prophase (118). Thus, polyploidization results in the early association of centromeres. This is consistent with the early studies by researchers who, when treating hexaploid wheat anthers with colchicine prior to their meiocytes being in meiotic prophase, observed an effect on chromosome pairing at metaphase I (17, 52–54, 169). The observation that centromeres are associated in pairs during floral development in hexaploid wheat and wild polyploid relatives but not in their diploid progenitors suggests that the chromosome-sorting process was initially taking place at the centromeres in polyploids. Pairs of homologous chromosomes were fluorescently labeled in hexaploid wheat and their behavior followed from early floral (anther development) through to meiotic prophase (118). These studies were undertaken using anther sections and confocal microscopy so that intact cells could be analyzed (7). The sections enabled the cells to be clearly classified. The study showed that early in anther development prior to the meiocytes being clearly recognizable from tapetal cells, some seven days prior to meiotic prophase, the centromeres in the developing anther associate in pairs (8, 10, 118). By five days prior to meiotic prophase, these associations are becoming homologous associations. Thus, the homologous chromosomes at this stage form a V-configuration. By three days prior to meiotic prophase, the stage at which meiocytes are in premeiotic interphase, 90% of the homologous chromosomes being visualized were associated via their centromeres (10). During premeiotic interphase, the homologous chromosomes colocalize along their length. However, the telomeres of the homologues do not associate as pairs (118). The meiocytes progress through S phase. The telomeres cluster to form a bouquet. The homologous chromosomes separate along their length and are associated by the centromeres and telomeres. The homologous chromosomes are visualized as sister chromatids at this stage (118). The sister chromatids then associate and the homologous chromosomes associate simultaneously. The homologous chromosomes are intimately aligned and then the telomere bouquet declusters (118). A schematic representation of the chromosome pairing process in polyploid and diploid *Triticeae* is shown in Figure 4 (see color plate).

Hexaploid and tetraploid wheats (69, 90) and rice have not suffered major rearrangements following polyploidization. The maize genome, on the other hand, evolved through breakage and fusion events of centromeres and telomeres, followed by tetraploidization and then by further rearrangements (Figure 3). It is not yet known whether maize centromeres associate premeiotically, as happens in hexaploid wheat and polyploid *Aegilops* species. However, visualizing the maize heterochromatin knobs, Cande and colleagues showed no premeiotic association of these sites (46). This implies that maize chromosomes do not align along their length during premeiotic interphase. During the leptotene and zygotene transition of meiotic prophase, maize telomeres cluster to form a bouquet, and then the heterochromatin knobs associate (16, 46). The homologues intimately associate along their length with the telomeres still clustered (75, 76). If centromeres in maize associate premeiotically, chromosome behavior in maize and the timing of the telomere bouquet would resemble that described for hexaploid wheat lines lacking the *Ph1* locus.

The *Ph1* Locus

The requirement for breeding purposes to introgress chromosome segments carrying beneficial traits from wild relatives into polyploids such as wheat encouraged researchers to study the genomic relationships of wheat (*Triticum*) and its wild relatives (*Aegilops*) through chromosome pairing. F1 hybrids between diploid progenitors of hexaploid wheat or wild relatives are capable of pairing and recombining (14, 181). However, polyploidization and the subsequent premeiotic association of centromeres promotes homologous pairing and thereby reduces the ability to introgress chromosome segments from wild species. Both Riley and Okamoto observed that by deleting both 5B chromosomes, a level of homoeologous pairing could be induced (140, 146). Sears identified a deletion (*ph1b*) of the 5B chromosome that also produced the same effect (155). A deletion of the same region of 5B causes a similar effect in tetraploid wheat. Researchers termed the deleted locus *Ph1*. Sears noted that the fertility of the line was around 30% of that of the wild-type wheat (155). This implied that the majority of pairing configurations observed at metaphase I in this line resulted in infertility, possibly through gamete abortion. A number of studies have been undertaken to characterize the effect of the *Ph1* locus. Analysis of anther sections during floral development revealed that, in common with hexaploid wheat (and its polyploid relatives), centromeres associated during floral development in the absence of the *Ph1* locus (118). However, the centromere structure is affected when the *Ph1* locus is deleted (9). The nature of the premeiotic centromere associations has been determined. In similarly staged sections, when 90% of the homologues were associated via their centromeres in the presence of the *Ph1* locus, only 30% of the homologues were in the absence of the locus (10). This is also the level of fertility of the *ph1b* line (155). Analysis of the wild relatives of wheat reveals that none of the chromosomes tested carry loci that can compensate for the loss of the 5B chromosome in hexaploid wheat (14). In the absence of the *Ph1* locus in wheat, centromeres still associate premeiotically as they do in polyploid *Aegilops* relatives (117, 118). The *Ph1* locus on chromosome 5B raises the fertility of wheat. It is quite possible therefore that the current 5B allele was not present in the original hybridization but was a mutation that arose and was then selected because it conferred increased fertility.

The timing of formation of the telomeric bouquet in the *ph1b* line is also delayed in meiotic prophase, thereby lengthening the whole stage (19, 118). The level of association between homologues via the centromeres observed during premeiotic interphase may increase further during meiotic prophase until the telomeric bouquet is formed. The change in timing of the telomeric bouquet will change the timing of synapsis, which ultimately could affect the recombination between chromosomes in the presence and absence of the *Ph1* locus. The *Ph1* locus appears to indirectly affect synapsis (78). Gillies noted that it was difficult to prepare synaptonemal complex spreads from wheat carrying the *Ph1* locus (78). Most preparations revealed short fragments of associated lateral elements. This is consistent with the observation that sister chromatids and homologues are associating at the same time, implying that lateral element formation is not complete before the central element formation occurs. In wheat hybrids in which only homoeologous chromosome pairing could occur, the chromosomes were correctly synapsed in the presence or absence of the *Ph1* locus (78). The homoeologous chromosomes differ in size, yet synapse via lateral elements of similar length. This implies that the chromosomes are associated prior to synapsis, which is consistent with the occurrence of centromeres in pairs and telomeres in the bouquet. Analysis of pairing in autotetraploids described above indicates that centromeres can be involved in multivalent formation (118). The observation that centromeres are in pairs in the *Ph1* mutant is consistent with a low level of chromosomes present as multivalents at metaphase I. In the *ph1b* line, only four chromosomes per nucleus on average are engaged in higher-order associations; the rest are bivalents or univalents.

Recombination and the *Ph1* Locus

Deleting one of the telomere regions of the pair of homologous chromosomes results in the failure of the pair of homologous chromosomes to recombine in hexaploid wheat and therefore to be found associated at metaphase I (43, 113). Deleting the telomere regions of both chromosomes to the same extent does not reduce recombination between homologues in hexaploid wheat (113) nor does the possession of nonhomologous centromeres, provided the telomere/subtelomeric regions exhibited homology (43). The *Ph1* locus affects recombination (55, 61, 78, 114). In wheat hybrids in which homoeologous chromosomes had synapsed in the presence of the *Ph1* locus, the chromosomes failed to recombine (78). Even homoeologous interstitial segments within a homologous chromosome fail to recombine despite the occurrence of recombination within the homologous segments (113).

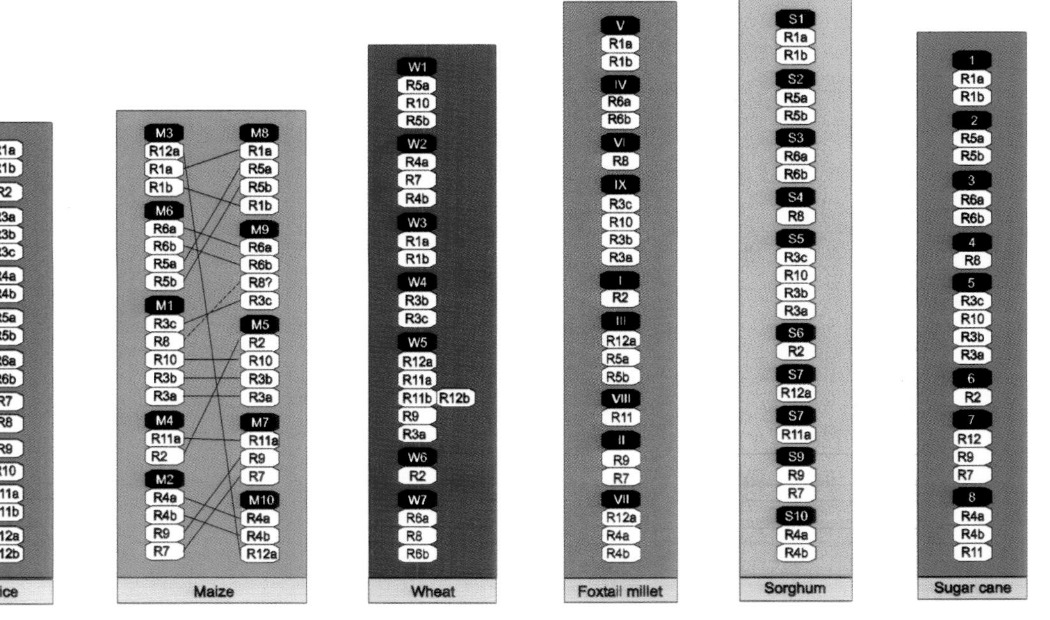

Figure 2 Comparisons of the cereal genome evolution based on rice linkage segments. (R1–R2) rice chromosomes dissected into linkage blocks (R1a, R1b, etc); 10 (M1–M10) maize; 7 (W1–W7) wheat, based on the linkage map of the D genome; 9 (I–IX) foxtail millet; 10 (S1–S10) sorghum; and 8 (1–8) basic sugarcane chromosomes represented as linkage segments on the basis of the conservation of gene order. Connecting lines indicate duplicate segments within the maize chromosomes. The designation of the sorghum and sugarcane linkage groups has varied between laboratories and publications.

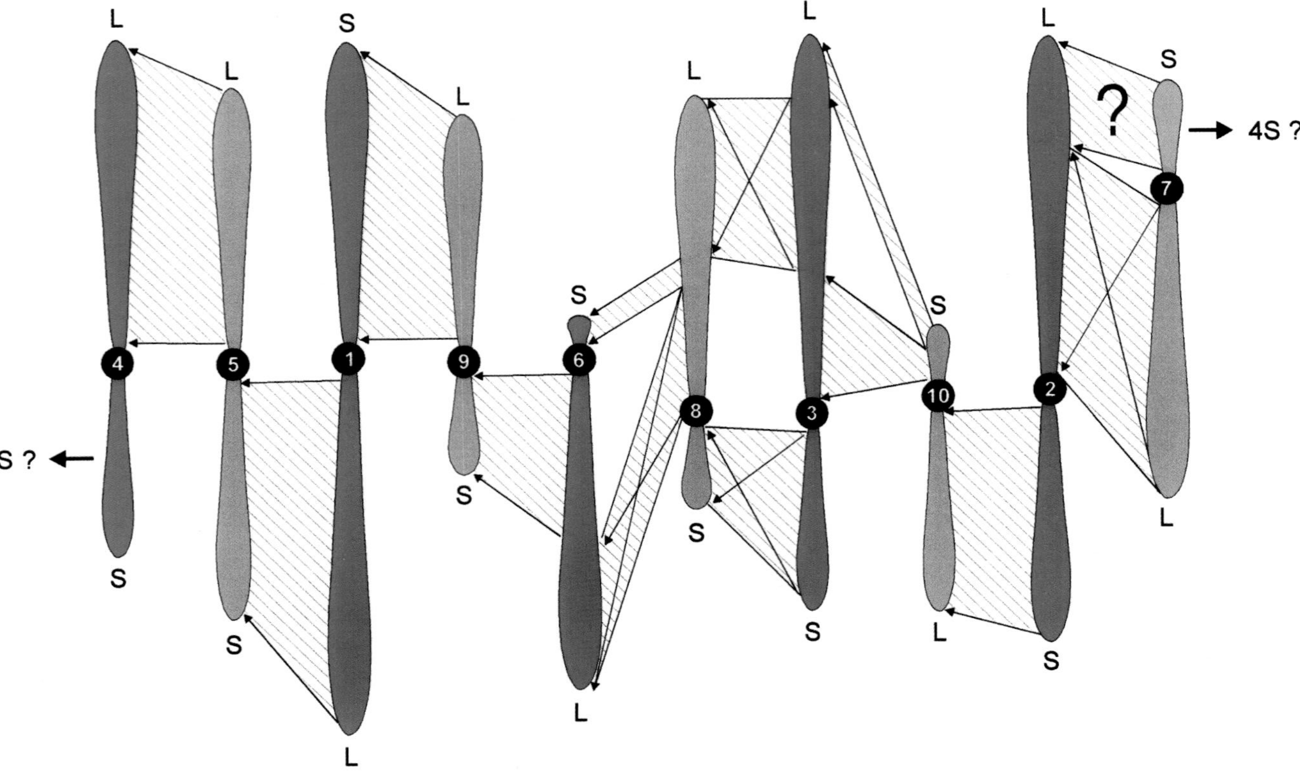

Figure 3 The homoeologous relationships of the two parental sets of maize (interphase) chromosomes forming a "translocation heterozygote." Chromosomes 4, 1, 6, 3, 2 are derived from one parental set and the rest from the other.

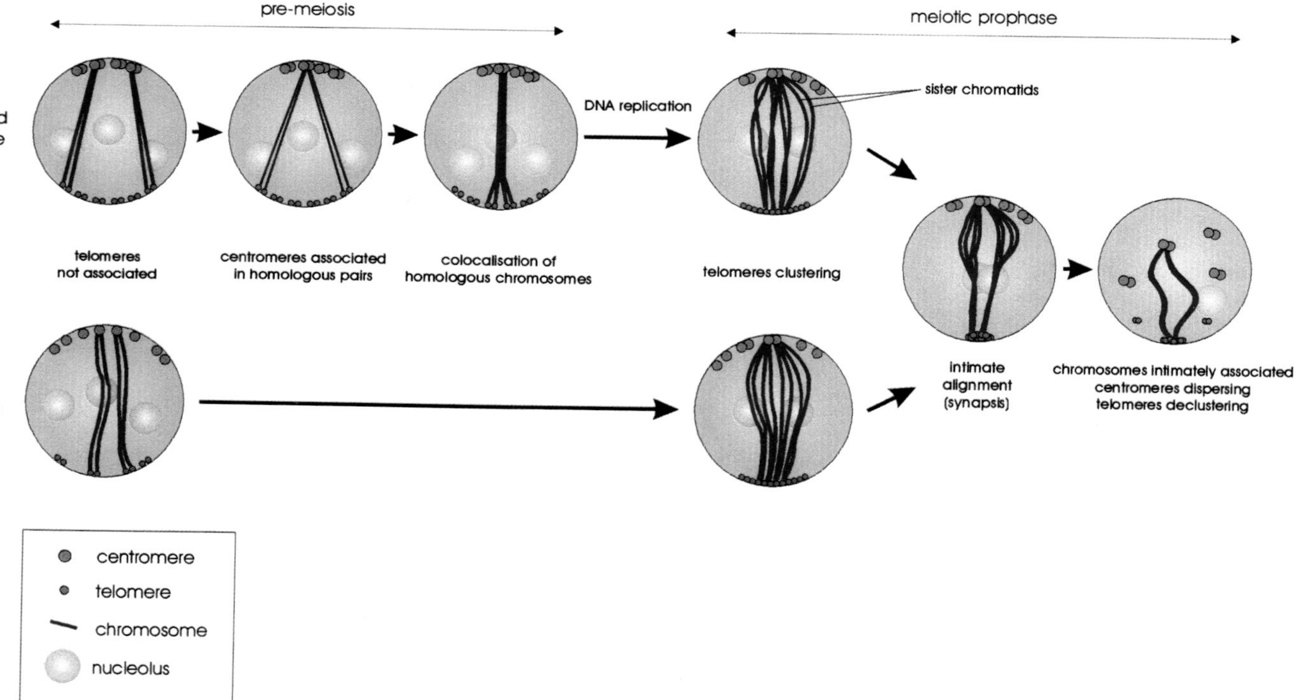

Figure 4 Diagram of chromosome-pairing events before and during meiosis in polyploid and diploid wheats. Polyploid and diploid *Aegilops* species within the *Triticeae* also exhibit similar pairing to the *Triticum* species.

However, in the absence of *Ph1* locus, homoeologous chromosomes and interstitial homoeologous segments within homologous chromosomes recombine. This implies that either the *Ph1* locus is a complex containing genes that affect the premeiotic alignment through centromere association and other genes affecting recombination, or that the high level of association of homologues during floral development leads to an early association of telomeres and synapsis. The expression levels of genes preventing homoeologous recombination may be comparable in similar staged floral tissues of wheat with and without the *Ph1* locus. However, at the time the telomeric bouquet and chromosome synapsis occur in wheat lacking the *Ph1* locus, the expression levels may be in decline, resulting in homoeologous recombination.

Pairing Models

Three hypotheses have long dominated the field of cereal chromosome pairing. How do these hypotheses stand in the light of recent data?

As researchers studying pairing in autotetraploid plants were to show, the chromosomes associate in these species as multivalents that later resolve as bivalents by metaphase I. Researchers at Carlsberg performed some of the initial studies using Bombyx (silkworms), and observed the initial multivalent formation and later bivalents. They proposed that a similar mechanism operated in the case of allopolyploids, such as hexaploid (bread) wheat. The initial studies seem to support this proposal (88, 93). However, a more detailed analysis by Holm revealed mostly bivalent formation (89). Although Holm concluded that "most chromosomes are at mid zygotene present as partially paired bivalents and only few have engaged in multiple associations," he nevertheless argued that multivalent formation and their resolution were important (89). Thus it is generally perceived in the literature that chromosomes in hexaploid wheat are involved in a searching process at synapsis through associating at a number of sites along their chromosome arms. This hypothesis is, however, problematic. First, most multivalents claimed from analyses of synaptonemal complexes were based on a single site association between two chromosomes that were synapsed as partially paired bivalents with another partner. If there is generalized searching operating, it would be expected that a number of sites would be involved along their chromosome arms visualized as multiple interactions. This is not the case. Moreover, the process of squashing and spreading to analyze the synaptonemal complex structures will result in some chance associations between partially paired bivalents. Clearly, these two types of association need to be differentiated. Autotetraploids of the progenitors of wheat do not exhibit interstitial multivalent associations (as described above) between identical chromosomes (79). Thus, why would multivalent associations at synapsis occur between nonidentical and not between identical chromosomes? Subsequent synapsis studies of an autotetraploid of rye (152) and *Triticum monococcum* (79) show that one site is engaged in multivalent associations. Recent data indicate that such sites are now likely to be the centromere regions. However, centromere sites are not engaged in multiple associations at meiotic prophase in allopolyploids such as hexaploid wheat. They associate mostly in pairs early in floral development and remain in pairs through meiotic prophase (118). Importantly, a number of spreads of partially synapsing wheat chromosomes show no associations that can be interpreted as multivalent formation (89). If all chromosomes had to go through the multivalent formation as part of synapsis, multivalents would be observed in all spreads. The close polyploid relatives of hexaploid wheat, *Aegilops* and *Avena*, have recently been shown not to exhibit multivalent formation at synapsis but simply bivalent formation (41, 42). Moreover, as stated previously, homoeologous chromosomes are synapsed by lateral elements of similar lengths, which implies that these chromosomes were already committed to pair prior to completion of lateral element formation (78). All data indicate that chromosomes in hexaploid wheat synapse as bivalents, as is in the case for its wild polyploid (*Aegilops*) relatives.

Early studies on *Drosophila* observed that chromosomes associate in somatic tissues during embryo development (123). Based on these *Drosophila* observations, Feldman proposed that most plant species associate their chromosomes premeiotically. Because he had observed the presence of univalents, multivalents and interlocking of bivalents in hexaploid wheat with increased doses of the 5BL arm, Feldman argued that the presence of the *Ph1* locus on 5B suppresses this premeiotic chromosome pairing "causing random distribution of chromosomes in the premeiotic nucleus." Premeiotic association was concluded to be partially suppressed at two doses of the *Ph1*, which eliminated pairing between homoeologues and led exclusively to homologous pairing at meiosis. The presence of extra *Ph1* doses even suppressed homologous chromosome pairing. Feldman and colleagues therefore used squashed preparations to assess whether there was somatic chromosome association in root cells (64, 65). This strategy would provide evidence of somatic association leading up to premeiosis, which then would be disrupted premeiotically by the *Ph1* locus. From the squashed preparations, Feldman and colleagues concluded that there was somatic chromosome association in roots. From further studies using colchicine treatment, they concluded that microtubule interactions with the centromeres were involved in this somatic pairing (12, 13). The cloning of sequences at centromeres now permits the question of whether centromeres associate in roots to be addressed (8). There is no evidence for centromere association in roots nor association of homologues as proposed (87, 118). Furthermore, homologous chromosomes are not randomly organized in the premeiotic nucleus but homologues are associated via their centromeres (117, 118). Therefore, although the principle that association of homologues involved centromeres proved to be correct, current data do not support the experimental evidence on which this proposal was based.

Finally, Watanabe proposed from studies on polyploid Chrysanthemum that obligate bivalent formation in polyploids is achieved by initiating chromosome pairing at two sites (174). He concluded that there would be two sites under independent and fundamentally different genic control and that the pairing of one site always precedes the pairing at the other. These proposals are entirely consistent with chromosome behavior of the *Triticum* and *Aegilops* (10, 117, 118). The analyses described above of autopolyploids and allopolyploids suggest that centromeres and telomeric regions are these two sites.

CONCLUDING REMARKS

In summary, centromeres and telomeres have been important in influencing cereal chromosome evolution and the pairing of cereal chromosomes. The development of new vector systems, particularly cereal artificial chromosomes for the biotech industry, is likely to promote centromere- and telomere-based studies. The better understanding of the structure of this regions and the proteins that interact with them will, in turn, help chromosome-pairing studies. The future characterization of the *Ph1* locus itself may well identify some of the key proteins/factors that interact at these sites.

ACKNOWLEDGMENTS

The author wishes to thank Tracie Foote, Simon Griffiths, and Enrique Martinez-Perez for help in preparing this manuscript.

LITERATURE CITED

1. Abranches R, Beven AF, Aragon-Alcaide L, Shaw PJ. 1998. Transcription sites are not correlated with chromosome domains in wheat nuclei. *J. Cell. Biol.* 143:5–12
2. Ahn S, Anderson JA, Sorrells ME, Tanksley SD. 1993. Homoeologous relationships of rice, wheat, and maize chromosomes. *Mol. Gen. Genet.* 241:483–90
3. Ahn S, Tanksley SD. 1993. Comparative linkage maps of the rice and maize genomes. *Proc. Natl. Acad. Sci. USA* 90:7980–84
4. Alfenito MR, Birchler JA. 1993. Molecular characterisation of a maize B chromosome centric sequence. *Genetics* 135:589–97
5. Ananiev EV, Phillips RL, Rines HW. 1998. Chromosome-specific molecular organisation of maize (*Zea mays* L) centromeric regions. *Proc. Natl. Acad. Sci. USA* 95:13073–78
6. Ananiev EV, Phillips R, Rines HW. 1998. Complex structure of knob DNA on maize chromosome 9. Retrotransposon invasion into heterochromatin. *Genetics* 149:2025–37
7. Aragon-Alcaide L, Bevan A, Moore G, Shaw P. 1997. The use of a vibratome sections of whole cereal sections to study anther development and meiosis. *Plant J.* 14:503–8
8. Aragon-Alcaide L, Miller TE, Schwarzacher T, Reader SM, Moore G. 1996. A cereal centomeric sequence. *Chromosoma* 105:261–68
9. Aragon-Alcaide L, Reader SM, Miller TE, Moore G. 1997. Centromere behaviour in wheat with high and low homoeologous chromosome pairing. *Chromosoma* 5:327–33

10. Aragon-Alcaide L, Reader SM, Shaw PJ, Beven AF, Miller TE, Moore G. 1997. Association of homologous chromosomes during floral development. *Curr. Biol.* 7:905–8
11. Ashikawa I, Kurata N, Naguramura Y, Minobe Y. 1994. Cloning and mapping of telomeric-associated sequences from rice. *DNA Res.* 1:67–76
12. Avivi L, Feldman M, Bushuk W. 1969. The mechanism of somatic association in common wheat. *Triticum aestivum* (L). The suppression of somatic association by colchicine. *Genetics* 62:745–52
13. Avivi L, Feldman M, Bushuk W. 1970. The mechanism of somatic association in common wheat, *Triticum aestivum* (L). Differential affinity for colchicine of spindle microtubules of plants having different doses of somatic-association suppressor. *Genetics* 65:585–92
14. Bakar MA, Kimber G. 1982. Chromosome pairing regulators in the former genus *Aegilops*. *Z. Pflanzenzuecht.* 89:130–38
15. Balcell C, Swinburne J, Coupland G. 1991. Transposons as tools for the isolation of plant genes. *Trends Biotechnol.* 9:31–36
16. Bass HW, Marshall WF, Sedat JW, Agard DA, Cande WZ. 1997. Telomeres cluster *de novo* before the initiation of synapsis: a three-dimensional spatial analysis of telomere positions before and during meiotic prophase. *J. Cell Biol.* 137:5–18
17. Bayliss MW, Riley R. 1972. Evidence of pre-meiotic control of chromosome pairing in *Triticum aestivum*. *Genet. Res.* 20:205–12
18. Bedbrook JR, Jones J, O'Dell M, Thompson RD, Flavell RB. 1980. A molecular description of telomeric heterochromatin in *Secale* species. *Cell* 19:545–60
19. Bennett MD, Dover GA, Riley R. 1974. Meiotic duration in wheat genotypes with and without homoeologous meiotic chromosomes. *Proc. R. Soc. London Ser. B* 187:191–207
20. Bennett MD, Smith JB. 1972. The effect of polyploidy on meiotic duration and pollen development in cereal anthers. *Proc. R. Soc. London Ser. B* 181:81–107
21. Bennett MD, Smith JB. 1991. Nuclear DNA amounts in angiosperms. *Philos. Trans. R. Soc. London Ser. B* 334:309–45
22. Bennetzen JL, Freeling M. 1993. Grasses as a single genetic system: genomic composition, collinearity and compatibility. *Trends Genet.* 9:259–61
23. Bennetzen JL, Kellogg EA. 1997. Do plants have a one-way ticket to genomic obesity? *Plant Cell* 9:1509–14
24. Bennetzen JL, San-Miguel P, Liu C-N, Chen M, Tikhonov A, et al. 1995. Microcollinearity and segmental duplication in the evolution of grass nuclear genomes. In *Unifying Plant Genomes*, ed. JS Heslop-Harrison, 50:1–3. Cambridge: Co. Biol.
25. Bennetzen JL, Schnick K, Springer PS, Brown WE, San-Miguel P. 1994. Active maize genes are unmodified and flanked by diverse classes of modified highly repetitive DNA. *Genome* 37:565–76
26. Bird AP. 1987. CpG islands as markers in the vertebrate nucleus. *Trends Genet.* 3:342–47
27. Bowan JG, Rajhathy T. 1977. Fusion of chromocenters in premeiotic interphase of *Secale cereale* and its possible relationship to chromosome pairing. *Can. J. Genet. Cytol.* 19:313–21
28. Brink RA, Cooper DC. 1932. Chromosome rings in maize and Oenothera. *Proc. Natl. Acad. Sci. USA* 18:447–55
29. Brown WRA. 1988. A physical map of the human pseudoautosomal region. *EMBO J.* 7:2377–85
30. Bureau TE. Wessler SR. 1992. Tourist—a large family of small inverted repeat elements frequently associated with maize genes. *Plant Cell* 4:1283–94
31. Bureau TE, Wessler SR. 1994. Mobile inverted-repeat elements of the tourist family are associated with the genes of many cereal grasses. *Proc. Natl. Acad. Sci. USA* 91:1411–15

32. Burr B, Burr FA, Matz EC. 1993. Maize molecular map (*Zea mays*) 2N = 20. In *Genetic Maps*, ed. SJ O'Brien, pp. 190–203. Cold Spring Harbor: Cold Spring Harbor Lab. Press
33. Burr B, Burr FA, Thompson K, Albertson MC, Stuber CW. 1988. Gene mapping with recombinant inbreeds in maize. *Genetics* 118:519–26
34. Causse M, Fulton TM, Cho YG, Ahn SN, Wu K, et al. 1994. Saturated molecular map of the rice genome based on an interspecific backcross population. *Genetics* 138:1251–74
35. Chen M, San-Miguel P, De Oliveira AC, Woo S-S, Zhang H, et al. 1997. Microcolinearity in sh2-homologous regions of maize, rice and sorghum genomes. *Proc. Natl. Acad. Sci. USA* 94:3431–35
36. Cheung WY, Chao S, Gale MD. 1991. Long range mapping of the α-amylase (α-Amy-l) loci on homoeologous group 6 chromosomes of wheat. *Mol. Gen. Genet.* 229:373–79
37. Cheung WY, Money TA, Abbo S, Devos KM, Gale MD, Moore G. 1994. A family of related sequences associated with (TTTAGGG)n repeats are located in interstitial regions of the wheat chromosomes. *Mol. Gen. Genet.* 245:349–54
38. Chittenden LM, Schertz KF, Lin Y-R, Wing RA, Paterson AH. 1994. A detailed RFLP map of *Sorghum bicolor* × *S. propinguum* suggests ancestral duplication of the sorghum chromosomes or chromosomal segments. *Theor. Appl. Genet.* 87:925–33
39. Coleman LC. 1941. The relation of chromocenters to the differential segments in *Rheo discolor*. *Am. J. Bot.* 28:742–48
40. Cook PR. 1997. The transcriptional basis of chromosome pairing. *J. Cell. Sci.* 110:1033–40
41. Cunado N, Callejas S, Garcia MJ, Fernandez A, Santos JL. 1996. The pattern of zygotene and pachytene pairing in allotetraploid *Aegilops* species sharing the U genome. *Theor. Appl. Genet.* 93:1152–55
42. Cunado N, Garcia MJ, Callejas S, Fernandez A, Santos L. 1996. The pattern of zygotene and pachytene pairing in allotetraploid *Aegilops* species sharing the D genome. *Theor. Appl. Genet.* 93:1175–79
43. Curtis CA, Lukaszewski AJ, Chrzastek M. 1991. Metaphase I pairing of deficient chromosomes and genetic mapping of deficiency breakpoints in common wheat. *Genome* 34:553–60
44. Darlington CD. 1929. Cytological theory of inheritance in Oenothera. *J. Genet.* 24:405–74
45. Davis BM. 1943. An amphidiploid in the F1 generation from a cross *Oenothera franciscana* × *Oenothera biennis* and its progeny. *Genetics* 28:275–85
46. Dawe RK, Sedat JW, Agard DA, Cande WZ. 1994. Meiotic chromosome pairing in maize is associated with a novel chromatin organisation. *Cell* 76:901–12
47. Devos KM, Atkinson MD, Chinoy CN, Harcourt RL, Koebner RMD, et al. 1993. Chromosomal rearrangements in the rye genome relative to that of wheat. *Theor. Appl. Genet.* 85:673–80
48. Devos KM, Gale MD. 1992. The genetic maps of wheat and their potential in plant breeding. *Outl. Agric.* 22:93–99
49. Devos KM, Millan T, Gale MD. 1993. Comparative RFLP maps of homoeologous chromosomes of wheat, rye and barley. *Theor. Appl. Genet.* 85:784–92
50. Devos KM, Wang ZM, Beales J, Sasaki Y, Gale MD. 1998 Comparative genetic maps of foxtail millet (*Setaria italica*) and rice (*Oryza sativa*). *Theor. Appl. Genet.* 96:63–68
51. Dong FJ, Miller T, Jackson SA, Wang GL, Ronald PC, et al. 1998. Rice (*Oryza sativa*) centromeric consist of complex DNA. *Proc. Natl. Acad. Sci. USA* 95:8135–40
52. Dover GA, Riley R. 1973. The effect of spindle inhibitors applied before meiosis on meiotic chromosome pairing. *J. Cell Sci.* 12:143–61

53. Driscoll CJ, Darvey NL. 1970. Chromosome pairing: effect of colchicine on isochromosome. *Science* 169:290–91
54. Driscoll CJ, Darvey NL, Barber HN. 1967. Effect of colchicine on meiosis of hexaploid wheat. *Nature* 216:687–88
55. Dubcovsky J, Luo MC, Dvorak J. 1995. Differentiation between homoeologous chromosomes 1A of wheat and 1A (M) of *Triticum monococcum* and its recognition by the wheat *Ph1* locus. *Proc. Natl. Acad. Sci. USA* 92:6645–47
56. Dubcovsky J, Luo M-C, Zhong G-Y, Kilian A, Kleinhofs A, Dvorak J. 1996. Genetic map of diploid wheat, *Triticum monococcum* L and its comparison with maps of *Hordeum vulgare* L. *Genetics* 143:983–99
57. Dufour P. 1996. *Cartographie moléculaire du genome du sorgho* (Sorghum bicolor *L. Moench*): *application en sélection variétale: cartographie comparé chez les andropogonées*. PhD thesis, Univ. Paris-Sud, France
58. Dufour P, Deu M, Grivet L, D'Hont A, Paulet F, et al. 1997. Construction of a composite sorghum genome map and comparison with sugarcane, a related complex polyploid. *Theor. Appl. Genet.* 94:409–18
59. Dufour P, Grivet L, D'Hont A, Deu M, Trouche G, et al. 1996. Comparative genetic mapping between duplicated segments on maize chromosomes 3 and 8 and homoeologous regions in sorghum and sugarcane. *Theor. Appl. Genet.* 92:1024–30
60. Dunford RP, Kurata N, Laurie DA, Money TA, Minobe Y, Moore G. 1995. Conservation of fine-scale DNA marker order in the genome of rice and the Triticeae. *Nucleic Acids Res.* 23:2724–28
61. Dvorak J, Dubcovsky J, Luo M-C, Devos KM, Gale MD. 1995. Differentiation between wheat chromosomes 4B and 4D. *Genome* 38:1139–47
62. Endo TR, Gill BS. 1996. The deletion stocks of common wheat. *J. Hered.* 87:295–307
63. Feldman M. 1966. The effect of chromosomes 5A, 5D, and 5D on chromosomal pairing in *Triticum aestivum*. *Proc. Natl. Acad. Sci. USA* 55:1447–53
64. Feldman M, Mello Sampavo T, Sears ER. 1966. Somatic association of *Triticum aestivum*. *Proc. Natl. Acad. Sci. USA* 56:1447–53
65. Flavell RB. 1986. Repetitive DNA and chromosome evolution in plants. *Philos. Trans. R. Soc. London Ser. B* 312:227–42
66. Flavell RB, Bennett MD, Hutchinson-Brace J. 1987. Chromosome structure and organisation. In *Wheat*, ed. F Lupton, pp. 211–68. London: Chapman & Hall
67. Flavell AJ, Dunbar E, Anderson R, Pearce SR, Hartley R, Kumar A. 1992. Ty1-copia group retrotransposons are ubiquitous and heterogenous in higher plants. *Nucleic Acids Res.* 20:3639–44
68. Foote T, Roberts M, Kurata N, Sasaki T, Moore G. 1997. Detailed comparative mapping of cereal chromosome regions corresponding to the Ph1 locus in wheat. *Genetics* 147:801–7
69. Gale MD, Atkinson MD, Chinoy CN, Harcourt RL, Jia J, et al. 1998. Genetic maps of the hexaploid wheat. In *Proc. Int. Wheat Genet. Symp., 8th*, ed. ZS Li, ZY Xin, pp. 29–40. Beijing: China Agric. Scientech Press
70. Gale MD, Devos KM. 1998. Comparative genetics in the grasses. *Proc. Natl. Acad. Sci. USA* 95:1971–74
71. Gardiner-Garden M, Frommer M. 1987. CpG islands in vertebrate genomes. *J. Mol. Biol.* 196:261–82
72. Gaut BS, Doebley JF. 1997. DNA sequences evidence for the segmental allotetraploid origin of maize. *Proc. Natl. Acad. Sci. USA* 94:6809–14
73. Gerlach WL, Bedbrook JR. 1979. Cloning and characterisation of the ribosomal RNA genes from wheat and barley. *Nucleic Acids Res* 7:1869–85
74. Gill KS, Gill BS, Endo TR, Boyko EV. 1996. Identification and high density

mapping of gene-rich regions in chromosome group 5 of wheat. *Genetics* 143:1001–12

75. Gillies CB. 1975. An ultrastructural analysis of chromosome pairing in maize. *C. R. Trav. Lab. Carlsberg* 40:135–61
76. Gillies CB. 1983. Ultrastructural studies of the association of homologous and nonhomologous parts of chromosomes in the mid-prophase of meiosis in *Zea mays*. *Maydica* 28:265–87
77. Gillies CB. 1985. An electron-microscopic study of synaptonemal complex-formation at zygotene in rye. *Chromosoma* 92:165–75
78. Gillies CB. 1987. The effect of *Ph1* gene alleles on synaptonemal complex formation in *Triticum aestivum* × *T kotschyi hybrids*. *Theor. Appl. Genet.* 74:430–38
79. Gillies CB, Kuspira J, Bhambhani RN. 1987. Genetic and cytogenetic analyses of the A genome of *Triticum monococcum*. IV. Synaptonemal complex formation in autotetraploids. *Genome* 29:309–18
80. Gillies CB, Lukaszewski AJ. 1989. Synaptonemal complex formation in rye (*Secale cereale*) heterozygous for telomeric C bands. *Genome* 32:901–7
81. Gomez MI, Islam-Faridi MN, Zwick MS, Czeschin DG, Hart GE, et al. 1998. Tetraploid nature of *Sorghum bicolor* (L) Moench. *J. Hered.* 89:188–90
82. Graner A, Jahoor A, Schondelmaier J, Siedler H, Pillen K, et al. 1991. Construction of an RFLP map of barley. *Theor. Appl. Genet.* 83:250–56
83. Gruenbaum Y, Naveh-Many T, Cedar H, Razin A. 1981. Sequence specificity of methylation in higher plant DNA. *Nature* 292:860–62
84. Guimaraes CT, Sills GR, Sobral BW. 1997. Comparative mapping of *Andropogoneae*: *Saccharum* L (sugarcane) and its relation to sorghum and maize. *Proc. Natl. Acad. Sci. USA* 94:14261–66
85. Han F, Kleinhofs A, Ullrich SE, Kilian A, Yano M, Sasaki T. 1998. Synteny with rice, analysis of barley malting quality QTLs and *rpg4* chromosome regions. *Genome* 41:373–80
86. Helentjaris T. 1995. Atlas of duplicated sequences. *Maize Genet. Coop. Newsl.* 69:67–81
87. Heslop-Harrison JS, Smith JB, Bennett MD. 1988. The absence of somatic association of centromeres of homologous chromosomes in grass mitotic metaphases. *Chromosoma* 96:119–31
88. Hobolth P. 1981. Chromosome pairing in allohexaploid wheat var. Chinese Spring. Transformation of multivalents into bivalents, a mechanism for exclusively bivalent formation. *Carlsberg Res. Commun.* 46:129–73
89. Holm PB. 1986. Chromosome pairing and chiasma formation in allohexaploid wheat. *Triticum aestivum* analysed by spreading of meiotic nuclei. *Carlsberg Res. Commun.* 51:239–94
90. Huang HC, Kochert G. 1994. Comparative RFLP mapping of an allotetraploid rice species and cultivated rice. *Plant Mol. Biol.* 25:633–48
91. Hutchinson J, Lonsdale DM. 1982. The chromosomal distribution of cloned highly repetitive sequences from hexaploid wheat. *Heredity* 48:371–76
92. Jauhar PP, Rieralizarazu O, Dewey WG, Gill BS, Crane CF, Bennett JH. 1991. Chromosome pairing relationships among A genome, B genome and D genome of bread wheat. *Theor. Appl. Genet.* 82:441–49
93. Jenkins G. 1983. Chromosome pairing in *Triticum aestivum* cv Chinese Spring. *Carlsberg Res. Commun.* 48:255–83
94. Jenkins G. 1985. Synaptonemal complex formation in hybrids of *Lolium temulentum* × *Lolium perenne* (L). *Chromosoma* 92:81–88
95. Jiang J, Nasuda S, Dong F, Scherrer CW, Woo SS, et al. 1996. A conserved repetitive DNA element located in the centromeres of cereal chromosomes. *Proc. Natl. Acad. Sci. USA* 93:14210–13

96. Jones M, Rees H, Jenkins G. 1989. Synaptonemal complex formation in *Avena* polyploids. *Heredity* 63:209–19
97. Kan YW, Dozy AM. 1978. Polymorphism of DNA sequence adjacent to human β globin structural gene: relationship to sickle mutation. *Proc. Natl. Acad. Sci. USA* 75:5631–35
98. Kellogg EA. 1998. Relationships of cereal crops and other grasses. *Proc. Natl Acad. Sci. USA* 95:2005–10
99. Kilian A, Kleinhofs A. 1992. Cloning and mapping of telomere-associated sequences in *Hordeum vulgare* L. *Mol. Gen. Genet.* 235:153–56
100. Kilian A, Chen J, Han F, Steffenson B, Kleinhofs A. 1997. Towards map-based cloning of the barley stem rust resistance genes *Rpg1* and *rpg4* using rice as an intergenomic cloning vehicle. *Plant Mol. Biol.* 35:187–95
101. Kilian A, Kudna D, Kleinhofs A. 1999. Genetic and molecular characterisation of barley chromosome telomeres. *Genome* 42:412–19
102. Kilian A, Kudna DA, Kleinhofs A, Yano M, Kurata N, et al. 1995. Rice-barley synteny and its application to saturation mapping of the barley *Rpg1* region. *Nucleic Acids Res.* 23:2729–33
103. Kleinhofs A, Kilian A, Saghai-Maroof MA, Biyashev RM, Hayes P, et al. 1993. A molecular isozyme and morphological map of barley (*Hordeum vulgare*) genome. *Theor. Appl. Genet.* 86:705–12
104. Kurata N, Moore G, Nagamura Y, Foote T, Yano M, et al. 1994 Conservation of genome structure between rice and wheat. *Nat. Biotechnol.* 12:276–78
105. Kurata N, Nagamura Y, Yamamoto K, Harushima Y, Sue N, et al. 1994. A 300 kilobase interval genetic map of rice including 883 expressed sequences. *Nat. Genet.* 8:365–75
106. Laurie DA, Pratchett N, Devos KM, Leitch IJ, Gale MD. 1993. Non-homoeologous translocations between group 4, 5 and 7 chromosomes in wheat and rye. *Theor. Appl. Genet.* 83:305–12
107. Lebo RG, Carrano AV, Burkhart-Schultz K, Dozy AM, L-C Yu, Kan YW. 1979. Assignment of human α, β and δ globin genes to the short arm of chromosome 11 by chromosome sorting and DNA restriction enzyme. *Proc. Natl. Acad. Sci. USA* 76:5804–8
108. Leeton PRJ, Smyth DR. 1993 An abundant LINE-like element amplified in the genome of *Lilium speciosum*. *Mol. Gen. Genet.* 237:97–104
109. Leister D, Kurth J, Laurie DA, Yano M, Sasaki T, et al. 1997. Rapid reorganisation of resistance gene homologues in cereal genomes. *Proc. Natl. Acad. Sci. USA* 95:370–75
110. Lilienfield FA. 1951. H Kihara: genome analysis in *Triticum* and *Aegilops* X. Concluding review. *Cytologia* 16:101–23
111. Linde-Lauren I. 1982. Linkage map of the long arm of barley chromosome 3 using C-bands and markers genes. *Heredity* 49:27–35
112. Liu CJ, Devos KM, Chinoy CN, Atkinson MD, Gale MD. 1992. Non-homoeologous translocations between 4, 5 and 7 chromosomes in wheat and rye. *Theor. Appl. Genet.* 83:305–12
113. Lukaszweski AJ. 1997. The development and meiotic behaviour of asymmetrical isochromosomes in wheat. *Genetics* 145:1155–60
114. Luo MC, Dubcovsky J, Dvorak J. 1996. Recognition of homeology by wheat *Ph1* locus. *Genetics* 144:1195–203
115. Maestra B, Naranjo T. 1999. Homoeologous relationships of *Aegilops speltoides* chromosomes to bread wheat. *Theor. Appl. Genet.* 17:181–86
116. Malheiros-Gards N, Garde A. 1950. Fragmentation as a possible evolutionary process in the genus *Luzula* DC. *Genet. Iberica* 2:257–62
117. Martinez-Perez E, Shaw P, Moore G. 1999. Homologous chromosome pairing

in wheat: chromosome pairing in wheat, and centromere behaviour in its diploid and polyploid relatives. *Int. Bot. Congr., 9th.* St Louis, MO

118. Martinez-Perez E, Shaw P, Reader S, Aragon-Alcaide, Miller T, Moore G. 1999. Homologous chromosome pairing in wheat. *J. Cell Sci.* 112:1761–69
119. McClintock B. 1930. A cytological demonstration of the location of an interchange between two non-homologous chromosomes in *Zea mays*. *Proc. Natl. Acad. Sci. USA* 16:791–96
120. McCouch SR, Chen X, Panaud O, Temnykh S, Xu Y, et al. 1997. Microsatellite marker development, mapping and applications in rice genetics and breeding. *Plant Mol. Biol.* 35:89–99
121. Meinke DW, Cherry JM, Dean C, Rounsley SD, Koornneef M. 1998. *Arabidopsis thaliana*: a model plant for genome analysis. *Science* 282:662–67
122. Melake Berhan A, Hulbert SH, Bulter LG, Bennetzen JL. 1993. Structure and evolution of the genomes of *Sorghum bicolor* and *Zea mays*. *Theor. Appl. Genet.* 86:598–604
123. Metz CW. 1916. Chromosome studies on the *Diptera* II. The paired association of chromosomes in the *Diptera* and its significance. *J. Exp. Zool.* 21:213–79
124. Miller JT, Dong F, Jackson SA, Song J, Jiang J. 1998. Retrotransposon-related DNA sequences in the centromeres of grass chromosomes. *Genetics* 150:1615–23
125. Ming R, Liu SC, Lin YR, daSilva J, Wilson W, et al. 1998. Detailed alignment of saccharum and sorghum chromosomes: comparative organisation of closely related diploid and polyploid genomes. *Genetics* 150:1663–82
126. Montefalcone G, Tempesta S, Rocchi M, Archidiacono N. 1999. Jumping centromeres: an additional oddness. *Eur. Cytogenet. Conf. 2nd*, Vienna, Aust.
127. Moore G. 1995. Cereal genome evolution: pastoral pursuits with lego genomes. *Curr. Opin. Genet. Dev.* 5:717–24
128. Moore G, Abbo S, Cheung W, Foote T, Gale M, et al., 1993. Key features of Cereal Genome Organisation as revealed by the use of cytosine methylation-sensitive restriction endonucleases. *Genomics* 15:472–82
129. Moore G, Cheung W, Schwaracher T, Flavell R. 1991. BIS1, a major component of the cereal genome and a tool for studying genomic organisation. *Genomics* 10:469–76
130. Moore G, Devos KM, Wang Z, Gale MD. 1995. Cereal Genome Evolution, Grasses, line up and form a circle. *Curr. Biol.* 5:737–739
131. Moore G, Foote T, Helentjaris T, Devos K, Kurata N, Gale M. 1995. Was there a single ancestral cereal chromosome? *Trends Genet.* 11:81–82
132. Moore G, Gale MD, Kurata N, Flavell RB. 1993. Molecular characterisation of small grain cereal genomes: current status and prospects. *Nat. Biotechnol.* 11:584–89
133. Moore G, Roberts M, Aragon-Alcaide L, Foote T. 1997. Centromeric sites and cereal chromosome evolution. *Chromosoma* 35:17–23
134. Namuth DM, Lapitan NLV, Gill KS, Gill BS. 1994. Comparative RFLP maps of *Hordeum vulgare* and *Triticum tauschii*. *Theor. Appl. Genet.* 89:865–72
135. Naranjo T. 1995. Chromosome structure of *Triticum longissimum* relative to wheat. *Theor. Appl. Genet.* 91:105–109
136. Naranjo T, Roca A, Goicoechea PG, Giraldez R. 1988. Chromosome structure of common wheat. Genome reassignment of chromosomes 4A and 4B. *Proc. Int. Wheat Genet. Symp., 7th, Cambridge* 1:115–20
137. Noma K, Obtsubo E, Ohtsubo H. 1999. Non-LTR retrotransposon (LINEs) are ubiquitous components of plant genomes. *Mol. Gen. Genet.* 261:71–79

138. Nonomura K-I, Kurata N. 1999. Organisation of the 1.9 Kb repeat unit RCE1 in the centromeric region of rice chromosomes. *Mol. Gen. Genet.* 261:1–10

139. Nordenskiold H. 1961. Tetrad analysis and course of meiosis in three hybrids of *Luzula campestris*. *Hereditas* 47:203–38

140. Okamoto M. 1957. A synaptic effect on chromosome V. *Wheat Inf. Serv.* 5:6–7

141. Panstruga R, Buschges R, Piffanelli P, Schulze-Lefert P. 1998. A contiguous 60 Kb genomic stretch from barley reveals molecular evidence for gene islands in a monocot genome. *Nucleic Acids Res.* 26:1056–62

142. Peacock WJ, Dennis ES, Rhoads MM, Pryov AJ. 1981. Highly repeated DNA sequences limited to knob heterochromatin in maize. *Proc. Natl. Acad. Sci. USA* 78:4490–94

143. Periera MG, Lee M, Bramel-Cox P, Woodman N, Doebley J, Whitkus R. 1993. Construction of an RFLP map in sorghum and comparative mapping in maize. *Genome* 37:236–43

144. Presting GG, Malysheva L, Fuchs J, Schubert I. 1998. A TY3/GYPSY retrotransposon-like sequence localises to the centromeric regions of cereal chromosomes. *Plant J.* 16:721–28

144a. Rabl C. 1885. Veber Zelltheilung. *Morphol. Jahrb.* 10:214–330

145. Ramsey L, Macaulay M, Cardle L, Morgante M, Invaissevich SD, et al. 1999. Intimate association of microsatellite repeats with retrotransposons and other dispersed repetitive elements in barley. *Plant J.* 17:415– 25

146. Riley R, Chapman V. 1958. Genetic control of the cytological diploid behaviour of hexaploid wheat. *Nature* 182:712–15

147. Roder MS, Lapitan NLV, Sorrells ME, Tanksley SD. 1993. Genetic and physical mapping of barley telomeres. *Mol. Gen. Genet.* 238:294–303

148. Russell GJ, Follet EAC, Subak-Sharpe JH, Harrison BD. 1971. The double strand DNA of cauliflower mosaic virus. *J. Gen. Virol.* 11:129–38

149. Saghai Maroof MA, Yang GP, Biyashev RM, Maughan PJ, Zhang Q. 1996. Analysis of barley and rice genomes by comparative RFLP linkage mapping. *Theor. Appl. Genet.* 92:541–51

150. San-Miguel P, Gaut BS, Tikhonov A, Nakajima Y, Bennetzen JL. 1998. The paleontology of intergene retrotransposons of maize. *Nat. Genet.* 20:43–45

151. San-Miguel P, Tikhonov A, Jin Y-K, Melake-Berham A, Springer PS, et al. 1996. Nested retrotransposons in the intergenic regions of the maize genome. *Science* 274:765–68

152. Santos JL, Cuadrado, Diez M, Romera C, Cunado N, et al. 1995. Further insights on chromosomal pairing of autopolyploids a triploid and tetraploids of rye. *Chromosoma* 104:298–307

153. Schwartz-Sommer Z, Leclercq L, Gobel E, Saedler H. 1987. *Cin4*, an insert altering the structure of the A1 gene in *Zea mays*, exhibits properties of nonviral retrotransposons. *EMBO J.* 6:3873–80

154. Schwarzacher T, Heslop-Harrison JS. 1991. *In situ* hybridisation to telomeres using synthetic oligomers. *Genome* 34:317–23

155. Sears ER. 1977. An induced mutant with homoeologous pairing in common wheat. *Can. J. Genet. Cytogenet.* 19:585–93

156. Singh K, Ishii T, Parco A, Huang N, Brar D, Khush GS. 1996. Centromere mapping and orientations of the molecular linkage map of rice (*Oryza sativa* L). *Proc. Natl. Acad. Sci. USA* 93:6163–68

157. Snape JW, Flavell RB, O'Dell M, Hughes WG, Payne PI. 1985. Intrachromosomal mapping of nucleolar organiser region relative to three marker loci on chromosome 1B of wheat (*Triticum aestivum*). *Theor. Appl. Genet.* 69:263–70

158. Solomon E, Bodmer W. 1979. Evolution of sickle variant genes. *Lancet* 1:923

159. Springer PS, Edwards KJ, Bennetzen JL. 1994. DNA class organisation on maize *Adh1* yeast artificial chromosomes. *Proc. Natl. Acad. Sci. USA* 91:863–67
160. Suoniemi A, Anamthuwat-Jonsson K, Arna T, Schulman AH. 1996. The retrotransposon BARE1 is a major dispersed component of the barley (*Hordeum vulgare*) genome. *Plant Mol. Biol.* 30:1321–29
161. Suoniemi A, Narvanto A, Schulman AH. 1996. The BARE1 retrotransposon is transcribed in barley from an LTR promoter active in transient assays. *Plant Mol. Biol.* 31:295–306
162. Suoniemi A, Tanskanen J, Schulman AH. 1998. Gypsy-like retrotransposon are widespread in the plant kingdom. *Plant J.* 13:699–705
163. Swartz MN, Trauter TA, Kornberg A. 1962. Enzymatic synthesis of deoxyribonucleic acid. *J. Biol. Chem.* 237:1961–67
164. Thomas JB, Kaltsikes PJ. 1976. A bouquet-like attachment plate for telomeres in leptotene of rye revealed by heterochromatin staining. *Heredity* 36:155–62
165. Timopheyeva LP, Kolomiyets OL, Vorontsova NI, Bogdanov YF. 1988. An electron microscope study of the synaptonemal complex in common wheat. Initiation of synapsis. *Tsitologiva* 30:390–94
166. Van Deynze AE, Dubcovsky J, Gill KS, Nelson JC, Sorrells ME, et al. 1995. Molecular-genetic maps for group 1 chromosomes of Triticeae species and their relation to chromosomes in rice and oat. *Genome* 38:45–59
167. Van Deynze AE, Nelson JC, Yglesias ES, Harrington SE, Braga DP, et al. 1995. Comparative mapping in grasses: wheat relationships. *Mol. Gen. Genet.* 248:744–54
168. Van Deynze AE, Nelson JC, O'Donoughue LS, Ahn SN, Siripoonwiwat W, et al. 1996. Comparative mapping in grass:oats relationships. *Mol. Gen. Genet.* 249:349–56
169. Vega JM, Feldman M. 1998. Effect of the paring gene Ph1 and premeiotic colchicine treatment on intra and interchromosomal pairing of isochromosomes in common wheat. *Genetics* 150:1199–208
170. Vershinin AV, Schwarzacher T, Heslop-Harrison JS. 1995. The large scale genome organisation of repetitive DNA families at the telomeres of chromosomes of rye chromosomes. *Plant Cell* 7:1823–33
171. Vershinin A, Svitashev S, Gummenson P-O, Salomon B, von Bothmer R, Bryngellsson T. 1994. Characterisation of a family of tandemly repeated DNA sequences in the Triticeae. *Theor. Appl. Genet.* 89:217–25
172. Walters MS. 1970. Evidence on the time of chromosome pairing from the preleptotene spiral stage in *Lilium longiflorum* "Croft". *Chromosoma* 29:375–418
173. Walters MS. 1972. Preleptotene chromosome contraction in *Lilium longiflorum* "Croft". *Chromosoma* 39:311–32
174. Wang, Z, Weber JL, Zhong G, Tanksley SD. 1994. Survey of plant short tandem DNA repeats. *Theor. Appl. Genet.* 88:1–6
175. Watanabe K. 1983 Studies on control of diploid-like meiosis in polyploid taxa of *Chrysanthemum*. *Theor. Appl. Genet.* 66:9–14
176. Werner JE, Kota RS, Gill BS, Endo TR. 1992. Distribution of telomeric repeats and their role in healing of broken chromosome ends in wheat. *Genetics* 35:844–48
177. Wessler SR, Bureau TE, White SE. 1995. LTR-retrotransposons and MITES-important players in the evolution of plant genomes. *Curr. Opin. Genet. Dev.* 5:814–21
177a. Westergaard M, Von Wettstein D. 1972. The synaptonemal complex. *Annu. Rev. Genetics* 6:71–110

178. Whitkus R, Doebley J, Lee M. 1992. Comparative genome mapping of sorghum and maize. *Genetics* 132:1119–30

179. Wilson WA, Harrington SE, Woodman WL, Lee M, Sorrells ME, McCouch SR. 1999. Can we infer the genome structure of progenitor maize through comparative analysis of rice, maize and the domesticated panicoids? *Genetics.* 153:453–73

180. Wolf KH, Gouy M, Yang Y-W, Sharp PM, Li WH. 1989. Date of the monocot-dicot divergence estimated from chloroplast DNA sequence data. *Proc. Natl. Acad. Sci. USA* 86:6201–5

181. Yen Y, Kimber G. 1990. Meiotic behaviour of induced autotetraploids in Triticum. *Genome* 33:302–7

182. Zakian VA. 1989. Structure and fusion of telomeres. *Annu. Rev. Genet.* 23:579–604

183. Zhang H, Jia J, Gale MD, Devos KM. 1998. Relationship between the chromosomes of *Aegilops umbellulata* and wheat. *Theor. Appl. Genet.* 96:69–75

Annu. Rev. Plant Physiol. Plant Mol. Biol. 2000. 51:223–56

AGROBACTERIUM AND PLANT GENES INVOLVED IN T-DNA TRANSFER AND INTEGRATION

Stanton B. Gelvin
Department of Biological Sciences, Purdue University, West Lafayette, Indiana 47907-1392; e-mail: gelvin@bilbo.bio.purdue.edu

Key Words crown gall tumorigenesis, plant genetic transformation, plant genetic engineering, interkingdom genetic exchange, protein and nucleic acid transfer

■ **Abstract** The phytopathogenic bacterium *Agrobacterium tumefaciens* genetically transforms plants by transferring a portion of the resident Ti-plasmid, the T-DNA, to the plant. Accompanying the T-DNA into the plant cell is a number of virulence (Vir) proteins. These proteins may aid in T-DNA transfer, nuclear targeting, and integration into the plant genome. Other virulence proteins on the bacterial surface form a pilus through which the T-DNA and the transferred proteins may translocate. Although the roles of these virulence proteins within the bacterium are relatively well understood, less is known about their roles in the plant cell. In addition, the role of plant-encoded proteins in the transformation process is virtually unknown. In this article, I review what is currently known about the functions of virulence and plant proteins in several aspects of the *Agrobacterium* transformation process.

CONTENTS

1040-2519/00/0601-0223$14.00

INTRODUCTION

The interaction between *Agrobacterium tumefaciens* (and the related "species" *A. rhizogenes, A. rubi,* and *A. vitis*) and plants involves a complex series of chemical signals communicated between the pathogen and the host. These signals include neutral and acidic sugars, phenolic compounds, opines (crown gall-specific molecules synthesized by transformed plants), Vir (virulence) proteins, and the T-(transferred) DNA that is ultimately transferred from the bacterium to the plant cell. A number of excellent review articles have discussed chemical signaling (9, 64, 225), and therefore I do not consider this aspect of the plant-microbe interaction at length. Although we currently have detailed knowledge of the molecular and genetic events leading to plant transformation that occur within *Agrobacterium* (30, 82, 98, 112, 226), at present we understand relatively little about the role that plant genes and proteins play in this process. Several recent reviews have begun to address this topic (65, 83, 105, 166, 182, 202, 203, 234, 237, 238), and Figure 1 (see color plate) depicts a summary of our current knowledge. The focus of this review is upon the roles of specific *Agrobacterium*- and plant-encoded proteins in the T-DNA transfer and integration events. Elucidation of these roles will further our knowledge of basic plant cell biology as well as promote the genetic manipulation of plants by *Agrobacterium*-mediated transformation. It is the belief of this author that scientists may have approached the limit of extending the host range of *Agrobacterium* by manipulation of the bacterium, and that further advances, if any, will come from manipulation of the plant host. Such manipulation will necessarily involve detailed understanding of the role that plant genes play in the transformation process, and how plant-encoded proteins interact with Vir proteins associated with the incoming T-DNA.

INITIATION OF THE T-DNA TRANSFER PROCESS

The T-DNA transfer process initiates when *Agrobacterium* perceives certain phenolic and sugar compounds from wounded plant cells (reviewed in 28, 82, 225). These phenolic compounds serve as inducers (or coinducers) of the bacterial *vir* genes, but may normally be involved in phytoalexin and lignin biosynthesis by the plant. Thus, *Agrobacterium* subverts part of the plant's defense mechanism and uses these compounds to signal the presence of a potentially susceptible plant. Phenolic chemicals such as acetosyringone and related compounds (18, 53, 135, 140, 185, 187, 188, 190, 208, 229) are perceived via the VirA sensory protein (3, 49, 94, 109, 133, 194, 207, 224), although there is some evidence for participation of a chromosomally encoded protein in the phenolic recognition process (79). Autophosphorylation of VirA protein and the subsequent transphosphorylation of VirG protein (92, 93) result in the activation of *vir* gene transcription. Most of the induced Vir proteins are directly involved in T-DNA processing from the Ti-(tumor inducing) plasmid and the subsequent transfer of T-DNA from the bacterium to the plant. Table 1 describes the functions of Ti-plasmid encoded virulence proteins in *Agrobacterium* and in the plant cell.

VirD2 protein is directly involved in processing the T-DNA from the Ti-plasmid. VirD2 nicks the Ti-plasmid at 25-bp directly repeated sequences, called T-DNA borders, that flank the T-DNA (47, 62, 90, 163, 191, 197, 214, 215, 230). Following nicking, VirD2 associates strongly, probably covalently, with the 5′ end of the resulting DNA molecule (52, 62, 77, 84, 223, 232) through tyrosine[29] (221). Although purified recombinant VirD2 protein can nick a single-stranded T-DNA border in vitro (89, 160), nicking at double-stranded borders requires both VirD1 and VirD2 proteins in vivo (90) and in vitro (177). T-DNA borders resemble sequences found at origins of transfer (oriT) of some conjugative plasmids (39, 111, 158, 159), and the nicking process carried out by VirD1 and VirD2 has been likened to that carried out by other conjugal DNA relaxing enzymes (112, 158). In a manner similar to that of bacterial conjugation systems (193), T-DNA is transferred to plants as a single-stranded DNA molecule called the T-strand (204, 233). Thus, T-DNA enters the plant as a protein/nucleic acid complex composed of a single VirD2 molecule attached to a single-stranded T-DNA.

WHICH VIR PROTEINS ACCOMPANY THE T-STRAND INTO THE PLANT?

VirD2 likely serves as a pilot protein to guide the T-strand from the bacterium into the plant cell. This transfer may take place through a pilus, composed of numerous proteins encoded by the *virB* operon, and VirD4 (12, 30, 97, 98, 226). An elucidation of the structure and assembly of the "T-pilus" (108) remains a critical area of research in *Agrobacterium* transformation. A processed form of VirB2 is the pilin protein that helps form the pilus (54, 95, 98, 108). Presumably, both the T-strand and VirE2 protein are transferred from *Agrobacterium* to the plant cell through this pilus, although the precise role of pili in DNA transfer is not clear in any conjugal system. Although the T-pilus may simply serve as a hook to bring the bacterium and plant cell into close proximity, it is clear that the pilus is required for transformation: All *virB* and *virD4* mutants (except nonpolar mutations in *virB1,* which are extremely attenuated in virulence) are avirulent (14, 141, 189, 194). It is therefore likely that the pilus, and thus VirB2 protein, interacts directly with the plant cell.

VirE2 protein plays an important role in the transformation process. VirE2 is a DNA binding protein that cooperatively associates with any single-stranded DNA sequence (31, 32, 35, 68, 180). *virE2* mutants are highly attenuated in virulence, although on some plant species such mutants incite tumors at low efficiency (48, 189). VirE2 is also the most abundant Vir protein accumulated in acetosyringone-induced *Agrobacterium* cells (55). However, it is likely that VirE2 protein functions primarily in the plant cell because *virE2* mutants accumulate normal levels of T-strands and other processed forms of the T-DNA in acetosyringone-induced bacterial cells (192, 196, 215).

The abundance of VirE2, its single-stranded DNA binding activity in vitro, and the single-stranded nature of the T-strand led to the proposal that the T-DNA was transferred from the bacterium to the plant cell as a "T-complex" composed of single-stranded DNA capped at the 5′ end with VirD2, and the entire molecule coated with VirE2 protein (83). Whereas this was an attractive (and still viable)

TABLE 1 Functions of Ti-plasmid–encoded virulence proteins in *Agrobacterium* and in plants

Virulence protein	Function in *Agrobacterium*	Function in the plant[a]	References
VirA	Phenolic sensor of a two-component regulatory system	N/A	3, 49, 94, 109, 133, 194, 207, 224
VirG	Phenolic response regulator of a two-component regulatory system	N/A	92, 115a, 194
VirB1-11	Synthesis and assembly of the T-pilus (VirB2 encodes a prepropilin)	N/A	12, 30, 54, 95, 97, 98, 108, 226
VirC1	Putative "overdrive" binding protein; enhancement of T-DNA transfer (?)	N/A	206a
VirD1	Required for T-DNA processing in vivo, and for double-strand T-DNA border nicking in vitro	N/A	90, 177
VirD2	1. T-DNA border-specific endonuclease		47, 62, 90, 163, 191, 197, 214, 215, 230
	2. Putative "pilot protein" that leads the T-strand through the transfer apparatus and into the plant		52, 62, 77, 84, 221, 223, 232
		1. Nuclear targeting of the T-strand	34, 78, 85, 105, 106, 142, 206
		2. Protection of the T-strand from 5′ exonucleolytic degradation	52
		3. T-stand integration into the plant genome	142, 147, 205
VirE1	1. Required for VirE2 export from *Agrobacterium*	N/A	198
	2. Chaperone for VirE2	N/A	45a, 198a, 234a

(continued)

TABLE 1 (*Continued*)

Virulence protein	Function in *Agrobacterium*	Function in the plant[a]	References
VirE2	1. Formation of a putative "T-complex" in *Agrobacterium*		83
		1. Formation of a putative "T-complex" in the plant	66
		2. Protection of the T-strand from nucleolytic degradation	175, 233
		3. Nuclear targeting of the T-strand	34, 36, 326
		4. Passage of the T-strand through the nuclear pore complex	33
VirF	?	1. Host range factor	134, 167
		2. Possible interaction with Skpl proteins to regulate plant cell division cycle	Paul Hooykaas, personal communication
VirH (PinF)	1. Putative cytochrome P450 enzyme	N/A	100a
VirJ/AcvB	1. Putative T-strand binding protein; T-strand export from *Agrobacterium* (?)	N/A	228

[a]N/A, not applicable.

model, recent research suggests that the T-strand/VirD2 complex may be transferred from the bacterium separate from VirE2 protein. This refinement of the model derives from several lines of experimentation.

In 1984, Otten et al (155) described a phenomenon termed extracellular complementation. They inoculated plants with two *Agrobacterium* strains, each individually avirulent. The first strain, termed the T-DNA donor, contained a wild-type T-DNA but a mutant *virE* gene. The second strain, termed the VirE2 donor, contained wild-type *vir* genes but lacked a T-DNA. However, when the strains were mixed, tumors developed on infected wound sites. The authors demonstrated that DNA transfer between the two strains did not occur, and that bacterial binding to the plant cell was necessary to effect tumorigenesis. These experiments suggested that VirE2 protein could be transferred to a plant cell independent of T-DNA transfer.

More recently, several lines of experimentation have confirmed the ability of *Agrobacterium* to transfer VirE2 protein and T-DNA separately. The first group of experiments relies upon a phenomenon called oncogenic suppression. Certain plasmids of the IncQ and IncW incompatibility groups, such as RSF1010 and pSa, respectively, can inhibit tumorigenesis when coresident with a Ti-plasmid in *Agrobacterium* (17, 59, 116). If present in the T-DNA donor strain (lacking a functional *virE2* gene), oncogenic suppression is either not effected [in the case of pSa (110)] or is only partially effected [in the case of RSF1010 (17)]. However, when either of these plasmids is present in the VirE2 donor strain, oncogenic suppression is relatively strong (17, 110). Thus, the transfer of T-DNA to a plant can occur under conditions that prevent VirE2 protein transfer. Second, *virE1* mutants can still deliver T-DNA, but not VirE2 protein, in extracellular complementation assays (198). Third, a *virE2* mutant *Agrobacterium* strain can deliver T-DNA to regenerating tobacco protoplasts, although the amount of T-DNA accumulating in the cytoplasm is five- to tenfold less than that detected (using PCR) when T-DNA is delivered from a wild-type strain (233). Fourth, tumors can be incited on a VirE2-producing transgenic tobacco plant by a *virE Agrobacterium* mutant, whereas tumorigenesis does not occur when wild-type tobacco plants are used (36). This latter experiment demonstrates that VirE2 protein functions in the plant cell and is not necessary in the bacterium for transformation to occur. These experiments did not indicate whether VirE2 interacts with the T-strand in the plant cytoplasm or in the nucleus. However, Gelvin (66) showed that a *virE2, virD2*ΔNLS (lacking the nuclear targeting sequence of VirD2 protein; see below) *Agrobacterium* double mutant could incite tumors on a VirE2-producing transgenic tobacco plant, but not on a wild-type plant. Thus, VirE2 protein can form a complex with the incoming T-DNA in the plant cytoplasm.

Although not proof, these experiments strongly suggest that *Agrobacterium* exports VirE2 and the T-DNA/VirD2 complex separately. This model corresponds with current models of conjugal DNA transfer between bacteria, which assert that free "T-strands" (single-stranded conjugal intermediates) do not exist in the bacterial cytoplasm, but rather that T-strand generation (by "unwinding" from the conjugal plasmid) occurs at the membrane surface concomitantly with T-strand export (112). Thus, in the case of T-DNA processing in *Agrobacterium*, free T-strands (in this case, representing the T-DNA) would not be available in the bacterium to bind VirE2 protein. The detection of T-strands in *Agrobacterium* may thus represent an experimental artifact derived from the "overinduction" of the vir genes by high concentrations of acetosyringone, and the use of proteases and strong detergents during T-strand isolation. Indeed, our laboratory could never detect T-strands in the supernatant of acetosyringone-induced *Agrobacterium* cells lysed in the absence of proteases and ionic detergents. However, we could identify T-strands fractionating in the membrane pellet of the centrifuged lysate (RK Jayaswal & SB Gelvin, unpublished observations).

The AcvB (VirJ) protein may play an important role in the T-DNA transfer process. In nopaline-type *Agrobacterium* strains that lack a functional *virJ* gene on the Ti-plasmid, chromosomal *acvB* mutants are avirulent (227). Octopine-type strains encode additionally an *acvB* homologue, *virJ*, on the Ti-plasmid. Thus in these strains, virulence is eliminated only in *acvB/virJ* double mutants (99, 157). Wirawan & Kojima (228) suggested that AcvB is a single-stranded DNA binding protein that interacts with the T-strand and helps in its export from the bacterium. If this model were proven, it would explain how the T-strand could be transferred from *Agrobacterium* as a T-strand/protein complex independent of VirE2 export.

A third Vir protein, VirF, is also likely exported to the plant. *Agrobacterium virF* mutants were originally described as host range mutants. Wild-type octopine-type *Agrobacterium* strains that contain the *virF* gene incite tumors on a wide variety of dicotyledonous plants, including many members of the genus *Nicotiana*. Nopaline-type strains that lack a functional *virF* gene are either avirulent or incite very small tumors on *Nicotiana glauca* (134). A similar, more limited host range is shown by *virF* mutant octopine-type strains, and addition of the octopine-type *virF* gene to nopaline-type bacterial strains extends the host range. However, nopaline-type strains are able to incite tumors on transgenic *N. glauca* plants that express the *virF* gene (167). These data strongly suggest that VirF protein functions in the plant rather than in the bacterium. At the 1999 meeting of the International Society of Molecular Plant-Microbe Interactions (IS-MPMI; Amsterdam) Dr. Paul Hooykaas described experiments that, for the first time, may shed light on the role of VirF in plant transformation. Using a yeast two-hybrid system, he showed interaction of VirF protein with *Arabidopsis* Skp1 proteins. Because Skp1 proteins are part of complexes involved in targeted proteolysis and entry of the cell into S phase, he suggested that VirF might help stimulate plant cells to divide and become more susceptible to transformation.

ATTACHMENT OF *AGROBACTERIUM* TO PLANT CELLS

Agrobacterium attaches to plant cells in a polar manner in a two-step process. The first step is likely mediated by a cell-associated acetylated, acidic capsular polysaccharide (169). The synthesis of this polysaccharide requires an intact *attR* locus; mutations in *attR* abolish virulence, bacterial attachment to carrot suspension cells, and the synthesis of the acetylated polysaccharide. This step in attachment is reversible because sheer forces such as those generated by vortexing or washing with a stream of water are sufficient to dislodge the bacteria.

The second step in attachment involves the elaboration of cellulose fibrils by the bacterium, which enmeshes large numbers of bacteria at the wound surface (41, 122, 123, 125–128). Although cellulose synthesis by the bacterium is not essential for tumorigenesis under laboratory inoculation conditions, it may help in the initial binding of the bacteria to the plant in nature because cellulose synthesis mutants are more easily washed off the plant surface than are wild-type bacteria (121).

The chromosomal virulence genes *chvA*, *chvB*, and *pscA* (*exoC*) are involved in the synthesis, processing, and export of cyclic β-1,2-glucans and other sugars

(4, 25, 26, 50, 51, 86, 100, 119, 150, 165, 201, 209) and may be involved indirectly in bacterial attachment. *Agrobacterium* strains mutant in *chvA* or *chvB* show attachment deficiencies and are either avirulent or highly attenuated in virulence under many inoculation conditions (50, 51). However, *chvB* mutants partially regain virulence if grown and inoculated at 19°C (11). Other *Agrobacterium* mutants, termed *att* mutants, are deficient in bacterial attachment (122, 129). All nonattaching bacterial mutants are avirulent under normal inoculation conditions. However, T-DNA can be delivered from nonattaching mutant bacterial cells when they are microinjected directly into the plant cytoplasm (57, 58). Thus, T-DNA (and probably VirE2 protein) export from *Agrobacterium* is not attachment dependent.

Agrobacterium can inefficiently infect plant cells in the absence of a wound site, most likely by entry through stomata (56). However, efficient infection normally requires wounding and/or a rapidly dividing cell suspension culture (5). Attachment most likely occurs at a cell wall surface of a wounded plant. Although one study reported that a plant cell wall is not required for infection (63), it is likely that small patches of cell wall material were still present on the cells used in that study (16). Bacterial attachment to the plant cell wall is saturable and is likely mediated by a protease-sensitive molecule on the plant cell surface (74, 124, 148, 149). Two plant cell wall proteins have been proposed to mediate bacterial attachment, a vitronectin-like protein (222) and a rhicadhesin-binding protein (199). However, the possible role of these plant proteins in bacterial attachment has not been shown by genetic analysis.

T-DNA transfer from *Agrobacterium* to plant cells may rapidly follow bacterial attachment. Sykes & Matthysse (200) used a cellulose-minus *Agrobacterium* strain that was easily washed off the plant surface to show that only 2–4 hours of cocultivation of the bacteria on wound sites of *Kalanchoe* leaves was sufficient to produce tumors. Similarly, Virts & Gelvin (220) showed that T-DNA transfer from *Agrobacterium* to regenerating *Petunia* protoplasts required only 2–6 hours of cocultivation. Narasimhulu et al (147) also showed that *Agrobacterium* that had previously been induced with acetosyringone required only 2 hours to transfer T-DNA to tobacco BY2 cells, as indicated by the expression of a T-DNA-encoded *gusA* gene 24 hours after inoculation.

Recently, several studies have been conducted to explore the roles of plant proteins in *Agrobacterium* attachment. Nam et al (144) identified several *Arabidopsis* ecotypes that were highly recalcitrant to *Agrobacterium*-mediated transformation. Two of these ecotypes, Bl-1 and Petergof, were blocked at an early step of the transformation process. Microscopic analysis revealed that these ecotypes bound fewer bacteria to the root surface than did a highly transformable ecotype such as Aa-0. Nam and colleagues later identified *Arabidopsis* T-DNA insertion mutants in the ecotype Ws that are resistant to Agrobacterium transformation (*rat* mutants). Table 2 describes the *rat* mutants initially isolated. Two of these mutants, *rat1* and *rat3,* are deficient in binding *Agrobacterium* to cut root surfaces (146). Plasmid rescue experiments identified T-DNA/plant DNA junction fragments from these

TABLE 2 Characterization of *Arabidopsis rat* mutants

Line	Transient transformation[a] % Root bundles showing GUS staining	Stable transformation[b] % Root bundles showing: ppt-Resistance[c]	Tumorigenicity	Probable function of *RAT* gene[d]
Wild-type	92 ± 6	87 ± 10	86 ± 15	—
rat1	22 ± 4	5 ± 2	7 ± 1	Arabinogalactan protein
rat3	31 ± 2	9 ± 2	10 ± 4	Likely cell wall protein
rat4	10 ± 4	14 ± 4	19 ± 8	Cellulose synthase-like protein
rat5	86 ± 2	15 ± 5	8 ± 3	Histone H2A
rat6	9 ± 2	12 ± 10	16 ± 5	Not sequenced
rat7	18 ± 10	10 ± 5	18 ± 10	Unknown
rat8	30 ± 8	40 ± 5	36 ± 5	Not sequenced
rat9	20 ± 4	6 ± 4	4 ± 5	Unknown
rat10	26 ± 12	19 ± 5	3 ± 4	Not sequenced
rat11	24 ± 4	14 ± 3	4 ± 5	Not sequenced
rat12	19 ± 5	10 ± 5	5 ± 6	Not sequenced
rat13	11 ± 2	20 ± 10	5 ± 8	Not sequenced
rat14	13 ± 1	9 ± 5	6 ± 5	Unknown
rat15	18 ± 2	11 ± 6	13 ± 5	Not sequenced
rat16	28 ± 7	33 ± 11	21 ± 10	Not sequenced
rat17	90 ± 6	20 ± 12	18 ± 12	myb-like transcription factor
rat18	88 ± 5	27 ± 8	21 ± 7	Not sequenced
rat19	18 ± 3	9 ± 4	4 ± 3	Not sequenced
rat20	83 ± 10	31 ± 10	10 ± 6	Not sequenced
rat21	20 ± 8	21 ± 8	8 ± 9	Not sequenced
rat22	70 ± 5	29 ± 15	17 ± 8	Unknown

[a]Three plants for each mutant and >100 root segments for each plant were examined.

[b]Five plants were tested for each mutant and 40–50 root bundles were tested for each plant.

[c]Resistance to the herbicide phosphinothricin.

[d]The proposed gene function is based on sequence homology of cDNA and genomic clones identified by the T-DNA-plant DNA junctions. For further details, see Reference 146.

mutants (J Nam, CT Ranjith Kumar, S Chattopadhyay & SB Gelvin, in preparation). DNA sequence analysis of these junction fragments, as well as wild-type cDNA and genomic clones corresponding to these junctions, indicated that *rat1* encodes an arabinogalactan protein (AGP) and *rat3* encodes a small protein that is likely secreted to the apoplast. The involvement of AGPs in Agrobacterium transformation was confirmed using β-glucosyl Yariv reagent, which binds AGPs specifically. When *Arabidopsis* root segments were incubated with an active Yariv reagent prior to inoculation with *Agrobacterium,* transformation was blocked. An inactive β-mannosyl Yariv reagent did not block transformation. Control experiments indicated that β-glucosyl Yariv reagent did not affect the viability of *Arabidopsis* root segments or *Agrobacterium* cells. Another *rat* gene, *Rat4,* encodes a cellulose synthase (CelA)-like protein (J Nam, unpublished observations). The *celA* and *celA*-like genes comprise a large multigene family in *Arabidopsis* composed of over 40 members (T Richmond, P Villand, S Cutler, & C Somerville, poster presented at the 9th International Conference on *Arabidopsis* Research, Madison, WI, 1998). Note that mutation of just this one member of such a large multigene family can abolish Agrobacterium-mediated transformation. Somewhat surprisingly, *Agrobacterium* cells bind to the cut surfaces of *rat4 Arabidopsis* roots as well as they do to wild-type roots (A Matthysse, unpublished observations).

Although the *Arabidopsis rat1* and *rat3* mutants and the Bl-1 and Petergof ecotypes display extremely attenuated virulence when they are inoculated on cut root surfaces with wild-type bacteria, they are transformed as well as are the corresponding wild-type plants and highly transformable ecotypes using the flower vacuum infiltration method. However, the attachment-deficient *Agrobacterium* mutants *chvA* and *chvB* fail to transform *Arabidopsis* using either inoculation method (142a). These results suggest that transformation of the female gametophyte by the flower vacuum infiltration method does not require plant cell surface proteins needed for transformation of somatic tissues.

NUCLEAR TARGETING OF THE T-DNA IN PLANT CELLS

Once inside the plant cell, the T-DNA must target to the nucleus. Several *Agrobacterium* virulence (Vir) proteins, as well as a number of plant proteins, are likely involved in this process.

The proteins VirD2 and VirE2 contain plant-active nuclear localization signal (NLS) sequences. VirD2, which is tightly associated with the 5′ end of the T-strand, contains two NLS regions. The first, a monopartite NLS resembling somewhat the NLS found in the SV40 large T-antigen, is located in the amino-terminal region of the protein. When fused to β-galactosidase, a peptide containing this NLS can target the chimeric protein to plant nuclei (78). However, it is unlikely that this NLS participates in nuclear localization of the T-strand (106, 174, 183). Tyrosine^{29}, located next to this NLS, is the site for VirD2 linkage to the T-strand (221) and it is therefore likely that this NLS is occluded by the connected T-strand.

A second, bipartite NLS resembling that of *Xenopus laevis* nucleoplasmin resides in the carboxy-terminal region of VirD2. Fusions of peptides containing this NLS to reporter proteins direct these chimeric proteins to plant and yeast nuclei (34, 85, 105, 106, 142, 206). In addition, VirD2 protein can localize to the nucleus of animal cells (73, 168).

VirE2 protein contains two separate bipartite NLS regions that can target linked reporter proteins to plant cell nuclei (34, 36). Fluorescently labeled single-stranded DNA coated with VirE2 and microinjected into plant cells localizes to the nucleus, whereas naked single-stranded DNA remains in the cytoplasm (236). Interestingly, VirE2 cannot direct single-stranded DNA into the nuclei of cells of *Xenopus laevis* or *Drosophila melanogaster*. When the VirE2 NLS amino acids are altered slightly to resemble more closely animal NLS sequences, the modified VirE2 can now target DNA to animal cell nuclei (73). These results suggests that nuclear targeting signals in plant and animal cells may differ slightly.

Considering the close association of the T-strand with both VirD2 and VirE2 proteins, it would be interesting to determine the relative roles these proteins play in T-strand nuclear localization. The literature on this subject is somewhat contradictory. Rossi et al (174) showed that deletion of the VirD2 bipartite NLS resulted in almost complete loss of transformation, which suggests that VirE2 NLS domains could not compensate for loss of the VirD2 NLS. Ziemienowicz (235) further showed that both VirD2 and VirE2 proteins were necessary for nuclear targeting of in vitro-synthesized T-complexes introduced into permeabilized HeLa cells, and that the presence of VirE2 protein could not compensate for deletion of the VirD2 NLS. These authors did not, however, rule out the possibility that VirE2 might participate in nuclear targeting in a non-NLS-dependent manner. For example, VirE2 may aid in transfer of the T-strand through the nuclear pore by maintaining a narrow, extended DNA conformation (33, 35). Several other groups demonstrated that *Agrobacterium* strains containing a *virD2* gene with a NLS deletion retained almost full virulence and transformation capability (142, 183). This result suggests that VirE2 may provide nuclear targeting capabilities in the absence of the VirD2 NLS. This latter result was confirmed by the observation that VirE2-producing transgenic tobacco plants can complement a *virE2,virD2*ΔNLS double mutant *Agrobacterium* strain (66). The T-complex delivered to the plant cell by this mutant bacterium lacks any known nuclear targeting signals, which again suggests that VirE2 can function in the plant to provide nuclear targeting capabilities to the incoming T-strand. The discrepancies between these groups may be traced to the types of NLS deletions examined: Shurvinton et al (183) and Mysore et al (142) used a VirD2 NLS deletion that removed only the critical basic domains of the VirD2 NLS, whereas Rossi et al (174) and Ziemienowicz et al (235) utilized a more extensive deletion of the NLS region. Deletions within this latter region may have severe consequences on VirD2 function in addition to affecting nuclear targeting (142, 147). More experimentation will be necessary to decipher the relative roles of the VirD2 and VirE2 NLSs in nuclear targeting of the T-strands.

Recent attention has turned to the role of plant proteins in the T-DNA nuclear targeting process. Two groups have used the yeast interaction trap (two-hybrid) approach (69) to identify plant-encoded proteins that interact with the bipartite NLS region of VirD2. Ballas & Citovsky (10) identified an *Arabidopsis* importin-α (α-karoypherin) that binds specifically to this NLS. This protein is a member of a multigene family (10, 80). In other species, importin-α has been shown to bind to NLS regions in karyophilic proteins and assist in nuclear targeting (70, 71). It would be interesting to determine whether only this member, or also other members, of the importin-α protein family participate in VirD2 and VirE2 nuclear targeting.

Tao et al (Y Tao, P Rao & SB Gelvin, submitted) used a similar approach to identify proteins encoded by a cDNA library derived from tomato that interact specifically with the VirD2 NLS. In addition to importin-α, they identified a cDNA, DIG3, that encodes a type 2C serine/threonine protein phosphatase (PP2C). This PP2C negatively affected nuclear import of a β-glucuronidase (GUS)-VirD2 NLS fusion protein: When a GUS-VirD2 NLS fusion gene was electroporated into tobacco BY2 protoplasts, more than 80% of the resulting GUS activity localized exclusively to nuclei. However, coelectroporation of the GUS-VirD2 NLS gene with the PP2C gene resulted in cytoplasmic localization of GUS activity in the majority of cells. The role of phosphorylation in nuclear targeting of VirD2 became more evident when these authors demonstrated phosphorylation of a VirD2 serine residue close to the bipartite NLS, both in vitro and in BY2 cells. Alteration of this serine to an alanine resulted in predominantly cytoplasmic localization of a GUS-VirD2 (mutant) NLS fusion protein, even in the absence of overexpression of the PP2C. Finally, this group showed that an *Arabidopsis abi1* mutant was two- to fourfold more transformation proficient than was the wild-type progenitor *Arabidopsis* line. The *Abi1* gene encodes a PP2C homologue of the DIG3 gene product from tomato. This result indicates that *Abi1* is a negative effector of *Agrobacterium* transformation. Taken together, these experiments suggest that phosphorylation of a serine residue near the VirD2 NLS potentiates nuclear import of VirD2, whereas dephosphorylation decreases nuclear import of VirD2, and hence the VirD2/T-DNA complex, in plant cells.

The precise roles of this PP2C and other proteins (such as the cognate kinase) in the T-DNA nuclear targeting process await development of a defined nuclear import system using purified plant nuclei and fractionated cellular extracts. Although several plant nuclear import assays have been described (75, 80, 136, 137), none is dependent upon the addition of defined proteins.

A third plant protein may play some role in T-DNA transport through the plant cell. Deng et al (46) recently identified an *Arabidopsis* cyclophilin that interacts in a yeast two-hybrid assay with a central domain of VirD2 protein. Because some cyclophilins have peptidyl-prolyl isomerase activity, the authors speculated that this protein might serve as a chaperonin to hold VirD2 in a transfer-competent conformation during T-strand trafficking through the plant cell.

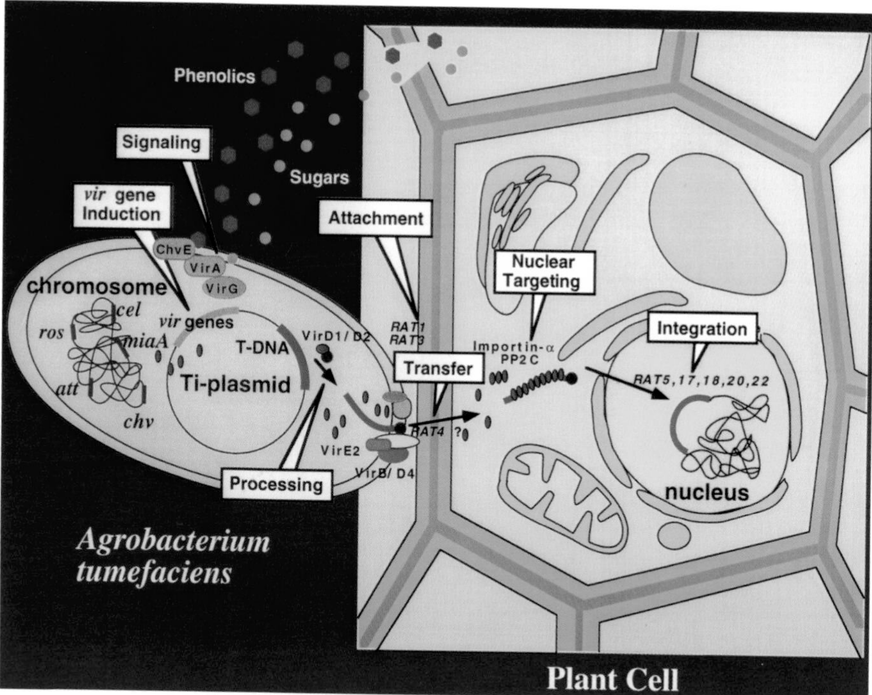

Figure 1 Schematic diagram of the Agrobacterium infection process. Critical steps that occur to or within the bacterium (chemical signaling, *vir* gene induction, and T-DNA processing) and within the plant cell (bacterial attachment, T-DNA transfer, nuclear targeting, and T-DNA integration) are highlighted, along with genes and/or proteins known to mediate these events.

ROLES OF VIR PROTEINS IN T-DNA INTEGRATION

Relatively little is known about the precise mechanism of T-DNA integration into the plant genome, or the role specific proteins play in this process. T-DNA integrates into plant chromosomes by illegitimate recombination (67, 120, 131, 154), the major mode of foreign DNA integration in plants (151, 161). It is likely that most T-DNA transferred to the plant cell nucleus does not stably integrate. This has been shown indirectly by measuring the relative amounts of transient versus stable expression of reporter gene activity in *Agrobacterium*-infected plant cells. Generally, the level of transient expression of T-DNA genes is much greater than is expression from stably integrated T-DNA (27, 88, 142, 144, 147). T-DNA enters the nucleus as a single-stranded molecule (204, 233). However, it is not yet clear whether T-DNA integrates via strand invasion of the locally denatured plant DNA by single-stranded T-strand, followed by second strand repair synthesis (107, 175, 202, 203, 205), or whether the T-strand is converted to an extrachromosomal double-stranded form prior to integration (45). Much of the extrachromosomal T-DNA that initially enters the nucleus likely becomes double-stranded, because the conversion of the T-strand to a transcriptionally competent form requires the synthesis of the DNA strand complementary to the T-strand (147). However, it is not clear whether this extrachromosomal double-stranded T-DNA is the substrate for integration, or whether it is a "dead end" molecule that is eventually lost from the dividing plant cells, either by dilution or by nucleolytic degradation.

Because VirD2 is covalently linked to the T-strand, it probably plays some role in the integration process. In vitro, VirD2 can catalyze the ligation of a T-DNA border "donor" sequence [containing VirD2 linked to the cleaved border (nucleotides 1-3) with a cleaved T-DNA border "acceptor" sequence (nucleotides 4-25)] (160). However, the in vitro ligation of the 5′ T-strand end linked to VirD2 with an acceptor sequence not containing a T-DNA border region requires either plant extracts or T4 DNA ligase (A Ziemienowitz & B Hohn, personal communication).

Two lines of evidence indicate that VirD2 participates in the T-DNA integration process in plants. The integration of the 5′ end of the T-strand into plant DNA is generally precise: Usually a few 5′ nucleotides at most are deleted when T-DNA integrates into the plant genome (175, 205). This may result from the protection from exonucleases that VirD2 offers to the capped 5′ T-strand end (52). Mutation of a conserved arginine residue in a domain of VirD2 that is common to domain III of the TraI protein of RP4 results in imprecise ligation of the 5′ end of the T-strand to plant DNA, frequently resulting in deletion of the 5′ end of the T-DNA (205). However, this mutation does not greatly reduce the efficiency of T-DNA integration.

Mysore et al (142) showed that a different mutation in VirD2 has the opposite effect. Shurvinton et al (183) demonstrated that a deletion/substitution mutation of a conserved domain near the carboxy-terminal end of VirD2, called the ω domain, resulted in an approximate two orders of magnitude decrease in tumorigenesis. This same mutation resulted in only a three- to fivefold decrease in transient T-DNA expression in tobacco and *Arabidopsis* cells (142, 147). These results suggested that this ω deletion/substitution mutation affected T-DNA integration to a much greater extent than it affected T-DNA transfer and nuclear targeting. This hypothesis was confirmed by Mysore et al (142) who directly demonstrated that an *Agrobacterium* strain harboring this ω mutation was deficient in T-DNA integration. However, the few tumors resulting from transformation of tobacco with this mutant strain contained relatively intact integrated T-DNA 5′ ends. Thus, this VirD2 mutation affected the efficiency but not the precision of T-DNA integration. It is unlikely, however, that the ω domain of VirD2 directly participates in T-DNA integration because precise deletion of this domain did not affect the frequency of stable Agrobacterium-mediated transformation (23). It is therefore likely that the ω deletion/substitution mutation (removing the conserved ω amino acids DDGR and replacing them with two serine residues) resulted in an altered structure of VirD2 that was no longer able to integrate T-DNA into the plant genome efficiently. The elucidation of the crystal structure of VirD2 may help to unravel the role of this virulence protein in T-DNA integration.

VirE2 may play only an indirect role in T-DNA integration. VirE2 mutants are extremely attenuated in virulence (48, 189). This attenuation likely results from nucleolytic degradation of the unprotected T-strand (31–33, 35, 68, 180) in either the plant cytoplasm (233), nucleus, or both. Integrated T-DNA molecules transferred from *virE2* mutant *Agrobacterium* strains exhibit extensive deletions corresponding to the 3′ ends of the T-strand (175). The 5′ ends of these T-DNAs, however, remain relatively intact, probably because of "capping" by VirD2 protein. It is also possible, however, that the severe attenuation of *virE2* mutant *Agrobacterium* strains results additionally from the inefficient nuclear targeting of the T-DNA by VirD2 protein alone.

ROLES OF PLANT PROTEINS IN T-DNA INTEGRATION

The roles of plant proteins in the T-DNA integration process are only now beginning to be defined. Because T-DNA integrates into the plant genome by illegitimate recombination (67, 120, 131, 154), plants deficient in DNA repair and recombination may be deficient in T-DNA integration. Such DNA metabolism mutants are likely to be hypersensitive to DNA damaging agents such as UV and γ-radiation, and the drug bleomycin. Sonti et al (186) investigated a number of radiation-sensitive *Arabidopsis* mutants for transient and stable Agrobacterium-mediated transformation. Although most mutants examined demonstrated wild-type levels of transformation, two mutants, *uvh1* and *rad5,* appeared to be resistant to transformation to the stable phenotypes of crown gall tumor formation and kanamycin resistance. However, these mutants were susceptible to transient transformation, as demonstrated by the expression of GUS activity directed by a *gusA* gene on the incoming T-strand. As noted previously, transient expression of GUS activity does not necessarily require T-DNA integration (142, 144, 146, 147). The authors therefore concluded that these *Arabidopsis* mutants were deficient specifically in the process of T-DNA integration. However, the results of Sonti et al (186) were questioned recently by Nam et al (145). These authors re-examined the transformation proficiency of the *Arabidopsis* mutants *uvh1* and *rad5*. They found, using both root inoculation and flower vacuum infiltration assays, that *uvh1* was as transformation proficient as was the progenitor wild-type plant, and further demonstrated directly that T-DNA integration was not impaired in this mutant. Jiang et al (91) confirmed the transformation proficiency of a different *Arabidopsis uvh1* mutant by vacuum infiltration. Nam et al (145) additionally confirmed, using a root inoculation assay, that the mutant *rad5* was deficient in stable transformation (transformation was 15% of that of wild-type plants). However, quantitative determinations of transient GUS activity using MUG fluorimetric assays indicated that this aspect of transformation was also 15% of that of the wild-type plants. Nam et al (145) therefore concluded that the *Arabidopsis* mutant *rad5* is blocked at a transformation step prior to T-DNA integration.

In addition to the *rad5* radiation hypersensitive *Arabidopsis* mutant, some radiation-sensitive *Arabidopsis* ecotypes are also recalcitrant to *Agrobacterium*-mediated transformation. Nam et al (144) examined almost 40 *Arabidopsis* ecotypes for susceptibility to root transformation by *Agrobacterium*. One ecotype, UE-1, was both slightly radiation hypersensitive and transformation deficient. These authors showed by transient and stable GUS activity assays that UE-1 was likely deficient in T-DNA integration: Transient GUS activity directed by an incoming T-DNA was actually twice as great in ecotype UE-1 than in the highly transformable ecotype Aa-0. However, stable GUS activity in UE-1 was very low compared to a high level in Aa-0. Using a T-DNA integration assay in which the amount of T-DNA integrated into high-molecular-weight plant DNA was assessed, these authors further directly showed that considerably less T-DNA integrated into the genome of UE-1 compared to Aa-0.

More recently, Nam et al (146) used a genetic approach to investigate *Arabidopsis rat* (resistant to Agrobacterium transformation) mutants, including T-DNA integration-deficient mutants. These mutants were identified in a T-DNA insertion library of the ecotype Ws (7, 60, 61). Five of the initial 21 *rat* mutants tested appeared to be integration deficient, as suggested by a high transient but low stable transformation efficiency (Table 2). Two of these putative integration-deficient mutants were investigated more thoroughly. The mutant *rat17* contains a T-DNA disruption of a gene encoding a myb-like transcription factor (CT Ranjith Kumar, unpublished). It is not yet clear whether this transcription factor interacts directly with the (yet undefined) T-DNA integration apparatus, or whether this factor is required for transcription of another gene(s) encoding proteins required for T-DNA integration.

rat5 is a more thoroughly characterized *Arabidopsis* mutant (142b, 146). *rat5* contains two tandem copies of T-DNA integrated into the 3′ untranslated region of a histone H2A gene (142b). Although highly recalcitrant to stable transformation by root inoculation, *rat5* is efficiently transformed by flower vacuum infiltration. These results suggest that some factor(s) required for efficient transformation are present in the female gametophyte [the tissue transformed by flower vacuum infiltration (231)] but absent in root somatic tissue (142a). In crosses between *rat5* and wild-type Ws plants, the *rat* phenotype appears dominant. However, the observation that *rat5* plants can be complemented to transformation-proficiency with the wild-type *RAT5* histone H2A gene suggests that the *rat5* mutant is haploinsufficient. Thus, the *rat5* phenotype is dosage dependent, and we have speculated that in this mutant the level of RAT5 histone H2A protein falls below the threshold level required for *Agrobacterium*-mediated transformation. *RAT5* is one of a six-member histone H2A multigene family (K Mysore & H Yi, unpublished), and it is still uncertain how disruption of one family member affects transformation. The role of a histone H2A in the T-DNA integration process is also unclear. It is possible that chromatin conformation of the T-DNA target site in the plant genome may influence the efficiency of T-DNA integration. T-DNA appears to integrate preferentially into transcriptionally active regions of the plant genome (76, 103, 104), and these regions may be comprised of a more "open" chromatin conformation. A possible role of the RAT5 histone H2A in T-DNA integration may be inferred by our recent finding that the RAT5 histone H2A protein interacts with VirD2 protein in a yeast two-hybrid assay system (CT Ranjith Kumar, unpublished). Perhaps this histone interacts directly with the incoming T-strand in a putative "integration complex." However, more experimentation is needed to confirm this role for the RAT5 histone H2A.

IS T-DNA INTEGRATION A LIMITING FACTOR IN *AGROBACTERIUM* TRANSFORMATION?

As described above, a number of steps in the *Agrobacterium* transformation process can limit transformation of a particular plant (Figure 1, see color plate). These include the synthesis of phenolic *vir* gene inducers by the plant, bacterial attachment, T-DNA transfer into the plant cytoplasm, T-DNA nuclear translocation, and T-DNA integration. Many plant species or cultivars/ecotypes that are recalcitrant to transformation still synthesize *vir* gene-inducing compounds (138, 144, 208, 218), and the addition of *vir* gene-inducing compounds to the bacteria prior to or during cocultivation often does not enhance transformation (172). However, there are numerous examples of transformation protocols that require the addition of *vir* gene inducers for efficient transformation (for examples, see 40, 87, 181). Bacterial attachment may also limit transformation, as demonstrated by the transformation-deficient *Arabidopsis* ecotypes Bl-1 and Petergof (144).

However, for many plant species or cultivars/ecotypes, T-DNA integration is clearly the limiting step for many transformation protocols. This can be demonstrated by relatively efficient transient transformation (defined by the expression of reporter genes encoded by the T-DNA early after infection) but poor stable transformation. Numerous examples of such recalcitrance to stable transformation, but susceptibility to transient transformation, exist for species including maize (172, 181), rice (113, 114), and apple (132). Maximova et al (132) estimated that the stable transformation of apple decreased approximately 10,000-fold from the initial transfer event. Nam et al (144) identified an *Arabidopsis* ecotype, UE-1, that was highly recalcitrant to Agrobacterium transformation because of a T-DNA integration, but not T-DNA transfer, deficiency.

If T-DNA integration were limiting for transformation of particular species or cultivars/ecotypes, it might be possible to increase the efficiency of transformation by overexpressing in plants genes involved in T-DNA integration. Mysore and colleagues (142b) recently identified an *Arabidopsis* histone H2A gene, *RAT5*, that was important for T-DNA integration but not transfer to *Arabidopsis*. The *rat5* mutant can be transiently transformed with an efficiency similar to that of the wild-type progenitor ecotype Ws. However, stable transformation to several phenotypes (crown gall tumorigenesis, phosphinothricin resistance, and stable GUS expression) is very inefficient (142b, 146). Overexpression of an additional copy of a *RAT5* gene in transgenic *Arabidopsis* plants already containing two wild-type copies of *RAT5* increased stable transformation frequencies two- to threefold. Indeed, the *RAT5* gene did not have to be integrated into the host genome to effect higher stable transformation rates. Expression of the *RAT5* gene from the incoming T-DNA of a binary vector also increased the stable transformation of *Arabidopsis* by two- to threefold. This increased transformation was seen both for genes harbored by the T-DNA of the binary vector, and also for genes encoded by the T-DNA of the Ti-plasmid coresident within the same *Agrobacterium* strain (142b). Although these experiments were conducted using the *Arabidopsis* ecotype Ws, which is susceptible to transformation, incorporation of the *RAT5* gene into the T-DNA also increased the stable transformation of several highly recalcitrant *Arabidopsis* ecotypes (L-Y Lee & SB Gelvin, unpublished). The effectiveness of the *RAT5* histone H2A gene on Agrobacterium transformation of several recalcitrant crop species is being tested presently.

Finally, it should be noted that the particular tissue transformed by *Agrobacterium* might affect the extent or pattern of T-DNA integration. For example, Grevelding et al (72) found that when root explants of *Arabidopsis* were the target tissue, the majority of T-DNA insertions were single-copy. However, multicopy T-DNA insertion was most frequent when leaf tissue was the target tissue. As noted above, Mysore et al (142a) found that many *Arabidopsis* ecotypes and mutants that were highly recalcitrant to stable transformation of root tissues were easily transformed by a flower vacuum infiltration method. Thus, certain factors that are important for T-DNA integration may be found in the female gametophyte [the tissue transformed by the vacuum infiltration procedure (231)] but may be lacking or limiting in somatic root tissue.

PROPERTIES OF PLANTS THAT AFFECT TRANSFORMATION

Various plant species differ greatly in their susceptibility to infection by *Agrobacterium tumefaciens* or *Agrobacterium rhizogenes* (6, 42, 162, 213). Even within a species, different cultivars or ecotypes may show vastly different degrees of susceptibility to tumorigenesis by particular *Agrobacterium* strains. These differences have been noted in rice (113), maize (172, 178), various legumes (81, 156, 179), aspen (13), cucurbits (184), *Pinus* species (15), tomato (212), *Arabidopsis* (144), grape (117), and other species. Although some differences in transformation frequency may be attributed to environmental or physiological factors, a genetic basis for susceptibility has clearly been established in a few plant species (8, 130, 144, 173).

Plant species may differ in their temporal competence for transformation following wounding. Armin Braun first noted this window of competence in *Vinca rosea* (19) and later in *Kalanchoe daigremontiana* (22). Bacteria were applied to cut plant surfaces various times after wounding. If the plants were inoculated within three days of wounding, tumor induction was relatively efficient. Inoculation four days after wounding resulted in only a few percent of the plants developing tumors, and after five days, tumorigenesis was absent. However, tomato remained susceptible to tumorigenesis up to two weeks after wounding (20, 21). Davis et al (40) showed that six days after wounding, tomato plants still retained approximately 25% of the susceptibility of plants inoculated immediately after wounding. Although susceptibility could be increased at later times by the addition of acetosyringone, a phenolic *vir* gene inducer synthesized specifically by wounded plant tissues (188), these treated plants never achieved susceptibility equal to untreated plants inoculated directly after wounding. Davis et al (40) also concluded that although suberization of the cell walls, which may present a physical barrier to transformation, did occur four days after wounding, suberized cells still retained a high transformation susceptibility. The authors thus concluded that factors additional to suberization and lack of *vir* gene inducers must play a role in temporal competence for transformation.

Several investigators have shown that various tissues, organs, and cell types within a plant may differ in their susceptibility to *Agrobacterium* transformation. Using transient expression of GUS activity as a reporter of transient transformation, Ritchie et al (172) showed that in maize, transformation occurred in mesocotyl segments originating from the intercalary meristem region. Histochemical analysis of sections from this region indicated that GUS activity occurred only in the vascular cylinder. Some GUS activity also localized to leaf and coleoptile tissues of shoot segments, but not in the shoot apex. Shen et al (181) similarly found that transient GUS expression was localized mainly to leaves and coleoptile regions of maize shoots. Schläppi & Hohn (178) demonstrated, using agroinoculation of maize streak virus as in indicator of transformation, that maize embryos showed a window of competence for transformation: Only embryos in which the shoot apical meristem had begun to differentiate showed competence, and the timing of this window differed among the three maize cultivars examined.

De Kathen & Jacobsen (43) showed that only dedifferentiating cells near the vascular system of cotyledon and epicotyl regions of *Pisum sativum* were susceptible to *Agrobacterium* transformation. In addition, the further the epicotyl section was from the shoot apex, the lower the transformation frequency became. In *Arabidopsis* cotyledon and leaf tissues, only dedifferentiating mesophyll cells were competent for transformation. In root tissue, competent cells were found in dedifferentiating pericycle. These cells had a common morphology: They were small, isodiametric, and had prominent nuclei and a dense cytoplasm (176).

A common feature of these cells is the induction of competence in response to wounding or phytohormone treatment. A number of plant species, including *Pinus radiata* (15), grape (117), *Arabidopsis thaliana* (176), and *Petunia hybrida* (219), show an increased transformation efficiency following hormone pretreatment. *Arabidopsis* leaf, root, hypocotyl, and cotyledon transformation protocols routinely include phytohormone pretreatment of the tissues prior to cocultivation with *Agrobacterium* (1, 2, 29, 37, 115, 118, 210).

The stimulation of cell division by phytohormones suggests that efficient *Agrobacterium* transformation may occur at a particular stage of the plant cell cycle. Villemont et al (219) investigated the role of the plant cell cycle in *Agrobacterium*-mediated transformation of *Petunia* mesophyll cells that had been synchronized with cell cycle phase-specific inhibitors. Noncycling cells that had not been treated with phytohormones, and were shown by flow-cytometric analysis of nuclear DNA and incorporation of bromodeoxyuridine to be blocked at G0-G1 phase, could not be transformed either transiently or stably to express a T-DNA-encoded *gusA* transgene. Cells treated with mimosine, which blocks the cell cycle in late G1 phase, similarly could not be transformed. However, cells blocked at M phase by colchicine treatment could express GUS activity transiently, although they could not be stably transformed. In addition, the cycling cells that showed the highest transformation competence were those cells that showed a very high S and G2 phase: M phase ratio. The authors concluded that T-DNA could be taken up, translocated to the nucleus, and expressed in cells conducting DNA synthesis but in the absence of cell division, and thus that *Agrobacterium*-mediated transient transformation required S phase DNA synthesis. Subsequent cell division was required for T-DNA integration and stabilization of transformation. However, an alternative explanation for these results is that cells can take up and translocate the T-DNA to the nucleus in the absence of S phase, but that S phase is required to convert the single-stranded T-strand to a transcription-competent double-stranded form. The development of a direct assay for T-strand uptake and nuclear translocation that does not depend upon T-DNA-encoded gene expression is required to resolve these two alternatives. Note that the generation of transgenic plants by direct DNA uptake protocols may be influenced by the stage of the cell cycle (143). In addition, the stage of the cell cycle may influence the complexity and pattern of integration of double-stranded DNA directly taken up from the medium (101).

FUTURE PROSPECTS

The molecular mechanism of plant transformation by *Agrobacterium* was first elucidated in the mid-1970s. By the early 1980s, scientists were already beginning to harness the natural genetic engineering ability of *Agrobacterium* to generate transgenic plants, and by the mid-1990s, genetically modified crops, engineered by *Agrobacterium*-mediated transformation, were becoming commonplace in many countries. Despite the use of *Agrobacterium* to transform an increasing number of agronomically and horticulturally important plant species, there remain many mysteries of the transformation process. We know relatively little about the events in the plant that mediate *Agrobacterium* attachment, T-DNA transfer to the host, nuclear targeting, and T-DNA integration into the host genome. Because viruses and bacterial pathogens take advantage of cellular machinery, by studying the mechanism of pathogenesis or tumorigenesis, we can learn about fundamental processes in plant cells. What practical applications may derive from such knowledge? I envisage at least three categories of technology that will benefit.

First, manipulation of plant genes will likely result in an increased frequency, and an increase in the host range, of *Agrobacterium*-mediated transformation. Many important plant species, or elite cultivars of particular species, currently remain recalcitrant to *Agrobacterium* transformation. For many species, this recalcitrance does not result from a lack of T-DNA transfer or nuclear targeting; transient transformation in many instances is relatively efficient. Rather, T-DNA integration into the genome of regenerable cells appears to be limiting. In the future, it may be possible to overexpress endogenous genes involved in the integration process, or to introduce homologous genes from other species, and thereby effect higher rates of stable transformation. The possibility of such future success is suggested by our recent ability to increase the stable transformation efficiency of various *Arabidopsis* ecotypes by overexpressing the *RAT5* histone H2A gene, which encodes a protein involved in T-DNA integration, during the transformation process (142b; L-Y Lee & SB Gelvin, unpublished). Furthermore, better control of the T-DNA integration process may mitigate the multicopy, inverted repeat T-DNA integration events frequently associated with homology-dependent gene silencing (38, 96, 195, 211).

Second, the generation of plants with a deficiency in T-DNA integration by illegitimate recombination may enhance our ability to effect homologous recombination, and thus gene targeting, in plants. Presently, illegitimate recombination events occur approximately 100,000-fold more frequently than do homologous recombination events (44, 102, 139, 152, 153, 164, 170, 171, 216, 217). A reduction in random T-DNA integration may allow scientists to select homologous integration events with a much-reduced random integration background.

Third, an understanding of plant genes required for Agrobacterium-mediated transformation may allow us to engineer plants for crown gall resistance. Crown gall can be an economically important disease on many plants including grapes, walnuts, apples, stonefruits, and certain ornamentals, such as roses and chrysanthemums (24). It may be possible to inhibit the expression of specific plant genes involved in *Agrobacterium* transformation to such an extent that transformation is greatly reduced but the plant remains phenotypically normal. The recent identification of *rat* (resistant to Agrobacterium transformation) mutants of *Arabidopsis* that appear normal in growth and development (146) suggests that such an approach may be feasible in crop plants.

ACKNOWLEDGMENTS

I thank Drs Barbara Hohn, Susan Karcher, Ann Matthysse, and Walt Ream for critical reading of this review, and Dr Lan-Ying Lee for help with the figure. Research in the author's laboratory is supported by grants from the US Department of Agriculture, the National Science Foundation, the Department of Energy (through the Consortium for Plant Biotechnology Research), the Biotechnology Research and Development Corporation, Novartis Agribusiness Biotechnology Research, Pioneer Hi-bred International, and Dow Agrosciences.

LITERATURE CITED

1. Akama K, Puchta H, Hohn, B. 1995. Efficient Agrobacterium-mediated transformation of *Arabidopsis thaliana* using the *bar* gene as selectable marker. *Plant Cell Rep.* 14:450–54
2. Akama K, Shiraishi H, Ohta S, Nakamura K, Okada K, et al. 1992. Efficient transformation of *Arabidopsis thaliana*: comparison of the efficiencies with various organs, plant ecotypes and *Agrobacterium* strains. *Plant Cell Rep.* 12:7–11
3. Albright LM, Huala E, Ausubel FM. 1989. Prokaryotic signal transduction mediated by sensor and regulator protein pairs. *Annu. Rev. Genet.* 23:311–36
4. Altabe S, de Iannino NI, de Mendoza D, Ugalde RA. 1990. Expression of the *Agrobacterium tumefaciens chvB* virulence region in *Azospirillum* spp. *J. Bacteriol.* 172:2563–67
5. An G. 1985. High efficiency transformation of cultured tobacco cells. *Plant Physiol.* 79:568–70
6. Anderson A, Moore L. 1979. Host specificity in the genus *Agrobacterium. Phytopathology* 69:320–23
7. Azpiroz-Leehan R, Feldmann KA. 1997. T-DNA insertion mutagenesis in *Arabidopsis*: going back and forth. *Trends Genet.* 13:152–56
8. Bailey M, Boerma HR, Parrott WA. 1994. Inheritance of *Agrobacterium tumefaciens*-induced tumorigenesis of soybean. *Crop Sci.* 34:514–19
9. Baker B, Zambryski P, Staskawicz B, Dinesh-Kumar SP. 1997. Signaling in plant-microbe interactions. *Science* 276: 726–33

10. Ballas N, Citovsky V. 1997. Nuclear localization signal binding protein from *Arabidopsis* mediates nuclear import of *Agrobacterium* VirD2 protein. *Proc. Natl. Acad. Sci. USA* 94:10723–28
11. Banta LM, Bohne J, Lovejoy SD, Dostal K. 1998. Stability of the *Agrobacterium tumefaciens* VirB10 protein is modulated by growth temperature and periplasmic osmoadaption. *J. Bacteriol.* 180:6597–606
12. Baron C, Zambryski PC. 1996.Plant transformation: a pilus in *Agrobacterium* T-DNA transfer. *Curr. Biol.* 6:1567–69
13. Beneddra T, Picard C, Nesme X. 1996. Correlation between susceptibility to crown gall and sensitivity to cytokinin in aspen cultivars. *Phytopathology* 86:225–31
14. Berger BR, Christie PJ. 1994. Genetic complementation analysis of the *Agrobacterium tumefaciens virB* operon: *virB2* through *virB11* are essential virulence genes. *J. Bacteriol.* 176:3646–60
15. Bergmann B, Stomp A-M. 1992. Effect of host plant genotype and growth rate on *Agrobacterium tumefaciens*-mediated gall formation in *Pinus radiata*. *Phytopathology* 82:1456–62
16. Binns AN. 1991. Transformation of wall deficient cultured tobacco protoplasts by *Agrobacterium tumefaciens*. *Plant Physiol.* 96:498–506
17. Binns AN, Beaupre CE, Dale EM. 1995. Inhibition of VirB-mediated transfer of diverse substrates from *Agrobacterium tumefaciens* by the IncQ plasmid RSF1010. *J. Bacteriol.* 177:4890–99
18. Bolton GW, Nester EW, Gordon MP. 1986. Plant phenolic compounds induce expression of the *Agrobacterium tumefaciens* loci needed for virulence. *Science* 232:983–85
19. Braun A. 1947. Thermal studies on the factors responsible for tumour induction in crown gall. *Am. J. Bot.* 34:234–40
20. Braun A. 1954. The physiology of plant tumors. *Annu. Rev. Plant Physiol.* 5:133–62
21. Braun A. 1978. Plant tumours. *Biochim. Biophys. Acta* 516:167–91
22. Braun A, Mandle RJ. 1948. Studies on the inactivation of the tumour-inducing principle in crown gall. *Growth* 12:255–69
23. Bravo-Angel AM, Hohn B, Tinland B. 1997. The omega sequence of VirD2 is important but not essential for efficient transfer of T-DNA by *Agrobacterium tumefaciens*. *Mol. Plant-Microbe Interact.* 11:57–63
24. Burr TJ, Bazzi C, Sule S, Otten L. 1998. Crown gall of grape: biology of *Agrobacterium vitis* and the development of disease control strategies. *Plant Dis.* 82:1288–97
25. Cangelosi GA, Hung L, Puvanesarajah V, Stacey G, Ozga DA, et al. 1987. Common loci for *Agrobacterium tumefaciens* and *Rhizobium meliloti* exopolysaccharide synthesis and their roles in plant interactions. *J. Bacteriol.* 169:2086–91
26. Cangelosi GA, Martinetti G, Leigh JA, Lee CC, Theines C, et al. 1989. Role of *Agrobacterium tumefaciens chvA* protein in export of β-1,2-glucan. *J. Bacteriol.* 171:1609–15
27. Castle LA, Morris RO. 1990. A method for early detection of T-DNA transfer. *Plant Mol. Biol. Rep.* 8:28–39
28. Charles TC, Jin S, Nester EW. 1992. Two-component sensory transduction systems in phytobacteria. *Annu. Rev. Phytopathol.* 30:463–84
29. Chaudhury AM, Signer ER. 1989. Non-destructive transformation of *Arabidopsis*. *Plant Mol. Biol. Rep.* 7:258–65
30. Christie PJ. 1997. *Agrobacterium tumefaciens* T-complex transport apparatus: A paradigm for a new family of multifunctional transporters in eubacteria. *J. Bacteriol.* 179:3085–94
31. Christie PJ, Ward JE, Winans SC, Nester EW. 1988. The *Agrobacterium tumefaciens virE2* gene product is a single-stranded-DNA-binding protein that associates with T-DNA. *J. Bacteriol.* 170:2659–67
32. Citovsky V, De Vos G, Zambryski P. 1988.

Single-stranded DNA binding protein encoded by the *virE* locus of *Agrobacterium tumefaciens*. *Science* 240:501–4
33. Citovsky V, Guralnick B, Simon MN, Wall JS. 1997. The molecular structure of *Agrobacterium* VirE2-single stranded DNA complexes involved in nuclear import. *J. Mol. Biol.* 271:718–27
34. Citovsky V, Warnick D, Zambryski P. 1994. Nuclear import of *Agrobacterium* VirD2 and VirE2 proteins in maize and tobacco. *Proc. Natl. Acad. Sci. USA* 91:3210–14
35. Citovsky V, Wong ML, Zambryski P. 1989. Cooperative interaction of *Agrobacterium* VirE2 protein with single-stranded DNA: implications for the T-DNA transfer process. *Proc. Natl. Acad. Sci. USA* 86:1193–97
36. Citovsky V, Zupan J, Warnick D, Zambryski P. 1992. Nuclear localization of *Agrobacterium* VirE2 protein in plant cells. *Science* 256:1802–5
37. Clarke MC, Wei W, Lindsey K. 1992. High-frequency transformation of *Arabi dopsis thaliana* by *Agrobacterium tumefaciens*. *Plant Mol. Biol. Rep.* 10:178–89
38. Cluster PD, O'Dell M, Metzlaff M, Flavell RB. 1996. Details of T-DNA structural organization from a transgenic *Petunia* population exhibiting co-suppression. *Plant Mol. Biol.* 32:1197–203
39. Cook DM, Farrand SK. 1992. The oriT region of the *Agrobacterium tumefaciens* Ti plasmid pTiC58 shares DNA sequence identity with the transfer origins of RSF1010 and RK2/RP4 and with T-region borders. *J. Bacteriol.* 174:6238–46
40. Davis ME, Miller AR, Lineberger RD. 1991. Temporal competence for transformation of *Lycopersicon esculentum* (L. Mill.) cotyledons by *Agrobacterium tumefaciens*: relation to wound-healing and soluble plant factors. *J. Exp. Bot.* 42:359–64
41. Deasey MC, Matthysse AG. 1984. Interactions of wild-type and a cellulose-minus mutant of *Agrobacterium tumefaciens* with tobacco mesophyll and tobacco tissue culture cells. *Phytopathology* 74:991–94
42. DeCleene M, DeLey J. 1976. The host range of crown gall. *Bot. Rev.* 42:389–466
43. de Kathen A, Jacobsen H-J. 1995. Cell competence for Agrobacterium-mediated DNA transfer in *Pisum sativum* L. *Transgen. Res.* 4:184–91
44. de Groot MJA, Offringa R, Groet J, Does MP, Hooykaas PJJ, et al. 1994. Non-recombinant background in gene targeting: Illegitimate recombination between a *hpt* gene and a defective 5′ deleted *nptII* gene can restore a Kmr phenotype in tobacco. *Plant Mol. Biol.* 25:721–33
45. De Neve M, De Buck S, Jacobs A, Van Montagu M, Depicker A. 1997. T-DNA integration patterns in co-transformed plant cells suggest that T-DNA repeats originate from co-integration of separate T-DNAs. *Plant J.* 11:15–29
45a. Deng W, Chen L, Peng W-T, Liang X, Sekiguchi S, et al. 1999. VirE1 is a specific molecular chaperone for the exported single-stranded DNA-binding protein VirE2 in *Agrobacterium*. *Mol. Microbiol.* 31:1795–807
46. Deng W, Chen L, Wood DW, Metcalfe T, Liang X, et al. 1998. *Agrobacterium* VirD2 protein interacts with plant host cyclophilins. *Proc. Natl. Acad. Sci. USA* 95:7040–45
47. De Vos G, Zambryski P. 1989. Expression of *Agrobacterium* nopaline-specific VirD1, VirD2, and VirC1 proteins and their requirement for T-strand production in *E. coli*. *Mol. Plant-Microbe Interact.* 2:43–52
48. Dombek P, Ream W. 1997. Functional domains of *Agrobacterium tumefaciens* single-stranded DNA-binding protein VirE2. *J. Bacteriol.* 179:1165–73
49. Doty SL, Yu MC, Lundin JI, Heath JD, Nester EW. 1996. Mutational analysis of the input domain of the VirA protein

of *Agrobacterium tumefaciens*. *J. Bacteriol.* 178:961–70
50. Douglas CJ, Halperin W, Nester EW. 1982. *Agrobacterium tumefaciens* mutants affected in attachment to plant cells. *J. Bacteriol.* 152:1265–75
51. Douglas CJ, Staneloni RJ, Rubin RA, Nester EW. 1985. Identification and genetic analysis of an *Agrobacterium tumefaciens* chromosomal virulence region. *J. Bacteriol.* 161:850–60
52. Durrenberger F, Crameri A, Hohn B, Koukolikova-Nicola Z. 1989. Covalently bound VirD2 protein of *Agrobacterium tumefaciens* protects the T-DNA from exonucleolytic degradation. *Proc. Natl. Acad. Sci. USA* 86:9154–58
53. Dye F, Berthelot K, Griffon B, Delay D, Delmotte FM. 1997. Alkylsyringamides, new inducers of *Agrobacterium tumefaciens* virulence genes. *Biochimie* 79:3–6
54. Eisenbrandt R, Kalkum M, Lai E-M, Lurz R, Kado CI, et al. 1999. Conjugative pili of IncP plasmids, and the Ti plasmid T pilus are composed of cyclic subunits. *J. Biol. Chem.* 274:22548–55
55. Engstrom P, Zambryski P, Van Montagu M, Stachel S. 1987. Characterization of *Agrobacterium tumefaciens* virulence proteins induced by the plant factor acetosyringone. *J. Mol. Biol.* 197:635–45
56. Escudero J, Hohn B. 1997. Transfer and integration of T-DNA without cell injury in the host plant. *Plant Cell* 9:2135–42
57. Escudero J, Neuhaus G, Hohn B. 1995. Intracellular *Agrobacterium* can transfer DNA to the cell nucleus of the host plant. *Proc. Natl. Acad. Sci. USA* 92:230–34
58. Escudero J, Neuhaus G, Schläppi M, Hohn B. 1996. T-DNA transfer in meristematic cells of maize provided with intracellular *Agrobacterium*. *Plant J.* 10:355–60
59. Farrand S, Kado CI, Ireland CR. 1981. Suppression of tumorigenicity by the IncW R plasmid pSa in *Agrobacterium tumefaciens*. *Mol. Gen. Genet.* 181:44–51
60. Feldmann KA. 1991. T-DNA insertion mutagenesis in *Arabidopsis*: mutational spectrum. *Plant J.* 1:71–82
61. Feldmann KA, Marks MD. 1987. Agrobacterium-mediated transformation of germinating seeds of *Arabidopsis thaliana*: a non-tissue culture approach. *Mol. Gen. Genet.* 208:1–9
62. Filichkin SA, Gelvin SB. 1993. Formation of a putative relaxation intermediate during T-DNA processing directed by the *Agrobacterium tumefaciens* VirD1, D2 endonuclease. *Mol. Microbiol.* 8:915–26
63. Firoozabady E, Galbraith DW. 1984. Presence of a plant cell wall is not required for transformation of *Nicotiana* by *Agrobacterium tumefaciens*. *Plant Cell Tissue Organ Cult.* 3:175–88
64. Gelvin SB. 1992. Chemical signaling between *Agrobacterium* and its plant host. In *Molecular Signals in Plant-microbe Communications,* ed. DPS Verma, 138–67. Boca Raton: CRC Press
65. Gelvin SB. 1993. Molecular genetics of T-DNA transfer from *Agrobacterium* to plants. In *Transgenic Plants,* ed. S-D Kung, R Wu, pp. 49–87. San Diego: Academic
66. Gelvin SB. 1998. *Agrobacterium* VirE2 proteins can form a complex with T strands in the plant cytoplasm. *J. Bacteriol.* 180:4300–2
67. Gheysen G, Villarroel R, Van Montagu M. 1991. Illegitimate recombination in plants: a model for T-DNA integration. *Genes Dev.* 5:287–97
68. Gietl C, Koukolikova-Nicola Z, Hohn B. 1987. Mobilization of T-DNA from *Agrobacterium* to plant cells involves a protein that binds single-stranded DNA. *Proc. Natl. Acad. Sci. USA* 84:9006–10
69. Golemis EA, Gyuris J, Brent R. 1994. Interaction trap/two-hybrid system to identify interacting proteins. In *Current Protocols in Molecular Biology*, ed. FM Ausubel, R Brent, RE Kingston, DD Moore, JG Seidman, JA Smith, K Struhl, pp. 13.14.1–13.14.17. New York: Wiley

70. Gorlich D, Kostka S, Kraft R, Dingwall C, Laskey RA, et al. 1995. Two different subunits of importin cooperate to recognize nuclear localization signals and bind them to the nuclear envelope. *Curr. Biol.* 5:383–92
71. Gorlich D, Mattaj IW. 1996. Nucleoplasmic transport. *Science* 271:1513–18
72. Grevelding C, Fantes V, Kemper E, Schell J, Masterson R. 1993. Single-copy T-DNA insertions in *Arabidopsis* are the predominant form of integration in root-derived transgenics, whereas multiple insertions are found in leaf discs. *Plant Mol. Biol.* 23:847–60
73. Guralnick B, Thomsen G, Citovsky V. 1996. Transport of DNA into the nuclei of *Xenopus* oocytes by a modified VirE2 protein of *Agrobacterium*. *Plant Cell* 8:363–73
74. Gurlitz RHG, Lamb PW, Matthysse AG. 1987. Involvement of carrot cell surface proteins in attachment of *Agrobacterium tumefaciens*. *Plant Physiol.* 83:564–68
75. Harter K, Kircher S, Frohnmeyer H, Krenz M, Nagy F, et al. 1994. Light-regulated modification and nuclear translocation of cytosolic G-box binding factors in parsley. *Plant Cell* 6:545–59
76. Herman L, Jacobs A, Van Montagu M, Depicker A. 1990. Plant chromosome/marker gene fusion assay for study of normal and truncated T-DNA integration events. *Mol. Gen. Genet.* 224:248–56
77. Herrera-Estrella A, Chen Z-M, Van Montagu M, Wang K. 1988. VirD proteins of *Agrobacterium tumefaciens* are required for the formation of a covalent DNA-protein complex at the 5′ terminus of T-strand molecules. *EMBO J.* 7:4055–62
78. Herrera-Estrella A, Van Montagu M, Wang K. 1990. A bacterial peptide acting as a plant nuclear targeting signal: The amino-terminal portion of *Agrobacterium* VirD2 protein directs a β-galactosidase fusion protein into tobacco nuclei. *Proc. Natl. Acad. Sci. USA* 87:9534–37
79. Hess KM, Dudley MW, Lynn DG, Joerger RD, Binns AN. 1991. Mechanism of phenolic activation of *Agrobacterium* virulence genes: development of a specific inhibitor of bacterial sensor/response systems. *Proc. Natl. Acad. Sci. USA* 88:7854–58
80. Hicks GR, Smith HMS, Lobreaux S, Raikhel NV. 1996. Nuclear import in permeabilized protoplasts from higher plants has unique features. *Plant Cell* 8:1337–52
81. Hood EE, Fraley RT, Chilton M-D. 1987. Virulence of *Agrobacterium tumefaciens* strain A281 on legumes. *Plant Physiol.* 83:529–34
82. Hooykaas PJJ, Beijersbergen AGM. 1994. The virulence system of *Agrobacterium tumefaciens*. *Annu. Rev. Phytopathol.* 32:157–79
83. Howard E, Citovsky V. 1990. The emerging structure of the *Agrobacterium* T-DNA transfer complex. *BioEssays* 12:103–8
84. Howard EA, Winsor BA, De Vos G, Zambryski P. 1989. Activation of the T-DNA transfer process in *Agrobacterium* results in the generation of a T-strand-protein complex: tight association of VirD2 with the 5′ ends of T-strands. *Proc. Natl. Acad. Sci. USA* 86:4017–21
85. Howard EA, Zupan JR, Citovsky V, Zambryski PC. 1992. The VirD2 protein of *A. tumefaciens* contains a C-terminal bipartite nuclear localization signal: implications for nuclear uptake of DNA in plant cells. *Cell* 68:109–18
86. Inon de Iannino N, Ugalde RA. 1989. Biochemical characterization of avirulent *Agrobacterium tumefaciens chvA* mutants: synthesis and excretion of β-(1,2)glucan. *J. Bacteriol.* 171:2842–49
87. Jacq B, Lesobre O, Sangwan RS, Sangwan-Norreel BS. 1993. Factors influencing T-DNA transfer in Agrobacterium-mediated transformation of sugarbeet. *Plant Cell Rep.* 12:621–24
88. Janssen B-J, Gardner RC. 1990. Localized transient expression of GUS in

leaf discs following cocultivation with *Agrobacterium. Plant Mol. Biol.* 14:61–72
89. Jasper F, Koncz C, Schell J, Steinbiss H-H. 1994. *Agrobacterium* T-strand production in vitro: Sequence-specific cleavage and 5′ protection of single-stranded DNA templates by purified VirD2 protein. *Proc. Natl. Acad. Sci. USA* 91:694–98
90. Jayaswal RK, Veluthambi K, Gelvin SB, Slightom JL. 1987. Double-stranded cleavage of T-DNA and generation of single-stranded T-DNA molecules in *Escherichia coli* by a *virD*-encoded border-specific endonuclease from *Agrobacterium tumefaciens*. *J. Bacteriol.* 169:5035–45
91. Jiang C-Z, Yen C-N, Cronin K, Mitchell D, Britt AB. 1997. UV- and gamma-radiation sensitive mutants of *Arabidopsis thaliana*. *Genetics* 147:1401–9
92. Jin S, Prusti RK, Roitsch T, Ankenbauer RG, Nester EW. 1990. Phosphorylation of the VirG protein of *Agrobacterium tumefaciens* by the autophosphorylated VirA protein: essential role in biological activity of VirG. *J. Bacteriol.* 172:4945–50
93. Jin S, Roitsch T, Ankenbauer RG, Gordon MP, Nester EW. 1990. The VirA protein of *Agrobacterium tumefaciens* is autophosphorylated and is essential for *vir* gene regulation. *J. Bacteriol.* 172:525–30
94. Jin S, Song Y-N, Deng W-Y, Gordon MP, Nester EW. 1993. The regulatory VirA protein of *Agrobacterium tumefaciens* does not function at elevated temperatures. *J. Bacteriol.* 175:6830–35
95. Jones AL, Lai E-M, Shirasu K, Kado CI. 1996. VirB2 is a processed pilin-like protein encoded by the *Agrobacterium tumefaciens* Ti plasmid. *J. Bacteriol.* 178:5706–11
96. Jorgensen RA, Cluster PD, English J, Que Q, Napoli CA. 1996. Chalcone synthase cosuppression phenotypes in petunia flowers: comparison of sense vs antisense constructs and single-copy vs complex T-DNA sequences. *Plant Mol. Biol.* 31:957–73
97. Kado CI. 1993. Promiscuous DNA transfer mechanism of *Agrobacterium tumefaciens* is mediated by a potential pilus-like structure encoded by the Ti plasmid *virB* operon. In *Pesticides/Environment: Molecular Biological Approaches*, ed. TMM Mitsui, I Yamaguchi, p. 17–26. Tokyo: Pestic. Soc. Jpn.
98. Kado CI. 1994. Promiscuous DNA transfer system of *Agrobacterium tumefaciens*: role of the *virB* operon in sex pilus assembly and synthesis. *Mol. Microbiol.* 12:17–22
99. Kalogeraki VS, Winans SC. 1995. The octopine-type Ti plasmid pTiA6 of *Agrobacterium tumefaciens* contains a gene homologous to the chromosomal virulence gene *acvB*. *J. Bacteriol.* 177:892–97
100. Kamoun S, Cooley MB, Rogowsky PM, Kado CI. 1989. Two chromosomal loci involved in production of exopolysaccharide in *Agrobacterium tumefaciens*. *J. Bacteriol.* 171:1755–59
100a. Kanemoto RH, Powell AT, Akiyoshi DE, Regier DA, Kerstetter RA, et al. 1989. Nucleotide sequence and analysis of the plant-inducible locus *pinF* from *Agrobacterium tumefaciens*. *J. Bacteriol.* 171:2506–12
101. Kartzke S, Saedler H, Meyer P. 1990. Molecular analysis of transgenic plants derived from transformations of protoplasts at various stages of the cell cycle. *Plant Sci.* 67:63–72
102. Kempin SA, Liljegren SJ, Block LM, Roundsley SD, Yanofsky MF, et al. 1997. Targeted disruption in *Arabidopsis*. *Nature* 389:802–3
103. Koncz C, Martini N, Mayerhofer R, Koncz-Kalman Z, Korber H, et al. 1989. High-frequency T-DNA-mediated gene tagging in plants. *Proc. Natl. Acad. Sci. USA* 86:8467–71
104. Koncz C, Nemeth N, Redei GP, Schell J. 1992. T-DNA insertional mutagenesis

in *Arabidopsis*. *Plant Mol. Biol.* 20:963–76

105. Koukolikova-Nicola Z, Hohn B. 1993. How does the T-DNA of *Agrobacterium tumefaciens* find its way into the plant cell nucleus? *Biochimie* 75:635–38
106. Koukolikova-Nicola Z, Raineri D, Stephens K, Ramos C, Tinland B, et al. 1993. Genetic analysis of the *virD* operon of *Agrobacterium tumefaciens*: a search for functions involved in transport of T-DNA into the plant cell nucleus and in T-DNA integration. *J. Bacteriol.* 175:723–31
107. Krizkova L, Hrouda M. 1998. Direct repeats of T-DNA integrated in tobacco chromosome: characterization of junction regions. *Plant J.* 16:673–80
108. Lai, E-M, Kado CI. 1998. Processed VirB2 is the major subunit of the promiscuous pilus of *Agrobacterium tumefaciens*. *J. Bacteriol.* 180:2711–17
109. Lee Y-W, Jin S, Sim W-S, Nester EW. 1995. Genetic evidence for direct sensing of phenolic compounds by the VirA protein of *Agrobacterium tumefaciens*. *Proc. Natl. Acad. Sci. USA* 92:12245–49
110. Lee L-Y, Gelvin SB, Kado CI. 1999. pSa causes oncogenic suppression of *Agrobacterium* by inhibiting VirE2 protein export. *J. Bacteriol.* 181:186–96
111. Lessl M, Balzler D, Pansegrau W, Lanka E. 1992. Sequence similarities between the RP4 Tra2 and the Ti *virB* region strongly support the conjugation model for T-DNA transfer. *J. Biol. Chem.* 267:20471–80
112. Lessl M, Lanka E. 1994. Common mechanisms in bacterial conjugation and Ti-mediated T-DNA transfer to plant cells. *Cell* 77:321–24
113. Li X-Q, Liu C-N, Ritchie SW, Peng J-Y, Gelvin SB, et al. 1992. High efficiency transient transformation of rice by *Agrobacterium tumefacien*. *Plant Mol. Biol.* 20:1037–48
114. Liu C-N, Li X-Q, Gelvin SB. 1992. Multiple copies of *virG* enhance the transient transformation of celery, carrot, and rice tissues by *Agrobacterium tumefaciens*. *Plant Mol. Biol.* 20:1071–87
115. Lloyd AM, Barnason AR, Rogers SG, Byrne MC, Fraley RT, et al. 1986. Transformation of *Arabidopsis thaliana* with *Agrobacterium tumefaciens*. *Science* 234:464–66

115a. Lohrke SM, Nechaev S, Yang H, Severinov K, Jin SJ.1999. Transcriptional activation of *Agrobacterium tumefaciens* virulence gene promoters in *Escherichia coli* requires *A. tumefaciens rpoA* gene, encoding the alpha subunit of RNA polymerase. *J. Bacteriol.* 181:4533–39

116. Loper JE, Kado CI. 1979. Host range conferred by the virulence-specifying plasmid of *Agrobacterium tumefaciens*. *J. Bacteriol.* 139:591–96
117. Lowe BA, Krul WR. 1991. Physical, chemical, developmental, and genetic factors that modulate the *Agrobacterium-Vitis* interaction. *Plant Physiol.* 96:121–29
118. Mandal A, Lang V, Orcyk W, Palva ET. 1993. Improved efficiency for T-DNA-mediated transformation and plasmid rescue in *Arabidopsis thaliana*. *Theor. Appl. Genet.* 86:621–28
119. Marks JR, Lynch TJ, Karlinsey JE, Thomashow MF. 1987. *Agrobacterium tumefaciens* virulence locus *pscA* is related to the *Rhizobium meliloti exoC* locus. *J. Bacteriol.* 169:5835–37
120. Matsumoto S, Ito Y, Hosoi T, Takahashi Y, Machida Y. 1990. Integration of *Agrobacterium* T-DNA into a tobacco chromosome: Possible involvement of DNA homology between T-DNA and plant DNA. *Mol. Gen. Genet.* 224:309–16
121. Matthysse AG. 1983. Role of bacterial cellulose fibrils in *Agrobacterium tumefaciens* infection.*J. Bacteriol.* 154:906–15
122. Matthysse AG. 1987. Characterization of

nonattaching mutants of *Agrobacterium tumefaciens*. *J. Bacteriol.* 169:313–23

123. Matthysse AG. 1995. Genes required for cellulose synthesis in *Agrobacterium tumefaciens*. *J. Bacteriol.* 177:1069–75
124. Matthysse AG, Gurlitz RHG. 1982. Plant cell range for attachment of *Agrobacterium tumefaciens* to tissue culture cells. *Physiol. Plant Pathol.* 21:381–87
125. Matthysse AG, Holmes KV, Gurlitz RHG. 1981. Elaboration of cellulose fibrils by *Agrobacterium tumefaciens* during attachment to carrot cells. *J. Bacteriol.* 145:583–95
126. Matthysse AG, Holmes KV, Gurlitz RHG. 1982. Binding of *Agrobacterium tumefaciens* to carrot protoplasts. *Physiol. Plant Pathol.* 20:27–33
127. Matthysse AG, McMahan S. 1998. Root colonization by *Agrobacterium tumefaciens* is reduced in *cel, attB, attD,* and *attR* mutants. *Appl. Environ. Microbiol.* 64:2341–45
128. Matthysse AG, Thomas DL, White AR. 1995. Mechanism of cellulose synthesis in *Agrobacterium tumefaciens*. *J. Bacteriol.* 177:1076–81
129. Matthysse AG, Yarnall HA, Young N. 1996. Requirement for genes with homology to ABC transport systems for attachment and virulence of *Agrobacterium tumefaciens*. *J. Bacteriol.* 178:5302–8
130. Mauro AO, Pfeiffer TW, Collins GB. 1995. Inheritance of soybean susceptibility to *Agrobacterium tumefaciens* and its relationship to transformation. *Crop Sci.* 35:1152–56
131. Mayerhofer RZ, Koncz-Kalman C, Nawrath G, Bakkeren A, Cramer, et al. 1991. T-DNA integration: a mode of illegitimate recombination in plants. *EMBO J.* 10:697–704
132. Maximova SN, Dandekar AM, Guiltinan MJ. 1998. Investigation of Agrobacterium-mediated transformation of apple using green fluorescent protein: high transient expression and low stable transformation suggest that factors other than T-DNA transfer are rate-limiting. *Plant Mol. Biol.* 37:549–59
133. McLean BG, Greene EA, Zambryski PC. 1994. Mutants of *Agrobacterium* VirA that activate the *vir* gene expression in the absence of the inducer acetosyringone. *J. Biol. Chem.* 269:2645–51
134. Melchers LS, Maroney MJ, den Dulk-Ras A, Thompson DV, van Vuuren HAJ, et al. 1990. Octopine and nopaline strains of *Agrobacterium tumefaciens* differ in virulence; molecular characterization of the *virF* locus. *Plant Mol. Biol.* 14:249–59
135. Melchers LS, Regensburg-Tuink AJA, Schilperoort RA, Hooykaas PJJ. 1989. Specificity of signal molecules in the activation of *Agrobacterium* virulence gene expression. *Mol. Microbiol.* 3:969–77
136. Merkle T, Leclerc D, Marshallsay C, Nagy F. 1996. A plant in vitro system for the nuclear import of proteins. *Plant J.* 10:1177–86
137. Merkle T, Nagy F. 1997. Nuclear import of proteins: putative import factors and development of in vitro import systems in higher plants. *Trends Plant Sci.* 2:458–64
138. Messens E, Dekeyser R, Stachel SE. 1990. A nontransformable *Triticum monococcum* monocotyledonous culture produces the potent *Agrobacterium vir*-inducing compound ethyl ferulate. *Proc. Natl. Acad. Sci. USA* 87:4368–72
139. Miao Z-H, Lam E. 1995. Targeted disruption of the *TGA3* locus in *Arabidopsis thaliana*. *Plant J.* 7:359–65
140. Morris JW, Morris RO. 1990. Identification of an *Agrobacterium tumefaciens* virulence gene inducer from the pinaceous gymnosperm *Pseudotsuga menziesii*. *Proc. Natl. Acad. Sci. USA* 87:3614–18
141. Mushegian AR, Fullner KJ, Koonin EV, Nester EW. 1996. A family of lysozyme-like virulence factors in bacterial

pathogens of plants and animals. *Proc. Natl. Acad. Sci. USA* 93:7321–26

142. Mysore KS, Bassuner B, Deng X-b, Darbinian NS, Motchoulski A, et al. 1998. Role of the *Agrobacterium tumefaciens* VirD2 protein in T-DNA transfer and integration. *Mol. Plant-Microbe Interact.* 11:668–83

142a. Mysore KS, Kumar CTR, Gelvin SB. 2000. *Arabidopsis* ecotypes and mutants that are recalcitrant to *Agrobacterium* root transformation are susceptible to germ-line transformation. *Plant J.* 21:9–16

142b. Mysore KS, Nam J, Gelvin SB. 2000. An *Arabidopsis* histonne H2A mutant is deficient in *Agrobacterium* T-DNA integration. *Proc. Natl. Acad. Sci. USA* 97:948–53

143. Nagata Z, Okada K, Takebe I. 1986. Strong dependency of the transformation of plant protoplast on cell cycle. In *Fallen Leaf Lake Conf. Agrobacterium Crown Gall,* ed. C Kado, 9. Univ. Calif., Davis

144. Nam J, Matthysse AG, Gelvin SB. 1997. Differences in susceptibility of *Arabidopsis* ecotypes to crown gall disease may result from a deficiency in T-DNA integration. *Plant Cell* 9:317–33

145. Nam J, Mysore KS, Gelvin SB. 1998. *Agrobacterium tumefaciens* transformation of the radiation hypersensitive *Arabidopsis thaliana* mutants *uvh1* and *rad5*. *Mol. Plant-Microbe Interact.* 11:1136–41

146. Nam J, Mysore KS, Zheng C, Knue MK, Matthysse AG, et al. 1999. Identification of T-DNA tagged *Arabidopsis* mutants that are resistant to transformation by *Agrobacterium*. *Mol. Gen. Genet.* 261:429–38

147. Narasimhulu SB, Deng X-b, Sarria R, Gelvin SB. 1996. Early transcription of *Agrobacterium* T-DNA genes in tobacco and maize. *Plant Cell* 8:873–86

148. Neff NT, Binns AN. 1985. *Agrobacterium tumefaciens* interaction with suspension-cultured tomato cells. *Plant Physiol.* 77:35–42

149. Neff NT, Binns AN, Brandt C. 1987. Inhibitory effects of a pectin-enriched tomato cell wall fraction on *Agrobacterium tumefaciens* binding and tumor formation. *Plant Physiol.* 83:525–28

150. O'Connell KP, Handelsman J. 1989. *chvA* locus may be involved in export of neutral cyclic β-1,2 linked D-glucan from *Agrobacterium tumefaciens*. *Mol. Plant-Microbe Interact.* 2:11–16

151. Offringa R, De Groot MJ, Haagsman HJ, Does MP, Van den Elzen PJM, et al. 1990. Extrachromosomal homologous recombination and gene targeting in plant cells after Agrobacterium mediated transformation. *EMBO J.* 9:3077–84

152. Offringa R, Franke-van Dijk MEI, De Groot MJA, Van Den Elzen PJM, Hooykaas PJJ. 1993. Nonreciprocal homologous recombination between *Agrobacterium* transferred DNA and a plant chromosomal locus. *Proc. Natl. Acad. Sci. USA* 90:7346–50

153. Offringa R, van den Elzen PJM, Hooykaas PJJ. 1992. Gene targeting in plants using the *Agrobacterium* vector system. *Transgenic Res.* 1:114–23

154. Ohba T, Yoshioka Y, Machida C, Machida Y. 1995. DNA rearrangement associated with the integration of T-DNA in tobacco: an example for multiple duplications of DNA around the integration target. *Plant J.* 7:157–64

155. Otten L, DeGreve H, Leemans J, Hain R, Hooykaas P, et al. 1984. Restoration of virulence of *vir* region mutants of *Agrobacterium tumefaciens* strain B6S3 by coinfection with normal and mutant *Agrobacterium* strains. *Mol. Gen. Genet.* 195:159–63

156. Owens LD, Cress DE. 1984. Genotypic variability of soybean response to

Agrobacterium strains harboring the Ti or Ri plasmids. *Plant Physiol.* 77:87–94

157. Pan SQ, Jin S, Boulton MI, Hawes M, Gordon MP, et al. 1995. An *Agrobacterium* virulence factor encoded by a Ti plasmid gene or a chromosomal gene is required for T-DNA transfer into plants. *Mol. Microbiol.* 17:259–69
158. Pansegrau W, Lanka E. 1991. Common sequence motifs in DNA relaxases and nick regions from a variety of DNA transfer systems. *Nucleic Acids Res.* 19:34–55
159. Pansegrau W, Lanka E, Barth PT, Figurski DH, Guiney DG, et al. 1994. Complete nucleotide sequence of Birmingham IncPα plasmids: compilation and comparative analysis. *J. Mol. Biol.* 239: 623–63
160. Pansegrau W, Schoumacher F, Hohn B, Lanka E. 1993. Site-specific cleavage and joining of single-stranded DNA by VirD2 protein of *Agrobacterium tumefaciens* Ti plasmids: analogy to bacterial conjugation. *Proc. Natl. Acad. Sci. USA* 90:11538–42
161. Paszkowski J, Baur M, Bogucki A, Potrykus I. 1998. Gene targeting in plants. *EMBO J.* 7:4021–26
162. Porter JR. 1991. Host range and implications of plant infection by *Agrobacterium rhizogenes*. *Crit. Rev. Plant Sci.* 10:387–421
163. Porter SG, Yanofsky MF, Nester EW. 1987. Molecular characterization of the *virD* operon from *Agrobacterium tumefaciens*. *Nucleic Acids Res.* 15:7503–17
164. Puchta H. 1997. Towards targeted transformation in plants. *Trends Plant Sci.* 3:77–78
165. Puvanesarajah V, Schell FM, Stacey G, Douglas, CJ, Nester EW. 1985. Role for 2-linked-α-D-glucan in the virulence of *Agrobacterium tumefaciens*. *J. Bacteriol.* 164:102–6
166. Ream W. 1989. *Agrobacterium tumefaciens* and interkingdom genetic exchange. *Annu. Rev. Phytopathol.* 27:583–86
167. Regensburg-Tuink AJG, Hooykaas PJJ. 1993. Transgenic *N. glauca* plants expressing bacterial virulence gene *virF* are converted into hosts for nopaline strains of *A. tumefaciens*. *Nature* 363:69–71
168. Relic B, Andjelkovic M, Rossi L, Nagamine Y, Hohn B. 1998. Interaction of the DNA modifying proteins VirD1 and VirD2 of *Agrobacterium tumefaciens*: analysis by subcellular localization in mammalian cells. *Proc. Natl. Acad. Sci. USA* 95:9105–10
169. Reuhs BL, Kim JS, Matthysse AG. 1997. Attachment of *Agrobacterium tumefaciens* to carrot cells and *Arabidopsis* wound sites is correlated with the presence of cell-associated, acidic polysaccharide. *J. Bacteriol.* 179:5372–79
170. Risseeuw E, Franke-van Dijk MEI, Hooykaas PJJ. 1997. Gene targeting and instability of *Agrobacterium* T-DNA loci in the plant genome. *Plant J.* 11:717–28
171. Risseeuw E, Offringa R, Franke-van Dijk MEI, Hooykaas PJJ. 1995. Targeted recombination in plants using *Agrobacterium* coincides with additional rearrangements at the target locus. *Plant J.* 7:109–19
172. Ritchie SW, Liu C-N, Sellmer JC, Kononowicz H, Hodges TK, et al. 1993. *Agrobacterium tumefaciens*-mediated expression of *gusA* in maize tissues. *Transgenic Res.* 2:252–65
173. Robbs SL, Hawes MC, Lin H-J, Pueppke SG, Smith LY. 1991. Inheritance of resistance to crown gall in *Pisum sativum*. *Plant Physiol.* 95:52–57
174. Rossi L, Hohn B, Tinland B. 1993. The VirD2 protein of *Agrobacterium tumefaciens* carries nuclear localization signals important for transfer of T-DNA to plants. *Mol. Gen. Genet.* 239:345–53
175. Rossi L, Hohn B, Tinland B. 1996. Integration of complete transferred DNA units is dependent on the activity of virulence E2 protein of *Agrobacterium*

tumefaciens. *Proc. Natl. Acad. Sci. USA* 93:126–30

176. Sangwan RS, Bourgeois Y, Brown S, Vasseur G, Sangwan-Norreel B. 1992. Characterization of competent cells and early events of Agrobacterium-mediated genetic transformation in *Arabidopsis thaliana*. *Planta* 188:439–56
177. Scheiffele P, Pansegrau W, Lanka E. 1995. Initiation of *Agrobacterium tumefaciens* T-DNA processing: purified protein VirD1 and VirD2 catalyze site- and strand-specific cleavage of superhelical T-border DNA in vitro. *J. Biol. Chem.* 270:1269–76
178. Schläppi M, Hohn B. 1992. Competence of immature maize embryos for Agrobacterium-mediated gene transfer. *Plant Cell* 4:7–16
179. Schroeder HE, Schotz AH, Wardley-Richardson T, Spencer D, Higgins TJV. 1993. Transformation and regeneration of two cultivars of pea (*Pisum sativum* L.). *Plant Physiol.* 101:751–57
180. Sen P, Pazour GJ, Anderson D, Das A. 1989. Cooperative binding of *Agrobacterium tumefaciens* VirE2 protein to single-stranded DNA. *J. Bacteriol.* 171:2573–80
181. Shen W-H, Escudero J, Schläppi M, Ramos C, Hohn B, et al. 1993. T-DNA transfer to maize cells: histochemical investigation of β-glucuronidase activity in maize tissues. *Proc. Natl. Acad. Sci. USA* 90:1488–92
182. Sheng J, Citovsky V. 1996. *Agrobacterium*-plant cell DNA transport: Have virulence proteins, will travel. *Plant Cell* 8:1699–710
183. Shurvinton CE, Hodges L, Ream W. 1992. A nuclear localization signal and the C-terminal omega sequence in the *Agrobacterium tumefaciens* VirD2 endonuclease are important for tumor formation. *Proc. Natl. Acad. Sci. USA* 89:11837–41
184. Smarrelli J, Watters MT, Diba LH. 1986. Response of various cucurbits to infection by plasmid-harboring strains of *Agrobacterium*. *Plant Physiol.* 82:622–24
185. Song Y-N, Shibuta M, Ebizuka Y, Sankawa U. 1990. Hydroxyacetosyringone is the major virulence gene activating factor in Belladonna hairy root cultures, and inositol enhances its activity. *Chem. Pharm. Bull.* 38:2063–65
186. Sonti RV, Chiruazzi M, Wong D, Davies CS, Harlow GR, et al. 1995. *Arabidopsis* mutants deficient in T-DNA integration. *Proc. Natl. Acad. Sci. USA* 92:11786–90
187. Spencer PA, Towers GHN. 1988. Specificity of signal compounds detected by *Agrobacterium tumefaciens*. *Phytochemistry* 27:2781–85
188. Stachel SE, Messens E, Van Montagu M, Zambryski P. 1985. Identification of the signal molecules produced by wounded plant cells that activate T-DNA transfer in *Agrobacterium tumefaciens*. *Nature* 318:624–29
189. Stachel SE, Nester EW. 1986. The genetic and transcriptional organization of the *vir* region of the A6 Ti plasmid of *Agrobacterium tumefaciens*. *EMBO J.* 5:1445–54
190. Stachel SE, Nester EW, Zambryski PC. 1986. A plant cell factor induces *Agrobacterium tumefaciens vir* gene expression. *Proc. Natl. Acad. Sci. USA* 83:379–83
191. Stachel SE, Timmerman B, Zambryski P. 1986. Generation of single-stranded T-DNA molecules during the initial stages of T-DNA transfer from *Agrobacterium tumefaciens* to plant cells. *Nature* 322:706–12
192. Stachel SE, Timmerman B, Zambryski P. 1987. Activation of *Agrobacterium tumefaciens vir* gene expression generates multiple single-stranded T-strand molecules from the pTiA6 T-region: requirement for 5′ *virD* gene products. *EMBO J.* 6:857–63
193. Stachel SE, Zambryski PC. 1986. *Agrobacterium tumefaciens* and the susceptible plant cell: A novel adaptation

of extracellular recognition and DNA conjugation. *Cell* 47:155–57

194. Stachel SE, Zambryski PC. 1986. *virA* and *virG* control the plant-induced activation of the T-DNA transfer process of *A. tumefaciens*. *Cell* 46:325–33
195. Stam M, de Bruijn R, Kenter S, van der Hoorn RAL, van Blokland R, et al. 1997. Post-transcriptional silencing of chalcone synthase in *Petunia* by inverted transgene repeats. *Plant J.* 12:63–82
196. Steck TR, Close TJ, Kado CI. 1989. High levels of double-stranded transferred DNA (T-DNA) processing from an intact nopaline Ti plasmid. *Proc. Natl. Acad. Sci. USA* 86:2133–37
197. Steck TR, Lin T-S, Kado CI. 1990. VirD2 gene product from the nopaline plasmid pTiC58 has at least two activities required for virulence. *Nucleic Acids Res.* 18:6953–58
198. Sundberg C, Meek L, Carroll K, Das A, Ream W. 1996. VirE1 protein mediates export of the single-stranded DNA-binding protein VirE2 from *Agrobacterium tumefaciens* into plant cells. *J. Bacteriol.* 178:1207–12

198a. Sundberg CD, Ream W. 1999. The *Agrobacterium tumefaciens* chaperone-like protein, VirE1, interacts with VirE2 at domains required for single-stranded DNA binding and cooperative interaction. *J. Bacteriol.* 181:6850–55

199. Swart S, Logman TJJ, Smit G, Lugtenberg BJJ, Kijne, JW. 1994. Purification and partial characterization of a glycoprotein from pea (*Pisum sativum*) with receptor activity for rhicadhesin, an attachment protein of *Rhizobiaceae*. *Plant Mol. Biol.* 24:171–83
200. Sykes LC, Matthysse AG. 1986. Time required for tumor induction by *Agrobacterium tumefaciens*. *Appl. Environ. Microbiol.* 52:597–98
201. Thomashow MF, Karlinsey JE, Marks JR, Hurlbert RE. 1987. Identification of a new virulence locus in *Agrobacterium tumefaciens* that affects polysaccharide composition and plant cell attachment. *J. Bacteriol.* 169:3209–16
202. Tinland B. 1996. The integration of T-DNA into plant genomes. *Trends Plant Sci.* 1:178–84
203. Tinland B, Hohn B. 1995. Recombination between prokaryotic and eukaryotic DNA: Integration of *Agrobacterium tumefaciens* T-DNA into the plant genome. In *Genetic Engineering*, ed. JK Setlow. New York: Plenum
204. Tinland B, Hohn B, Puchta H. 1994. *Agrobacterium tumefaciens* transfers single-stranded transferred DNA (T-DNA) into the plant cell nucleus. *Proc. Natl. Acad. Sci. USA* 91:8000–4
205. Tinland B, Schoumacher F, Gloeckler V, Bravo-Angel AM, Hohn B. 1995. The *Agrobacterium tumefaciens* virulence D2 protein is responsible for precise integration of T-DNA into the plant genome. *EMBO J.* 14:3585–95
206. Tinland B, Koukolikova-Nicola Z, Hall MN, Hohn B. 1992. The T-DNA-linked VirD2 protein contains two distinct functional nuclear localization signals. *Proc. Natl. Acad. Sci. USA* 89:7442–46

206a. Toro N, Datta A, Carmi OA, Young C, Prusti RK, Nester EW. 1989. The *Agrobacterium tumefaciens virC1* gene product binds to overdrive, a T-DNA transfer enhancer. *J. Bacteriol.* 171:6845–49

207. Turk SCHJ, van Lange RP, Regensburg-Tuink TJG, Hooykaas PJJ. 1994. Localization of the VirA domain involved in acetosyringone-mediated *vir* gene induction in *Agrobacterium tumefaciens*. *Plant Mol. Biol.* 25:899–907
208. Usami S, Okamoto S, Takebe I, Machida Y. 1988. Factor inducing *Agrobacterium tumefaciens vir* gene expression is present in monocotyledonous plants. *Proc. Natl. Acad. Sci. USA* 85:3748–52
209. Uttaro AD, Cangelosi GA, Geremia RA,

Nester EW, Ugalde RA. 1990. Biochemical characterization of avirulent *exoC* mutants of *Agrobacterium tumefaciens*. *J. Bacteriol.* 172:1640–46
210. Valvekens D, Van Montagu M, Van Lijsebettens M. 1988. *Agrobacterium tumefaciens*-mediated transformation of *Arabidopsis thaliana* root explants by using kanamycin selection. *Proc. Natl. Acad. Sci. USA* 85:5536–40
211. Van Houdt H, Ingelbrecht I, Van Montagu M, Depicker A. 1997. Post-transcriptional silencing of a neomycin phosphotransferase II transgene correlates with the accumulation of unproductive RNAs and with increased cytosine methylation of 3′ flanking regions. *Plant J.* 12:379–92
212. van Roekel JSC, Damm B, Melchers LS, Hoekema A. 1993. Factors influencing transformation frequency of tomato (*Lycopersicon esculentum*). *Plant Cell Rep.* 12:644–47
213. van Wordragen MF, Dons HJM. 1992. *Agrobacterium tumefaciens*-mediated transformation of recalcitrant crops. *Plant Mol. Biol. Rep.* 10:12–36
214. Veluthambi K, Jayaswal RK, Gelvin SB. 1987. Virulence genes A, G, and D mediate the double-stranded border cleavage of T-DNA from the *Agrobacterium* Ti plasmid. *Proc. Natl. Acad. Sci. USA* 84:1881–85
215. Veluthambi K, Ream W, Gelvin SB. 1988. Virulence genes, borders, and overdrive generate single-stranded T-DNA molecules from the A6 Ti plasmid of *Agrobacterium tumefaciens*. *J. Bacteriol.* 170:1523–32
216. Vergunst AC, Hooykaas PJJ. 1998. Cre/*lox*-mediated site-specific integration of *Agrobacterium* T-DNA in *Arabidopsis thaliana* by transient expression of *cre*. *Plant Mol. Biol.* 38:393–406
217. Vergunst AC, Jansen LET, Hooykaas PJJ. 1998. Site-specific integration of *Agrobacterium* T-DNA in *Arabidopsis thaliana* mediated by Cre recombinase. *Nucleic Acids Res.* 26:2729–34
218. Vijayachandra K, Palanichelvam K, Veluthambi K. 1995. Rice scutellum induces *Agrobacterium tumefaciens vir* genes and T-strand generation. *Plant Mol. Biol.* 29:125–33
219. Villemont E, Dubois F, Sangwan RS, Vasseur G, Bourgeois Y, et al. 1997. Role of the host cell cycle in the Agrobacterium-mediated genetic transformation of *Petunia*: evidence of an S-phase control mechanism for T-DNA transfer. *Planta* 201:160–72
220. Virts EL, Gelvin SB. 1985. Analysis of transfer of tumor-inducing plasmids from *Agrobacterium tumefaciens* to *Petunia* protooplasts. *J. Bacteriol.* 162:1030–38
221. Vogel AM, Das A. 1992. Mutational analysis of *Agrobacterium tumefaciens virD2*: tyrosine 29 is essential for endonuclease activity. *J. Bacteriol.* 174:303–8
222. Wagner VT, Matthysse AG. 1992. Involvement of a vitronectin-like protein in attachment of *Agrobacterium tumefaciens* to carrot suspension culture cells. *J. Bacteriol.* 174:5999–6003
223. Ward ER, Barnes WM. 1988. VirD2 protein of *Agrobacterium tumefaciens* very tightly linked to the 5′ end of T-strand DNA. *Science* 242:927–30
224. Winans SC. 1991. An *Agrobacterium* two-component regulatory system for the detection of chemicals released from plant wounds. *Mol. Microbiol.* 5:2345–50
225. Winans SC 1992. Two-way chemical signaling in *Agrobacterium*-plant interactions. *Microbiol. Rev.* 56:12–31
226. Winans SC, Burns DL, Christie PJ. 1996. Adaptation of a conjugal transfer system for the export of pathogenic macromolecules. *Trends Microbiol.* 4:64–68
227. Wirawan IGP, Kang HW, Kojima M. 1993. Isolation and characterization of a new chromosomal virulence gene of

Agrobacterium tumefaciens. J. Bacteriol. 175:3208–12

228. Wirawan IGP, Kojima M. 1996. A chromosomal virulence gene (*acvB*) product of *Agrobacterium tumefaciens* that binds to a T-strand to mediate its transfer to host plant *cells. Biosci. Biotechnol. Biochem.* 60:44–49
229. Xu Y, Bu W, Li B. 1993. Metabolic factors capable of inducing *Agrobacterium vir* gene expression are present in rice (*Oryza sativa L.*). *Plant Cell Rep.* 12:160–64
230. Yanofsky MF, Porter SG, Young C, Albright LM, Gordon MP, et al. 1986. The *virD* operon of *Agrobacterium tumefaciens* encodes a site-specific endonuclease. *Cell* 47:471–77
231. Ye G-N, Stone D, Pang S-Z, Creely W, Gonzalez K, Hinchee M. 1999. *Arabidopsis* ovule is the target for *Agrobacterium in planta* vacuum infiltration transformation. *Plant J.* 19:249–57
232. Young C, Nester EW. 1988. Association of the virD2 protein with the 5′ end of T strands in *Agrobacterium tumefaciens. J. Bacteriol.* 170:3367–74
233. Yusibov VM, Steck TR, Gupta V, Gelvin SB. 1994. Association of single-stranded transferred DNA from *Agrobacterium tumefaciens* with tobacco cells. *Proc. Natl. Acad. Sci. USA* 91:2994–98
234. Zambryski PC. 1992. Chronicles from the *Agrobacterium*-plant cell DNA transfer story. *Annu. Rev. Plant Physiol. Plant Mol. Biol.* 43:465–90

234a. Zhou X-R, Christie PJ. 1999. Mutagenesis of the *Agrobacterium* VirE2 single-stranded DNA-binding protein identifies regions required for self-association and interaction with VirE1 and a permissive site for hybrid protein construction. *J. Bacteriol.* 181:4342–52

235. Ziemienowicz A, Gorlich D, Lanka E, Hohn B, Rossi L. 1999. Import of DNA into mammalian nuclei by proteins originating from a plant pathogenic bacterium. *Proc. Natl. Acad. Sci. USA* 96:3729–33
236. Zupan JR, Citovsky V, Zambryski P. 1996. *Agrobacterium* VirE2 protein mediates nuclear uptake of single-stranded DNA in plant cells. *Proc. Natl. Acad. Sci. USA* 93:2392–97
237. Zupan JR, Zambryski P. 1995. Transfer of T-DNA from *Agrobacterium* to the plant cell. *Plant Physiol.* 107:1041–47
238. Zupan J, Zambryski P. 1997. The *Agrobacterium* DNA transfer complex. *Crit. Rev. Plant Sci.* 16:279–95

Annu. Rev. Plant Physiol. Plant Mol. Biol. 2000. 51:257–88

SIGNALING TO THE ACTIN CYTOSKELETON IN PLANTS

Chris J. Staiger
Department of Biological Sciences, Purdue University, West Lafayette, Indiana 47907–1392; e-mail: cstaiger@bilbo.bio.purdue.edu

Key Words actin-binding protein, signal transduction, tip growth, pathogen attack, pollen

■ **Abstract** Plants have developed finely tuned, cellular mechanisms to respond to a variety of intrinsic and extrinsic stimuli. In several examples, these responses necessitate rearrangements of the cytoplasm that are coordinated by a network of actin microfilaments and microtubules, dynamic polymers collectively known as the cytoskeleton. This review focuses on five different cellular responses in which the actin cytoskeleton redistributes following extracellular stimulation: pollen tube tip growth and the self-incompatibility response; root hair responses to bacterial nodulation factors; light-mediated plastid positioning; nonhost resistance to fungal attack; and guard cell shape and turgor changes. For each of these systems, there is reasonable knowledge about what signals induce the plant response and the function(s) of the actin rearrangement. This review aims to build beyond a description of cytoskeletal changes and look at specific actin-binding proteins that have been implicated as effectors of each response, as sites of action for second messengers, and as fundamental coordinators of actin dynamics.

CONTENTS

1040-2519/00/0601-0257$14.00

INTRODUCTION

Plant cells respond to a variety of internal and external stimuli with rapid and dramatic rearrangements of their cytoplasm. These changes are often mediated by a dynamic cytoskeleton, the network of microtubules, actin microfilaments, and their associated proteins. There is growing evidence that signal transduction cascades converge on these cytoskeletal proteins during many fundamental cellular responses. However, no full cascade of events or intermediates, from signal to cytoskeletal response, has been elucidated for any system. This review describes several stimulus-mediated cellular responses in which cytoskeletal rearrangements are known, makes some educated guesses about the accessory proteins likely to be responsive to the signaling intermediates, and discusses the potential function of each cytoskeletal rearrangement. The focus is exclusively on recent findings regarding the actin-based cytoskeleton, and microtubules are mentioned only when it is obvious that the two cytoskeletal systems act in concert to achieve a common cellular response. The reader is referred to several excellent reviews that cover additional aspects of signaling to the plant cytoskeleton (28, 44, 95, 124–126, 151, 177).

ACTIN-BINDING PROTEINS AS SENSORS OF THE CELLULAR ENVIRONMENT

Since its discovery and purification from skeletal muscle nearly half a century ago, actin has received the intense scrutiny of biochemists, cell biologists, and biophysicists. Its structure has been described at atomic resolution, and its physico-chemical properties have been mapped with great precision (150). Actin, a globular protein (G-actin) of approximately 42,000 Da, comprises two large domains between which is a cleft containing an adenine nucleotide- and divalent cation-binding site. Under physiological salt conditions, the actin monomers self-assemble into polymeric structures called filamentous actin (F-actin) or actin microfilaments, which resemble a double-stranded string of pearls, with an average diameter of 7 nm and a right-handed helical twist of 13–14 subunits per half turn. Polymerization is normally coupled to hydrolysis of ATP, but does not depend on this event. The intrinsic asymmetry of each subunit and a lag between subunit addition and ATP hydrolysis give actin filaments a distinct functional polarity. For example, subunit addition and loss are different at the two ends of a filament, with a fast-growing, plus (or barbed) end and a slow-growing, minus (or pointed) end that act as the sites for growth and shrinkage at equilibrium.

In all eukaryotic cells, actin exists in both polymeric and monomeric forms, with the ratio between the two populations varying with cell type, differentiation, and physiological status. Usually, F-actin is considered to have a particular function in the cytoplasm, for example as the molecular railroad tracks upon which plant cytoplasmic streaming runs (184). However, monomeric actin might play a unique role under certain circumstances in animal systems, especially in the nucleus as part of a chromatin remodeling complex (193) or transcriptional activation machinery (159).

The dynamic nature of the actin cytoskeleton depends upon the unique constellation of actin-binding proteins present in the cytoplasm of each cell, as well as their spatial distribution and local activation. In non-plant systems, a plethora of actin-binding proteins modulate the organization and functional properties of the cytoskeleton (5, 114, 133, 134, 146). Actin-binding proteins control the size and activity of the subunit pool, the subcellular sites for nucleation, the rate of filament turnover, and the organization of microfilaments into higher-order structures. These can be grouped into several categories based on their biochemical activities: Monomer-binding proteins regulate polymerization and depolymerization; side-binding proteins stabilize filaments, allow the formation of orthogonal networks, or bundle F-actin into highly ordered arrays; capping and severing factors control the availability of filament ends for subunit addition or loss; nucleation factors form a seed of actin subunits that catalyzes polymerization; and motor molecules (myosins) convert ATP hydrolysis into movement along filaments. Although members for each of these classes are likely to exist in plants, only monomer-binding proteins (profilin & ADF/cofilin), side-binding proteins (fimbrin, villin, EF-1α) and myosins have been isolated and characterized to date [reviewed in (2, 30, 161)]. Nevertheless, the rapid growth of plant expressed sequence tag and genome databases, as well as the application of novel biochemical approaches, will certainly lead to an explosion of discoveries in the next few years.

Many actin cytoskeletal elements are located near the plasma membrane (26, 107), where they are uniquely positioned to receive and transduce information. Actin-binding proteins can act as efficient sensors of the intracellular environment and likely serve at the crossroads between signaling events and rearrangement of the cytoplasmic architecture (76, 111, 147, 148). Because the properties and activity of many actin-binding proteins are remarkably sensitive to fluctuations in signaling intermediates and second messengers, this seems to be an acceptable generalization. For example, non-plant actin-binding proteins are regulated by phosphorylation, changes in cellular calcium or pH, or association with membrane polyphosphoinositides. The limited data available for plant actin-binding proteins confirm that they too are regulated. Plant profilins and actin depolymerizing factors (ADF) bind to PtdIns(4,5)P_2 and this inhibits their ability to interact with actin (38, 61). The activity of maize and *Arabidopsis* ADFs is negatively regulated through phosphorylation by a calcium-dependent protein kinase (137, 156). The first definitive proof that a eukaryotic profilin could be phosphorylated in vivo comes from a recent study on bean seedlings (60). Surprisingly, phosphorylation occurs on tyrosine residues, which is normally a rare event in plants. However, the consequences of profilin phosphorylation on ligand binding activity have not been described. A challenge for the future in the plant systems described below will be to demonstrate that posttranslational modifications or changes in the activity of actin-binding proteins occur in response to specific intracellular or environmental stimuli.

The implication of actin in a specific cellular response is traditionally inferred from the results of studies with cytoskeletal inhibitors. Actin inhibitors group into two broad classes (160): stabilizing agents (phalloidin, jasplakinolide) and depolymerizing compounds or disrupters of actin polymerization (cytochalasins, macrolides). Although phalloidin and cytochalasins have been most popular historically, other agents are being used more frequently because of their membrane permeability properties (jasplakinolide) or distinct modes of action (macrolides). Use of these novel inhibitors has already provided unique insights into actin function in plant systems (10, 54, 144, 152, 194). When both a stabilizing agent and a disrupter of actin polymerization inhibit a particular cellular response, it is usually assumed that actin dynamics or turnover are required for the process.

POLLEN TUBE TIP GROWTH AND THE SELF-INCOMPATIBILITY RESPONSE

Plant reproduction is exquisitely regulated through the precise coordination of pollen tube behavior (183). The male gametophyte, or pollen grain, functions to deliver two sperm to an embryo sac that is embedded within the ovule and other sporophytic tissues. It does so in a dramatic example of cellular morphogenesis, by formation of a pollen tube through which the sperm travel. Upon reaching the embryo sac, the tip of the pollen tube bursts and releases the sperm, which then migrate to the egg and central cell to effect double-fertilization. After pollen lands upon a receptive surface, and during the growth of the pollen tube through the transmitting tissue of the style, a multitude of signals modulate the rate and directionality of cell growth. The nature of these chemical and mechanical signals is only just being discovered (110, 135). Perhaps best understood are the genetic mechanisms, called self-incompatibility (SI) responses, that block pollen germination or tube growth on pistils with the same *S*-genotype. As described later, in at least one system, the signal transduction cascade that prevents self-fertilization and inbreeding seems to result in the destruction of the actin cytoskeleton and cessation of tip growth.

Pollen tubes extend by a highly specialized and localized form of wall and membrane deposition called tip growth. Pollen tubes have a characteristic polarization of their cytoplasm and cytoskeletal components, similar to other tip-growing cells such as root hairs and fungal hyphae (15, 168). The extreme apex is often referred to as the clear zone, where Golgi-derived, secretory vesicles accumulate in a distinct, inverted-cone arrangement. Secretory vesicles are presumed to fuse within a very restricted zone at the apex, although even the best electron micrographs from rapid-freeze/freeze-substitution experiments fail to show a significant number of vesicle fusion events (97). On the lateral shanks of the pollen tube ~5–15 μm distal to the apex, excess membrane is recovered, as indicated by the presence of clathrin-coated pits and vesicles (11, 33). Although the extreme apex contains some endoplasmic reticulum, the majority of cellular organelles are prevented from entering and accumulating in this area.

Polarized ionic gradients are also characteristic of tip-growing cells. For example, pollen tubes have a fluctuating, tip-high gradient of free Ca^{2+} (73, 116, 119, 132, 136). The shank of the pollen tube has a Ca^{2+} concentration of 100–500 nM, whereas the extreme apex may see pulses that reach 1–10 μM (73, 116). More recently, evidence for cytosolic pH gradients has been reported (42, 117, 118), with the tip showing a fluctuating acidic domain. Proton efflux across the plasma membrane and a corresponding cytosolic alkaline band occur in a zone distal to the tip that corresponds roughly to the area of vesicle recovery (42).

A variety of approaches, including immunocytochemistry of cryofixed pollen (97–99), microinjection of live cells with fluorescent-phalloidin (122), and transient expression of a GFP–talin fusion protein (94), allow a consensus view of the actin cytoskeleton in growing pollen tubes to be constructed. The shank of the pollen tube contains thick, axial cables of actin filaments that end 5–10 μm from the tip (Figure 1*a, b*). It is reasonable to assume that these are the tracks that form the basis of reverse-fountain streaming of organelles, because streaming stops and the cables are disrupted when pollen is treated with cytochalasins (45, 71, 112, 129, 131, 167, 170). Because myosins move unidirectionally, toward the plus-end of actin filaments [see (182) for an important exception], it can be inferred that filaments in the cortical cytoplasm have a polarity opposite to those in the subcortical cytoplasm. A dense "collar" of fine actin filaments is often observed in the zone just distal to the tip, and the extreme apex has a reduced amount of fine F-actin. The collar-like region observed in fixed cells seems to be the same as the zone in which axial cables flare out and make contact with the plasma membrane in fluorescent phalloidin injected pollen tubes (122). It also coincides with the alkaline band (42) and the region where microfilaments contact clathrin-coated pits (33). In addition to a role in exocytosis, it is becoming increasingly clear that actin microfilaments play a major role during endocytosis in yeast and mammalian cells (8). One interesting, but untested, possibility is that the collar-like region localizes the endocytotic machinery to a precise subdomain of the pollen tube plasma membrane. Alternatively, the collar may define the tip region by restricting the exit of calcium channels, secretory vesicles, and other components necessary for regulated exocytosis. These speculative models need to be evaluated further, especially as additional molecular and cytological markers for exocytosis and endocytosis become available.

New data about the importance of actin organization at the pollen tube apex come from the use of macrolide toxins called latrunculins (54). Latrunculin B (LatB) treatment demonstrates that cytoplasmic streaming is not the sole, or even the most important, function for F-actin during pollen tube growth, since this process continues after tip growth has ceased. Even at concentrations of 10 nM, this agent causes rapid reorganization of the apical cytoplasm. The collar of fine actin filaments disappears within 5 min and the long axial bundles of actin extend into the

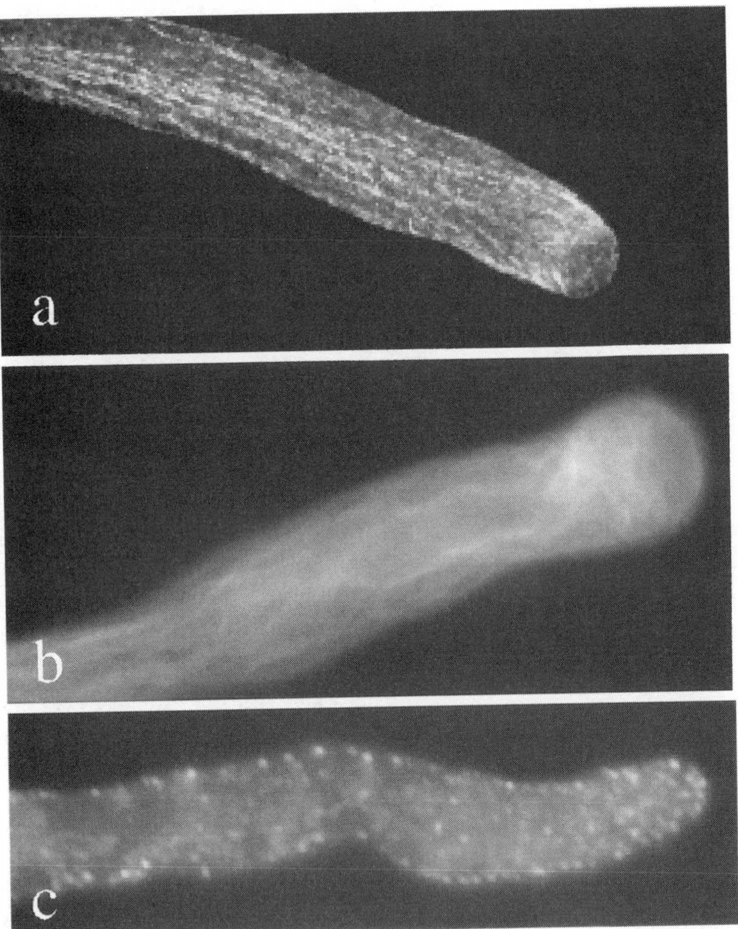

Figure 1 Actin organization in pollen tubes. (*a–b*) Actin organization in untreated pollen tubes. (*a*) Maize pollen (BC Gibbon & CJ Staiger, unpublished data) and (*b*) poppy pollen are shown, respectively. Both have long, axial cables of actin that end about 5–10 μm from the apex, in a region that contains a dense collar or meshwork of fine actin filaments. The extreme tip has relatively few short, fine filaments. (*c*) Actin distribution changes dramatically during the self-incompatibility response. A poppy pollen tube challenged for 20 min with incompatible S-protein shows punctate foci of phalloidin staining throughout the cortex (BN Snowman & VE Franklin-Tong, unpublished data).

apex. Surprisingly, disruption of tip growth occurs at LatB concentrations that have only a negligible impact on F-actin levels. From this, the authors infer that pollen maintains precise control over polymer levels and that the tip cytoskeleton may represent a unique population of filaments that is differentially sensitive to LatB. A correlate may be that the tip actin array is highly dynamic and thus sensitive to depolymerization, whereas the large cables in the shank of the tube are stabilized against depolymerization by barbed-end capping factors or bundling proteins, or both.

Actin-binding proteins are potent sensors of the pollen tube ionic and signaling environment and may dictate the characteristic organization of actin at the tip. Although the monomer-binding protein profilin is distributed uniformly in the cytoplasm of pollen tubes (59, 175), changes in its activity in response to local fluxes in Ca^{2+} could modulate actin organization. Profilin is a better actin-sequestering protein at elevated Ca^{2+} concentrations in vitro (100, 143), presumably because of the reduced ability of the profilin–CaATP-actin complex to add onto filament ends (62a). Native maize pollen profilin shows a 3.5-fold increase in sequestering activity for pollen actin when free Ca^{2+} is raised from 25 nM to 1 μM (DR Kovar, BK Drøbak & CJ Staiger, manuscript submitted). This increased profilin sequestering activity could, therefore, contribute to the reduced levels of F-actin at the extreme apex of actively growing pollen tubes. Actin-depolymerizing factor (ADF), another pollen actin-binding protein (85, 109), is sensitive to changes in pH and could assist in organizing the tip cytoskeleton. ADF binds both F-actin and G-actin and is thought to accelerate filament dynamics by facilitating the loss of subunits from the pointed end (19). It may also destabilize filaments through severing activity or altering the twist of the actin helix (7). At alkaline pH, maize ADF shows preferential binding to subunits rather than polymeric actin (61). It has been predicted that the organization of actin filaments in the alkaline band region could be a consequence of ADF activity (42).

Evidence for a number of signaling components that are implicated in linking cellular polarization and tip growth to cytoskeletal reorganization in other systems has accumulated in recent years (47, 188). Of particular interest is the Rho family of monomeric GTPases. In yeast and vertebrate cells, Rho GTPases are members of multiprotein complexes that establish cellular polarity and organization of the cytoskeleton. Pollen has several Rac/Rop GTPases that are localized to the apical plasma membrane (93, 104) and which may perform functions similar to those described for yeast (188). These were implicated initially in pollen tip growth by elegant studies in which antibodies were injected into the cytoplasm of living pollen tubes, resulting in the cessation of tip growth (105). More recently, overexpression and dominant mutants of Rac/Rop implicate the pollen GTPases in actin filament organization (93, 101). Constitutive active Rac mutants, or transient overexpression, result in a loss of cellular polarity, swelling of the tip and rearrangement of the actin cytoskeleton into transverse arrays (93). Antisense expression and dominant negative mutants lead to cessation of tip growth and perhaps a decrease in filament bundling, although the latter was not convincingly depicted (93). Rac/Rop signaling may be linked to cytoskeletal organization in pollen through modulation of PtdIns-kinase activity and membrane phospholipid pools (93) or through regulation of Ca^{2+} fluxes at the tip (101).

The field poppy, *Papaver rhoeas*, is rapidly becoming a powerful model system to study signaling to the actin cytoskeleton (47, 157). During the self-incompatibility (SI) response, pollen on the stigmatic surface is recognized by the products of a multi-allelic *S*-gene and results in the discriminative inhibition of "self", but not "non-self" pollen growth (46). Several *P. rhoeas S*-alleles have been cloned and the recombinant proteins have biological activity in an in vitro pollen growth assay (43, 96, 181). Treatment of pollen with S-proteins triggers a signaling pathway, with cytosolic Ca^{2+} acting as a second messenger, that mediates pollen tube inhibition. Within minutes of applying incompatible S-proteins, a Ca^{2+}-wave of greater than 1.5 μM floods the shank of the tube and the tip-high gradient dissipates (49–51). Inhibition of poppy pollen tube growth can be mimicked by artificial increases in cytosolic calcium, through photolysis of caged-Ca^{2+} (48). In lily pollen tubes, when Ca^{2+} is forced to concentrations greater than 10 μM experimentally, actin filaments become fragmented and cytoplasmic streaming stops (90, 92). Indeed, the actin cytoskeleton in poppy pollen tubes also undergoes rapid, dramatic and S-specific reorganization (Figure 1*b, c*) (A Geitmann, BN Snowman, VE Franklin-Tong & AMC Emons, unpublished results). Within 1–2 min of incompatible S-protein challenge, the dense collar of actin filaments fills the extreme apical cytoplasm of the pollen tube tip. During the subsequent 5 min, a destruction of axial actin cables and uniform phalloidin labeling at the cortex occurs. From 10 min onwards, S-protein challenge leaves punctate foci of actin filaments that are uniformly distributed in the cortex along the shank and in the apex of the pollen tube (Figure 1*c*). Challenge with compatible S-proteins, by comparison, does not alter actin organization, tip growth, or cytoplasmic streaming. Quantitative analysis reveals an ~80% reduction in F-actin levels in the first 5–10 min after incompatible S-protein challenge (BN Snowman, VE Franklin-Tong & CJ Staiger, unpublished data). This finding is consistent with the dramatic alterations in actin organization observed by light microscopy and suggests that the fragmented appearance of the cytoskeleton is due to more than just an activation of calcium-dependent severing activity or inhibition of myosin activity. A simple and testable model is that the flood of Ca^{2+} promotes the ability of profilin to sequester monomeric actin, thereby resulting in net depolymerization of F-actin.

DYNAMIC CHANGES IN ROOT HAIR GROWTH IN RESPONSE TO *RHIZOBIUM* NOD FACTORS

Symbiotic bacteria belonging to the genus *Rhizobium* can subvert the host cell machinery of leguminous plants to allow for their cellular uptake and proliferation in nitrogen-fixing nodules (83, 128, 130, 138). Numerous examples in which pathogenic bacteria pirate vertebrate host-cell components, in particular the actin cytoskeleton, are now known (37, 55). It should not be surprising, therefore, that *Rhizobium* spp. have learned to use host-cell signaling pathways and cytoskeletal responses for their own purposes.

Bacteria from the rhizosphere are attracted to the root by the release of wound induced flavonoids. *Rhizobium* gains access to root tissues through the tip-growing epidermal hair cells or through the intercellular spaces at wound sites (83, 138). In the former case, infection occurs on actively growing hairs, or even on the early bulge-like protuberance of trichoblasts (83). The process of bacterial uptake often begins with root hair swelling, curling, and other distortions of the actively growing hairs. Subsequently, localized cell wall loosening, cytoplasmic aggregation and nuclear migration to the site of infection are initiated (142). The bacteria gain entry to the hair by redirecting plasma membrane and cell wall deposition inward, forming an invagination. As they enter the cell, the bacteria are surrounded at all times by a host-cell membrane and plant and bacterial cell wall material. Ultimately, this "inside-out" tip growth (138), or modified, local endocytosis (83), leads to the formation of infection threads that carry the bacteria through the root hair, the epidermal cell, and finally into the cortical cells. Intriguingly, these cells anticipate the arrival of the infection thread, and form a radial path for its travel through the tissue (174).

Early ultrastructural analysis provided preliminary evidence for cytoskeletal rearrangements during root hair deformation and infection thread formation (139, 140). In *Vicia hirsuta* roots, there appears to be a loss of microfilaments from the tip region and an accumulation of microtubules at the penetration site (138). Intriguingly, the root hair nucleus tracks the progression of the tip of the infection thread, much as it maintains a constant distance from the actively growing root hair during normal tip growth (108). Based on these observations, Ridge presented a model for how *Rhizobium* redirects the host-cell cytoskeleton to produce infection threads (138).

Not only do *Rhizobium* spp. receive signals from the plant root, they transmit their own information in the form or lipochito-oligosaccharide Nod factors. When applied to legume roots and in the absence of bacteria, these species-specific signaling factors are sufficient to elicit many of the normal responses associated with rhizobial infection, including root hair swelling, membrane depolarization, ion mobilization, cell wall loosening, pre-infection thread formation, and host nodulation-gene expression (83). Current research designed to dissect the signal transduction cascades during early cytoplasmic responses typically makes use of Nod factors rather than direct infection by bacterial cells (39, 69, 70, 174).

Like pollen tubes, root hairs maintain a polar organization of their cytoplasm, a specific organization of the cytoarchitecture and cytoskeleton at the tip, and a tip-high gradient of cytoplasmic calcium (121). Several groups have studied Nod factor–induced changes in actin organization in the root hairs of legumes (18, 29, 120). Differences in the details of the plant response have arisen probably because distinct populations of cells along the root axis are sensitive to Nod factor between the two systems studied. In the work of Cárdenas et al, only rapidly growing hairs are responsive (17, 18), whereas in the vetch root system, as used by Emons and coworkers, hairs that are terminating growth are responsive (31, 69, 120). However, both groups concur in finding that rearrangements of the actin cytoskeleton and alterations to the tip-growth machinery are among the earliest cellular responses. Coincident with and perhaps causal for these cytoskeletal rearrangements, dramatic ion fluxes in the cytoplasm have also been reported.

The effects of Nod factor in living root hair cells that were loaded with fluorescent-phalloidin to mark the actin microfilaments have been examined (18). In *Phaseolus vulgaris* root hairs treated with *R. etli* Nod factors, the large axial cables begin to disappear within 5–10 min and an abundance of fluorescent material accumulates near the plasma membrane at the extreme apex of the tip. The large axial cables recover after 1 h, but the high fluorescence at the tip remains. By 4 h, distortion of root hair morphology has occurred and foci of actin are present at the tip. The rapid reorganization of actin has been attributed to a calcium influx that occurs within 5 min of Nod factor application. Ratiometric imaging demonstrates that actively growing root hairs have a gradient of cytosolic calcium, with a high of 400 nM at the tip and dropping to 80–100 nM in the shank (17). Nod factor treatment elevates calcium at the tip to 1500 nM. After 10–15 min, hot spots of calcium are also observed in the vicinity of the nucleus, which are qualitatively similar to the calcium spikes reported by Ehrhardt et al (39). The authors prefer fragmentation, perhaps through activation of calcium-regulated actin-severing proteins like villin or gelsolin, as the explanation for the origin of the short actin filaments (17). The recent identification of a villin-related protein from lily pollen (176) suggests that such a mechanism could occur in tip-growing plant cells. Cárdenas et al speculate further that these Nod factor–induced actin rearrangements may be a necessary prerequisite for infection thread formation (18).

When *Vicia sativa* (vetch) roots are grown in Fåhraeus slides to observe the effects of *R. leguminosarum* Nod factors, only a small region of the root axis is responsive (69). This so-called zone II contains root hairs that are terminating active tip growth. Nod factor causes the tip of these hairs to swell within an hour, and then to determine a new growth axis and return to active tip growth. Actively growing root hairs of zone I have a tip-high cytosolic calcium gradient, ranging from 1–2 μM at the apex to ~160 nM near the base, and accumulate a spectrin-like protein at the extreme apex (31). In a detailed study of actin organization during *V. sativa* root hair development, Miller et al show that actively growing root hairs have long axial cables of thick filaments that end 5–10 μm from the tip and a dense array of fine filamentous actin, which they term fine bundles or FB-actin (120). The extreme apex is devoid of actin, or contains very short dynamic actin filaments. The importance of this characteristic organization is demonstrated by treatment with low concentrations of cytochalasin D, which disrupts tip growth but allows cytoplasmic streaming to continue. Cessation of growth coincides with loss of the FB-actin network, which is inferred to regulate the delivery of cell wall–containing vesicles to the tip zone.

Zone II hairs have diminished calcium and spectrin levels at the tip (31), and a slightly different actin organization (120). Although they still show streaming and have a polarized tip cytoplasm, the vacuole is often closer to the apex and the vesicle-rich clear zone is smaller. When actin architecture is examined, the FB-actin has diminished and actin cables often extend into the tip. Nod factors

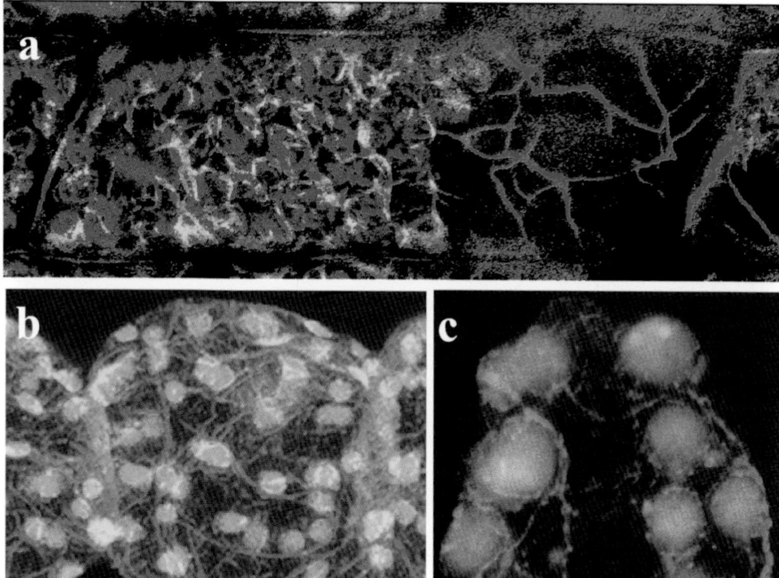

Figure 2 Light-induced chloroplast aggregration and cytoskeleton reorganization. (*a*) The micrograph shows the outer periclinal wall (OPW) of a *Vallisneria* epidermal cell, with chlorophyll fluorescence in red and Alexa-phalloidin stained actin filaments in green. The cell was first irradiated with dim red light for 4 h to induce accumulation of chloroplasts on the OPW. Subsequently, the right half of the cell was irradiated with strong blue light for 2 h. The blue-light induced half of the cell is devoid of chloroplasts and has thick actin bundles, whereas the unexposed half has chloroplasts surrounded by a 'honeycomb array' of actin filaments (S Takagi, unpublished data). (*b*) Actin arrays in cryofixed *Arabidopsis* mesophyll cells labeled with anti-actin antibody. Chloroplasts (*red*) are in close association with large cables and an interconnected array of fine filaments (*green*). (*c*) Higher magnification view from a portion of an *Arabidopsis* mesophyll cell. The chloroplasts are surrounded by a basket-like array of microfilaments (M Kandasamy & RB Meagher, unpublished data)

result in swelling at the tip of zone II hairs and a return of the spectrin-like antigen and increased cytosolic calcium (31). This occurs initially in a general fashion throughout the swelling, and is localized subsequently to a region that is the presumptive site for new tip growth. Technical limitations prevented the authors from examining early responses to Nod factor, or from studying dynamic redistribution of spectrin and calcium in individual hairs. Actin in the Nod factor–treated swellings takes the form of numerous, but disordered, actin cables near the plasma membrane. Foci of actin filaments eventually accumulate at a particular location on the swelling, again the presumed site of future outgrowth. The authors argue that Nod factor elicits a calcium influx necessary to disorder the actin array prior to resumption of tip growth. How a localized site for polar growth is selected, and how the unique cytoarchitecture and actin organization of an actively growing tip are regenerated remain as unresolved issues. In an attempt to characterize the earliest cytoskeletal responses, de Ruijter et al (29) have shown that Nod factor increases the FB-actin in all classes of root hairs within 3–15 min. In the future it will be useful to develop systems for examining dynamic changes in calcium, spectrin, and actin reorganization, and to characterize root hair responses to the presence of single or few bacteria.

LIGHT-INDUCED CHLOROPLAST ANCHORING AND MOVEMENT

The orientation and motility of chloroplasts, whether stationary or streaming in the cytoplasm, are precisely regulated in response to the angle, wavelength, and fluence rate of light [for reviews, see (64, 65, 124, 163, 178, 184)]. Light-mediated chloroplast behavior is best characterized in several algae, ferns, and aquatic angiosperms, but probably also occurs in some terrestrial angiosperms. The conventional wisdom holds that plant cells, by maintaining control over plastid position, optimize the use of light for photosynthesis. Conversely, when fluence rates are too high the plastids can be positioned to avoid or minimize photodamage. Following its detection by photoreceptors in the cytoplasm, the light signal is rapidly translated into changes in the behavior and localization of the actin cytoskeleton, through as yet poorly defined signal transduction events.

A blue light receptor, or cryptochrome, is the most common sensor of the light stimulus for plastid orientation [reviewed in (64, 178)]. In some plants, however, phytochrome seems to play a major role and in a few systems there is cooperation between phytochrome and blue-light receptors or between phytochrome and photosynthetic pigments (36). In *Adiantum*, where both red and blue light contribute to plastid positioning (189), the identification of a hybrid molecule with features of both phytochrome and blue light receptors (127) allows the attractive hypothesis that a single sensor mediates the response to different wavelengths of light to be tested. Alternatively, as supported by recent genetic and biochemical data for angiosperm photoreceptors, separate phytochrome and cryptochrome molecules might interact directly in a common signal transduction pathway (20).

During light-induced chloroplast positioning, new and unique actin arrays form around the plastids and most probably anchor the plastids in the cytoplasm. In *Selaginella helvetica* epidermal cells examined by high-voltage EM, plastid orientation on the anticlinal walls in high-intensity conditions correlates with the proliferation of a meshwork of 8-nm filaments surrounding each chloroplast (27). Chloroplasts in the coenocytic alga *Vaucheria sessilis* participate in the longitudinal flow of cytoplasmic streaming, but following partial irradiation of a cell with low-fluence rate blue light, they become trapped or immobilized in the illuminated region of cytoplasm (12, 13). This trapping is preceded by a local reticulation of the cytoplasmic fibers, which elsewhere form the longitudinal tracks upon which streaming occurs. The tentative identification of these fibers as actin filaments was made on the basis that phalloidin blocked formation of the reticulum, and that filaments extruded from the cytoplasm could be decorated with myosin-S1 fragment (13). These early observations indicated that spatial control over actomyosin organization and activity could regulate the subcellular distribution of chloroplasts.

The three-dimensional arrangement of actin filaments has been examined in detail with fluorescent-phalloidin or antibody labeling methods. Protonemal cells of the fern *Adiantum capillus-veneris*, when maintained in the dark, have thick axial actin cables in the subcortical cytoplasm and a fine actin meshwork in the cortex (77, 78). Following partial irradiation of a cell with a microbeam of polarized light at a high- or low-fluence rate, the plastids gather in the shaded or the illuminated region, respectively. Once the plastids move to their new location, a novel circular actin array forms on the plasma membrane side of each organelle, and the fine filament network becomes less dense. This response is localized; circular structures do not form on plastids outside of the influence of the microbeam. Because the circular actin arrays form only after the chloroplasts have gathered in the illuminated region (low-fluence response) and disappear prior to the plastid movement out of this area, this actin was proposed to function in anchoring organelles at the cortex. Photo-orientation in *Adiantum* is blocked by cytochalasins, but not by colchicine (79).

In leaf epidermal cells of the aquatic angiosperm *Vallisneria*, under low or moderate fluences of unidirectional light, chloroplasts maintain a rather stable position on the periclinal face. When irradiated with high-fluence rate blue light, chloroplasts move away from the periclinal face and participate in the rotational cytoplasmic streaming along the anticlinal walls. In dark-adapted cells, the actin filaments on the periclinal face form a uniformly distributed random network of thick cables (34, 35). Irradiation with as little as 5 min of red light causes the chloroplasts to accumulate some 40 min later on the periclinal face, and this repositioning coincides with the formation of honeycomb-like networks of actin filaments (Figure 2*a*, left side; see color plate). Similarly, Liebe & Menzel (103) report that the immobilized chloroplasts of *Vallisneria* mesophyll cells have short actin filaments arranged in a circular pattern. The unique actin rearrangement can be prevented by subsequent far-red light or by photosynthetic inhibitors, and is reversed by strong blue-light treatment (Figure 2*a*, right side). The increased stability of chloroplasts on the periclinal face against centrifugal force (164), and sensitivity of photo-orientation to cytochalasins (34), supports a role for the honeycomb array in plastid anchoring.

Even the terrestrial angiosperm *Arabidopsis* undergoes light-mediated chloroplast rearrangements (173), which should facilitate future genetic approaches to identify components of the signaling cascade and effector system. A recent study of the actin cytoskeleton, using cryofixation of leaf tissues, provides an exquisite picture of the chloroplast-associated array (80). *Arabidopsis* chloroplasts are aligned along the axial, thick actin bundles and also associate with the thin actin filaments, which extend from the major cables (Figure 2*b*). In addition, each chloroplast appears to have its own circumferential basket of fine actin microfilaments (Figure 2*c*). When these actin arrays are abolished by latrunculin B treatment, chloroplasts in the mesophyll and bundle sheath cells aggregate into aberrant patterns, demonstrating their importance in plastid positioning. Reticulate patches of actin filaments that lie between the chloroplasts and the plasma membrane in *Zinnia* mesophyll cells are also predicted to perform an anchoring function (52).

In what is perhaps the most dramatic example of chloroplast orientation, a novel and transient actin array assists in the rotational movement of individual chloroplasts in the green alga, *Mougeotia*. Each cell has just one large, ribbon-shaped chloroplast that can rotate such that its large face or its thin-edge is exposed to the light source. In the high-fluence rate response, the chloroplast becomes oriented within 10–15 min, with its thin edge parallel to the irradiation. The stimulus for this movement is perceived by phytochrome (and possibly cryptochrome) and is thought to be transmitted through calcium to an actomyosin response system (179, 180). Mineyuki and coworkers (123) were the first to demonstrate that fibrous elements in the cytoplasm, previously observed by interference-contrast microscopy, indeed contained actin filaments extending from the margins of the chloroplast to make contact with the cortex. The appearance of these actin cables precedes or is coincident with the initiation of movement, and they disappear immediately upon the cessation of rotation. A working model holds that phytochrome (P_{FR}) negatively regulates the availability of cortical attachment sites for the contractile actin filament cables (178, 180). Following the recent demonstration that a related green alga, *Mesotaenium caldariorum*, has a phytochrome with serine/threonine protein kinase activity (190), one enticing possibility is that the photoreceptor directly targets and modifies the activity of cytoskeletal components, such as a myosin light chain, light chain kinase, or the yet-to-be-discovered plasma membrane microfilament anchors, through protein phosphorylation events.

Because of the diversity of actin rearrangements and the different photoreceptors employed, it seems unlikely that a single mechanism will explain all aspects of light-mediated plastid positioning. However, several theoretical models for how signaling cascades alter the activity and organization of the actomyosin response system should be considered. These include activation or inactivation of chloroplast-associated myosins, myosin binding to or release from the plastid, establishment of cortical anchorage sites, rearrangement of the cortical actin filament network, stimulation of chloroplast-actin interactions, or formation of new actin arrays on the plastid, and various combinations of these.

Cytoplasmic streaming of a multitude of organelles, including chloroplasts, can also be stimulated by light (124, 184). This is called secondary cytoplasmic streaming or photodinesis, and it reportedly occurs even in certain root hairs (82). Induction of streaming probably does not require the generation of new actin microfilament tracks, but rather is thought to depend upon the activation or recruitment of myosin onto organelles. Light-induced changes in the motility of plastids have been observed in *Vallisneria* epidermal cells (35). Red-light treatment leads to an almost immediate increase in the mobility of chloroplasts, which then subsides some 10 min later (34). Takagi and coworkers have partially purified a myosin-like motor from *Vallisneria* (163), which, like other higher plant myosins (90, 91, 191, 192), is inhibited when calcium concentrations are greater than 1 μM. Thus, one way to modulate streaming in response to light might be to lower cytosolic calcium concentration and thereby activate chloroplast-associated motor molecules. Another, perhaps cooperative, mechanism would be to recruit myosin to chloroplasts in response to light. Immunofluorescence data show that this may indeed be the case (103). And finally, a more direct model in which the phytochrome (P_{FR}) kinase activity alters motor activity by phosphorylation should also be considered. Phosphorylation of non-muscle myosin light chain activates myosin ATPase, whereas evidence from plants seems to indicate that phosphorylation inhibits cytoplasmic streaming (171). Further investigation of the kinase(s) responsible for myosin phosphorylation (113, 141), and the functional impact of posttranslational modification, is urgently needed.

PLANT CELL RESPONSES TO FUNGAL INVASION

Plant cells are bombarded continuously with the spores from fungi, yet fungal invasion of plant tissues is the exception rather than the rule. In numerous cases, it appears that plant cells use a cytoskeleton-based, nonhost resistance mechanism to fend off attack from foreign invaders prior to penetration, or to prevent spread of the fungus beyond the initial infection site [for reviews see (56, 63, 67, 158)].

The fungal spore or conidium, upon landing on a receptive surface, germinates and forms a tip-growing germ tube. Subsequently, a specialized infection structure, the appressorium, is delimited at the end of the germ tube (Figure 3*d*). Penetration is achieved by digestion of the plant cell wall, by direct puncture, or both. In certain cases, the appressorium develops a specialized wall and accumulates solutes that allow it to develop very high turgor pressures. The rice blast fungus, *Magnaporthe grisea*, develops cell pressures of 6–8 MPa, which it directs through a 2-μm diameter appendage, the penetration peg (74). The penetration peg rams its way through the plant cuticle, exerting forces as high as 17 μN, to effect intracellular invasion (9). In other systems, for example the flax rust fungus (86),

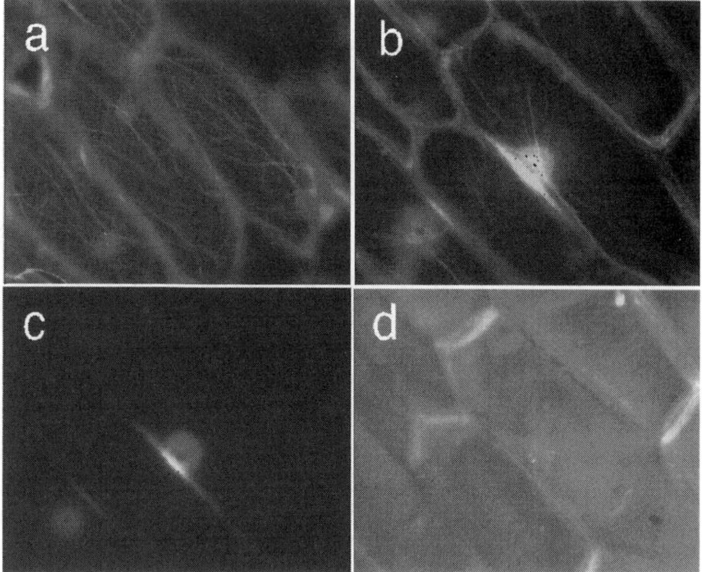

Figure 3 Actin rearrangement in epidermal cells that are under attack by a non-pathogenic fungus. (*a*) Uninfected onion epidermal cells stained with rhodamine-phalloidin. Actin filaments form thick bundles that are oriented transverse or oblique to the long axis of the cell. (*b–d*) Onion epidermal tissue inoculated with the rice blast fungus, *Magnaporthe grisea*. (*b*) The large microfilament cables have rearranged to focus on a plasma membrane site subtending the fungal appressorium. (*c*) Autofluorescent compounds in the plant cell wall and cytoplasm subtending the appressorium are observed. (*d*) DIC (differential interference contrast) image of the plant cell surface shows the location of the *Magnaporthe* appressorium. The germ tube and conidium have broken off during fixation and handling (CJ Staiger & FA Tenjo, unpublished data).

the appressorium develops over a stoma on the leaf surface and infectious hyphae gain access to the intercellular space between mesophyll cells.

Several studies provide compelling evidence that plant cells sense the developing fungus on their surface and take preventative measures to inhibit subsequent invasion. Following the observation that fungal attack leads to accumulation of cytoplasm and sometimes even the nucleus at the site of attack, precise rearrangements of the plant cytoskeleton in cells which are under attack have been reported in six systems: barley–*Erysiphe* (6, 87, 88); parsley–*Phytophthora* (58); flax–*Melampsora* (86); cowpea–*Uromyces* (154); onion–*Magnaporthe* (187); and onion–*Botrytis* (115). The local cytoskeletal rearrangement probably functions to focus the recruitment of cell-wall strengthening and antimicrobial activities to the site of attack.

The first report of the cytoskeletal basis for plant cell defense response was made by Kobayashi and coworkers who examined barley coleoptiles inoculated with the powdery mildew fungus, *Erysiphe* (87). Uninfected epidermal cells have a characteristic arrangement of actin filaments with axial or oblique thick cables, a fine actin meshwork in the cortical cytoplasm, and a cage of actin surrounding the nucleus (Figure 3*a*). On barley coleoptiles inoculated with a non-pathogenic strain of *Erysiphe*, the actin cytoskeleton of the epidermal cell rearranges during the formation and maturation of the appressorium on the cell surface (87, 88). Typically, actin cables focus at a site on the plasma membrane that subtends the fungal appressorium. Some 3–4 h later, the microtubule cytoskeleton also focuses on the site of infection. However, there seems to be considerable variation in the timing and nature of host cell microtubule rearrangements. The dramatic redistribution of actin in flax mesophyll cells inoculated with the flax rust fungus *Melampsora* occurs prior to penetration, but microtubules rearrange at the same time (86). In parsley cells challenged with *Phytophthora*, microfilament rearrangement coincides with infection, but the microtubules completely depolymerize at the infection site (58).

Pathogenic fungi can suppress the plant cytoskeletal responses, suggesting that pathogens harbor the ability to control plant actin dynamics and rearrangements. This has been examined in detail for the barley–*Erysiphe* interaction where the incidence of actin foci is 70% under appressoria of the non-pathogen *E. pisi*, but only 30% in the presence of the pathogen *E. graminis* (87). Similarly, microtubules radiate toward the fungal contact site in 72% of flax mesophyll cells inoculated with an incompatible strain of *Melampsora*, whereas only 5–6% of cells inoculated with a compatible (pathogenic) strain showed a response (86).

Limited evidence hints that the nonhost resistance response involves not only rearrangement of filaments, but also net actin polymerization. In potato protoplasts, *Phytophthora* hyphal cell-wall fractions stimulate cytoplasmic aggregation and a coincident increase in actin polymerization of 40%. Furuse et al demonstrated that these responses were sensitive to actin, protein kinase, Ca-CaM, and phospholipase inhibitors (53). Moreover, the same group has preliminary evidence that two of the defense response proteins, osmotin and chitinase, associate with the actin cytoskeleton (165), which could position them at the site of fungal attack or contribute to the stabilization of dynamic actin filaments.

It is not clear at present whether these cytoskeletal responses are stimulated by fungal elicitors, by plant cell wall degradation products, or by the physical pressure of appressorium adhesion or penetration peg protrusion. Perhaps the cytoskeletal response is sufficiently robust to sense any or all of these potential stimuli. Physical pressure exerted on the plant plasma membrane by a microneedle is sufficient in some cases, to stimulate cytoplasmic aggregation, nuclear migration, and generation of reactive oxygen intermediates, as well as to elicit activation of certain defense-response genes (24, 62, 81). In addition, wounding of the cell wall and application of hemicellulase stimulates nuclear migration in cowpea cells (68). All of these results support a wound-response model for cytoplasmic rearrangements. However, the timing of the cytoplasmic response in several systems provides circumstantial evidence that the plant cell perceives something other than wounding. In barley coleoptile and flax cells, actin rearrangement occurs 3–4 h prior to penetration by *Erysiphe* or *Melampsora*, respectively. Onion epidermal cells inoculated with *Botrytis* or *Magnaporthe* develop actin foci roughly coincident with the timing of penetration peg formation (115, 187) (Figure 3*b*). Convincing evidence that this is not a wound response comes from a mutational analysis of *Magnaporthe*. The mutant, *mps1*, is defective for MAP-kinase function and is unable to form a penetration peg on artificial surfaces or plant epidermal cells. Nonetheless, this mutant fungal strain still elicits rearrangement of the actin cytoskeleton, thus demonstrating that a signal must emanate from the appressorium on the cell surface or from an earlier stage of fungal development (187).

Although the precise function of actin rearrangements following fungal contact is not fully understood, the importance of an actin-based response in preventing fungal proliferation in plant tissues is revealed by pharmacological studies. It has been proposed that the actin cytoskeleton is involved in transmitting the fungal contact signal to the nucleus for changes in gene expression (87). Alternatively, actin could assist the formation of a mechanical and chemical barrier by recruiting defense-related components to the site of fungal attack (87). Several investigations concur in finding that the inability to rearrange the plant cell cytoskeleton allows non-pathogenic fungi an enhanced ability to penetrate and allows fungal spread by blocking hypersensitive cell death. Cytochalasins inhibit or delay hypersensitive cell death in barley coleoptiles (66), potato (166, 172), and cowpea (154), presumably leading to increased spread of the invading fungus to neighboring plant cells. This was confirmed and extended by Kobayashi and colleagues who measured the cytoplasmic responses of barley and other nonhost plants following inoculation with *Erysiphe* (88, 89). Cytochalasins A and E, at concentrations that did not inhibit fungal growth, increased the penetration efficiency for *Erysiphe* on barley as well as on the cells of several dicot species. Furthermore, cytochalasin A increased the ability of a wide range of nonpathogenic fungi to penetrate barley and tobacco cells.

In addition to accumulation of cytoplasm and actin filaments at the site of fungal contact, plant cells also generate elaborate wall appositions, or papillae. The generation of wall fortifications includes deposition of callose, proteins, phenolics, quanidine compounds, and other autofluorescent substances (Figure 3*c*). The cytoskeleton could be involved in directing secretory vesicles to the site of wall fortification, or could recruit the necessary synthetic machinery to a specific region of the plasma membrane. Cytochalasins inhibit callose deposition and papilla formation in barley (88, 89) and cowpea (153, 155). Although a wealth of circumstantial evidence points toward their existence, the direct physical links between the plasma membrane-associated callose/cellulose synthase complex and the cytoskeleton remain undiscovered (32). Perhaps a step in this direction was taken when Winter et al (185) provided biochemical evidence that sucrose synthase (SuSy) could associate with the actin cytoskeleton. Following on from Delmer and coworkers' demonstration that SuSy might channel carbon to the membrane glucan synthase (1), this attractive hypothesis deserves further attention.

Generation of hydrogen peroxide and active oxygen species (AOS) has been proposed to cross-link cell wall components and to strengthen the wall against penetration (14). In addition, AOS may act as antimicrobial agents directly. By analogy with phagocytic vertebrate cells, it is thought that a membrane-associated

NADPH oxidase is at least partially responsible for AOS generation during fungal attack. NADPH oxidase is a multiprotein complex that transfers electrons from NADPH at the cytosolic face of the plasma membrane to molecular oxygen on the extracellular face. The complex is activated by recruitment of several cytosolic proteins to the membrane. Specifically, p67-phox (*ph*agocyte *ox*idase), p47-phox, and the monomeric GTPase Rac move from the cytosol to associate with a membrane-associated cytochrome b composed of gp91-phox and p22-phox. In an elegant study, Xing et al show that the activation of NADPH-oxidase activity in tomato cells challenged with fungal elicitors from an incompatible race, but not a compatible race, correlates with an accumulation of p67, p47, and rac2 proteins at the plasma membrane (186). Even more interesting, the complex associates with a detergent-resistant fraction assumed to be the membrane skeleton. DNase I treatment leads to release of p67, p47, and rac2 from the membrane skeleton. Because DNase I binds to monomeric actin in vitro (150) and causes depolymerization of filamentous actin in plant cells (162), this line of investigation provides circumstantial evidence for association of the phox proteins, directly or indirectly, with actin filaments.

It is tempting to speculate that cytoskeletal rearrangements could function to restrict AOS generation to the vicinity of fungal attack. Indeed, Thordal-Christensen et al demonstrate that hydrogen peroxide production is spatially restricted at sites in the barley epidermal cell wall that subtend the developing fungus (169). Future efforts might be directed toward determining whether the activated NADPH oxidase is localized precisely at the site of fungal attack or whether it is more generally distributed on epidermal cell plasma membranes. Immunocytochemical studies will be facilitated by the demonstration that several components of the plant oxidase complex are recognized by antibodies raised against mammalian proteins, and by the cloning of the gp91-phox homologs (14). If localization of the oxidase complex at membrane sites where actin filaments are focused prior to fungal attack can be demonstrated, a search for coronin-like actin-binding proteins from plants would be justified. Vertebrate coronin apparently docks the oxidase complex onto the membrane cytoskeleton in neutrophils (57). There is also circumstantial evidence that ADF/cofilin participates in the oscillatory pattern of AOS productin in neutrophils, by regulating polymerization and depolymerization of actin at ruffled membranes (72).

STOMATAL OPENING AND CLOSING: Links Between Actin Filaments and Ion Channels

One key function of the epidermal layer of leaves and stems is to regulate the rate of gas exchange and transpiration between underlying plant tissues and the environment. Pairs of specialized epidermal cells form an aperture, the stoma, that can open or close in a controlled manner (40). Guard cells are usually kidney-shaped (many dicots) or dumbell-shaped (grasses) and are joined at their ends, with the inner anticlinal walls defining the edges of the stoma. The stomatal aperture can dilate or close in response to CO_2 concentration, light quality and intensity, humidity, or temperature through turgor-mediated changes in shape of the surrounding guard cells. The various stimuli are integrated through signal-transduction cascades—involving calcium and proton fluxes, changes in phosphoinositide levels, protein phosphorylation and dephosphorylation, and GTP-binding proteins—and ultimately focus on the regulation of several types of ion channels (3, 4, 149). A decrease in guard cell turgor leads to stomatal closing in response to the stress hormone ABA for example, and is mediated by K^+ and anion efflux as well as malate metabolism. Guard cells also undergo increases in turgor pressure that lead to stomatal opening, as in the case of light stimulation. Increases in turgor are thought to be driven primarily by activation of inward K^+ channels, which are probably the products of *KAT1*-like genes (102, 145).

The actin cytoskeleton has a unique organization in mature guard cells and this pattern changes during stimulus-mediated stomatal opening and closing. Cortical actin microfilaments usually mirror the pattern of cortical microtubules and cellulose microfibril wall thickenings (41). The first description of actin microfilament arrangement in guard cells was by Cho & Wick, who examined the developing stomatal complex of rye (21, 22). In the subcortical cytoplasm of young dumbell-shaped guard cells, large longitudinal bundles of microfilaments are observed, whereas in the cortical cytoplasm, a meshwork with occasional transverse filaments is present (22). The mature and functional kidney-shaped guard cells of *Selaginella* (23), *Tradescantia* (25), *Commelina* (41, 84), and *Arabidopsis* (94) have fine cortical microfilaments that radiate outward from the inner anticlinal wall.

Youngsook Lee and colleagues made the interesting observation that the number and distribution of actin filaments in *Commelina communis* guard cells are different when the aperture is open than when it is closed (41). When stomata are open, for example in the light, microfilaments fan outward from the anticlinal wall, as described above. In the dark and during ABA-induced closing, however, the microfilaments depolymerize or become fragmented. Short, randomly oriented microfilaments are seen throughout the cortical cytoplasm, with some concentration near the ventral side of the guard cell. Infrequently, long filaments aligned parallel to the long axis of the cell are found. These actin-based rearrangements are among the fastest known cytoskeletal responses to extracellular stimuli, with ABA treatment leading to changes within 3–10 min.

A pharmacological approach was used to examine whether microfilament depolymerization or polymerization and stabilization are necessary for aperture closing or opening, respectively (75, 84). Phalloidin blocks ABA-induced stomatal closing, consistent with a model in which filament breakdown is necessary for loss of turgor and stomatal closing. However, phalloidin also inhibits light-induced stomatal opening. Cytochalasins have the opposite effect, enhancing both light-induced opening (75, 84) and ABA-induced closing (J-U Hwang & Y Lee, unpublished results). The combined results are not consistent with a simple model in which actin filament polymerization and depolymerization are necessary for opening and closing of stomata, and thus led the authors to propose that microfilaments are involved in modulating ion channels or signal transduction cascades (84).

Significant new insight into the role of actin in shape and volume changes was provided by patch-clamp analysis of ion channel activity on *Vicia faba* guard cell membrane patches. Lee and coworkers showed that cytochalasins potentiate the activity of an inward-rectified K^+ channel, whereas phalloidin inhibits the inward K^+ current (75). Cytochalasin presumably enhances guard cell opening in the light by activating K^+ channels and thereby accelerates increases in osmotic turgor pressure. These findings provide the first evidence that actin filaments in plant cells can modulate ion channel activity and influence physiological responses.

Liu & Luan propose that the actin filament network in guard cells is part of an osmosensor that functions in a positive-feedback mechanism to accelerate cell shape and volume changes (106). When patch-clamp analysis of *Vicia* guard cell protoplasts in a whole-cell configuration was performed, it was found that inward K^+ currents were activated by hypotonic conditions and inhibited by hypertonic conditions. Ion channel regulation was shown to correlate positively with volume changes of the guard cell protoplasts; hypertonic bathing solution leads to cell shrinking and hypotonic solution to cell swelling. Actin microfilament arrays are also sensitive to changes in extracellular osmolarity; protoplasts in hypertonic conditions have a filament network that ramifies throughout the cytoplasm, whereas hypotonic conditions lead to filament depolymerization and collapse around the nucleus. A cautionary note should be injected here, because the actin filament arrays were quite scant and did not resemble those found in intact guard cells. Cytochalasin treatment increases the inward K^+ current, but only under hypertonic conditions, i.e. when a filament network is expected to be intact. Through extrapolation, it is assumed that depolymerization of actin filaments links hypotonic swelling to activation of inward K^+ channels. These observations must now be extended to studies with intact, walled guard cells.

Data from animal cell model systems suggest several ways in which ion channels might be linked to the cytoskeleton (16). It is thought that during cell volume regulation of vertebrate cells a deformation of the cytoskeleton transmits information by direct activation or inactivation of ion channels, or by altering the availability of signal transduction components. Connection of ion channels to the cortical cytoskeleton could be through direct interaction with F-actin or, more likely, through a membrane skeleton and specific anchoring proteins like spectrin and ankyrin. Evidence for a membrane skeleton in plant cells is quite limited; however, spectrin-like molecules have been immunolocalized to the cortical cytoplasm in several cell types (30). Pharmacological data and quantification of changes in F-actin levels suggest that microfilament length and the degree of crosslinking, in other words the gel properties of the actin cytoskeleton, are important for osmotic regulation of cell volume (16). Thus, complementary actin inhibitors and careful measurements of actin cytoskeletal components must be applied to the guard cell system to further elucidate regulation of ion channel activity.

Recently, Youngsook Lee's group has provided further evidence that actin is involved in signal-mediated changes in stomatal aperture (S-O Eun & Y Lee, personal communication). If actin simply responds to opening and closing signals by polymerizing and depolymerizing, respectively, then its configuration should be the same regardless of the stimulus. Fusicoccin, a stimulator of H^+-ATPases and a potent stomatal opening agent, leads to disruption of actin arrays and hyper-promotes opening even in the light. Stabilization of actin filaments with phalloidin delayed the course of fusicoccin-induced opening. From these data, they predict that radial actin arrays do not determine the direction of aperture change, but rather that intact filaments slow the process of opening and closing. The converse is also thought to be true; the depolymerization of actin filaments enhances stomatal movements, and the direction of movement is determined by the type of stimulus that leads to activation of various ion transporters. Additional experimentation will be necessary to confirm this interesting model. For example, investigation of real-time actin dynamics during responses of intact guard cells under different environmental conditions might be productive. This should soon be possible, since Kost et al (94) have provided a preliminary view of actin arrays in *Arabidopsis* guard cells with the GFP–talin reporter.

FUTURE PROSPECTS

This survey of the recent exciting developments from five systems in which cytoplasmic rearrangements occur in response to external stimuli makes it clear that in each case we have a reasonable idea about the nature of the signal or stimulus. Furthermore, at the output end of the signaling cascade, it is also clear that changes in the cytoplasmic architecture are mediated by an actin-based effector response system. A major goal must, therefore, be the elucidation of the molecular components in the black box signal transduction cascades, which connect perception of the stimulus with the actin-based response system. The current rapid progress in identifying plant actin-binding proteins, many of which are likely to be regulated by second messengers, will facilitate working backwards from the response. Direct evidence for the involvement of a particular actin-binding protein will, however, require advanced cell biological approaches and mutational analyses. Additional insight into each response system could also be gained through a careful analysis of actin dynamics and/or examination of cytoskeletal redistribution in real time. The use of GFP-reporter proteins to study these processes in live cells (93, 94) provides one powerful tool to accomplish this goal. The future promises additional fascinating discoveries about how the dynamic framework of the cytoskeleton coordinates responses to external and intrinsic stimuli in non-motile plant cells.

ACKNOWLEDGMENTS

I thank current and past members of my laboratory and the Cytoskeletal Research Group at Purdue for fruitful and stimulating discussions. I am also extremely grateful to Dave Collings, Noni Franklin-Tong, Youngsook Lee, Dan Szymanski, and Shingo Takagi for constructive comments on the manuscript. Thanks to Shingo Takagi, Muthugapatti Kandasamy, Rich Meagher, Ben Snowman, and Noni Franklin-Tong for providing images and sharing unpublished information. Research in my laboratory is supported by the US Department of Agriculture, the US Department of Energy, and the Showalter Fund of the Purdue Research Foundation.

Visit the Annual Reviews home page at www.AnnualReviews.org

LITERATURE CITED

1. Amor Y, Haigler CH, Johnson S, Wainscott M, Delmer D. 1995. A membrane-associated form of sucrose synthase and its potential role in synthesis of cellulose and callose in plants. *Proc. Natl. Acad. Sci. USA* 92:9353–57
2. Asada T, Collings D. 1997. Molecular motors in higher plants. *Trends Plant Sci.* 2:29–37
3. Assmann SM. 1993. Signal transduction in guard cells. *Annu. Rev. Cell Biol.* 9:345–75
4. Assmann SM, Shimazaki K-i. 1999. The multisensory guard cell. Stomatal responses to blue light and abscisic acid. *Plant Physiol.* 119:809–15
5. Ayscough KR. 1998. *In vivo* functions of actin-binding proteins. *Curr. Opin. Cell Biol.* 10:102–11
6. Baluška F, Bacigálová K, Oud JL, Hauskrecht M, Kubica Š. 1995. Rapid reorganization of microtubular cytoskeleton accompanies early changes in nuclear ploidy and chromatin structure in postmitotic cells of barley leaves infected with powdery mildew. *Protoplasma* 185:140–51
7. Bamburg JR, McGough A, Ono S. 1999. Putting a new twist on actin: ADF/cofilins modulate actin dynamics. *Trends Cell Biol.* 9:364–70
8. Battey NH, James NC, Greenland AJ, Brownlee C. 1999. Exocytosis and endocytosis. *Plant Cell* 11:643–59
9. Bechinger C, Giebel K-F, Schnell M, Leiderer P, Deising HB, Bastmeyer M. 1999. Optical measurements of invasive forces exerted by appressoria of a plant pathogenic fungus. *Science* 285:1896–99
10. Bibikova TN, Blancaflor EB, Gilroy S. 1999. Microtubules regulate tip growth and orientation in root hairs of *Arabidopsis thaliana*. *Plant J.* 17:657–65
11. Blackbourn HD, Jackson AP. 1996. Plant clathrin heavy chain: sequence analysis and restricted localisation in growing pollen tubes. *J. Cell Sci.* 109:777–86
12. Blatt MR, Briggs WR. 1980. Blue-light-induced cortical fiber reticulation concomitant with chloroplast aggregation in the alga *Vaucheria sessilis*. *Planta* 147:355–62
13. Blatt MR, Wessels NK, Briggs WR. 1980. Actin and cortical fiber reticulation in the siphonaceous alga *Vaucheria sessilis*. *Planta* 147:363–75
14. Blumwald E, Aharon GS, Lam BC-H. 1998. Early signal transduction pathways in plant—pathogen interactions. *Trends Plant Sci.* 3:342–46
15. Cai G, Moscatelli A, Cresti M. 1997.

Cytoskeletal organization and pollen tube growth. *Trends Plant Sci.* 2:86–91
16. Cantiello HF. 1997. Role of actin filament organization in cell volume and ion channel regulation. *J. Exp. Zool.* 279:425–35
17. Cárdenas L, Feijó JA, Kunkel JG, Sánchez F, Holdaway-Clarke T, et al. 1999. *Rhizobium* Nod factors induce increases in intracellular free calcium and extracellular calcium influxes in bean root hairs. *Plant J.* 19:347–52
18. Cárdenas L, Vidali L, Domínguez J, Pérez H, Sánchez F, et al. 1998. Rearrangement of actin microfilaments in plant root hairs responding to *Rhizobium etli* nodulation signals. *Plant Physiol.* 116:871–77
19. Carlier M-F, Laurent V, Santolini J, Melki R, Didry D, et al. 1997. Actin depolymerizing factor (ADF/cofilin) enhances the rate of filament turnover: implication in actin-based motility. *J. Cell Biol.* 136:1307–22
20. Cashmore AR, Jarillo JA, Wu Y-J, Liu D. 1999. Cryptochromes: blue light receptors for plants and animals. *Science* 284:760–65
21. Cho S-O, Wick SM. 1990. Distribution and function of actin in the developing stomatal complex of winter rye (*Secale cereale* cv. Puma). *Protoplasma* 157:154–64
22. Cho S-O, Wick SM. 1991. Actin in the developing stomatal complex of winter rye: a comparison of actin antibodies and Rh-phalloidin labelling of control and CB-treated tissues. *Cell Motil. Cytoskelet.* 19:25–36
23. Cleary AL, Brown RC, Lemmon BE. 1993. Organisation of microtubules and actin filaments in the cortex of differentiating *Selaginella* guard cells. *Protoplasma* 177:37–44
24. Cleary AL, Hardham AR. 1993. Pressure induced reorientation of cortical microtubules in epidermal cells of *Lolium rigidum* leaves. *Plant Cell Physiol.* 34:1003–8
25. Cleary AL, Mathesius U. 1996. Rearrangements of F-actin during stomatogenesis visualised by confocal microscopy in fixed and permeabilised *Tradescantia* leaf epidermis. *Bot. Acta* 109:15–24
26. Collings DA, Asada T, Allen NS, Shibaoka H. 1998. Plasma membrane-associated actin in Bright Yellow 2 tobacco cells. Evidence for interaction with microtubules. *Plant Physiol.* 118:917–28
27. Cox G, Hawes CR, van der Lubbe L, Juniper BE. 1987. High-voltage electron microscopy of whole, critical-point dried plant cells. 2. Cytoskeletal structures and plastid motility in *Selaginella. Protoplasma* 140:173–86
28. Cyr RJ, Palevitz BA. 1995. Organization of cortical microtubules in plant cells. *Curr. Opin. Cell Biol.* 7:65–71
29. de Ruijter NCA, Bisseling T, Emons AMC. 1999. *Rhizobium* Nod factors induce an increase in sub-apical fine bundles of actin filaments in *Vicia sativa* root hairs within minutes. *Mol. Plant-Microbe Interact.* 12:829–32
30. de Ruijter NCA, Emons AMC. 1999. Actin-binding proteins in plant cells. *Plant Biol.* 1:26–35
31. de Ruijter NCA, Rook MB, Bisseling T, Emons AMC. 1998. Lipochito-oligosaccharides re-initiate root hair tip growth in *Vicia sativa* with high calcium and spectrin-like antigen at the tip. *Plant J.* 13:341–50
32. Delmer DP. 1999. Cellulose biosynthesis: exciting times for a difficult field of study. *Annu. Rev. Plant Physiol. Plant Mol. Biol.* 50:245–76
33. Derksen J, Rutten T, Lichtscheidl IK, de Win AHN, Pierson ES, Rongen G. 1995. Quantitative analysis of the distribution of organelles in tobacco pollen tubes: implications for exocytosis and endocytosis. *Protoplasma* 188:267–76
34. Dong X-J, Nagai R, Takagi S. 1998. Microfilaments anchor chloroplasts along the outer periclinal wall in *Vallisneria* epidermal cells through cooperation of PFR

and photosynthesis. *Plant Cell Physiol.* 39:1299–306

35. Dong X-J, Ryu J-H, Takagi S, Nagai R. 1996. Dynamic changes in the organization of microfilaments associated with the photocontrolled motility of chloroplasts in epidermal cells of *Vallisneria. Protoplasma* 195:18–24
36. Dong X-J, Takagi S, Nagai R. 1995. Regulation of the orientation movement of chloroplasts in epidermal cells of *Vallisneria:* cooperation of phytochrome with photosynthetic pigment under low-fluence-rate light. *Planta* 197:257–63
37. Dramsi S, Cossart P. 1998. Intracellular pathogens and the actin cytoskeleton. *Annu. Rev. Cell Dev. Biol.* 14:137–66
38. Drøbak BK, Watkins PAC, Valenta R, Dove SK, Lloyd CW, Staiger CJ. 1994. Inhibition of plant plasma membrane phosphoinositide phospholipase C by the actin-binding protein, profilin. *Plant J.* 6:389–400
39. Ehrhardt DW, Wais R, Long SR. 1996. Calcium spiking in plant root hairs responding to Rhizobium nodulation signals. *Cell* 85:673–81
40. Esau K. 1977. *Anatomy of Seed Plants.* New York: Wiley & Sons. 550 pp. 2nd ed.
41. Eun S-O, Lee Y. 1997. Actin filaments of guard cells are reorganized in response to light and abscisic acid. *Plant Physiol.* 115:1491–98
42. Feijó JA, Sainhas J, Hackett GR, Kunkel JG, Hepler PK. 1999. Growing pollen tubes possess a constitutive alkaline band in the clear zone and a growth-dependent acidic tip. *J. Cell Biol.* 144:483–96
43. Foote HCC, Ride JP, Franklin-Tong VE, Walker EA, Lawrence MJ, Franklin FCH. 1994. Cloning and expression of a distinctive class of self-incompatibility (S) gene from *Papaver rhoeas* L. *Proc. Natl. Acad. Sci. USA* 91:2265–69
44. Fowler JE, Quatrano RS. 1997. Plant cell morphogenesis: plasma membrane interactions with the cytoskeleton and cell wall. *Annu. Rev. Cell Dev. Biol.* 13:697–743
45. Franke WW, Herth W, Van der Woude J Morré DJ. 1972. Tubular and filamentou structures in pollen tubes: possible involvement as guide elements in protoplasmic streaming and vectorial migration of secretory vesicles. *Planta* 105:317–41
46. Franklin FCH, Lawrence MJ, Franklin Tong VE. 1995. Cell and molecular biology of self-incompatibility in flowering plants *Int. Rev. Cytol.* 158:1–64
47. Franklin-Tong VE. 1999. Signaling and the modulation of pollen tube growth. *Plan Cell* 11:727–38
48. Franklin-Tong VE, Drøbak BK, Allan AC Watkins PAC, Trewavas AJ. 1996. Growth of pollen tubes of Papaver rhoeas is regulated by a slow-moving calcium wave propagated by inositol 1,4,5-trisphosphate *Plant Cell* 8:1305–21
49. Franklin-Tong VE, Hackett G, Hepler PK 1997. Ratio-imaging of $Ca^{2+}{}_i$ in the self incompatibility response in pollen tubes o *Papaver rhoeas. Plant J.* 12:1375–86
50. Franklin-Tong VE, Ride JP, Franklin FCH. 1995. Recombinant stigmatic self incompatibility (S-) protein elicits a Ca^{2+} transient in pollen of *Papaver rhoeas. Plan J.* 8:299–307
51. Franklin-Tong VE, Ride JP, Read ND Trewavas AJ, Franklin FCH. 1993. The self-incompatibility response in *Papaver rhoeas* is mediated by cytosolic free calcium. *Plant J.* 4:163–77
52. Fukuda H, Kobayashi H. 1989. Dynamic organization of the cytoskeleton during tracheary-element differentiation. *Dev Growth Differ.* 31:9–16
53. Furuse K, Takemoto D, Doke N, Kawakita K. 1999. Involvement of actin filament association in hypersensitive reactions in potato cells. *Physiol. Mol. Plant Pathol* 54:51–61
54. Gibbon BC, Kovar DR, Staiger CJ. 1999 Latrunculin B has different effects on maize pollen germination and tube growth *Plant Cell* 11:2349–63
55. Goosney DL, de Grado M, Finlay BB

1999. Putting *E. coli* on a pedestal: a unique system to study signal transduction and the actin cytoskeleton. *Trends Cell Biol.* 9:11–14

56. Grant M, Mansfield J. 1999. Early events in host-pathogen interactions. *Curr. Opin. Plant Biol.* 2:312–19
57. Grogan A, Reeves E, Keep N, Wientjes F, Totty NF, et al. 1997. Cytosolic *phox* proteins interact with and regulate the assembly of coronin in neutrophils. *J. Cell Sci.* 110:3071–81
58. Gross P, Julius C, Schmelzer E, Hahlbrock K. 1993. Translocation of cytoplasm and nucleus to fungal penetration sites is associated with depolymerization of microtubules and defence gene activation in infected, cultured parsley cells. *EMBO J.* 12:1735–44
59. Grote M, Swoboda I, Meagher RB, Valenta R. 1995. Localization of profilin- and actin-like immunoreactivity in in vitro-germinated tobacco pollen tubes by electron microscopy after special water-free fixation techniques. *Sex. Plant Reprod.* 8:180–86
60. Guillén G, Valdés-López V, Noguez R, Olivares J, Rodríguez-Zapata LC, et al. 1999. Profilin in *Phaseolus vulgaris* is encoded by two genes (only one expressed in root nodules) but multiple isoforms are generated in vivo by phosphorylation on tyrosine residues. *Plant J.* 19:497–508
61. Gungabissoon RA, Jiang C-J, Drøbak BK, Maciver SK, Hussey PJ. 1998. Interaction of maize actin-depolymerising factor with actin and phosphoinositides and its inhibition of plant phospholipase C. *Plant J.* 16:689–96
62. Gus-Mayer S, Naton B, Hahlbrock K, Schmelzer E. 1998. Local mechanical stimulation induces components of the pathogen defense response in parsley. *Proc. Natl. Acad. Sci. USA* 95:8398–403

62a. Gutsche-Perelroizen I, Lepault J, Ott A, Carlier M-F. 1999. Filament assembly from profilin-actin. *J. Biol. Chem.* 274:6234–43

63. Hardham AR, Mitchell HJ. 1998. Use of molecular cytology to study the structure and biology of phytopathogenic and mycorrhizal fungi. *Fungal Gen. Biol.* 24:252–84
64. Haupt W. 1982. Light-mediated movement of chloroplasts. *Annu. Rev. Plant Physiol.* 33:205–33
65. Haupt W. 1998. Chloroplast movement: from phenomenology to molecular biology. *Prog. Bot.* 60:3–36
66. Hazen BE, Bushnell WR. 1983. Inhibition of the hypersensitive reaction in barley to powdery mildew by heat shock and cytochalasin B. *Physiol. Plant Pathol.* 23:421–38
67. Heath MC. 1997. Signalling between pathogenic rust fungi and resistant or susceptible host plants. *Ann. Bot.* 80:713–20
68. Heath MC, Nimchuk ZL, Xu H. 1997. Plant nuclear migrations as indicators of critical interactions between resistant or susceptible cowpea epidermal cells and invasion hyphae of the cowpea rust fungus. *New Phytol.* 135:689–700
69. Heidstra R, Geurts R, Franssen H, Spaink HP, van Kammen A, Bisseling T. 1994. Root hair deformation activity of nodulation factors and their fate on *Vicia sativa*. *Plant Physiol.* 105:787–97
70. Heidstra R, Yang WC, Yalcin Y, Peck S, Emons AM, et al. 1997. Ethylene provides positional information on cortical cell division but is not involved in Nod factor-induced root hair tip growth in *Rhizobium*-legume interaction. *Development* 124:1781–87
71. Herth W, Franke WW, Van der Woude WJ. 1972. Cytochalasin stops tip growth in plants. *Naturwissenschaften* 59:38–39
72. Heyworth PG, Robinson JM, Ding J, Ellis BA, Badwey JA. 1997. Cofilin undergoes rapid dephosphorylation in stimulated neutrophils and translocates to ruffled membranes enriched in products of the NADPH

oxidase complex. Evidence for a novel cycle of phosphorylation and dephosphorylation. *Histochem. Cell Biol.* 108:221–33

73. Holdaway-Clarke TL, Feijó JA, Hackett GR, Kunkel JG, Hepler PK. 1997. Pollen tube growth and the intracellular cytosolic calcium gradient oscillate in phase while extracellular calcium influx is delayed. *Plant Cell* 9:1999–2010
74. Howard RJ, Ferrari MA, Roach DH, Money NP. 1991. Penetration of hard substances by a fungus employing enormous turgor pressures. *Proc. Natl. Acad. Sci. USA* 88:11281–84
75. Hwang J-U, Suh S, Yi H, Kim J, Lee Y. 1997. Actin filaments modulate both stomatal opening and inward K^+-channel activities in guard cells of *Vicia faba* L. *Plant Physiol.* 115:335–42
76. Janmey PA. 1994. Phosphoinositides and calcium as regulators of cellular actin assembly and disassembly. *Annu. Rev. Physiol.* 56:169–91
77. Kadota A, Wada M. 1989. Photoinduction of circular F-actin on chloroplast in a fern protonemal cell. *Protoplasma* 151:171–74
78. Kadota A, Wada M. 1992. Photoinduction of formation of circular structures by microfilaments on chloroplasts during intracellular orientation in protonemal cells of the fern *Adiantum capillus-veneris. Protoplasma* 167:97–107
79. Kadota A, Wada M. 1992. Photoorientation of chloroplasts in protonemal cells of the fern *Adiantum* as analyzed by use of a video-tracking system. *Bot. Mag. Tokyo* 105:265–79
80. Kandasamy MK, Meagher RB. 1999. Actin-organelle interaction: Association with chloroplast in *Arabidopsis* leaf mesophyll cells. *Cell Motil. Cytoskelet.* 44:110–18
81. Kennard JL, Cleary AL. 1997. Pre-mitotic nuclear migration in subsidiary mother cells of Tradescantia occurs in G1 of the cell cycle and requires F-actin. *Cell Motil. Cytoskelet.* 36:55–67
82. Keul M, Soran V, Lazar-Keul G. 1969. The chemical and photodynamic action of neutral red on rotational streaming in barley (*Hordeum vulgare* L.) root hairs. *Protoplasma* 67:279–93
83. Kijne JW. 1992, The Rhizobium infection process. In *Biological Nitrogen Fixation*, ed. G Stacey, RH Burris, HJ Evans, pp. 349–98. New York: Chapman & Hall
84. Kim M, Hepler PK, Eun S-O, Ha KS, Lee Y. 1995. Actin filaments in mature guard cells are radially distributed and involved in stomatal movement. *Plant Physiol.* 109:1077–84
85. Kim S-R, Kim Y, An G. 1993. Molecular cloning and characterization of anther-preferential cDNA encoding a putative actin-depolymerizing factor. *Plant Mol. Biol* 21:39–45
86. Kobayashi I, Kobayashi Y, Hardham AR. 1994. Dynamic reorganization of microtubules and microfilaments in flax cells during the resistance response to flax rust infection. *Planta* 195:237–47
87. Kobayashi I, Kobayashi Y, Yamaoka N, Kunoh H. 1992. Recognition of a pathogen and a nonpathogen by barley coleoptile cells. III. Responses of microtubules and actin filaments in barley coleoptile cells to penetration attempts. *Can. J. Bot.* 70:1815–23
88. Kobayashi Y, Kobayashi I, Funaki Y, Fujimoto S, Takemoto T, Kunoh H. 1997. Dynamic reorganization of microfilaments and microtubules is necessary for the expression of non-host resistance in barley coleoptile cells. *Plant J.* 11:525–37
89. Kobayashi Y, Yamada M, Kobayashi I, Kunoh H. 1997. Actin microfilaments are required for the expression of nonhost resistance in higher plants. *Plant Cell Physiol.* 38:725–33
90. Kohno T, Shimmen T. 1987. Ca^{2+}-induced fragmentation of actin filaments in pollen tubes. *Protoplasma* 141:177–79

91. Kohno T, Shimmen T. 1988. Accelerated sliding of pollen tube organelles along *Characeae* actin bundles regulated by Ca^{2+}. *J. Cell Biol.* 106:1539–43
92. Kohno T, Shimmen T. 1988. Mechanism of Ca^{2+} inhibition of cytoplasmic streaming in lily pollen tubes. *J. Cell Sci.* 91:501–09
93. Kost B, Lemichez E, Spielhofer P, Hong Y, Tolias K, et al. 1999. Rac homologues and compartmentalized phosphatidylinositol 4,5-bisphosphate act in a common pathway to regulate polar pollen tube growth. *J. Cell Biol.* 145:317–30
94. Kost B, Spielhofer P, Chua N-H. 1998. A GFP-mouse talin fusion protein labels plant actin filaments *in vivo* and visualizes the actin cytoskeleton in growing pollen tubes. *Plant J.* 16:393–401
95. Kropf DL, Bisgrove SR, Hable WE. 1998. Cytoskeletal control of polar growth in plant cells. *Curr. Opin. Cell Biol.* 10:117–22
96. Kurup S, Ride JP, Jordan N, Fletcher G, Franklin-Tong VE, Franklin FCH. 1998. Identification and cloning of related self-incompatibility S-genes in *Papaver rhoeas* and *Papaver nudicaule*. *Sex. Plant Reprod.* 11:192–98
97. Lancelle SA, Cresti M, Hepler PK. 1987. Ultrastructure of the cytoskeleton in freeze-substituted pollen tubes of *Nicotiana alata*. *Protoplasma* 140:141–50
98. Lancelle SA, Hepler PK. 1988. Cytochalasin-induced ultrastructural alterations in *Nicotiana* pollen tubes. *Protoplasma* (Suppl.) 2:65–75
99. Lancelle SA, Hepler PK. 1989. Immunogold labelling of actin on sections of freeze-substituted plant cells. *Protoplasma* 150:72–74
100. Larsson H, Lindberg U. 1988. The effect of divalent cations on the interaction between calf spleen profilin and different actins. *Biochim. Biophys. Acta* 953:95–105
101. Li H, Lin Y, Heath RM, Zhu MX, Yang Z. 1999. Control of pollen tube tip growth by a Rop GTPase—dependent pathway that leads to tip-localized calcium influx. *Plant Cell* 11:1731–42
102. Li J, Lee Y-RJ, Assmann SM. 1998. Guard cells possess a calcium-dependent protein kinase that phosphorylates the KAT1 potassium channel. *Plant Physiol.* 116:785–95
103. Liebe S, Menzel D. 1995. Actomyosin-based motility of endoplasmic reticulum and chloroplasts in *Vallisneria* mesophyll cells. *Biol. Cell* 85:207–22
104. Lin Y, Wang Y, Zhu J-K, Yang Z. 1996. Localization of a Rho GTPase implies a role in tip growth and movement of the generative cell in pollen tubes. *Plant Cell* 8:293–303
105. Lin Y, Yang Z. 1997. Inhibition of pollen tube elongation by microinjected anti-Rop1Ps antibodies suggests a crucial role for Rho-type GTPases in the control of tip growth. *Plant Cell* 9:1647–59
106. Liu K, Luan S. 1998. Voltage-dependent K^+ channels as targets of osmosensing in guard cells. *Plant Cell* 10:1957–70
107. Lloyd CW, Drøbak BK, Dove SK, Staiger CJ. 1996. Interactions between the plasma membrane and the cytoskeleton in plants. In *Membranes: Specialized Functions in Plants*, ed. M Smallwood, JP Knox, DJ Bowles, pp. 1–20. Oxford: BIOS Sci.
108. Lloyd CW, Pearce KJ, Rawlins DJ, Ridge RW, Shaw PJ. 1987. Endoplasmic microtubules connect the advancing nucleus to the tip of legume root hairs, but F-actin is involved in basipetal migration. *Cell Motil. Cytoskelet.* 8:27–36
109. Lopez I, Anthony RG, Maciver SK, Jiang C-J, Khan S, et al. 1996. Pollen specific expression of maize genes encoding actin depolymerizing factor-like proteins. *Proc. Natl. Acad. Sci. USA* 93:7415–20
110. Lush WM. 1999. Whither chemotropism

and pollen tube guidance? *Trends Plant Sci.* 4:413–18

111. Machesky LM, Insall RH. 1999. Signaling to actin dynamics. *J. Cell Biol.* 146:267–72
112. Mascarenhas JP, LaFountain J. 1972. Protoplasmic streaming, cytochalasin B, and growth of the pollen tube. *Tissue Cell* 4:11–14
113. McCurdy DW, Harmon AC. 1992. Phosphorylation of a putative myosin light chain in *Chara* by calcium-dependent protein kinase. *Protoplasma* 171:85–88
114. McGough A. 1998. F-actin-binding proteins. *Curr. Opin. Struct. Biol.* 8:166–76
115. McLusky SR, Bennett MH, Beale MH, Lewis MJ, Gaskin P, Mansfield JW. 1999. Cell wall alterations and localized accumulation of feruloyl-3′-methoxytramine in onion epidermis at sites of attempted penetration by *Botrytis allii* are associated with actin polarisation, peroxidase activity and suppression of flavonoid biosynthesis. *Plant J.* 17:523–34
116. Messerli M, Robinson KR. 1997. Tip localized Ca^{2+} pulses are coincident with peak pulsatile growth rates in pollen tubes of *Lilium longiflorum. J. Cell Sci.* 110:1269–78
117. Messerli MA, Danuser G, Robinson KP. 1999. Pulsatile influxes of H^+, K^+ and Ca^{2+} lag growth pulses of *Lilium longiflorum* pollen tubes. *J. Cell Sci.* 112:1497–509
118. Messerli MA, Robinson KR. 1998. Cytoplasmic acidification and current influx follow growth pulses of *Lilium longiflorum* pollen tubes. *Plant J.* 16:87–91
119. Miller DD, Callaham DA, Gross DJ, Hepler PK. 1992. Free Ca^{2+} gradient in growing pollen tubes of *Lilium. J. Cell Sci.* 101:7–12
120. Miller DD, de Ruijter NCA, Bisseling T, Emons AMC. 1999. The role of actin in root hair morphogenesis: studies with lipochito-oligosaccharide as a growth stimulator and cytochalasin as an actin perturbing drug. *Plant J.* 17:141–54
121. Miller DD, de Ruijter NCA, Emons AMC. 1997. From signal to form: aspects of the cytoskeleton—plasma membrane—cell wall continuum in root hair tips. *J. Exp. Bot.* 48:1881–96
122. Miller DD, Lancelle SA, Hepler PK. 1996. Actin microfilaments do not form a dense meshwork in *Lilium longiflorum* pollen tube tips. *Protoplasma* 195:123–32
123. Mineyuki Y, Kataoka H, Masuda Y, Nagai R. 1995. Dynamic changes in the actin cytoskeleton during the high-fluence rate response of the *Mougeotia* chloroplast. *Protoplasma* 185:222–29
124. Nagai R. 1993. Regulation of intracellular movements in plant cells by environmental stimuli. *Int. Rev. Cytol.* 145:251–310
125. Nick P. 1998. Signaling to the microtubular cytoskeleton in plants. *Int. Rev. Cytol.* 184:33–80
126. Nick P. 1999. Signals, motors, morphogenesis—the cytoskeleton in plant development. *Plant Biol.* 1:169–79
127. Nozue K, Kanegae T, Imaizumi T, Fukud S, Okamoto H, et al. 1998. A phytochrome from the fern *Adiantum* with features of the putative photoreceptor NPH1 *Proc. Natl. Acad. Sci. USA* 95:15826–30
128. Pawlowski K, Bisseling T. 1996. Rhizobial and actinorhizal symbioses: What ar the shared features? *Plant Cell* 8:1899-913
129. Perdue TD, Parthasarathy MV. 1985. *I situ* localization of F-actin in pollen tubes *Eur. J. Cell Biol.* 39:13–20
130. Peterson RL, Farquhar ML. 1996. Roo hairs: specialized tubular cells extendin root surfaces. *Bot. Rev.* 62:1–40
131. Picton JM, Steer MW. 1981. Determination of secretory vesicle production rate by dictyosomes in pollen tubes of *Tradescantia* using cytochalasin D. *J. Cell Sc* 49:261–72
132. Pierson ES, Miller DD, Callaham DA van Aken J, Hackett G, Hepler PK

1996. Tip-localized calcium entry fluctuates during pollen tube growth. *Dev. Biol.* 174:160–73

133. Pollard TD, Almo S, Quirk S, Vinson V, Lattman EE. 1994. Structure of actin binding proteins: insights about function at atomic resolution. *Annu. Rev. Cell Biol.* 10:207–49
134. Pollard TD, Cooper JA. 1986. Actin and actin-binding proteins. A critical evaluation of mechanisms and functions. *Annu. Rev. Biochem.* 55:987–1035
135. Pruitt RE. 1999. Complex sexual signals for the male gametophyte. *Curr. Opin. Plant Biol.* 2:419–22
136. Rathore KS, Cork RJ, Robinson KR. 1991. A cytoplasmic gradient of Ca^{2+} is correlated with the growth of lily pollen tubes. *Dev. Biol.* 148:612–19
137. Ressad F, Didry D, Xia G-X, Hong Y, Chua N-H, et al. 1998. Kinetic analysis of the interaction of actin-depolymerizing factor (ADF)/cofilin with G- and F-actins. Comparison of plant and human ADFs and effect of phosphorylation. *J. Biol. Chem.* 273:20894–902
138. Ridge RW. 1992. A model of legume root hair growth and *Rhizobium* infection. *Symbiosis* 14:359–73
139. Ridge RW, Rolfe BG. 1985. *Rhizobium* sp. degradation of legume root hair cell wall at the site of infection thread origin. *Appl. Env. Microbiol.* 50:717–20
140. Ridge RW, Rolfe BG. 1986. Sequence of events during the infection of the tropical legume *Macroptilium atropurpureum* Urb. by broad-host-range, fast-growing *Rhizobium* ANU240. *J. Plant Physiol.* 122:121–37
141. Roberts DM. 1989. Detection of a calcium-activated protein kinase in *Mougeotia* by using synthetic peptide substrates. *Plant Physiol.* 91:1613–19
142. Rolfe BG, Redmond JW, Batley M, Chen H, Djordjevic SP, et al, 1986. Intercellular communication and recognition in the rhizobium-legume symbiosis. In *Recognition in Microbe Plant Symbiotic and Pathogenic Interactions*. NATO ASI Series. Ser. H, Cell Biol., ed. B Lugtenberg, 4:9–54 Berlin: Springer-Verlag.
143. Ruhlandt G, Lange U, Grolig F. 1994. Profilins purified from higher plants bind to actin from cardiac muscle and to actin from a green alga. *Plant Cell Physiol.* 35:849–54
144. Sawitsky H, Liebe S, Willingale-Theune J, Menzel D. 1999. The anti-proliferative agent jasplakinolide rearranges the actin cytoskeleton of plant cells. *Eur. J. Cell Biol.* 78:424–33
145. Schachtman DP, Schroeder JI, Lucas WJ, Anderson JA, Gaber RF. 1992. Expression of an inward-rectifying potassium channel by the *Arabidopsis KAT1* cDNA. *Science* 258:1654–58
146. Schleicher M, André B, Andréoli C, Eichinger L, Haugwitz M, et al. 1995. Structure/function studies on cytoskeletal proteins in *Dictyostelium* amoeba as a paradigm. *FEBS Lett.* 369:38–42
147. Schlüter K, Jockusch BM, Rothkegel M. 1997. Profilins as regulators of actin dynamics. *Biochim. Biophys. Acta* 1359:97–109
148. Schmidt A, Hall MN. 1998. Signaling to the actin cytoskeleton. *Annu. Rev. Cell Dev. Biol.* 14:305–38
149. Schroeder JI, Schwarz M, Pei Z-M, 1998, Protein kinase and phosphatase regulation during abscisic acid signaling and ion channel regulation in guard cells. In *Cellular Integration of Signalling Pathways in Plant Development*, ed. F Lo Schiavo, RF Last, G Morelli, NV Raikhel, pp. 59–69. Berlin: Springer-Verlag
150. Sheterline P, Clayton J, Sparrow JC. 1998. Actin. *Protein Profile* 4:1–272
151. Shibaoka H. 1994. Plant hormone-induced changes in the orientation of cortical microtubules: alterations in the cross-linking between microtubules and the plasma membrane. *Annu. Rev. Plant Physiol. Plant Mol. Biol.* 45:527–44

152. Shimmen T, Hamatani M, Saito S, Yokota E, Mimura T, et al. 1995. Roles of actin filaments in cytoplasmic streaming and organization of transvacuolar strands in root hair cells of *Hydrocharis*. *Protoplasma* 185:188–93

153. Škalamera D, Heath MC. 1996. Cellular mechanisms of callose deposition in response to fungal infection or chemical damage. *Can. J. Bot.* 74:1236–42

154. Škalamera D, Heath MC. 1998. Changes in the cytoskeleton accompanying infection-induced nuclear movements and the hypersensitive response in plant cells invaded by rust fungi. *Plant J.* 16:191–200

155. Škalamera D, Jibodh S, Heath MC. 1997. Callose deposition during the interaction between cowpea (*Vigna unguiculata*) and the monokaryotic stage of the cowpea rust fungus (*Uromyces vignae*). *New Phytol.* 136:511–24

156. Smertenko AP, Jiang C-J, Simmons NJ, Weeds AG, Davies DR, Hussey PJ. 1998. Ser6 in the maize actin-depolymerizing factor, ZmADF3, is phosphorylated by a calcium-stimulated protein kinase and is essential for the control of functional activity. *Plant J.* 14:187–94

157. Snowman BN, Geitmann A, Clarke SR, Staiger CJ, Franklin FCH, et al. 1999. Signalling and the cytoskeleton of pollen tubes of *Papaver rhoeas*. *Ann. Bot.* In press

158. Somssich IE, Hahlbrock K. 1998. Pathogen defence in plants—a paradigm of biological complexity. *Trends Plant Sci.* 3:86–90

159. Sotiropoulos A, Gineitis D, Copeland J, Treisman R. 1999. Signal-regulated activation of serum response factor is mediated by changes in actin dynamics. *Cell* 98:159–69

160. Spector I, Braet F, Shochet NR, Bubb MR. 1999. New anti-actin drugs in the study of the organization and function of the actin cytoskeleton. *Microsc. Res. Tech.* 47:18–37

161. Staiger CJ, Gibbon BC, Kovar DR, Zonia LE. 1997. Profilin and actin depolymerizing factor: modulators of actin organization in plants. *Trends Plant Sci.* 2:275–81

162. Staiger CJ, Yuan M, Valenta R, Shaw PJ, Warn R, Lloyd CW. 1994. Microinjected profilin affects cytoplasmic streaming in plant cells by rapidly depolymerizing actin microfilaments. *Curr. Biol.* 4:215–19

163. Takagi S. 1997. Photoregulation of cytoplasmic streaming: cell biological dissection of signal transduction pathway. *J. Plant Res.* 110:299–303

164. Takagi S, Kamitsubo E, Nagai R. 1991. Light-induced changes in the behavior of chloroplasts under centrifugation in *Vallisneria* epidermal cells. *J. Plant Physiol.* 138:257–62

165. Takemoto D, Furuse K, Doke N, Kawakita K. 1997. Identification of chitinase and osmotin-like protein as actin-binding proteins in suspension-cultured potato cells. *Plant Cell Physiol.* 38:441–48

166. Takemoto D, Maeda H, Yoshioka H, Doke N, Kawakita K. 1999. Effect of cytochalasin D on defense responses of potato tuber discs treated with hyphal wall components of *Phytophthora infestans*. *Plant Sci.* 141:219–26

167. Tang X, Lancelle SA, Hepler PK. 1989. Fluorescence microscopic localization of actin in pollen tubes: comparison of actin antibody and phalloidin staining. *Cell Motil. Cytoskelet.* 12:216–24

168. Taylor LP, Hepler PK. 1997. Pollen germination and tube growth. *Annu. Rev. Plant Physiol. Plant Mol. Biol.* 48:461–91

169. Thordal-Christensen H, Zhang Z, Wei Y, Collinge DB. 1997. Subcellular localization of H_2O_2 in plants. H_2O_2 accumulation in papillae and hypersensitive response during the barley—powdery

mildew interaction. *Plant J.* 11:1187–94
170. Tiwari SC, Polito VS. 1990. An analysis of the role of actin during pollen activation leading to germination in pear (*Pyrus communis* L.): treatment with cytochalasin D. *Sex. Plant Reprod.* 3:121–29
171. Tominaga Y, Wayne R, Tung HYL, Tazawa M. 1987. Phosphorylation-dephosphorylation is involved in Ca^{2+}-controlled cytoplasmic streaming of Characean cells. *Protoplasma* 136:161–69
172. Tomiyama K, Sato K, Doke N. 1982. Effect of cytochalasin B and colchicine on hypersensitive death of potato cells infected by incompatible race of *Phytophthora infestans. Ann. Phytopathol. Soc. Jpn.* 48:228–30
173. Trojan A, Gabryś H. 1996. Chloroplast distribution in *Arabidopsis thaliana* (L.) depends on light conditions during growth. *Plant Physiol.* 111:419–25
174. van Brussel AAN, Bakhuizen R, van Spronsen PC, Spaink HP, Tak T, et al. 1992. Induction of pre-infection thread structures in the leguminous host plant by mitogenic lipo-oligosaccharides of *Rhizobium. Science* 257:70–72
175. Vidali L, Hepler PK. 1997. Characterization and localization of profilin in pollen grains and tubes of *Lilium longiflorum. Cell Motil. Cytoskelet.* 36:323–38
176. Vidali L, Yokota E, Cheung AG, Shimmen T, Hepler PK. 1999. The 135 kDA actin-binding protien from *Lilium longiflorum* pollen is the plant homologue of villin. *Protoplasma* 209:283–91
177. Volkmann D, Baluška F. 1999. Actin cytoskeleton in plants: from transport networks to signaling networks. *Microsc. Res. Tech.* 47:135–47
178. Wada M, Grolig F, Haupt W. 1993. Light-oriented chloroplast positioning. Contribution to progress in photobiology. *J. Photochem. Photobiol.* 17:3–25
179. Wagner G, Haupt W, Laux A. 1972. Reversible inhibition of chloroplast movement by cytochalasin B in the green alga Mougeotia. *Science* 176:808–9
180. Wagner G, Klein K. 1981. Mechanism of chloroplast movement in *Mougeotia. Protoplasma* 109:169–85
181. Walker EA, Ride JP, Kurup S, Franklin-Tong VE, Lawrence MJ, Franklin FCH. 1996. Molecular analysis of two functional homologues of the S_3 allele of the *Papaver rhoeas* self-incompatibility gene isolated from different populations. *Plant Mol. Biol* 30:983–94
182. Wells AL, Lin AW, Chen LQ, Safer D, Cain SM, et al. 1999. Myosin VI is an actin-based motor that moves backwards. *Nature.* 401:505–8
183. Wilhelmi LK, Preuss D. 1999. The mating game: pollination and fertilization in flowering plants. *Curr. Opin. Plant Biol.* 2:18–22
184. Williamson RE. 1993. Organelle movements. *Annu. Rev. Plant Physiol. Plant Mol. Biol.* 44:181–202
185. Winter H, Huber JL, Huber SC. 1998. Identification of sucrose synthase as an actin-binding protein. *FEBS Lett.* 430:205–8
186. Xing T, Higgins VJ, Blumwald E. 1997. Race-specific elicitors of *Cladosporium fulvum* promote translocation of cytosolic components of NADPH oxidase to the plasma membrane of tomato cells. *Plant Cell* 9:249–59
187. Xu J-R, Staiger CJ, Hamer JE. 1998. Inactivation of the mitogen-activated protein kinase Mps1 from the rice blast fungus prevents penetration of host cells but allows activation of plant defense responses. *Proc. Natl. Acad. Sci. USA* 95:12713–18
188. Yang Z. 1998. Signaling tip growth in plants. *Curr. Opin. Plant Biol.* 1:525–30
189. Yatsuhashi H, Kadota A, Wada M. 1985. Blue- and red-light action in photoorientation of chloroplasts in *Adiantum protonemata. Planta* 165:43–50

190. Yeh K-C, Lagarias JC. 1998. Eukaryotic phytochromes: light-regulated serine/threonine protein kinases with histidine kinase ancestry. *Proc. Natl. Acad. Sci. USA* 95:13976–81

191. Yokota E, Muto S, Shimmen T. 1999. Inhibitory regulation of higher-plant myosin by Ca^{2+} ions. *Plant Physiol.* 119:231–39

192. Yokota E, Yukawa C, Muto S, Sonobe S, Shimmen T. 1999. Biochemical and immunocytochemical characterization of two types of myosins in cultured tobacco Bright Yellow-2 cells. *Plant Physiol.* 121:525–34

193. Zhao K, Wang W, Rando OJ, Xue Y, Swiderek K, et al. 1998. Rapid and phosphoinositol-dependent binding of the SWI/SNF-like BAF complex to chromatin after T lymphocyte receptor signaling. *Cell* 95:625–36

194. Zonia L, Tupý J, Staiger CJ. 1999. Unique actin and microtubule arrays co-ordinate the differentiation of microspores to mature pollen in *Nicotiana tabacum*. *J. Exp. Bot.* 50:581–94

Annu. Rev. Plant Physiol. Plant Mol. Biol. 2000. 51:289–322

CYTOSKELETAL PERSPECTIVES ON ROOT GROWTH AND MORPHOGENESIS

Peter W. Barlow
IACR–Long Ashton Research Station, Department of Agricultural Sciences, University of Bristol, Long Ashton, Bristol BS41 9AF, United Kingdom; e-mail: peter.barlow@bbsrc.ac.uk

František Baluška
Botanisches Institut, Rheinische Friedrich-Wilhelms-Universität Bonn, Kirschallee 1, D-53115 Bonn, Germany; e-mail: baluska@uni-bonn.de

Key Words actin, microfilaments, microtubules, tubulin

■ **Abstract** Growth and development of all plant cells and organs relies on a fully functional cytoskeleton comprised principally of microtubules and microfilaments. These two polymeric macromolecules, because of their location within the cell, confer structure upon, and convey information to, the peripheral regions of the cytoplasm where much of cellular growth is controlled and the formation of cellular identity takes place. Other ancillary molecules, such as motor proteins, are also important in assisting the cytoskeleton to participate in this front-line work of cellular development.

Roots provide not only a ready source of cells for fundamental analyses of the cytoskeleton, but the formative zone at their apices also provides a locale whereby experimental studies can be made of how the cytoskeleton permits cells to communicate between themselves and to cooperate with growth-regulating information supplied from the apoplasm.

CONTENTS

040-2519/00/0601-0289$14.00

INTRODUCTION

Microtubules (MTs), polymers of the tubulin protein, are generally held to be responsible for the orientation of cellulose microfibrils within plant cell walls. The microfibrils both provide a scaffold for the assembly of other wall components and influence the orientation of cell growth. The latter process is driven by an internal hydrostatic pressure which shows no preferential direction in the application of its force upon the cell periphery. However, anisotropic expansion of cells is possible if different wall facets, or portions of a facet, have different yield thresholds to the internal pressure. Where they exist, such anisotropies come about because of differential depositions or modifications of wall materials. Again, this could be a consequence of an MT-directed process since MTs in the peripheral (cortical) zone of the cytoplasm can help shape the interior surface of the wall, thus bringing about the characteristic microanatomy of plant tissues and their cells (3, 105, 122, 174).

Given the dual function of MTs in individual cells—helping to define not only cell growth orientation but also cellular microanatomy—it is a challenge to understand how MTs participate in the more large-scale development of multicellular organs. Both aspects of MT function can be appreciated within multicellular callus systems (213). Often, what is lacking in callus is a signal for turning haphazard growth into orderly organogenetic growth. Usually such a cue is supplied by growth-regulating substances directed to target zones via the apoplasm or symplasm. Thus, MTs and actin microfilaments (MFs) might collectively be part of a sensory system for capturing and transducing information contained within the cellular milieu and then converting it into a coherent growth response which includes further cell differentiation (23, 175, 200, 222).

Microtubules and MFs are generally considered to be major components of an intracellular system, broadly known as the plant cytoskeleton (77). The limits of the cytoskeleton, in terms of what types of cytoplasmic structures are, or are not part of it, are hard to define; the suffix skeleton might be regarded as being relevant only to a support structure. Nevertheless, it is fairly clear that MTs and MFs fulfil a cytoskeletal role in the sense that they confer structural order and stability on the interior of the cell and these, in turn, permit the orderly unfolding of cell growth and, consequently, organ growth.

Research into root biology continues to be central to plant sciences on account of the practical implications of the subject. Moreover, roots are a convenient source of tissues for fundamental studies of tissue differentiation and the physiological responses to environmental perturbations. Roots provide a ready source of cells for the examination of MTs and MFs in both the electron and fluorescence microscopes (for MTs see 22, 83; for MFs see 184, 187). In fact, the now classical relationship between the orientations of MTs in the cell cortex and the cellulose microfibrils in the cell wall, as well as the very existence of MTs in plants, were discovered in root meristems of *Juniperus chinensis* and *Phleum pratense* (133). In conjunction with fluorescence microscopy, where extensive use is made of fluorescent antibodies to cytoskeletal proteins (140), pharmacological agents, to which roots can easily be exposed, help dissect the relationship between cellular chemistry and cytoskeletal substructure (102, 171, 232).

This article draws upon the long history of various aspects of root research, as well as the notable advances in knowledge of the plant cytoskeleton, and combines them to appraise root development and morphogenesis from a cytoskeletal perspective. Although many other articles on roots have appeared in this series of Annual Reviews, the present one seems to be the first to examine their development from such a point of view. Observations from various systems other than roots are also mentioned because these give useful clues as to how the cytoskeleton–root development concept can be furthered. We would claim—if we may be permitted to paraphrase the dictum of the famous geneticist, Theodosius Dobzhansky—that many aspects of root growth and development make sense only in the light of cytoskeletal behavior, particularly in the way cytoskeletal MTs are deployed.

TYPES OF MTS IN ROOTS

Cortical MTs

The first observation of plant MTs in root meristem cells by Ledbetter & Porter (133) arose from the utilization of glutaraldehyde as a fixative for use in electron microscopy. Hitherto, the popular use of $KMnO_4$ as a fixative had rendered MTs unobservable (and many other structures as well). The MTs revealed in this study were of the type now known as cortical MTs due to their presence in the cell cortex underlying the cell periphery (56). Their conspicuous coalignment with the microfibrillar constituents of the adjacent cell wall at once made sense of Green's conjecture in the previous year (81) that cytoplasmic fibers or elements, similar in some of their properties to those of the mitotic spindle, lay within the cortical cytoplasm from where they somehow directed cell wall biosynthesis in a manner which influenced the orientation of cell growth.

Three of Ledbetter & Porter's initial observations (133) on cortical microtubules in root meristems are still relevant today, and are still without satisfactory explanation: (*a*) Adjacent, parallel MTs were never less than 35 nm apart (center to center).

This agrees with numerically more detailed observations (212) on radish roots that cortical MTs were mostly about 90 nm apart. Here, it is worth recalling observations on insect ovarian cells which had been caused to express two different mammalian MT-associated proteins, MAP2 and tau (46). When the cells expressed MAP2, the distance (as seen in cross-sections) between MTs in the induced cytoplasmic processes was ca 65 nm, but when tau was expressed the inter-MT distance was 20 nm. Deep-etch microscopy corroborated these findings and revealed molecules of correspondingly larger or smaller size linking the MTs. The two sets of inter-MT distances were within the range found in dendrites (60–70 nm) and axons (20–30 nm) of rat neuronal tissue. Thus, where different arrangements of MTs are associated with root tissue differentiation, they may have been defined by different complements of MT-associated proteins. (*b*) The zone of the meristematic cell cortex inhabited by MTs was enriched with ribosomes, so much so that Ledbetter & Porter thought this suggested an interrelationship between the two structures. More recently, preparations of pea roots were reported to show an association between polysomes and actin which cosedimented with the cytoskeleton fraction (236). Actin MFs are also components of the cell cortex (49, 55, 138, 149, 183), and it may be because of their presence (often undetectable in the electron microscope) that ribosomes are intermingled with the cortical MTs. Recent summaries of views about the plant cell cortex (23a, 98) indicates a complex set of structural and functional interactions between membranes, MFs, MTs, and even genes (31, 154). (*c*) MTs were aligned in as many as three layers beneath the plasma membrane. This observation is intriguing since current ideas suggest that the outermost layer of MTs is associated with wall biosynthesis; so, do MTs in the other two inner layers have a function? Or do the different layers of MTs reflect a sequence of recruitment from a more internal zone of cytoplasm, where they are assembled, to the outer zone, where the MTs are putatively active? The fact that the MT arrays in the innermost layer were less well ordered (i.e. nonparallel MTs) than those of the outer layers (which showed parallel MTs) suggests this might be the case; but an opposite sequence, of MT return to the inner cytoplasm, is also possible. Studies of the sequence of MT reorganization following treatment of tobacco BY-2 cells with the anti-MT agent, propyzamide, showed that nonparallel arrangements of cortical MTs reappeared a few minutes before parallel arrays (89). As for the lengths of cortical MTs in meristematic cells, these were estimated as being mostly between 2–4 μm long (86). The values were estimated from the cortices underlying longitudinal walls of interphase cells of *Azolla* and maize roots; cortical MT lengths in *Impatiens* roots were 4–6 μm long. In radish roots, mean cortical MT lengths were less in meristematic cells (means for individual root varying between 0.9–1.3 μm) than in expanded cells of the root hair zone (2.6–6.7 μm) (212). Here, however, most cortical MTs were short; the difference between the mean lengths in the two cell types was due to an increased proportion of longer (upto 14 μm) MTs in the expanded cells. In all cases, MT lengths corresponded to the cross-sectional width of a cell wall facet, suggesting the possibility for local cytoskeletal control of the wall properties between neighboring cells.

It has often been speculated that the cortical MTs have some connection with the so-called rosettes embedded in the internal face of the plasma membrane and the terminal globules in the exterior face. These rosettes and globules seem to be two halves of a common structure associated with cellulose microfibril synthesis at the inner surface of the cell wall (172). The cortical MTs do not lie directly beneath the rosettes, but are located to one side of them: in developing xylem cells of cress roots lateral connections were found between MTs and rosettes (99), and evidence from freeze-fractured plasma membranes of the alga, *Closterium*, suggested this too (76). Unidentified proteins link the MTs to the plasma membrane (1, 156; see 114 for a review), but whether any of these proteins are part of the rosette protein is not known. A further question is whether the rosettes determine the above-mentioned ~90 nm spacing between parallel, cortical MTs (212), and whether they account for the images of bridges between these MTs seen in the electron microscope.

Concerning MT-associated proteins and the order which they might confer upon MT arrays (101), one needs to distinguish between associations that bring about bridging between MTs, and which would thus favor MT bundling, from associations that connect the MTs to the plasma membrane, as well as from those which anneal the free ends of MTs. Another category consists of the MT-organizing proteins which facilitate MT polymerization (195, 218), the main sites for this being the nuclear envelope (208), the cell cortex including the preprophase band, the spindle and phragmoplast (90), and the centromeres of mitotic chromosomes (32). And in this regard, attention should be paid to the 120-kDa protein isolated by Chan et al (45) and to γ-tubulin (137, 155), especially since this last-mentioned protein has significance for MT organization in animal (238) and fungal (150) cells. As far as MT–MT bridging is concerned, a 65-kDa protein with this property was extracted from tobacco BY-2 cells (117), and a 76-kDa protein isolated from suspension-cultured carrot cells also caused bundling of MTs (58). The MTs assayed in this last-mentioned work, and also in another study (217) in which an 83-kDa protein was isolated from maize suspension culture cells, were from animal brains, though a similar bundling response was shown when the 83-kDa protein was added to native plant MTs. The bundled MTs showed a center-to-center spacing of <350 nm. In view of the much closer spacing of cortical MTs in fixed cells (mentioned earlier), the question is whether these observations have relevance for MT bundling in the cell. A MT-annealing-type protein has been isolated which increased the rate of MT elongation (109). Concomitant bundling of MTs was also noticed. Thus, MT conformation may play a part in regulating MT dynamics, or vice versa. Also significant for modulation of plant MT arrays is elongation factor-1α (Ef-1α). This ubiquitous, ribosome-bound protein not only serves as a protein translation factor in eukaryotic cells, but also interacts with MTs. It, too, encourages MT-bundling, this property being negatively regulated by calcium and calmodulin (68). Ef-1α may also help re-establish the perinuclear MT complex following cytokinesis (130). Both these effects may involve interactions with actin and, hence, could be responsible for the stability of complexes between

MTs, actin MFs, and ribosomes. In this way, Ef-1α could assist in the intracellular compartmentation of protein synthesis (53).

Specialized bands of cortical MTs which help shape the secondary walls of plant cells are well known (174) even though the conditions which bring about these MT distributions are obscure; some perhaps, could involve self-organizing processes dependent upon reaction-diffusion mechanisms (208a). Less well known, however, are the small rings of MTs (2–3 μm diameter) which develop at peripheral sites of primary and secondary vascular cells. The MTs rings are responsible for defining pit fields, simple pits, and bordered pits (44, 103). They begin to form following a clearing of MTs from sites (4 μm diameter) in the cytoplasmic cortex. How this comes about is not known—it may relate to local changes in the plasma membrane—but such areas have also been recorded following various experimental treatments (9, 16, 33). Given the stiffness of MTs, the rings probably consist of short MTs linked together in some way, perhaps with the participation of actin. Even larger MT rings are features of the end walls of developing vessels (44). They involve nearly the whole perimeter of these walls and are thought to participate in their removal, thus allowing vessel–vessel continuity. Interestingly, a ring of MTs with a similar function is responsible for fashioning the lid of cyst cells of the alga *Acetabularia* (161). Whether the nucleus participates in forming the MT ring of vessels, as it does in *Acetabularia,* is not known.

A variant of the cortical class of MTs is the preprophase band (PPB), a transient example of MT bundling (163). The PPB begins to develop in cells which have reached late interphase of the mitotic cycle. It is recognizable at this stage partly because all the other cortical MTs are becoming disassembled in preparation for redeployment in the mitotic spindle. Those MTs which remain, i.e. those of the incipient PPB become crowded together in the cell cortex and hence appear particularly bright in the fluorescence microscope. With the passage from early to late prophase, the cortical MTs of the PPB of onion root cells become more numerous (50 MTs in cross-section, rising to 250 MTs), are arranged in more layers (3 layers, rising to 10 layers), and come closer together (center-to-center distances decreasing from 40 nm to <30 nm) (179). Treatment with the protein synthesis inhibitor, cycloheximide (36 μM for 2 h), diminished all these trends so that the PPB remained broader than usual (on average, the PPB was 4.5 μm wide in contrast to 3.2 μm in untreated roots) and contained 70% fewer MTs.

These observations on the microtubular PPB have to be considered in the light of the participation of actin MFs in its structure (65). The actin-disrupter, cytochalasin D, like cycloheximide, also prevented the narrowing of the PPB in prophase cells of onion roots (69, 167). Therefore, it is reasonable to suggest that the synthesis of a prophase-specific protein, or proteins, is required for the cell cycle-linked evolution of PPB structure. Such a protein may also protect the PPB MTs from whatever conditions cause the cortical MTs elsewhere in the cell to dissassemble. One protein suggested for this role is the p34^{cdc2} homologue from maize, known to be associated with the PPB (54). The corresponding maize cdc2 antibody, however, recognized only about 10% of PPBs, these being late, not early, PPBs (162). On the basis of results using the conserved PSTAIR sequence of p34^{cdc2}, it was suggested that some nonstaining of PPBs was a technical problem rather than being of any biological significance (164). Probably, many other proteins associated with the PPB (reviewed in 163) could be considered as possible regulators of its function.

Even more intriguing is the significance that the PPB has for cell division and morphogenesis because the position at which the cell plate is inserted into the wall of a dividing cell is intimately linked with the position of the PPB during the preceding late interphase. On the basis of electron microscope evidence from onion roots, it was suggested (181) that the PPB continues to support incorporation of precursors into the underlying cell wall. The absence of cortical MTs elsewhere in the dividing cell would make this the only site of wall synthesis at this stage of the cell cycle. The PPB might therefore prepare a site at which wall precursor material contained within the expanding cell plate can adhere. The cell plate is also attracted to the location of the former PPB by long-range mechanisms, as centrifugation experiments have shown (75). This attraction is unlikely to involve the actin MFs associated with the PPB since the MF structures at this site disperse during prophase (182). Some kind of "negative" imprinting at the PPB site has also been suggested (51), but of what this might consist is unclear.

Circumstantial evidence about whether or not the parental wall plays an active role in cell plate insertion also comes from regenerating protoplasts and suggests that a minimal external wall, as well as a minimal cortical MT network, are required for cell division (85, 196, 202). Unfortunately, in these studies, no observations were made to determine the presence or absence of PPBs, only about whether or not there were division walls. Nor do the observations indicate why certain protoplasts lack a PPB (202), a finding which could be relevant for explaining how some higher plant cell types, such as cambium fusiform initials, lack a PPB (72) yet divide satisfactorily. It may be that such elongated cambial cells (cf. 42) simply lack sufficient numbers of MTs and tubulin gene transcripts to form a PPB.

If wall deposition does occur at the site of the PPB (181), this could help explain how some type of division wall growth can occur even in the absence of a phragmoplast. For instance, following caffeine treatment, stubs of what might normally have been part of the new division wall were found attached to parental cell walls (118, 193). Such wall stubs are not uncommon in other circumstances: they have been found, for example, in nematode-induced syncytia of *Impatiens* roots (119). These observations suggest that cytokinesis consists of two processes: centrifugal growth and maturation of the cell plate within the phragmoplast, and centripetal growth of the new division wall from the PPB site. Since the latter process is slow relative to the former, centripetal wall growth is not usually appreciated unless phragmoplast and cell plate are destroyed.

Endoplasmic MTs and Mitotic Spindle

Although Ledbetter & Porter (133) did not demonstrate MTs in the interior of interphase cells, they expected that MTs would be found there, even if identifiable only with difficulty. Fortunately, the anticipated difficulty disappeared with the advent of the immunofluorescence marking of MTs (231); hence, endoplasmic MT arrays were identified (129). However, the frequent failure to appreciate endoplasmic MTs in squashes of root cells is understandable as they are easily masked by numerous over- and underlying cortical MTs. Laser scanning confocal microscopy removes this constraint to their identification (71, 84, 166). Tissue sections also provide an excellent means of revealing endoplasmic MTs (15) and for examining the relationships between their conformation and the differentiation status of the cell and, in the case of meristematic cells, the phase of the cell cycle (JS Parker & PW Barlow, unpublished data).

Endoplasmic MTs do not form the clustered associations that are characteristic of cortical MTs. They are more usually visualized as sparse populations of single or branched tubules traversing the cytoplasmic space, though a superabundance of endoplasmic MTs form in onion and maize root cells following exposure to cycloheximide (9, 166). Close observation has often suggested that one end of these endoplasmic MTs is attached to the nuclear surface and the other reaches into the cell cortex, a configuration which suggests the potential for communication between these two zones of the cell (21, 28).

The plant mitotic spindle may be regarded as a transformed set of endoplasmic MTs (10). In general, the mitotic spindle is a conservative structure and the behavior of its MTs has been reviewed (30). Endoplasmic and spindle MTs are also related through their common property of being organized upon the nuclear envelope. A good deal of evidence suggests that the nucleus itself is the site of synthesis of MT-organizing material, and that this material continually emerges from the nucleus to overlay the surface of its external membrane (21). Whereas during most of the interphase of the cell cycle endoplasmic MTs radiate from all over the nuclear surface, late in interphase putative motor proteins (4, 170) bring about the segregation of the MT-organizing material into two groups on opposite sides of the nuclear surface. Also segregated at this time is γ-tubulin (155). All the while, the organizing material continues to assemble endoplasmic MTs, some of them making contact with the PPB (179). When fully segregated, the two opposite groups of organizing material serve as the sole foci for MT assembly. At pro-metaphase, fine endoplasmic-like MTs radiate from each of the two half-spindle cones toward the end-walls and side-walls of the mitotic cell (84). These associations may ensure a suitable position for the nucleus and spindle prior to, and during, mitosis. When the nuclear envelope breaks down and the chromosomes condense, these foci (which may consist of a small number of subfoci) serve as the two poles of the mitotic spindle. At the same time, MT-organizing material, such as the 49-kDa protein of Hasezawa & Nagata (90), latches onto the centromeric regions of the prophase chromosomes, enabling their association with spindle MTs.

An essential feature of mitotically cycling cells is the sequential transformation of their MT arrays, the timing of which has been estimated in onion root meristems (216). Spindles exist only briefly, for about 1.5 h within a total cycle of 34 h (at 15°C); PPBs and phragmoplasts persist for 2.3 h and 2.0 h, respectively. These transformations are based on the continually changing equilibrium between free tubulin and MTs (153, 233) and the preferential activation of MT-organizing centres. How the timescale of these events is determined remains an open question

Phragmoplast

Root meristem cells are usually devoid of any large vacuole, so there is no need of a phragmosome to support either the dividing nucleus or the forming phragmoplast and cell plate—unless one regards the whole of the cell interior as a type of phragmosome! But this does not mean that some of the cytoskeletal components and properties characteristic of phragmosomes, such as the tension that exists within cytoskeletal filaments (78, 80), are absent from meristematic cells. For example, actin filaments may radiate out from the edge of the phragmoplast and secure the attachment of the expanding cell plate with the parental wall (141) even if, in meristematic cells, there is no definite phragmosome structure within which this could occur. Indeed, disruption, by latrunculin A, of actin in meristematic cells leads to twisted phragmoplasts (17).

The density of MTs within the phragmoplast suggests the presence of many MT-organizing sites, as well as proteins which link MTs laterally and confer dynamic movements upon them (5). The rapidity with which the phragmoplast grows at telophase, and the coincidence of its formation with spindle disassembly, suggest a movement of tubulin dimers from one structure to the other. MT-organizing materials may be similarly redeployed at this time, materials formerly at the poles of the spindle and at the centromeres being relocated to the zone between the two telophase sister-nuclei (10). Possibly, the reformation of the pair of nuclear envelopes, together with an affinity of motor-proteins for the free ends of the MTs (6) which are polymerized on the nuclear surfaces, assist in relocating MT-organizing material toward the mid-zone. Besides MTs, actin MFs also provide an important component of the phragmoplast (197). Moreover, once cytokinesis is complete, actin remains within the plasmodesmata which traverse the new division wall (227), as well as heavily decorating this region of the cell periphery (17, 18). Myosin is also associated with the plasmodesmata (190, 191).

The role of the phragmoplast is to attract into itself membrane-bound vesicles bearing precursors for the cell plate. Studies with low doses of colchicine have shown that this process principally requires the participation of phragmoplast MTs (131). Also necessary are motor proteins such as the dynamin-like protein, phragmoplastin (82), and the filamentous protein, centrin, which colocates with the phragmoplast vesicles (62). When the component molecules become assembled into a cell plate, the MT-organizing material of the phragmoplast is displaced toward its edge. What permits the cell plate to expand as a flat disc and not as a spheroid in the mid-zone [as it does in the *pilz* mutants of *Arabidopsis* (157)] is not firmly established. It seems that as material is added to the edges of the growing cell plate by the phragmoplast, so these edges are pushed toward sites on the parental cell wall already prepared by the PPB to accept them. Whether this is a form of self-assembly (cell-plate crystallization), in much the same way that cellulose-forming rosettes are thought to be pushed forward by the crystallization of the new cellulose fibrils, is not known. Actin and vinculin filaments radiating from the edges of each sister nucleus may provide this pushing force (70). These molecules, as well as centrin (62), may also be responsible for straightening out the undulations in the new division wall (165) which, at this stage, is rich in callose. Further details of cell plate formation are mentioned in the final section.

MT Cables

A little-understood fourth type of microtubular structure comprises the fluorescent MT cables seen following immunostaining with antitubulin. Long-lived tissues, such as the xylem ray parenchyma cells of roots of *Aesculus* and shoots of *Populus* (NJ Chaffey & PW Barlow, in preparation) clearly show such MT cables. Generally, they seem to exist in cells which have completed their growth and, hence, may be considered as mature and fully functional. The cables consist of 3–4 MTs (in cross-section) and are similar in this respect to the MT bundles found in cortical cells in mature regions of hyacinth roots (57, 136). We speculate that such cables are concerned with intra- and intercellular transport processes and with the general polarity of tissues, properties not so strongly expressed in association with younger cells with their more usual arrays of transverse or reticulate MTs. The cables are presumably not a degenerate type of MT because the final stages of cell differentiation in root and other tissues are usually marked by bright fluorescent spots of tubulin which supercede the MT population and which have no apparent order within the cell (88).

CYTOSKELETAL PROTEIN VARIABILITY AND ITS RELATION TO ROOT TISSUE IDENTITY

Tubulin Genes and Isotypes

Whatever their role in plant cell development, the MTs associated with the four types of array mentioned previously are all constructed of α- and β-tubulin dimers. The correlation between MTs and the developmental program of root cells suggests that genes for α- and β-tubulins, but with different coding sequences, are functional in different cell types. This leads to the identification of tubulin isotypes; the isotypes can, however, also be the result of posttranslational and postpolymerizational modification of one given type of tubulin protein (37, 77, 144, 145). It is not known to what extent the isotypes are interchangeable. In human HeLa cells, for example, four different β-tubulin isotypes can all be found within interphase and spindle MT arrays (134). But there is other, firm evidence that isotype substitution leads to developmental abnormalities if it occurs in a cell which does not normally support that isotype (106). Isotype interchangeability cannot therefore be generally acceptable, perhaps because of the different proteins with which the MTs associate and the consequences this has for cell differentiation (46). A second hypothesis is that tubulin isotypes differentially regulate the turnover of the MTs which contain them and that this variation of MT half-life might have some adaptive significance. Plant organisms, with a range of tissues, as well as a range of environmental conditions to contend with, may increase the options for the regulation of MT dynamics by possessing multiple isotypes. However, examination of root tissue in relation to tubulin isotypes suggests that they also participate in specialized activities of cell differentiation. It seems, therefore, that there is a default system of tubulin deployment which operates in a standard, or optimal, growth environment, whereas another system, involving alternative tubulin isotypes, is evoked when growth is challenged by nonstandard environments.

c-DNA prepared from different tissues of maize revealed at least six α-tubulin sequences (221), some of which were similar to those found in the dicot *Arabidopsis thaliana* (126, 146). The $\alpha 2$ isotype of maize was identified as a product of the tubα5 gene (121). Specifically examining root tips, four α- and four β-tubulins were identified in *Phaseolus vulgaris* (111), and six β-tubulins were identified in both carrot (112) and maize (121). Although seven different β-tubulin genes were expressed in *Arabidopsis* roots (207), only three of them (TUB1, TUB6, and TUB8) showed notable amounts of transcript. Later work with Arabidopsis root (48) showed that the TUB1 gene product (as identified by GUS transgene reaction) was localized in the epidermis and cortex tissues, whereas TUB8 was confined to endodermis and phloem. In both cases, a strong GUS reaction was found in the zone of rapid elongation but none was evident in meristem or root cap. Of the β-tubulin isotypes described for carrot, the most strongly expressed was $\beta 1$; some differences were encountered between seedling and mature tap roots with respect to the expression of isotypes $\beta 5$ and $\beta 6$. Among the c-DNA sequences identified from maize were those of the tubα1, tubα2, and tubα3 genes (169). In maize roots, these genes showed a tissue-dependent pattern of expression (121, 214); tubα1 was specifically expressed in roots (and pollen), but not in other organs (169, 192). In situ mRNA hybridization to root tissue sections revealed tubα1 to be strongly expressed in the meristematic cells of the root cortex and root cap, but to a lesser degree in the vascular cylinder and quiescent center. Similar results were obtained when the promoter of tubα1 was inserted into the genome of tobacco. Here, root meristem cells, but not the quiescent center, showed promoter expression (192). One difficulty in interpreting such observations in terms of MT function and of MT gene-switching is that different regions within root tissues have different concentrations of RNA in their cells (24). Differences in the level of tubα expression may therefore be a function of more general regional differences of RNA metabolism. Another problem is that it is by no means certain that all the tubulin RNA is translated into tubulin protein. Thus, on the basis of in situ hybridization, it may go beyond the evidence to imply (214) a link between the distribution of tubα1 and the variously oriented divisions associated with the generation of cell files (formative divisions) within the root apex, although in such a zone of the meristem a relatively rapid deployment of tubulin might be expected to associated with the high frequency of cell division. The gene tubα3 was not so strongly expressed in meristematic regions; the amount of tubα3 mRNA was about 100-fold less than tubα1 (168). Nevertheless, in situ hybridization with the tubα3 probe gave a strong signal in the meristem, except for cortical cells which seemed to be relatively weakly labeled (214). The pattern of tubα2 is especially interesting since it was expressed only in epidermal cells, whereas this tissue was not particularly marked by tubα1 (214)—which is unexpected if the product of tubα1 is associated with cell division. Epidermis was also marked by tubα3, but only in the older regions of root.

By separating the maize root into a tip portion and a mature portion 1–2 cm from the tip, and then further dissecting the mature portion into vascular cylinder and cortex, the α1 and α4 isotypes were seen to predominate in the tip while mature tissue expressed an abundance of α2 and α3 isotypes (121). β4 and β5 isotypes were features of vascular tissue, whereas β1 and β2 isotypes were abundant in the cortex; β1 isotype was absent from the root tip. When the transcripts of the six tubα genes were considered individually (121), those of tubα4 were particularly evident in vascular tissue (see also 66), suggesting that this gene is active when MTs are required to regulate secondary wall synthesis in either xylem or phloem (or both). The general situation for the maize root (see also 71a) has similarities with tubulin isotype distributions in barley leaves (92). Along a developmental gradient, from 0 mm to 35 mm from the leaf base, the α- and β-isotypes changed in frequency. This was paralleled by a change from random to bundled MTs in the mesophyll cells, just as occurs to the MT arrays in maturing maize root cells (15).

An evolutionary aspect of tubulin diversity is indicated by the discovery of a species biotype (R biotype) of the grass, *Eleusine indica,* which is resistant to anti-MT chemicals, such as dinitroaniline, and at the same time displays sensitivity to the MT-stabilizing compound, taxol (220). The R biotype possesses a novel β-tubulin isotype and its tubulin can polymerize in vitro in the presence of oryzalin (219). Although these results from roots of R-biotype *Eleusine* could not be confirmed by Waldin et al (224), it is possible that there was a change in the tubulin could have occurred which was undetectable by electrophoresis. Further investigation (233a) showed the R biotype to possess missense mutations in the TUA1 gene for α-tubulin.

MTs in roots of freeze-tolerant rye usually disassemble following exposure to freezing (0°C or less) temperatures, but this effect can be offset by a short period of acclimation at a slightly warmer temperature (4°C). In this rye-root test system, during the acclimation period, the freezing-sensitive population of MTs is exchanged for one that is more resistant (124). It was also found that, as a consequence of acclimation, two α- and one β-tubulin isotypes disappeared, while one new β-isotype appeared (125). Taxol rendered the rye root MTs more resistant to cold (47), tending to confirm that the cold-induced pattern of tubulin isotypes had increased their stability. Taxol treatment also stabilized actin MFs in the root cells toward cold, whereas MT disruption by amiprophos-methyl increased the cold-susceptibility of the MFs. These results suggest a structural link between MTs and MFs. Treatment of rye roots with hypertonic solutions of sorbitol simulated some of the physiological effects of chilling, but did not induce such a complete disintegration of the cortical MTs (188). As plant meristems often have to withstand long periods of cold during the winter months, and utilize their dormancy mechanism to do so, it would be of interest to compare isotypes from dormant and nondormant apices, with and without chilling. Dormant vascular cambium in *Aesculus* roots showed conspicuous cortical MTs, even when taken for fixation from frozen surroundings (43).

Another environmental challenge to maize roots is infection by arbuscular mycorrhizal fungi. Following infection with *Glomus versiforme* or *Gigaspora margarita,* tubα3 expression increased in the cortical cells containing the fungal arbuscules (36). Similarly, in ectomycorrhizal formation (short roots) induced by *Suillus bovinus* in Scots pine, new α-tubulin isotypes were detected, whereas there were no alterations to β-tubulin patterns (178). However, no particular changes in MT organization were noted which might correlate with the modified α-tubulin complement (177). Increased α-tubulin transcripts were also detected in the developing ectomycorrhizal root system of *Eucalyptus glomus* (41), but this may have been due simply to increased numbers of highly proliferative cells contained in lateral root primordia.

Posttranslational Modifications of Tubulin

Posttranslational modifications of α- or β-tubulins in plant cell are now becoming better understood as a means to regulate MT dynamics. Although a large number of modifications are possible, the best known are phosphorylation, acetylation, tyrosination, and polyglycylation (145), and all these, with the exception of polyglycylation, have been found in plants (113). Different modifications may exist even along a single microtubule, giving tremendous scope for MT polymorphism. The β-tubulin of tobacco (204, 205) was modified by polyglutamylation, whereas several other types of posttranslational modifications were found for the α-tubulins, all of which could be detected by immunofluorescence in the MTs of interphase and dividing root cells (76a, 76b, 204, 226).

Actin Genes and Isotypes

Actin genes are more diverse than tubulin genes (158, 159), many dozens having been discovered in *Petunia,* for example (8). The findings concerning actin isotypes are similar to those of tubulin. Although many of the actin isotypes are common to certain plant organs, some are specific to roots (115). In soybean roots, tissue-specific patterns have been found (152). Antibodies were raised to peptides of κ- and λ-actin and conjugated with gold particles. Particular cells of the cap flank, as well as some older statocyte cells, were marked by the λ-actin antibody, whereas the κ-actin antibody showed neglible reactivity toward the cap. Application of chicken anti-actin antibody, N350, also failed to react with soybean root cap. A lack of reaction of maize root caps toward another chick actin antibody (14, 18) is in keeping with this negative result. By contrast, using GUS constructs, two actin genes of *Arabidopsis, ACT1* and *ACT2,* were both found to be active in the *Arabidopsis* root cap and root meristem (2).

Four actin isotypes were identified in rice roots and the expression of their respective RNAs followed over a 35-day period following germination (151). The amounts of mRNA of two isotypes (Rac2 and Rac3) declined by approximately 80%, whereas Rac7 isotype remained constant; Rac1 transcripts showed a slight decrease. The significance of these findings remains obscure until the respective proteins are colocalized to the cellular structure of the roots. In roots of *Phaseolus vulgaris,* where two main actin isotypes have been found (186), only one was present in nodules induced by *Rhizobium.* Whether this was a new isotype or one of the original isotype complement of uninfected roots, the second one being suppressed, was not mentioned. Complex rearrangements of actin MFs and MTs occurred during development of the *Bradyrhizobium*-induced nodules of soybean roots (229). The actin MFs formed an unusual honeycomb pattern, the significance of which may be to establish and maintain the distribution of symbiosomes in the infected cells. They may also take part in delivering vesicles for the elaboration of additional membranous structures.

EARLIEST DEVELOPMENT OF ROOTS, THEIR MT ARRAYS, AND HOW MTS HELP GENERATE CELL FILES

The first root of most dicot plants differentiates in the early embryo. In the case of *Arabidopsis thaliana,* root differentiation begins after the first 5 or 6 cell divisions have established a proembryo (123). Microtubules have been examined during the early embryogenic stages of a few species, including *Arabidopsis* (107, 225, 235). Their behavior shares many of the features which continue to be seen in the subsequent phases of organogenesis. Thus, early embryos, just like adult meristems, have MT-based mechanisms that establish the planes of cell division and subsequent cell growth.

A comprehensive description of MTs in early *Arabidopsis* embryos is due to Webb & Gunning (225). Cortical MTs appeared before the first division of the zygote and were aligned perpendicular to the direction of cell elongation. A more dense, transverse array of MTs occurred at the distal end of the zygote and may mark a zone of localized wall extension, although an alternative idea is that the closely packed MTs restrict lateral expansion and hence reinforce the cell periphery (235). Later, another transverse band of MTs, which corresponded to the PPB, made its appearance. The first-mentioned distal array is of interest because similar rings of cortical MTs, which look like (and could be mistaken for) PPBs, have been seen in other systems (79, 173, 223) and also in growing root hairs of the fern, *Azolla* (AL Cleary, cited in Reference 225). Wherever these MT rings have been reported, the cells are free of lateral contact with other cells. However, the correlation between cortical MTs and localized growth rate of the wall is not clear. In the giant internodal cells of the alga, *Nitella flexilis*, for example, bands of growth and nongrowth alternate along the cell wall, yet no corresponding variation in cortical MT patterns could be found (128).

Following the first division of the proembryo, cortical MTs became oriented randomly and growth entered an isotropic phase (225). Endoplasmic MTs were prominent and closely associated with the nuclei and probably position them in the center of the cells. During the isotropic growth phase, nuclei divided in each of the three available orthogonal planes. The ability to divide successively in three planes is probably set not by chance, but is due to a property of the MT-organizing material located on the nuclear surface. This material partitions into two equal groups which repel each other (see 59). When repulsion is maximal, the material lies in two sites on opposite sides of the nucleus where it serves as organizers for the two poles of the forthcoming mitotic spindle (21). How the nucleus, following mitosis, senses in which plane to set up the future spindle is not known. It may be that some imprint on the nuclear surface survives mitosis; or each pole of the spindle may mark out a domain at the cell periphery which then does not permit a spindle pole of the next division to form in proximity to it. This last-mentioned system would require communication between cell periphery and spindle, a feature which, in animal cells, is mediated by dynactin (38).

Eventually, both the sequence of early orthogonal divisions and the symmetry of globular growth are broken. This may have to do with the fact that after three zygotic divisions there are now four internal division walls. Opportunities arise for cortical MTs to form transverse arrays around the perimeter of these cells and their growth can become polarized; for this to occur, a specification of an apical-basal axis is required (e.g. a reference point, such as the suspensor, may mark the basal end). This would be the prelude for cell file formation parallel to the embryonic axis, and hence would establish a bipolar, torpedo-shaped embryo containing domains where root and shoot structures can form. Meanwhile, a group of structural initial cells (26) establishes a series of divisions within the new root apex.

The later stages of embryogenesis see the completion of the primary root meristems in terms of the number of dividing cells. The ensuing dormant period seems to be accompanied by the loss of the microtubular cytoskeleton in the embryonic primary root meristem—at least within the radicle of tomato seeds. In this system, three β-tubulin isotypes become strongly evident at the time of the first wave of DNA synthesis during germination (61). In seasonally dormant, two-year-old taproots of horse chestnut, cortical MTs persisted in the presumptive vascular cambial cells (43). Here, all the MTs adopted a helical mode, whereas MTs in active cambium were more randomly arranged.

ROOT GROWTH AND THE CYTOSKELETON

Rectilinear Root Growth

The reactivation of cell growth and division during germination builds up the primary root meristem to a maximal length. Later, the meristem shortens and root growth rate slows. At the same time, the number of cell files may be reduced and the root becomes thinner (194). Germination also enables cells to enter a phase of rapid growth, a growth step which was missing in the embryonic radicle. This rapid mode of growth is assumed to depend on the development of the vacuolar compartment in cells immediately behind the meristem, but whether it is entirely driven by turgor is an open question in view of the fact that retardation of root growth follows on from treatments with cytochalasins (189, 209) or latrunculin (17; F Baluška & D Volkmann, submitted) and the consequent disassembly of actin. Likewise, whether the level of tubulin gene transcripts and cortical MT numbers are directly related to the rate of cell or root growth—as seems to be the case in some shoot tissues, for example (39, 160)—is not known. Increased numbers of cortical MTs and rates of their interpolation within root cells of *Azolla pinnata* occurred during their transition from meristematic to elongation growth and differentiation (87). That these increases in MTs were related to increased rates of secondary wall deposition is clear, and it may follow that they were also related to changing cellular growth rates. The dramatic disappearance of *Tub B1* gene transcripts which occurs in soybean roots older than 6 days postgermination (120) was, unfortunately, not considered in relation to any root growth characteristics.

Unlike the cells of the meristem, where cortical MTs have a variable but mainly transverse orientation, cells of the elongation zone of maize roots have strictly transverse MTs (15). This new orientation, which can be accompanied by a bundling of MTs in the cortical cells, develops in the transition zone at the base of the meristem (19). A similar maintenance of transverse orientation was noted for radish roots, though, as the cells elongated, the angle of the cortical MTs with respect to the cell axis became more variable (212). The MT-switch in the transition zone might depend upon the activity of a new set of tubulin genes or the utilization of a new population of MT-associated proteins, notably those which favor MT bundling. Presumably, these bundles must not be fixed in their position; they need to equalize their association with the longitudinal walls in order to produce a uniform thickness of secondary wall. However, some locational stability of MTs along the longitudinal walls must also occur because small areas from which cortical MTs are excluded correspond to regions where pit fields will form (33). These areas need to remain free of cortical MTs long enough to allow this wall feature to develop.

A similar sequence of cellular development—meristem, elongation, maturation—also exists in the root cap. The corresponding arrays of cortical MTs have been examined in root caps of maize (15, 28), radish (63), and cress (93). In the first two species, cortical MTs rearranged from transverse to random as the cells progressed from meristem to the cap flanks. By the time the cap cells detached, their MTs were either random or absent. No longitudinal cortical MTs were found in mature cap cells and in this respect MT reorganization differed from that occurring in derivatives of the proximal root meristem. Equally noteworthy is the contrast between the cells of the cap and of the root proper with respect to the disposition of actin MFs (18). Conspicuous MFs were largely absent from the cap, whereas cable-like MFs were abundant in root cells, especially within those of the vascular cylinder. In the central cap cells, actin is thought to exist as fine filaments whose structure is modified to assist in the gravisensing role of these cells (discussed later). The difficulty of observing these filaments by immunofluorescence may, in part, because the process involves chemical fixation. Freeze-substituted tobacco root caps sectioned for electron microscopy (64) revealed various categories of MFs. Both techniques, however, showed that actin MFs were absent in the quiescent center (14, 18, 64).

The rearrangement of cortical MTs along the growth zone of roots has been mapped in maize, radish, and pea roots. For pea, Hogetsu & Oshima (104) noted that the zone where the MTs switched from transversal to oblique orientation was located in a 1-mm zone where growth had recently ceased; it was estimated that the reorientation required 2 h to complete. Often the cortical MTs reoriented longitudinally and then made a meandering course between each end of the cell. The transverse-to-longitudinal orientation is paralleled in other systems: it occurs in the single cells of cotton fibers as they grow (198), and also in files of isolated BY-2 cells growing and maturing in culture (88). It is as though MT reorientation is an inherent feature of a cellular growth cycle, irrespective of whether division occurs or not. Because MTs are sensitive to electrical stimuli (110), it is tempting to speculate that the changing pattern of electrical flux along the length of growing roots (210) has some relationship with the reorientations of their MTs. Similar transverse-to-oblique reorientations of MTs were also found in onion root meristem cells exposed to inhibitors of RNA synthesis (215), suggesting a metabolic basis for this effect.

In the maturing cortex of both pea (104) and hyacinth roots (136), it was noticed that the cortical MTs existed in a criss-cross arrangement against the longitudinal walls of neighboring cells. A more detailed study of cortical and epidermal cells of maize and *Arabidopsis* roots revealed that the cortical MTs had a particular chirality (135). By observing MTs against the radial longitudinal walls, arrays of MTs in an S helix were observed toward the end of the elongation zone. Just beyond the elongation zone, however, cortical MT arrays displayed a Z helix arrangement. In the intervening zone, the cortical MTs were longitudinal. The S → Z helix transition is most simply achieved by a rotation of the MT array in a clockwise direction. The consistency of the S → Z helix transition (over 100 maize roots and 13 *Arabidopsis* roots were examined) raises the question, at least for *Arabidopsis,* whether this microtubular chirality has any relationship with the natural chirality of root growth (203). Not only can there be consistent handedness of the MT arrays, but cortical MT bundles in adjacent cells sometimes also appear to have co-alignment (e.g. 230), as though the MTs had been aligned by a common factor. If so, this would be an indication that some type of intercellular communication influences the cytoskeleton.

Root Contraction

If root cells were able to redirect their growth in the transverse plane and also to shrink their length, root contraction would result. The longitudinal hoops of cortical MTs found in the elongated, mature zone of cortical tissue of hyacinth roots (206) could assist this natural shrinkage process, especially if helped by some type of cytoskeletal contraction mechanism. Longitudinal MTs could also bring about the expansion of the previously unwidened transversal end walls. Tubulin levels more than doubled in the contracting zone of the hyacinth root and, in the electron microscope, the cortical MTs were surrounded by additional amounts of electron-dense material which was suggested to be MT-associated proteins (57).

In the root cortex of hyacinth and other species, the shift in cortical MT orientation commenced at different distances from the tip, with the outer cortical cell files showing MT reorientation before the inner files (116). Usually, the more internal cell files, including pericycle (which belongs to the stele) remain juvenile, with random MTs, longer than outer cortical files (15). Moreover, cortical MTs of the inner root cortex of maize showed increased instability toward ethylene (11). This probably accounts for the ethylene-induced swelling of these roots, a process that enables them to overcome mechanical impedence.

Curvilinear Root Growth

The curvilinear growth of roots, which is associated with tropisms, results from differential extension rates of cells on opposite sides of the root in the distal part of the zone of rapid elongation. The question is whether there is a contribution from the cytoskeleton to the growth differential. In the case of gravitropic coleoptiles, alterations to their cortical MTs (100) and wall microfibrils (74) provide an affirmative answer. But in roots the answer seems to be negative, if only because of the simple observation that gravitropism occurs in roots which had been exposed to oryzalin or colchicine, and hence lacked MTs, before they were gravistimulated (13). Nevertheless, altered arrangements of cortical MTs have been seen, and complex rearrangements of growth did occur, within gravireacting maize roots. However, the stresses and strains associated with the bending (237) and the gravity-induced alterations in auxin levels could have had rapid effects on cortical MTs levels (20, 34). Accordingly, it is not easy to disentangle which are the primary physiological and biophysical perturbations to the cytoskeleton resulting from the root graviresponse, and which are secondary effects induced by the bending reaction.

The positive graviresponse involves alterations to the timing of cellular development in upper and lower portions of the root. Thus, when longitudinal arrays of MTs have been seen on the upper side of a horizontal root tip 2 h into its gravireaction (13), it could be an indication that cell maturation has been advanced during this period, especially since a transverse-to-longitudinal MT reorientation normally accompanies the maturation process (15, 104). Other criteria also suggest premature cell maturation on the upper side of gravireacting roots (27, 60). The longitudinal orientation of MTs, whether found on the upper or lower sides of the root, would have the effect of diminishing the contribution of the affected cells to forward elongation growth and, hence, would initiate the required growth differential.

Careful examination of F-actin networks in gravibending maize roots did not reveal any difference between the cells on the upper and the lower sides (35). However, it is timely to reconsider this observation in the light of the proposal that actin plays an active role in cell elongation (209). The critical location for a contribution from actin may be in the transition zone, at the changeover between meristematic and rapid elongation growth. If an actin-triggered switch to rapid elongation were advanced in the upper portion of the root relative to the lower side, a gravireaction could be initiated without any dramatic change in actin configuration

Cytoskeleton, Cell Structure, and Graviperception

A more certain area where the cytoskeleton impinges on root gravitropism is in graviperception. There are three aspects to consider: (*a*) the structure of the graviperceptive cells, generally held to be the statocytes of the central root cap (*b*) the capture of information relating to root orientation with respect to the gravity vector, and (*c*) the subsequent transduction of graviperception into a signal for a graviresponse.

A crucial element in relation to the role of statocyte cells in graviperception is their asymmetry. In the cap cells of vertical roots of cress or lentil, where ultrastructural organization has been examined in some detail (95, 96, 185), endoplasmic reticulum (ER) and amyloplasts gather at the distal end of the cell, whereas the nucleus is found at the proximal end. This asymmetry is developed during germination and involves actin-dependent movement of the ER (96). Observations on roots exposed to cytochalasin (95, 143) or to colchicine (96) suggest that the position of the nucleus in the statocytes is regulated by actin MFs, whereas positioning of the distal ER is mediated by cortical MTs. Interestingly, MTs associated with the ER tend to be more sensitive to dissassembly by colchicine than are the MTs elsewhere in the same statocyte (94).

Comparison of results of experiments performed in the microgravity (1×10^{-4} **g**) of spaceflights with those done on Earth (at 1 **g**) reveals that statocyte asymmetry is, in part, regulated by the cytoskeleton (142). A similar conclusion was also reached from simulated microgravity experiments performed with the clinostat (142). In microgravity, the nucleus is displaced distally from its usual position and the amyloplasts move basally. Thus, these two organelles normally (i.e. in 1 **g**) assume a position within the cell that is the result of the restraints imposed upon them by the cytoskeleton and by their tendency to displacement due to their mass.

The relationship between the amyloplasts and the cytoskeleton is becoming clearer. Schemes have been proposed whereby the amyloplasts are supported by, or impinge upon, delicate transcellular filaments of actin (12). Evidence for this comes from an experiment which showed that, when the actin was dissassembled by cytochalasin, amyloplast sedimentation was at least three times quicker than usual (201). Less significant was the alteration of sedimentation rate following dissassembly of the MTs (14). Unfortunately, it has been difficult to visualize the putative actin strands in the statocytes by immunofluorescence (14, 97, 127, 228, but see 64). It is probable that the actin turns over rapidly and that it exists only as short filamentous elements, or even as G actin. The actin filaments within the statocytes probably attach to receptors in the plasma membrane and/or ER. During graviperception they trigger the asymmetric outflow of information from the cells which is crucial for initiating the graviresponse (12). Mutation at the gene locus *ARG1* (Altered Response to Gravity) in *Arabidopsis* brings about a slower rate of root gravitropic bending. The protein encoded by *ARG1* is a DnaJ-like protein (199) and it is possible that it transduces gravity signals in the statocytes by its interaction with the actin cytoskeleton.

MUTATIONS AFFECTING THE CYTOSKELETON AND ROOT DEVELOPMENT

Although mutations affecting root systems have been know for a long time, it is only recently that those affecting the cellular behavior of roots have come to the fore. There are at least two types of mutation which affect the cytoskeleton and, hence, the morphogenesis of the root apex. They involve: (*a*) impairment to the orientation of cortical MTs, and (*b*) the formation of the cell plate.

Screens of mutagenized *Arabidopsis* seedlings revealed stunted individuals with depolarized cell growth in their roots (211). Meristematic cells of the *ton1* and *ton2* mutants were characterized by random orientations of their cortical MTs. Significantly, PPBs were consistently absent, although mitotic spindles and phragmoplasts were present and normal. The *fass* mutant presented a similar phenotype to *ton* and may be allelic to it. Electron microcopy showed the cortical MTs to be more sparsely spaced along the plasma membranes and to have a more haphazard orientation (148). One possible basis of the cytoskeletal lesion is the failure to regulate endogenonous hormonal levels. *fass* seedlings contained 2.6-times more free auxin (IAA) than did wild-type, though these levels varied considerably between samples (73). The elevated auxin may be responsible for a two- to threefold rise in ethylene production which, in turn, could have had an impact upon the cortical MT arrays: earlier we mentioned that ethylene and auxin tend to disturb cortical MT orientation (11, 20).

More subtle effects on cortical MTs were associated with altered levels of gibberellins. The respective *d5* and *gib-1* mutants of maize and tomato, with impaired gibberellin biosynthesis, had slightly thicker roots and, in the meristematic cells, the cortical MTs also tended to deviate from the usual transverse orientation (16). This, in turn, led to altered division patterns in the formative zone of the root cortex (23, 25). These effects on the MTs can be phenocopied in wild-type roots by exposing them to the gibberellin biosynthesis-inhibitor, paclobutrazol, and corrected by addition of gibberellic acid. Although there is a large conceptual gap between hormones and MT orientation, one factor which could provide a link is the posttranslational modification of tubulin isotypes. Internodal epidermal cells of a pea mutant, dwarfed as a result of the *le* gene which depresses gibberellin levels, had cortical MTs which tended to be longitudinally oriented (67). Within 2 h of its application, gibberellic acid had promoted transverse MTs. This alteration to the MTs was associated with an inability of the α-1 tubulin isotype to react with YL1/2 antibody which specifically probes tyrosinated tubulin. Since the α-1 isotype continued to be present, the implication is that the α-1 tubulin of the *le* mutant was detyrosinated when gibberellin levels were high, and that this led to MT reorientation. In another system (protoplasts from maize cell suspension), addition of gibberellic acid resulted in a stimulation of α-tubulin acetylation (as judged by affinity for 6-11B-1 antibody), more organized cortical MTs, and greater resistance to freezing temperatures (108).

A contrasting situation is found in the roots of certain conditional mutants of *Arabidopsis* (91). The mutations *cobra, quill,* and some others, showed altered polarity of cell growth, but it is uncertain whether this was due to altered cortical MT orientation. Presumably, PPBs were normal since there was no mention of aberrant cell divisions. A deeper analysis of cell growth seems in prospect, given the numerous mutations that influence cell shape in the root (29, 91). The *tangled* mutant of maize, for example, is a good candidate for further study, given its effect on MT orientation in leaves (52). In fact, root tissues of the *pygmy* mutant of maize (40), which is cognate with *tangled,* also show disturbed cortical MT arrangements (PW Barlow & JS Parker, unpublished data), and these almost certainly account for the stunted root growth and irregular cell files (40). Nevertheless, it is possible that such mutants have their basis in cytoskeletal components other than the MTs. Involvement of actin filaments is a possibility.

A second class of mutants, again in *Arabidopsis*, affected in the division process, yield information about the molecules involved in phragmoplast structure and function. The *knolle* mutant presents embryos with both large and small cells with incomplete division walls (147). It was said that endomitotic cycles were present, but the evidence suggests that the nuclei in question are mitotic and polyploid as a result of nuclear fusion following incomplete cytokinesis. Further characterization revealed (132) that the *KNOLLE* gene encodes for a syntaxin protein involved in the fusion of the vesicles from which the cell plate is assembled at late telophase. The defect is not totally effective since walls do form, though a fraction of them are incomplete in their central portion. Wall stubs are often present. As mentioned earlier, these stubs may develop by a complementary pathway which becomes apparent only when cell plate formation is defective. The incompleteness of the walls can also result in a failure to develop cell layers. Only when such layers have been constructed can the positional information inherent within the developing embryo be interpreted correctly. Thus, in *knolle*, anthocyanin accumulates in epidermal cells instead of subepidermal cells, an error which probably occurs because the walls separating these two layers are incomplete. The *keule* mutant also shows incomplete division walls due to defective cytokinesis (7), but these are seen mainly in meristematic rather than mature cells. It may be that the defective division wall is eventually completed by continued growth of the wall stubs. Another cytokinesis-defective mutant of *Arabidopsis, cyt-1,* also shows defective division walls (wall stubs were present) (176). This mutant seems to be affected in a way distinct from *keulle* or *knolle* since it also showed an altered pattern of callose deposition not shared by the other mutants. Callose is a component of the cell plate and early division wall, so it is possible that it is not deployed correctly within the phragmoplast, and it is this that causes the cell plate to fail.

Mutations in the *PILZ* group of genes of *Arabidopsis* appear to abolish the functional assembly of all classes of MTs (157). Although some aberrant cell divisions can nevertheless occur (again with poorly differentiated wall stubs), no root organ forms in such mutant embryos.

A mutant, *cyd,* discovered in pea, has some similarities with *keulle* and *knolle* of *Arabidopsis* in that multinucleate cells occur (139). Such cells were more frequent in embryonic cotyledons (73%) than they were in roots (28%). However, it is not clear where in the root tip the abnormal cells were produced since meristematic cells were not multinucleate. This led to the presumption that they had been formed in cells outside the meristem. This unsatisfactory conclusion may indicate that the gene shows penetrance only during the early stages of embryo axis development and that multinucleate cells were formed at this time, but not later on. This would agree with the finding that seedlings derived from cultured embryos had a lower frequency of multinucleate cells than did the embryos. It could be that the mutation has a penetrance regulated by the type of meristem or the stage of its development. Penetrance effects are clearly seen in the multinucleate *MUN* mutants of *Arabidopsis:* defects of cytokinesis are found only in the roots and not elsewhere (180). By contrast, the *cyd1* mutation results in aberrant cytokineses in all dividing cells except those of the root meristems (234). Is it possible that such differential effects are regulated by organ-specific isoforms of cytoskeletal proteins?

CONCLUDING REMARKS

The identification of mutations which impair cell division and disturb the usual orientation and arrangement of cell growth in roots will continue to unravel the molecular mechanism by which cells in general reproduce and attain identities within tissues. The role of the cytoskeleton in differentiation could also be revealed in a more specific way if the affected cells—i.e. those which utilize particular cytoskeletal arrays for their development—could be identified in mutagenized plant populations. Close analysis of tissue differentiation, with the concept of positional information as a context for interpretation, might reveal the role of tissue compartmentation (resulting from cytokinesis) in the differentiation process. It might even be possible to test whether there are local genetic controls over cytokinesis, invoking specific orientations of cell division at precise locations within the root meristem.

ACKNOWLEDGMENTS

We are grateful to the many colleagues who have helped us in studying the cytoskeleton over many years, but particular thanks are due to NJ Chaffey, JS Parker, and D Volkmann. Much of our own work has been supported by the Alexander von Humboldt-Stiftung, the Deutsche Agentur für Raumfahrtangelegenheiten, and the Ministerium für Wissenschaft und Forschung. IACR receives grant-aided support from the Biotechnology and Biological Sciences Research Council of the UK.

LITERATURE CITED

1. Akashi T, Shibaoka H. 1991. Involvement of transmembrane proteins in the association of cortical microtubules with the plasma membrane in tobacco BY-2 cells. *J. Cell Sci.* 98:169–74
2. An Y-Q, Huang S, McDowell JM, McKinney EC, Meagher RB. 1996. Conserved expression of the *Arabidopsis* ACT1 and ACT3 actin subclass in organ primordia and mature pollen. *Plant Cell* 8:15–30
3. Apostolakos P, Galatis B, Panteris E. 1991. Microtubules in cell morphogenesis and

intercellular formation in *Zea mays* leaf mesophyll and *Pilea cadierei* epithem. *J. Plant Physiol.* 137:591–601
4. Asada T, Collings D. 1997. Molecular motors in higher plants. *Trends Plant Sci.* 2:29–37
5. Asada T, Kuriyama R, Shibaoka H. 1997. TKRP125, a kinesis-related protein involved in the centrosome-independent organization of the cytokinetic apparatus in tobacco BY-2 cells. *J. Cell Sci.* 110:179–89
6. Asada T, Shibaoka H. 1994. Isolation of polypeptides with microtubule-translocating activity from phragmoplasts of tobacco BY-2 cells. *J. Cell Sci.* 107:2249–57
7. Assaad FF, Mayer U, Wanner G, Jürgens G. 1996. The *KEULE* gene is involved in cytokinesis in *Arabidopsis. Mol. Gen. Genet.* 253:267–77
8. Baird WV, Meagher RB. 1987. A complex gene superfamily encodes actin in petunia. *EMBO J.* 6:3223–31
9. Baluška F, Barlow PW, Hauskrecht M, Kubica Š, Parker JS, Volkmann D. 1995. Microtubule arrays in maize root cells. Interplay between the cytoskeleton, nuclear organization and post-mitotic cellular growth patterns. *New Phytol.* 130:177–92
10. Baluška F, Barlow PW, Lichtscheidl IK, Volkmann D. 1998. The plant cell body: a cytoskeletal tool for cellular development and morphogenesis. *Protoplasma* 202:1–10
11. Baluška F, Brailsford RW, Hauskrecht M, Jackson MB, Barlow PW. 1993. Cellular dimorphism in the maize root cortex: involvement of microtubules, ethylene and gibberellin in the differentiation of cellular behaviour in postmitotic gowth zones. *Bot. Acta* 106:394–403
12. Baluška F, Hasenstein KH. 1997. Root cytoskeleton: its role in perception of and response to gravity. *Planta* 303:S69–S78
13. Baluška F, Hauskrecht M, Barlow PW, Sievers A. 1996. Gravitropism of the primary root of maize: a complex pattern of differential cellular growth in the cortex independent of the microtubular cytoskeleton. *Planta* 198:310–18
14. Baluška F, Kreibaum A, Vitha S, Parker JS, Barlow PW, Sievers A. 1997. Central root cap cells are depleted of endoplasmic microtubules and actin microfilament bundles: implications for their role as gravity-sensing statocytes. *Protoplasma* 196:212–23
15. Baluška F, Parker JS, Barlow PW. 1992. Specific patterns of cortical and endoplasmic microtubules associated with cell growth and tissue differentiation in roots of maize (*Zea mays* L.). *J. Cell Sci.* 103:191–200
16. Baluška F, Parker JS, Barlow PW. 1993. A role for gibberellic acid in orienting microtubules and cell growth polarity in the maize root cortex. *Planta* 191:149–57
17. Baluška F, Šamaj J, Kendrick-Jones J, Barlow PW, Staiger CJ, Volkmann. 1997. Tissue- and domain-specific distributions and re-distributions of actin microfilaments, myosins, and profilin isoforms in cells of root apices. *Cell Biol. Int.* 21:852–54
18. Baluška F, Vitha S, Barlow PW, Volkmann D. 1997. Rearrangements of F-actin in growing cells of intact maize root tissue: A major developmental switch occurs in the postmitotic transition region. *Eur. J. Cell Biol.* 72:113–21
19. Baluška F, Volkmann D, Barlow PW. 1996. Specialized zones of development in roots: view from the cellular level. *Plant Physiol.* 112:3–4
20. Baluška F, Volkmann D, Barlow PW. 1996. Complete disintegration of the microtubular cytoskeleton precedes auxin-mediated reconstruction in post-mitotic maize root cells. *Plant Cell Physiol.* 37:1013–21
21. Baluška F, Volkmann D, Barlow PW. 1997. Nuclear components with microtubule-organizing properties in multicellular eukaryotes: functional and evolutionary considerations. *Int. Rev. Cytol.* 175:91–135

22. Baluška F, Volkmann D, Barlow PW. 1998. Tissue- and development-specific distributions of cytoskeletal elements in growing cells of the maize root apex. *Plant Biosyst.* 132:251–65
23. Baluška F, Volkmann D, Barlow PW. 1999. Hormone-cytoskeleton interactions in plant cells. In *Biochemistry and Molecular Biology of Plant Hormones*, ed. PJJ Hooykaas, MA Hall, KR Libbenga, pp. 363–90. Amsterdam: Elsevier

23a. Baluška F, Volkmann D, Barlow PW. 2000. Actin-based cell cortex domains and their association with polarized 'plant-cell bodies' in higher plants. *Plant. Biol.* In press

24. Barlow PW. 1971. Properties of cells in the root apex. *Rev. Fac. Agron. La Plata* 47:275–301
25. Barlow PW. 1995. The cytoskeleton and its role in determining the cellular architecture of roots. *G. Bot. Ital.* 129:863–72
26. Barlow PW. 1997. Stem cells and founder zones in plants, particularly their roots. In *Stem Cells*, ed. CS Potten, pp. 29–57. London: Academic
27. Barlow PW, Hofer R-M. 1982. Mitotic activity and cell elongation in geostimulated roots of *Zea mays*. *Physiol. Plant.* 54:137–41
28. Barlow PW, Parker JS. 1996. Microtubular cytoskeleton and root morphogenesis. *Plant Soil* 187:23–36
29. Baskin TI, Betzner AS, Hoggart R, Cork A, Williamson RE. 1992. Root morphology mutants in *Arabidopsis thaliana*. *Aust. J. Plant Physiol.* 10:427–37
30. Baskin TI, Cande WZ. 1990. The structure and function of the mitotic spindle in flowering plants. *Annu. Rev. Plant Physiol. Plant Mol. Biol.* 41:277–315
31. Bassell G, Singer RH. 1997. mRNA and cytoskeletal filaments. *Curr. Opin. Cell Biol.* 9:109–15
32. Binarova P, Rennie P, Fowke L. 1994. Probing microtubule organizing centres with MPM-2 in dividing cells of higher plants using immunofluorescence and immunogold techniques. *Protoplasma* 180: 106–17
33. Blancaflor EB, Hasenstein KH. 1993. Organization of cortical microtubules in graviresponding maize roots. *Planta* 191: 231–37
34. Blancaflor EB, Hasenstein KH. 1995. Time course and auxin sensitivity of cortical microtubule reorientation in maize roots. *Protoplasma* 185:72–82
35. Blancaflor EB, Hasenstein KH. 1997. The organization of the actin cytosksleton in vertical and graviresponding primary roots of maize. *Plant Physiol.* 113:1447–55
36. Bonfante P, Bergero R, Uribe X, Romera C, Rigau J, Puigdomènech P. 1996. Transcriptional activation of maize α-tubulin gene in mycorrhizal maize and transgenic tobacco plants. *Plant J.* 9:737–43
37. Bulinski JC, Gundersen GG. 1991. Stabilization and post-translational modification of microtubules during cellular morphogenesis. *BioEssays* 13:285–93
38. Busson S, Dujardin D, Moreau A, Dompierre J, De May JR. 1998. Dynein and dynactin are localized to astral microtubules and at cortical sites in mitotic epithelial cells. *Curr. Biol.* 8:541–44
39. Bustos MM, Guiltinan MF, Cyr RJ, Ahdoot D, Fosket DE. 1989. Light regulation of β-tubulin gene expression during internode development in soybean (*Glycine max* [L.] Merr.). *Plant Physiol.* 91:1157–61
40. Byer M. 1957–58. Cytohistological abnormalities associated with the gene "Pigmy" (Py-1) in the primary root tip of corn (*Zea mays* L.). *Proc. Minn. Acad. Sci.* 25/26:1–10
41. Carnero Diaz E, Martin F, Tagu D. 1996 Eucalypt α-tubulin: cDNA cloning and increased level of transcript in ectomycorrhizal root system. *Plant Mol. Biol* 31:905–10
42. Chaffey NJ, Barlow PW, Barnett JR. 1997

Cortical microtubules rearrange during differentiation of vascular cambial derivatives, microfilaments do not. *Trees* 11:333–41
43. Chaffey NJ, Barlow PW, Barnett JR. 1998. A seasonal cycle of cell wall structure is accompanied by a cyclical rearrangement of cortical microtubules in fusiform cambial cells within taproots of *Aesculus hippocastanum* (Hippocastanaceae). *New Phytol.* 139:623–35
44. Chaffey NJ, Barnett JR, Barlow PW. 1999. A cytoskeletal basis for wood formation in angiosperm trees: the involvement of cortical microtubules. *Planta* 208:19–30
45. Chan J, Rutten T, Lloyd C. 1996. Isolation of microtubule-associated proteins from carrot cytoskeletons: A 120 kDa MAP decorates all four microtubule arrays and the nucleus. *Plant J.* 10:251–59
46. Chen J, Kanai Y, Cowan NJ, Hirokawa N. 1992. Projection domains of MAP2 and tau determine spacing between microtubules in dendrites and axons. *Nature* 360:674–77
47. Chu B, Kerr GP, Carter JV. 1993. Stabilizing microtubules with taxol increases microfilament stability during freezing of rye root tips. *Plant Cell Environ.* 16:883–89
48. Chu B, Wilson TJ, McCune-Zierath C, Snustad DP, Carter JV. 1998. Two β-tubulin genes, TUB1 and TUB8, of *Arabidopsis* exhibit largely nonoverlapping patterns of expression. *Plant Mol. Biol.* 37:785–90
49. Clayton L, Lloyd CW. 1985. Actin organization during the cell cycle in meristematic plant cells. *Exp. Cell* Res. 156:231–38
50. Deleted in proof
51. Cleary AL, Gunning BES, Wasteneys GO, Hepler PK. 1992. Microtubules and F-actin dynamics at the division site in living *Tradescantia* stamen hair cells. *J. Cell Sci.* 103:977–88
52. Cleary AL, Smith LC. 1998. The *Tangled1* gene is required for spatial control of cytoskeletal arrays associated with cell division during maize leaf development. *Plant Cell* 10:1875–88
53. Clore AM, Dannehafer JM, Larkins BM. 1996. Ef-1α is associated with a cytoskeletal network surrounding protein bodies in maize endosperm cells. *Plant Cell* 8:2003–14
54. Colasanti J, Cho S-O, Wick S, Sundaresan V. 1993. Localization of the functional p34^{cdc2} homolog of maize in root tip and stomatal complex cells: association with predicted division sites. *Plant Cell* 5:1101–11
55. Collings DA, Asada T, Allen NS, Shibaoka H. 1998. Plasma membrane-associated actin in Bright Yellow 2 tobacco cells. Evidence for interaction with microtubules. *Plant Physiol.* 118:917–28
56. Cyr RJ. 1994. Microtubules in plant morphogenesis. Role of the cortical array. *Annu. Rev. Cell Biol.* 10:153–80
57. Cyr RJ, Lin B-L, Jernstedt JA. 1988. Root contraction in hyacinth. II. Changes in tubulin levels, microtubule number and orientation associated with differential cell expansion. *Planta* 174:446–52
58. Cyr RJ, Palevitz BA. 1989. Microtubule-binding proteins from carrot I. Initial characterization and microtubule binding. *Planta* 177:245–60
59. Czihak G, Kojima M, Linhart J, Vogel H. 1991. Multipolar mitosis in procaine-treated polyspermic sea urchin eggs and in eggs fertilized with UV-irradiated spermatozoa with a computer model to simulate the positioning of centrosomes. *Eur. J. Cell Biol.* 55:255–61
60. Darbelley N, Perbal G. 1984. Gravité et différentiation des cellules corticales dans la racine de lentille. *Biol. Cell* 50:93–98
61. De Castro R, Zheng X, Bergervoet JHW, De Vos CHR, Bino RJ. 1995. β-tubulin accumulation and DNA replication in

imbibing tomato seeds. *Plant Physiol.* 109:499–504

62. Del Vecchio AJ, Harper JDI, Vaughn KC, Baron AT, Salisbury JL, Overall RL. 1997. Centrin homologues in higher plants are prominently associated with the developing cell plate. *Protoplasma* 196:224–34
63. Derksen J, Jeucken G, Traas JA, Van Lammeren AAM. 1986. The microtubular cytoskeleton in differentiating root tips of *Raphanus sativus* L. *Acta Bot. Neerl.* 35:223–31
64. Ding B, Turgeon R, Parthasarathy MV. 1991. Microfilament organization and distribution in freeze substituted tobacco plant tissues. *Protoplasma* 165:96–105
65. Ding B, Turgeon R, Parthasarathy MV. 1991. Microfilaments in the preprophase band of freeze substituted tobacco root cells. *Protoplasma* 165:209–11
66. Dolfini S, Consonni G, Mereghetti M, Tonelli C. 1993. Antiparallel expression of the sense and antisense transcripts of maize α-tubulin genes. *Mol. Gen. Genet.* 241:161–69
67. Duckett CM, Lloyd, CW. 1994. Gibberellic acid-induced microtubule reorientation in dwarf peas is accompanied by rapid modification of an α-tubulin isotype. *Plant J.* 5:363–72
68. Durso NA, Cyr RJ. 1994. Beyond translation: elongation factor-1α and the cytoskeleton. *Protoplasma* 180:99–105
69. Eleftheriou EP, Palevitz BA. 1992. The effect of cytochalsin D on preprophase band organization in root tip cells of *Allium. J. Cell Sci.* 103:989–98
70. Endlé M-C, Stoppin V, Lambert A-M, Schmit A-C. 1998. The growing cell plate of higher plants is a site of both actin assembly and vinculin-like antigen recruitment. *Eur. J. Cell Biol.* 77:10–18
71. Ericson ME, Carter JV. 1996. Immunolabeled microtubules and microfilaments are visible in multiple cell layers of rye root tip sections. *Protoplasma* 191:215–19

71a. Eun S-O, Wick SM. 1998. Tubulin isoform usage in maize microtubules. *Protoplasma* 204:235–44

72. Farrar JJ, Evert RF. 1997. Ultrastructure of cell division in the fusiform cells of the vascular cambium of *Robinia pseudoacacia. Trees* 11:203–15
73. Fisher RH, Barton MK, Cohen JD, Cooke TJ. 1996. Hormonal studies of *fass,* an *Arabidopsis* mutant that is altered in organ elongation. *Plant Physiol.* 110:1109–21
74. Folsom DB, Brown RM Jr. 1987. Changes in cellulose microfibril orientation during differential growth in oat coleoptiles. In *Physiology of Cell Expansion during Plant Growth,* ed. DJ Cosgrove, DP Knievel, pp. 58–73. Rockville: Am. Soc. Plant Physiol.
75. Galatis B, Apostolakos P, Katsaros C. 1984. Experimental studies on the function of the cortical cytoplasmic zone of the preprophase microtubule band. *Protoplasma* 122:11–26
76. Giddings TH Jr, Staehelin LA. 1988. Spatial relationship between microtubules and plasma-membrane rosettes during the deposition of primary wall microfibrils in *Closterium* sp. *Planta* 173:22–30

76a. Glimer S, Clay P, MacRae TH, Fowke LC. 1999. Acetylated tubulin is found in all microtubule arrays of two species of pine. *Protoplasma* 207:174–85

76b. Glimer S, Clay P, MacRae TH, Fowke LC. 1999. Tyrosinated, but not detyrosinated, α-tubulin is present in root tip cells. *Protoplasma* 210:92–98

77. Goddard RH, Wick SM, Silflow CD, Snustad DP. 1994. Microtubule components of the plant cytoskeleton. *Plant Physiol.* 104:1–6
78. Goodbody KC, Venverloo CJ, Lloyd CW. 1991. Laser microsurgery demonstrates that cytoplasmic strands anchoring the nucleus across the vacuole of

premitotic plant cells are under tension. Implications for division plane alignment. *Development* 113:931–39
79. Goode JA, Alfano F, Stead AD, Duckett JG. 1993. The formation of aplastidic abscission (tmema) cells and protonemal disruption in the moss *Bryum tenuisetum* Limpr. is associated with transverse arrays of microtubules and microfilaments. *Protoplasma* 174:158–72
80. Grabski S, Xie XG, Holland JF, Schindler M. 1994. Lipids trigger changes in the elasticity of the cytoskeleton in plant cells: a cell optical displacement assay for live cell measurements. *J. Cell Biol.* 126:713–26
81. Green PB. 1962. Mechanism for plant cellular morphogenesis. *Science* 138:1404–5
82. Gu X, Verma DPS. 1997. Dynamics of phragmoplastin in living cells during cell plate formation and microscopy of elongation from the plane of cell division. *Plant Cell* 9:157–69
83. Gunning BES. 1982. The root of the water fern *Azolla:* cellular basis of development and multiple roles for cortical microtubules. In *Developmental Order: Its Origin and Regulation,* ed. S Subtelny, pp. 379–421. New York: Liss
84. Gunning BES. 1992. Use of confocal microscopy to examine transitions between successive microtubule arrays in the plant cell division cycle. In *Proc. Int. Jpn. Prize Symp. Cell. Basis Growth Dev. Plants*, 7th, ed. H Shibaoka, pp. 145–55. Osaka: Osaka Univ. Press
85. Hahne G, Hoffmann F. 1985. Cortical microtubular lattices: absent from mature mesophyll and necessary for division? *Planta* 166:309–13
86. Hardham AR, Gunning BES. 1978. Structure of cortical microtubule arrays in plant cells. *J. Cell Biol.* 77:14–34
87. Hardham AR, Gunning BES. 1979. Interpolation of microtubules into cortical arrays during cell elongation and differentiation in roots of *Azolla pinnata. J. Cell Sci.* 37:411–42
88. Hasezawa S, Hogetsu T, Syono K. 1988. Rearrangement of cortical microtubules in elongating cells derived from tobacco protoplasts—a time-course observation by immunofluorescence microscopy. *J. Plant Physiol.* 133:46–51
89. Hasezawa S, Kumagai F, Nagata T. 1997. Sites of microtubule reorganization in tobacco BY-2 cells during cell-cycle progression. *Protoplasma* 198:202–9
90. Hasezawa S, Nagata T. 1993. Microtubule organizing centers in plant cells: localization of a 49 kDa protein that is immunologically cross-reactive to a 51 kDa protein from sea urchin centrosomes in synchronized tobacco BY-2 cells. *Protoplasma* 176:64–74
91. Hauser M-T, Morikami A, Benfy PN. 1995. Conditional root expansion mutants of *Arabidopsis. Development* 121:1237–52
92. Hellmann A, Wernicke W. 1998. Changes in tubulin protein expression accompanying reorganization of microtubular arrays during cell shaping in barley leaves. *Planta* 204:220–25
93. Hensel W. 1984. Microtubules in statocytes from roots of cress (*Lepidium sativum* L.). *Protoplasma* 119:121–34
94. Hensel W. 1984. A role for microtubules in the polarity of statocytes from roots of *Lepidium sativum* L. *Planta* 162:404–14
95. Hensel W. 1985. Cytochalasin B affects the structural polarity of statocytes from cress roots (*Lepidium sativum* L.). *Protoplasma* 129:178–87
96. Hensel W. 1986. Cytodifferentiation of polar plant cells. Use of anti-microtubular agents during differentiation of statocytes from cress roots (*Lepidium sativum* L.). *Planta* 169:293–303
97. Hensel W. 1989. Tissue slices from living root caps as a model system in which to study cytodifferentiation of polar cells. *Planta* 177:296–303
98. Hepler PK, Palevitz BA, Lancelle SA, McCauley MM, Lichtscheidl IK. 1990.

Cortical endoplasm in plants. *J.Cell Sci.* 96:355–73

99. Herth W. 1985. Plasma-membrane rosettes involved in localized wall thickening during xylem vessel formation of *Lepidium sativum* L. *Planta* 164:12–21
100. Himmelspach R, Wymer CL, Lloyd CW, Nick P. 1999. Gravity-induced reorientation of cortical microtubules observed in vivo. *Plant J.* 18:449–53
101. Hirokawa N. 1994. Microtubule organization and dynamics dependent on microtubule-associated proteins. *Curr. Opin. Cell Biol.* 6:74–81
102. Hoffmann JC, Vaughn KC. 1994. Mitotic disrupter herbicides act by a single mechanism but vary in efficacy. *Protoplasma* 179:16–25
103. Hogetsu T. 1990. Mechanism for formation of the secondary wall thickening in tracheary elements: microtubules and microfibrils of tracheary elements of *Pisum sativum* L. and *Commelina communis* L. and the effects of amiprophos-methyl. *Planta* 185:190–200
104. Hogetsu T, Oshima Y. 1986. Immunofluorescence microscopy of microtubule arrangement in root cells of *Pisum sativum* L. var Alaska. *Plant Cell Physiol.* 27:939–45
105. Hoss S, Wernicke W. 1995. Microtubules and the establishment of apparent cell wall invaginations in mesophyll cells of *Pinus silvestris* L. *J. Plant Physiol.* 147:474–76
106. Hoyle HD, Raff EC. 1990. Two *Drosophila* beta tubulin isoforms are not functionally equivalent. *J. Cell Biol.* 111:1009–26
107. Huang B-Q, Ye X-L, Yeung EC, Zee SY. 1998. Embryology of *Cymbidium sinense:* the microtubule organization of early embryos. *Ann. Bot.* 81:741–50
108. Huang RF, Lloyd CW. 1999. Gibberellic acid stabilizes microtubules in maize suspension cells to cold and stimulates acetylation of α-tubulin. *FEBS Letts.* 443:317–20
109. Hugdahl JD, Bokros CL, Morejohn LC. 1995. End-to-end annealing of plant microtubules by the p86 subunit of eukaryotic initiation factor–(iso)4F. *Plant Cell* 7:2129–38
110. Hush JM, Overall RL. 1991. Electrical and mechanical fields orient cortical microtubules in higher plant tissues. *Cell Biol. Int. Rep.* 15:551–60
111. Hussey PJ, Gull K. 1985. Multiple isotypes of alpha-tubulin and beta-tubulin in the plant *Phaseolus vulgaris. FEBS Letts.* 181:113–18
112. Hussey PJ, Lloyd CW, Gull K. 1988. Differential and developmental expression of β-tubulins in a higher plant. *J. Biol. Chem.* 263:5474–79
113. Hussey PJ, Snustad DP, Silflow CD. 1991. Tubulin gene expression in higher plants. See Ref. 140a, pp. 15–27

113a. Hyams JS, Lloyd CW, eds. 1994. *Microtubules.* New York: Liss

114. Isenberg G, Niggli V. 1998. Interaction of cytoskeletal proteins with membrane lipids. *Int. Rev. Cytol.* 178:73–125
115. Janssen M, Hunte C, Schulz M, Schnabl H. 1996. Tissue specification and intracellular distribution of actin isoforms in *Vicia faba* L. *Protoplasma* 191:158–63
116. Jernstedt JA. 1984. Root contraction in Hyacinth. I. Effects of IAA on differential cell expansion. *Am. J. Bot.* 71:1080–89
117. Jiang C-L, Sonobe S. 1993. Identification and preliminary characterization of a 65 kDa higher-plant microtubule-associated protein. *J. Cell Sci.* 105:891–901
118. Jones MGK, Payne HL. 1977. Cytokinesis in *Impatiens balsamina* and the effect of caffeine. *Cytobios* 20:79–91
119. Jones MGK, Payne HL. 1977. Early stages of nematode-induced giant cell formation in roots of *Impatiens*

balsamina. J. Nematol. 10:70–84

120. Jongewaard I, Colon A, Fosket DE, 1994. Distribution of transcripts of the *tub B1* β-tubulin gene in developing soybean (*Glycine max* [L.] Merr.) seedling organs. *Protoplasma* 183:77–85
121. Joyce CM, Villemur R, Snustad DP, Silflow CD. 1992. Tubulin gene expression in maize (*Zea mays* L.). Change in isotype expression along the developmental axis of seedling root. *J. Mol. Biol.* 227:97–107
122. Jung G, Wernicke W. 1990. Cell shaping and microtubules in developing mesophyll of wheat (*Triticum aestivum* L.). *Protoplasma* 153:141–48
123. Jürgens G, Mayer U. 1994. Arabidopsis. In *Embryos. A Colour Atlas of Development*, ed. J Bard, pp. 7–21. London: Wolfe
124. Kerr GP, Carter JV. 1990. Relationship between freezing tolerance of root-tip cells and cold stability of microtubules in rye (*Secale cereale* L. cv Puma). *Plant Physiol.* 93:77–82
125. Kerr GP, Carter JV. 1990. Tubulin isotypes in rye roots are altered during cold acclimation. *Plant Physiol.* 93:83–88
126. Kopczak SD, Haas NA, Hussey PJ, Silflow CD, Snustad DP. 1992. The small genome of *Arabidopsis* contains at least six expressed α-tubulin genes. *Plant Cell* 4:539–47
127. Koropp K, Volkmann D. 1994. Monoclonal antibody CRA against a fraction of actin from cress roots recognizes its antigen in different plant species. *Eur. J. Cell Biol.* 64:116–26
128. Kropf DL, Williamson RE, Wasteneys GO. 1997. Microtubule orientation and dynamics in elongating characean internodal cells following cytosolic acidification, induction of pH bands, or premature growth arrest. *Protoplasma* 197:188–98
129. Kubiak JZ, Tarkowska JA. 1987. Evidence for two sets of cytoplasmic microtubules in interphase and prophase cells of onion root. An immunofluorescence study. *Cytologia* 52:781–86
130. Kumagai F, Hasezawa S, Takahashi Y, Nagata T. 1995. The involvement of protein synthesis elongation factor 1α in the organization of microtubules on the perinuclear region during the cell cycle transition from M phase to G_1 phase in tobacco BY-2 cells. *Bot. Acta* 108:467–73
131. Lasselain M-J, Deysson G. 1979. Maturation et cheminement des vésicules de cytodiérèse: étude cytochimique sur les phragmoplastes incomplets. *C. R. Acad. Sci. Paris, Ser. D* 288:879–81
132. Lauber M, Waizenegger I, Steinmann T, Schwarz H, Mayer U, et al. 1997. The *Arabidopsis KNOLLE* protein is a cytokinesis-specific syntaxin. *J. Cell Biol.* 139:1485–93
133. Ledbetter MC, Porter KR. 1963. A "microtubule" in plant cell fine structure. *J. Cell Biol.* 19:239–50
134. Lewis SA, Gu W, Cowan NJ. 1987. Free intermingling of mammalian β-tubulin isotypes among functionally distinct microtubules. *Cell* 49:539–48
135. Liang BM, Dennings AM, Sharp RE, Baskin TI. 1996. Consistent handedness of microtubule helical arrays in maize and *Arabidopsis* primary roots. *Protoplasma* 190:8–15
136. Lin B-L, Jernstedt JA. 1987. Microtubule organization in root cortical cells of *Hyacinthus orientalis*. *Protoplasma* 141:13–23
137. Liu B, Marc J, Joshi HC, Palevitz BA. 1993. A γ-tubulin-related protein associated with the microtubule arrays of higher plants in a cell cycle-dependent manner. *J. Cell Sci.* 104:1217–28
138. Liu B, Palevitz BA. 1992. Organization of cortical microfilaments in dividing root cells. *Cell Motil. Cytoskel.* 23:252–64
139. Liu C-M, Johnson S, Wang TL. 1995. *cyd*, a mutant of pea that alters embryo morphology is defective in cytokinesis. *Dev. Genet.* 16:321–31
140. Lloyd CW. 1987. The plant cytoskeleton:

the impact of fluorescence microscopy. *Annu. Rev. Plant Physiol.* 38:119–39

140a. Lloyd CW, ed. 1991. *The Cytoskeletal Basis of Plant Growth and Form.* London: Academic

141. Lloyd CW, Traas JA. 1988. The role of F-actin in determining the division plane in carrot suspension cells. Drug studies. *Development* 102:211–21

142. Lorenzi G, Perbal G. 1990. Root growth and statocyte polarity in lentil seedling roots grown in microgravity or on a slowly rotating clinostat. *Physiol. Plant.* 78:532–37

143. Lorenzi G, Perbal G. 1990. Actin filaments responsible for the location of the nucleus in the lentil statocyte are sensitive to gravity. *Biol. Cell* 68:259–63

144. Ludueña RF. 1993. Are tubulin isotypes functionally significant? *Mol. Biol. Cell* 4:445–57

145. Ludueña RF. 1998. Multiple forms of tubulin: different gene products and covalent modifications. *Int. Rev. Cytol.* 178:207–75

146. Ludwig SR, Oppenheimer DG, Silflow CD, Snustad DP. 1987. Characterization of the α-tubulin gene family in *Arabidopsis thaliana. Proc. Natl. Acad. Sci. USA* 84:5833–37

147. Lukowitz W, Mayer U, Jürgens G. 1996. Cytokinesis in the *Arabidopsis* embryo involves the syntaxin-related *KNOLLE* gene product. *Cell* 84:61–71

148. McClinton RS, Sung ZR. 1997. Organization of cortical microtubules at the plasma membrane in *Arabidopsis. Planta* 201:252–60

149. McCurdy, DW, Sammut M, Gunning BES. 1988. Immunofluorescent visualization of arrays of transverse cortical actin microfilaments in wheat root-tip cells. *Protoplasma* 147:204–6

150. McDaniel DP, Roberson RW. 1998. γ-tubulin is a component of the Spitzenkörper and centrosomes in hyphal-tip cells of *Allomyces macrogynus. Protoplasma* 203:118–23

151. McElroy D, Rothenberg M, Wu R. 1990. Structural characterization of a rice actin gene. *Plant Mol. Biol.* 14:163–77

152. McLean BG, Eubanks S, Meagher RB 1990. Tissue-specific expression of divergent actins in soybean root. *Plant Cell* 2:335–44

153. McNally FJ. 1996. Modulation of microtubule dynamics during the cell cycle *Curr. Opin. Cell Biol.* 8:23–29

154. Maniotis AJ, Chen CS, Ingber DE. 1997 Demonstration of mechanical connections between integrins, cytoskeletal filaments, and nucleoplasm that stabilize nuclear structure. *Proc. Natl. Acad. Sci. USA* 94:849–54

155. Marc J. 1997. Microtubule-organizing centres in plants. *Trends Plant Sci.* 2:223–30

156. Marc J, Sharkey DE, Durso NA, Zhang M, Cyr RJ. 1996. Isolation of a 90-kD microtubule-associated protein from tobacco membranes. *Plant Cell* 8:2127–38

157. Mayer U, Herzog U, Berger F, Inzé D, Jürgens G. 1999. Mutations in the *PILZ* group genes disrupt the microtubule cytoskeleton and uncouple cell cycle progression from cell division in *Arabidopsis* embryo and endosperm. *Eur. J. Cell Biol.* 78:100–8

158. Meagher RB. 1991. Divergence and differential expression of actin gene families in higher plants. *Int. Rev. Cytol.* 125:139–163

159. Meagher RB, McKinney EC, Kandasamy MK. 1999. Isovariant dynamics expand and buffer the responses of complex systems: the diverse plant actin gene family. *Plant Cell* 11:995–1005

160. Mendu N, Silflow CD. 1993. Elevated levels of tubulin transcripts accompany the GA3-induced elongation of oat internode segments. *Plant Cell Physiol.* 34:973–83

161. Menzel D, Elsner-Menzel C. 1990. The microtubule cytoskeleton in developing

cysts of the green alga *Acetabularia:* involvement in cell wall differentiation. *Protoplasma* 157:52–63
162. Mews M, Sek FJ, Moore R, Volkmann D, Gunning BES, John PCL. 1997. Mitotic cyclin distribution during maize cell division: implications for the sequence diversity and function of cyclins in plants. *Protoplasma* 200:128–45
163. Mineyuki Y. 1999. The preprophase band of microtubules: its function as a cytokinetic apparatus in higher plants. *Int. Rev. Cytol.* 187:1–49
164. Mineyuki Y, Aioi H, Yamashita M, Nagahama Y. 1996. A comparative study of stainability of preprophase bands by the PSTAIR antibody. *J. Plant Res.* 109:185–92
165. Mineyuki Y, Gunning BES. 1990. A role for preprophase bands of microtubules in maturation of new cell walls, and a general proposal on the function of preprophase band sites in cell division in higher plants. *J. Cell Sci.* 97:527–37
166. Mineyuki Y, Iida H, Anraku Y. 1994. Loss of microtubules in the interphase cells of onion (*Allium cepa* L.) root tips from the cell cortex and their appearance in the cytoplasm after treatement with cycloheximide. *Plant Physiol.* 104:281–84
167. Mineyuki Y, Palevitz BA. 1990. Relationship between preprophase band organization, F-actin and the division site in *Allium.* Fluorescence and morphometric studies on cytochalasin-treated cells. *J. Cell Sci.* 97:283–95
168. Montoliu Ll, Puigdomènech P, Rigau J. 1990. The *Tubα3* gene from *Zea mays*: structure and expression in dividing plant tissues. *Gene* 94:201–7
169. Montoliu Ll, Rigau J, Puigdomènech P. 1989. A tandem of α-tubulin genes preferentially expressed in radicular tissue from *Zea mays. Plant Mol. Biol.* 14:1–15
170. Moore JD, Endow SA. 1996. Kinesin proteins: a phylum of motors for microtubule-based motility. *BioEssays* 18:207–19
171. Morejohn LC. 1991. The molecular pharmacology of plant tubulin and microtubules. See Ref. 140a, pp. 29–43
172. Mueller SC, Brown RM Jr. 1980. Evidence for an intramembrane component associated with a cellulose microfibril-synthesizing complex in higher plants. *J. Cell Biol.* 84:315–26
173. Murata T, Kadota A, Hogetsu T, Wada M. 1987. Circular arrangement of cortical microtubules around the subapical part of a tip-growing fern protonema. *Protoplasma* 141:135–38
174. Newcomb EH. 1969. Plant microtubules. *Annu. Rev. Plant Physiol.* 20:253–88
175. Nick P. 1998. Signalling to the microtubular cytoskeleton in plants. *Int. Rev. Cytol.* 184:33–80
176. Nickle TC, Meinke DW. 1998. A cytokinesis-defective mutant of *Arabidopsis (cyt1)* characterized by embryonic lethality, incomplete cell walls, and excessive callose accumulation. *Plant J.* 15:321–32
177. Niini S, Raudaskoski M. 1998. Growth patterns in non-mycorrhizal and mycorrhizal short shoots of *Pinus sylvestris. Symbiosis* 25:101–14
178. Niini SS, Tarkka MT, Raudaskoski M. 1996. Tubulin and actin protein patterns in Scots pine (*Pinus sylvestris*) roots and developing ectomycorrhiza with *Suillus bovinus. Physiol. Plant.* 96:186–92
179. Nogami A, Suzaki T, Shigenaka Y, Nagahama Y, Mineyuki Y. 1996. Effects of cycloheximide on preprophase bands and prophase spindles in onion (*Allium cepa* L.) root tip cells. *Protoplasma* 192:109–21
180. Oveka M, iamporová M, Hauser M-T. 1996. Root specific mutation of *Arabidopsis thaliana* affecting cytokinesis: production and fate of multinucleate cells. *J. Comput.-Assist. Microsc.* 8:277–78

181. Packard MJ, Stack SM. 1976. The preprophase band: possible involvement in the formation of the cell wall. *J. Cell Sci.* 22:403–11

182. Palevitz BA. 1987. Actin in the preprophase band of *Allium cepa*. *J. Cell Biol.* 104:1515–19

183. Panteris E, Apostolakos P, Galatis B. 1992. The organization of F-actin in root tip cells of *Adiantum capillus veneris* throughout the cell cycle. A double label fluorescence microscopy study. *Protoplasma* 170:128–37

184. Parthasarathy MV, Pesacreta TC. 1980. Microfilaments in plant vascular cells. *Can. J. Bot.* 58:807–15

185. Perbal G. 1978. The mechanism of geoperception in lentil roots. *J. Exp. Bot.* 29:631–38.

186. Pérez HE, Sánchez N, Vidali L, Hernández JM, Lara M, Sánchez F. 1994. Actin isoforms in non-infected roots and symbiotic root nodules of *Phaseolus vulgaris* L. *Planta* 193:51–56

187. Pesacreta TC, Carley WW, Webb WW, Parthasarathy MV. 1982. F-actin in conifer roots. *Proc. Natl. Acad. Sci. USA* 79:2898–901

188. Pihakaski-Maunsbach K, Puhakainen T. 1995. Effect of cold exposure on cortical microtubules of rye (*Secale cereale*) as observed by immunocytochemistry. *Physiol. Plant.* 93:563–71

189. Pope DG, Thorpe JR, Al-Azzawi MJ, Hall JL. 1979. The effect of cytochalasin B on the rate of growth and ultrastructure in wheat coleoptiles and maize roots. *Planta* 144:373–83

190. Radford JE, White RG. 1998. Localization of a myosin-like protein to plasmodesmata. *Plant J.* 14:743–50

191. Reichelt S, Knight AE, Hodge TP, Baluška F, Šamaj J, et al. 1999. Characterization of the unconventional myosin VIII in plant cells and its localisation at the post-cytokinetic wall. *Plant J.* 19:555–69

192. Rigau J, Capellades M, Montoliu Ll, Torres MA, Romera C, et al. 1993. An analysis of a maize α-tubulin gene promoter by transient expression and in transgenic tobacco plants. *Plant J.* 4:1043–50

193. Röper W, Röper S. 1977. Centripetal wall formation in roots of *Vicia faba* after caffeine treatment. *Protoplasma* 93:89–100

194. Rost TL, Baum S. 1988. On the correlation of primary root length, meristem size and protoxylem tracheary element position in pea seedlings. *Am. J. Bot.* 75:414–24

195. Schellenbaum P, Vantard M, Lambert A-M. 1992. Higher plant microtubule-associated proteins (MAPs): a survey. *Biol. Cell* 76:359–64

196. Schilde-Rentschler L. 1977. Role of the cell wall in the ability of tobacco protoplasts to form callus. *Planta* 135:177–81

197. Schmit A-C, Lambert A-M. 1990. Microinjected fluorescent phalloidin in vivo reveals the F-actin dynamics and assembly in higher plant mitotic cells. *Plant Cell* 2:129–38

198. Seagull RW. 1992. A quantitative electron microscopic study of changes in microtubule arrays and wall microfibril orientation during in vitro cotton fiber development. *J. Cell Sci.* 101:561–77

199. Sedbrook JC, Chen R, Masson PH. 1999. *ARG1* (Altered Response to Gravity) encodes a DnaJ-like protein that potentially interacts with the cytoskeleton. *Proc. Natl. Acad. Sci. USA* 96:1140–45

200. Shibaoka H. 1994. Plant hormone-induced changes in the orientation of cortical microtubules: alterations in the cross-linking between microtubules and the plasma membrane. *Annu. Rev. Plant Physiol. Plant Mol. Biol.* 45:527–44

201. Sievers A, Kruse S, Kuo-Huang L.-L., Wendt M. 1989. Statoliths and microfilaments in plants. *Planta* 179:275–78

202. Simmonds DH. 1992. Plant cell wall removal. Cause for microtubular instability and division abnormalities in protoplast cultures. *Physiol. Plant.* 85:387–90

203. Simmons C, Söll D, Migliaccio F. 1995. Circumnutation and gravitropism cause root waving in *Arabidopsis thaliana. J. Exp. Bot.* 46:143–50

204. Smertenko A, Blume Y, Viklický V, Opatrný Z, Dráber P. 1997. Post-translational modifications and multiple tubulin isoforms in *Nicotiana tabacum* L. *cells. Planta* 201:349–58

205. Smertenko AP, Lawrence SL, Hussey PJ. 1998. Immunological homologues of the *Arabidopsis thaliana* β1 tubulin are polyglutamylated in *Nicotiana tabacum. Protoplasma* 203:138–43

206. Smith-Huerta NL, Jernstedt JA. 1989. Root contraction in hyacinth III. Orientation of cortical microtubules visualized by immunofluorescence microscopy. *Protoplasma* 151:1–10

207. Snustad DP, Haas NA, Kopczak SD, Silflow CD. 1992. The small genome of *Arabidopsis* contains at least nine expressed β-tubulin genes. *Plant Cell* 4:549–56

208. Stoppin V, Lambert A-M, Vantard M. 1996. Plant microtubule-associated proteins (MAPs) affect microtubule nucleation and growth at plant nuclei and mammalian centrosomes. *Eur. J. Cell Sci.* 69:11–23

208a. Tabony J. 1996. Self-organization in a simple biological system through chemically dissipative processes. *Nanobiology* 4:117–37

209. Thimann KV, Reese K, Nachmias VT. 1992. Actin and the elongation of plant cells. *Protoplasma* 171:153–61

210. Toko K, Iiyama S, Tanaka C, Hayashi K, Yamafuji K, Yamafuji K. 1987. Relation of growth process to spatial patterns of electric potential and enzyme activity in bean roots. *Biophys. Chem.* 27:39–58

211. Traas J, Bellini C, Nacry P, Kronenberger J, Bouchez D, Caboche M. 1995. Normal differentiation patterns in plants lacking microtubular preprophase bands. *Nature* 375:676–77

212. Traas JA, Braat P, Derksen JW. 1984. Changes in microtubule arrays during the differentiation of cortical root cells of *Raphanus sativus. Eur. J. Cell Biol.* 34:229–38

213. Traas JA, Renaudin JP, Teyssendier de la Serve B. 1990. Changes in microtubular organization mark the transition to organized growth during organogenesis in *Petunia* hybrid. *Plant Sci.* 68:249–56

214. Uribe X, Torres MA, Capellades M, Puigdomènech P, Rigau J. 1998. Maize α-tubulin genes are expressed according to specific patterns of cell differentiation. *Plant Mol. Biol.* 37:1069–78

215. Utrilla L, De la Torre C. 1991. Loss of microtubular orientation and impaired development of prophase bands upon inhibition of RNA synthesis in root meristems. *Plant Cell Rep.* 9:492–95

216. Utrilla L, Giménez-Abián MI, De la Torre C. 1993. Timing the phases of microtubule cycles involved in cytoplasmic and nuclear division in cells of undisturbed onion root meristems. *Biol. Cell* 78:235–41

217. Vantard M, Schellenbaum P, Fellous A, Lambert A-M. 1991. Characterization of maize microtubule-associated proteins, one of which is immunologically related to tau. *Biochemistry* 38:9334–46

218. Vaughn KC, Harper JDI. 1998. Microtubule-organizing centers and nucleating sites in land plants. *Int. Rev. Cytol.* 181:75–149

219. Vaughn KC, Vaughan MA. 1990. Structural and biochemical characterization of dinitroaniline-resistant *Eleusine.* In *Managing Resistance to Agrochemicals. From Fundamental Research to Practical Strategies*, ed. MB Gree, HM LeBaron, WK Moberg, Am. Chem. Soc. Symp. Ser. 421:364–75. Washington, DC: Am. Chem. Soc.

220. Vaughn KC, Vaughan MA. 1991. Dinitroaniline resistance in *Eleusine indica* may be due to hyperstabilized

microtubules. In *Herbicide Resistance in Weeds and Crops*, ed. JC Caseley, GW Cussans, RK Atkin, pp. 177–86. Oxford: Butterworth/Heinemann

221. Villemur R, Joyce CM, Haas NA, Goddard RH, Kopczak SD, et al. 1992. α-Tubulin gene family of maize (*Zea mays* L.). Evidence for two ancient α-tubulin genes in plants. *J. Mol. Biol.* 227:81–96
222. Volkmann D, Baluška F. 1999. The actin cytoskeleton in plants: from transport networks to signaling networks. *Microsc. Res. Tech.* 47:135–54
223. Wada M, Nozue K, Kadota A. 1998. Cytoskeletal pattern change during branch formation in a centrifuged *Adiantum* protonema. *J. Plant Res.* 111:53–58
224. Waldin TR, Ellis JR, Hussey PJ. 1992. Tubulin-isotype analysis of two grass species resistant to dinitroaniline herbicides. *Planta* 188:258–64
225. Webb MC, Gunning BES. 1991. The microtubular cytoskeleton during development of the zygote, proembryo and free nuclear endosperm in *Arabidopsis thaliana* (L.) Heynh. *Planta* 184:187–95
226. Wehland J, Schroeder M, Weber K. 1984. Organization of microtubules in stabilized meristematic plant cells revealed by a rat monclonal antibody reacting only with the tyrosinated form of α-tubulin. *Cell Biol. Int. Rep.* 8:147–50
227. White RG, Badelt K, Overall RL, Vesk M. 1994. Actin associated with plasmodesmata. *Protoplasma* 180:169–84
228. White RG, Sack FD. 1990. Actin microfilaments in presumptive statocytes of root caps and coleoptiles. *Am. J. Bot.* 77:17–26
229. Whitehead LF, Day DA, Hardham AR. 1998. Cytoskeletal arrays in the cells of soybean root nodules: the role of actin microfilaments in the organization of symbiosomes. *Protoplasma* 203:194–205
230. Wick SM. 1985. Immunofluorescence microscopy of tubulin and microtubule arrays in plant cells. III. Transitions between mitotic/cytokinetic and interphase microtubule arrays. *Cell Biol. Int. Rep.* 9:357–71
231. Wick SM, Seagull RM, Osborn M, Weber K, Gunning BES. 1981. Immunofluorescence microscopy of organized microtubule arrays in structurally-stabilized meristematic plant cells. *J. Cell Biol.* 89:685–90
232. Wilson L, Jordan MA. 1994. Pharmacological probes of microtubule function. See Ref. 113a, pp. 59–83
233. Wordeman L, Mitchison TJ. 1994. Dynamics of microtubule assembly in vivo. See Ref. 113a, pp. 287–301

233a. Yamamoto E, Zeng L, Baird WV. 1998. α-Tubulin missense mutations correlate with antimicrotubule drug resistance in *Eleusine indica. Plant Cell* 10:297–308

234. Yang M, Nadeau JA, Zhao L, Sack FD. 1999. Characterization of a *cytokinesis defective (cyd1)* mutant of *Arabidopsis. J. Exp. Bot.* 50:1437–46
235. Ye XL, Zee SY, Yeung EC. 1997. Suspensor development in the nun orchid, *Phaius tankervilliae. Int. J. Plant Sci.* 158:704–12
236. You W, Abe S, Davies E. 1992. Cosedimentation of pea root polysomes with the cytoskeleton. *Cell Biol. Int. Rep.* 16:663–73
237. Zandomeni K, Schopfer P. 1994. Mechanosensory microtubule reorientation in the epidermis of maize coleoptiles subjected to bending stress. *Protoplasma* 182:96–101
238. Zheng Y, Jung MK, Oakley BR. 1991. γ-tubulin is present in *Drosophila melanogaster* and *Homo sapiens* and is associated with the centrosome. *Cell* 65:817–23

Annu. Rev. Plant Physiol. Plant Mol. Biol. 2000. 51:323–47

THE GREAT ESCAPE: Phloem Transport and Unloading of Macromolecules[1]

Karl J. Oparka and Simon Santa Cruz
Unit of Cell Biology, Scottish Crop Research Institute, Invergowrie, Dundee DD2 5DA, United Kingdom; e-mail: kopark@scri.sari.ac.uk

Key Words companion cells, macromolecules, plasmodesmata, phloem transport, sieve elements

■ **Abstract** The phloem of higher plants translocates a diverse range of macromolecules including proteins, RNAs, and pathogens. This review considers the origin and destination of such macromolecules. A survey of the literature reveals that the majority of phloem-mobile macromolecules are synthesized within companion cells and enter the sieve elements through the branched plasmodesmata that connect these cells. Examples of systemic macromolecules that originate outside the companion cell are rare and are restricted to viral and subviral pathogens and putative RNA gene-silencing signals, all of which involve a relay system in which the macromolecule is amplified in each successive cell along the pathway to companion cells. Evidence is presented that xenobiotic macromolecules may enter the sieve element by a default pathway as they do not possess the necessary signals for retention in the sieve element–companion cell complex. Several sink tissues possess plasmodesmata with a high-molecular-size exclusion limit, potentially allowing the nonspecific escape of a wide range of small (<50-kDa) macromolecules from the phloem. Larger macromolecules and systemic mRNAs appear to require facilitated transport through sink plasmodesmata. The fate of phloem-mobile macromolecules is considered in relation to current models of long-distance signaling in plants.

CONTENTS

[1]Dedicated to Don Fisher on his retirement.

1040-2519/00/0601-0323$14.00

INTRODUCTION

The phloem of higher plants forms an extensive conduit for the long-distance transport of a diverse range of compounds. The elongated conducting elements of the phloem, the sieve elements (SEs), are joined by perforated end walls known as sieve plates to form sieve tubes, a functional continuum of cells that is closely connected with adjoining, nucleate companion cells (CCs) and a range of associated parenchyma elements (2, 3, 58, 60, 80, 82). During differentiation, SEs lose many of their organelles and at maturity, function as enucleate pipes through which solutes are transported from source (net carbon exporting) to sink (net carbon importing) regions of the plant. During maturation of SEs, partial autolysis of the protoplast results in a reduced component of organelles comprising the plasma membrane (PM), smooth endoplasmic reticulum (ER), mitochondria, and starch-storing plastids (2, 3, 58, 60, 80, 82). This thin layer of organelles, appressed against the SE wall, is referred to as the parietal layer and appears to provide the only means of membrane continuity between the SE and CC (63, 82). Because of the intimate structural and functional connections between SE and CC, the SE-CC complex is frequently viewed as a single functional entity within the phloem (63, 82, 92–94). Although considerable research has been devoted to unraveling the mechanism(s) of long-distance solute movement through the phloem, the pressure-flow hypothesis of Münch (55) is now commonly accepted (63, 82, 92, 93). According to this model, solute movement through the sieve tubes occurs by mass flow and is driven by a pressure gradient between source and sink regions of the plant (55). In the 1980s, attention turned to the mechanisms by which solutes, particularly sucrose, were loaded and unloaded from the phloem (for review, see 80). This sharp focus on sucrose movement to a large extent detracted from the fact that the phloem carries a wide range of macromolecules, in addition to solutes, throughout the plant. These include proteins, systemic wound signals and pathogens (6, 10, 14, 27, 50, 58, 75, 80, 86, 103). Studies of phloem exudate collected from aphid stylets, highlighted the enormous diversity of materials moving in the translocation stream (23, 25, 76, 78, 103), but unfortunately offered little insight as to their role.

Recent research has depicted the phloem as a dynamic transport system capable of carrying several macromolecular signals, including RNA information molecules, from source to sink (14, 27, 41, 53, 70, 75, 78, 79, 87). For example, during systemic posttranscriptional gene silencing, small RNA signal molecules are believed to be transported from their sites of generation in source tissues to induce gene silencing in distant sink tissues (31, 41, 64, 95, 96). Research on phloem exudates has also demonstrated the presence of numerous proteins (25, 28, 29, 32, 34, 76, 79, 103) and endogenous mRNAs (70, 78, 101) in sieve tube exudate. Together with a large body of evidence that viral genomes, either as nucleoprotein complexes or intact virions, may be translocated in the phloem (4, 10, 27, 50, 58, 65, 69, 72, 75, 77, 81, 91), the picture that emerges is one in which the transport phloem functions as an information superhighway (41, 70), translocating a wide range of macromolecular traffic throughout the plant. What is the role of the many diverse macromolecules in the phloem and what is their final destination? The following discussion focuses attention on how macromolecules enter and exit the translocation stream and examines their potential fate within sink tissues. In particular, we examine critically the evidence for macromolecular signaling through the phloem and address the problems inherent in unloading macromolecules from the terminal phloem of sink organs. Space constraints restrict us only to case histories in which specific studies have begun to shed light on the origin and destination of phloem-mobile macromolecules.

ENTERING THE SE-CC COMPLEX

Before macromolecules gain access to the plant's long-distance transport pathway, they must enter the SE-CC complexes in source tissues. In exporting leaves this is assumed to occur in the phloem of minor veins, where the bulk of solute loading is thought to occur (46, 69, 80, 89, 92) (Figure 1*a*). However, larger vein classes could also potentially be entry points for systemic macromolecules, and there is no a priori case as to why macromolecules should enter the phloem at the same sites as solutes (58). Minor veins of dicotyledons usually have a characteristic architecture, comprising two or three mature sieve elements, associated companion cells, and phloem parenchyma elements (4, 19, 26, 46, 58, 69, 75, 76, 92) (Figure 1*a*). Prior to the sink-source transition these minor vein complexes are immature (69, 89) and symplastically coupled to the mesophyll. Some macromolecules may enter the SE-CC at this early stage of differentiation and subsequently be translocated out of the leaf when the SE-CC matures. Early invasion of the SE-CC complex can occur with systemic viruses, which may enter the immature phloem and replicate prior to maturation of the SE (58, 69) (Figure 1*b*). At this early stage of SE-CC differentiation, macromolecules could enter SEs either from the CC or directly from phloem parenchyma elements, as at this stage all cell types are symplastically coupled to the mesophyll (58, 69, 92).

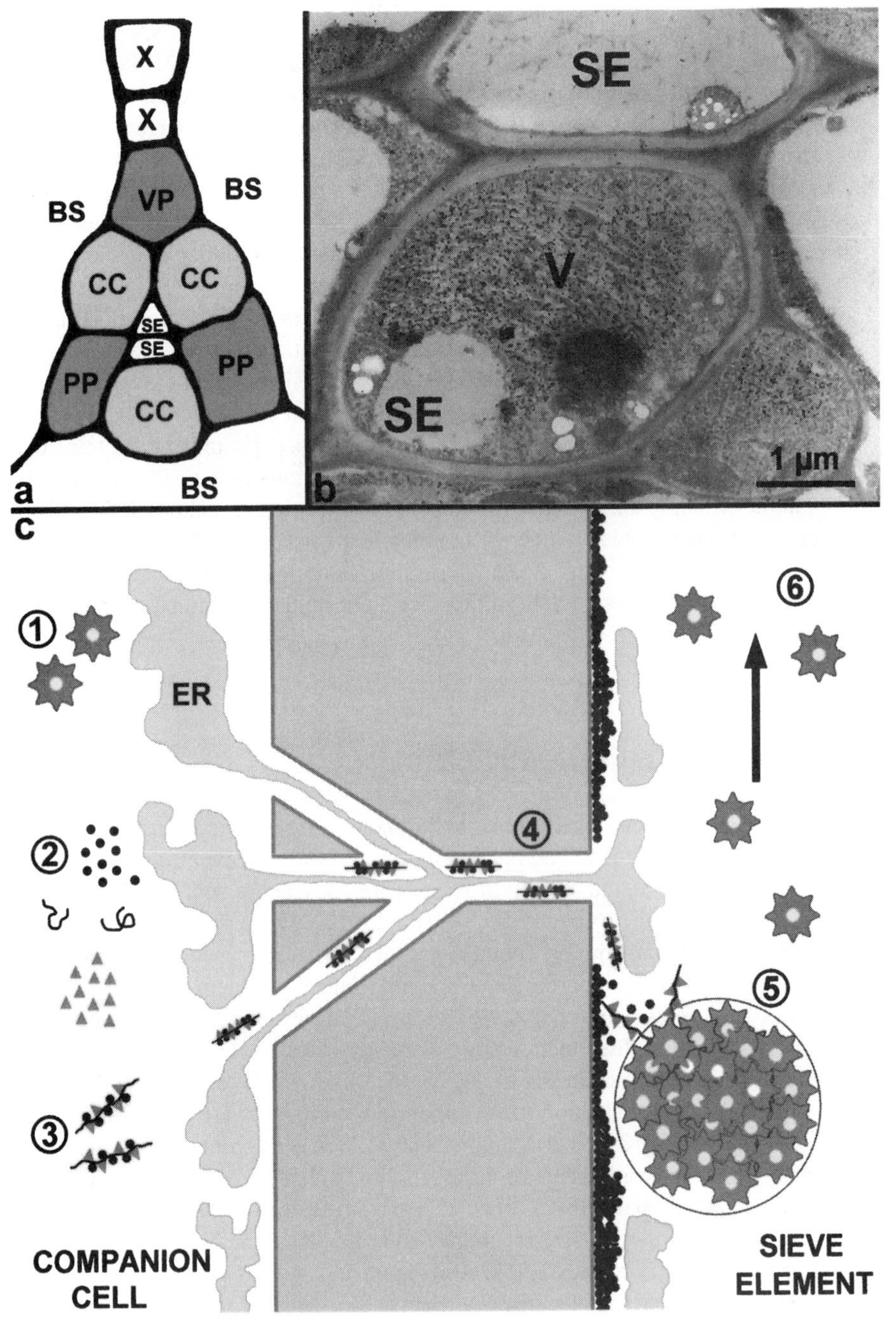
X
X
VP
BS
BS
CC
CC
SE
SE
PP
PP
CC
a
BS
SE
V
SE
1 µm
b
c
①
ER
②
③
④
⑤
⑥
COMPANION
CELL
SIEVE
ELEMENT

SYMPLASTIC VERSUS APOPLASTIC LOADING

The mechanism of phloem loading appears to differ among plant species (26, 58, 80, 92), suggesting that the pathways of macromolecular trafficking into the SE might also vary. To date, two main pathways of phloem loading have been identified, apoplastic and symplastic (80, 92). In putative apoplastic loading species, sucrose is loaded into the SE-CC actively by transmembrane carriers (80, 92), and the numbers of plasmodesmata around the SE-CC complex become progressively reduced during development (92). However, a few plasmodesmata remain at this interface (58, 69, 92), and these become potential routes through which systemic macromolecules may enter the SE-CC complex. During maturation of the SE-CC complex, the plasmodesmata that connect the SE and CC become branched on the CC-side of the wall and open up into the sieve element via a single pore (2–4, 45, 58, 80, 82, 92–94) (Figure 1*c*). Callose is associated with the entrances of these plasmodesmata (Figure 2*b*, see color plate), and they have been termed pore-plasmodesma units (PPUs) to reflect their unique architecture (94). In putative symplastic loading species, solutes enter the SE via specialized CCs known as an intermediary cells (ICs; 80, 89, 90, 92). In symplastically loading Cucurbitaceae, branched plasmodesmata increase in numbers between the ICs and bundle sheath cells prior to the sink-source transition (97). However, the PPUs that form between SE and CC are similar in architecture regardless of whether the mode of loading is apoplastic or symplastic (94). Intuitively, one might expect that apoplastic loaders would be less susceptible to invasion by macromolecular pathogens such as viruses. However, both apoplastic and symplastic loaders can be infected by a plethora of host-specific viruses (10, 19, 58, 75), suggesting that the absolute number of plasmodesmata around the SE-CC complex is not the sole determinant of virus invasion. Macromolecules that enter the translocation stream may originate within or outside the SE-CC complex (Figure 3, see color plate). In the former,

←

Figure 1 Sieve element-companion cell complexes. (*a*) Minor vein architecture of tobacco, comprising two sieve elements (SE) and three companion cells (CC). Two phloem parenchyma elements (PP) abut the lower SE-CC complex. BS, bundle sheath; VP, vascular parenchyma; X, xylem. (From Reference 69, courtesy of *The Plant Cell*.) (*b*) Early viral invasion of an immature minor-vein SE. Immunogold labeling of potato virus X reveals abundant virus particles (V) in the cytoplasm of the lower SE. (From Reference 69, courtesy of *The Plant Cell*.) (*c*) Putative trafficking mechanism of the genome of cucumber mosaic virus (CMV) between CC and SE. **1.** Intact virus particles are formed in the CC cytoplasm. **2.** The virus disassembles into RNA (strands) and coat protein subunits (circles). **3.** A linear transport complex comprising the viral RNA, coat protein and movement protein (MP, triangles) is formed in the CC. **4.** The ribonucleoprotein complex is trafficked by the MP through the pore-plasmodesma units. **5.** Virus assembly occurs in the SE parietal layer within membrane bound assembly sites, while the MP is deposited close to the pore entrance. **6.** Assembled virions are subsequently released into the translocation stream. (From Reference 4, courtesy of *The Plant Cell*.)

they are synthesized specifically within companion cells (or ICs) and enter the SE via the PPUs that connect the two cell types, whereas in the latter they are synthesized in cells outside the SE-CC complex (e.g. in mesophyll cells) and enter the SE-CC complex by the plasmodesmata that connect the SE-CC complexes with the mesophyll. For a given systemic macromolecule, it is important to establish which of the above routes is taken. As discussed below, the site of origin of macromolecules will to a large extent determine their ability to enter the translocation stream.

SYNTHESIS OF MACROMOLECULES WITHIN THE CC

Amino acids may enter the CC symplastically; alternatively, they may be transported into the CC by transmembrane carriers (Figure 3). In the case of proteins, synthesized on the ER of the CC, at least three fates are possible: (*a*) The protein may be retained within the CC cytoplasm or targeted to specific CC organelles, (*b*) it may be targeted to the SE parietal layer, or (*c*) it may be exported in the translocation stream (Figure 3). Over the past decade, a substantial body of evidence has accumulated to suggest that macromolecules synthesized within CCs may be transported into enucleate SEs. The ability of a given protein to enter the SE from the CC depends on a number of features unique to the PPUs. One of these is the unusually high size exclusion limit (SEL) of these plasmodesmata, which have the ability to traffic macromolecules that are normally too large to pass between source-leaf mesophyll cells (25, 32, 43, 58). The SEL of PPUs has been probed in a number of ways. First, microinjected fluorescent dextrans of at least 10 kDa have been shown to move between SE and CC (43). Second, labeling of wheat leaves with radioactive methionine established that wheat phloem sap, collected as exudate from aphid stylets, contained more than 200 proteins within the size range 10–40 kDa, as well as a number of larger proteins (60–79 kDa; 25). A similar range of proteins has been detected in sieve-tube exudate of rice (34) and castor bean (77). An important feature noted in the wheat study was a constant exchange of proteins between SEs and CCs along the translocation pathway (43). A third demonstration of macromolecular exchange between CC and SE was provided recently by Imlau et al (32). These authors used the CC-specific promoter of the *Arabidopsis* sucrose transporter, SUC2, to drive the production of the jellyfish green fluorescent protein (GFP) within source CCs. The 27-kDa GFP moved into the SE and was translocated to sink regions of the plant (Figure 4*c*, see color plate). In the case of xenobiotic macromolecules such as microinjected dextrans and GFP, one could argue that such macromolecules are not normal constituents of the CC and may not reflect the normal trafficking behavior of proteins between SE and CC. However, recent evidence suggests that several SE-specific proteins are also trafficked from the CC to the SE. The following section catalogues some examples of this phenomenon.

SUCROSE CARRIER PROTEINS

Given the role of the SE-CC complex in apoplastic phloem loading, it is perhaps not surprising to find that proteins responsible for carrier-mediated sucrose transport are trafficked between CC and SE. If such proteins are destined for the SE parietal layer, targeting signals are envisaged that direct the protein to the correct location within the SE (63). Although the sucrose transport protein, SUC2, of *Arabidopsis* is localized exclusively in CCs (32, 84), the SUT1 protein of tobacco is synthesized in CCs and subsequently moves into SEs (46, 45). It is unclear whether both the SUT1 protein and its mRNA move into SEs. In situ hybridization showed that SUT1 mRNA was localized to both SEs and CCs and was associated strongly with the orifices of PPUs, leading to the suggestion that the protein may be translated in the SEs (45). However, this conflicts with the dogma that enucleate SEs contain only parietal sheets of smooth ER (2, 3). Polyribosomes have been detected in enucleate protophloem SEs of rice (59), and ribosomes have also been detected on the outermost (lumen-facing) cisternae of the stacked ER in the parietal layer in recently matured SEs (2, 3). However, ribosomes are not thought to persist in mature, translocating SEs (2, 3, 25, 80). The correct insertion of SUT1 to the SE plasma membrane clearly requires a targeting mechanism that prevents loss of this protein to the translocation stream. This could occur by transport along the plasma membrane or ER, both of which are continuous through the PPUs (53, 63, 80, 82). Whatever the means by which SUT1 reaches SEs, its function in sucrose retrieval requires that the protein is retained at the SE plasma membrane and that it is not lost to the translocation stream. Oparka & Turgeon (63) have suggested that some proteins destined for the SE parietal layer may be transported beneath the blanket of smooth ER that lines the SE plasma membrane. This ER is separated from the PM by an electron-dense material, as are the stacked sheets of ER cisternae (2, 3, 63, 82). Conceivably, these canals may provide a transport pathway for proteins along the parietal layer of the SE.

P-PROTEINS

The SEs of most plant species contain phloem-specific proteins (P-proteins) that occur in a variety of forms including filaments, tubules, and crystalline aggregates (2, 3, 15, 44, 74, 80). The biochemistry of P-proteins is best characterized in the Cucurbitaceae, where phloem exudate is easily collected from several species. *Cucurbita maxima* exudate contains two prominent polypetides: phloem protein 1 (PP1; 96 kDa), which forms the P-protein filaments seen in electron micrographs (2, 3, 15, 74, 82); and PP2 (a 48-kDa dimer), which functions as a lectin (74, 83). PP1 and PP2 transcription and translation occur in CCs (5, 11, 28, 29) (Figure 3), and therefore the subunits of these proteins are probably transported through the PPUs prior to final protein assembly in the SE parietal layer. It has been well established that one of the roles of P-proteins is to seal the sieve-plate pores upon wounding (2, 3, 15, 44, 74, 80). Traditionally, it has been thought that in intact, translocating SEs, P-proteins are anchored to the parietal layer (2, 3, 44, 74), with the phloem lectin protein perhaps attaching the P-protein filaments to the SE reticulum or plasma membrane (83). However, recent evidence suggests that both PP1 and PP2 may be translocated across both interspecific and intergeneric grafts of the Cucurbitaceae (28, 29, 70, 87). Aside from their function in sealing damaged SEs, P-proteins might also function as part of a long-distance signaling system (87), although such a role has yet to be confirmed. Immunolocalization studies of PP1 and PP2 revealed the exclusive localization of these proteins in the SE-CC complexes and showed that the proteins, but not their respective transcripts, were detectable in the scions of interspecific grafts (28, 29). The presence of PP1, but not PP2, in scion CCs demonstrated that the SEL of the PPUs was sufficiently large to permit the passage of PP2 dimers (or monomers) between SE and CC while limiting the larger 96-kDa filament protein to the SE (29) (Figure 3). A significant finding was that neither PP1 nor PP2 was detected outside SE-CC complexes by either light or electron microscopy. Thus, while some macromolecules, such as viruses, can exit the phloem (see below), P-proteins do not exit the SE-CC complex in sink tissues.

Curiously, both PP1 and PP2 increase the SEL of mesophyll cells in *C. maxima* cotyledons. FITC-labeled PP2 moved extensively while unlabeled PP2, co-injected with dextrans, facilitated the cell-cell movement of 20-kDa but not 40-kDa dextrans (1). Other phloem exudate proteins, such as RPP-13-1, a thioredoxin h protein, also modify the SEL of mesophyll plasmodesmata (33). Given that PP1 and PP2 are synthesized exlusively in CCs and do not leave the SE-CC complex in sink tissues, the biological significance of these observations is obscure. One possibility is that these proteins possess motifs that normally modify only PPUs and that their behavior in mesophyll cells mimics this role (53). It appears, however, that P-proteins are incapable of modifying the plasmodesmata between the SE-CC complex and surrounding bundle sheath cells to mediate their escape to the mesophyll.

CmPP16

Recently, Lucas and co-workers (102) isolated a 16-kDa phloem protein from *C. maxima* (CmPP16) that appears to share several features with viral movement proteins. CmPP16 was isolated using antibodies to the MP of red clover necrotic mosaic virus (RCNMV) and has been shown to bind to and mediate the cell-cell trafficking of nonsequence-specific RNA between mesophyll cells (102). Like other phloem proteins, CmPP16 has been immunolocalized to the periphery of sieve elements (102). Significantly, high-resolution in situ PCR detected CmPP16 mRNA in CCs and also in mature SEs in petioles and stems, indicating that the mRNA as well as the protein may be translocated in the phloem (Figure 3). Movement of CmPP16 protein and its mRNA across interspecific grafts confirmed this mobility (102). Although injection experiments demonstrate that CmPP16 can modify mesophyll plasmodesmata, the extent to which CmPP16 is found naturally in the mesophyll has not been determined, nor has its ability to traffic mRNA molecules into and out of the phloem. Thus, the suggestion that CmPP16 is a plant paralog of a viral MP (102) requires unequivocal demonstration.

SYNTHESIS OF MACROMOLECULES OUTSIDE THE CC

Reports of proteins synthesized outside the SE-CC complex that are capable of subsequent entry into the translocation stream are scant in comparison with those that document synthesis within the CC. To date, no direct evidence has been produced to suggest that endogenous plant proteins traffic from mesophyll into the SE-CC complex. If macromolecules are synthesized in tissues outside the phloem, the problem faced is one of entering the SE-CC complex to gain phloem mobility. Such putative trafficking molecules require not only the capacity to move from cell-to cell through mesophyll tissues but also the ability to breach the plasmodesmata around the SE-CC complex (63).

VIRUSES

Plant viruses have been studied extensively with respect to their ability to move from cell to cell and to enter the phloem. Most viruses encode movement proteins (MPs) that facilitate cell-cell movement of the viral genome, either as virions or some form of nucleoprotein complex, through plasmodesmata (10, 17, 27, 48, 58, 75). A number of viral MPs bind nucleic acid (98), target plasmodesmata (4, 35, 61, 71, 76, 88) (Figure 2*c*), and gate the plasmodesmal pore to a higher than normal exclusion limit (35, 61, 76, 101). Despite extensive evidence for modification of mesophyll plasmodesmata by viral MPs, only recently has it been shown that a viral MP can traffic directly into SEs. Blackman et al (4) made a fusion of the MP of cucumber mosaic virus (CMV) to GFP and expressed this fusion protein from a potato virus X vector. The MP-GFP fusion entered minor vein SEs (Figure 2*a*) and became associated with a reticular structure in the SE parietal layer (Figure 2*c*, *d*). Bright punctate accumulations of MP-GFP were probably associated with the entrances of PPUs (Figure 2*c*, *d*). Immunogold cytochemistry of MP distribution during a wild-type CMV infection confirmed that the MP entered the SE (4). Much of the MP detected within the SE remained associated with the parietal layer, close to the point of MP entry (Figure 1*c*). Given that viral MPs may enter SEs, some of the MP could be translocated to distant sites in the plant and exert an influence on plasmodesmata in these regions. Given the small size (>50 kDa) of many viral MPs, and the known ability of these proteins to traffic between cells, large quantities of virus movement protein may be exported from infected CCs in source tissues and exit into the postphloem pathway of sink tissues. Although undemonstrated, such an efflux of MPs could prime sink tissue for the subsequent export and transport of virus.

An extensive study of virus accumulation in minor veins suggested that the route taken to the SE invariably involves phloem parenchyma elements (19), as these cell often directly abut minor-vein SEs (Figure 1*a*). In all cases in which CCs became infected, the phloem parenchyma elements were also infected (19). Furthermore, in apoplastic loading species in which the CCs are specialized transfer cells (92), the symplastically isolated CCs did not become infected, suggesting that these cells are circumvented in establishing phloem-mediated infection (19). Thus, not all viruses may enter SEs across PPUs, and the plasmodesmata that link phloem parenchyma directly to SEs may provide a potential Achilles' heel for SE invasion.

For many viruses, encapsidated virions appear to represent the functional long-distance movement complex (4, 27, 57, 58, 81, 91) (Figure 1*c*). However, the preceding cell-cell movement steps through mesophyll cells may occur as a ribonucleoprotein complex (4, 18, 27, 58, 71) or as an intact virion (58, 76), depending on the virus. In CMV, virions have not been detected in mesophyll plasmodesmata nor in the PPUs that connect the SE and CC (4). For CMV, it appears that final viral assembly prior to long-distance transport occurs in the SE parietal layer, subsequent to cell-to-cell transport of complexes of viral nucleic acid, MP, and coat protein through the PPUs (4) (Figure 1*c*). In contrast, other spherical viruses have been detected within PPUs in an encapsidated form (56), suggesting considerable plasticity of the PPUs in accommodating viral trafficking. The exact site of encapsidation may differ among different viral groups. For example, tobacco mosaic virus (TMV), type member of the tobamoviruses, can move cell to cell without its CP. However, CP is an absolute requirement for long-distance movement of TMV, suggesting that phloem transport of TMV involves virions (reviewed in 10, 27, 58). A related tobamovirus, cucumber green mottle mosaic virus (CGMMV), has been detected as intact particles in *Cucurbita* sieve tube exudate (81). In this study, no evidence of free viral RNA or other CGMMV-related structures was found in exudate, suggesting that movement through the phloem occurs exclusively as virus particles. Some infectious viral agents do not encode a CP and yet move long distances through the phloem. In the case of viroids, small pathogenic RNAs with no protein-coding capacity, both cell-to-cell (18) and long-distance movement (65) occur, indicating that any protein(s) involved in viroid movement must be encoded by the host. The umbravirus, groundnut rosette virus (GRV), also moves via the phloem without encoding a CP (71, 72). The ORF3 protein of this virus encodes a protein that is essential for long-distance movement (72), whereas a second protein (ORF4), which shares sequence homology with several known viral MPs (71), is essential for cell-cell movement. In GRV, the respective short- and long-distance movement components appear to be under the control of separate gene products. Significantly, the ORF3 protein can replace the CP of TMV for long-distance movement, providing evidence that the ORF3 product represents a class of *trans*-acting long-distance movement factors that can facilitate trafficking of an unrelated viral RNA (72). It will be interesting to determine if this viral protein can traffic endogenous plant RNAs over long distances.

TRANSCRIPTION FACTORS

Although the subject of this review concerns the systemic transport of macromolecules, it is worth considering the available evidence for intercellular transport of endogenous plant proteins, peptides, and nucleic acids in order to discriminate between macromolecules capable only of intercellular movement and those capable of combined intercellular and systemic movement.

The first class of plant proteins to be ascribed a cell-to-cell transport function were transcription factors involved in plant meristem identity. Experiments using periclinal chimeras between wild-type and mutant plants demonstrated that several transcription factors functioned in a non-cell-autonomous fashion (9). Subsequently, elegant studies, based on in situ hybridization of mRNA and proteins, suggested that this nonautonomous behavior was due to an ability of these transcription factors to traffic from cell to cell (37, 38, 67). Notably, the intercellular transport of transcription factors was directional, restricted to meristems, and occurred only between one or two cell layers (36, 67). Surprisingly, however, a recombinant form of the maize transcription factor KNOTTED1 (KN1; 25 kDa), purified from *Escherichia coli*, moved between both maize and tobacco mesophyll cells following microinjection of fluorescently labeled protein (52). Moreover, microinjected KN1 increased the plasmodesmatal SEL in mesophyll cells and also mediated the selective plasmodesmal trafficking of *kn1* sense RNA (52). Curiously, in microinjected mesophyll tissue, KN1 showed no affinity for nuclei, the predicted site of transcription factor localization (52). Furthermore, despite the ability of recombinant KN1 to traffic its own mRNA, previous in situ hybridization studies had indicated that only KN1 and not KN1 transcripts were present in the L1 layer of the maize epidermis. These discrepancies between the behavior of KN1 in mesophyll and meristematic tissue suggest that additional mechanisms exist to regulate the trafficking of KN1 and other transcription factors when expressed in their native context. A curious paradox emerges; endogenous proteins of nonphloem origin, such as KN1, which can modify mesophyll plasmodesmata, remain restricted to limited cellular domains, whereas some CC-synthesized proteins, such as PP2, which can also modify mesophyll plasmodesmata, do not exit the phloem.

SYSTEMIN

The ability of viruses to move from mesophyll to phloem tissue is undisputed. In contrast, the evidence that endogenous proteins, synthesized in the mesophyll, enter the phloem is less compelling. The 18-amino acid polypeptide, systemin, is a powerful inducer of defense-related genes and has been suggested to be the primary systemic signal of gene induction (6, 20, 73). Whole-leaf autoradiographs (57) showed that when ^{14}C systemin was applied to fresh wound it was delivered to upper leaves within 2–4 h, showing a distribution pattern similar to ^{14}C-labeled sucrose (57). Over short time periods, the labeled systemin was found in both xylem and phloem of leaf veins, suggesting that systemin may move initially in the apoplast and from here be loaded into the phloem (57, 73). However, as pointed out by Bowles (6), there is no direct evidence that endogenous systemin is mobile in the wounded plant. The initial site of systemin synthesis in wounded tissues has not been clearly demonstrated. For example, it has yet to be established whether systemin moves from cell to cell through mesophyll plasmodesmata, and it is possible that phloem-mobile sytemin is manufactured in CCs prior to movement into SEs for long-distance transport. Using expression of a reporter gene (GUS) driven by the prosystemin promoter, activity was found to be located only in the CCs and associated parenchyma (36). When up-regulated upon wounding, this high cell specificity was maintained. This evidence strongly suggests that other signals, originating outside the SE-CC complex, give rise to systemin synthesis specifically within CCs.

SYSTEMIC RNA SIGNALING

Recent studies have provided evidence that the systemic signal(s) involved in gene silencing can enter the translocation stream and be transported and unloaded in sink regions of the plant (Figure 4*a*, *b*). Using grafts, Vaucheret and coworkers demonstrated that signals specifying silencing of both nitrate reductase and nitrite reductase could be transmitted from silenced stocks to nonsilenced scions (64). In studies performed in the Baulcombe laboratory, a stably integrated *gfp* transgene was silenced by infiltration with *A. tumefaciens* carrying *gfp* in the T-DNA of a binary vector (95). At 18 days after infiltration of lower leaves, silencing (indicated by loss of GFP fluorescence) was observed in young developing leaves and shoot tips (95). When a single source leaf was similarly inoculated, silencing of the *gfp* transgene was observed on the stem one month after treatment and was restricted to shoots that emerged from the same side of the stem as the inoculated leaf (96). Although interpreted as evidence for phloem transport of the systemic signal (96), the time scale for systemic silencing to occur is particularly long when compared with the movement of photoassimilates (minutes to hours; 23, 44, 50, 54, 69) and systemic viruses (3–10 days; 10, 19, 50, 58, 69, 72). The fact that several viruses trigger a response in their hosts that leads to a recovery from virus infection, together with the observation that some viruses possess mechanisms to suppress gene silencing, have led to the suggestion that systemic gene silencing may have evolved as a host response to viral attack (8, 13, 42, 68). If so, then it might be expected that systemic gene silencing would occur at a rate greater than that observed for virus movement, a feature not observed. Clearly, further characterization of the long-distance signal is required before functional relationships between gene silencing and virus movement can be established.

Despite the long time scale for the establishment of systemic gene silencing, the progression of silencing in young sink leaves mirrors the pattern of phloem unloading of GFP and viruses (75, 96) (Figure 4*b*, *c*, *d*). Furthermore, the systemic signal moved through a threeway graft in which the middle section did not contain a GFP transgene (96), strongly suggesting that transport of the silencing signal was truly phloem dependent. The case for phloem-transmission of gene silencing will be strengthened substantially if the systemic signal responsible for gene silencing can be isolated from phloem exudates.

A further unresolved question regarding gene silencing is whether the systemic signals are initiated in CCs or originate outside the phloem prior to entering the SE-CC complex. Although cell-to-cell propagation of gene silencing clearly occurs in sink tissues (95, 96), evidence for cell-to-cell transmission of silencing in source tissues is less clear. In studies where silencing is initiated by either stably integrated transgenes or agroinfiltration, the trigger responsible for gene silencing is likely be present in both CCs and mesophyll cells (Figure 3). Evidence for intercellular transmission of gene silencing in source tissue is provided by agroinfiltration experiments in which the silencing of an integrated *gfp* transgene by infiltration with *Agrobacterium* carrying the *gfp* gene was seen to extend beyond the margin of the infiltrated tissue (95). In addition, these authors demonstrated systemic silencing of a *gfp* transgene following biolistic bombardment of *gfp*-carrying plasmids into single leaf cells. Cobombardment of seedlings with a 35S-GUS plasmid revealed, on average, less than eight randomly distributed individual cells that exhibited blue staining (96). Thus, very localized events can apparently initiate production and spread of the sequence-specific signal of gene silencing, and at least limited cell-to-cell movement of the silencing signal may occur in source tissue.

What is the nature of the transported signal? RNA molecules seem the most likely candidate for transmission of the cosuppressed state between cells (40) (Figure 3). Recent studies examining both transgene and viral-induced posttranscriptional gene silencing have identified small (25-nucleotide) RNA molecules whose accumulation required either transgene sense transcription or RNA virus replication (31). Note that endogenous and viral RNA movement both involve a relay system in which the signal is produced and amplified in each successive invaded cell (10, 68, 75, 95, 96). One problem faced by the plant in utilizing an RNA-based signaling system is the potential for RNA degradation as it moves between cells and within the phloem. It has been suggested that the systemic RNA signal may be protected in transit by a host protein that also facilitates its movement, similar to the long-distance movement of viral RNA (41, 53, 70).

TRANSLOCATION OF mRNAS

The presence of endogenous RNAs in the phloem was first reported in the literature in 1975 (103). Recent reports have noted the presence of thioredoxin h, oryzacystatin-I, and actin mRNAs in rice phloem sap collected by an insect laser method (78). Thioredoxin h mRNA has been immunolocalized to the CCs of the leaf sheath of rice (34); hence, the most likely origin of these mRNAs in phloem sap is the CC (Figure 3). An analysis of the phloem sap of *Cucurbita maxima* has disclosed several mRNA species, some with putative roles in meristem identity (70). Using RT-PCR, Ruiz-Medrano et al (70) showed that *NACP*, a member of the

NAC domain gene family involved in apical meristem development, was present in sieve elements and companion cells of stem and root phloem. Significantly, longitudinal sections of root and shoot apices showed transcript continuity between meristems and sieve elements of the protophloem, the presumed exit point of the mRNA, suggesting that *NACP* was transported over long distances and accumulated subsequently in vegetative and floral meristems. In grafting experiments in which *Cucurbita maxima* acted as the stock and *Cucumis sativus* as the scion, the *C. maxima NACP* mRNA moved through the phloem and accumulated in apical tissues of *C. sativus*. An additional significant observation was that only *NACP* mRNA, and not NACP protein, entered the translocation stream, consistent with the view that the transcript may have a role in long-distance signaling (41, 53). Ruiz-Medrano et al (70) have suggested that many phloem-specific transcripts or their proteins may play a general role in physiological events within developing leaves, as well as developmental events taking place in meristems.

Do all the mRNAs detected in the phloem have a signaling function? Conceivably, some translocated mRNAs are translated in sink tissues or act as signals to regulate the transcription of related genes (53). The signal that induces flowering has been well documented to be translocated to the vegetative apex via the phloem (12, 47) and, although not yet identified, speculation continues to grow that this signal may be RNA or an RNA-protein complex (70). At present, a direct functional link is lacking between the presence of mRNA species in the phloem and the regulation of specific cellular functions within meristems. In particular, if mRNAs are unloaded from the phloem, what factors mediate their selective transport and targeting to apical meristems?

SUPERHIGHWAY OR SEWAGE SYSTEM?

The phloem provides an ideal conduit for long-distance signals. However, very few of the 200 soluble proteins detected in sieve tube exudate have been identified, much less attributed a function, and of the RNA species identified in exudate, not all have an obvious function in sink tissues (70, 78). Some of the macromolecules present in the translocation stream may indeed have entered by default rather than design. The indiscriminate movement of large dextrans (43) and proteins such as GFP (32, 62) between SE and CC provides clear evidence that at least some macromolecules may enter the SE freely through the PPUs. This raises the possibility that unless a protein has a retention signal for the CC, or a targeting signal that directs it the SE parietal layer (see above), it will be exported in the translocation stream (63). Fisher et al (25) provided compelling evidence for protein turnover by CCs along the transport pathway following the radiolabeling of wheat leaves with amino acids. These authors proposed highly selective regulation of protein removal from SE in the transport phloem and nonselective protein removal from the SEs in sink tissues (25). As the SEs alone do not possess the machinery to degrade proteins in the translocation stream (see 87), many of the proteins detected in sieve tube exudate may reflect the flotsam produced by CCs along the phloem transport pathway. In the above model, loss of small proteins (and possibly nucleic acids) to the SE would be an inevitable consequence of the intimate symplastic continuity between SE and CC (43, 94). Distinguishing between macromolecules that enter SEs for a signaling purpose and those that enter by a default pathway may prove to be a difficult task for the future.

An additional problem in sampling the phloem is whether sieve-tube exudate represents an accurate reflection of the moving translocation stream. When the phloem is severed, the sudden loss of turgor pressure from the sieve tubes can lead to the indiscriminate movement of macromolecules between CCs and SEs (63). This seems less likely to be a problem during aphid feeding (23) but may give rise to potential artefacts when collecting exudate from direct incisions made into the phloem (63). In the case of CC-specific enzymes such as dehydrogenase (49), this can result in rapid displacement of the enzyme into the SE. Given the high natural SEL between SE and CC, one is left wondering if a little bit of everything enters SEs during the collection of phloem exudate from cut tissues. Clearly, stringent controls are required to ensure that the macromolecules present in sieve-tube exudate are normal constituents of the translocation stream.

SE UNLOADING

A growing body of evidence suggests that in rapidly growing sink tissues the pathway of unloading from SE-CC complexes is symplastic (22, 24, 66, 80). Apoplastic unloading, involving the loss of solutes across the SE-CC membranes (24, 66, 80), appears to be restricted to pathway phloem, where routine solute retrieval occurs into the SE-CC complexes by carrier-mediated transport (54, 63, 66, 80, 93). In terminal sinks, such as root tips, fruits, and seeds, symplastic unloading of the SE-CC complexes appears to be almost universal (63, 66, 80). Teleologically, to achieve efficient and rapid unloading, the simplest solution is that the postphloem symplast does not place major constraints on the exit of solutes from the SE-CC complex (63). This could be achieved by increasing the number and permeability of the plasmodesmata in the postphloem pathway (22, 24, 66). For symplastic phloem unloading, it is commonly assumed that the limiting path cross-sectional area is set by the contiguous walls containing the least number of plasmodesmatal connections (66). Such studies have assumed an almost universal plasmodesmal SEL of <1 kDa for sink tissues (66). For example, it has been calculated that plasmodesmata in unloading root tips of corn may not accommodate the observed fluxes of sucrose, necessitating apoplastic postphloem transfer (7). However, recent evidence from a range of species suggests that plasmodesmata in the postphloem pathways of sink tissues may have a higher than normal SEL. For example, Fisher & Wu (25) applied fluorescent dextrans to the postphloem pathway in wheat grains and found the SEL of plasmodesmata to be in excess of 10 kDa (25) (Figure 2*d*). As discussed below, such a large SEL of plasmodesmata in sink tissues forces a re-evaluation of the pathways and mechanisms of postphloem transport.

MOLECULAR MASS VERSUS STOKES RADIUS

Most studies of plasmodesmal function cite the SEL as a function of molecular mass (M_r; see 16, 30, 43, 51, 53, 61, 101). Early studies, utilizing microinjected fluorescent probes, estimated the SEL of plasmodesmata to be about 0.8–1 kDa (30, 86). Terry & Robards (86) emphasized the role of the Stokes radius (R_s) determining molecular mobility through plasmodesmata. The R_s is the molecular dimension of an equivalent sphere with the same hydrodynamic drag as the molecule in question (22, 40). The larger molecules (~1 kDa) that pass through plasmodesmata had an R_s of 0.9 nm, suggesting functional channel diameters in the plasmodesmal pore of about 3 nm (86). This value has now been shown to be closer to 4 nm (21). Unfortunately, as pointed out by Fisher & Cash-Clark (22), there is no unique relationship between M_r and molecular dimensions. The R_s of transported molecules assumes greater importance with the demonstration that some macromolecules may be transported through sink plasmodesmata (22, 24, 32, 62, 99). It should be stressed that fluorescent dextrans, which have been used extensively in microinjection experiments, do not have the same R_s of proteins of identical M_r (22, 40). In considering the estimated functional SEL of a plasmodesmal pore, it should be noted that a 20-kDa dextran has the same R_s as a 51-kDa globular protein (22, 40). By infiltrating fluorescent dextrans into the postphloem pathway of wheat grains, Wang & Fisher (99) demonstrated the passage of 10-kDa dextran through plasmodesmata, equating to a functional channel diameter of 7 nm (Figure 2*d*). The effect of channel dimensions on the conductance for diffusive transport of low-molecular-weight solutes is very high (21, 22, 24, 66). For example, in the case of sucrose, the per channel conductance of a 7-nm channel would be 12 times that of a standard 3-nm channel (24). Although evidence favors diffusive transport in the postphloem pathway of wheat (22, 99), the high conductance of sink plasmodesmata could also facilitate bulk (convective) flow, rather than diffusion, of solutes from the phloem in cases where import rates are exceptionally high (66).

UNLOADING OF MACROMOLECULES

The diversity of macromolecules moving in the phloem poses a potential problem for the phloem located at the terminus of the translocation stream: How do the SE-CC complexes located in sink tissues discriminate between the different solutes and macromolecules arriving in the phloem? The simplest answer is that they do not. As pointed out by Oparka & Turgeon (63), apoplastic SE unloading would necessitate the presence of carriers for an enormous range of low-molecular-weight solutes. Furthermore, proteins moving in the translocation stream would have to be rapidly degraded into smaller moieties in sink CCs so that these could be unloaded by membrane carriers. To achieve efficient and rapid unloading, the simplest scenario is that the postphloem symplast does not place major constraints on the exit of solutes or small proteins from the SE-CC complex. An analysis of unloading patterns in sink leaves shows a remarkably similar pattern for a diverse range of

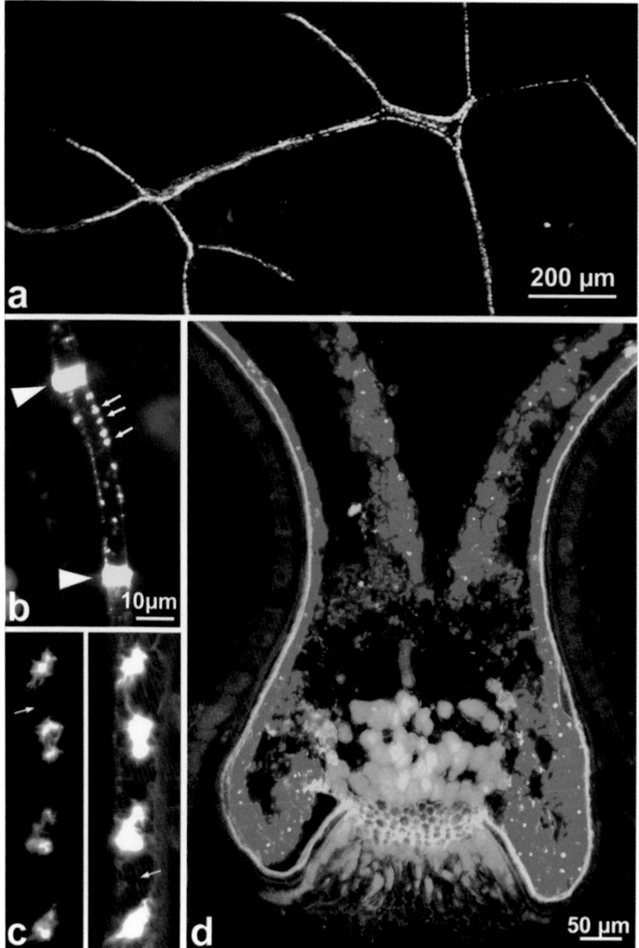

Figure 2 (*a*) Targeting of the MP of CMV (expressed as a MP-GFP fusion) to minor vein sieve elements. Note the punctate fluorescence arising from plasmodesmata. (From Reference 4, courtesy of *The Plant Cell*). (*b*) Sieve element stained with aniline blue to reveal callose associated with the sieve plates (darts) and pore-plasmodesma units (*arrows*). (*c*) Targeting of pore-plasmodesma units by a CMV MP-GFP fusion. The left panel shows the plasmodesmal entry sites of the MP-GFP, imaged under a confocal microscope. The right panel, taken at higher gain, reveals a labeled reticular structure within the SE parietal layer (*arrows*) (From Reference 4, courtesy of *The Plant Cell*). (*d*) The crease region of a developing wheat caryopsis showing the cell-cell transport of fluorescein dextran through chalazal tissues. Red autofluorescence is generated from chlorophyll-containing pericarp tissues. (From Reference 99, courtesy of *The American Society of Plant Physiologists.*)

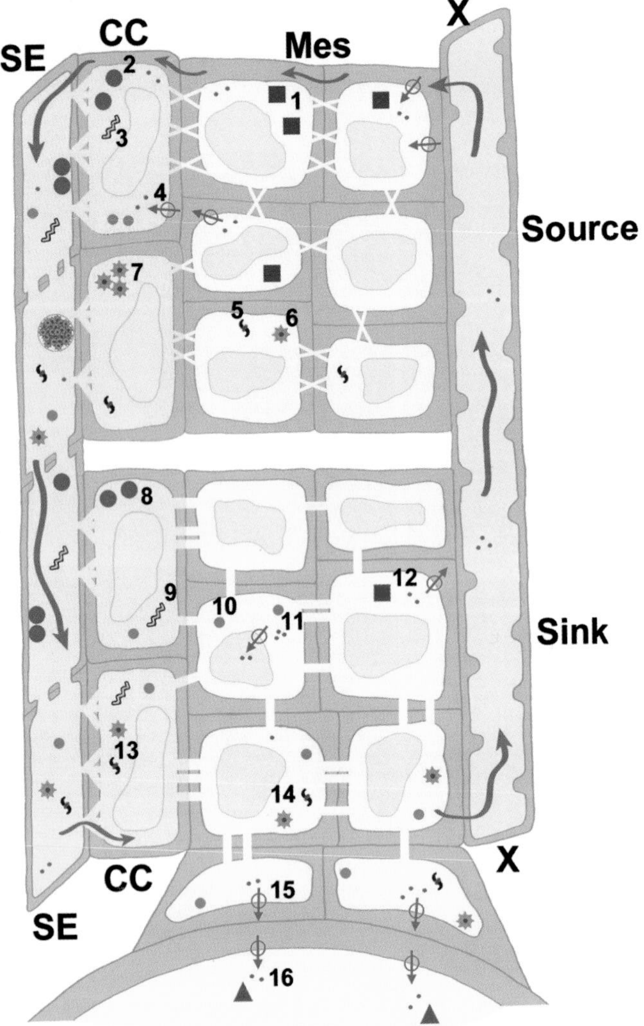

Figure 3 Schematic representation of macromolecular trafficking between source and sink regions of the plant. In source tissues: 1. Most mesophyll-synthesized proteins (*purple squares*) remain in this cellular domain and do not traffic through the plasmodesmata that connect the mesophyll with the SE-CC complexes. 2. CC-synthesized proteins, such as P-proteins (*blue circles*), may traffic into the SE. Some of these remain anchored to the parietal layer although a proportion may enter the translocation stream. 3. Some CC-specific mRNAs (*yellow strands*) may also enter the SE and be translocated to sink tissues. 4. Some CC-specific proteins may be synthesized from amino acids that enter the CC by carrier-mediated transport from the mesophyll. Xenobiotic proteins, such as GFP (*green circles*), synthesized within the CC may also enter the translocation stream. 5. Systemic RNA signals (*blue strands*) that originate outside the SE-CC complex are amplified as they move between cells. (Legend continues on the next page.)

6. In the case of systemic viruses, also originating outside the SE-CC complex, the viral genome is trafficked into the SE-CC by specific movement proteins. 7. Viral replication occurs in the CC, prior to transport of the viral genome (probably as particles) through the phloem. In sink tissues: 8. Some large proteins, such as P-proteins, may enter the CCs of sink tissues but are too large to exit the plasmodesmata around the SE-CC complex. 9. Some mRNA molecules may enter the sink CC only. Other mRNAs, with putative roles in signaling, may be able to leave the SE-CC complex to target meristematic tissues. 10. Small proteins such as GFP exit the SE-CC complex and traffic through sink tissues due to the high size exclusion limit of sink plasmodesmata. 11. Many of these small proteins may be degraded, the amino acid derivatives subsequently being transported into the vacuole for storage. 12. Some amino acids may be resynthesized into sink-specific proteins (*red squares*) or actively transported into the xylem for recycling to source tissues. 13. Both small systemic RNA signals and viruses escape the SE and recommence a replication (amplification) cycle in the CC. 14. These RNAs may then be transported and amplified further in sink tissues. 15. In cases where a symplastic discontinuity occurs (e.g. in seeds) carrier-mediated transport of amino acids is essential. Such barriers provide an important safeguard to prevent macromolecules (including viral RNA) from entering filial tissues of the seed. 16. Amino acids, transported into the seed by carriers, may be synthesized into seed-specific proteins (*red triangles*).

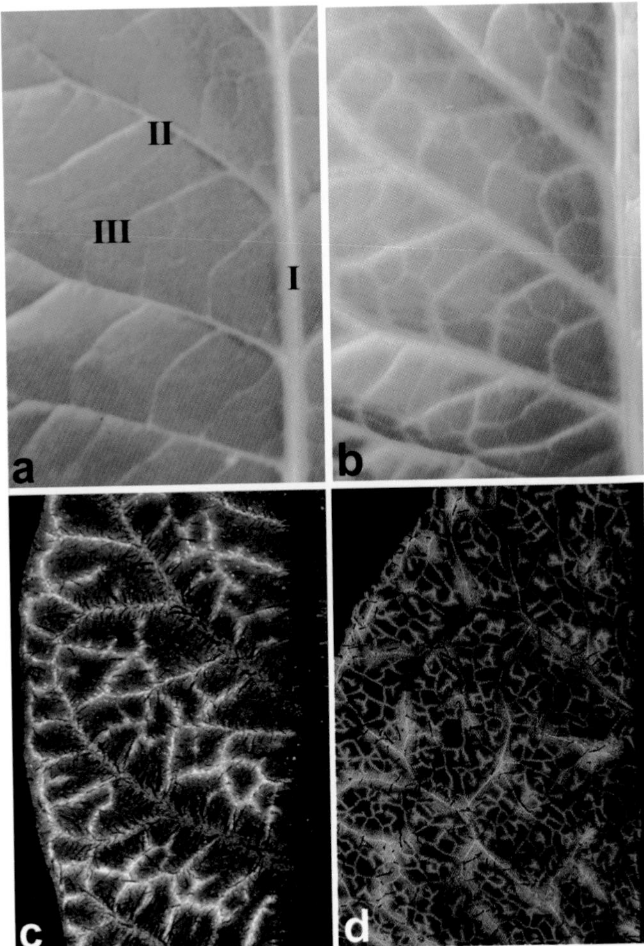

Figure 4 Phloem unloading of systemic macromolecules is restricted to major veins. (*a*) Control leaf of tobacco showing three distinct vein classes. (*b*) Systemic silencing of the nitrite reductase gene is first seen as chlorosis around the class III veinal network. (Figure courtesy of Hervé Vaucheret.). (*c*) Phloem unloading of GFP from class III veins in a tobacco leaf. The GFP was expressed in source leaves under the CC-specific promoter SUC2, and was subsequently translocated to, and unloaded within, the sink leaf (See Reference 62). (*d*) Phloem unloading of a tobacco etch virus (TEV) construct expressing GFP. Virus exit (seen as green fluorescence) is first seen from the class III veinal network. The xylem of this leaf was allowed to transpire Texas red in order to reveal the minor vein networks. Note that virus unloading does not occur from minor veins. (Figure courtesy of Sophie Haupt.)

phloem-transported compounds. For example, radioactive solutes (50, 89), fluorescent solutes (69, 75), GFP (32, 62), and systemic RNA signals (62, 75, 96) all exit the phloem from major veins in similar patterns (Figure 4*b*, *c*, *d*), suggesting that the same vein classes utilized in solute unloading are also involved in the unloading of some phloem-mobile macromolecules (75). Given the relatively small size of RNA molecules involved in systemic posttranscriptional gene silencing (31), it would appear that the postphloem pathway could accommodate the diffusional transport of these signals through the symplast.

Recently, Fisher & Cash-Clark (22) injected a range of fluorescent proteins and dextrans into wheat grains through severed aphid stylets and observed the SE unloading and postphloem transport of dextrans up to M_r 16 kDa (R_s 2.6 nm), predicting aqueous channel diameters in the plasmodesmata of 8–9 nm. To probe the SEL of sink-leaf tissues of tobacco, Oparka et al (62) biolistically bombarded sink and source leaves with plasmids encoding GFP-fusion proteins. Free GFP and a GFP-sporamin fusion (47 kDa; R_s 2.21) moved freely through sink (but not source) leaf plasmodesmata whereas larger fusion proteins (61 kDa; R_s 2.6 and 67 kDa; R_s 2.8) failed to move in both sink and source leaves. The decrease in plasmodesmal permeability during the sink-source transition was correlated with a change in architecture from simple to branched plasmodesmata (62). When considered in terms of R_s, the estimates of aqueous channel diameters in postphloem tissues of wheat grains and in sink tobacco leaves give similar values of about 8–9 nm, considerably larger than previously published values for source tissues (30, 51, 61, 86, 101). Since many of the proteins detected in sieve tube exudate fall in the range 10–40 kDa, the postphloem pathway in sink leaves could potentially accommodate the symplastic transport of many of the small proteins present in the translocation stream.

As pointed out, some of the large macromolecules present within the SE-CC complexes, such as the P-proteins PP1 and PP2, do not exit the SE-CC complex in sink tissues (29) (Figure 3). It remains to be demonstrated whether such molecules are retained within the SE-CC complex by specific targeting mechanisms, or whether they are simply too large to pass through the plasmodesmata that surround the SE-CC complex. Fisher & Cash-Clark (22) have suggested that the larger proteins present in sieve tube exudate (up to 70 kDa in wheat; 25) may require recognition factors that facilitate a specific interaction with plasmodesmata if they are to move through the postphloem pathway. Similarly, if sink plasmodesmata impose an SEL of <50 kDa (62), the mRNA molecules transported from the phloem to meristematic tissues (70) would also require a facilitated transport mechanism.

UNLOADING OF VIRUSES

Some systemic viruses facilitate their escape in sink tissues, although the mechanism of exit from the SE-CC complex remains unknown. It is perhaps significant that very few viruses fail to be unloaded from the SE-CC complex. In phloem-limited viruses, restriction of virus movement usually occurs at some point outside the SE-CC complex (10, 58, 85), suggesting that the limiting steps in tissue invasion lie in cell-to-cell postphloem movement rather than in the ability to escape the SE-CC complex. Since many viruses appear to move in an encapsidated form, one possibility is that viral disassembly occurs within the SE prior to movement of the viral RNA into the CC. How this might occur in the translocation stream in the absence of viral MPs is unclear, unless MP is also translocated in some form to permit RNA trafficking into the sink CC. A second possibility is that the SEL of PPUs, in common with the majority of sink plasmodesmata (22, 62, 99), is considerably larger than in source tissues, allowing intact virions to pass freely between SE and CC. In the CC, disassembly could occur followed by reinitiation of the viral replication and movement cycles (Figure 3). Oparka et al (62) examined the possibility that, due to the high SEL of sink-leaf tissues, viral MPs may not be required for postphloem movement of the viral genome through plasmodesmata. A GFP-expressing PVX mutant was constructed that lacked the 25-kDa protein of the triple gene block, a gene product shown to be essential for cell-cell movement of PVX (76).This mutant virus was restricted to single cells on sink as well as source tissues, indicating that an interaction is still required between the viral MP and sink plasmodesmata before the viral genome can pass between cells.

POSTPHLOEM SORTING OF MACROMOLECULES

The symplastic unloading of solutes from the phloem in sink tissues may be a mechanism that ensures the efficient delivery of carbon to rapidly importing sinks (24, 66). Implicit in this unloading mechanism is the accompanying exit of a range of low-molecular-weight macromolecules into the postphloem pathway. By allowing both solutes and macromolecules to exit the phloem freely the plant overcomes the immediate problem of fouling the terminal transport pathway, but shifts the sorting problem to the cells of the postphloem pathway. In several types of reproductive sinks, specialized parenchyma elements surround the terminal phloem elements (24, 60, 66, 99, 100). In cereal caryopses, the terminal sieve elements are separated from the endosperm by specialized chalazal tissues. These tissues have very high frequencies of simple plasmodesmata (60, 100) and accumulate a wide range of compounds during grain development (60, 100). Such postphloem tissues are most likely sites of degradation and recycling of macromolecules within the postphloem pathway, breaking down and/or storing those materials that are unable to be used by the seed and passing on only those that can be utilized (63) (Figure 3).The postphloem pathway in several sinks may thus have a dual function; facilitation of symplastic transport and retrieval/degradation of unloaded macromolecules.

A common feature of symplastic phloem unloading in reproductive structures is a complete symplastic barrier between maternal and filial generations of the caryopsis, necessitating the exclusive apoplastic transfer of solutes by carrier-mediated transport at this interface (24, 32, 60, 66, 99, 100) (Figure 3). The judicious placement of this symplastic barrier at a location distant to the SE-CC complexes may be a mechanism that allows the postphloem sink tissues both time and space in which to process the cargo being unloaded from the translocation stream, as well as a mechanism for preventing the movement of phloem-unloaded pathogens into the seed (Figure 3). In the case of GFP unloaded into the seed coat of *Arabidopsis*, the protein is translocated extensively around the seed coat without entry into the endosperm (32). Presumably, with time GFP is degraded by these tissues and recycled, as without continued turnover the protein appears to have a relatively short biological half-life (32, 62). Several small solutes, particularly amino acids, produced as degradation products from unloaded macromolecules, could be recycled to the shoot (39), along with phloem-unloaded water, via the xylem (Figure 3).

CONCLUSION

Of the several macromolecules detected in the phloem exudates of plants, it appears that most originate in the CC, with subsequent movement into the SE occurring by the PPUs that connect these cell types. Unusually, several phloem-mobile proteins possess the capacity to traffic through mesophyll cells, although such proteins are not normally present in this cellular domain. Conversely, endogenous mobile proteins such as the transcription factor KN1 appear to be able to modify nonphloem plasmodesmata but have not been detected in the translocation stream. These observations may point to the selective operation of endogenous plant mobile proteins within specific tissue domains. Identifying those phloem-mobile macromolecules that act as long-distance signaling agents is likely to prove a difficult task for the future. The ability of xenobiotic macromolecules to enter the SE and be subsequently unloaded in sink tissues suggests that some macromolecules may be present in the translocation stream by default rather than design. To assess the role of macromolecules in long-distance signaling will thus require more than their mere presence in sieve-tube exudate. For a given phloem-mobile macromolecule, number of criteria must be met before a signaling role can be attributed.

1. The macromolecule must be demonstrated to move from source to sink in the translocation stream. In this respect, grafting experiments are likely to play a major role in determining uniqueness of the macromolecule in sink tissues.
2. It must be demonstrated that the macromolecule leaves the SE-CC complex in sink tissues. As detailed above, several sink tissues possess plasmodesmata with a SEL sufficiently large to accommodate the bulk of proteins that have been detected in sieve-tube exudate. On the other hand, several large proteins, including P-proteins, appear to be phloem-mobile across graft unions but are not unloaded from the SE-CC complex. Some of these macromolecules may initiate a signal cascade from within the sink CC, involving a second signal that transmits information through the

postphloem pathway (53), but such a mechanism remains hypothetical. It seems equally possible that several of the macomolecules unloaded from the phloem are targeted for degradation in postphloem tissues.

3. The macromolecule must be shown unequivocally to target and modify specific cell(s) following unloading from the phloem. To assess this capacity, it will be essential to follow the putative phloem-unloaded signal to its destination and to demonstrate an unequivocal and unique effect on cellular function.

Of the above criteria, the last is clearly the most important (and probably the most difficult) to demonstrate. The observation that several macromolecules are present in phloem exudate is an intriguing discovery that warrants further detailed study. However, it may be premature to attribute a signaling function to each new species of macromolecule identified in the translocation stream. In this respect, it is essential to move from a situation of guilt by association to one in which long-distance macromolecular signals can be identified and characterized in detail. Such experiments may prove conceptually challenging but should provide important insights into long-distance communication in plants.

ACKNOWLEDGMENTS

We are grateful to Don Fisher for critically reading the manuscript. We also thank Jim Carrington for supplying us with tobacco etch virus (TEV)-GFP, Norbert Sauer for the gift of AtSUC2-GFP tobacco plants, and Hervé Vaucheret for sharing unpublished data. We are also grateful to Alison Roberts for preparation of the figures. Work in the authors' laboratories was funded by the Scottish Executive Rural Affairs Department (SERAD).

LITERATURE CITED

1. Balachandran S, Xiang Y, Schobert C, Thompson G, Lucas WJ. 1997. Phloem sap proteins from *Cucurbita maxima* and *Ricinus communis* have the capacity to traffic cell to cell through plasmodesmata. *Proc. Natl. Acad. Sci. USA* 94:14150–55
2. Behnke H-D. 1989. Structure of the phloem. In *Transport of Photoassimilates*, ed. DA Baker, JA Milburn, pp. 79–137. Harlow, UK: Longman
3. Behnke H-D, Sjolund RD. 1990. *Sieve Elements: Comparative Structure, Isolation and Development*. Berlin: Springer-Verlag
4. Blackman LM, Boevink P, Santa Cruz S, Palukaitis P, Oparka KJ. 1998. The movement protein of cucumber mosaic virus traffics into sieve elements in minor veins of *Nicotiana clevelandii*. *Plant Cell* 10:525–37
5. Bostwick DE, Dannehoffer JM, Skaggs MI, Liser RM, Larkins BA, Thomson GA. 1992. Pumpkin phloem lectin genes are specifically expressed in companion cells. *Plant Cell* 4:1539–48
6. Bowles D. 1998. Signal transduction in the wound response of tomato plants. *Philos.*

Trans. R. Soc. London Ser. B 353:1495–510
7. Bret-Hart MS, Silk WK. 1994. Nonvascular, symplasmic diffusion of sucrose cannot satisfy the carbon demands of growth in the primary root tip of *Zea mays* L. *Plant Physiol.* 105:19–33
8. Brignetti G, Voinnet O, Li W-X, Ji L-H, Ding S-W, Baulcombe DC. 1998. Viral pathogenicity determinants are suppressors of transgene silencing in *Nicotiana benthamiana*. *EMBO J.* 17:6739–46
9. Carpenter R, Coen ES. 1995. Transposon induced chimera show that *floricula*, a meristem identity gene, acts non-autonomously between cell layers. *Development* 121:19–26
10. Carrington JC, Kasschau KD, Mahajan SK, Schaad MC. 1996. Cell-to-cell and long-distance transport of viruses in plants. *Plant Cell* 8:1669–81
11. Clark AM, Jacobsen KR, Bostwick DE, Dannehoffer JM, Skaggs MI, Thompson GA. 1997. Molecular characterization of a phloem-specific gene encoding the filament protein, phloem protein 1 (PP1), from *Cucurbita maxima*. *Plant J.* 12:49–61
12. Colasanti J, Yuan Z, Sundaresan V. 1998. The *indeterminate* gene encodes a zinc finger protein and regulates a leaf-generated signal required for the transition to flowering in maize. *Cell* 93:593–603
13. Covey SN, Al-Kaff NS, Langara A, Turner DS. 1997. Plants combat infection by gene silencing. *Nature* 385:781–82
14. Crawford KM, Zambryski PC. 1999. Phloem transport: Are you chaperoned? *Curr. Biol.* 9:R281–85
15. Cronshaw J, Esau K. 1967. Tubular and fibrillar components of mature and differentiating sieve elements. *J. Ultrastruct. Res.* 34:244–259
16. Ding B. 1997. Cell-to-cell transport of macromolecules through plasmodesmata: a novel signalling pathway in plants. *Trends Cell Biol.* 7:5–9
17. Ding B, Itaya A, Woo Y-M. 1999. Plasmodesmata and cell-to-cell communication in plants. *Int. Rev. Cytol.* 190:251–316
18. Ding B, Kwon M-O, Hammond R, Owens R. 1997. Cell-to cell movement of potato spindle tuber viroid. *Plant J.* 12:931–36
19. Ding XS, Carter SA, Deom CM, Nelson RS. 1998. Tobamovirus and potyvirus accumulation in minor veins of inoculated leaves from representatives of the Solanaceae and Fabaceae. *Plant Physiol.* 116:125–36
20. Enyedi AJ, Yalpani N, Silverman P, Raskin I. 1992. Signal molecules in systemic plant resistance to pathogens and pests. *Cell* 70:879–86
21. Fisher DB. 1999. The estimated pore diameter for plasmodesmal channels in the *Abutilon* nectary should be about 4 nm, rather than 3 nm. *Planta* 208:299–300
22. Fisher DB, Cash-Clark CE. 2000. Sieve tube unloading and post-phloem transport of fluorescent tracers and proteins injected into sieve tubes via severed aphid stylets. *Plant Physiol.* In press
23. Fisher DB, Frame JM. 1984. A guide to the use of the exuding stylet technique in phloem physiology. *Planta* 161:385–93
24. Fisher DB, Oparka KJ. 1996. Postphloem transport. Principles and problems. *J. Exp. Bot.* 47:1141–54
25. Fisher DB, Wu Y, Ku MSB. 1992. Turnover of soluble proteins in the wheat sieve tube. *Plant Physiol.* 100:1433–41
26. Gamalei Y. 1989. Structure and function of leaf minor veins in trees and herbs. A taxonomic review. *Trees* 3:96–110
27. Gilbertson RL, Lucas WJ. 1996. How do viruses traffic on the vascular highway? *Trends Plant Sci.* 1:260–67
28. Golecki B, Schulz A, Carstens-Behrens U, Kollmann R. 1998. Evidence for graft transmission of structural phloem proteins or their precursors in heterografts of Cucurbitaceae. *Planta* 206:630–40
29. Golecki B, Schulz A, Thompson GA. 1999. Translocation of structural P-proteins in the phloem. *Plant Cell* 11:127–40

30. Goodwin PB. 1983. Molecular size exclusion limit for movement through the symplast of the *Elodea* leaf. *Planta* 157:124–30
31. Hamilton AJ, Baulcombe DC. 1999. A species of small antisense RNA in posttranscriptional gene silencing in plants. *Science* 286:950–52
32. Imlau A, Truernit E, Sauer N. 1999. Cell-to-cell and long-distance trafficking of the green fluorescent protein in the phloem and symplastic unloading of the protein into sink tissues. *Plant Cell* 11:309–22
33. Ishiwatari Y, Fujiwara T, McFarland KC, Nemoto K, Hayashi H, et al. 1998. Rice phloem thioredoxin h has the capacity to mediate its own cell-to-cell transport through plasmodesmata. *Planta* 205:12–22
34. Ishiwatari Y, Honda C, Kawashima I, Nakamura S, Hirano H, et al. 1995. Thioredoxin h is one of the major proteins in rice phloem sap. *Planta* 195:456–63
35. Itaya A, Woo Y-M, Masuta C, Bao Y, Nelson RS, Ding B. 1998. Developmental regulation of intercellular protein trafficking through plasmodesmata in tobacco leaf epidermis. *Plant Physiol.* 118:373–85
36. Jacinto T, McGurl B, Franceschi V, Delano-Freier J, Ryan CA. 1997. Tomato prosystemin promoter confers wound-inducible, vascular bundle-specific expression of the β-glucoronidase gene in transgenic tomato plants. *Planta* 203:406–12
37. Jackson D, Hake S. 1997. Morphogenesis on the move: cell-to-cell trafficking of plant regulatory proteins. *Curr. Opin. Genet. Dev.* 7:495–500
38. Jackson D, Veit B, Hake S. 1994. Expression of maize *KNOTTED 1* related homeobox genes in the shoot apical meristem predicts patterns of morphogenesis in the vegetative shoot. *Development* 120:405–13
39. Jeschke WD, Wolf O, Pate JS. 1991. Solute exchanges from xylem to phloem in the leaf and from the phloem to the xylem in the root. In *Recent Advances in Phloem Transport and Assimilate Compartmentation*, ed. J-L Bonnemain, S Delrot, WJ Lucas, J Dainty, pp. 96–105. Nantes, France: Ouest Editions
40. Jorgensen KE, Moller JV, 1979. Use of flexible polymers as probes of glomelular pore size. *Am. J. Physiol.* 236:F103–11
41. Jorgensen RA, Atkinson RG, Forster RL, Lucas WJ. 1998. An RNA-based information superhighway in plants. *Science* 279:1486–87
42. Kasschau KD, Carrington. JC. 1998. A counterdefense strategy of plant viruses: suppression of postranscriptional gene silencing. *Cell* 95:461–70
43. Kempers R, van Bel AJE. 1997. Symplasmic connections between sieve element and companion cell in the stem phloem of *Vicia faba* L. have a molecular exclusion limit of at least 10 kD. *Planta* 201:195–201
44. Knoblauch M, van Bel AJE. 1998. Sieve tubes in action. *Plant Cell* 10:35–50
45. Kuhn C, Franceschi VR, Schulz A, Lemoine R, Frommer WB. 1997. Macromolecular trafficking indicated by localization and turnover of sucrose transporters in enucleate sieve elements. *Science* 275:1298–300
46. Kuhn C, Quick WP, Schulz A, Riesmeier JW, Sonnewald U, Frommer WB. 1996. Companion cell-specific inhibition of the potato sucrose transporter SUT1. *Plant Cell Environ.* 19:1115–23
47. Lang A. 1965. Physiology of flower initiation, differentiation and development. In *Encyclopedia of Plant Physiology*, ed. W Ruhland, pp. 1380–536. Berlin: Springer
48. Lazarowitz SG, Beachy RN. 1999. Viral movement proteins as probes for intracellular and intercellular trafficking in plants. *Plant Cell* 11:535–48
49. Lehmann J. 1973. Zur Lokalisation von Dehydrogenase des Energeistoffwechsels im Phloem von *Cucurbita maxima* L. *Planta* 111:187–98
50. Leisner SM, Turgeon R. 1993. Movement

of virus and photoassimilate in the phloem. *BioEssays* 15:741–48

51. Lucas WJ. 1995. Plasmodesmata: intercellular channels for macromolecular transport in plants. *Curr. Opin. Cell Biol.* 7:673–80
52. Lucas WJ, Bouche-Pillon S, Jackson DP, Nguyen L, Baker L, et al. 1995. Selective trafficking of KNOTTED 1 homeodomain protein and its mRNA through plasmodesmata. *Science* 270:1980–83
53. Mezitt LA, Lucas WJ. 1996. Plasmodesmal and cell-to-cell transport of proteins and nucleic acids. *Plant Mol. Biol.* 32:251–73
54. Minchin PEH, Thorpe MR. 1987. Measurement of unloading and reloading of photo-assimilate within the stem of bean. *J. Exp. Bot.* 38:211–20
55. Münch E. 1930. *Die Stoffwebegungen in der Pflanze*. Jena, Germany: G Fischer
56. Murant AF, Roberts IM. 1979. Virus-like particles in phloem tissues of chervil (*Anthriscus cerefolium*) infected with carrot red leaf virus. *Ann. App. Biol.* 92:343–46
57. Narvaez-Vasquez J, Orozco-Cardenas ML, Franceschi VR, Ryan CA. 1995. Autoradiographic and biochemical evidence for the systemic translocation of systemin in tomato plants. *Planta* 195:593–600
58. Nelson RS, van Bel AJE. 1998. The mystery of virus trafficking into, through and out of the vascular tissue. *Prog. Bot.* 59:476–533
59. Oparka KJ. 1980. Polysomes and intracisternal accumulations in enucleate sieve elements of rice (*Oryza sativa* L.). *Planta* 150:249–54
60. Oparka KJ, Gates P. 1981. Transport of assimilates in the developing caryopsis of rice (*Oryza sativa* L.). Ultrastructure of the pricarp vascular bundle and its connections with the aleurone layer. *Planta* 151:561–73
61. Oparka KJ, Prior DAM, Santa Cruz S, Padgett HS, Beachy RN. 1997. Gating of epidermal plasmodesmata is restricted to the leading edge of expanding infection sites of tobacco mosaic virus (TMV). *Plant J.* 12:781–89
62. Oparka KJ, Roberts AG, Boevink P, Santa Cruz S, Roberts IM, et al. 1999. Simple, but not branched, plasmodesmata allow the nonspecific trafficking of proteins in developing tobacco leaves. *Cell* 97:743–54
63. Oparka KJ, Turgeon R. 1999. Sieve elements and companion cells—traffic control centers of the phloem. *Plant Cell* 11:739–50
64. Palauqui J-C, Elmayan T, Pollien J-M, Vaucheret H. 1997. Systemic acquired silencing: transgene-specific post-translational silencing is transmitted by grafting from silenced to non-silenced scions. *EMBO J.* 16:4738–45
65. Palukaitis P. 1987. Potato tuber spindle viroid: investigations of the long-distance, intra-plant transport route. *Virology* 158:239–41
66. Patrick JW. 1997. Phloem unloading: sieve element unloading and post-phloem transport. *Annu. Rev. Plant Physiol Plant Mol. Biol.* 48:191–222
67. Perbal M-C, Haughn G, Saedler H, Schwarz-Sommer Z. 1996. Non-cell autonomous function of *Antirrhinum* floral homeotic proteins *DEFICIENS* and *GLOBOSA* is exerted by their polar cell-to-cell trafficking. *Development* 122:3433–41
68. Ratcliff F, Harrison BD, Baulcombe DC. 1997. A similarity between viral defense and gene silencing in plants. *Science* 276:1558–60
69. Roberts AG, Santa Cruz S, Roberts IM, Prior DAM, Turgeon R, Oparka KJ. 1997. Phloem unloading in sink leaves of *Nicotiana benthamiana*: comparison of a fluorescent solute with a fluorescent virus. *Plant Cell* 9:1381–96
70. Ruiz-Medrano R, Xonocostle-Cazares B, Lucas WJ. 1999. Phloem long-distance transport of *CmNACP* mRNA: implications for supracellular regulation in plants. *Development* 126:4405–19
71. Ryabov EV, Oparka KJ, Santa Cruz S,

Robinson DJ, Taliansky ME. 1998. Intracellular location of two groundnut rosette umbravirus proteins delivered by PVX and TMV vectors. *Virology* 242:303–13
72. Ryabov EV, Robinson DJ, Taliansky M. 1999. A plant-virus encoded protein facilitates long-distance movement of heterologous viral RNA. *Proc. Natl. Acad. Sci. USA* 96:1212–17
73. Ryan CA, Pearce G. 1998. SYSTEMIN: a polypeptide signal for plant defensive genes. *Annu. Rev. Cell Dev. Biol.* 14:1–17
74. Sabnis DD, Sabnis HM. 1995. Phloem proteins: structure, biochemistry and function. In *The Cambial Derivatives. Encyclopedia of Plant Anatomy*, ed. M Iqbal, 9/4:271–92. Berlin: Bortraeger
75. Sakuth T, Schobert C, Pecsvaradi A, Eicholz A, Komor E. 1993. Specific proteins in the sieve-tube exudate of *Ricinus comunis* L. seedlings: separation, characterisation, and in-vivo labeling. *Planta* 191:207–13
76. Santa Cruz S. 1999. Phloem transport of viruses and macromolecules—what goes in must come out. *Trends Microbiol.* 6:237–41
77. Santa Cruz S, Roberts AG, Prior DAM, Chapman S, Oparka KJ. 1998. Cell-to-cell and phloem-mediated transport of potato virus X: the role of virions. *Plant Cell* 10:495–510
78. Sasaki T, Chino M, Hayashi H. Fujiwara T. 1998. Detection of several mRNA species in rice phloem sap. *Plant Cell Physiol.* 39:895–97
79. Schobert C, Grossmann P, Gottschalk M, Komor E, Pecsvaradi A, Nieden UZ. 1995. Sieve-tube exudate from *Ricinus communis* L. seedlings contain ubiquitin and chaperones. *Planta* 196:205–10
80. Schulz A. 1998. Phloem. Structure related to function. *Prog. Bot.* 59:431–75
81. Simon-Buelo L, Garcia-Arenal F. 1999. Virus particles of cucumber green mottle mosaic tobamovirus move systemically in the phloem of infected cucumber plants. *Mol. Plant Microbe Interact.* 12:112–18
82. Sjolund RD. 1997. The phloem sieve element: a river runs through it. *Plant Cell* 9:1137–46
83. Smith LM, Sabnis DD, Johnson RPC. 1987. Immunolocalization of phloem lectin from *Cucurbita maxima* using peroxidase and colloidal gold labels. *Planta* 170:461–70
84. Stadler R, Sauer N. 1996. The *Arabidopsis thaliana* AtSUC2 gene is specifically expressed in companion cells. *Bot. Acta* 109:299–306
85. Sudarshana MR, Wang HL, Lucas WJ, Gilbertson RL. 1998. Dynamics of bean dwarf mosaic geminivirus cell-to-cell and long-distance movement in *Phaseolus vulgaris* revealed, using the green fluorescent protein. *Mol. Plant Microbe. Interact.* 11:277–91
86. Terry BR, Robards AW. 1987. Hydrodynamic radius alone governs the mobility of molecules through plasmodesmata. *Planta* 171:145–57
87. Thompson GA, Schulz A. 1999. Macromolecular trafficking in the phloem. *Trends Plant Sci.* 4:354–61
88. Tomenius K, Clapham D, Meshi T. 1987. Localization by immunogold cytochemistry of the virus-encoded 30 K protein in plasmodesmata of leaves infected with tobacco mosaic virus. *Virology* 160:363–71
89. Turgeon R. 1989. The sink-source transition in leaves. *Annu. Rev. Plant Physiol. Plant Mol. Biol.* 40:119–38
90. Turgeon R, Beebe DU, Gowan E. 1993. The intermediary cell: minor vein anatomy and raffinose oligosaccharide synthesis in the Scrophulariaceae. *Planta* 191:446–56
91. Vaewhongs AA, Lommel SA. 1995. Virion formation is required for the long-distance movement of red clover necrotic mosaic virus in movement protein transgenic plants. *Virology* 212:607–13
92. Van Bel AJE. 1993. Strategies of phloem loading. *Annu. Rev. Plant Physiol. Plant Mol. Biol.* 44:253–81

Annu. Rev. Plant Physiol. Plant Mol. Biol. 2000. 51:349–70

DEVELOPMENT OF SYMMETRY IN PLANTS

A. Hudson

Institute of Cell and Molecular Biology, University of Edinburgh, King's Buildings, Mayfield Road, Edinburgh EH9 3JH United Kingdom; e-mail: Andrew.Hudson@ed.ac.uk

Key Words asymmetry, axes, embryogenesis, lateral organs, flowers

■ **Abstract** Plant development involves specification and elaboration of axes of asymmetry. The apical-basal and inside-outside axes arise in embryogenesis, and are probably oriented maternally. They are maintained during growth post-germination and interact to establish novel axes of asymmetry in flowers and lateral organs (such as leaves). Whereas the genetic control of axis elaboration is now partially understood in embryos, floral meristems, and organs, the underlying mechanisms of axis specification remain largely obscure. Less functionally significant aspects of plant asymmetry (e.g. the handedness of spiral phyllotaxy) may originate in random events and therefore have no genetic control.

CONTENTS

WHAT IS SYMMETRY?

An object can be said to show symmetry if it appears the same after a transformation—for example, a rotation, a shift along a straight line to a new position or on reflection about a plane running through it (for more extensive, nonmathematical considerations of symmetry, see 21, 91). Symmetry therefore implies uniformity. The most symmetrical form possible for an organism is a sphere—a sphere remains the same on rotation or reflection in any axis passing through its midpoint and therefore shows infinite planes of rotational and reflectional symmetry. The early

1040-2519/00/0601-0349$14.00

stage embryos of higher plants and many animals approximate morphologically to spheres. In contrast, mature plant organs tend to show reduced symmetry. A leaf, for example, usually has only one vertical plane of reflectional symmetry passing through its midrib (a form of symmetry termed *bilateral* by biologists). Development of a plant therefore involves loss of uniformity as a consequence of the formation and elaboration of axes of asymmetry. Many of these axes can be traced back to asymmetry specified in early embryogenesis.

In many model animal species, development of asymmetry is understood well enough to allow it to be divided into discrete phases. First, an axis is specified—often by long-range signals or unequal partitioning of a determinant within a cell before division. Second, specification of different identities occurs in domains along the axis. This usually results from activation of different transcription factors that may interact to reinforce the distinctions between domains. Third, short-range interactions can then elaborate the pattern, a process that may be coupled with growth.

This review considers how axes of asymmetry might be formed during higher plant embryogenesis and how they can be maintained and elaborated during growth to maturity. Of particular interest is how the mechanisms that specify different axes are co-ordinated and interact to ensure a normally patterned, functional plant.

ASYMMETRY IN EMBRYOGENESIS

Apical-Basal (A-B) Asymmetry

The* Fucus *Paradigm Plant A-B axis formation has been extensively characterized in brown algae, mainly members of the genus *Fucus*. Although they share no common multicellular ancestor with angiosperms and are therefore likely to have adopted different solutions to axis specification, fucoid algae provide insights into mechanisms that might operate in angiosperms. The subject has been reviewed recently (15, 29) and is summarized only briefly here. Fucoid algae produce spherical free-floating eggs lacking all signs of polarity. After fertilization, polar environmental cues (e.g. directional light or gravity) establish an axis of asymmetry that is initially labile and can reorient in response to a change in the direction of environmental cues. Polarity subsequently becomes fixed and organelles are asymmetrically distributed along the A-B axis. In the absence of polar environmental cues, zygotes are able to form A-B axes apparently at random, although the site of sperm entry may have an influence. After axis fixation, the cell wall at the basal pole subsequently bulges and a new cell wall is laid down perpendicular to the A-B axis, dividing the zygote asymmetrically into a larger apical cell, from which the shoot-like thallus forms, and a smaller basal cell, which gives rise to the root-like rhizoid. The cell wall is implicated in maintaining A-B asymmetry in the early embryo because a cell isolated with its walls intact can regenerate an embryo retaining the original polarity. In contrast, isolated protoplasts regain spherical symmetry and sensitivity to environmental cues. Ablation experiments suggest that signaling between apical and basal cells is not necessary to reinforce cell fates (e.g. an apical cell retains its identity when the rhizoid cell is ablated, and vice versa), but that cell walls may continue to have a role in this process.

A-B Asymmetry in Angiosperms Asymmetry along the apical basal (A-B) axis is usually apparent in mature angiosperm embryos in the position of the shoot meristem apically and the root meristem basally separated by embryonic stem (hypocotyl) and root. The A-B axis of asymmetry is the first to become apparent in embryogenesis, and it forms the major axis of growth of roots and shoots after germination.

As in fucoid algae, initial division of the angiosperm zygote is usually asymmetric, frequently resulting in a large, vacuolated basal cell and a smaller, cytoplasmically dense apical cell (Figure 1, see color plate) (71). Subsequent embryonic development has been best characterized in *Arabidopsis* by observation and fate mapping of genetically marked cells (81), and involves a relatively invariant pattern of cell division. The basal cell of *Arabidopsis* divides horizontally to form a filamentous suspensor of 6 to 9 cells through which metabolites can pass to the developing embryo, and may be a source of factors necessary for normal embryonic growth (108, 109). The most apical derivative of the basal cell, the hypophysis, follows a different fate and divides asymmetrically to give a smaller, apical daughter cell that forms the quiescent center and a larger basal cell that gives rise to the central (columella) root cap initials (25, 80). The remainder of the embryo is derived from the original apical cell, which divides twice parallel to the A-B axis and subsequently perpendicular to it to form an embryo with two four-celled tiers. The upper tier gives rise to the most apical structure of the embryo—the shoot apical meristem and most of the cotyledons, whereas the lower tier contributes the basal part of the cotyledon, the hypocotyl, embryonic root, and the remaining initials of the root apical meristem.

Maternal Orientation of the A-B Axis The mechanisms of A-B axis specification in angiosperm embryos remain enigmatic. In contrast to fucoid algae, the zygotes of most angiosperms are derived from egg cells that show A-B asymmetry before fertilization. They are usually elongated and show an asymmetrical distribution of organelles along the future A-B axis (71). They are also oriented relative to maternal tissues and the female gametophyte, so that the apical pole of the embryo faces the proximal part of the ovule. Polarity of the gametophyte can in turn be traced to early stages of ovule development when the archesporial cell (the diploid precursor of the female gametophyte) elongates along the future embryonic A-B axis (20, 55, 76), and this asymmetry is maintained through subsequent development of the gametophyte in the polarity of meiotic and mitotic divisions and the fate and migration of haploid nuclei (84).

The invariant orientation of the embryonic A-B axis implies that it is determined by asymmetry of the ovule. However, embryos can form an A-B axis in the absence of ovules—for example, from cultured protoplasts (33, 51), zygotes (33, 38, 43), or zygotes formed by fertilization of spherical egg cell protoplasts in vitro (13). Ectopic embryos can also be formed from epidermal cells of *Arabidopsis* plants misexpressing the embryo-specific gene *LEAFY COTYLEDONS1* (46) or naturally from the leaves of certain species (94). Therefore, A-B axis formation might be regarded as an intrinsic property of embryogenic cells, analogous to that of *Fucus* zygotes which can form an axis in the absence of asymmetric environmental cues. In the same way that directional light or gravity sets the direction of the *Fucus* axis, asymmetry of the ovule might influence the orientation of the embryonic A-B axis in angiosperms. Somatic embryos produced in culture often orient their A-B axis with reference to existing asymmetry of the embryonic tissue or isolated initial. For example, somatic embryos can be induced to form from pollen grains of *Hyoscyamus niger*, which first undergo an asymmetric division to produce a larger vegetative cell and a smaller generative cell, as in normal pollen development. Embryos develop from the vegetative cell with their basal pole toward the generative cell, which itself divides to form a suspensor-like structure (70). These observations favor a mechanism in which the axis is oriented by existing internal asymmetry of the embryogenic cell, rather than by signals from other tissues.

In genetically tractable animal species, specification of at least one primary axis of asymmetry involves maternally expressed gene products deposited in the egg [e.g. anterior-posterior axis formation in *Drosophila* embryos; summarized in (14)]. These genes and others needed for their expression and deposition of their products are identified by maternal-effect mutations disrupting embryonic pattern. Genetic analysis in *Arabidopsis* has identified the *SHORT INTEGUMENT (SIN)* gene as a potential component of the maternal contribution to embryonic asymmetry (72). Homozygous *sin* mutants produce defective embryos that often lack A-B asymmetry, even when fertilized with pollen from wild-type plants. This suggests that *SIN* is needed maternally for embryo polarity (72). In addition, *SIN* is required for at least one aspect of ovule asymmetry—cells of *sin* mutant integuments (maternal tissues around the gametophyte) show reduced elongation along the long axis of the ovule, corresponding to the embryonic A-B axis (74). Therefore, the embryonic effect of *sin* mutations might result from polarity defects in the ovule.

The Role of Asymmetric Cell Division The first asymmetric cell division of the angiosperm zygote appears to be necessary for elaboration of a normal A-B axis. *Arabidopsis* zygotes lacking *GNOM/EMB30* (*GN*) gene activity show reduced elongation along the A-B axis, undergo a more symmetric first division, and form embryos with variable defects in A-B polarity (57, 59, 82). The most severely affected embryos develop into spherically symmetrical balls of tissue, whereas in others, A-B polarity is reversed with respect to maternal tissues (102). Similarity of the *GN* gene product to a yeast protein required for membrane vesicle formation suggested a role in export of material to growing cell walls (either existing walls that are growing or new walls being formed at cell division). This view is supported by postembryonic expression of *GN* in wild-type plants and the requirement for *GN* activity in cell elongation, division, and adhesion in a variety of tissues (17, 82). Because the primary defect in *gn* mutant zygotes appears to be failure of the first asymmetric division, this division appears necessary for correct A-B asymmetry.

Genes Elaborating the A-B Axis Mutations in other *Arabidopsis* genes affect development of specific elements of the A-B pattern. For example, *gurke* (*gk*) mutant embryos often fail to form cotyledons and a shoot apical meristem (97); *hobbit* (*hbt*) mutant embryos lack root meristems because they failed to undergo a horizontal asymmetric division in the hypophysis at the eight-cell stage (Figure 1, see color plate) (80, 106), and *fackel* (*fk*) mutant embryos lack hypocotyl tissue between apical and basal structures (59). The *gn* mutation is epistatic to *gk*, *fk*, and *hbt* mutations, affects embryos earlier in development, and disrupts A-B patterning globally. Therefore, *GK*, *FK*, and *HBT* may be expressed in domains along the A-B axis in response to *GN*-dependent polarity. The domains are proposed to correspond to apical or basal tiers of the eight-cell embryo (*GK* and *FK*, respectively) or the hypophysis (*HBT*; Figure 1, see color plate). To explain why each mutation has an effect first in one domain, but then subsequently affects neighboring cells, interaction between domains is proposed to refine and elaborate the A-B pattern. For example, lower tier cells inhibit suspensor fate in the hypophysis and allow cells in both domains to interact at their boundary to organize a functional root meristem (reviewed in 39a, 58). Although the patterns of *GK*, *FK*, and *HBT* expression have yet to be reported, other genes show early asymmetric expression along the A-B axis. These include *AtML1*, a homeobox gene of unknown developmental function, which is first expressed in the apical cell after the first embryonic division (47). This indicates that an A-B pre-pattern capable of directing asymmetric gene expression exists at this stage.

A Role for Auxin in A-B Asymmetry? Analysis of another class of *Arabidopsis* mutants has suggested that the phytohormone auxin is involved in A-B axis development. Several lines of evidence suggest that the *MONOPTEROS* (*MP*) and *BODENLOS* (*BDL*) loci act in response to auxin signaling: *MP* encodes a transcription factor capable of binding auxin-responsive promoter elements (37) and *mp* mutant plants produced via tissue culture show phenotypes characteristic of disrupted auxin activity (69). Similarly, *bdl* mutants are insensitive to auxin and share other phenotypes with the auxin-insensitive mutant *axr1* (35). Mutations in either *MP* or *BDL* block formation of the central and basal regions of the embryo. The defects can be traced back to the two-cell stage, where the apical cell may divide horizontally, rather than vertically, resulting in an embryo with four cell tiers expressing apical characters (10, 35). Together, these findings implicate auxin signaling in specification of basal identity early in embryonic A-B axis formation. However, the expression of *MP* and *BDL* provides no clues as to where the potential auxin signal might come from. *MP* mRNA appears to accumulate uniformly along the A-B axis of early wild-type embryos (37), which suggests that

MP activity might confer sensitivity to auxin on all these cells. The observation that the *bdl* mutation affects development of apical as well as basal regions in an *mp* mutant background suggests that *BDL* is also expressed apically (35).

Signals Reinforcing A-B Asymmetry Signaling between the products of the apical and basal daughter cells is also implicated in maintaining A-B asymmetry. Mutations in a number of different *Arabidopsis* genes, including two *RASPBERRY* loci (107), three *ABNORMAL SUSPENSOR* loci (85), and *TWIN2* (*TWN2*) (110), lead to arrest of embryo development. Suspensor cells are then able to assume embryonic patterns of division and marker gene expression, suggesting that a viable embryo is needed to suppress embryo fate in the suspensor. At least one of these genes, *TWN2*, encodes a housekeeping function (valyl-tRNA synthetase) necessary for continued growth and division of the original embryo (110). In contrast, *twn1* mutations allow formation of ectopic embryos without arresting development of the original embryo, suggesting that *TWN1* might have a more specific role in the inhibition of embryo fate in the suspensor, from as early as the two-cell stage (101).

Inside-Outside Asymmetry of the Embryo

An histological distinction between inner and outer parts of the *Arabidopsis* embryo becomes apparent from the eight-cell stage (Figure 1, see color plate) (59). Previous divisions of the original apical cell have been anticlinal, producing new cell walls perpendicular to the surface of the embryo. The next round of cell division is periclinal and asymmetric, as new cell walls are formed approximately parallel to the surface. Larger daughter cells on the outside of the embryo form the protoderm. They subsequently divide only anticlinally to give rise to the epidermis of the embryo, including the outer cell layer of the shoot apical meristem. The behavior of the smaller, inner daughter cells depends on their position along the A-B axis. In the basal region, they are split by vertical, periclinal divisions to give an inner cell, which will give rise to vasculature and pericycle, and an external cell, which will form ground tissue (light-green and darker-green cells in Figure 1, see color plate). These cells are then subdivided by further vertical, periclinal divisions. Vertical anticlinal divisions increase the number of cells in the circumference of each layer, and horizontal divisions subdivide the basal tier of the embryo (79). The result is a hypocotyl and root consisting of concentric cell layers that appear radially symmetrical in transverse section (Transition Stage embryo in Figure 1, see color plate). Division of the hypophysis appears similarly ordered, whereas those of internal cells in the apical region are more randomly oriented.

Origins of Inside-Outside Asymmetry In addition to having A-B asymmetry, the zygote is also inherently asymmetric along an axis from its inside to its outside (or vice versa). Therefore, the first periclinal divisions in the eight-cell embryo might be oriented by signals from outside the embryo, or polarity within the cells

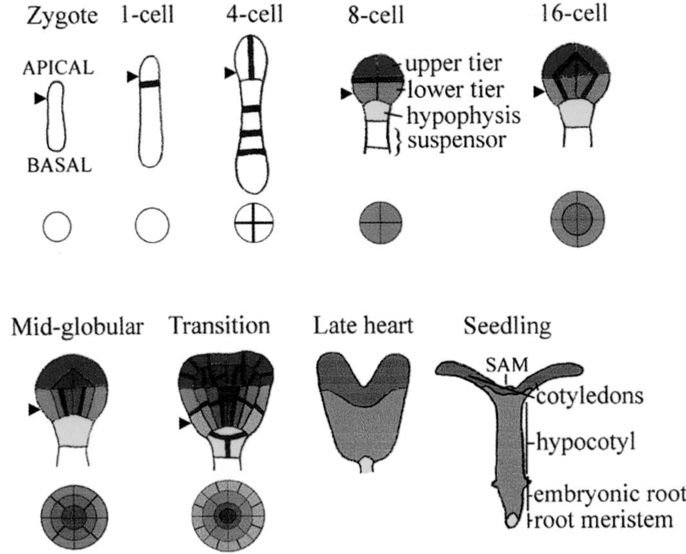

Figure 1 Axes of asymmetry in *Arabidopsis* embryogenesis. Embryos are shown in longitudinal sections along the apical-basal axis above, with transverse sections at the positions of arrowheads below. Cell walls formed since the previous stage are shown by thicker lines and derivative of the upper and lower tiers and hypophysis in blue, green, and yellow, respectively. [Redrawn from (79, 59).] Not to scale.

themselves. Removal of the protoderm of lemon embryos at the globular stage has provided no evidence that external or internal signals play a part in specifying protoderm cell fate: Exposed underlying cells proliferated to form undifferentiated callus (16). One speculation is that inside-out polarity is set by the zygotic cell wall. Internal cell walls of the embryo are produced de novo, whereas all external walls are descended from that of the zygote and may therefore inherit a determinant of outside identity. After germination, the outer cell wall grows to cover the entire shoot and primary root (lateral roots organize a new epidermis from internal cells of the cortex; 54). This proposal is consistent with the fate of cells produced by aberrant periclinal divisions in the shoot protoderm of dicots. The inner daughter cell, which lacks derivatives of the external wall, is displaced internally and assumes a non-protoderm fate, while the outside daughter cell retains the external wall and protoderm identity. Similarly, inner daughter cells of the protoderm are able to contribute to internal tissue of monocot organs and dicot petals during normal development (discussed in 93).

The Role of Asymmetric Cell Division The periclinal divisions in the eight-cell *Arabidopsis* embryo appear necessary for correct inside-out patterning. Mutations in a number of genes, including *KNOLLE* (*KN*), disrupt this division and thereby prevent formation of a normal protoderm (26, 48, 59, 96, 107). *KN* encodes a syntaxin required for membrane vesicle fusion during cell wall synthesis, and *kn* mutant embryos frequently form misoriented or incomplete cell walls that allow cytoplasmic continuity between daughter cells along the inside-outside axis (42). Defects in patterning along this axis are therefore consistent with a failure of cell division to partition a determinant of inner or outer cell fate. However, *kn* and other mutants including *fass*, which shows random orientation of cell divisions (59, 98), retain some aspects of inside-outside asymmetry. They produce vascular tissue centrally, surrounded by ground tissue, although the number and arrangement of cells in each layer may be disrupted. Cells toward the outside of *kn* mutant embryos also express the lipid transfer protein gene *AtLTP1*, which is normally restricted to the single layer of protoderm cells (102). Periclinal cell divisions therefore appear necessary to maintain or elaborate correct inside-outside polarity, rather than to form it.

Genes Elaborating Inside-Outside Asymmetry Genes that may respond to radial asymmetry in the lower part of the embryo have been identified by mutations that affect formation of functional root meristems (79). These include *SCARECROW* (*SCR*), which encodes a transcription factor necessary for the asymmetric periclinal division that splits outer cortical initials from those of the endodermis internally (23). *SCR* is expressed in the common cortical-endodermal initial before division, which suggests that it responds to an inside-outside pre-pattern. Subsequent expression is confined to the endodermal daughter cell, which suggests that the periclinal division further refines the pattern along the inside-outside axis. In the absence of the periclinal division, progeny of the initial cell express both cortical and endodermal characters, consistent with this view. In the apical region of the embryo, expression of *ATML1* becomes confined to the protoderm following the first asymmetric division (47), while that of *WUSCHEL* (*WUS*), required for apical meristem function, is established in the internal daughters of the most apical cells (56). This indicates that a pre-pattern along the inside-outside axis can also direct gene expression in this region.

Interaction Between Axes of Asymmetry The observation that inside-outside symmetry has different consequences along the apical-basal axis (e.g. in patterns of gene expression and cell division) implies that the two axes interact to specify cell fate. Expression of *WUS* in internal apical cells indicates that interaction occurs at least as early as the 16-cell stage. Responses to the two axes can be uncoupled genetically. For example, *gn* mutants, which show defects in A-B polarity soon after fertilization, may produce a normal protoderm layer expressing the marker *AtLTP1* (102), which implies that inside-out asymmetry is not dependent on A-B asymmetry. Similarly, *kn* mutants show a disrupted inside-outside pattern but retain aspects of A-B asymmetry, including asymmetry of the first zygotic division.

Bilateral Symmetry of the Embryo

Development of Bilateral Symmetry The combination of A-B and inside-out axes of asymmetry results in a globular embryo that shows full rotational symmetry in the arrangement of tissues, i.e. the embryo appears the same after any rotation about its A-B axis (although the degrees of rotational symmetry are fewer if the radial walls separating cells in each layer are taken into account; Figure 1, see color plate). At this stage, a further axis of asymmetry becomes imposed on the embryo, which in most dicots results in formation of a pair of cotyledons. Viewed in transverse section, apical cells at the same distance from the center are now only equivalent if diametrically opposite each other. The embryo can therefore be said to show twofold rotational symmetry about its A-B axis. Viewed from the side, the heart-shaped embryo has only one plane of reflectional symmetry (along its A-B axis) and therefore is often said to show bilaterally symmetry—a term usually reserved for onefold rotational symmetry. For convenience, the term bilateral symmetry is used here.

Genes needed to impose the additional asymmetry might therefore be identified by mutations that block the globular to heart transition. However, this must be treated with caution. First, formation of cotyledons is an apical character and can be affected by mutations in A-B pattern genes [e.g. *FK* (59)]. Second, a block in globular-heart transition is among the most common class of embryonic mutant phenotype in *Arabidopsis*, which suggest that reduced activity of any one of over 200 genes can become limiting at this stage (19, 30). In at least two cases, the affected genes appear to be required for housekeeping functions (83, 100). Similarly, mutations that block bilateral symmetry of maize and rice embryos (which first becomes apparent in flattening of the scutellum) are equally frequent (40, 88).

Identification of genes that specify bilateral symmetry before cotyledon formation will therefore require detailed genetic and phenotypic characterization of a large number of mutants.

Auxin and Bilateral Symmetry A class of more informative *Arabidopsis* mutations allows formation of cotyledons with full rotational symmetry. The result is an embryo with a single, cup-shaped cotyledon that resembles a golf tee. Such cotyledons are produced occasionally by *gn* mutants with apical defects (82), which suggests that asymmetry might be dependent on correct specification of apical identity. However, mutations in two additional genes, *PIN-FORMED1* (*PIN1*) and *PINOID* (*PID*), give similar phenotypes and implicate polar auxin transport in imposition of bilateral symmetry (6, 44, 66). *PIN1* encodes a component of the auxin efflux carrier that functions in directional transport of auxin (32, 67). *PID* has a similar mutant phenotype to *PIN1*, is required for normal activity of polar auxin transport in mature tissues, and acts redundantly with other genes involved in auxin responses, which suggests that it has a similar role (6). Involvement of auxin transport is further supported by the finding that globular embryos of black mustard, *Brassica juncea*, form cup-shaped cotyledons when cultured in the presence of auxin transport inhibitors (34, 44). The question remains as to how movement of auxin is involved in specifying cell fate at the apex. One proposal is that auxin is removed from cells adjacent to cotyledon initials and low concentrations are necessary for non-cotyledon fates (34). The view is supported by the finding that embryos of *B. juncea* also produce cup-shaped cotyledons when cultured in the presence of high concentrations of exogenous auxin (34). However, involvement of auxin transport in another direction, e.g. along the A-B axis, cannot be ruled out.

Genes Elaborating Bilateral Symmetry Formation of cup-shaped cotyledons may result from failure to specify non-cotyledon fate in cells which would normally lie in the sinus between cotyledons [termed the *embryonic peripheral region*, EPR, by Long & Barton (45)]. Therefore cotyledon identity might be a default state and its inhibition in the EPR necessary for bilateral symmetry. This hypothesis is supported by the expression pattern of the transcription factor gene *AINTEGUMENTA* (*ANT*). In postembryonic growth, *ANT* expression is confined to lateral organ initials and primordia (27) and acts redundantly in formation of floral organs (27, 41). It is expressed in developing cotyledons, and might therefore have a similar redundant role in their formation. However, at an earlier stage of embryo development (late globular), *ANT* is expressed in a radially symmetric ring of cells around the apex, which includes the EPR and only becomes confined to cotyledon initials before the transition to the heart stage (45). This shift from radial to bilateral expression of *ANT* occurs after two other genes have begun to show bilateral expression. The *CUP-SHAPED COTYLEDON1* (*CUC1*) gene, encoding a NAC protein of unknown biochemical function, is expressed in a stripe of cells running across the apex of the late globular embryo from which the EPR and SAM will form (2). Embryos lacking activity of *CUC1* and a second gene, *CUC2*, often produce cup-shaped cotyledons and lack SAMs, which suggests that *CUC* activity is required to repress cotyledon fate (1, 2). The homeobox gene *SHOOT MERISTEMLESS*, *STM*, is expressed in a similar domain to *CUC1* in the late globular embryo (45). Strong *stm* mutations prevent formation of a functional SAM, but also allow production of cotyledons that are united at their bases (5). Therefore, *STM* also acts to inhibit cotyledon fate in the EPR, although it does not appear to do this by repressing *ANT* expression (45). *CUC* activity is required for *STM* expression in the late globular embryo, which suggests that bilateral *CUC* expression is responsible for the bilateral expression of *STM*. However, *STM* expression can be detected in embryos at earlier developmental stages than those reported to express *CUC1*, which suggests that interactions between these two genes might be more complex, or that earlier *CUC1* expression has yet to be detected.

Although *STM* is not necessary for bilateral symmetry, its early pattern of expression reveals how this symmetry might arise. It is first detected in one or two cells on the flank of the midglobular embryo and subsequently becomes established in a second domain diametrically opposite. Intermediate cells then express *STM* to form the stripe seen at the late globular stage (45). If the midglobular embryo possesses only A-B and radial symmetry, cells at the same position along the A-B axis and at the same horizontal distance from it should be equivalent and express the same complement of genes (as observed for the ring of *ANT* expression). The finding that a group of asymmetrically placed cells express *STM* suggests that transcription is not activated in response to existing polarity, but is determined at random. Because it is unlikely that a random event would also establish the second domain of *STM* expression exactly opposite the first, the first domain is likely to determine the position of the second by interaction (e.g. by inhibiting EPR identity locally and therefore promoting cotyledon fate).

Unlike the A-B and inside-out axes of the embryo, orientation of the axis of bilateral symmetry appears to have no functional consequences for the plant, and might therefore be left to chance. Alternatively, it may be specified by maternal tissues. The ovule itself is bilaterally symmetrical. It imposes additional asymmetry on the late-stage embryo, which has to fold its cotyledons back along its A-B axis as it grows to fill the ovule. The axis of bilateral symmetry in the mature embryo coincides with that of the ovule and, in addition, one cotyledon becomes longer than the other (although both contain the same number of cells). Because heart-stage embryos appear randomly oriented, the ovule would appear to position the late-stage embryo by forcing it to rotate about its A-B axis to fit the seed coat, rather than by specifying its axis of bilateral symmetry earlier in development.

Although reduction in rotational symmetry has been discussed in the context of the apical region, it also becomes apparent in the central parts of the dicot embryo. In *Arabidopsis*, vascular tissue of the root and hypocotyl shows a bilateral arrangement, with two opposite xylem poles extending radially from the stele (31). The positions of these poles correspond to those of the cotyledons, but it is not yet known whether cotyledons might induce xylem formation, xylem induce cotyledons, or whether both respond independently to bilateral symmetry. A reduction in the degree of rotational symmetry is also apparent in the epidermal cells of the mature embryo in the alternate arrangement of two cell types around the circumference [e.g. hair-cell and non-hair-cell progenitors in the embryonic root, protruding and nonprotruding cells in the hypocotyl (9, 25)] Signaling from underlying cells or the radial walls separating them is implicated in specifying the fates of different cells, and interaction between epidermal cells in reinforcing them (8).

ELABORATION OF ASYMMETRY AFTER EMBRYOGENESIS

Phyllotaxy and Translational Symmetry in the Shoot

By maturation, most dicots show A-B asymmetry, with root meristem and SAM separated by hypocotyl and embryonic root tissues. After germination, the SAM gives rise to shoot tissue in the A-B axis and lateral organs (e.g. leaves), which elaborate their own axes of asymmetry (see below). The relative positioning of organ and stem components (termed phyllotaxy) is usually regular, and can therefore be described in terms of symmetry. For example, leaves at successive positions (nodes) along the A-B stem axis may occur immediately above each other and therefore one node can be made to resemble the node above or below by a shift along the A-B axis (referred to mathematically as a *translation*). Alternatively, a node may need to be both translated and rotated to resemble an adjacent node, in which case the stem shows spiral symmetry. However, the term *spiral phyllotaxy* is traditionally reserved for one specific example of spiral symmetry in which plants form a single organ at each node and successive organs are offset by less than 180° [often approximately 137° (73)].

Phyllotaxy results from specification of organ and stem fate in cells of the SAM. Although the mechanisms involved remain poorly characterized, most available evidence suggests that organ initials specify the position of future organs by a process involving local inhibition of organ fate. For example, surgical ablation of one group of leaf initials in the fern, *Dryopteris*, allowed subsequent leaves to be initiated closer to the site of ablation (105). Similarly, premature leaf initiation from the tomato SAM, induced by localized application of expansin protein, could disrupt the positions at which subsequent leaves were formed (28). However, the positions of future leaves in *Arabidopsis* are marked by expression of the *PINHEAD/ZWILLE* gene in vascular initials below future organ initials. This occurs before changes in expression of other genes that mark organ fate become apparent (e.g. down-regulation of *STM* in the SAM), which suggests the possible involvement of inductive signals originating basally (53).

Spiral phyllotaxy involves an increase in asymmetry. Shoots with spiral phyllotaxy usually develop from a seedling with an opposite pair of cotyledons (i.e. two-fold rotational symmetry). Maintenance of the phyllotaxy shown in cotyledons would result in a shoot with an opposite pair of leaves at each node, and, because leaves inhibit leaf formation locally, with alternate nodes rotated by 90°. This form of decussate phyllotaxy is seen in many species, including *Antirrhinum*. However, in most plants that show spiral phyllotaxy, symmetry is reduced at nodes with a single leaf (they now have only one plane of reflectional symmetry) and also by the inherent chirality of a spiral (it is either left- or right-handed). Species with spiral phyllotaxy tend to produce equal proportions of left- and right-handed forms, and no genetic basis for the handedness of the spiral has been detected, which suggests that it is specified at random (3). In *Arabidopsis*, the first two leaves are initiated together, opposite each other and at 90° to the cotyledons. A single-leaf primordium is then formed nearer to one or other of the initial leaves, to either its left or right side. If it is closer to one side, the resulting spiral phyllotaxy is left-handed; if it is closer to the other, the spiral is right-handed. Therefore, the handedness of the spiral seems likely to represent random positioning of the third-leaf primordium.

Many other randomly determined inequalities may operate to reduce symmetry in plant development. For example, ovules in one chamber of an *Arabidopsis* carpel develop later than in the other, and the pattern of vascular tissue or lobes on one side of a leaf is not an exact reflection of the other side. Inequality between cotyledons has been suggested to result in the unorthodox, asymmetric morphology of *Monophyllea horsfieldii*. This species, like other members of the Gesneriaceae, produces embryos with a pair of equal-sized cotyledons but no SAM. One of the cotyledons grows to produce a large photosynthetic organ that then forms reproductive inflorescences from its midrib. The other cotyledon appears to be capable of growth but is repressed by its partner, because all embryos in which one cotyledon has been removed at random can produce a normal plant (99).

Formation of New Axes of Asymmetry in Organs

In contrast to the stem, which maintains the original A-B and radial axes established early in embryogenesis, formation of lateral organs involves initiation and elaboration of new axes of asymmetry. Organ initiation first becomes apparent as a group of cells grow out from the shoot apical meristem to form the proximal-distal (P-D) organ axis. Further asymmetry along the P-D axis may become apparent later, for example, in monocot leaves between proximal sheath and distal blade tissue, or in simple dicot leaves between the proximal petiole and distal blade. A mature organ usually also shows asymmetry along its dorsal-ventral (D-V, or adaxial-abaxial) axis. The dorsal epidermal cells of leaves often differ from ventral, and asymmetry is often seen in the arrangement of internal cell layers. Leaves and other lateral organs also appear to have a medial-lateral axis (M-L, from midline to edge). Leaf primordia are either flattened along this axis at emergence from the SAM or become flattened by growth of the primordium soon after initiation. Further M-L asymmetry is reflected in the position of the midrib medially and blade tissue laterally.

Proximal-Distal Asymmetry Because initiation of an organ primordium from the SAM involves growth along the P-D axis, this axis is the first to become apparent (52). Mutations that prevent specification of the P-D axis are therefore likely to block organ initiation, which makes them difficult to distinguish from mutations in genes needed to specify organ fate. The mechanisms of P-D asymmetry are, as a consequence, poorly understood.

Inflorescence meristems of *Arabidopsis* plants mutant for either the *PIN1* or *PID* genes produce stem-like tissue without lateral organs. However, the inflorescence meristems of both *pid* and *pin1* mutants produce lateral bulges that resemble organ primordia but fail to elaborate a P-D axis (6, 66), which suggests that the two genes are needed for P-D outgrowth. *PIN1* and *PID* are necessary for normal polar auxin transport, which suggests that auxin is involved in P-D axis formation. Homeobox genes of the *knotted1*-like family (*knox* genes) are also implicated in formation or elaboration of the P-D axis. Expression of these genes is normally confined to the SAM and excluded from organ initials from before primordium initiation (reviewed in 90). Ectopic *knox* gene expression in organs causes them to develop proximal identities (e.g. sheath) at a more distal position. Although these results indicate that *knox* gene activity can influence the P-D axis of leaves, their implications for the role of *knox* genes in wild-type organs are not clear (discussed in 31, 39). In the *Arabidopsis* carpel, the *ETTIN* (*ET*) gene is required to prevent proximal development of distal stylar tissue (87). At least part of *ET* function appears to be repression of TSL2, a nuclear protein kinase needed for formation of the distal carpel (75, 86). *ET* encodes a transcription factor similar to the *MP* gene product and is therefore likely to act in auxin-mediated responses. It is transcribed uniformly along the P-D axis of carpel initials and primordia, but only in a ventral domain. Again, the significance of these findings for P-D axis elaboration is not yet clear (discussed in 65).

Dorsal-Ventral Asymmetry In contrast, mutations in a number of different genes disrupt D-V asymmetry of lateral organs. For example, loss of *PHANTASTICA* (*PHAN*) activity of *Antirrhinum* allows production of leaves and petal lobes consisting only of ventral cell types (104), which suggests that the MYB transcription factor encoded by *PHAN* is needed for specification of dorsal identity and that ventral identity occurs by default (Figure 2) (103, 104). The semidominant *phabulosa-1d* (*phb-1d*) mutation in *Arabidopsis* has the opposite effect of dorsalizing all lateral organs in a dose-dependent manner, although semidominance does not clarify whether the mutation involves loss of a ventralizing activity, or a gain-of-function of a dorsalising gene (Figure 2) (60).

Although *PHAN* is required for dorsal cell identity, it is expressed in both dorsal and ventral organ initials; this suggests that other, spatially restricted factors are also needed for asymmetry. A number of candidates have been identified in *Arabidopsis*. Loss-of-function mutations in the *ARGONAUTE* (*AGO*) gene, which encodes a potential translation initiation factor, cause partial ventralization of organs, whereas constitutive *AGO* overexpression partially dorsalizes them (11, 53).

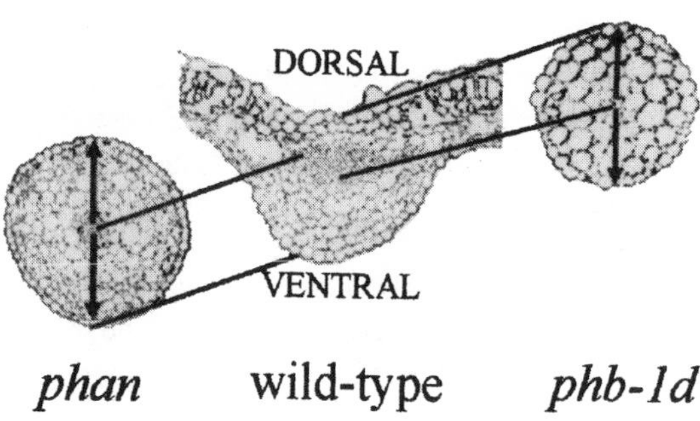

Figure 2 Effects of *phantastica* and *phabulosa-1d* mutations on dorsal-ventral asymmetry of leaves. Sections were made perpendicular to the proximal-distal axes *phan* mutant leaf and the midrib of a wild-type leaves of *Antirrhinum* and a *phab-1d* mutant leaf of *Arabidopsis*. Not to scale.

Although this suggests that *AGO* specifies dorsal identity, it is expressed ubiquitously in organ initials and primordia of wild-type plants. A structurally similar gene, *PINHEAD/ZWILLE* (*PNH/ZLL*), is expressed in an apical region of the SAM including organ initials but confined to a dorsal domain of organs at the time of primordium initiation (53, 63). Whereas *pnh* mutations allow formation of organs with normal D-V asymmetry, *ago*; *pnh* double mutants (or *ago* mutants carrying a single wild-type copy of *PNH*) show more severely ventralized leaves, which suggests that *PNH/ZLL* and *AGO* act redundantly to specify dorsal fate (53).

Conversely, several members of the YABBY transcription factor gene family, *FILAMENTOUS FLOWER* (*FIL*), *YABBY2* (*YAB2*), and *YAB3*, are involved in specifying ventral identity in *Arabidopsis* organs (77, 89). Expression of *FIL*, *YAB2*, and *YAB3* is uniform in organ initials within the SAM but becomes limited to ventral regions of primordia (including cotyledons) at initiation. *FIL* acts redundantly with *YAB3* to specify ventral fate. Constitutitve expression of either *YAB3* or *FIL* appears sufficient to confer ventral characters on the dorsal regions of leaves (89).

Dorsal expression of *ZLL/PNH* and ventral expression of *YAB2*, *YAB3*, and *FIL* presumably occurs in response to D-V asymmetry present within the SAM (where patterned expression in organ initials is first seen). The three YABBY genes are expressed independently of each other, but regulated by *PHB* (89). Isolation of *PHB* should therefore provide a clue to the mechanisms acting upstream of ventral fate specification. However, because the D-V axis of the organ corresponds to the A-B axis of the stem, it seems likely that the D-V asymmetric pre-pattern is established in organs in response to the A-B axis of the SAM. This view is supported by surgical experiments in which incisions were made above leaf initials in potato and sesame, isolating them from contact with more apical SAM cells. This resulted in formation of ventralized organs (36, 92), and suggests that signals originating apically are necessary for dorsal fate. The origin of the P-D organ axis is less clear. At primordium initiation it corresponds to the inside-outside axis of the meristem and may therefore be derived from this. However, the inside-outside axis of organs appears to have another role, discussed below.

Interaction Between Axes The two major axes of organ growth (P-D and M-L) lie at right angles to each other and to the D-V axis, which implies that they are not specified independently (otherwise at least one of them should be random). They are also oriented with respect to the remainder of the plant, suggesting that at least two of them are derived from asymmetry present in the SAM. This connection is a functional necessity—shoots, for example, tend to grow toward light and therefore leaves, which are flattened in two axes perpendicular to the stem, expose their largest photosynthetic area to incident light.

Antirrhinum phan and *Arabidopsis phb-1d* mutant phenotypes suggest a connection between M-L growth and D-V asymmetry. Extreme *phan* or *phb-1d* mutant leaves consist of only ventral or dorsal cell types, respectively, and have no lateral blade; this suggests that both dorsal and ventral identities are required for M-L growth (60, 104). The mutants can also produce leaves that are mosaics of normal tissue and either ventralized tissue (in *phan*) or dorsalized tissue (in *phb-1d*). In both cases, juxtaposition of dorsal and ventral identities results in an ectopic leaf blade. Therefore, interaction between dorsal and ventral cells might direct M-L growth of the wild-type blade. The extreme *phan* or *phb-1d* mutant leaves also suggest that the M-L axis, although recognizable morphologically, might not be specified directly. In the absence of D-V asymmetry, mutant leaves are needle-like and show full rotational symmetry in the arrangement of tissue layers; this indicates that inside-out and P-D axes of asymmetry within the organ can be specified independently of the D-V axis. The fate of a cell along the M-L axis of a wild-type leaf might therefore be specified by is its coordinates along the inside-outside and D-V axes (a polar system), rather than along D-V and M-L axes (a Cartesian system). Organs are formed from all cell layers of the meristem. Therefore a mechanism of inside-outside asymmetry that involved cell layering could be inherited by organs from the SAM.

Dicot leaves tend to have few markers for M-L asymmetry (often only the distinctions between midrib, blade, and leaf edge). In contrast, the presence of a distinct marginal domain in the maize leaf is suggested by the *narrow sheath* (*ns*) mutant phenotype. In *ns* mutants, SAM cells that would normally become initials of the leaf margins fail to participate in primordium initiation, which results in narrower leaves lacking marginal characters (78). Maize leaves differ from those of dicots in showing marked flattening in the M-L axis at primordium emergence, whereas in dicots flattening usually occurs later. However, the maize *leafbladeless1* (*lbl*) mutation, which conditions a phenotype similar to *phan* in *Antirrhinum*, suggests that M-L flattening results from D-V interactions (95). Severely affected *lbl* mutant leaves lack dorsal cell types and are radially symmetrical, whereas other leaves produce ectopic M-L axes where dorsal and ectopic ventral identities are juxtaposed.

Asymmetry of the Flower

Apical-Basal (Radial) Asymmetry The flowers of most species, including *Antirrhinum* and *Arabidopsis*, show asymmetry in the identity of organs produced in concentric whorls. The axis of asymmetry corresponds to the A-B axis of the flower. Because the axis is usually highly compressed, it is more usually termed radial. The function and regulation of floral homeotic genes that specify the distinctions between organs in different whorls have been characterized extensively and are the subject of many recent reviews (e.g. 61). However, the mechanisms that specify A-B/radial asymmetry in the floral meristem, leading to activation of floral homeotic genes in specific domains, are less well understood [see, for example, (68)].

Dorsal-Ventral Asymmetry Flowers of many species, including *Antirrhinum*, show an additional, dorsal-ventral, axis of asymmetry coincident with the A-B axis of the inflorescence stem. Such flowers are termed zygomorphic or bilaterally symmetrical [the ecological significance and evolution of this character are discussed in (64)]. The *Antirrhinum* corolla consists of five distal petal lobes (two dorsal, two lateral, and one ventral) that differ in identity along the D-V axis and, with the exception of the ventral petal, also show internal D-V asymmetry. D-V asymmetry is also seen in the adjacent whorl, in which development of the dorsal stamen arrests and the remaining pair of lateral stamens develop differently from the ventral pair. Three genes, *CYCLOIDEA* (*CYC*), *DICHOTOMA* (*DICH*), and *RADIALIS* (*RAD*), are required for dorsal identity (18, 50). In *cyc* mutants, lateral petals adopt ventral identities and dorsal petals become partly lateralized. The remaining dorsal/lateral characters in *cyc* mutants can be attributed to *DICH* activity, because all petals are ventralized in *cyc;dich* double mutants. In wild-type flowers, *CYC* expression occurs in dorsal petal initials and in the dorsal-most part of lateral petals. However, *CYC* is required for normal development of lateral petals as a whole, which suggests that it acts non-cell-autonomously in their more ventral regions (18). *RAD* is likely to be involved in this process, because homozygous *rad* or *cyc* mutations condition similar phenotypes, and *rad/+*; *cyc/+trans*-heterozygotes show slight ventralization of lateral petals, which suggests that the two genes act sequentially in the same developmental pathway.

In contrast, loss of *DIVARICATA* (*DIV*) activity allows the ventral petal to assume lateral identity, which suggests that *DIV* acts to specify ventral fate (4). *DIV* activity is limited to the ventral petal by the action of the dorsalizing genes *CYC* and *DICH*, with *CYC* alone repressing it in lateral petals. Further evidence for antagonism of *DIV* and *CYC* is provided by the observation that three wild-type doses of *DIV* can partially overcome the dorsalizing effect of *CYC* in lateral petals. However, the range over which *CYC* acts non-cell-autonomously is increased by *DIV* activity; more complex interactions may thus be possible. Because the ventralizing effects of *DIV* are dose-dependent, one model for D-V axis specification is that a decreasing gradient of *DIV* activity is formed from the ventral to the dorsal part of the flower, as a result of *CYC* and *DICH* repressing *DIV* dorsally. Different levels of *DIV* activity along the D-V axis could then specify regional identities (4).

One effect of *CYC* expression is to retard growth of the dorsal part of the floral meristem. A related gene, *teosinte branched*, has a similar effect on growth of axillary meristems in maize (24), which suggests a common role for members of this gene family in growth control (22). To what extent growth-repression is necessary for the dorsalizing effects of *CYC* in the *Antirrhinum* flower is not yet known.

CYC expression is established in a dorsal domain of the early floral meristem, which suggests that it responds to A-B asymmetry of the inflorescence axis. Two observations support this view. First, *centroradialis* mutants of *Antirrhinum* produce ectopic flowers from the inflorescence apex (12). The flowers lack D-V asymmetry, which indicates that A-B asymmetry in the flanks of the inflorescence meristem may be needed. Second, ablation of parts of the inflorescence meristem dorsal to developing floral meristems or localized application of auxin can result in the formation of flowers with radial asymmetry (7).

PERSPECTIVES

Most of the genes known to influence asymmetry in plants appear either to act indirectly in the process (e.g. in cell division) or, because they have a restricted domain of expression or action, to be required for axis elaboration. The mechanisms of axis specification, which establish the fundamental asymmetry, remain poorly understood. A distinction between axis specification and elaboration may be partly artificial because new axes can be elaborated from existing ones in postembryonic development. However, the question of what regulates asymmetry of gene expression or action remains. Identification of upstream regulators of asymmetrically expressed genes like the YABBIES, may provide part of the answer. Axis specification in animals can result from a chemical gradient, which specifies position along the axis in a concentration-dependent manner. The finding that auxin transport and response mutations affect plant asymmetry has provided a hint that auxin might be involved in such a process, and its role should quickly become clearer. Further insights should also come from molecular analysis of *PHB* and *DIV* genes, which act dose-dependently and might therefore form gradients of activity which specify asymmetry along the D-V axis of lateral organs and flowers.

ACKNOWLEDGMENTS

I thank John Bowman for making results available before publication, and John Golz for his comments on the manuscript.

LITERATURE CITED

1. Aida M, Ishida T, Fukaki H, Fujisawa H, Tasaka M. 1997. Genes involved in organ separation in *Arabidopsis*: an analysis of the *cup-shaped cotyledon* mutant. *Plant Cell* 9:841–57
2. Aida M, Ishida T, Tasaka M. 1999. Shoot apical meristem and cotyledon formation during *Arabidopsis* embryogenesis: interaction among the *CUP-SHAPED COTYLEDON* and *SHOOT MERISTEMLESS* genes. *Development* 126:1563–70

3. Allard HA. 1945. Clockwise and counterclockwise spirality in the phyllotaxy of tobacco. *J. Agric. Res.* 71:237–42
4. Almeida J, Rocheta M, Galego L. 1997. Genetic control of flower shape in *Antirrhinum majus*. *Development* 124:1387–92
5. Barton MK, Poethig RS. 1993. Formation of the shoot apical meristem in *Arabidopsis thaliana*—an analysis of development in the wild-type and in the *shootmeristemless* mutant. *Development* 119:823–31
6. Bennett SRM, Alvarez J, Bossinger G, Smyth D. 1995. Morphogenesis in *pinoid* mutants of *Arabidopsis thaliana*. *Plant J.* 8:505–20
7. Bergbusch VL. 1999. A note on the manipulation of flower symmetry in *Antirrhinum majus*. *Ann. Bot.* 83:483–88
8. Berger F, Haseloff J, Schiefelbein J, Dolan L. 1998. Positional information in root epidermis is defined during embryogenesis and acts in domains with strict boundaries. *Curr. Biol.* 8:421–30
9. Berger F, Linstead P, Dolan L, Haseloff J. 1998. Stomatal patterning on the hypocotyl of *Arabidopsis thaliana* is controlled by genes involved in the control of root epidermis patterning. *Dev. Biol.* 194:116–34
10. Berleth T, Jürgens G. 1993. The role of the *monopteros* gene in organising the basal body region of the *Arabidopsis* embryo. *Development* 118:575–87
11. Bohmert K, Camus I, Bellini C, Bouchez D, Caboche M, Benning C. 1997. *AGO1* defines a novel locus of *Arabidopsis* controlling leaf development. *EMBO J.* 17:17–80
12. Bradley D, Carpenter R, Copsey L, Vincent C, Rothstein S, Coen E. 1996. Control of inflorescence architecture in *Antirrhinum*. *Nature* 379:791–97
13. Breton C, Faure JE, Dumas C. 1995. From in vitro fertilisation to early embryogenesis in maize. *Protoplasma* 187:3–12
14. Brody TB. 1999. *The Interactive Fly*. http://sdb.bio.purdue.edu/fly/aimain/1aahome.htm az
15. Brownlee C, Bouget F-Y. 1998. Polarity determination in *Fucus*: from zygote to multicellular embryo. *Semin. Cell Dev. Biol.* 9:179–85
16. Bruck DK, Walker DB. 1985. Cell determination during embryogenesis in *Citrus jambhiri*. II Epidermal differentiation as a one-time event. *Am. J. Bot.* 72:1602–9
17. Busch M, Mayer U, Jürgens G. 1996. Molecular analysis of the *Arabidopsis* pattern formation gene *GNOM*: gene structure and intragenic complementation. *Mol. Gen. Genet.* 250:681–91
18. Carpenter R, Coen ES. 1990. Floral homeotic mutations produced by transposon-mutagenesis in *Antirrhinum majus*. *Genes Dev.* 4:1483–93
19. Castel LA, Errampalli D, Atherton TL, Franzmann LH, Yoon ES, Meinke DW. 1993. Genetic and molecular characterisation of embryonic mutants identified following seed transformation in *Arabidopsis*. *Mol. Gen. Genet.* 241:504–14
20. Christensen CA, King EJ, Jordan JR, Drews GN. 1997. Megagametogenesis in *Arabidopsis* wild-type and the *Gf* mutant. *Sex. Plant Reprod.* 10:49–64
21. Coen E. 1999. *The Art of Genes—How Organisms Make Themselves*. Oxford: Oxford Univ. Press
22. Cubas P, Lauter N, Doebley J, Coen E. 1999. The TCP domain: a motif found in proteins regulating plant growth and development. *Plant J.* 18:215–22
23. Di Laurenzio L, Wysocka-Diler J, Malmy JE, Pysh L, Helariutta Y, et al. 1996. The *SCARECROW* gene regulates an asymmetric cell division that is essential for generating radial organisation of the Arabidopsis root. *Cell* 86:423–33
24. Doebley J, Stec A, Hubbard L. 1997. The evolution of apical dominance in maize. *Nature* 386:485–88
25. Dolan L, Janmaat K, Willemsen V, Linstead P, Schiefelbein JW, et al. 1993. Cellular organisation of the *Arabidopsis thaliana* root. *Development* 119:71–84

26. Dunn SM, Drews GN, Fischer RL, Harada JJ, Goldberg RB, Koltunow AM. 1997. *fist:* an Arabidopsis mutant with altered cell division planes and radial pattern disruptions during embryogenesis. *Sex. Plant. Reprod.* 10:358–67
27. Elliott RC, Betzner AS, Huttner E, Oakes MP, Tucker WQ, et al. 1996. *AINTEGUMENTA*, an *APETALA2*-like gene of *Arabidopsis* with pleiotropic roles in ovule development and floral organ growth. *Plant Cell* 8:155–68
28. Fleming AJ, McQueen-Mason S, Mandel T, Kuhlemeier C. 1997. Induction of leaf primordia by the cell wall protein expansin. *Science* 276:1425–18
29. Fowler JE, Quatrano RS. 1997. Plant cell morphogenesis: plasma membrane interactions with the cytoskeleton and cell wall. *Annu. Rev. Cell Dev. Biol.* 13:697–43
30. Franzmann LH, Yoon ES, Meinke DW. 1995. Saturating the genetic map of *Arabidopsis thaliana* with embryonic mutations. *Plant J.* 7:341–50
31. Freeling M. 1992. A conceptual framework for maize leaf development. *Dev. Biol.* 153:44–58
32. Gälweiler L, Guan CM, Müller A, Wisman E, Mendgen K, et al. 1998. Regulation of polar auxin transport by AtPIN1 in *Arabidopsis* vascular tissue. *Science* 282:2226–30
33. Grambow HJ, Kao KN, MILLER RA, Gamborg OL. 1972. Cell division and plant development from protoplasts of carrot suspension cultures. *Planta* 103:348–55
34. Hadfi K, Speth V, Neuhaus G. 1998. Auxin-induced developmental patterns in *Brassica juncea* embryos. *Development* 125:879–87
35. Hamann T, Mayer U, Jürgens G. 1999. The auxin-insensitive *bodenlos* mutation affects primary root formation and apical-basal patterning in the *Arabidopsis* embryo. *Development* 126:1387–95
36. Hanawa J. 1961. Experimental studies of leaf dorsiventrality in *Sesamum indicum* L. *Bot. Mag. Tokyo* 74:303–9
37. Hardtke CS, Berleth T. 1998. The Arabidopsis gene *MONOPTEROS* encodes a transcription factor mediating embryo axis formation and vascular development. *EMBO J* 17:1405–11
38. Holm PB, Knudsen S, Mouritzen P, Negri D, Olsen FL, Rové C. 1994. Regeneration of fertile barley plants from mechanically isolated protoplasts of fertilized egg cells. *Plant Cell* 6:531–43
39. Hudson A. Axes and axioms in leaf formation? *Curr. Opin. Plant Biol.* 2:56–60
39a. Jürgens G. 1995. Axis formation in plant embryogenesis: cues and clues. *Cell* 81:467–70
40. Kitano H, Tamura Y, Satoh H, Nagato Y. 1993. Hierarchical regulation of organ differentiation during maize embryogenesis. *Plant J.* 3:607–10
41. Klucher KM, Chow H, Reiser L, Fischer RL. 1996. The *AINTEGUMENTA* gene of *Arabidopsis* required for ovule and female gametophyte development is related to the floral homeotic gene *APETALA2*. *Plant Cell* 8:137–53
42. Lauber MH, Waizenegger I, Steinmann T, Schwartz H, Mayer U, et al. 1997. The *Arabidopsis* KNOLLE protein is a cytokinesis-specific syntaxin. *J. Cell Biol.* 139:1485–93
43. Leduc N, Matthy S, Rochon E, Rougier M, Morgensen L, et al. 1996. Isolated maize zygotes mimic *in vivo* embryonic development and express microinjected genes when cultured in vitro. *Dev. Biol.* 177:190–203
44. Liu C-M, Xu Z-H, Chua N-H. 1993. Auxin polar transport is essential for the establishment of bilateral symmetry during early plant embryogenesis. *Plant Cell* 5:621–30
45. Long JA, Barton K. 1998. The development of apical embryonic pattern in *Arabidopsis*. *Development* 125:3027–35
46. Lotan T, Ohto M, Yee KM, West MAL, Lo R, et al. 1998. *Arabidopsis* LEAFY

COTYLEDON1 is sufficient to induce embryo development in vegetative cells. *Cell* 93:1195–105

47. Lu P, Porat R, Nadeau J, O'Neill S. 1996. Identification of a meristem L1 layer specific gene in Arabidopsis that is expressed during embryonic pattern formation and defines a new class of homeobox genes. *Plant Cell* 8:2155–68
48. Lukowitz W, Mayer U, Jürgens G. 1996. Cytokinesis in the Arabidopsis embryo involved the syntaxin-related KNOLLE protein. *Cell* 84:61–71
49. Deleted in proof
50. Luo D, Carpenter R, Vincent C, Copsey L, Coen ES. 1996. Origin of floral asymmetry in *Antirrhinum*. *Nature* 383:794–99
51. Luo Y, Koop H-U. 1997. Somatic embryogenesis in cultured immature zygotic embryos and leaf protoplasts of *Arabidopsis thaliana* ecotypes. *Planta* 202:387–96
52. Lyndon RF. 1990. *Plant Development—The Cellular Basis*. London: Unwin Hyman
53. Lynn K, Fernandez A, Aida M, Sedbrook J, Tasaka M, et al. 1999. The *PINHEAD/ZWILLE* gene acts pleiotropically in Arabidopsis development and has overlapping functions with the *ARGONAUTE1* gene. *Development* 126:469–81
54. Malamy JE, Benfey PN. 1997. Organization and cell differentiation in lateral roots of *Arabidopsis thaliana*. *Development* 124:33–44
55. Mansfield SG, Briarty LG, Erni S. 1991. Early embryogenesis in *Arabidopsis thaliana*. I The mature embryo sac. *Can. J. Bot.* 69:447–60
56. Mayer KFX, Schoof H, Haecker A, Lenhard M, Jürgens G, Laux T. 1998. Role of *WUSCHEL* in regulating stem cell fate in the *Arabidopsis* shoot meristem. *Cell* 95:805–15
57. Mayer U, Büttner G, Jürgens G. 1993. Apical-basal pattern formation in the *Arabidopsis* embryo: studies on the role of the *GNOM* gene. *Development* 117:149–62
58. Mayer U, Jürgens G. 1998. Pattern formation in plant embryogenesis: a reassessment. *Semin. Cell Dev. Biol.* 9:187–93
59. Mayer U, Torres Ruiz R, Beleth T, Misera S, Jürgens G. 1991. Mutations affecting body organisation in the *Arabidopsis* embryo. *Nature* 353:402–7
60. McConnell JR, Barton K. 1998. Leaf polarity and meristem formation in *Arabidopsis*. *Development* 125:2935–42
61. Meyerowitz EM. 1998. Genetic and molecular mechanisms of pattern formation in *Arabidopsis* flower development. *J. Plant Res.* 111:233–42
62. Deleted in proof
63. Moussian B, Schoof H, Haecker A, Jürgens G, Laux T. 1998. Role of the *ZWILLE* gene in the regulation of central shoot meristem cell fate during *Arabidopsis* embryogenesis. *EMBO J.* 17:1799–809
64. Neal PR, Dafni A, Giurfa M. 1998. Floral symmetry and its role in plant-pollinator systems: terminology, distribution, and hypotheses. *Annu. Rev. Ecol. System.* 29:345–73
65. Nemhauser JL, Zambryski PC, Roe JL. 1998. Auxin signalling in *Arabidopsis* flower development? *Curr. Opin. Plant Biol.* 1:531–35
66. Okada K, Ueda J, Komai MK, Bell CJ, Shimuray Y. 1991. Requirement of the polar auxin transport system in early stages of *Arabidopsis* bud formation. *Plant Cell* 3:677–84
67. Palme K. 2000. Polar auxin transport. *Annu. Rev. Plant Phys. Mol. Biol.* 51:000–00. In press
68. Parcy F, Nilsson O, Busch MA, Lee I, Weigel D. 1998. A genetic framework for floral patterning. *Nature* 395:561–66
69. Przemeck GKH, Mattsson J, Hardtke CS, Sung ZR, Berleth T. 1996. Studies on the role of the *Arabidopsis* gene *MONOPTEROS* in vascular development and plant cell axialisation. *Planta* 200:229–37

70. Raghavan V. 1976. Role of the generative cell in androgenesis in henbane. *Science* 191:388–89
71. Raghavan V. 1997. *Molecular Embryology of Flowering Plants*. Cambridge, UK: Cambridge Univ. Press
72. Ray S, Golden T, Ray A. 1996. Maternal effects of the *short integuments* mutation on embryo development in *Arabidopsis*. *Dev. Biol.* 180:365–69
73. Richards FJ. 1948. The geometry of phyllotaxis and its origin. *Symp. Soc. Exp. Biol.* 2:217–45
74. Robinson-Beers K, Pruitt RE, Gasser CS. 1992. Ovule development in wild-type *Arabidopsis* and in two female-sterile mutants. *Plant Cell* 4:1237–49
75. Roe JL, Nemhauser JL, Zambryski PC. 1997. *TOUSLED* participates in apical tissue formation during gynoecium development in *Arabidopsis*. *Plant Cell* 9:335–53
76. Russel SD. 1979. Fine structure of gametophyte development in *Zea mays*. *Can. J. Bot.* 57:1093–110
77. Sawa S, Watanabe K, Goto K, Kanaya E, Morita EH, Okada K. 1999. *FILAMENTOUS FLOWER*, a meristem and organ identity gene of *Arabidopsis*, encodes a protein with a zinc finger and HMG-related domains. *Genes Dev.* 13:1079–88
78. Scanlon MJ, Schneeberger RG, Freeling M. 1996. The maize mutant *narrow sheath* fails to establish leaf margin identity in a meristematic domain. *Development* 122:1683–91
79. Scheres B, Di Laurenzio L, Willemsen V, Hauser M-T, Janmaat K, et al. 1995. Mutations affecting the radial organisation of the *Arabidopsis* root display specific defects throughout the embryonic axis. *Development* 121:53–62
80. Scheres B, Di Laurenzio L, Willemsen V, Hauser M-T, Janmaat K, et al. 1995. Mutations affecting the radial organisation of the *Arabidopsis* root display specific defects throughout the embryonic axis. *Development* 121:53–62
81. Scheres B, Wolkenfelt H, Willemsen V, Terlouw M, Lawson E, et al. 1994. Embryonic origin of the *Arabidopsis* primary root and root meristem initials. *Development* 120:2475–87
82. Schevell DE, Leu W-M, Gillmor CS, Xia G, Feldmann KA, Chua N-H. 1994. *EMB30* is essential for normal cell division, cell expansion and cell adhesion in Arabidopsis and encodes a protein that has similarity to Sec7. *Cell* 77:1051–62
83. Schneider T, Dinkins R, Robinson K, Shellhammer J, Meinke DW. 1989. An embryo-lethal mutant of *Arabidopsis thaliana* is a biotin auxotroph. *Dev. Biol.* 131:161–67
84. Schneitz K. 1999. The molecular and genetic control of ovule development. *Curr. Opin. Plant Biol.* 2:13–17
85. Schwartz BW, Yeung EC, Meinke DW. 1994. Disruption of morphogenesis and transformation of the suspensor in abnormal suspensor mutants of *Arabidopsis*. *Development* 120:3235–45
86. Sessions A, Nemhauser JL, McColl A, Roe JL, Feldmann KA, Zambryski PC. 1997. *ETTIN* patterns the *Arabidopsis* floral meristem and reproductive organs. *Development* 124:4481–91
87. Sessions RA, Zambryski PC. 1995. *Arabidopsis* gynoecium structure in the wild-type and in *ettin* mutants. *Development* 121:1519–32
88. Sheridan WF, Clark JK. 1993. Mutational analysis of morphogenesis of the maize embryo. *Plant J.* 3:347–58
89. Siegfried KR, Eshed Y, Baum SF, Otsuga D, Drews GN, Bowman JL. 1999. Members of the YABBY gene family specify abaxial cell fate in *Arabidopsis*. *Development*. In press
90. Sinha N. 1999. Leaf development in angiosperms. *Ann. Rev. Plant Phyiol. Mol. Biol.* 50:129–46
91. Stewart I, Golubitsky M. 1993. *Fearful Symmetry—Is God a Geometer?* London: Penguin

92. Sussex IM. 1955. Morphogenesis in *Solanum tuberosum* L.: experimental investigation of leaf dorsiventrality and orientation in the juvenile shoot. *Phytomorphology* 5:286–300
93. Szymkowiack E. 1996. Is the extent of the proliferation of component cell lineages critical during organ morphogenesis? *Semin. Cell. Dev. Biol.* 7:849–56
94. Taylor LR. 1967. The foliar embryos of *Malaxis paludosa*. *Can. J. Bot.* 45:1553–56
95. Timmermans MCP, Schultes NP, Jankovsky JP, Nelson T. 1998. *Leafbladeless1* is required for dorsoventrality of lateral organs in maize. *Development* 125:2813–23
96. Topping JF, May VT, Muskett PR, Linsay K. 1997. Mutations in the *HYDRA1* gene of Arabidopsis perturb shape and disrupt embryonic and seedling morphogenesis. *Development* 124:4415–24
97. Torrez-Ruiz RA, Lohner A, Jürgens G. 1996. The *GURKE* gene is required for normal organisation of the apical region in the Arabidopsis embryo. *Plant J.* 10:1005–16
98. Traas J, Bellini C, Nacry C, Kronenberger J, Bouchez D, Caboche M. 1995. Normal differentiation patterns in plants lacking microtubular preprophase bands. *Nature* 375:676–77
99. Tsukaya H. 1997. Determination of the unequal fate of cotyledons. *Development* 124:1275–80
100. Uwer U, Willmitzer T, Altmann T. 1998. Inactivation of a glycyl-tRNA synthetase leads to an arrest in plant embryonic development. *Plant Cell* 10:1277–94
101. Vernon DM, Meinke DW. 1994. Embryonic transformation of the suspensor in *twin*, a polyembryonic mutant of *Arabidopsis*. *Dev. Biol.* 165:566–73
102. Vroeman C, Langeveld S, Mayer U, Ripper G, Jürgens G, et al. 1996. Pattern formation in the Arabidopsis embryo revealed by position-specific lipid transfer protein gene expression. *Plant Cell* 8:783–91
103. Waites R, Selvadurai HRN, Oliver IR, Hudson A. 1988. The *PHANTASTICA* gene encodes a MYB transcription factor involved in growth and dorsoventrality of lateral organs in *Antirrhinum*. *Cell* 93:779–89
104. Waites R, Hudson A. 1995. *phantastica:* a gene required for dorsoventrality in leaves of *Antirrhinum majus*. *Development* 121:2143–54
105. Wardlaw CW. 1948. Experiments in organogenesis in ferns. *Symp. Soc. Exp. Biol.* 2:93–131
106. Willemsen V, Wolkenfelt H, De Vrieze G, Welsbee P, Scheres B. 1998. The *HOBBIT* gene is required for formation of the root meristem in the *Arabidopsis* embryo. *Development* 125:521–31
107. Yadegari R, De Paiva GR, Laux T, Koltunow AM, Apuya N, et al. 1994. Cell differentiation and morphogenesis are uncoupled in Arabidopsis *raspberry* embryos. *Plant Cell* 6:1713–29
108. Yeung EC, Meinke DW. 1993. Embryogenesis in angiosperms: development of the suspensor. *Plant Cell* 5:1371–81
109. Yeung EC, Sussex IM. 1979. Embryogeny of *Phaseolus coccineus*: the suspensor and growth of the embryo-proper *in vitro*. *Z. Pflantzenphysiol.* 91:423–33
110. Zhang JZ, Somerville CR. 1997. Suspensor-derived polyembryony caused by altered expression of a valyl-tRNA synthetase in the *twn2* mutant of *Arabidopsis*. *Proc. Natl. Acad. Sci. USA* 94:7349–55

Annu. Rev. Plant Physiol. Plant Mol. Biol. 2000. 51:371–400

Plant Thioredoxin Systems Revisited

P. Schürmann
Laboratoire de Biochimie Végétale, Université de Neuchâtel, Rue Emile-Argand 11, CH-2007 Neuchâtel, Switzerland; e-mail: Peter.Schurmann@bota.unine.ch

J.-P. Jacquot
Laboratoire de Biologie Forestière, Associé INRA, Biochimie et Biologie Moléculaire Végétale, Université de Nancy 1, F-54506 Vandoeuvre Cedex, France; e-mail: Jean-Pierre.Jacquot@scbiol.uhp-nancy.fr

Key Words NADPH:thioredoxin reductase, ferredoxin:thioredoxin reductase, target enzymes, regulatory sites, redox potentials, crystal structures

■ **Abstract** Thioredoxins, the ubiquitous small proteins with a redox active disulfide bridge, are important regulatory elements in plant metabolism. Initially recognized as regulatory proteins in the reversible light activation of key photosynthetic enzymes, they have subsequently been found in the cytoplasm and in mitochondria. The various plant thioredoxins are different in structure and function. Depending on their intracellular location they are reduced enzymatically by an NADP-dependent or by a ferredoxin (light)-dependent reductase and transmit the regulatory signal to selected target enzymes through disulfide/dithiol interchange reactions. In this review we summarize recent developments that have provided new insights into the structures of several components and into the mechanism of action of the thioredoxin systems in plants.

CONTENTS

1040-2519/00/0601-0371$14.00

INTRODUCTION

Thioredoxins are small proteins with a redox active disulfide bridge present in the characteristic active site sequence-Trp-**Cys**-Gly-Pro-**Cys**-. They have a molecular mass of approximately 12 kDa and are universally distributed in animal, plant, and bacterial cells. In their reduced form they constitute very efficient protein disulfide oxido-reductases. Initially described as hydrogen carriers in ribonucleotide reduction in *Escherichia coli*, they have been found to serve as electron donors in a variety of cellular redox reactions (65). In oxygenic photosynthetic cells thioredoxins were recognized as important regulatory proteins in carbon assimilation (16, 22).

Only two types of thioredoxins have been found in bacteria and animal tissues, but plant tissues contain multiple forms: two in chloroplasts, one in the cytoplasm, and one in mitochondria. Although all thioredoxins are of comparable size and appear to have very similar redox properties, they fulfill specific functions. This specificity is due to structural complementarity, which allows specific interaction between the different thioredoxins and their respective target proteins.

Depending on their intracellular location, thioredoxins are reduced by a different electron donor system. Thioredoxins in nonphotosynthetic tissue and in the cytosol of photosynthetic cells are reduced with electrons from NADPH via the NADP/thioredoxin system, whereas the chloroplast thioredoxins of plants and eukaryotic algae and the thioredoxins of oxygenic photosynthetic prokaryotes are reduced via the ferredoxin/thioredoxin system with elecrons provided by photosynthetic electron transport.

In this review we describe recent developments, including new structural information, that provide a better understanding of the structure and function of thioredoxin systems in plants and present some information on new developments and on proteins with structural or functional similarity to thioredoxins. Different aspects of the thioredoxin systems in plants have been discussed in several recent publications (7, 17, 18, 20, 21, 44, 49, 49a, 73, 95a, 128, 135a, 136, 140, 166a). For detailed information on earlier developments we refer the reader to former reviews in this series (16, 65).

CYTOPLASMIC AND MITOCHONDRIAL NADP-DEPENDENT THIOREDOXIN SYSTEMS

Plant cells, like bacterial and animal cells, contain a cytosolic thioredoxin system that is dependent on NADPH for its reduction (48, 151). In this cytosolic system the reducing power of NADPH is transferred to thioredoxin via a flavoprotein—the NADPH:thioredoxin reductase (NTR).[1]

Genomic Organization, Structure, and Functions of NADPH:Thioredoxin Reductase

A cDNA coding for this enzyme has been isolated from *Arabidopsis thaliana*. The deduced primary structure of the protein displays significant homology to *E. coli* NTR (75). Southern blot experiments indicate that at least two different genes coding for NTR are present in *Arabidopsis*, with the total number of genes being low. Following the isolation of cDNAs coding for NTR, the crystal structure of the recombinant protein has been elucidated. The enzyme is a homodimer with subunits of about 35 kDa. Each of the subunits possesses two subdomains, a central NADPH binding domain, and a domain constituted by the N and C termini, which binds FAD. In addition, each subunit contains a redox active disulfide in a **CATC** motif (35). Much as in the *E. coli* enzyme, the redox active site is facing the isoalloxazin ring of the flavin bringing the redox entities in close contact, thereby facilitating electron transfer from FAD to the disulfide. Comparison of the crystal structures of the *E. coli* and *Arabidopsis* proteins suggests that in order to accommodate a thioredoxin molecule, the reductase must undergo a structural change, likely to occur at a hinge situated at the junction between the FAD and NADPH subdomains. Plant NTR appears to be much more closely related to the prokaryotic type enzyme (prototype *E. coli*) than to the mammalian enzyme that contains an N-terminal extension as well as an essential selenocysteine (53). Like many other flavoproteins, the *Arabidopsis* NTR also possesses diaphorase activity and it is able to reduce quinone and nitrocompounds (11, 101).

Genomic Organization, Structure of the Thioredoxin *h* Component

The cytosolic thioredoxin has been termed thioredoxin *h*, as heterotrophic, since it was originally found in nonphotosynthetic tissues (79, 151). A number of thioredoxin *h* cDNA sequences have been isolated from both nonphotosynthetic and photosynthetic tissues, most notably in *Arabidopsis* (125). All deduced sequences share a rather high degree of homology, except that the canonical active-site motif (**WCGPC**) is modified to **WCPPC** in several members. At leastfive genes coding

[1]Abbreviations: DTT, dithiothreitol; FBPase, fructose 1,6-bisphosphatase; Fd, ferredoxin; FTR, ferredoxin:thioredoxin reductase; GSH, reduced glutathione; G6PDH, glucose 6-phosphate dehydrogenase; GAPDH, glyceraldehyde 3-phosphate dehydrogenase; NADP-MDH, NADP-dependent malate dehydrogenase; NTR, NADPH:thioredoxin reductase; PRK, phosphoribulokinase; Rubisco, ribulose 1,5-bisphosphate carboxylase/oxygenase; SBPase, sedoheptulose 1,7-bisphosphatase; Trx, thioredoxin.

for thioredoxin *h* are present in *Arabidopsis*. Primary structure analyses indicate that all these proteins are closely related from an evolutionary standpoint (73). Thioredoxin *h* sequences are slightly longer than the one of *E. coli*—the prototype prokaryotic thioredoxin—mostly owing to N- and/or C-terminal extensions. So far, the NMR solution structure of oxidized thioredoxin *h* from the photosynthetic green alga *Chlamydomonas reinhardtii* is the only one available (102). In agreement with primary structure analyses, the model indicates that plant thioredoxin *h* is much closer to human thioredoxin than to the *E. coli* type. Nevertheless, all thioredoxins share a common fold with a succession of the following secondary structural elements: $\beta1$, $\alpha1$, $\beta2$, $\alpha2$, $\beta3$, $\alpha3$, $\beta4$, $\beta5$, $\alpha4$.

The thioredoxin *h* structure presents only one major difference compared to the models of other thioredoxins, namely an elongated $\alpha1$ helix. The isolated protein has two specific properties that allow its identification: first, the presence of a conserved Trp residue (Trp13 in *Chlamydomonas*) that imparts peculiar UV spectral characteristics; and second, a marked instability when subjected to high temperature and slightly alkaline pH (150). Analysis of the intron positions suggests that there is a common ancestor for plant and green algal thioredoxin *h* genes (130).

Expression and Targets of Thioredoxin *h*

Although this area is not as well documented by in vitro studies as the one dealing with chloroplast thioredoxins, a number of functions have been proposed for the cytosolic thioredoxin system. It has been suggested to act as a reducing system participating in the mobilization of protein reserves during seed germination, and these reactions may possibly be linked to the reduction of storage proteins and enzyme inhibitors and also mediated by a redox-dependent protease, thiocalsin (8, 82). It has also been proposed that thioredoxin *h* may play a role in sulfate assimilation and in conferring resistance to hydrogen peroxide, similar to what occurs in yeast or mammalian cells since some of the *Arabidopsis* isoforms complement yeast mutants in these two processes (105).

Several recent reports indicate that in cyanobacteria and plants thioredoxin is involved in the detoxification of H_2O_2 via the enzyme thioredoxin peroxidase, or peroxiredoxin. In cyanobacteria, the reduction of H_2O_2, alkyl hydroperoxides and *t*-butyl hydroperoxide was shown to be significantly increased by dithiol reagents in vitro or by light in vivo, which suggests that in cyanobacteria the activity of thioredoxin peroxidase is coupled to the photosynthetic electron transport system (155, 168). In the cytosol, peroxiredoxin might be a target of thioredoxin *h*, since this enzyme has been isolated as an in vivo target of plant thioredoxin *h* in yeast by stabilizing a mixed disulfide intermediate (158). Structural and functional equivalents of the yeast and mammalian peroxiredoxin are present in plant cells. Peroxiredoxins of the two-Cys type, believed to function as antioxidants, have also been characterized in plants such as *Hordeum* and *Spinacia* (5).

Although no 3-D structure of plant peroxiredoxins is available yet, the crystal structure of the human enzyme has been solved. Figure 1 (see color section) shows a structural composite of the thioredoxin cytosolic system comprised of NADPH, the *Arabidopsis thaliana* NTR monomer, *Chlamydomonas reinhardtii* thioredoxin (Trx) *h*, and human peroxiredoxin.

Thioredoxin *h* is also a potent regulator of membrane-bound, receptor-like kinases in plants (13). Expression studies (essentially northern and western blots) indicate that thioredoxin *h* proteins are present in many plant organs and that their expression is developmentally regulated. Transcription of the gene in *Chlamydomonas* was found to be under the control of light and a circadian rhythm (87). Furthermore, it is strongly activated after addition of heavy metals, such as Cd^{2+}, to algal cells (86).

Thioredoxin *h* as a Messenger in Plants?

Several experiments suggest that thioredoxin *h* could act as a messenger protein in plants. First, thioredoxin *h* was found to be one of the major proteins in the phloem sap of rice (72). Furthermore, it is detected not only in the phloem sap of monocots such as rice and wheat, but also in dicots such as *Ricinus* and *Cucurbita* (138). Since mature phloem cells are enucleated, the site of mRNA synthesis was located to companion cells, and transcripts coding for thioredoxin *h* were transferred to the sieve tubes (132). In addition, microinjection of recombinant thioredoxin *h* demonstrated that the protein itself has the capacity to mediate its own cell-to-cell transport through plasmodesmata (71). It thus appears that thioredoxin *h* is a mobile element in plants and could act as a message carrier, presumably by regulating membrane receptors, as described for the membrane-bound, receptor-like kinase of *Brassica* (13).

Thioredoxin Systems in Mitochondria

It is now well established that animal cells contain the full complement for a functional thioredoxin system (selenium-containing NTR and thioredoxin) in mitochondria (100, 111). Early reports indicated that plant mitochondria also contain at least one thioredoxin (12, 92), but no cDNA sequence has been unambiguously shown to code for a transit peptide that targets the mitochondrial compartment. Similarly, no plant mitochondrial NTR sequence has yet been isolated. Although more solid evidence for the presence of a functional thioredoxin system in plant mitochondria is still lacking, these thioredoxins can be reduced in vitro by lipoic acid generated via the 2-oxoacid dehydrogenase complex, as in animal mitochondria (24). A possible role for plant mitochondrial thioredoxins could be to activate/stabilize the 2-oxoacid dehydrogenase complex present in this organelle (23). Another possible role for mitochondrial thioredoxin could be the redox regulation of the cyanide-resistant alternative oxidase in the plant electron transport chain (124, 157a).

93. Van Bel AJE. 1993. The transport phloem. Specifics of its functioning. *Prog. Bot.* 54:134–50

94. Van Bel AJE, Kempers R. 1996. The pore/plasmodesm unit: key elements in the interplay between sieve element and companion cell. *Prog. Bot.* 58:278–91

95. Voinnet O, Baulcombe DC. 1997. Systemic signalling in gene silencing. *Nature* 389:553

96. Voinnet O, Vain P, Angell S, Baulcombe DC. 1998. Systemic spread of sequence-specific transgene RNA degradation in plants is initiated by localized introduction of ectopic promoterless DNA. *Cell* 95:177–87

97. Volk GM, Turgeon R, Beebe DU. 1996. Secondary plasmodesmata formation in the minor-vein phloem of *Cucumis melo* L. and *Cucurbita pepo* L. *Planta* 199:425–32

98. Waigmann E, Lucas WJ, Citovsky V, Zambryski P. 1994. Direct functional assay for tobacco mosaic virus cell-to-cell movement protein and identification of a domain involved in increasing plasmodesmal permeability. *Proc. Natl. Acad. Sci. USA* 91:1433–37

99. Wang N, Fisher DB. 1994. The use of fluorescent tracers to characterize the post-phloem transport pathway in maternal tissues of developing wheat grains. *Plant Physiol.* 104:17–27

100. Wang HL, Offler CE, Patrick JW. 1995. The cellular pathway of photosynthate transfer in the developing wheat grain. II. A structural analysis and histochemical studies of the pathway from the crease phloem to the endosperm cavity. *Plant Cell Environ.* 18:373–88

101. Wolf S, Deom CM, Beachy RN, Lucas WJ. 1989. Movement protein of tobacco mosaic virus modifies plasmodesmatal size exclusion limit. *Science* 246:377–79

102. Xoncostle-Cazares B, Xiang Y, Ruiz-Medrano R, Wang H-L, Monzer J, et al. 1999. Plant paralog to viral movement protein potentiates transport of mRNA into the phloem. *Science* 283:94–98

103. Ziegler H. 1975. Nature of transported substances in the phloem. In *Encyclopedia of Plant Physiology*, Vol. 1, *Transport in Plants: Phloem Transport*, ed. MH Zimmermann, JA Milburn, pp. 59–100. Berlin: Springer-Verlag

THE CHLOROPLASTIC, FERREDOXIN-DEPENDENT THIOREDOXIN SYSTEM

In chloroplasts and cyanobacterial cells thioredoxins are reduced by way of light. Light-driven photosynthetic electron transport produces reduced ferredoxin (Fd), which serves as electron donor for ferredoxin:thioredoxin reductase (FTR). This enzyme in turn reduces thioredoxins, which interact with target proteins to reduce and activate biosynthetic, or deactivate catabolic pathways. This system of enzyme regulation is the ferredoxin/thioredoxin system (16, 17). Significant progress has been made in understanding this system with respect to function and especially to structure, inasmuch as the crystal structure of each component protein member is now known (Figure 2; see color section).

Ferredoxin

Fd, the soluble, stromal [2Fe-2S] protein, is a negatively charged electron donor that interacts quite strongly with its electron acceptor proteins to form electrostatically stabilized complexes. It was demonstrated that spinach ferredoxin and FTR form a high-affinity, 1:1 complex (61). Negatively charged residues involved in this complex formation have been located in spinach and *Chlamydomonas reinhardtii* ferredoxin (37, 77). A crystal structure is now available for spinach ferredoxin at 1.7 Å resolution (10). Although there are positively charged residues in the putative ferredoxin interaction domain of spinach FTR, no experiments have yet shown their involvement in protein-protein interaction.

Ferredoxin:Thioredoxin Reductase

FTR is the central enzyme of the ferredoxin/thioredoxin system; it converts a photosynthetic "electron signal" received from reduced ferredoxin to a "thiol signal" that is transmitted to thioredoxin. Unlike cytoplasmic NTR, this chloroplast stromal enzyme is a yellowish-brown [4Fe-4S] protein, composed of two dissimilar subunits, designated catalytic and variable. The catalytic subunit of the enzyme from different photosynthetic organisms has a constant size of about 13 kDa, whereas the variable subunit ranges between 8 and 13 kDa (41, 69).

Gene and Protein Sequences In plants both subunits of FTR are nucleus-encoded. Two slightly different cDNAs have been sequenced for both subunits of spinach FTR, suggesting the presence of two genes (46, 50). In *Porphyra purpurea*, a red alga (121), and in *Guillardia theta*, a cryptomonad (40), a single gene for the catalytic subunit was found in the chloroplast genome.

The variable subunits show pronounced sequence diversity, with only 46% to 60% identity within the eukaryotes and 33% to 40% between eukaryote and prokaryotes. The most striking difference is the extended N terminus, present in all three known eukaryotic subunits, but absent in the prokaryotic ones. In spinach

FTR, this N-terminal extension was found to be unstable; it was degraded to discrete shorter peptides (157) that exhibit no difference in functional properties.

By contrast, the catalytic subunits have a highly conserved primary structure. Among the strictly conserved residues are seven Cys, six of which are organized into two CPC and one CHC motifs. These six Cys are the functionally essential residues that constitute the redox active disulfide bridge and ligate the Fe-S cluster. Cluster ligation does not follow known consensus motifs, but shows a new arrangement with the following fingerprint: **C**PCX_{16}**C**P**C**X_8**C**HC (cluster ligands are in bold). In spinach FTR Cys54 and Cys84 form the active-site disulfide. Cys54 is accessible to solvent, whereas Cys84 is protected. The four remaining cysteines, Cys52, Cys71, Cys73, and Cys82, are ligands to the iron center. This arrangement places the redox-active disulfide bridge adjacent to the [4Fe-4S] cluster (30).

The genomes of two archaebacteria, *Archaeoglobus fulgidus* (81) and *Methanobacterium thermoautotrophicum* (146), both contain a gene coding for a protein with a striking resemblance to the catalytic subunit of FTR. Although the overall identities between the archaebacterial proteins and the photosynthetic FTR are rather low (25%–35%), the residues found to be essential for the functioning of FTR, the CXC motifs, are conserved. However, no functions are known for these proteins in the archaea. Nonetheless, the striking structural similarities suggest that the catalytic subunit of photosynthetic FTR might be derived from an ancient precursor protein whose function has been adapted during evolution.

The genes coding for the two subunits of FTR from spinach (50) and *Synechocystis* (142) have been introduced into expression vectors. These constructs yielded soluble, heterodimeric recombinant enzymes containing the correctly inserted Fe-S cluster, demonstrated by spectral properties and enzyme activity.

Crystal Structure Recombinant *Synechocystis* FTR has recently been crystallized and its structure solved (34, 35a). The variable subunit has an open β-barrel structure and the catalytic subunit is essentially α-helical. The two subunits, with the catalytic subunit sitting on top of the variable subunit, form an unusually thin molecule, with the shape of a concave disc measuring only 10 Å across its center. In this center are located the functionally important structures, the cubane [4Fe-4S] cluster and, in close proximity, the redox active disulfide bridge. The Fe-S cluster is accessible from one side of the concave disk, the ferredoxin docking side, whereas the disulfide bridge is accessible from the opposite side, the thioredoxin docking side. This allows simultaneous docking of the electron donor and electron acceptor proteins—a prerequisite for the proposed reaction-mechanism.

The docking sites are also well adapted for their function: The ferredoxin docking site contains four positively charged residues for specific interaction with negatively charged ferredoxin, whereas the thioredoxin docking site contains mainly hydrophobic residues, making it compatible for interaction with different thioredoxins.

Surprisingly, the shape of the variable subunit shows striking similarities to PsaE, the ferredoxin-binding protein of photosystem 1, with the SH3 domain and with GroES, although there are no sequence similarities with either protein (34).

Chloroplast Thioredoxins

Chloroplasts of plants and algae contain two different types of thioredoxins, an *f*- and an *m*-type. They have different phylogenetic origins (58) and different target-enzyme specificities. Thioredoxin *f* was originally described as the activator protein for spinach stromal fructose 1,6-bisphosphatase (FBPase) and thioredoxin *m* for NADP-malate dehydrogenase (NADP-MDH) (16, 17, 73). Although a clear distinction between the two Trx types has not always been obtained in experiments under in vitro conditions, these two types can clearly be distinguished by their primary structures.

Structure of Thioredoxin m This thioredoxin, present in oxygenic prokaryotes, algae, monocots, and dicots, strongly resembles the thioredoxin from prokaryotes, both heterotrophic and anoxygenic photosynthetic. In plants and green algae this Trx is nuclear-encoded, but in red algae its gene was found in the chloroplast genome (121, 123), which supports its bacterial endosymbiotic origin. Functionally and structurally, it strongly resembles bacterial thioredoxins, which can be used interchangeably. Thioredoxins of the *m*-type are, therefore, considered bacterial-type thioredoxins. A comparison of the primary structures of all *m*-type thioredoxins available in the databases shows that they are clearly related. However, there is less sequence similarity than for the *f*-type proteins. Recombinant *m*-type thioredoxins have been reported from *Chlamydomonas* (84), spinach (139), and pea (90). For the *Chlamydomonas* protein, an NMR analysis showed that its secondary structure and general protein folding is similar to its *E. coli* counterpart (84). For the recombinant spinach thioredoxin *m*, the crystal structure has been solved in the oxidized and reduced state at 2.1 and 2.3 Å resolution, respectively (27). As already found for *Chlamydomonas* thioredoxin *m*, its structure is very similar to that of *E. coli* thioredoxin (80), with nearly identical secondary structure. The surface around the active-site Cys residues largely resembles its *E. coli* counterpart (Figure 3, center and bottom panels; see color section). This corroborates biochemical evidence that the proteins are functionally interchangeable. There is little difference in conformation of the active site between oxidized and reduced thioredoxin *m*; however, some slight structural changes in the main chain conformation of the active site render the solvent-exposed Cys (Cys37) more accessible upon reduction (G Capitani, Z Markovic-Housley, G del Val, M Morris, JN Jansonius, P Schürmann, Crystal structures of two functionally different thioredoxins in spinach chloroplasts. In preparation).

Structure of Thioredoxin f There are fewer sequences known for this Trx, which is considered to be restricted to photosynthetic eukaryotic organisms. Thioredoxin *f* is nuclear-encoded and apparently derived from the eukaryotic host during evolution. *f*-Type thioredoxins are highly conserved, with 75% to 90% residue identity in the mature protein from dicots and with 60% to 67% residue identity between dicots and monocots (140). Owing to additional amino acids at the N terminus, they are slightly longer than other Trx types. Interestingly, the C-terminal part of the sequence resembles classical animal thioredoxin in containing a third, strictly conserved Cys (Cys73 in spinach). The crystal structure of two different forms of recombinant thioredoxin *f* has been solved (G Capitani, Z Markovic-Housley, G del Val, M Morris, JN Jansonius, P Schürmann, Crystal structures of two functionally different thioredoxins in spinach chloroplasts. In preparation), a long form closely resembling the in vivo form (1) and an N terminally truncated short form (38). Both structures are essentially identical aside from the N terminus, which contains an additional α-helix in the long form. Although the overall structure of thioredoxin *f* does not differ markedly from a typical thioredoxin, its surface topography is distinct from that of other Trxs (27) (Figure 3). Positive charges surrounding the active site may be instrumental in orienting thioredoxin *f* correctly with its target proteins. Hydrophobic residues, also prominent in this area, may be more important in the nonspecific interaction with FTR that reduces various thioredoxins. A striking difference is the presence of the third Cys exposed on the surface (Cys73 in spinach), 9.7 Å away from the accessible Cys of the active-site dithiol (Cys46). As mentioned, this third Cys is conserved in all *f*-type thioredoxins. The structural analysis also shows that the active-site Cys closest to the N terminus (Cys46 in spinach) is exposed, whereas its partner (Cys49 in spinach) is buried, confirming experiments which indicated that Cys46 is the attacking nucleophile in the reduction of the target protein's disulfide (14).

Specificity of Thioredoxins Chloroplast thioredoxins show selectivity in their interaction with target enzymes when tested under conditions approaching their in vivo situation. The stromal Calvin cycle enzymes FBPase (51), SBPase (166), PRK (165), as well as Rubisco activase (169) and CF_1 (141), are exclusively or most efficiently activated by thioredoxin *f*. NADP-MDH, although originally reported to specifically interact with thioredoxin *m*, was shown to be even more efficiently activated by thioredoxin *f* (51, 64). G6PDH (glucose 6-phosphate dehydrogenase) of the stromal oxidative pentose phosphate pathway, on the contrary, interacts specifically with thioredoxin *m* (162). These observations suggest that, at least as far as carbohydrate metabolism is concerned, reduced thioredoxin *f* functions primarily in enzyme activation (i.e. enhancing the rate of biosynthesis), whereas reduced thioredoxin *m* acts mainly in enzyme deactivation (i.e. decreasing the rate of degradation).

The specificity of thioredoxin *f* raises the question of which structural features are possibly responsible for it. An answer has been sought using site-directed mutagenesis of thioredoxins. The results confirm the importance of the third Cys (38) and of positive charges on the contact surface of thioredoxin *f* (51, 83, 103), and can essentially be reconciled with the recent structural model of Trx *f* (27).

Mechanism of Thioredoxin Reduction by FTR

There is clear evidence from experiments with isolated chloroplasts (33) and purified protein components (42) that FTR converts the photosynthetic electron transport signal, provided by reduced ferredoxin, to a thiol signal, detected with the fluorescent label mono-bromobimane, which then appears sequentially in thioredoxin and the target enzyme. Based on recent spectroscopic (148, 149) and structural (30, 34) analyses, it has become possible to propose a reaction mechanism of how FTR might achieve this conversion. To obtain the two electrons needed for reduction of the active-site disulfide bridge FTR mediates two consecutive one-electron transfers from reduced ferredoxin and stabilizes the one electron–reduced reaction intermediate. Characterization of the Fe-S cluster in native and chemically modified FTR (148, 149) provided evidence for a cubane $[4Fe\text{-}4S]^{2+}$ cluster in the oxidized enzyme, which cannot be reduced, but oxidized (redox potential for the couple $[4Fe\text{-}4S]^{3+/2+} = +420$ mV). The postulated reaction mechanism features a new biological role for this Fe-S cluster that catalyzes the reduction of a disulfide bond by establishing a transient $[4Fe\text{-}4S]^{3+}$ cluster coordinated by five cysteinates (149), with a cluster Fe atom as the fifth ligand (34, 35a).

Another aspect of the mechanism of thioredoxin-mediated enzyme regulation concerns the redox potentials of the component proteins in the regulatory chain. The oxidation-reduction properties of the different regulatory disulfides influence the activation state of the individual target enzymes. To understand the kinetics of activation, one needs to know the oxidation/reduction midpoint potentials (E_m) of the disulfide/sulfhydryl couples involved. Over the past several years, midpoint potentials of the members of the regulatory chain and several target enzymes have been obtained by cyclic voltammetry (131), the fluorescent probe monobromobimane, and activity measurements (60, 62, 63, 70, 120). Despite some relatively small differences for the E_m values determined by the different groups, there emerges a general agreement as to which members of the regulatory chain have the most positive E_m values and which the most negative. The more negative midpoint potential of the disulfide of spinach FTR ($E_m = -320$ mV, pH 7) allows both thioredoxins (−290 mV for *f*; −300 mV for *m*) to become reduced. Target enzymes involved indirectly or directly in regeneration of the CO_2 acceptor, CF_1 (−280 mV) and PRK (−295 mV), have the most positive potentials. FBPase [−305 mV, pH 7, spinach, P Schürmann, M Hirasawa & DB Knaff, unpublished data, replacing a former erroneous E_m value of −330 mV (63)], and SBPase (−300 mV, pH 7, tomato), which together control the entry into the regenerative phase of the Calvin cycle and into starch synthesis, have a slightly more negative potential than thioredoxin *f*, whereas NADP-MDH has a significantly more negative potential (−330 mV, sorghum) that keeps this latter enzyme oxidized/inactive as long as there is no surplus of reducing equivalents (NADPH). The differences in redox potential suggest a sequential order in the activation of chloroplast enzymes in the light, with the most critical metabolic processes given priority. They further suggest that the redox equilibria reached will result in different degrees of activation.

These equilibria may then be modified by enzyme effectors (47) that would allow for fine-tuning of activity after the target enzymes are switched on by light.

Chloroplast Target Enzymes

The fact that the light-dependent increase in enzyme activity could be mimicked in vitro by sulfhydryl compounds like dithiothreitol (DTT) was traditionally taken as evidence of an involvement of reduced thioredoxin. However, final proof of a redox event involving thioredoxin needs experimental evidence for an absolute requirement and the demonstration of an accessible disulfide bridge on the target structure. The presence of a redox-active disulfide has been confirmed for most of the light-regulated enzymes and its involvement verified by site-directed mutagenesis. For some of the enzymes, three-dimensional models, based either on X-ray crystallography or homology modeling, have provided further insights into possible structural changes occurring due to reduction.

Comparison of primary structures reveals that there is no Cys-containing consensus motif present in most of the light-regulated target enzymes, although in some a **CXXXXC** sequence is the responsible element or part of it (Table 1). However, the two active cyst(e)ines can also be separated by many residues. For certain enzymes, particularly those also occurring as cytosolic isoforms (e.g. FBPase, MDH), the regulatory disulfide structures are located on extra loops or extensions, indicating that they were added during adaptation to photosynthetic function. These observations suggest that the adaptation of enzymes to light-mediated redox regulation arose multiple times during evolution.

Reduction of the regulatory disulfides of the target enzymes by thioredoxin proceeds, like thioredoxin reduction, with the formationof a transient heterodisulfide

TABLE 1 Regulatory site sequences of chloroplast target enzymes and their principal activating thioredoxin

Target enzyme	Plant	Regulatory site[a]	Activator
FBPase	Spinach	**Cys155**X_{18}**Cys174**Val *Val*Asn*Val*Cys179	Trx *f*
SBPase	Wheat	**Cys52**Gly*GlyThr*Ala**Cys57**	Trx *f*
PRK	Spinach	**Cys16**X_{38}**Cys55**	Trx *f*
ATP synthase (γ-subunit of CF_1)	Spinach	**Cys199***Asp*IleAsn *Gly*Lys**Cys205**	Trx *f*
NADP-MDH	Sorghum	**Cys24**Phe*Gly*Val*Phe***Cys29** **Cys365**X_{11}**Cys377**	Trx *f* (*m*)
G6PDH	Potato	**Cys149***ArgIleAsp*Lys *Arg*GluAsp**Cys157**	Trx *m*
Rubisco activase	*Arabidopsis*	**Cys392**X_{18}**Cys411**	Trx *f*

[a]The regulatory Cys are in bold and additional conserved residues are in italics.

complex between the two reaction partners. The reactive Cys, which is the solvent-accessible residue closest to the N terminus of thioredoxin, cleaves the target disulfide by nucleophilic attack, thereby forming a covalently linked mixed disulfide. In a rapid second step, the second sulfhydryl, which is inaccessible to solvent, attacks the mixed disulfide to produce oxidized thioredoxin and reduced target enzyme.

Fructose 1,6-Bisphosphatase Stromal FBPase is one of the first enzymes in which activity was demonstrated to be clearly dependent on reduced thioredoxin *f* (16). Sequencing revealed the presence of an insert in the middle of the primary structure compared to its cytosolic isoform (118). This insert contains three Cys—**Cys**155X_{18} **Cys**174ValValAsnVal**Cys**179 in spinach—two separated by four hydrophobic residues and the third, Cys155, by 18 residues upstream toward the N terminus (Table 1). Cys174 and Cys179, present in a **CXXXXC** motif, were thought to constitute the redox active disulfide bridge (93). However, site-directed mutagenesis experiments indicated that all three Cys present in the insert are involved in regulation. The replacement of Cys155 results in a constitutively fully active enzyme, while the replacement of either of the remaining Cys (C174, C179) results in a partially active enzyme that still requires reduction by thioredoxin for full activity. These results suggest that Cys155 is an obligatory part of the regulatory disulfide, whereas the remaining two Cys (C174, C179) act interchangeably in constituting its bonding partner (74, 76, 126). Cys155 also forms the transient heterodisulfide bond during reduction by thioredoxin *f*, which indicates that it must be surface exposed (Y Balmer, P Schürmann, unpublished data). Interestingly, the region situated between Cys155 and Cys174 contains several negative charges (seven Asp and Glu in the pea sequence, with three of them highly conserved). Mutagenesis of this regulatory region suggests that these anionic charges are important for thioredoxin docking, presumably by establishing electrostatic interactions with positively charged, surface residues of thioredoxin *f* (129).

The structure of spinach chloroplast FBPase has been solved by X-ray crystallography at 2.8 Å resolution (159). The model shows that the three regulatory Cys are on a loop extending out of the core structure of the enzyme. Unfortunately, the definition of its crystal structure, probably due to its flexibility, was not sufficient to locate a disulfide bridge. Recent crystallographic analysis of the oxidized recombinant pea enzyme reveals a disulfide bond between Cys153 (Cys155 in spinach) and Cys 173 (Cys174 in spinach), whereas the third Cys in the loop is present as a free sulfhydryl, which is sufficiently close to form an alternate disulfide with Cys153 when Cys173 is altered by site-directed mutagenesis (29a).

Sedoheptulose 1,7-Bisphosphatase A substrate-specific SBPase is found only in oxygenic photosynthetic eukaryotes. It is unique to the Calvin cycle and has no cytosolic counterpart (117). Primary structures for this 76-kDa, homodimeric, nucleus-encoded chloroplast enzyme are known from *Arabidopsis* (164), *Chlamydomonas* (56), spinach (94), and wheat (119). In cyanobacteria, SBPase activity is contributed by a bifunctional FBPase/SBPase enzyme (153, 154).

Eukaryotic SBPase shows considerable overall sequence similarity with stromal FBPase; however, it lacks the regulatory Cys insert. The number of Cys varies, but four are found at strictly conserved positions in the N-terminal domain. The two most N-terminal Cys, in wheat arranged as a **Cys**52GlyGlyThrAla**Cys**57 motif (Table 1), have been shown to be involved in redox regulation. Mutation of these two Cys to Ser resulted in an active, redox-insensitive SBPase (43). As in stromal FBPase, the regulatory Cys residues of SBPase are not part of the catalytic site. Based on structural similarities between the two enzymes, the regulatory disulfide could be located on a flexible loop near the junction between the two SBPase monomers. Oxidation or reduction of the regulatory disulfide may alter the conformation of the homodimer and thereby change the activity of the enzyme (117). Unlike FBPase, this enzyme exhibits an absolute requirement for redox activation (106) and once reduced, its activity is modified further by stromal factors such as pH, [Mg^{2+}], and metabolites (25, 118, 137).

Phosphoribulokinase A redox-regulated PRK is present as a 80-kDa homodimer in photosynthetic eukaryotes (113) and as a homotetramer in cyanobacteria (160). Each subunit contains four Cys residues at conserved positions, two near the N terminus and two near the C terminus. For the spinach enzyme the N-terminal pair, Cys16 and Cys55, were identified by chemical modification (115) and site-directed mutagenesis (98) as the regulatory cysteines forming an intramolecular disulfide bond in the oxidized/inactive enzyme. Both are located in the nucleotide binding domain of the active site where Cys55 may play a facilitative role in catalysis (112, 114) by binding the sugar phosphate substrate (98). This Cys forms the transient heterodisulfide with Cys46 of thioredoxin *f* during reductive activation (15). PRK is the only example of a thioredoxin-linked enzyme with a regulatory Cys as part of its active site. Since one of the reduced Cys residues appears to be involved in substrate binding, the formation of a disulfide bridge very effectively blocks catalytic activity, a situation found also in NADP-MDH.

A crystal structure of the allosterically regulated PRK from *Rhodobacter sphaeroides* is now available (57). Although this enzyme is not redox regulated, the structural data will be useful for modeling.

ATP Synthase The chloroplast ATP synthase complex is composed of the integral thylakoid membrane portion CF_0 and the hydrophilic CF_1, which is composed of five different subunit-types. The γ subunit of the latter contains the structural element, which allows for thiol modulation of the enzyme by Trx *f* (108, 141). The regulatory motif— **Cys**199AspIleAsnGlyLys**Cys**205 in spinach (Table 1)— is present in the subunit from plants (97) and green algae (32), but is absent in the enzyme of cyanobacteria (32, 95, 163) and diatoms (109) or mitochondrial F_1. The disulfide bridge between the two Cys residues seems to be inaccessible in the inactive enzyme in the dark and becomes exposed upon activation by the transmembrane electrochemical proton gradient (35b). No information is available on possible structural changes brought about by reduction of the regulatory disulfide. However, the main purpose of reduction does not seem to be the modulation of enzyme activity, but rather to permit a higher rate of ATP formation at limiting electrochemical potentials. Regulation linked to thioredoxin also allows the enzyme to be switched off in the dark to avoid wasteful hydrolysis of ATP.

NADP-Dependent Malate Dehydrogenase NADP-MDH is the most extensively studied thioredoxin-regulated enzyme. It appears to have a rather complex regulatory mechanism, involving more than one disulfide bridge, which has recently been described in detail (96, 128). The homodimeric enzyme of 85 kDa differs from its cytoplasmic, NAD-dependent homologue by the presence of N- and C-terminal extensions and by containing eight Cys residues at strictly conserved positions. Two are located in the N-terminal extension in a **Cys**24PheGlyValPhe**Cys**29 motif (C_4 sorghum), and another two, Cys365 and Cys377, separated by 11 residues, are in the C-terminal region (Table 1). The four remaining Cys are located in the core part of the polypeptide. Systematic functional analysis of the Cys residues by chemical modification and site-directed mutagenesis provided evidence that one regulatory disulfide is formed between the two N-terminal Cys. A second regulatory disulfide bridge is present in the C terminus. These two regulatory sites are not functionally equivalent but have distinct effects on activation. Removal of the N-terminal disulfide yields an inactive, oxidized enzyme still in need of thioredoxin activation. However, this activation is almost instantaneous upon addition of reduced thioredoxin, which contrasts with the slow activation seen with the native enzyme containing both regulatory disulfides. Removal of the C-terminal disulfide produces a mutant enzyme with activation properties very similar to wild type, but the activation process is no longer inhibited by NADP. In addition, this mutant enzyme exhibits a slight constitutive activity (i.e. independent of reduction), but with a very high K_m for oxaloacetate. Removal of both N- and C-terminal disulfides results in a thioredoxin-insensitive, permanently active enzyme.

A fifth, internal Cys residue might also be involved in these regulatory events by forming a transient disulfide bridge with one of the N-terminal Cys (127). This internal Cys207 (in C_4 sorghum) becomes accessible after removal of the N-terminal disulfide by mutation and can form a heterodisulfide with thioredoxin (54).

Because the crystal structures of C_4 NADP-MDH from *Flaveria* (29) and sorghum (78) have recently been solved, the biochemical results can now be placed in a structural perspective. The models confirm the presence of two surface-exposed, thioredoxin-accessible disulfide bonds. The C-terminal disulfide holds down the C-terminal extension, which bends back over the surface and reaches with its tip into the active site, obstructing access of the C_4 acid substrate (82a). The tip of the C terminus carries two negative charges that interact with bound NADP, but not with NADPH. These interactions retard the release of the C-terminal extension and can explain the inhibition of reductive activation by NADP observed earlier (29). The N-terminal extension appears to be rather flexible, and the structural changes due to reduction of its disulfide bond are less clear. Based on the two structures, two possibilities are offered: The N-terminal disulfide might lock domains and thus maintain an unfavorable conformation (78) or the N-terminal residues might reach over the surface of the molecule toward the adenosine end of the active site and limit substrate access (29). The mobility of the N-terminal extension would also allow for the proposed formation of a transient disulfide bridge between an N-terminal Cys and the internal Cys207 (127). However, it is suggested that this transient disulfide is more likely formed between the two subunits of the homodimer than within a single subunit (29).

The rather complex regulation of NADP-MDH by redox equilibrium and the NADP/NADPH ratio appears to be important in C_3 and C_4 plants. In C_3 plants, where this stromal enzyme exports reducing equivalents in the form of malate from the chloroplast to the cytosol through the "malate valve" (135), it is essential that reducing equivalents are exported only when they are in excess in the chloroplast. In certain C_4 plants, where the enzyme functions as an essential catalyst of the CO_2 trapping and transport mechanism in mesophyll cells, it is equally important to turn off the C_4 pathway in the dark. The additional requirement of pyruvate, Pi dikinase and phosphoenolpyruvate carboxylase for reversible light activation helps to insure that the C_4 pathway is switched off completely in darkness.

Glucose 6-Phosphate Dehydrogenase G6PDH, catalyzing the first step of the oxidative pentose phosphate cycle, exists in plants in at least two isoforms, one in the cytosol and one in chloroplasts. Only the latter isoform is subject to redox regulation by thioredoxin. As this stromal enzyme is part of a dissimilatory pathway, it is regulated in the opposite way from biosynthetic enzymes, i.e. it is inactive in the light (reduced state) and active in the dark (oxidized state). The chloroplast isoform as well as the enzyme from cyanobacteria contain a number of Cys residues that are not present in the cytosolic counterpart. Mutational analysis of the six Cys of the recombinant potato enzyme, all situated in the N-terminal part of the polypeptide, revealed that Cys149 and Cys157 are engaged in a regulatory disulfide bridge (162). This disulfide, specifically reduced by thioredoxin *m*, is part of a regulatory sequence with no similarity to others. The cyanobacterial enzyme, although redox regulated (3, 31, 52), is quite different, with fewer (two to four) Cys residues at other positions. Two Cys (Cys188, Cys447 in *Synechococcus*) are at conserved positions within the cyanobacterial enzyme and might form the regulatory disulfide (52, 134), although there is no supportive experimental evidence to date.

Homology modeling using the crystal coordinates of the *Leuconostoc* enzyme locates the proposed regulatory cysteines of the chloroplast enzyme on an exposed loop, sufficiently close to permit a disulfide bridge, and freely accessible for interaction with thioredoxin (162). However, this model does not provide any insight into how disulfide reduction brings about the observed 30-fold increase in K_m for the substrate.

Rubisco Activase Until recently, no link between the observed light activation of Rubisco and thioredoxin could be made, the sole effect of light being indirect via a change in the stromal ADP/ATP ratio modulating activase activity (116). In certain plants, such as *Arabidopsis*, the activase polypeptide exists in two forms: short and long. The long form was recently found to be activated by reduced thioredoxin *f* in vitro at physiological ADP/ATP ratios and the regulatory disulfide, consisting of Cys392 and Cys411, was identified at the C-terminal end of the protein (169). However, in other plants, such as tobacco, where there is apparently no long-form activase, the mode of light-induced regulation (activation) remains to be clarified.

Other Possible Chloroplast Targets Three additional chloroplast enzymes—glyceraldehyde 3-phosphate dehydrogenase (GAPDH), acetyl CoA carboxylase, and Fd-dependent glutamate synthase (GOGAT)—are reported to be activated by reduced thioredoxin; however, no specific regulatory sites have yet been demonstrated. Although GAPDH activity can be stimulated by reduced thioredoxin *f*, no compelling evidence exists to date for an absolute requirement for thioredoxin as reductant. Indeed, recent experiments have shown that the same effects can be achieved by low concentrations of GSH (4). Acetyl CoA carboxylase, which catalyzes the first committed step in de novo fatty acid biosynthesis, appears to be activated by thioredoxin (68, 133). This enzyme exists in two isoforms in most plants, a eukaryotic, multifunctional form in the cytoplasm and a prokaryotic, multisubunit isoform in the chloroplast. The activity of the prokaryotic, pea plastid enzyme is significantly increased by the addition of dithiols, but not monothiols, and thioredoxin *f* proved to be the most efficient activator (133). Varying numbers of conserved Cys residues are present in the primary structure of the enzyme from different plant species, but no regulatory Cys are known. The activity of Fd-dependent GOGAT from spinach and soybean chloroplasts was reported to be significantly stimulated by DTT, but not GSH. Thioredoxin *m* was the most efficient activator of the spinach enzyme, which upon reduction exhibited an increased reaction velocity (89). Finally, the addition of DTT-reduced thioredoxin to *Chlamydomonas reinhardtii* preparations significantly enhanced translation of *psbA* RNA, which encodes the D1 protein of photosystem II. This provides a possible direct link between light and the replacement (via translation) of a reaction center protein known to be subject to photooxidative damage (36).

OTHER CATALYSTS WITH THE THIOREDOXIN FOLD IN PLANTS

Several catalysts contain in their polypeptide structure either a simplification of the "thioredoxin fold," e.g. glutaredoxin, or a thioredoxin module that is often present as an extension of the polypeptide.

E. coli glutaredoxin is prototypical of a simplified thioredoxin molecule, which lacks the two first structural units of thioredoxin, i.e. the $\beta 1$ and $\alpha 1$ modules. It is thus a shorter polypeptide ($\sim$85 residues) with an active site, typically -Cys-Pro-Phe/Tyr-Cys, that is readily reduced by GSH and kept reduced by NADPH and glutathione reductase. *E. coli* glutaredoxin is an excellent hydrogen donor for ribonucleotide reductase. Glutaredoxin has also been isolated from plant tissues (104, 144). It was found to be located in the cytosol and easily reduced by GSH. The isolation of a cDNA and gene from *Oryza sativa* (99, 145) established that

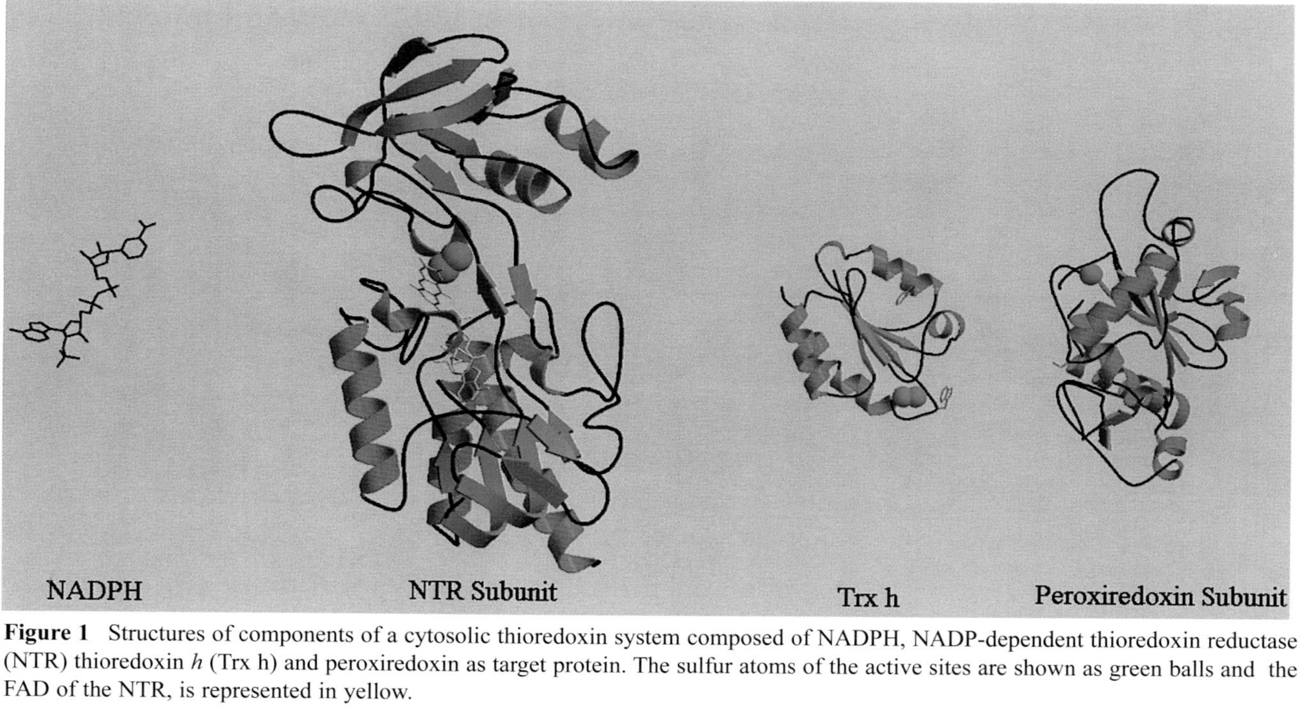

Figure 1 Structures of components of a cytosolic thioredoxin system composed of NADPH, NADP-dependent thioredoxin reductase (NTR) thioredoxin *h* (Trx h) and peroxiredoxin as target protein. The sulfur atoms of the active sites are shown as green balls and the FAD of the NTR, is represented in yellow.

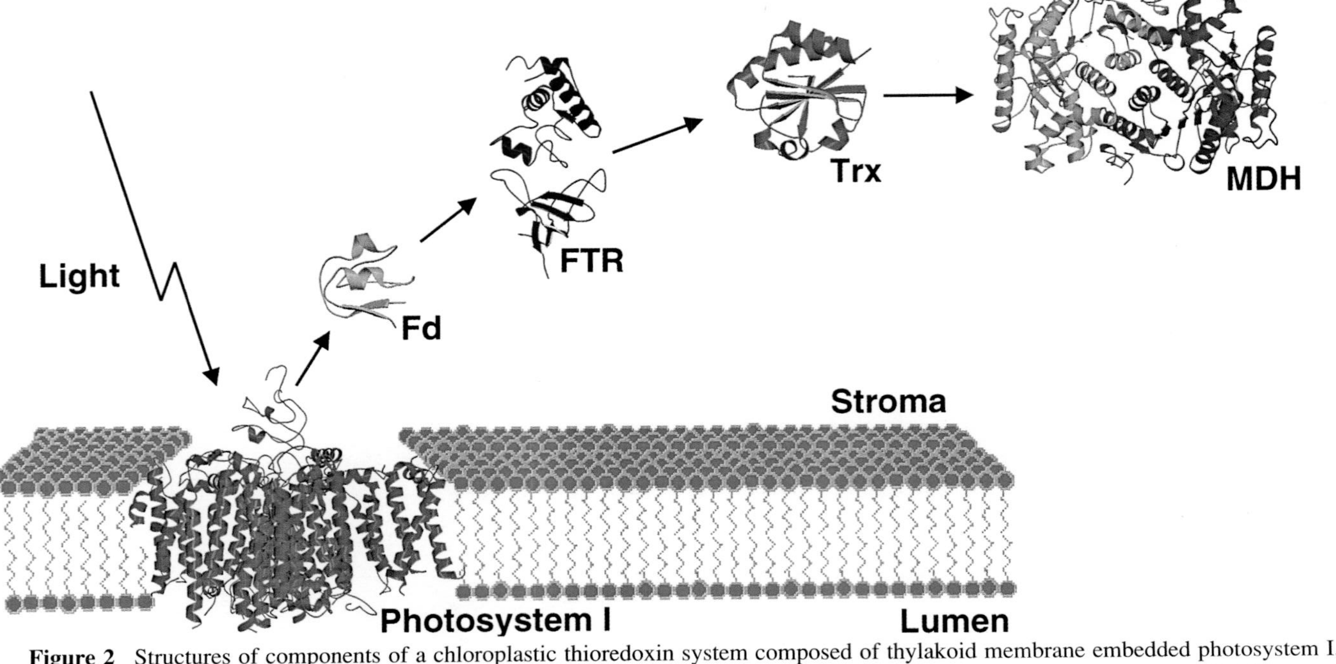

Figure 2 Structures of components of a chloroplastic thioredoxin system composed of thylakoid membrane embedded photosystem I, ferredoxin (Fd), ferredoxin:thioredoxin reductase (FTR), thioredoxin (Trx) and NADP-dependent malate dehydrogenase (MDH) as target protein. Modified figure reprinted with permission from (78). Copyright 1999 American Chemical Society.

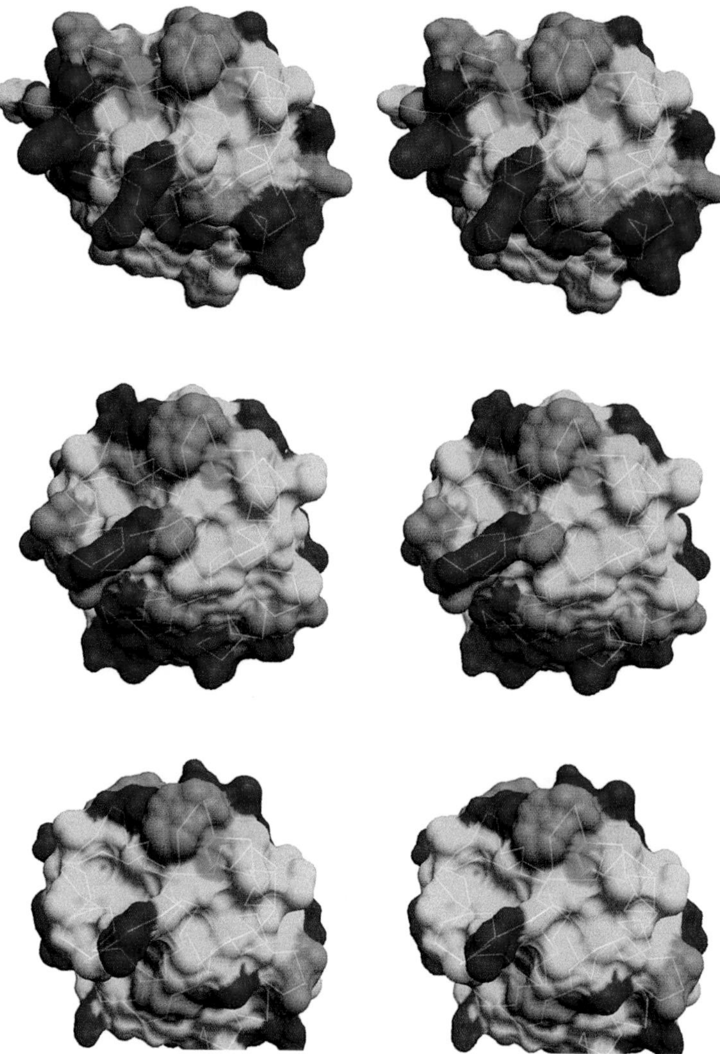

Figure 3 Surface stereo views of spinach chloroplast thioredoxin *f (top)*, thioredoxin *m (center)* and thioredoxin from *E. coli (bottom)*. The colors represent the following residue types: green–Cys; red–charged (+ or -); blue–polar; yellow–apolar; gray–backbone.

the protein is longer than its bacterial counterpart (~112 residues) and that there is a low number of gene copies (~2) in rice. Glutaredoxin was also found to be a major component of the phloem sap, together with its reductant glutathione, which suggests that it could potentially play a role in the redox regulation of sieve tube proteins or membrane-located receptors (152).

Among the plant enzymes that feature an extension comprised of either a thioredoxin or a glutaredoxin module is 5′ adenylsulfate reductase (APS reductase) (9, 55, 143, 151a, 167). In plants, this protein is the equivalent of the bacterial PAPS reductase and is involved in the assimilation of sulfur. Two types of cDNA sequences coding for APS reductase have been isolated that feature either a transit peptide or no transit peptide. APS reductase can be produced as a precursor of ~460 amino acids with an N-terminal extension (~70 amino acids) believed to target the protein to the chloroplast, a central domain equivalent to bacterial PAPS reductase, and a C-terminal domain of about 140 amino acids with a thioredoxin-like sequence and a glutaredoxin activity when isolated. Other cDNA clones do not show the putative transit sequence, which indicates that there may be several subcellular localizations for this enzyme. However, the thioredoxin-like module is consistently found in all analyzed clones. In bacteria, PAPS reductase is dependent on the addition of reduced thioredoxin for in vitro activity. In contrast to the bacterial enzyme, plant APS reductase is active independently of exogenous thioredoxin, presumably because of the presence of the thioredoxin module in its own polypeptide.

When plants such as *Solanum tuberosum* are subjected to water stress, they produce a variety of stress-related proteins among which is one stromal protein, CDSP 32, for *c*hloroplastic *d*rought-induced *s*tress *p*rotein (122). This nuclear-encoded, 32-kDa protein is comprised of two thioredoxin-like domains, with only the C-terminal domain containing a **CGPC** active site. CDSP 32 is proposed to play a role in the preservation of the thiol/disulfide redox potential of chloroplast proteins during water deficit.

A new protein of the thioredoxin family called nucleoredoxin (NRX) has recently been isolated from maize (85). This protein is located in the nucleus and contains three consecutive repeats of the thioredoxin domain. The first and third modules contain the active site **WCPPC**, which suggests a potential for catalyzing redox reactions, and the C-terminal part of the protein can form a Zn finger. The presence of this protein in developing kernels indicates that it could regulate the activity of transcription factors, presumably by altering their redox state.

NEW FUNCTIONS FOR THIOREDOXINS IN PLANTS

This topic has been extensively discussed in recent reviews (49, 73), so only the most recent developments are reported here. Thioredoxin has recently been implicated in the organization of the photosynthetic apparatus. A *Rhodobacter sphaeroides* thioredoxin minus mutant has been constructed and shown to be impaired in the formation of its photosynthetic apparatus (110). In this photosynthetic bacterium, thioredoxin appears to exert a dual role, activating aminolevulinic acid synthetase and regulating *puf* (an operon encoding pigment-binding proteins of the LHC and bacterial reaction center complexes) expression at the transcriptional level. In plants, phosphorylation of certain PSII proteins (D1, D2, and LHCII), although dependent on the redox state of plastoquinone, is also dependent on the presence of thiol-containing compounds (D1 and D2 phosphorylation is stimulated by reduced compounds and LHCII phosphorylation by oxidized compounds) (28). It is thus proposed that a second loop of redox regulation of thylakoid protein phosphorylation is under control of the ferredoxin/thioredoxin system.

ADP-glucose pyrophosphorylase, catalyzing the first committed step of starch biosynthesis, can be reduced by thioredoxins, thereby increasing its affinity for 3-phosphoglycerate, the principal activator (6). The reduction opens an intermolecular disulfide bridge between the two small subunits of the heterotetrameric enzyme, which probably imparts better access to 3-phosphoglycerate. It is proposed that in photosynthetic tissue the ferredoxin/thioredoxin system could be involved in this covalent mechanism of regulation of starch biosynthesis (6).

Three other enzymatic systems are also thioredoxin dependent in plants. The first one is enolase of *Arabidopsis* and ice plant, which is activated by oxidation, although not in other species such as tomato or maize (2). The second is carboxyarabinitol-1-phosphate (CA1P) phosphatase, an enzyme involved in the hydrolysis of CA1P, which is a naturally occurring nocturnal inhibitor of Rubisco. CA1P phosphatase is activated by either reduced thioredoxin or GSH (59). A third protein that could be redox regulated is CP12, a small nuclear-encoded protein capable of binding both PRK and NADP-GAPDH through conserved cysteine residues (161).

Finally, thioredoxin-like molecules have been found on extracellular pollen and detected in self-incompatibility reactions (88, 156). A thioredoxin activity linked to the C terminus of the S gene product seems to be responsible for self-incompatibility. In addition, thioredoxin *h*-like proteins have been shown to interact with the kinase domain of membrane-bound receptor-like protein kinases, which suggests a role in the self-incompatibility signal cascade (13).

TECHNOLOGICAL DEVELOPMENTS

This area has also been covered in earlier recent reviews (49, 73), and therefore only the latest developments are described. It was recently reported that thioredoxin has the capacity to decrease the allergenicity of wheat and milk proteins (19, 39). These findings may have considerable impact on human nutrition, since allergens have many undesirable effects.

Many expression systems have been devised for thioredoxin, which is readily produced with extremely high yields in bacteria such as *E. coli*. The high stability and solubility of thioredoxin has prompted several groups to use it as a stabilizer/solubilizer of peptides/proteins that are normally produced in inclusion bodies (45, 66, 67, 107, 147). With these expression systems, the protein to be solubilized is fused to an *E. coli* thioredoxin sequence, and, if necessary, to a poly His-tag. The resulting fusion protein can, in turn, be isolated by chromatography on a Ni^{2+}-containing affinity matrix and the thioredoxin module cleaved, if necessary, by a protease. The large number of successful reports attests to the power and efficiency of this technique.

A very elegant method to map protein-protein interactions by using thioredoxin has been reported by Lu et al (91). Briefly, the thioredoxin sequence is inserted into flagellin, a major structural component of the *E. coli* flagellum system. Small peptides can then be introduced into the loop of thioredoxin that contains the active site, a location known to be extremely permissive for peptide insertion. Bacteria engineered in this way will display the peptides in their pili, and this property can be used to build peptide libraries that allow, in turn, the identification of proteins binding with a high affinity. This method is thus an alternative to the well-developed yeast two-hybrid system.

CONCLUDING REMARKS

During the past several years, the use of recombinant technology has greatly contributed to advancements in the understanding of the thioredoxin systems. The availability of recombinant proteins has enabled crystallographers to solve the structures of all the members and even of a few target enzymes. Most of the regulatory sites have been located and the functions of their Cys elucidated by site directed mutagenesis. This same technique has also been employed to probe the protein-protein interaction surfaces for residues responsible for the specificity of these interactions. Not only has the structural field greatly advanced, the mechanism of thioredoxin reduction in chloroplasts has also been intensively studied. This has led to a comprehensive model involving a new role for a Fe-S protein in the reduction of a disulfide bridge, as well as to knowledge of the redox potentials of the regulatory chain and target enzymes.

A number of new functions for thioredoxin have recently been described in mammalian cells; these mostly relate to redox regulation of transcription factors, possibly mediated through the translocation of some components of these systems to the nucleus. It will be interesting to observe whether such developments are reported in plant systems in the future.

It is also interesting to comment on how differently the thioredoxin systems have evolved in plants compared with mammalian cells. In plants, evolution has brought an incredible diversification of the thioredoxin sequences in relation to the existence of another intracellular compartment, the chloroplast. Mammalian cells, on the other hand, have acquired a new NTR catalyst containing a selenocysteine at the active site required for catalytic activity. This modification has not yet been reported in plants or bacteria and seems to be specific to animal cells.

ACKNOWLEDGMENTS

The authors thank K Johansson and Drs G Capitani and G Mulliert-Carlin for providing figures used in this chapter; Professors H Eklund, DB Knaff, D Ort, and Drs M Miginiac-Maslow and AR Portis, Jr for access to material prior to publication; and Professor BB Buchanan for critically reading the manuscript. The authors would also like to acknowledge helpful discussions with Drs Capitani and S Dai, Professors Buchanan, Eklund, Knaff, and Y Meyer. Work in the authors' laboratories is supported by the Schweizerischer Nationalfonds (P.S.) and the INRA and CNRS, France (J.-P. J.)

LITERATURE CITED

1. Aguilar F, Brunner B, Gardet-Salvi L, Stutz E, Schürmann P. 1992. Biosynthesis of active spinach-chloroplast thioredoxin f in transformed *E. coli*. *Plant Mol. Biol.* 20:301–6
2. Anderson LE, Li AD, Stevens FJ. 1998. The enolases of ice plant and *Arabidopsis* contain a potential disulphide and are redox sensitive. *Phytochemistry* 47:707–13
3. Austin PA, Ross IS, Mills JD. 1996. Regulation of pigment content and enzyme activity in the cyanobacterium *Nostoc* sp Mac grown in continuous light, a light-dark photoperiod, or darkness. *Biochim. Biophys. Acta* 1277:141–49
4. Baalmann E, Backhausen JE, Rak C, Vetter S, Scheibe R. 1995. Reductive modification and nonreductive activation of purified spinach chloroplast NADP-dependent glyceraldehyde-3-phosphate dehydrogenase. *Arch. Biochem. Biophys.* 324:201–8
5. Baier M, Dietz KJ. 1996. Primary structure and expression of plant homologues of animal and fungal thioredoxin-dependent peroxide reductases and bacterial alkyl hydroperoxide reductases. *Plant Mol. Biol.* 31:553–64
6. Ballicora MA, Frueauf JB, Fu Y, Schürmann P, Preiss J. 1999. Activation of the potato tuber ADP-glucose pyrophosphatase by thioredoxins. *J. Biol. Chem.* 275:1315–20
7. Besse I, Buchanan BB. 1997. Thioredoxin-linked plant and animal processes: the new generation. *Bot. Bull. Acad. Sin. (Taipei)* 38:1–11
8. Besse I, Wong JH, Kobrehel K, Buchanan BB. 1996. Thiocalsin: a thioredoxin-linked, substrate-specific protease dependent on calcium. *Proc. Natl. Acad. Sci. USA* 93:3169–75
9. Bick JA, Åslund F, Chen YC, Leustek T. 1998. Glutaredoxin function for the carboxyl-terminal domain of the plant-type 5′-adenylylsulfate reductase. *Proc. Natl. Acad. Sci. USA* 95:8404–9
10. Binda C, Coda A, Aliverti A, Zanetti G, Mattevi A. 1998. Structure of mutant E92K of [2Fe-2S] ferredoxin I from *Spinacia oleracea* at 1.7 Å resolution. *Acta Cryst.* D54:1353–58
11. Bironaite D, Anusevicius Z, Jacquot JP, Cenas N. 1998. Interaction of quinones with *Arabidopsis thaliana* thioredoxin reductase. *Biochim. Biophys. Acta* 1383:82–92
12. Bodenstein-Lang J, Buch A, Follmann H. 1989. Animal and plant mitochondria contain specific thioredoxins. *FEBS Lett.* 258:22–26
13. Bower MS, Matias DD, Fernandes-Carvalho E, Mazzurco M, Gu TS, et al. 1996. Two members of the thioredoxin-h family interact with the kinase domain of

a *Brassica S* locus receptor kinase. *Plant Cell* 8:1641–50

14. Brandes HK, Larimer FW, Geck MK, Stringer CD, Schürmann P, Hartman FC. 1993. Direct identification of the primary nucleophile of thioredoxin f. *J. Biol. Chem.* 268:18411–14
15. Brandes HK, Larimer FW, Hartman FC. 1996. The molecular pathway for the regulation of phosphoribulokinase by thioredoxin *f*. *J. Biol. Chem.* 271:3333–35
16. Buchanan BB. 1980. Role of light in the regulation of chloroplast enzymes. *Annu. Rev. Plant Physiol.* 31:341–74
17. Buchanan BB. 1991. Regulation of CO_2 assimilation in oxygenic photosynthesis: the ferredoxin/thioredoxin system: perspective on its discovery, present status, and future development. *Arch. Biochem. Biophys.* 288:1–9
18. Buchanan BB. 1992. Carbon dioxide assimilation in oxygenic and anoxygenic photosynthesis. *Photosynth. Res.* 33:147–62
19. Buchanan BB, Adamidi C, Lozano RM, Yee BC, Momma M, et al. 1997. Thioredoxin-linked mitigation of allergic responses to wheat. *Proc. Natl. Acad. Sci. USA* 94:5372–77
20. Buchanan BB, Schürmann P, Decottignies P, Lozano RM. 1994. Thioredoxin: a multifunctional regulatory protein with a bright future in technology and medicine. *Arch. Biochem. Biophys.* 314:257–60
21. Buchanan BB, Schürmann P, Jacquot J-P. 1994. Thioredoxin and metabolic regulation. *Semin. Cell Biol.* 5:285–93
22. Buchanan BB, Wolosiuk RA, Schürmann P. 1979. Thioredoxin and enzyme regulation. *Trends Biochem. Sci.* 4:93–96
23. Bunik V, Raddatz G, Lemaire S, Meyer Y, Jacquot JP, Bisswanger H. 1999. Interaction of thioredoxins with target proteins: role of particular structural elements and electrostatic properties of thioredoxins in their interplay with 2-oxoacid dehydrogenase complexes. *Protein Sci.* 8:65–74
24. Bunik V, Shoubnikova A, Loeffelhardt S, Bisswanger H, Borbe HO, Follmann H. 1995. Using lipoate enantiomers and thioredoxin to study the mechanism of the 2-oxoacid-dependent dihydrolipoate production by the 2–oxoacid dehydrogenase complexes. *FEBS Lett.* 371:167–70
25. Cadet F, Meunier J-C. 1988. pH and kinetic studies of chloroplast sedoheptulose-1,7-bisphosphatase from spinach (*Spinacia oleracea*). *Biochem. J.* 253:249–54
26. Deleted in proof
27. Capitani G, Markovic-Housley Z, Jansonius JN, del Val G, Morris M, Schürmann P. 1998. Crystal structures of thioredoxins *f* and *m* from spinach chloroplasts. In *Photosynthesis: Mechanisms and Effects (Proc. Int. Congr. Photosynth., 11th, Budapest, Hungary)*, ed. G Garab, 3:1939–42. Dordrecht, The Netherlands: Kluwer
28. Carlberg I, Rintamäki E, Aro EM, Andersson B. 1999. Thylakoid protein phosphorylation and the thiol redox state. *Biochemistry* 38:3197–204
29. Carr PD, Verger D, Ashton AR, Ollis DL. 1999. Chloroplast NADP-malate dehydrogenase: structural basis of light-dependent regulation of activity by thiol oxidation and reduction. *Structure* 7:461–75

29a. Chiadmi M, Navaza A, Miginiac-Maslow M, Jacquot JP, Cherfils J. 1999. Redox signaling in the chloroplast: structure of the oxidized pea fructose-1,6-bisphosphatase. *EMBO J.* 18:6809–15

30. Chow L-P, Iwadate H, Yano K, Kamo M, Tsugita A, et al. 1995. Amino acid sequence of spinach ferredoxin:thioredoxin reductase catalytic subunit and identification of thiol groups constituting a redox active disulfide and a [4Fe-4S] cluster. *Eur. J. Biochem.* 231:149–56
31. Cossar JD, Rowell P, Stewart WDP. 1984. Thioredoxin as a modulator of

glucose-6-phosphate dehydrogenase in a N_2-fixing cyanobacterium. *J. Gen. Microbiol.* 130:991–98

32. Cozens AL, Walker JE. 1987. The organization and sequence of the genes for ATP synthase subunits in the cyanobacterium *Synechococcus* 6301. Support for an endosymbiotic origin of chloroplasts. *J. Mol. Biol.* 194:359–83
33. Crawford NA, Droux M, Kosower NS, Buchanan BB. 1989. Function of the ferredoxin/thioredoxin system in reductive activation of target enzymes of isolated intact chloroplasts. In *Photosynthesis*, ed. WR Briggs, pp. 425–36. New York: Liss
34. Dai S. 1998. *Structural and functional studies of NADPH and ferredoxin dependent thioredoxin reductases*. PhD thesis. Swedish Univ. Agric. Sci., Uppsala. 75 pp.
35. Dai S, Saarinen M, Ramaswamy S, Meyer Y, Jacquot JP, Eklund H. 1996. Crystal structure of *Arabidopsis thaliana* NADPH dependent thioredoxin reductase at 2.5 Å resolution. *J. Mol. Biol.* 264:1044–57

35a. Dai S, Schwendtmayer C, Schürmann P, Ramaswamy S, Eklund H. 2000. Redox signaling in chloroplasts: cleavage of disulfides by an iron-sulfur cluster. *Science* 287:655–58

35b. Dann MS, McCarty RE. 1992. Characterization of the activation of membrane-bound and soluble CF_1 by thioredoxin. *Plant Physiol.* 99:153–60

36. Danon A, Mayfield SP. 1994. Light-regulated translation of chloroplast messenger RNAs through redox potential. *Science* 266:1717–19
37. De Pascalis AR, Schürmann P, Bosshard HR. 1994. Comparison of the binding sites of plant ferredoxin for two ferredoxin-dependent enzymes. *FEBS Lett.* 337:217–20
38. del Val G, Maurer F, Stutz E, Schürmann P. 1999. Modification of the reactivity of spinach chloroplast thioredoxin *f* by site-directed mutagenesis. *Plant Sci.* 149:183–90
39. del Val G, Yee BC, Lozano RM, Buchanan BB, Ermel RW, et al. 1999. Thioredoxin treatment increases digestibility and lowers allergenicity of milk. *J. Allergy Clin. Immunol.* 103:690–97
40. Douglas SE, Penny SL. 1999. The plastid genome of the Cryptophyte alga, *Guillardia theta*: Complete sequence and conserved synteny groups confirm its common ancestry with red algae. *J. Mol. Evol.* 48:236–44
41. Droux M, Jacquot J-P, Miginiac-Maslow M, Gadal P, Huet JC, et al. 1987. Ferredoxin-thioredoxin reductase, an iron-sulfur enzyme linking light to enzyme regulation in oxygenic photosynthesis: purification and properties of the enzyme from C3, C4, and cyanobacterial species. *Arch. Biochem. Biophys.* 252:426–39
42. Droux M, Miginiac-Maslow M, Jacquot J-P, Gadal P, Crawford NA, et al. 1987. Ferredoxin-thioredoxin reductase: A catalytically active dithiol group links photoreduced ferredoxin to thioredoxin functional in photosynthetic enzyme regulation. *Arch. Biochem. Biophys.* 256:372–80
43. Dunford RP, Durrant MC, Catley MA, Dyer T. 1998. Location of the redox-active cysteines in chloroplast sedoheptulose-1,7-bisphosphatase indicates that its allosteric regulation is similar but not identical to that of fructose-1,6-bisphosphatase. *Photosynth. Res.* 58:221–30
44. Eklund H, Gleason FK, Holmgren A. 1991. Structural and functional relations among thioredoxins of different species. *Proteins* 11:13–28
45. Ems-McClung SC, Hainline BE. 1998. Expression of maize gamma zein C-terminus in *Escherichia coli*. *Protein Expr. Purif.* 13:1–8
46. Falkenstein E, von Schaewen A, Scheibe R. 1994. Full-length cDNA sequences for

both ferredoxin-thioredoxin reductase subunits from spinach (*Spinacia oleracea* L.). *Biochim. Biophys. Acta* 1185:252–54

47. Faske M, Holtgrefe S, Ocheretina O, Meister M, Backhausen JE, Scheibe R. 1995. Redox equilibria between the regulatory thiols of light/dark-modulated chloroplast enzymes and dithiothreitol: fine-tuning by metabolites. *Biochim. Biophys. Acta* 1247:135–42
48. Florencio FJ, Yee BC, Johnson TC, Buchanan BB. 1988. An NADP/thioredoxin system in leaves. Purification and characterization of NADP-thioredoxin reductase and thioredoxin h from spinach. *Arch. Biochem. Biophys.* 266:496–507
49. Follmann H, Häberlein I. 1996. Thioredoxins: universal, yet specific thiol-disulfide redox cofactors. *BioFactors* 5:147–56

49a. Fridlyand LE, Scheibe R. 1999. Regulation of the Calvin cycle for CO_2 fixation as an example for general control mechanisms in metabolic cycles. *Biosystems* 51:79–93

50. Gaymard E, Schürmann P. 1995. Cloning and expression of cDNAs coding for the spinach ferredoxin:thioredoxin reductase. In *Photosynthesis: From Light to Biosphere. (Proc. Int. Photosynth. Congr., 10th, Montpellier, France)*, ed. P Mathis, 2:761–64. Dordrecht, The Netherlands: Kluwer
51. Geck MK, Larimer FW, Hartman FC. 1996. Identification of residues of spinach thioredoxin *f* that influence interactions with target enzymes. *J. Biol. Chem.* 271:24736–40
52. Gleason FK. 1996. Glucose-6-phosphate dehydrogenase from the cyanobacterium, *Anabaena* sp PCC 7120: purification and kinetics of redox modulation. *Arch. Biochem. Biophys.* 334:277–83
53. Gorlatov SN, Stadtman TC. 1998. Human thioredoxin reductase from HeLa cells: Selective alkylation of selenocysteine in the protein inhibits enzyme activity and reduction with NADPH influences affinity to heparin. *Proc. Natl. Acad. Sci. USA* 95:8520–25
54. Goyer A, Decottignies P, Lemaire S, Ruelland E, Issakidis-Bourguet E, et al. 1999. The internal Cys207 of sorghum leaf NADP-malate dehydrogenase can form mixed disulfides with thioredoxin. *FEBS Lett.* 444:165–69
55. Gutierrez-Marcos JF, Roberts MA, Campbell EI, Wray JL. 1996. Three members of a novel small gene-family from *Arabidopsis thaliana* able to complement functionally an *Escherichia coli* mutant defective in PAPS reductase activity encode proteins with a thioredoxin-like domain and "APS reductase" activity. *Proc. Natl. Acad. Sci. USA* 93:13377–82
56. Hahn D, Kaltenbach C, Kück U. 1998. The Calvin cycle enzyme sedoheptulose-1, 7-bisphosphatase is encoded by a light-regulated gene in *Chlamydomonas reinhardtii*. *Plant Mol. Biol.* 36:929–34
57. Harrison DH, Runquist JA, Holub A, Miziorko HM. 1998. The crystal structure of phosphoribulokinase from *Rhodobacter sphaeroides* reveals a fold similar to that of adenylate kinase. *Biochemistry* 37:5074–85
58. Hartman H, Syvanen M, Buchanan BB. 1990. Contrasting evolutionary histories of chloroplast thioredoxins f and m. *Mol. Biol. Evol.* 7:247–54
59. Heo JY, Holbrook GP. 1999. Regulation of 2-carboxy-D-arabinitol 1-phosphate phosphatase: activation by glutathione and interaction with thiol reagents. *Biochem. J.* 338:409–16
60. Hirasawa M, Brandes HK, Hartman FC, Knaff DB. 1998. Oxidation-reduction properties of the regulatory site of spinach phosphoribulokinase. *Arch. Biochem. Biophys.* 350:127–31
61. Hirasawa M, Droux M, Gray KA, Boyer

JM, Davis DJ, et al. 1988. Ferredoxin-thioredoxin reductase: properties of its complex with ferredoxin. *Biochim. Biophys. Acta* 935:1–8

62. Hirasawa M, Ruelland E, Schepens I, Issakidis-Bourguet E, Miginiac-Maslow M, Knaff DB. 2000. Oxidation-reduction properties of the regulatory disulfides of sorghum chloroplast NADP-malate dehydrogenase. *Biochemistry.* 39:3344–50
63. Hirasawa M, Schürmann P, Jacquot J-P, Manieri W, Jacquot P, et al. 1999. Oxidation-reduction properties of chloroplast thioredoxins, ferredoxin:thioredoxin reductase and thioredoxin *f*-regulated enzymes. *Biochemistry* 38:5200–5
64. Hodges M, Miginiac-Maslow M, Decottignies P, Jacquot J-P, Stein M, et al. 1994. Purification and characterization of pea thioredoxin f expressed in *Escherichia coli. Plant Mol. Biol.* 26:225–34
65. Holmgren A. 1985. Thioredoxin. *Annu. Rev. Biochem.* 54:237–71
66. Huang EX, Huang QL, Wildung MR, Croteau R, Scott AI. 1998. Overproduction, in *Escherichia coli*, of soluble taxadiene synthase, a key enzyme in the taxol biosynthetic pathway. *Protein Expr. Purif.* 13:90–96
67. Huang KX, Huang QL, Scott AI. 1998. Overexpression, single-step purification, and site-directed mutagenetic analysis of casbene synthase. *Arch. Biochem. Biophys.* 352:144–52
68. Hunter SC, Ohlrogge JB. 1998. Regulation of spinach chloroplast acetyl-CoA carboxylase. *Arch. Biochem. Biophys.* 359:170–78
69. Huppe HC, Lamotte-Guéry Fd, Jacquot J-P, Buchanan BB. 1990. The ferredoxin-thioredoxin system of a green alga, *Chlamydomonas reinhardtii*. Identification and characterization of thioredoxins and ferredoxin-thioredoxin reductase components. *Planta* 180:341–51
70. Hutchison RS, Groom Q, Ort DR. 2000. Differential effects of low temperature induced oxidation on thioredoxin mediated activation of photosynthetic carbon reduction cycle enzymes. *Biochemistry* 39. In press
71. Ishiwatari Y, Fujiwara T, McFarland KC, Nemoto K, Hayashi H, et al. 1998. Rice phloem thioredoxin h has the capacity to mediate its own cell-to-cell transport through plasmodesmata. *Planta* 205:12–22
72. Ishiwatari Y, Honda C, Kawashima I, Nakamura S, Hirano H, et al. 1995. Thioredoxin h is one of the major proteins in rice phloem sap. *Planta* 195:456–63
73. Jacquot J-P, Lancelin J-M, Meyer Y. 1997. Thioredoxins: structure and function in plant cells. *New Phytol.* 136:543–70
74. Jacquot J-P, López-Jaramillo J, Chueca A, Cherfils J, Lemaire S, et al. 1995. High-level expression of recombinant pea chloroplast fructose-1,6-bisphosphatase and mutagenesis of its regulatory site. *Eur. J. Biochem.* 229:675–81
75. Jacquot J-P, Rivera-Madrid R, Marinho P, Kollarova M, Le Maréchal P, et al. 1994. *Arabidopsis thaliana* NADPH thioredoxin reductase cDNA characterization and expression of the recombinant protein in *Escherichia coli. J. Mol. Biol.* 235:1357–63
76. Jacquot JP, López-Jaramillo J, Miginiac-Maslow M, Lemaire S, Cherfils J, et al. 1997. Cysteine-153 is required for redox regulation of pea chloroplast fructose-1,6-bisphosphatase. *FEBS Lett.* 401:143–47
77. Jacquot JP, Stein M, Suzuki K, Liottet S, Sandoz G, Miginiac-Maslow M. 1997. Residue Glu-91 of *Chlamydomonas reinhardtii* ferredoxin is essential for electron transfer to ferredoxin-thioredoxin reductase. *FEBS Lett.* 400:293–96
78. Johansson K, Ramaswamy S, Saarinen M, Lemaire-Chamley M, Issakidis-Bourguet E, et al. 1999. Structural basis for light activation of a chloroplast enzyme. The

structure of sorghum NADP-malate dehydrogenase in its oxidized form. *Biochemistry* 38:4319–26

79. Johnson TC, Cao RQ, Kung JE, Buchanan BB. 1987. Thioredoxin and NADP-thioredoxin reductase from cultured carrot cells. *Planta* 171:321–31
80. Katti SK, LeMaster DM, Eklund H. 1990. Crystal structure of thioredoxin from *Escherichia coli* at 1.68 Å resolution. *J. Mol. Biol.* 212:167–84
81. Klenk HP, Clayton RA, Tomb JF, White O, Nelson KE, et al. 1997. The complete genome sequence of the hyperthermophilic, sulphate-reducing archaeon *Archaeoglobus fulgidus*. *Nature* 390:364–70
82. Kobrehel K, Wong JH, Balogh A, Kiss F, Yee BC, Buchanan BB. 1992. Specific reduction of wheat storage proteins by thioredoxin *h*. *Plant Physiol.* 99:919–24

82a. Krimm I, Goyer A, Issakidis-Bourguet E, Miginiac-Maslow M, Lancelin J-M. 1999. Direct NMR observation of the thioredoxin-mediated reduction of the chloroplast NADP-malate dehydrogenase provides a structural basis for the relief of auto-inhibition. *J. Biol. Chem.* 274:34539–42

83. Lamotte-Guéry Fd, Miginiac-Maslow M, Decottignies P, Stein M, Minard P, Jacquot J-P. 1991. Mutation of a negatively charged amino acid in thioredoxin modifies its reactivity with chloroplastic enzymes. *Eur. J. Biochem.* 196:287–94
84. Lancelin J-M, Stein M, Jacquot J-P. 1993. Secondary structure and protein folding of recombinant chloroplastic thioredoxin Ch2 from the green alga *Chlamydomonas reinhardtii* as determined by ^{1}H NMR. *J. Biochem.* 114:421–31
85. Laughner BJ, Sehnke PC, Ferl RJ. 1998. A novel nuclear member of the thioredoxin superfamily. *Plant Physiol.* 118:987–96
86. Lemaire S, Keryer E, Stein M, Schepens I, Issakidis-Bourguet E, et al. 1999. Heavy-metal regulation of thioredoxin gene expression in *Chlamydomonas reinhardtii*. *Plant Physiol.* 120:773–78
87. Lemaire S, Stein M, Issakidis-Bourguet E, Keryer E, Benoit V, et al. 1999. The complex regulation of ferredoxin/thioredoxin-related genes by light and the circadian clock. *Planta* 209:221–29
88. Li X, Nield J, Hayman D, Langridge P. 1996. A self-fertile mutant of *Phalaris* produces an S protein with reduced thioredoxin activity. *Plant J.* 10:505–13
89. Lichter A, Häberlein I. 1998. A light-dependent redox signal participates in the regulation of ammonia fixation in chloroplasts of higher plants—ferredoxin:glutamate synthase is a thioredoxin-dependent enzyme. *J. Plant Physiol.* 153:83–90
90. López-Jaramillo J, Chueca A, Jacquot JP, Hermoso R, Lázaro JJ, et al. 1997. High-yield expression of pea thioredoxin *m* and assessment of its efficiency in chloroplast fructose-1,6-bisphosphatase activation. *Plant Physiol.* 114:1169–75
91. Lu Z, Murray KS, Van Cleave V, LaVallie ER, Stahl ML, McCoy JM. 1995. Expression of thioredoxin random peptide libraries on the *Escherichia coli* cell surface as functional fusions to flagellin: a system designed for exploring protein-protein interactions. *Biotechnology* 13:366–72
92. Marcus F, Chamberlain SH, Chu C, Masiarz FR, Shin S, et al. 1991. Plant thioredoxin *h*: an animal-like thioredoxin occurring in multiple cell compartments. *Arch. Biochem. Biophys.* 287:195–98
93. Marcus F, Moberly L, Latshaw SP. 1988. Comparative amino acid sequence of fructose-1,6-bisphosphatases: identification of a region unique to the light-regulated chloroplast enzyme. *Proc. Natl. Acad. Sci. USA* 85:5379–83
94. Martin W, Mustafa AZ, Henze K,

Schnarrenberger C. 1996. Higher-plant chloroplast and cytosolic fructose-1, 6-bisphosphatase isoenzymes: origins via duplication rather than prokaryote-eukaryote divergence. *Plant Mol. Biol.* 32:485–91

95. McCarn DF, Whitaker RA, Alam J, Vrba J, Curtis SE. 1988. Genes encoding the alpha, gamma, delta, and four F0 subunits of ATP synthase constitute an operon in the cyanobacterium *Anabaena* sp. strain PCC 7120. *J. Bacteriol.* 170:3448–58

95a. Meyer Y, Verdoucq L, Vignols F. 1999. Plant thioredoxins and glutaredoxins: identity and putative roles. *Trends Plant Sci.* 4:388–94

96. Miginiac-Maslow M, Issakidis E, Lemaire M, Ruelland E, Jacquot JP, Decottignies P. 1997. Light-dependent activation of NADP-malate dehydrogenase: a complex process. *Aust. J. Plant Physiol.* 24:529–42

97. Miki J, Maeda M, Mukohata Y, Futai M. 1988. The y-subunit of ATP synthase from spinach chloroplasts primary structure deduced from the cloned cDNA sequence. *FEBS Lett.* 232:221–26

98. Milanez S, Mural RJ, Hartman FC. 1991. Roles of cysteinyl residues of phosphoribulokinase as examined by site-directed mutagenesis. *J. Biol. Chem.* 266:10694–99

99. Minakuchi K, Yabushita T, Masumura T, Ichihara K, Tanaka K. 1994. Cloning and sequence analysis of a cDNA encoding rice glutaredoxin. *FEBS Lett.* 337:157–60

100. Miranda-Vizuete A, Damdimopoulos AE, Pedrajas JR, Gustafsson JA, Spyrou G. 1999. Human mitochondrial thioredoxin reductase cDNA cloning, expression and genomic organization. *Eur. J. Biochem.* 261:405–12

101. Miskiniene V, Sarlauskas J, Jacquot J-P, Cenas N. 1998. Nitroreductase reactions of *Arabidopsis thaliana* thioredoxin reductase. *Biochim. Biophys. Acta* 1366:275–83

102. Mittard V, Blackledge MJ, Stein M, Jacquot JP, Marion D, Lancelin JM. 1997. NMR solution structure of an oxidised thioredoxin *h* from the eukaryotic green alga *Chlamydomonas reinhardtii. Eur. J. Biochem.* 243:374–83

103. Mora-García S, Rodriguez-Suárez RJ, Wolosiuk RA. 1998. Role of electrostatic interactions on the affinity of thioredoxin for target proteins. Recognition of chloroplast fructose-1,6-bisphosphatase by mutant *Escherichia coli* thioredoxins. *J. Biol. Chem.* 273:16273–80

104. Morell S, Follmann H, Häberlein I. 1995. Identification and localization of the first glutaredoxin in leaves of a higher plant. *FEBS Lett.* 369:149–52

105. Mouaheb N, Thomas D, Verdoucq L, Monfort P, Meyer Y. 1998. *In vivo* functional discrimination between plant thioredoxins by heterologous expression in the yeast *Saccharomyces cerevisiae. Proc. Natl. Acad. Sci. USA* 95:3312–17

106. Nishizawa AN, Buchanan BB. 1981. Enzyme regulation in C4 photosynthesis. Purification and properties of thioredoxin-linked fructose bisphosphatase and sedoheptulose bisphosphatase from corn leaves. *J. Biol. Chem.* 256:6119–26

107. Nowakowski JL, Courtney BC, Bing QA, Adler M. 1998. Production of an expression system for a synaptobrevin fragment to monitor cleavage by botulinum neurotoxin B. *J. Protein Chem.* 17:453–62

108. Ort DR, Oxborough K. 1992. In situ regulation of chloroplast coupling factor activity. *Annu. Rev. Plant Physiol. Plant Mol. Biol.* 43:269–91

109. Pancic PG, Strotmann H. 1993. Structure of the nuclear encoded gamma subunit of CF_0CF_1 of the diatom *Odontella sinensis* including its presequence. *FEBS Lett.* 320:61–66

110. Pasternak C, Haberzettl K, Klug G. 1999. Thioredoxin is involved in oxygen-regulated formation of the photosynthetic

apparatus of *Rhodobacter sphaeroides*. *J. Bacteriol.* 181:100–6

111. Pedrajas JR, Kosmidou E, Miranda-Vizuete A, Gustafsson JÅ, Wright AP, Spyrou G. 1999. Identification and functional characterization of a novel mitochondrial thioredoxin system in *Saccharomyces cerevisiae*. *J. Biol. Chem.* 274:6366–73
112. Porter MA, Hartman FC. 1990. Exploration of the function of a regulatory sulfhydryl of phosphoribulokinase from spinach. *Arch. Biochem. Biophys.* 281:330–34
113. Porter MA, Milanez S, Stringer CD, Hartman FC. 1986. Purification and characterization of ribulose-5-phosphate kinase from spinach. *Arch. Biochem. Biophys.* 245:14–23
114. Porter MA, Potter MD, Hartman FC. 1990. Affinity labeling of spinach phosphoribulokinase subsequent to *S*-methylation at Cys16. *J. Protein Chem.* 9:445–52
115. Porter MA, Stringer CD, Hartman FC. 1988. Characterization of the regulatory thioredoxin site of phosphoribulokinase. *J. Biol. Chem.* 263:123–29
116. Portis AR Jr. 1995. The regulation of Rubisco by Rubisco activase. *J. Exp. Bot.* 46:1285–91
117. Raines CA, Lloyd JC, Dyer TA. 1999. New insights into the structure and function of sedoheptulose-1,7-bisphosphatase; an important but neglected Calvin cycle enzyme. *J. Exp. Bot.* 50:1–8
118. Raines CA, Lloyd JC, Longstaff M, Bradley D, Dyer T. 1988. Chloroplast fructose-1,6-bisphosphatase: the product of a mosaic gene. *Nucleic Acids Res.* 16:7931–42
119. Raines CA, Lloyd JC, Willingham NM, Potts S, Dyer TA. 1992. cDNA and gene sequences of wheat chloroplast sedoheptulose-1,7-bisphosphatase reveal homology with fructose-1,6-bisphosphatases. *Eur. J. Biochem.* 205:1053–59
120. Rebeille F, Hatch MD. 1986. Regulation of NADP-malate dehydrogenase in C4 plants: effect of varying NADPH to NADP ratios and thioredoxin redox state on enzyme activity in reconstituted systems. *Arch. Biochem. Biophys.* 249:164–70
121. Reith M, Munholland J. 1997. Complete nucleotide sequence of the *Porphyra purpurea* chloroplast genome. *Plant Mol. Biol. Report.* 13:333–35
122. Rey P, Pruvot G, Becuwe N, Eymery F, Rumeau D, Peltier G. 1998. A novel thioredoxin-like protein located in the chloroplast is induced by water deficit in *Solanum tuberosum* L. plants. *Plant J.* 13:97–107
123. Reynolds AE, Chesnick JM, Woolford J, Cattolico RA. 1994. Chloroplast encoded thioredoxin genes in the red algae *Porphyra yezoensis* and *Griffithsia pacifica*: evolutionary implications. *Plant Mol. Biol.* 25:13–21
124. Rhoads DM, Umbach AL, Sweet CR, Lennon AM, Rauch GS, Siedow JN. 1998. Regulation of the cyanide-resistant alternative oxidase of plant mitochondria. Identification of the cysteine residue involved in α-keto acid stimulation and intersubunit disulfide bond formation. *J. Biol. Chem.* 273:30750–56
125. Rivera-Madrid R, Mestres D, Marinho P, Jacquot J-P, Decottignies P, et al. 1995. Evidence for five divergent thioredoxin *h* sequences in *Arabidopsis thaliana*. *Proc. Natl. Acad. Sci. USA* 92:5620–24
126. Rodriguez-Suárez RJ, Mora-García S, Wolosiuk RA. 1997. Characterization of cysteine residues involved in the reductive activation and the structural stability of rapeseed (*Brassica napus*) chloroplast fructose-1,6-bisphosphatase. *Biochem. Biophys. Res. Commun.* 232:388–93
127. Ruelland E, Lemaire-Chamley M, Le Maréchal P, Issakidis-Bourguet E, Djukic N, Miginiac-Maslow M. 1997. An

internal cysteine is involved in the thioredoxin-dependent activation of sorghum leaf NADP-malate dehydrogenase. *J. Biol. Chem.* 272:19851–57

128. Ruelland E, Miginiac-Maslow M. 1999. Regulation of chloroplast enzyme activities by thioredoxins: activation or relief from inhibition? *Trends Plant Sci.* 4:136–41
129. Sahrawy M, Chueca A, Hermoso R, Lázaro JJ, Gorgé JL. 1997. Directed mutagenesis shows that the preceding region of the chloroplast fructose-1, 6-bisphosphatase regulatory sequence is the thioredoxin docking site. *J. Mol. Biol.* 269:623–30
130. Sahrawy M, Hecht V, López-Jaramillo J, Chueca A, Chartier Y, Meyer Y. 1996. Intron position as an evolutionary marker of thioredoxins and thioredoxin domains. *J. Mol. Evol.* 42:422–31
131. Salamon Z, Tollin G, Hirasawa M, Knaff DB, Schürmann P. 1995. The oxidation-reduction properties of spinach thioredoxins f and m and of ferredoxin:thioredoxin reductase. *Biochim. Biophys. Acta* 1230:114–18
132. Sasaki T, Chino M, Hayashi H, Fujiwara T. 1998. Detection of several mRNA species in rice phloem sap. *Plant Cell Physiol.* 39:895–97
133. Sasaki Y, Kozaki A, Hatano M. 1997. Link between light and fatty acid synthesis: Thioredoxin-linked reductive activation of plastidic acetyl-CoA carboxylase. *Proc. Natl. Acad. Sci. USA* 94:11096–101
134. Scanlan DJ, Newman J, Sebaihia M, Mann NH, Carr NG. 1992. Cloning and sequence analysis of the glucose-6-phosphate dehydrogenase gene from the cyanobacterium *Synechococcus* PCC 7942. *Plant Mol. Biol.* 19:877–80
135. Scheibe R. 1987. NADP$^+$-malate dehydrogenase in C3-plants: regulation and role of a light-activated enzyme. *Physiol. Plant.* 71:393–400

135a. Scheibe R, 1999. Light/dark modulation: regulation of chloroplast metabolism in a new light. *Bot. Acta* 103:327–34

136. Scheibe R. 1994. Lichtregulation von Chloroplastenenzymen. *Naturwissenschaften* 81:443–48
137. Schimkat D, Heineke D, Heldt HW. 1990. Regulation of sedoheptulose-1,7-bisphosphatase by sedoheptulose-7-phosphate and glycerate, and of fructose-1,6-bisphosphatase by glycerate in spinach chloroplasts. *Planta* 181: 97–103
138. Schobert C, Baker L, Szederkenyi J, Grossmann P, Komor E, et al. 1998. Identification of immunologically related proteins in sieve-tube exudate collected from monocotyledonous and dicotyledonous plants. *Planta* 206:245–52
139. Schürmann P. 1995. The ferredoxin/thioredoxin system. In *Methods in Enzymology*, ed. L Packer, 252:274–83. Orlando, FL: Academic
140. Schürmann P, Buchanan BB. 2000. The structure and function of the ferredoxin/thioredoxin system. In *Advances in Photosynthesis*, ed. B Andersson, EM Aro. Dordrecht, The Netherlands: Kluwer. In press
141. Schwarz O, Schürmann P, Strotmann H. 1997. Kinetics and thioredoxin specificity of thiol modulation of the chloroplast H$^+$-ATPase. *J. Biol. Chem.* 272:16924–27
142. Schwendtmayer C, Manieri W, Hirasawa M, Knaff DB, Schürmann P. 1998. Cloning, expression and characterization of ferredoxin:thioredoxin reductase from *Synechocystis* sp. PCC6803. See Ref. 27, pp. 1927–30
143. Setya A, Murillo M, Leustek T. 1996. Sulfate reduction in higher plants: molecular evidence for a novel 5′-adenylylsulfate reductase. *Proc. Natl. Acad. Sci. USA* 93:13383–88
144. Sha S, Minakuchi K, Higaki N, Sato K, Ohtsuki K, et al. 1997. Purification and characterization of glutaredoxin

(thioltransferase) from rice (*Oryza sativa* L). *J. Biochem.* 121:842–48

145. Sha S, Yabushita T, Minakuchi K, Masumura T, Tanaka K. 1997. Structure of the rice glutaredoxin (thioltransferase) gene. *Gene* 188:23–28
146. Smith DR, Doucette-Stamm LA, Deloughery C, Lee H, Dubois J, et al. 1997. Complete genome sequence of *Methanobacterium thermoautotrophicum* deltaH: functional analysis and comparative genomics. *J. Bacteriol.* 179:7135–55
147. Smith PA, Tripp BC, DiBlasio-Smith EA, Lu ZJ, LaVallie ER, McCoy JM. 1998. A plasmid expression system for quantitative *in vivo* biotinylation of thioredoxin fusion proteins in *Escherichia coli*. *Nucleic Acids Res.* 26:1414–20
148. Staples CR, Ameyibor E, Fu W, Gardet-Salvi L, Stritt-Etter A-L, et al. 1996. The nature and properties of the iron-sulfur center in spinach ferredoxin:thioredoxin reductase: a new biological role for iron-sulfur clusters. *Biochemistry* 35:11425–34
149. Staples CR, Gaymard E, Stritt-Etter AL, Telser J, Hoffman BM, et al. 1998. Role of the $[Fe_4S_4]$ cluster in mediating disulfide reduction in spinach ferredoxin: thioredoxin reductase. *Biochemistry* 37:4612–20
150. Stein M, Jacquot J-P, Jeannette E, Decottignies P, Hodges M, et al. 1995. *Chlamydomonas reinhardtii* thioredoxins: structure of the genes coding for the chloroplastic *m* and cytosolic *h* isoforms; expression in *Escherichia coli* of the recombinant proteins, purification and biochemical properties. *Plant Mol. Biol.* 28:487–503
151. Suske G, Wagner W, Follmann H. 1979. NADP-dependent thioredoxin reductase and a new thioredoxin from wheat. *Z. Naturforsch.* 34c:214–21

151a. Suter M, von Ballmoos P, Kopriva S, Op den Camp R, Schaller J, et al. 2000. Adenosine 5′-phosphosulfate sulfotransferase and adenosine 5′-phosphosulfate reductase are identical enzymes. *J. Biol. Chem.* 275:930–36

152. Szederkényi J, Komor E, Schobert C. 1997. Cloning of the cDNA for glutaredoxin, an abundant sieve-tube exudate protein from *Ricinus communis* L and characterisation of the glutathione-dependent thiol-reduction system in sieve tubes. *Planta* 202:349–56
153. Tamoi M, Ishikawa T, Takeda T, Shigeoka S. 1996. Molecular characterization and resistance to hydrogen peroxide of two fructose-1, 6-bisphosphatases from *Synechococcus* PCC 7942. *Arch. Biochem. Biophys.* 334:27–36
154. Tamoi M, Murakami A, Takeda T, Shigeoka S. 1998. Acquisition of a new type of fructose-1,6-bisphosphatase with resistance to hydrogen peroxide in cyanobacteria: molecular characterization of the enzyme from *Synechocystis* PCC 6803. *Biochim. Biophys. Acta* 1383:232–44
155. Tichy M, Vermaas W. 1999. In vivo role of catalase-peroxidase in *Synechocystis* sp. strain PCC 6803. *J. Bacteriol.* 181:1875–82
156. Toriyama K, Hanaoka K, Okada T, Watanabe M. 1998. Molecular cloning of a cDNA encoding a pollen extracellular protein as a potential source of a pollen allergen in *Brassica rapa*. *FEBS Lett.* 424:234–38
157. Tsugita A, Yano K, Gardet-Salvi L, Schürmann P. 1991. Characterization of spinach ferredoxin-thioredoxin reductase. *Protein Seq. Data Anal.* 4:9–13

157a. Vanlerberghe GC, McIntosh L. 1997. Alternative oxidase: from gene to function. *Annu. Rev. Plant Physiol. Plant Mol. Biol.* 48:703–34

158. Verdoucq L, Jacquot J-P, Vignols F, Chartier Y, Meyer Y. 1999. In vivo characterization of a thioredoxin h target

protein defines a new peroxiredoxin family. *J. Biol. Chem.* 274:19714–22

159. Villeret V, Huang S, Zhang Y, Xue Y, Lipscomb WN. 1995. Crystal structure of spinach chloroplast fructose-1, 6-bisphosphatase at 2.8 Å resolution. *Biochemistry* 34:4299–306
160. Wadano A, Nishikawa K, Hirahashi T, Satoh R, Iwaki T. 1998. Reaction mechanism of phosphoribulokinase from a cyanobacterium, *Synechococcus* PCC7942. *Photosynth. Res.* 56:27–33
161. Wedel N, Soll J, Paap BK. 1997. CP12 provides a new mode of light regulation of Calvin cycle activity in higher plants. *Proc. Natl. Acad. Sci. USA* 94:10479–84
162. Wenderoth I, Scheibe R, von Schaewen A. 1997. Identification of the cysteine residues involved in redox modification of plant plastidic glucose-6-phosphate dehydrogenase. *J. Biol. Chem.* 272:26985–90
163. Werner S, Schumann J, Strotmann H. 1990. The primary structure of the gamma-subunit of the ATPase from *Synechocystis* 6803. *FEBS Lett.* 261:204–8
164. Willingham NM, Lloyd JC, Raines CA. 1994. Molecular cloning of the *Arabidopsis thaliana* sedoheptulose-1,7-biphosphatase gene and expression studies in wheat and *Arabidopsis thaliana*. *Plant Mol. Biol.* 26:1191–200
165. Wolosiuk RA, Crawford NA, Yee BC, Buchanan BB. 1979. Isolation of three thioredoxins from spinach leaves. *J. Biol. Chem.* 254:1627–32
166. Wolosiuk RA, Hertig CM, Nishizawa AN, Buchanan BB. 1982. Enzyme regulation in C4 photosynthesis. Role of Ca^+ in thioredoxin-linked activation of sedoheptulose bisphosphatase from corn leaves. *FEBS Lett.* 140:31–35

166a. Wolosiuk RA, Ballicora MA, Hagelin K. 1993. The reductive pentose phosphate cycle for photosynthetic CO_2 assimilation: enzyme modulation. *FASEB J.* 7:622–37

167. Wray JL, Campbell EI, Roberts MA, Gutierrez-Marcos JF. 1999. Redefining reductive sulfate assimilation in higher plants: a role for APS reductase, a new member of the thioredoxin superfamily? *Chem. Biol. Interact.* 109:153–67
168. Yamamoto H, Miyake C, Dietz KJ, Tomizawa KI, Murata N, Yokota A. 1999. Thioredoxin peroxidase in the cyanobacterium *Synechocystis* sp. PCC 6803. *FEBS Lett.* 447:269–73
169. Zhang N, Portis Jr AR. 1999. Mechanism of light regulation of Rubisco: a unique role for the larger Rubisco activase isoform involving reductive activation by thioredoxin-f. *Proc. Natl. Acad. Sci. USA* 96:9438–43

Annu. Rev. Plant Physiol. Plant Mol. Biol. 2000. 51:401–32

SELENIUM IN HIGHER PLANTS

N. Terry, A. M. Zayed, M. P. de Souza, and A. S. Tarun
Department of Plant and Microbial Biology, University of California Berkeley, Berkeley, California 94720-3102; e-mail: nterry@nature.berkeley.edu

Key Words phytoremediation, volatilization, tolerance, essentiality, toxicity

■ **Abstract** Plants vary considerably in their physiological response to selenium (Se). Some plant species growing on seleniferous soils are Se tolerant and accumulate very high concentrations of Se (Se accumulators), but most plants are Se nonaccumulators and are Se-sensitive. This review summarizes knowledge of the physiology and biochemistry of both types of plants, particularly with regard to Se uptake and transport, biochemical pathways of assimilation, volatilization and incorporation into proteins, and mechanisms of toxicity and tolerance. Molecular approaches are providing new insights into the role of sulfate transporters and sulfur assimilation enzymes in selenate uptake and metabolism, as well as the question of Se essentiality in plants. Recent advances in our understanding of the plant's ability to metabolize Se into volatile Se forms (phytovolatilization) are discussed, along with the application of phytoremediation for the cleanup of Se contaminated environments.

CONTENTS

1040-2519/00/0601-0401$14.00

INTRODUCTION

Interest in selenium (Se) has escalated in the past two decades. In trace amounts, Se is an essential micronutrient and has important benefits for animal and human nutrition. At high dosages, however, it may be toxic to animals (96, 114, 165) and to humans (162). The concentration range from trace element requirement to lethality is quite narrow; the minimal nutritional level for animals is about 0.05 to 0.10 mg Se kg^{-1} dry forage feed, while exposure to levels of 2 to 5 mg Se kg^{-1} dry forage causes toxicity (165, 168). The first report of the nutritional benefit of Se was published in 1957 (138). In 1973, Se was shown to form part of the important antioxidant enzyme, glutathione (GSH) peroxidase (133). Other health benefits include carcinoma suppression (87) and the relief of certain symptoms associated with AIDS (82).

The toxicity of Se has been known for many years (54). However, it was not until the Kesterson Reservoir controversy in the 1980s that scientists, regulators, politicians, and the general public of the United States were made acutely aware of the importance of Se as an environmental contaminant. Selenium present in the waters at the natural wildlife refuge at Kesterson Reservoir, California, was shown to be the agent responsible for mortality, developmental defects, and reproductive failure in migratory aquatic birds (116) and in fish (135). The contamination of the Kesterson Reservoir arose from Se-laden agricultural drainage water that had been allowed to flow into it from neighboring farms. Selenium toxicity is encountered in arid and semiarid regions of the world that have seleniferous, alkaline soils derived from the weathering of seleniferous rocks and shales. In the western United States, the leaching of soluble, oxidized forms of Se, especially selenate, from these seleniferous soils is accelerated by intensive irrigation, with selenate being accumulated in high concentrations in drainage water. Selenium is also released into the environment by various industrial activities; for example, oil refineries and electric utilities generate Se-contaminated aqueous discharges.

The narrow margin between the beneficial and harmful levels of Se has important implications for human health. Plants can play a pivotal role in this respect: For example, plants that accumulate Se may be useful as a "Se-delivery system" (in forage or crops) to supplement the mammalian diet in many areas that are deficient in Se. On the other hand, the abilities of plants to absorb and sequester Se can also be harnessed to manage environmental Se contamination by phytoremediation, whereby green plants are used to remove pollutants from contaminated soil or water (153). One special attribute of plants that has potential benefit for Se phytoremediation is their ability to convert inorganic Se to volatile forms, predominantly dimethylselenide (DMSe), a process called phytovolatilization.

The chemistry of Se has been reviewed extensively by several authors (e.g. 88, 91). It is of interest to note the following. Selenium was discovered by the Swedish chemist Jakob Berzelius in 1817. Its name comes from the Greek word for moon, selene. It is a group VIA metalloid with an atomic weight of 78.96. Selenium shares many similar chemical properties with sulfur (S), although the

Se atom is slightly larger; the radius of Se^{2+} is 0.5 Å whereas the radius of the S^{2+} is 0.37 Å. Like S, Se can exist in five valence states, selenide (2^-), elemental Se (0), thioselenate (2^+), selenite (4^+), and selenate (6^+) (91). The speciation of Se depends on redox conditions and pH: Selenate tends to be the major species in aerobic and neutral to alkaline environments, whereas selenide and elemental Se dominate in anaerobic environments. Selenium also exists in volatile forms other than DMSe, e.g. dimethyldiselenide (DMDSe), and probably dimethyl selenone, dimethyl selenylsulfide, and methaneselenol (65, 129).

In this review we have focused on the main conceptual developments in Se physiology and biochemistry in plants, including Se uptake and transport, the question of Se essentiality in plants, the assimilation and volatilization of Se, as well as mechanisms of toxicity and tolerance. We have also included sections on phytoremediation and phytovolatilization because of the important implications of these strategies in dealing with environmental Se contamination.

ACCUMULATORS AND NONACCUMULATORS

Plants differ in their ability to accumulate Se in their tissues. Certain native plants are able to hyperaccumulate Se in their shoots when they grow on seleniferous soils. These species are called Se accumulators and include a number of species of *Astragalus, Stanleya, Morinda, Neptunia, Oonopsis*, and *Xylorhiza* (29, 84, 132, 156). They can accumulate from hundreds to several thousand milligrams of Se kg^{-1} dry weight in their tissues. On the other hand, most forage and crop plants, as well as grasses, contain less than 25 mg Se kg^{-1} dry weight and do not accumulate Se much above a ceiling of 100 mg Se kg^{-1} dry weight when grown on seleniferous soils. These plants are referred to as Se nonaccumulators (29). Crop plants grown on nonseleniferous soils typically have Se concentrations ranging from 0.01 to 1.0 mg kg^{-1} dry weight (102). On soils containing moderate concentrations of Se, Burau et al (33) found that tissue Se levels in 17 different crops rarely exceeded 1 mg Se kg^{-1} dry weight, although salt-tolerant species of genera such as *Distichlis* and *Atriplex*, grown at similar soil Se levels, exhibited concentrations of 10 to 20 mg Se kg^{-1} dry weight (83, 120).

Although Se accumulators grow on seleniferous soils, not all plant species on seleniferous soils are Se accumulators: Some plants accumulate only a few milligrams of Se kg^{-1} dry weight. For example, the genus *Astragalus* contains both Se-accumulating species and nonaccumulating species and these can grow next to each other on the same soil (58). Trelease & Beath (156) observed that several species of plants growing on seleniferous soil of the Niobrara Formation had markedly different tissue Se concentrations, e.g. *Astragalus bisulcatus* contained 5530, *Stanleya pinnata* 1190, *Atriplex nuttallii* 300, and grasses 23, mg Se kg^{-1} dry weight.

A third category of plants, known as secondary Se accumulators (29), grow on soils of low-to-medium Se content and accumulate up to 1000 mg Se kg^{-1} dry

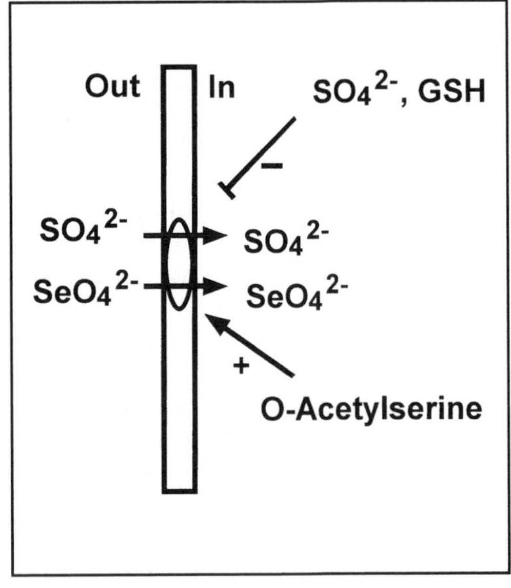

Figure 1 Selenate uptake across the root plasma membrane is mediated by the high-affinity sulfate transporter. The expression of the high-affinity sulfate transporter is regulated positively by O-acetylserine, and negatively by sulfate and glutathione. SO_4^{2-} = sulfate; SeO_4^{2-} = selenate; GSH = reduced glutathione.

weight. Examples of plants in this group are species of *Aster, Astragalus, Atriplex, Castilleja, Comandra, Grayia, Grindelia, Gutierrezia*, and *Machaeranthera* (120). Most recently, research by Bañuelos and his colleagues has identified the fast-growing *Brassica* species, Indian mustard (*Brassica juncea*) and canola (*B. napus*), as new secondary Se-accumulator plant species with a typical Se concentration of several hundred milligrams of Se kg^{-1} dry weight in their shoot tissues when grown on soils contaminated with moderate levels of Se (11).

UPTAKE AND TRANSPORT

Uptake of Selenate, Selenite, and Organic Se

Selenate is accumulated in plant cells against its likely electrochemical potential gradient through a process of active transport (29). Selenate readily competes with the uptake of sulfate and it has been proposed that both anions are taken up via a sulfate transporter in the root plasma membrane (Figure 1) (1, 5, 27, 39, 94). Selenate uptake in other organisms, including *Escherichia coli* (100) and yeast (38), is also mediated by a sulfate transporter.

Research with yeast enabled the first sulfate transporter genes to be cloned from plants. The approach was to select for resistance to high concentrations of selenate as a means of isolating yeast mutants defective in sulfate transport (27, 38, 146). Using functional complementation, three genes (*SHST1, SHST2*, and *SHST3*) encoding the sulfate transporter were isolated from the tropical legume

Stylosanthes hamata (146) and one gene (*HVST1*) was isolated from barley (*Hordeum vulgare*) (145). Sequence analysis predicts that the transporter contains 12 membrane-spanning domains. Furthermore, the transporter genes exhibit a high degree of sequence conservation with other sulfate transporters cloned from animals and microorganisms (44, 146). By using the sequence of these highly conserved regions, homologs of the sulfate transporter gene have now been cloned from many plants including *Arabidopsis* (149, 169), Indian mustard (80), and corn (23). Kinetic and expression studies have indicated that the sulfate transporters belong to two main classes, i.e. transporters that are either high affinity or low affinity for sulfate. The high-affinity transporter, which is likely to be the primary transporter involved in sulfate uptake, has a K_m for sulfate of ~7-10 μM and is expressed primarily in the roots (145, 146). The low-affinity transporter has a higher K_m for sulfate (100 μM) and is expressed in both shoots and roots; this transporter may be involved in the internal intercellular transport of sulfate (146).

The expression of the transporter genes is regulated by the S status of the plant, as well as by the regulators, glutathione (GSH) and O-acetylserine (44; see Figure 1). While high levels of sulfate and GSH decrease transcription, high levels of O-acetylserine increase transcription of the high-affinity transporter genes as well as sulfate uptake (44). Thus, increasing O-acetylserine levels can potentially increase selenate uptake: Our results show that application of O-acetylserine increased selenate accumulation in Indian mustard almost twofold compared to untreated plants (MP de Souza & N Terry, unpublished data). O-acetylserine, a precursor of cysteine (Cys) and a product of the nitrate assimilation pathway, may be of pivotal importance as a coregulator of the S and nitrogen metabolic pathways (44).

Because of the role of the sulfate transporter in selenate uptake, we have undertaken studies to determine whether overexpression of either the high-affinity (*SHST1*) or low-affinity (*SHST3*) transporter genes from *S. hamata* increases the uptake of Se in Indian mustard. Overexpression of *SHST1* increased selenate accumulation up to twofold in transgenic plants compared to wild type; on the other hand, transgenic plants overexpressing *SHST3* did not differ significantly from wild type in their accumulation of selenate (S Huang & N Terry, unpublished data). These data support the view that the high-affinity sulfate transporter is involved in selenate uptake.

Unlike selenate, there is no evidence that the uptake of selenite is mediated by membrane transporters (1, 5, 6, 143). Selenite uptake was inhibited by only 20% by the addition of a respiratory inhibitor, hydroxylamine, to nutrient solution; selenate uptake, however, was inhibited by 80% (5). Asher et al (6) showed that, although the Se concentration in the xylem exudate of selenate-supplied detopped roots exceeded that of the external medium by 6 to 13 times, Se concentration was always lower in the xylem exudate than outside when selenite was supplied. Plants can also take up organic forms of Se such as selenomethionine (SeMet) actively. Abrams et al (1) showed that SeMet uptake by wheat seedlings followed Michaelis Menten kinetics and that uptake was coupled to metabolism, as indicated by evidence of inhibition by metabolic inhibitors (e.g. dinitrophenol) and by anaerobic conditions (i.e. nitrogen bubbling).

Interaction with Salinity

As indicated above, sulfate competes with selenate for uptake by the sulfate transporter. It is hardly surprising therefore that sulfate salinity drastically inhibits plant uptake of selenate (19, 107, 141, 166, 171). Not all plant species are affected to the same extent by sulfate salinity. In Se-accumulators, selenate is taken up preferentially over sulfate. Calculation of the Se/S discrimination coefficient (the ratio of the plant Se/S ratio to the solution Se/S ratio) indicated that *A. bisulcatus*, rice, and Indian mustard are able to take up Se preferentially in the presence of a high sulfate supply (19). Other species (e.g. alfalfa, wheat, ryegrass, barley, broccoli) had discrimination coefficient values lower than 1, and selenate uptake was significantly inhibited by the increases in sulfate supply. Chloride salinity had much less effect on selenate uptake than sulfate salinity (107, 141, 166). Generally, there is a small decrease in shoot accumulation of Se with increasing salt levels (15).

Transport and Distribution

The translocation of Se from root to shoot is dependent on the form of Se supplied. Selenate is transported much more easily than selenite, or organic Se, such as SeMet. Zayed et al (171) showed that the shoot Se/root Se ratio ranged from 1.4 to 17.2 when selenate was supplied but was only 0.6 to 1 for plants supplied with SeMet and less than 0.5 for plants supplied with selenite. Arvy (5) demonstrated that within 3 h, 50% of the selenate taken up by bean plant roots moved to shoots, whereas in the case of selenite, most of the Se remained in the root and only a small fraction was found in the shoot. Time-dependent kinetics of Se uptake by Indian mustard showed that only 10% of the selenite taken up was transported from root to shoot, whereas selenate (which was taken up twofold faster than selenite) was rapidly transported into shoots (49). Thus, plants transport and accumulate substantial amounts of selenate in leaves but much less selenite or SeMet. The reason why selenite is poorly translocated to shoots may be because it is rapidly converted to organic forms of Se such as SeMet (171), which are retained in the roots.

The distribution of Se in various parts of the plant differs according to species, its phase of development, and its physiological condition (156). In Se accumulators, Se is accumulated in young leaves during the early vegetative stage of growth; during the reproductive stage, high levels of Se are found in seeds while the Se content in leaves is drastically reduced (58). Nonaccumulating cereal crop plants, when mature, often show about the same Se content in grain and in roots, with smaller amounts in the stems and leaves (17). Distribution of Se in plants also depends on the form and concentration of Se supplied to the roots and on the nature and concentration of other substances, especially sulfates, accompanying the Se (49, 171).

Plants can absorb volatile Se from the atmosphere via the leaf surface. Zieve & Peterson (174) demonstrated that *Agrostis tenuis, H. vulgare, Lycopersicon esculentum*, and *Raphanus sativus* accumulate ^{75}Se when fumigated with (^{75}Se)-DMSe in a sealed system. The Se absorbed by the leaves was accumulated in roots as inorganic selenite, selenoglutathione (SeGSH), SeMet, and protein-bound SeMet. Using an isotope dilution method, Haygarth et al (79) assessed the importance of soil and atmospheric Se inputs to ryegrass; with soil pH of 6, 47% of the ^{75}Se in ryegrass leaves was derived from the soil, whereas at pH 7, it was 70%. It was assumed that the remainder of the Se came from the atmosphere.

BIOCHEMISTRY OF Se

The Question of Se Essentiality in Plants

Although there is mounting evidence that Se is required for the growth of algae (126, 170), the question of the essentiality of Se as a micronutrient in higher plants is unresolved and remains controversial. There are indications that Se may be required for Se-accumulating plants, which are endemic to seleniferous soils. Trelease & Trelease (157) observed an increase in biomass production when the Se accumulator, *Astragalus pectinatus*, was treated with 0.38 mM Se. These results were challenged subsequently by Broyer et al (31), who attributed the growth stimulation in the *Astragalus* plants to the ability of Se in the nutrient solution to counteract phosphate toxicity; at low phosphate concentrations, growth was not stimulated by Se treatment. There is no evidence for a Se requirement in nonaccumulators (142). When alfalfa and clover plants were grown in hydroponic culture with highly purified salts, addition of selenite did not result in an enhancement of growth (30), However, this type of experiment does not prove that there is no requirement for Se; it merely shows that, if there is a requirement, it must be satisfied at an Se concentration lower than that obtained in the plant tissue under the conditions of the experiment (since there are always trace amounts of Se in plants coming from impurities in the nutrient salts or from the atmosphere).

In order to investigate the essentiality of Se in higher plants, attempts have been made to establish whether plants contain essential selenoproteins, such as those discovered for bacteria and animals. For example, several Se-dependent enzymes have been identified in which an integral selenocysteine (SeCys) residue is inserted in the catalytic site (67, 147). Selenoenzymes, such as formate dehydrogenase from bacteria and Archaea, and the GSH peroxidase (GPX) and type 1 iodothyronine deiodinase family of enzymes from animals, are involved in oxido-reduction reactions in which the Se (present in the reduced selenol form) functions as a redox center (148). Mutagenesis studies have shown that this SeCys residue plays a critical role because replacement of the active site SeCys by a Cys residue greatly reduced catalytic activity (9, 20).

The incorporation of the active site-SeCys into these essential selenoproteins is a cotranslational process directed by a UGA codon (22). UGA normally functions as a universal termination codon; in order for UGA to function as a SeCys codon, both specific secondary structural elements in the mRNA and a unique SeCys-charged $tRNA^{ser/sec}$ that contains the UGA anticodon are required (148). The biosynthesis of SeCys is itself a unique process because it occurs in a tRNA-bound state distinct from the S/Se assimilation pathway (see next section). A key reaction in this biosynthesis is the activation of selenide to form selenophosphate by the enzyme selenophosphate synthetase (148). Selenophosphate is the Se donor for the conversion of the serine-bound $tRNA^{ser}$ to the SeCys-bound $tRNA^{sec}$ (148).

Attempts have been made to establish whether plants also contain these essential selenoproteins. Radioactive labeling experiments have so far failed to detect essential selenoproteins in plants. For example, Anderson (2), using radioactively labeled [^{75}Se]selenite and [^{35}S]sulfate, was unable to detect proteins with significantly higher ^{75}Se/^{35}S ratios than other proteins. A second approach to detecting selenoproteins in plants is to test for the presence of a family of selenoproteins, the GSH peroxidases (GPX), which catalyze the reduction of hydrogen peroxide by reduced GSH in order to protect cells against oxidative damage (see 60). The presence of GPX is determined by GSH-dependent reduction of hydrogen peroxide. Based on this enzymatic assay, plant extracts with GPX activity and partially purified preparations of GPX were reported recently from various plants (reviewed in 60). Sabeh et al (134) found a 16-kd tetrameric protein in *Aloe vera*, which they conclude is a GPX selenoprotein similar to that found in mammals.

Molecular evidence has been obtained that suggests that, although there are GPX-like enzymes in higher plants, they may not be selenoproteins. For example, genes with significant sequence homology to animal GPX genes were recently isolated from a number of plants including *Nicotiana sylvestris* (42), *Citrus sinensis* (81), and *Arabidopsis* (Genbank accession no. X89866). Analysis of these sequences indicates that the plant genes showed homology to only one subgroup of animal GPX genes, the phospholipid hydroxyperoxide GSH peroxidases (60). However, in contrast to animal GPX genes that contain a SeCys codon, UGA, in the active site, the plant genes contain the Cys codon, UGU (60). The gene isolated from *Citrus sinensis* was expressed into protein and assayed; it exhibited only 0.2 to 0.8% activity of the animal GPX (18, 81). Thus, it would appear to have another function in plants (60). Furthermore, peptide sequencing of the purified protein has confirmed that Cys, and not SeCys, is at the active site of the enzyme (64). On the basis of all the available information published, we conclude that no essential higher plant selenoprotein has been clearly identified by either protein or DNA sequence analysis to date.

Although there is no evidence for an essential selenoprotein in higher plants, there is evidence that part of the machinery for synthesizing selenoproteins may be present, i.e. a UGA decoding-$tRNA^{sec}$ was demonstrated to be present in *Beta vulgaris* (sugar beet) (76). This $tRNA^{sec}$ was observed in bacterial (95) and mammalian cells (93) and has also been identified in the diatom *Thalassiosira pseudonana* (77), and in the Ascomycete *Gliocladium virens* (76). Hence, UGA appears to have the dual function of termination codon and codon for SeCys in all groups of organisms.

Se Assimilation, Volatilization, and Incorporation into Proteins

In addition to the possible incorporation of trace amounts of Se into specific or essential selenoproteins, higher plants metabolize Se via the S-assimilation pathway (172). This involves the nonspecific incorporation of Se into selenoamino acids and their proteins, as well as volatilization, which occurs when Se is supplied to plants in excess of any potential Se requirement. Most of the available information on Se assimilation and volatilization is for Se nonaccumulators (Figures 2–5); the pathway for Se accumulators is presented separately (Figure 6).

Role of ATP Sulfurylase After selenate is absorbed into the root via the sulfate transporter, it is translocated without chemical modification (49, 171) through the xylem to the leaves. Once inside the leaf, selenate enters chloroplasts where it is metabolized by the enzymes of sulfate assimilation. The first step in the reduction of selenate is its activation by ATP sulfurylase to adenosine phosphoselenate (APSe), an activated form of selenate (Figure 2). ATP sulfurylase has been shown to activate selenate, as well as sulfate, in vitro (35, 53, 140). Molecular studies in our laboratory provided the first in vivo evidence that ATP sulfurylase is responsible for selenate reduction, and that this enzyme is rate limiting for both selenate reduction and Se accumulation (125). Transgenic Indian mustard plants that overexpressed ATP sulfurylase were developed using a gene construct containing the *Arabidopsis thaliana aps1* gene (97), with its own chloroplast transit sequence, fused to the Cauliflower Mosaic Virus 35S promoter. X-ray absorption spectroscopy (XAS) analysis of wild-type Indian mustard plants supplied with selenate showed that selenate was accumulated in both roots and shoots, but when selenite was supplied, an organo-Se compound (similar to SeMet) accumulated (49). We concluded therefore that the reduction of selenate was rate limiting to selenate assimilation. This rate-limiting step was overcome in transgenic plants overexpressing ATP sulfurylase because these plants accumulated a SeMet-like compound when supplied with selenate (125). Furthermore, when the shoots of the transgenic plants were removed, and the detopped roots supplied with selenate, plants were unable to reduce the selenate, which remained as the principal form of Se in roots. This supports the view that the chloroplasts are the site for selenate reduction (125).

Reduction of APSe to Selenide There is evidence that APSe can be reduced nonenzymatically to GSH-conjugated selenite (GS-selenite) following the pathway outlined in Figure 2*A* (3, 53). The GS-selenite is then reduced via GSH to produce the intermediate selenodiglutathione (GS-Se-SG). When selenite is taken up by plants (instead of selenate), it reacts nonenzymatically with GSH to form

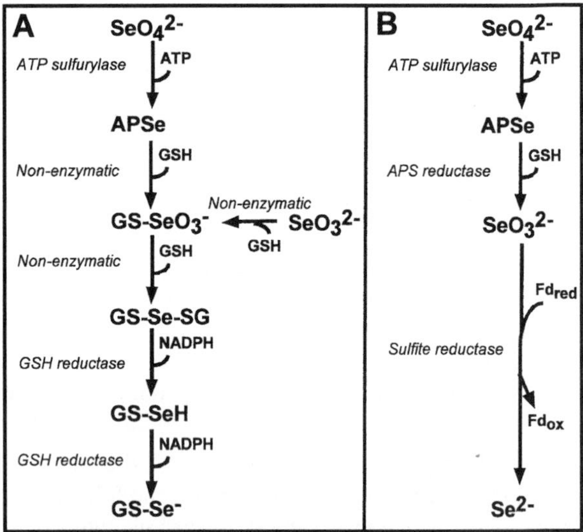

Figure 2 The activation of selenate by ATP sulfurylase, which is followed by (*A*) reduction to selenide (Se^{2-}) via nonenzymatic reactions and glutathione reductase, (*B*) reduction to selenide via APS reductase and sulfite reductase. Selenite (SeO_3^{2-}) can enter the pathway via a nonenzymatic reaction to selenodiglutathione (GS-Se-SG), which is reduced to the selenol, GS-SeH. $GS\text{-}Se^-$ is glutathione-conjugated selenide.

GS-Se-SG (2, 3). GS-Se-SG is reduced with NADPH to the corresponding selenol (GS-SeH), and subsequently to GSH-conjugated selenide ($GS\text{-}Se^-$) by the enzyme GSH reductase (113). Later, Anderson & Scarf (3) proposed that the reduction of GS-Se-SG to GS-SeH may also proceed nonenzymatically with GSH as a reductant.

In the S-assimilation pathway, the S analog of APSe, 5′-adenylylsulfate (APS), is at a branchpoint where it can be converted to either sulfite (72, 136, 139) or 3′-phosphoadenosine 5′-phosphosulfate (PAPS), which is used to form sulfated compounds (e.g. sulfolipids) (21). By analogy to the S-assimilation pathway, APSe could be reduced to selenite with GSH by APS reductase (21), and then to selenide by the enzyme, sulfite reductase (using ferredoxin as a reductant) (Figure 2*B*). Genes encoding the chloroplast enzymes, APS reductase and sulfite reductase, have been cloned from *A. thaliana* (32, 72, 139). However, there is no evidence that APS reductase and sulfite reductase are involved in selenate reduction in vivo, or that there is a Se analog for PAPS or selenolipids in plants (35).

Although GSH reductase is clearly an important enzyme in the reduction of selenate (Figure 2), we were unable to obtain molecular evidence that it is a rate-limiting enzyme. We transformed Indian mustard plants to overexpress a bacterial GSH reductase in the cytoplasm (cytGR) and in the chloroplast (cpGR) (EAH Pilon-Smits & N Terry, unpublished data). The GSH reductase gene was cloned

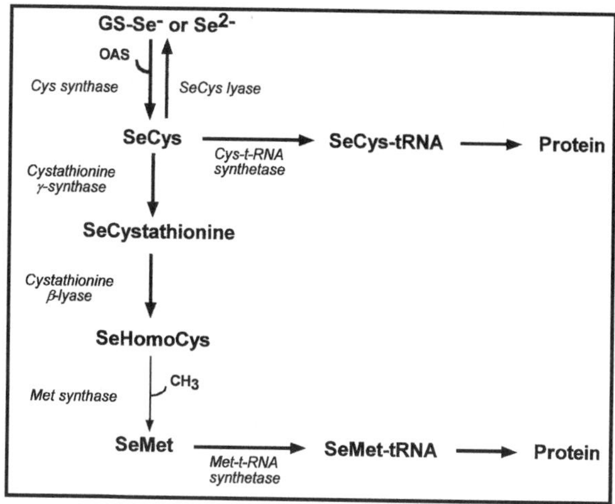

Figure 3 Incorporation of selenide into SeCys, SeMet, and proteins. SeCys lyase is a Se-specific enzyme, whereas all the other enzymes shown recognize both S and Se. The only exception is Met synthase, which is involved in Met synthesis, and is very likely to be involved in SeMet synthesis although there is no evidence supporting this (represented by the thin arrow).

from *E. coli* and adapted for expression in plants by Foyer et al (69). However, transgenic cpGR and cytGR plants treated with selenate or selenite did not show any differences from the wild type with respect to the form of Se accumulated, Se uptake, or Se volatilization rate.

Selenite can undergo other transformations besides its assimilation into selenide; for example, plants supplied with selenite have been shown to oxidize it to selenate (6; D Hansen, CM Lytle & N Terry, unpublished data). Although common in bacteria, the reduction of selenate or selenite to elemental Se (52, 101, 117, 118) has not been reported in plants.

Amino Acid and Protein Synthesis Plants are thought to assimilate SeCys in a manner similar to bacteria (2), where SeCys is metabolized to SeMet, both of which are nonspecifically incorporated into proteins. In plants, the formation of SeCys very likely takes place within the chloroplasts (113). SeCys is formed by the action of Cys synthase, which couples selenide with O-acetylserine (112) (Figure 3). GS-Se^- may be the physiological substrate of Cys synthase rather than free Se^{2-} (159). The activity of Cys synthase may be influenced by the ratio of sulfide to selenide; selenide inhibits the synthesis of Cys while excess sulfide inhibits the synthesis of SeCys (112). When the Cys synthase gene (cloned from spinach; 137) was overexpressed in Indian mustard, the transgenic plants treated with selenate or selenite did not show any differences from the wild type with respect to Se uptake, Se species accumulated, or Se volatilization (EAH Pilon-Smits, CM Lytle &

N Terry, unpublished data). The results from this overexpression study suggest that Cys synthase is not rate limiting for selenate or selenite assimilation to SeMet.

In mammals, SeCys can be reconverted to selenide by the Se-specific enzyme SeCys lyase, with the concomitant release of L-alanine (Figure 3). This enzyme, which was purified from pig liver, is specific for SeCys because it discriminates against Cys as a substrate (59). The same enzyme may also exist in plants, since an *Arabidopsis thaliana* homolog of the mammalian SeCys lyase gene was recently cloned (Genbank accession no. CAA16686).

By analogy to the methionine (Met) biosynthetic pathway (55, 128), SeMet may be produced from SeCys via SeCystathionine and SeHomoCys (Figure 3). There is some evidence for this from kinetic studies of in vitro enzymes. Cystathionine-γ-synthase, which catalyzes the condensation of phosphohomoserine with SeCys to form SeCystathionine, exhibited a preference for SeCys: It had a higher affinity for SeCys ($K_m = 70\ \mu M$) than for Cys ($K_m = 240\ \mu M$) (45). Cystathionine-β-lyase, which most likely cleaves selenocystathionine to selenohomoCys, did not differentiate between the Se and S forms of cystathionine, since the enzyme had a similar affinity for cystathionine ($K_m = 0.31$ mM) and selenocystathionine ($K_m = 0.35$ mM) (105).

The most likely enzyme for the synthesis of SeMet from SeHomoCys is the cytosolic enzyme, Met synthase (Figure 3), which has been cloned from *Arabidopsis, Catharanthus roseus, and Coleus blumei* (57, 122, 128). In plants, Met synthase uses methyl-tetrahydrofolate as a methyl donor (40). It is possible that SeMet is subject to the recycling pathways that have been described for Met (55).

Selenium is readily incorporated into proteins in nonaccumulator plants treated with Se (28). The incorporation into proteins occurs through the nonspecific substitution of SeCys and SeMet in place of Cys and Met, respectively (29) (Figure 3). With regard to Cys versus SeCys incorporation, Burnell & Shrift (36) showed that the mung bean cysteinyl-tRNA synthetase, the enzyme catalyzing the first step in the incorporation of Cys into protein, supported Cys- and SeCys-dependent PPi-ATP exchange at rates that were approximately equal. Similar studies showed that both Met and SeMet are substrates for the methionyl t-RNA synthetase (34, 62).

Selenomethionine to DMSe Lewis et al (99) showed that SeMet may be methylated to Se-methylSeMet, which was cleaved to DMSe by an enzyme in a crude extract of cabbage leaves (Figure 4). The most likely enzyme responsible for this reaction is S-methylMet hydrolase, which produces DMS from S-methylMet in higher plants (70, 71, 78). Another possible pathway for DMSe production is via the intermediate, dimethylselenoniopropionate (DMSeP) (Figure 4). Evidence for this was obtained by Ansede et al (4), who showed that selenate-supplied *Spartina alterniflora* plants accumulated the selenonium compounds, Se-methylSeMet and DMSeP, especially at high salinity and at Se concentrations greater than 50 μM. The production of DMSeP very likely will occur via the same biochemical pathway proposed for the synthesis of its S analog, dimethylsulfoniopropionate (DMSP) in higher plants (74, 75, 90). If so, the first step would be the methylation of SeMet by the cytosolic enzyme, Met methyltransferase (MMT) (25, 85). The enzymes

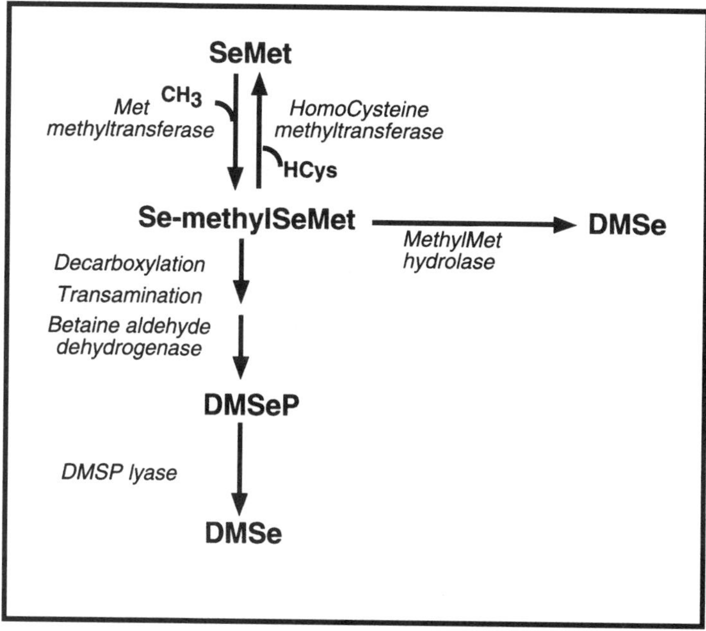

Figure 4 Production of DMSe from SeMet can take place directly from Se-methylSeMet, which is produced by the methylation of SeMet, or it can be produced after the conversion of SeMet to DMSeP.

for the formation of DMSP from S-methylMet are not known although DMSP-aldehyde and DMSP-amine have been identified as intermediates, depending on the plant species (86, 90). The biochemical steps involved are a decarboxylation, a transamination involving the conversion of methylMet to DMSP-aldehyde, and the chloroplastic enzyme betaine aldehyde dehydrogenase (BADH), which oxidizes DMSP-aldehyde to DMSP (131, 161). The formation of DMSe from DMSeP may proceed in plants as it does in bacteria, i.e. DMSeP may be volatilized to DMSe by the enzyme DMSP lyase (4), which is thought to exist in plants (43) (Figure 4).

A futile S-methylMet cycle occurs in plants. In this cycle, homoCys methyltransferase catalyzes the conversion of S-methylMet to Met by donating the methyl group to homoCys (71, 109). A similar reaction could exist to form SeMet from Se-methylSeMet.

Localization of the Pathway The movement of Se through the plant and the localization of enzymes involved in Se assimilation and volatilization may be summarized as follows. Selenate is transported to the chloroplasts where it is reduced. All the enzymes involved in this reduction, ATP sulfurylase, APS reductase, GSH reductase, sulfite reductase, Cys synthase, cystathionine γ-synthase, and cystathionine β-lyase, are chloroplastic (32, 68, 72, 89, 92, 112, 128, 139, 160, 163).

The production of SeMet from SeHomoCys and the methylation of SeMet to Se-methylSeMet by Met synthase and Met methyltransferase, respectively, most likely takes place in the cytosol because both of these enzymes are cytosolic (57, 85, 163). In shoots, Se could be volatilized through the action of methylMet hydrolase producing DMSe directly from methyl-SeMet (70, 71, 99). By analogy with dimethyl sulfide (DMS) production from DMSP that occurs in leaves (43), DMSeP synthesis from Se-methylSeMet could take place in the chloroplast (158), with DMSe being produced by DMSP lyase. However, since roots volatilize Se as much as 26 times more than shoots, it is clear that most of the DMSe is produced from the roots (173). For this to occur, the DMSe precursors, Se-methylSeMet and/or DMSeP, would have to be transported to roots in order to form DMSe. This presumes that methyl-Met hydrolase and DMSP lyase are also present in roots.

Other Possible Rate-Limiting Steps In addition to the rate limitation imposed by ATP sulfurylase, there is evidence of other possible rate-limiting steps for Se assimilation and volatilization. Since SeCys may move readily into other pathways (e.g. protein synthesis), one might speculate that the availability of SeCys might be rate limiting for Se volatilization. To test this hypothesis, de Souza et al (50) supplied Indian mustard plants with a crude preparation of SeCys (which contained impurities of selenocystine and dithiothreitol). The SeCys-supplied plants accumulated high concentrations of Se in their roots but did not volatilize Se significantly faster than selenate or selenite; thus, there was no evidence that SeCys was rate limiting (Figure 5). Plants supplied with SeMet on the other hand, volatilized Se at rates that were almost fivefold higher than the rates measured from SeCys, even though the plants accumulated slightly lower amounts of Se in roots. Thus, the synthesis of SeMet would appear to be rate limiting to Se volatilization. Furthermore, the rate of Se volatilization from DMSeP-supplied plants was fivefold higher than that measured from SeMet-supplied plants, even though the roots of plants supplied with SeMet accumulated eightfold more Se than DMSeP-supplied plants (Figure 5). This suggests that the conversion of SeMet to DMSeP is also rate limiting. In support of this view, XAS and HPLC data showed that a SeMet-like compound accumulates in selenite- and SeMet-supplied plants, but only relatively low levels of selenonium compounds (MP de Souza, CM Lyttle, MM Mulholland, ML Otte, N Terry, unpublished data). That DMSeP is volatilized at very high rates but does not accumulate in significant amounts in Indian mustard provides evidence that the production of DMSe from DMSeP is not likely to be rate limiting.

Pathway in Se Accumulators The pathway for the assimilation of inorganic forms of Se to SeCys in Se accumulators is believed to be the same as for nonaccumulators (29, 154) (Figure 6). However, Se accumulators differ from nonaccumulators in that they metabolize the SeCys primarily into various nonprotein selenoamino acids. The synthesis of these nonprotein selenoamino acids probably occurs along pathways associated with S metabolism (29). Three possible pathways are shown in Figure 6 for the conversion of SeCys to (I) Se-methylSeCys which has been found in many Se-accumulators (29); (II) Se-cystathionine, which

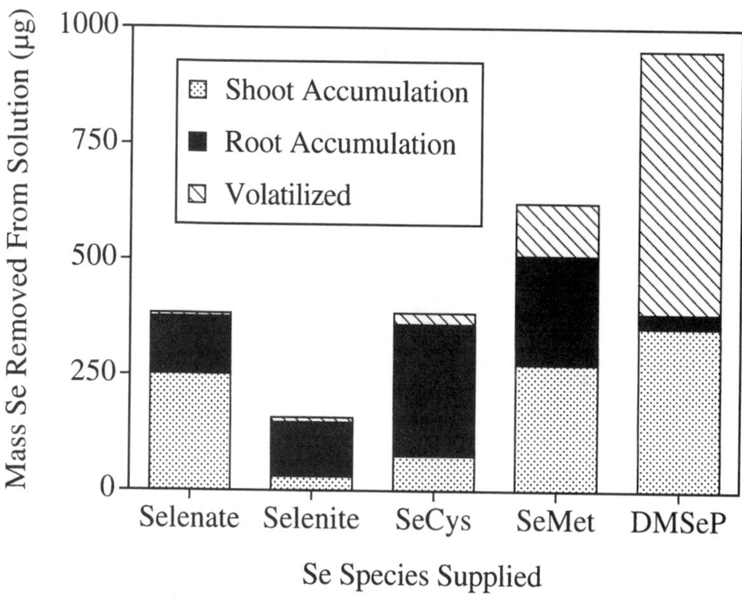

Figure 5 Total amounts of Se removed from hydroponic media per plant when Indian mustard plants were supplied with different forms of Se at 20 μM. The total amount of Se shown includes the amount of Se volatilized and the amounts accumulated in roots and shoots over an 8-day period under hydroponic culture conditions.

has been observed to accumulate in *Neptunia amplexicaulis* and *Morinda reticulata* (123); and (III) the dipeptide, γ-glutamyl-Se-methylSeCys, which has been observed in two *Astragalus* Se-accumulator species (114). Se-methylSeCys is the product of the enzyme SeCys methyltransferase, which has recently been purified and cloned from *Astragalus bisulcatus* (110, 111). SeCys methyltransferase methylates SeCys specifically (i.e. not Cys) into methylSeCys (110). It may be further methylated to produce DMDSe, which is volatilized (Figure 6) (63, 98). In (II), Se-cystathionine accumulates because the enzyme cystathionine β-lyase (which cleaves cystathionine (Figures 3, 6), is unable to cleave the Se analog (123).

Toxicity and Tolerance

When plants are exposed to high concentrations of Se in their root medium, they may exhibit symptoms of injury including stunting of growth, chlorosis, withering and drying of leaves, decreased protein synthesis, and premature death of the plant (106, 156). There are striking differences between the Se-accumulating plants and the nonaccumulators in the amount of Se they may absorb without showing symptoms of toxicity. In nonaccumulators, the threshold Se concentration in shoot tissue resulting in a 10% reduction in yield varied from 2 mg Se kg^{-1} in rice to 330 mg Se kg^{-1} in white clover (108). Se accumulators, on the other hand, may contain Se concentrations in excess of 4000 mg Se kg^{-1} without exhibiting any negative

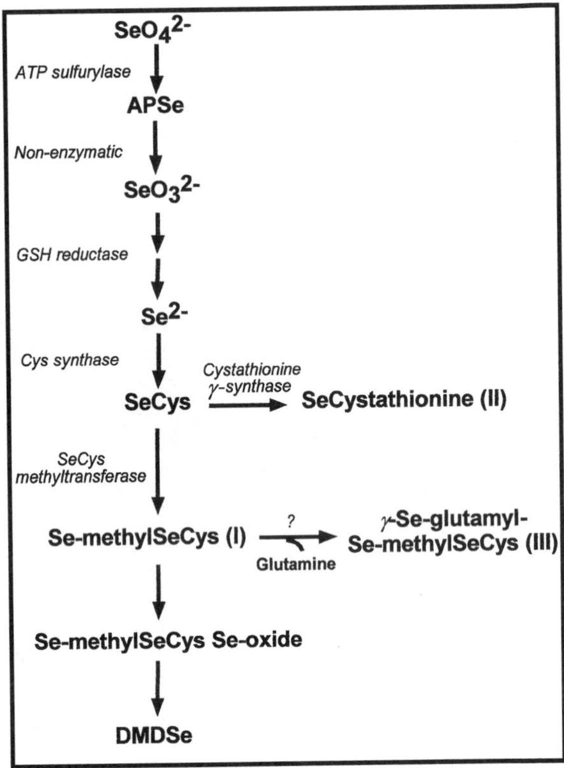

Figure 6 The assimilation of selenate by Se accumulators. The incorporation of selenate to SeCys takes place in a manner similar to Se nonaccumulators (Figures 1–3). In Se-accumulating *Astragalus* species, SeCys is methylated to Se-methylSeCys(I), which can be volatilized to DMDSe, and/or conjugated to glutamine to the dipeptide, γ-glutamyl-Se-methyl-SeCys(III). In other Se-accumulators SeCys may be converted to SeCystathionine, which is accumulated (II).

effects on growth (142). The threshold Se concentration may also vary with plant age and with sulfate supply. For example, Rosenfeld & Beath (132) observed that in the nonaccumulators, wheat and corn, younger plants were more susceptible and growth inhibition was greater than in mature plants. Tolerance to Se toxicity may increase with increasing sulfate supply so that threshold Se concentrations may not be the same at different sulfate concentrations in the root environment (108).

In nonaccumulators, the threshold toxicity concentration is also dependent on the form of Se accumulated. Selenate and selenite are the major forms that are toxic to plants because both are readily absorbed by the plant and assimilated to organic Se compounds. Some studies indicate that selenite is more toxic than selenate (e.g. 132). This may be due to the faster conversion of selenite to selenoamino acids (171), which may then be incorporated into plant proteins in replacement of S and cause toxicity to the plant (see below). Other studies have shown that selenate is more toxic than selenite: Wu et al (167) observed that selenate caused a greater growth inhibition than selenite with respect to the growth of tall fescue.

The major mechanism whereby high Se accumulation in plant tissues induces Se toxicity is almost certainly associated with the incorporation of SeCys and SeMet into proteins in place of Cys and Met, respectively (28, 61). The differences in size and ionization properties of S and Se may result in significant alterations in protein structure. The bond between two Se atoms is approximately one seventh longer and one fifth weaker than the disulfide bond (29). Therefore, the incorporation of SeCys in place of Cys into protein could interfere with the formation of disulfide bridges, resulting in a slightly altered tertiary structure of S-proteins and a negative effect on their catalytic activity (3, 29). Furthermore, Se may diminish the rate of protein synthesis; this is because the substitution of SeMet for Met into proteins may be less effective as a substrate for peptide bond formation during translation (62).

There are other ways by which Se can induce toxicity in plants. Se induces chlorosis, possibly through an adverse effect on the production of porphobilinogen synthetase, an enzyme required for chlorophyll biosynthesis (119). Selenate and selenite interfere with the in vivo reduction of nitrate in leaves (7). Selenate may interfere with the synthesis of GSH. The addition of selenate strongly reduced sulfate-induced GSH accumulation in spinach leaf disks (46). Also, incubation of spruce needles with selenate led to a substantial decrease in GSH content (24). Thus, interference of GSH synthesis in plants by selenate or other Se compounds may diminish plant defense against hydroxyl radicals and oxidative stress.

Tolerance Mechanisms Given the toxic effects of Se on plant function, how are Se-accumulators able to tolerate such high concentrations of Se in their cells? The main mechanism appears to be the reduction of the intracellular concentration of SeCys and SeMet, which otherwise would be incorporated into proteins with damaging effects on plant function. This may be achieved by accumulating Se in nonprotein seleno-amino acids, e.g. Se-methylSeCys and SeCystathionine, or as the dipeptide, γ-glutamyl-Se-methylSeCys, which accumulates in some species of *Astragalus* (114) (see discussion above). Since SeCys methyltransferase has threefold less activity with Cys compared to SeCys in the methylation reaction, Se detoxification by Se-methylSeCys accumulation can proceed without depleting the cell of Cys (110). Neuhierl et al (111) demonstrated that expression of the SeCys methyltransferase gene in *E. coli* could confer tolerance to high concentrations of Se and reduce the nonspecific incorporation of Se into proteins. The role of this enzyme in Se tolerance is further reinforced by the finding that moderate tolerance of the nonaccumulator *Astragalus cicer* to selenite coincided with the expression of an enzyme with SeCys methyltransferase activity (164).

There is evidence that a similar enzyme to SeCys methyltransferase exists in bacteria and in nonaccumulators but the enzyme has a different function. This was shown by the work of Neuhierl et al (111) who found genes that had sequence similarity to the *A. bisulcatus* SeCys methyltransferase gene in *E. coli (yagD), Arabidopsis*, and rice. Biochemical analysis of the enzyme encoded by *yagD* showed that it was involved in the catabolism of S-methylMet, i.e. it was a methyltransferase gene. These researchers postulated that the SeCys methyltransferase in Se accumulators may have developed under selective pressure from a S-methylMet-dependent thiol/selenol methyltransferase found in all plants.

An alternative mechanism for excluding Se from proteins may be through the ability of Se accumulators to discriminate against the incorporation of selenoaminoacids into proteins. For example, in the Se accumulator, *A. bisulcatus*, cysteinyl-tRNA synthetase is unable to attach SeCys to the Cys tRNA, thereby excluding SeCys from cellular incorporation into proteins (37). Selenium tolerance may also be achieved by compartmentation into the vacuole as selenate and, in the case of Se accumulators, as nonprotein seleno-amino acids.

PHYTOVOLATILIZATION

Variation Among Plant Species

The rate of Se volatilization varies substantially among plant species (56, 124, 151). Terry et al (151) measured the rate of Se volatilization from 15 crop species grown in solution culture in the presence of 20 μM selenate. Rice, broccoli, and cabbage volatilized Se at the highest rates, i.e. 200–350 μg Se m^{-2} leaf area day^{-1}, and sugar beet, bean, lettuce, and onion exhibited the lowest rates, >15 μg Se m^{-2} leaf area day^{-1}. Duckart et al (56) found that *Astragalus bisulcatus* and broccoli showed the highest rates of Se volatilization, followed by tomato, tall fescue, and alfalfa, respectively. For wetland plant species cultured hydroponically, there was a 50-fold variation in Se volatilization rates; the highest Se volatilization rate from selenate was attained by mare's tail, 1.0 mg Se kg^{-1} dry weight d^{-1}, and from selenite, 4.0 mg Se kg^{-1} dry weight d^{-1} (attained by *Azolla*) (124). Terry & Lin (152) compared the rate of volatilization of 11 different plant species, including pickleweed, saltgrass, cordgrass, Eucalyptus, cotton, and canola, under field conditions. They found that pickleweed (*Salicornia bigelovii*) had the highest average volatilization rate, 420 μg Se m^{-2} soil surface d^{-1}, a rate that was 10 to 100 times greater than rates measured for the other species.

Plant/Microbe Interactions

Bacteria, fungi, and algae can volatilize Se at high rates independently of plants. Pure cultures of bacteria, fungi, and algae tested under laboratory conditions volatilized Se at rates as high as or higher than plants when expressed per gram dry weight of tissue (26, 51, 65, 66, 127, 155). A complicating factor in the assessment of the contribution of Se volatilization from plants is the volatilization of Se by soil microorganisms (154). An important question to resolve is: To what extent is Se volatilization by plants dependent on the presence and activity of microbes in the rhizosphere?

One of the first indications that bacteria are involved in plant Se volatilization was obtained when detopped roots of broccoli were treated with penicillin-G and chlortetracycline (173). These antibiotics inhibited 95% of the Se volatilization activity from selenate-supplied detopped roots, supporting the view that rhizosphere bacteria may be essential for plant Se volatilization. Several attempts have been made to determine whether plants can volatilize Se independently of microorganisms. One approach to resolving this problem was to culture plants axenically, i.e. from seeds washed with ethanol and hypochlorite to remove microorganisms. When sterile and nonsterile plants were supplied with selenate, selenite, or SeMet, sterile plants volatilized SeMet at slightly lower rates than nonsterile plants but with selenate or selenite, sterile plants produced very little if any volatile Se (AM Zayed, N Terry, unpublished data). Thus, these data suggest that bacteria in the rhizosphere of Indian mustard are required for Se volatilization from selenate and selenite but not from SeMet.

What role do rhizosphere bacteria play in the phytovolatilization of selenate or selenite? De Souza et al (47, 48) obtained evidence that bacteria facilitate the uptake of selenate into root tissues. Two different approaches were used to inhibit microbial activity in roots: antibiotics and axenic plants. Using these approaches, rhizosphere bacteria were shown to increase selenate uptake in different plant species two- to threefold faster than in plants with inhibited rhizosphere activity (47, 48). De Souza et al (47) also obtained evidence that the role of bacteria in selenate uptake may be the production of a heat-labile compound(s) that was proteinaceous in nature. One possibility is that bacteria may stimulate selenate uptake in plants by producing O-acetylserine and serine in the rhizosphere (sterile plants inoculated with bacteria had twofold more O-acetylserine and tenfold more serine in their rhizosphere than sterile plants had). In support of this view, O-acetylserine and serine supplied to axenic plants were shown to stimulate selenate uptake two- and 1.5-fold, respectively (MP de Souza & N Terry, unpublished data).

The effect of rhizosphere microorganisms on Se uptake appeared to be specific for selenate; there was no such effect on selenite uptake. There is some evidence that the role of bacteria in selenite volatilization by plants may be in the production of organo-Se compounds such as SeMet, which is more readily volatilizable than selenite or selenate (171): XAS speciation of selenite-supplied plants showed that plants treated with ampicillin to inhibit rhizosphere bacteria accumulated SeCys, whereas plants not treated with ampicillin accumulated Se in the form of SeMet (MP de Souza, CM Lytle & N Terry, unpublished data).

Environmental Factors

Several studies indicate that the ability of plants to volatilize Se is influenced by the concentration of Se in the root medium and by the chemical form of Se supplied. The rate of Se volatilization was shown to be correlated linearly with the external Se concentration and with the internal plant tissue Se concentration for Indian mustard plants supplied with selenate or selenite over an increasing concentration range (49). Terry et al (151) showed the rate of Se volatilization by different plant species was strongly correlated with the plant tissue Se concentration in selenate-supplied plants. The dependence of Se volatilization rate on the form of Se supplied was demonstrated by Lewis et al (99), who showed that cabbage leaves from plants supplied with selenite released 10 to 16 times more volatile Se than those taken from plants supplied with selenate. Our research showed that plants of broccoli, Indian mustard, sugar beet, and rice volatilized SeMet at the highest rate, followed by selenite, then selenate (171), whereas Indian mustard plants supplied with DMSeP volatilized Se at a rate that was 6 times higher than that measured from plants supplied with SeMet, and 26 times higher than from SeCys (50). Under field conditions, Zijian et al (175) showed that when soils from a low-Se region in China were amended with SeMet, the rate of Se volatilization was enhanced 14-fold, while the addition of less reduced forms of Se resulted in only a fourfold increase. In experiments where SeMet and SeCys were added to soil, Martens & Suarez (103) showed that 50% to 80% of the added SeMet was volatilized, whereas little or no volatile Se was measured from SeCys-amended soil because it was converted mainly to selenite and selenide.

A very important factor influencing phytovolatilization is the concentration of sulfate compared to selenate in the substrate. Selenium volatilization by broccoli was inhibited strongly by increased sulfate supply: Zayed & Terry (173) showed that with increase in the sulfate concentration in nutrient solution from 0.25 to 10 mM, the rate of Se volatilization by broccoli plants decreased from 97 to 14 μg Se m^{-2} leaf area day^{-1}. The inhibition in volatilization with sulfate increased progressively with the increase in the ratio of S/Se in plant tissue. These results suggested that sulfate competed with selenate for uptake and that other S compounds out-competed their Se analogs for the active sites of enzymes responsible for the conversion of inorganic Se to volatile forms. The inhibitory effects of sulfate on Se volatilization by plants were less detectable when Se was supplied as selenite or SeMet (171).

Selenium volatilization rates vary enormously in the field, with very high rates occurring at certain times of the year, especially in spring and early summer (73, 152). Lin & Terry (unpublished data) examined the potential effects of 15 different environmental and plant factors on Se volatilization in a constructed wetland, including the Se concentrations in water, sediment, and plants, the microbial biomass in sediments, and the pH, salinity (EC), dissolved oxygen, standing depth, and temperature of water. Multiple-regression analysis indicated that water temperature, Se concentration in sediment (or in roots), and the level of microbial biomass (especially in the rhizosphere) were among the most important environmental factors influencing phytovolatilization.

PHYTOREMEDIATION

Plants have been shown to be highly effective in the phytoremediation of Se-contaminated soils. Bañuelos and his colleagues reduced the total Se inventory in the top 75 cm of soil from California's Central Valley by almost 50% over a period of 3 years (14). With their copious root systems, plants can scavenge large areas and volumes of soils, removing Se as selenate, selenite, and organic forms of selenium such as SeMet. Once absorbed by plant roots, Se is translocated to the shoot, where it may be harvested and removed from the site, a process called phytoextraction. One difficulty with phytoextraction is that Se is accumulated in plant tissues where it may become available to wildlife, especially birds.

Phytovolatilization, on the other hand, circumvents this problem because it removes Se completely from the local ecosystem into the atmosphere. This minimizes the entry of Se into the food chain, particularly as Se is volatilized by roots (173). Studies have shown that volatile Se entering the atmosphere is dispersed and diluted by air currents directly away from the contaminated areas, with deposition possibly occurring in Se-deficient areas (8).

Many species have been evaluated for their efficacy in phytoremediation (13, 19 104, 121, 130, 132, 167). Certain species of *Astragalus* were found to accumulate the most Se (e.g. *A. bisulcatus*), which led some researchers to suggest the use of the Se-accumulators, *A. bisulcatus* and *A. racemosus* (121). However, these are slow-growing plants and Se accumulated in *Astragalus* shoots is mostly soluble and can easily be leached from plant tissues back to the soil by rainfall (41).

The ideal plant species for phytoremediation of Se is one that can accumulate and volatilize large amounts of Se, grow rapidly and produce a large biomass on Se-contaminated soil, tolerate salinity and other toxic conditions, and provide a safe source of forage for Se-deficient livestock. Indian mustard (*Brassica juncea*), which typically contains 350 mg Se kg^{-1} dry weight, has most of the desired attributes (10, 13, 172). It is also easily genetically engineered (16, 172). EAH Pilon-Smits et al (125) transformed Indian mustard by overexpressing the gene encoding ATP sulfurylase; the transgenic plants exhibit much promise in the cleanup of Se from contaminated soils in that they accumulate two- to threefold more Se per plant than the wild-type plants.

Another major environmental problem is how to cleanup Se from contaminated water. One effective solution to this problem is to use constructed wetlands: up to 90% of the Se from oil refinery effluents was shown to be removed by this means (73). Although much of the Se is retained in the sediments, a substantial portion is taken up and immobilized into plant tissue, and a further portion (possibly as much as 10% to 30%) may be volatilized to the atmosphere (73). The choice of suitable wetland species is important in the cleanup of Se-contaminated wastewater by constructed wetlands because different species vary substantially in their ability to absorb, accumulate, and volatilize Se. Ten wetland cells were planted with different plant species in mono- and mixed cultures to evaluate the best plant species composition for maximum removal of Se from irrigation drainage water (150). This study, located at the Tulare Lake Drainage District at Corcoran, California, is now entering its third year; early results indicate that Se removal is greatly enhanced by the planting of wetland vegetation and that cattail (high biomass) and widgeon grass (high bioaccumulation) removed the greatest amount of Se per unit area of wetland (150).

One problem most often raised in connection with Se phytoremediation is how to dispose of Se-containing vegetation. Because Se is an essential trace element for adequate nutrition and health in humans and animals, one solution is to use seleniferous plant materials as a forage blend in Se-deficient regions to improve the Se status of animals. Another option is to add Se as a source of organic Se fertilizer to soils supporting forage crops (e.g. alfalfa; 12). Plants used for the phytoremediation of Se may generate other useful byproducts such as fibers for the production of paper and building materials, energy for heat production by the combustion of dried plant material or by fermentation to methane or ethanol. In addition, a large variety of chemical compounds (e.g. oil, sugars, fatty acids, proteins, pharmacological substances, vitamins, and detergents) are naturally produced by plants and may be useful byproducts of the phytoremediation process (115).

CONCLUSIONS AND FUTURE DIRECTIONS

Despite the fact that Se has been shown to be an essential micronutrient for animals, bacteria, and probably algae, the question of whether Se is required for the growth of higher plants is still controversial and unresolved. Clearly, one of the most important goals for future Se research is to continue to search for the presence of essential selenoproteins in higher plants using protein or DNA sequence analysis. Advances in the sequencing of genomes of higher plants such as *Arabidopsis* and rice may facilitate the discovery of genes containing in-frame UGA codons or homologs of the genes encoding SeCys incorporation enzymes (e.g. selenophosphate synthetase). This would provide strong (though indirect) evidence that SeCys-containing selenoproteins are present in plants and thus support the view of Se essentiality in plant growth.

Molecular studies have enabled significant strides to be made in other areas of Se physiology and biochemistry. Overexpression of genes encoding transporters and enzymes involved in the uptake, assimilation, and volatilization of sulfate has been highly successful in elucidating the role of these proteins in Se metabolism and in identifying rate-limiting steps in the assimilation of selenate to volatile Se. For example, it is now well established that selenate is taken up actively via a sulfate transporter in the root plasma membrane, that the S enzyme, ATP sulfurylase, mediates the reduction of selenate in plants, and that the reduction of selenate and the conversion of organic Se to volatile Se is rate limiting to Se uptake and assimilation.

One important area in need of more research is the role of rhizosphere bacteria in Se volatilization by plants. Plants volatilize relatively low amounts of selenate or selenite in the absence of bacteria, and it is not clear why this is so. Part of the reason may be that rhizosphere bacteria are required to facilitate the uptake of selenate into root tissues, possibly by providing key proteinaceous compounds. Furthermore, bacteria promote the conversion of SeCys to SeMet, thereby facilitating selenite volatilization.

The ability of plants to absorb, sequester, and volatilize Se has important implications in the management of environmental Se contamination by phytoremediation. Indian mustard has been identified as a plant species that has a superior ability to accumulate and volatilize Se, grow rapidly on Se-contaminated soil, tolerate salinity and other toxic conditions, and provide a safe source of forage for Se-deficient livestock. By overexpressing genes encoding key enzymes in the S assimilation pathway, it has been possible to genetically engineer plants to enhance their ability for Se phytoremediation.

ACKNOWLEDGMENTS

We gratefully acknowledge the generous financial support provided by the Electric Power Research Institute (contract nos. W04163-01 and W08021-30), and by the University of California Salinity/Drainage Task Force for much of the research described in this review. We also thank the United States Department of Energy and the Stanford Synchrotron Radiation Laboratory for beam-time and on-line support for x-ray absorption spectroscopy. We are especially grateful to Dr. Jean-Claude Davidian for the critical reading of the manuscript, and to Dr. Steve Whiting for preparing the manuscript for publication.

LITERATURE CITED

1. Abrams MM, Shennan C, Zazoski J, Burau RG. 1990. Selenomethionine uptake by wheat seedlings. *Agron. J.* 82:1127–30
2. Anderson JW. 1993. Selenium interactions in sulfur metabolism. In *Sulfur Nutrition and Assimilation in Higher Plants: Regulatory Agricultural and Environmental Aspects*, ed. LJ De Kok, I Stulen, H Rennenberg, C Brunold, WE Rauser, pp. 49–60. The Hague, The Netherlands: SPB Academic
3. Anderson JW, Scarf AR. 1983. Selenium and plant metabolism. In *Metals and Micronutrients: Uptake and Utilization by Plants*, ed. DA Robb, WS Pierpoint, pp. 241–75. London: Academic
4. Ansede JH, Yoch DC. 1997. Comparison of selenium and sulfur volatilization by dimethylsulfoniopropionate lyase (DMSP) in two marine bacteria and estuarine sediments. *FEMS Microbiol. Ecol.* 23:315–24
5. Arvy MP. 1993. Selenate and selenite uptake and translocation in bean plants (*Phaseolus vulgaris*). *J. Exp. Bot.* 44:1083–87
6. Asher CJ, Butler GW, Peterson PJ. 1977. Selenium transport in root systems of tomato. *J. Exp. Bot.* 23:279–91
7. Aslam M, Harbit KB, Huffaker RC. 1990. Comparative effects of selenite and selenate on nitrate assimilation in barley seedlings. *Plant Cell Environ.* 13:773–82
8. Atkinson R, Aschmann SM, Hasegawa D, Thompson-Eagle ET, Frankenberger WT Jr. 1990. Kinetics of the atmospherically important reactions of dimethylselenide. *Environ. Sci. Technol.* 24:1326–32
9. Axley MJ, Boeck A, Stadtman TC. 1991. Catalytic properties of an *Escherichia coli* formate dehydrogenase mutant in which sulfur replaces selenium. *Proc. Natl. Acad. Sci. USA* 88:8450–54
10. Bañuelos G, Schrale G. 1989. Plants that

remove selenium from soils. *Calif. Agric.* 43:19–20

11. Bañuelos GS, Ajwa HA, Mackey M, Wu L, Cook C, et al. 1997. Evaluation of different plant species used for phytoremediation of high soil selenium. *J. Environ. Qual.* 26:639–46
12. Bañuelos GS, Mead R, Akohoue S. 1991. Adding selenium-enriched plant tissue to soil causes the accumulation of selenium in alfalfa. *J. Plant Nutr.* 14:701–14
13. Bañuelos GS, Meek DW. 1990. Accumulation of selenium in plants grown on selenium-treated soil. *J. Environ. Qual.* 19:772–77
14. Bañuelos GS, Terry N, Zayed A, Wu L. 1995. *Managing high soil selenium with phytoremediation*. Presented at Proc. Annu. Natl. Meet. Am. Soc. Surf. Min. Reclam., 12th, Gillette, WY
15. Bañuelos GS, Zayed A, Terry N, Wu L, Akohoue S, et al. 1996. Accumulation of selenium by different plant species grown under increasing sodium and calcium chloride salinity. *Plant Soil* 183:49–59
16. Barfield DG, Pua EC. 1991. Gene transfer in plants of *Brassica juncea* using *Agrobacterium tumefaciens* mediated transformation. *Plant Cell Report.* 10:308–14
17. Beath OA. 1937. The occurence of selenium and seleniferous vegetation in Wyoming. II Seleniferous vegetation. *Wyoming Agric. Exper. Stn. Bull.* 221:29–64
18. Beeor-Tzahar T, Ben-Hayyim G, Holland D, Faltin Z, Eshdat Y. 1995. A stress-associated citrus protein is a distinct plant phospholipid hydroperoxide glutathione peroxidase. *FEBS Lett.* 366:151–55
19. Bell PF, Parker DR, Page AL. 1992. Contrasting selenate sulfate interactions in selenium-accumulating and nonaccumulating plant species. *Soil Sci. Soc. Am. J.* 56:1818–24
20. Berry MJ, Kieffer JD, Harney JW, Larsen PR. 1991. Selenocysteine confers the biochemical properties characteristic of the type I iodothyronine deiodinase. *J. Biol. Chem.* 266:14155–58
21. Bick JA, Leustek T. 1998. Plant sulfur metabolism—the reduction of sulfate to sulfite. *Curr. Opin. Plant Biol.* 1:240–44
22. Boeck A, Forchhammer K, Heider J, Baron C. 1991. Selenoprotein synthesis: an expansion of the genetic code. *Trends Biochem. Sci.* 16:463–67
23. Bolchi A, Petrucco S, Tenca PL, Foroni C, Ottonello S. 1999. Coordinate modulation of maize sulfate permease and ATP sulfurylase mRNAs in response to variations in sulfur nutritional status: stereospecific down-regulation by L-cysteine. *Plant Mol. Biol.* 39:527–37
24. Bosma W, Schupp R, De Kok LJ, Rennenberg H. 1991. Effect of selenate on assimilatory sulfate reduction and thiol content spruce needles. *Plant Physiol. Biochem.* 29:131–38
25. Bourgis F, Roje S, Nuccio ML, Fisher DB, Tarczynski MC, et al. 1999. S-methylmethionine plays a major role in phloem sulfur transport and is synthesized by a novel type of methyltransferase. *Plant Cell* 11:1485–97
26. Brady JM, Tobin JM, Gadd GM. 1996. Volatilization of selenite in aqueous medium by a *Penicillium* species. *Mycol. Res.* 100:955–61
27. Breton A, Surdin-Kerjan Y. 1977. Sulfate uptake in *Saccharomyces cerevisiae:* biochemical and genetic study. *J. Bacteriol.* 132: 224–32
28. Brown TA, Shrift A. 1981. Exclusion of selenium from proteins in selenium-tolerant *Astragalus* species. *Plant Physiol.* 67:1951–53
29. Brown TA, Shrift A. 1982. Selenium: toxicity and tolerance in higher plants. *Biol. Rev.* 57:59–84
30. Broyer TC, Lee DC, Asher CJ. 1966. Selenium nutrition of green plants. Effects of selenite supply on growth and selenium

content of alfalfa and subterranean clover. *Plant Physiol.* 41:1425–28

31. Broyer TC, Lee DC, Asher CJ. 1972. Selenium and nutrition of *Astragalus*. I. Effect of selenite or selenate supply on growth and selenium content. *Plant Soil* 36:635–49
32. Bruhl A, Haverkamp T, Gisselmann G, Schwenn JD. 1996. A cDNA clone from *Arabidopsis thaliana* encoding plastidic ferredoxin: sulfite reductase. *Biochim. Biophys. Acta* 1295:119–24
33. Burau RG, McDonald A, Jacobson A, May D, Grattan S, et al. 1988. Selenium in tissues of crops sampled from the west side of the San Joaquin Valley, California. In *Selenium Contents in Animal and Human Food Crops Grown in California*, ed. KK Tanji, L Valoppi, RC Woodring, pp. 61–67. Berkeley: Univ. Calif., Div. Agric. Natl. Resourc.
34. Burnell JN. 1981. Methionyl-tRNA synthetase from *Phaseolus aureus*: purification and properties. *Plant Physiol.* 67:325–29
35. Burnell JN. 1981. Selenium metabolism in *Neptunia amplexicaulis*. *Plant Physiol.* 67:316–24
36. Burnell JN, Shrift A. 1977. Cysteinyl tRNA synthetase from *Phaseolus aureus*. *Plant Physiol.* 60:670–74
37. Burnell JN, Shrift A. 1979. Cysteinyl-tRNA synthetase from *Astragalus* species. *Plant Physiol.* 63:1095–97
38. Cherest H, Davidian J-C, Thomas D, Benes V, Ansorge W, et al. 1997. Molecular characterization of two high affinity sulfate transporters in *Saccharomyces cerevisiae*. *Genetics* 145:627–35
39. Clarkson DT, Luttge U. 1991. II. Mineral nutrition: inducible and repressible nutrient transport systems. *Prog. Bot.* 52:61–83
40. Cossins EA, Chen L. 1997. Folates and one-carbon metabolism in plants and fungi. *Phytochemistry* 45:437–52
41. Cowgill UM. 1990. The selenium cycle in three species of *Astragalus*. *J. Plant Nutr.* 13:1309–18
42. Criqui MC, Jamet E, Parmentier Y, Marbach J, Durr A, et al. 1992. Isolation and characterization of a plant cDNA showing homology to animal glutathione peroxidases. *Plant Mol. Biol.* 18:623–27
43. Dacey JWH, King GM, Wakeham SG. 1987. Factors controlling emission of dimethylsulfide from salt marshes. *Nature* 330:643–45
44. Davidian J-C, Hatzfield Y, Cathala N, Tagmount A, Vidmar JJ. 2000. Sulfate uptake and transport in plants. In *Sulfur Nutrition and Sulfur Assimilation in Higher Plants: Molecular, Biochemical and Physiological Aspects*, ed. C Brunold, H Rennenberg, LJ De Kok, I Stuhlen, J-C Davidian. Bern: Paul Haupt. In press pp. 1–19
45. Dawson JC, Anderson JW. 1988. Incorporation of cysteine and selenocysteine into cystathionine and selenocystathionine by crude extracts of spinach. *Phytochemistry* 27:3453–60
46. De Kok LJ, Kuiper PJC. 1986. Effect of short-term dark incubation with sulfate, chloride and selenate on the glutathione content of spinach [*Spinacia oleracea* cultivar Estivato] leaf discs. *Physiol. Plant.* 68:477–82
47. De Souza MP, Chu D, Zhao M, Zayed AM, Ruzin SE, et al. 1999. Rhizosphere bacteria enhance selenium accumulation and volatilization by Indian mustard. *Plant Physiol.* 119:565–73
48. De Souza MP, Huang CPA, Chee N, Terry N. 1999. Rhizosphere bacteria enhance the accumulation of selenium and mercury in wetland plants. *Planta* 209:259-63
49. De Souza MP, Pilon-Smits EAH, Lytle CM, Hwang S, Tai J, et al. 1998. Rate-limiting steps in selenium assimilation and volatilization by Indian mustard. *Plant Physiol.* 117:1487–94
50. De Souza MP, Pilon-Smits EAH, Terry N. 1999. The physiology and biochemistry of Selenium volatilization by plants. In *Phytoremediation of Toxic Metals*, ed. I Raskin,

BD Ensley, pp. 171–90. New York: Wiley & Sons

51. De Souza MP, Terry N. 1997. Selenium volatilization by rhizosphere bacteria. *Abstr. Gen. Meet. Am. Soc. Microbiol.* 97:499
52. Demoll-Decker H, Macy JM. 1993. The periplasmic nitrite reductase of *Thauera selenatis* may catalyze the reduction of selenite to elemental selenium. *Arch. Microbiol.* 160:241–47
53. Dilworth GL, Bandurski RS. 1977. Activation of selenate by adenosine 5′-triphosphate sulfurylase from *Saccharmoyces cereviseae*. *Biochem. J.* 163:521–29
54. Draize JH, Beath OA. 1935. Observations on the pathology of "blind staggers" and "alkali disease". *Am. Vet. Med. Assoc. J.* 86:753–63
55. Droux M, Gakiere B, Denis L, Ravanel S, Tabe L, et al. 2000. Methionine biosynthesis in plants: biochemical and regulatory aspects. In *Sulfur Nutrition and Sulfur Assimilation in Higher Plants: Molecular, Biochemical and Physiological Aspects*, ed. C Brunold, H Rennenberg, LJ De Kok, I Stulen, J-C Davidian. Bern: Paul Haupt. In press
56. Duckart EC, Waldron LJ, Donner HE. 1992. Selenium uptake and volatilization from plants growing in soil. *Soil Sci.* 153:94–99
57. Eichel J, Gonzalez JC, Hotze M, Matthews RG, Schroeder J. 1995. Vitamin B-12-independent methionine synthase from a higher plant (*Catharanthus roseus*) molecular characterization, regulation, heterologous expression, and enzyme properties. *Eur. J. Biochem.* 230:1053–58
58. Ernst WHO. 1982. Selenpflanzen (Selenophyten). In *Pflanzenokologie und Mineralstoffwechsel*, ed. H Kinzel, pp. 511–19. Stuttgart: Verlag Eugen Ulmer
59. Esaki N, Nakamura T, Tanaka H, Soda K. 1982. Selenocysteine lyase, a novel enzyme that specifically acts on selenocysteine. Mammalian distribution and purification and properties of pig liver enzyme. *J. Biol. Chem.* 257:4386–91
60. Eshdat Y, Holland D, Faltin Z, Ben-Hayyim G. 1997. Plant glutathione peroxidases. *Physiol. Plant.* 100:234–40
61. Eustice DC, Foster I, Kull FJ, Shrift A. 1980. *In vitro* incorporation of selenomethionine into protein by *Vigna radiata* polysomes. *Plant Physiol.* 66:182–86
62. Eustice DC, Kull FJ, Shrift A. 1981. Selenium toxicity: aminoacylation and peptide bond formation with selenomethionine. *Plant Physiol.* 67:1054–58
63. Evans CS, Asher CJ, Johnson CM. 1968. Isolation of dimethyl diselenide and other volatile selenium compounds from *Astragalus racemosus* (Pursh.). *Aust. J. Biol. Sci.* 21:13–20
64. Faltin Z, Camoin L, Ben-Hayyim G, Perl A, Beeor-Tzahar T, et al. 1998. Cysteine is the presumed catalytic residue of *Citrus sinensis* phospholipid hydroperoxide glutathione peroxidase over-expressed under salt stress. *Physiol. Plant.* 104:741–46
65. Fan TW-M, Lane AN, Higashi RM. 1997. Selenium biotransformations by a euryhaline microalga isolated from a saline evaporation pond. *Environ. Sci. Technol.* 31:569–76
66. Fleming RW, Alexander M. 1972. Dimethylselenide and dimethyltelluride formation by a strain of *Penicillium*. *Appl. Microbiol.* 24:424–29
67. Forchhammer K, Boeck A. 1991. Biology and biochemistry of the element selenium. *Naturwissenschaften* 78:497–504
68. Foyer CH, Halliwell B. 1976. The presence of glutathione and glutathione reductase in chloroplasts: a proposed role in ascorbic acid metabolism. *Planta* 133:21–25
69. Foyer CH, Souriau N, Perret S, Lelandais M, Kunert KJ, et al. 1995. Overexpression of glutathione reductase but not glutathione synthetase leads to increases in antioxidant capacity and resistance to photoinhibition

in poplar trees. *Plant Physiol.* 109:1047–57

70. Gessler NN, Bezzubov AA. 1988. Study of the activity of S-methylmethionine sulfonium salt hydrolase in plant and animal tissues. *Prikl. Biokhim. Mikrobiol.* 24:240–46
71. Giovanelli J, Mudd SH, Datko AH. 1980. Sulfur amino acids in plants. In *Sulfur Amino Acids in Plants,* ed. BJ Miflin, pp. 453–505. New York: Academic
72. Gutierrez-Marcos JF, Roberts MA, Campbell EI, Wray JL. 1996. Three members of a novel small gene-family from *Arabidopsis thaliana* able to complement functionally an *Escherichia coli* mutant defective in PAPS reductase activity encode proteins with a thioredoxin-like domain and "APS reductase" activity. *Proc. Natl. Acad. Sci. USA* 93:13377–82
73. Hansen D, Duda PJ, Zayed A, Terry N. 1998. Selenium removal by constructed wetlands: role of biological volatilization. *Environ. Sci. Technol.* 32:591–97
74. Hanson AD, Rivoal J, Paquet L, Gage DA. 1994. Biosynthesis of 3-dimethylsulfoniopropionate in *Wollastonia biflora* (L.) DC: evidence that S-methylmethionine is an intermediate. *Plant Physiol.* 105:103–10
75. Hanson AD, Trossat C, Nolte KD, Gage DA. 1997. 3-Dimethylsulphonioproprionate biosynthesis in higher plants. In *Sulphur Metabolism in Higher Plants: Molecular, Ecophysical and Nutrition Aspects*, ed. WJ Cram, LJ De Kok, I Stulen, C Brunold, H Rennenberg, pp. 147–54. Leiden, The Netherlands: Backhuys
76. Hatfield D, Choi IS, Mischke S, Owens LD. 1992. Selenocysteinyl-tRNAs recognize UGA in *Beta vulgaris*, a higher plant, and in *Gliocladium virens*, a filamentous fungus. *Biochem. Biophys. Res. Commun.* 184:254–59
77. Hatfield DL, Lee BJ, Price NM, Stadtman TC. 1991. Selenocysteyl-transfer RNA occurs in the diatom *Thalassiosira* and in the ciliate *Tetrahymena. Mol. Microbiol.* 5:1183–86
78. Hattula T, Granroth B. 1974. Formation of dimethyl sulphide from S-methylmethionine in onion seedlings (*Allium cepa*). *J. Sci. Food Agric.* 25:1517–21
79. Haygarth PM, Harrison AF, Jones KC. 1995. Plant selenium from soil and the atmosphere. *J. Environ. Qual.* 24:768–71
80. Heiss S, Schaefer HJ, Haag-Kerwer A, Rausch T. 1999. Cloning sulfur assimilation genes of *Brassica juncea* L.: Cadmium differentially affects the expression of a putative low-affinity sulfate transporter and isoforms of ATP sulfurylase and APS reductase. *Plant Mol. Biol.* 39:847–57
81. Holland D, Ben-Hayyim G, Faltin Z, Camoin L, Strosberg AD, et al. 1993. Molecular characterization of salt-stress-associated protein in citrus protein and cDNA sequence homology to mammalian glutathione peroxidases. *Plant Mol. Biol.* 21:923–27
82. Hori K, Hatfield D, Maldarelli F, Lee BJ, Clause KA. 1997. Selenium supplementation supresses tumor necrosis factor alpha-induced human immunodeficiency virus type 1 replication in vitro. *AIDS Res. Hum. Retrovir.* 13:1325–32
83. Huang ZZ, Wu L. 1991. Species richness and selenium accumulation of plants in soils with elevated concentration of selenium and salinity. *Ecotoxicol. Environ. Saf.* 22:251–66
84. Ihnat M. 1989. *Occurrence and Distribution of Selenium*, p. 354. Boca Raton, FL: CRC Press
85. James F, Nolte KD, Hanson AD. 1995. Purification and properties of S-adenosyl-L-methionine: L-methionine S-methyltransferase from *Wollastonia biflora* leaves. *J. Biol. Chem.* 270:22344–50
86. James F, Paquet L, Sparace SA, Gage DA, Hanson AD. 1995. Evidence implicating dimethylsulfoniopropionaldehyde as an intermediate in dimethylsulfoniopropionate biosynthesis. *Plant Physiol.* 108:1439–48

87. Jansson B. 1980. The role of selenium as a cancer-protecting trace element. In *Carcinogenicity and Metal Ions*, ed. H Sigel, pp. 281–311. New York: Dekker
88. Kabata-Pendias AJ. 1998. Geochemistry of selenium. *J. Environ. Pathol. Toxicol. Oncol.* 17:173–77
89. Kim J, Leustek T. 1996. Cloning and analysis of the gene for cystathionine gamma-synthase from *Arabidopsis thaliana*. *Plant Mol. Biol.* 32:1117–24
90. Kocsis MG, Nolte KD, Rhodes D, Shen TL, Gage DA, et al. 1998. Dimethylsulfoniopropionate biosynthesis in *Spartina alterniflora*: Evidence that S-methylmethionine and dimethylsulfoniopropylamine are intermediates. *Plant Physiol.* 117:273–81
91. Lauchli A. 1993. Selenium in plants: uptake, functions, and environmental toxicity. *Bot. Acta* 106:455–68
92. Le Guen L, Thomas M, Kreis M. 1994. Gene density and organization in a small region of the *Arabidopsis thaliana* genome. *Mol. Gen. Genet.* 245:390–96
93. Lee BJ, Worland PJ, Davis JN, Stadtman TC, Hatfield DL. 1989. Identification of a selenocysteylseryl transfer RNA in mammalian cells that recognizes the nonsense codon, UGA. *J. Biol. Chem.* 264:9724–27
94. Leggett JE, Epstein E. 1956. Kinetics of sulfate absorption by barley roots. *Plant Physiol.* 31:222–26
95. Leinfelder W, Stadtman TC, Boeck A. 1989. Occurrence in vivo of selenocysteylseryl transfer RNAUCA in *Escherichia coli*: effect of sel mutations. *J. Biol. Chem.* 264:9720–23
96. Lemly AD. 1997. Environmental implications of excessive selenium: a review. *Biomed. Environ. Sci.* 10:415–35
97. Leustek T, Murillo M, Cervantes M. 1994. Cloning of a cDNA encoding ATP sulfurylase from *Arabidopsis thaliana* by functional expression in *Saccharomyces cerevisiae*. *Plant Physiol.* 105:897–902
98. Lewis BG. 1971. *Volatile selenium in higher plants*. PhD thesis, Univ. Calif., Berkeley
99. Lewis BG, Johnson CM, Broyer TC. 1974. Volatile selenium in higher plants. The production of dimethyl selenide in cabbage leaves by enzymatic cleavage of Se-methyl selenomethionine selenonium salt. *Plant Soil* 40:107–18
100. Lindblow-Kull C, Kull FJ, Shrift A. 1985. Single transporter for sulfate, selenate, and selenite in *Escherichia coli* K12. *J. Bacteriol.* 163:1267–69
101. Losi ME, Frankenberger WT Jr. 1997. Reduction of selenium oxyanions by *Enterobacter cloacae* SLD1a-1: isolation and growth of the bacterium and its expulsion of selenium particles. *Appl. Environ. Microbiol.* 63:3079–84
102. Marschner H. 1995. *Mineral Nutrition of Higher Plants*, pp. 430–33. London: Academic
103. Martens DA, Suarez DL. 1997. Mineralization of selenium-containing amino acids in two California soils. *Soil Sci. Soc. Am. J.* 61:1685–94
104. Mayland HF, James LF, Panter KE, Sonderegger JL. 1989. Selenium in seleniferous environments. In *Selenium in Agriculture and the Environment*, ed. LW Jacobs, pp. 15–50. Madison, WI: Am. Soc. Agron.
105. McCluskey TJ, Scarf AR, Anderson JW. 1986. Enzyme-catalyzed $\alpha\beta$-elimination of selenocystathionine and selenocystine and their sulfur isologues by plant extracts. *Phytochemistry* 25:2063-68
106. Mengel K, Kirkby EA, International Potash Institute. 1987. *Principles of Plant Nutrition*, p. 687. Bern: Int. Potash Inst.
107. Mikkelsen RL, Page AL, Haghnia GH 1988. Effect of salinity and its composition on the accumulation of selenium by alfalfa. *Plant Soil* 107:63–67
108. Mikkelson RL, Page AL, Bingham FT

1989. Factors affecting selenium accumulation by agricultural crops. *Soil Sci. Soc. Am. Spec. Publ.* 23:65–94
109. Mudd SH, Datko AH. 1990. The S-methylmethionine cycle in *Lemna paucicostata. Plant Physiol.* 93:623–30
110. Neuhierl B, Boeck A. 1996. On the mechanism of selenium tolerance in selenium-accumulating plants: purification and characterization of a specific selenocysteine methyltransferase from cultured cells of *Astragalus bisulcatus. Eur. J. Biochem.* 239:235–38
111. Neuhierl B, Thanbichler M, Lottspeich F, Boeck A. 1999. A family of S-methylmethionine-dependent thiol/selenol methyltransferases: role in selenium tolerance and evolutionary relation. *J. Biol. Chem.* 274:5407–14.
112. Ng BH, Anderson JW. 1978. Synthesis of selenocysteine by cysteine synthases from selenium accumulator and non-accumulator plants. *Phytochemistry* 17:2069–74
113. Ng BH, Anderson JW. 1979. Light-dependent incorporation of selenite and sulphite into selenocysteine and cysteine by isolated pea chloroplasts. *Phytochemistry* 18:573–80
114. Nigam SN, Tu J-I, McConnell WB. 1969. Distribution of selenomethylselenocysteine and some other amino acids in species of *Astragalus*, with special reference to their distribution during the growth of *A. bisulcatus. Phytochemistry* 8:1161–65
115. Nyberg S. 1991. Multiple use of plants: studies on selenium incorporation in some agricultural species for the production of organic selenium compounds. *Plant Foods Human Nutr.* 41:69–88
116. Ohlendorf HM, Hoffman DJ, Salki MK, Aldrich TW. 1986. Embryonic mortality and abnormalities of aquatic birds: apparent impacts of selenium from irrigation drain water. *Sci. Total Environ.* 52: 49–63
117. Oremland RS, Blum JS, Culbertson CW. 1994. Isolation, growth, and metabolism of an obligately anaerobic, selenate-respiring bacterium. *Abstr. Gen. Meet. Am. Soc. Microbiol.* 94:423
118. Oremland RS, Steinberg NA, Presser TS, Miller LG. 1991. In situ bacterial selenate reduction in the agricultural drainage systems of western Nevada [USA]. *Appl. Environ. Microbiol.* 57:615–17
119. Padmaja K, Prasad DDK, Prasad ARK. 1989. Effect of selenium on chlorophyll biosynthesis in mung bean seedlings. *Phytochemistry* 28:3321–24
120. Parker DR, Page AL. 1994. Vegetation management strategies for remediation of selenium-contaminated soils. In *Selenium in the Environment*, ed. WT Frankenberger Jr, S Benson, pp. 327–47. New York: Marcel Dekker
121. Parker DR, Page AL, Thomason DN. 1991. Salinity and boron tolerances of candidate plants for the removal of selenium from soils. *J. Environ. Qual.* 20:157–64
122. Petersen M, van der Straeten D, Bauw G. 1995. Full-length cDNA clone from *Coleus blumei* with high similarity to cobalamin-independent methionine synthase. *Plant Physiol.* 109:338
123. Peterson PJ, Robinson PJ. 1972. L-Cystathionine and its selenium analogue in *Neptunia amplexicaulis. Phytochemistry* 11:1837–39
124. Pilon-Smits EAH, de Souza MP, Hong G, Amini A, Bravo RC, et al. 1999. Selenium volatilization and accumulation by twenty aquatic plant species. *J. Environ. Qual.* 28:1011–18
125. Pilon-Smits EAH, Hwang S, Lytle CM, Zhu Y, Tai JC, et al. 1999. Overexpression of ATP sulfurylase in Indian mustard leads to increased selenate uptake, reduction, and tolerance. *Plant Physiol.* 119:123–32
126. Price NM, Thompson PA, Harrison PJ. 1987. Selenium: an essential element

for growth of the coastal marine diatom *Thalassiosira pseudonana* (Bacillariophyceae). *J. Phycol.* 23:1–9

127. Rael RM, Frankenberger WT Jr. 1996. Influence of pH, salinity, and selenium on the growth of *Aeromonas veronii* in evaporation agricultural drainage water. *Water Resour.* 30:422–30
128. Ravanel S, Gakiere B, Job D, Douce R. 1998. The specific features of methionine biosynthesis and metabolism in plants. *Proc. Natl. Acad. Sci. USA* 95:7805–12
129. Reamer DC, Zoller WH. 1980. Selenium biomethylation products from soil and sewage sludge. *Science* 208:500–2
130. Retana J, Parker DR, Amrhein C, Page AL. 1993. Growth and trace element concentrations of five plant species growth in a highly saline soil. *J. Environ. Qual.* 22:805–11
131. Rhodes D, Gage DA, Cooper AJL, Hanson AD. 1997. S-methylmethionine conversion to dimethylsulfoniopropionate: evidence for an unusual transamination reaction. *Plant Physiol.* 115:1541–48
132. Rosenfeld I, Beath OA. 1964. *Selenium, Geobotany, Biochemistry, Toxicity, and Nutrition*. New York: Academic. 411 pp.
133. Rotruck JT, Pope AL, Ganther HE, Swanson AB, Hafeman DG, et al. 1973. Selenium: biochemical role as a component of gluthathione peroxidase. *Science* 179:588–90
134. Sabeh F, Wright T, Norton SJ. 1993. Purification and characterization of a glutathione peroxidase from the *Aloe vera* plant. *Enzyme Prot.* 47:92–98
135. Saiki MK, Lowe TP. 1987. Selenium in aquatic organisms from subsurface agricultural drainage water, San Joaquin Valley, California. *Arch. Environ. Contam. Toxicol.* 19:496–99
136. Saito K, Takahashi H, Noji M, Inoue K, Hatzfeld Y. 2000. Molecular regulation of sulfur assimilation and cysteine synthesis. In *Sulfur Nutrition and Sulfur Assimilation in Higher Plants: Molecular, Biochemical and Physiological Aspects*, ed. C Brunold, H Rennenberg, LJ De Kok, I Stulen, J-C Davidian. Bern: Paul Haupt. In press
137. Saito K, Tatsuguchi K, Takagi Y, Murakoshi I. 1994. Isolation and characterization of cDNA that encodes a putative mitochondrion-localizing isoform of cysteine synthase (O-acetylserine(thiol)-lyase) from *Spinacia oleracea*. *J. Biol. Chem.* 269:28187–92
138. Schwarz K, Foltz CM. 1957. Selenium as an integral part of factor 3 against dietary necrotic liver degeneration. *J. Am. Chem. Soc.* 70:3292–93
139. Setya A, Murillo M, Leuster T. 1996. Sulfate reduction in higher plants: molecular evidence for a novel 5′-adenylylsulfate reductase. *Proc. Natl. Acad. Sci. USA* 93:13383–88
140. Shaw WH, Anderson JW. 1972. Purification, properties and substrate specificity of adenosine triphosphate sulphurylase from spinach leaf tissue. *Biochem. J.* 127:237–47
141. Shennan C, Schachtman DP, Cramer GR. 1990. Variation in selenium-75-labeled selenate uptake and partitioning among tomato cultivars and wild species. *New Phytol.* 115:523–30
142. Shrift A. 1969. Aspects of selenium metabolism in higher plants. *Annu. Rev. Plant Physiol.* 20:475–94
143. Shrift A, Ulrich JM. 1976. Transport of selenate and selenite into *Astragalus* roots. *Plant Physiol.* 44:893–96
144. Skorupa JP. 1998. Selenium poisoning of fish and wildlife in nature: lessons from twelve real-world examples. In *Environmental Chemistry of Selenium*, ed. WT Frankenberger Jr, RA Engberg, pp. 315–54. New York: Marcel Dekker
145. Smith FW, Hawkesford MJ, Ealing PM Clarkson DT, Berg PJV, et al. 1997. Regulation of expression of a cDNA from barley roots encoding a high affinity sulphate transporter. *Plant. J.* 12:875–84

146. Smith FW, Hawkesford MJ, Prosser IM, Clarkson DT. 1995. Isolation of a cDNA from *Saccharomyces cerevisiae* that encodes a high affinity sulphate transporter at the plasma membrane. *Mol. Gen. Genet.* 247:709–15
147. Stadtman TC. 1990. Selenium biochemistry. *Annu. Rev. Biochem.* 59:111–28
148. Stadtman TC. 1996. Selenocysteine. *Annu. Rev. Biochem.* 65:83–100
149. Takahashi H, Saskura N, Noji M, Saito K. 1996. Isolation and characterization of a cDNA encoding a sulfate transporter from *Arabidopsis thaliana. FEBS Lett.* 392:95–99
150. Terry N. 1998. Use of flow-through constructed wetlands for the remediation of selenium in agricultural tile drainage water. In *1997–1998 Tech. Progr. Rep., UC Salin./Drain. Res. Prog.*, Univ. Calif. Berkeley
151. Terry N, Carlson C, Raab TK, Zayed AM. 1992. Rates of selenium volatilization among crop species. *J. Environ. Qual.* 21:341–44
152. Terry N, Lin ZQ. 1999. *Managing High Selenium in Agricultural Drainage Water by Agroforestry Systems: Role of Selenium Volatilization.* Rep. Calif. State Dep. Water Resourc., Sacramento
153. Terry N, Zayed A. 1998. Phytoremediation of selenium. In *Environmental Chemistry of Selenium*, ed. WT Frankenberger Jr, R Engberg, pp. 633–56. New York: Marcel Dekker
154. Terry N, Zayed AM. 1994. Selenium volatilization by plants. *In Selenium in the Environment.* ed. WT Frankenberger Jr, S Benson, pp. 343–67. New York: Marcel Dekker
155. Thompson-Eagle ET, Frankenberger WT Jr, Karlson U. 1989. Volatilization of selenium by *Alternaria alternata. Appl. Environ. Microbiol.* 55:1406–13
156. Trelease SF, Beath OA. 1949. *Selenium, Its Geological Occurrence and Its Biological Effects in Relation to Botany, Chemistry, Agriculture, Nutrition, and Medicine*, p. 292. New York: Trelease & Beath
157. Trelease SF, Trelease HM. 1939. Physiological differentiation in *Astragalus* with reference to selenium. *Am. J. Bot.* 26:530–35
158. Trossat C, Nolte KD, Hanson AD. 1996. Evidence that the pathway of dimethylsulfoniopropionate biosynthesis begins in the cytosol and endos in the chloroplast. *Plant Physiol.* 111:965–73
159. Tsang MLS, Schiff JA. 1978. Studies of sulfate utilization by algae. 18. Identification of glutathione as a physiological carrier in assimilatory sulfate reduction by *Chlorella. Plant Sci. Lett.* 11:177–83
160. Turner WL, Pallett KE, Lea PJ. 1998. Cystathionine beta-lyase from *Echinochloa colonum* tissue culture. *Phytochemistry* 47:189–96
161. Vojtechova M, Hanson AD, Munoz-Clares RA. 1997. Betaine-aldehyde dehydrogenase from Amaranth leaves efficiently catalyzes the NAD-dependent oxidation of dimethylsulfoniopropionaldehyde to dimethylsulfoniopropionate. *Arch. Biochem. Biophys.* 337:81–88
162. von Vleet JF, Ferrans VJ. 1992. Etiological factors and pathologic alterations in selenium-vitamin E deficiency and excess in animals and humans. *Biol. Trace Elem. Res.* 33:1–21
163. Wallsgrove RM, Lea PJ, Miflin BJ. 1983. Intracellular localization of aspartate kinase and the enzymes of threonine and methionine biosynthesis in green leaves. *Plant Physiol.* 71:780–84
164. Wang Y, Boeck A, Neuhierl B. 1999. Acquisition of selenium tolerance by a selenium non-accumulating *Astragalus* species via selection. *Biofactors* 9:3–10
165. Wilber CG. 1980. Toxicology of selenium: a review. *Clin. Toxicol.* 17:171–230
166. Wu L, Huang ZZ. 1991. Chloride and sulfate salinity effects on selenium accumulation by tall fescue. *Crop Sci.* 31:114–18

167. Wu L, Huang ZZ, Burau RG. 1988. Selenium accumulation and selenium-salt cotolerance in five grass species. *Crop Sci.* 28:517–22

168. Wu L, Mantgem PJV, Guo X. 1996. Effects of forage plant and field legume species on soil selenium redistribution, leaching, and bioextraction in soils contaminated by agricultural drain water sediment. *Arch. Environ. Contam. Toxicol.* 31:329–38

169. Yamaguchi Y, Nakamura T, Harada E, Koizumi N, Sano H. 1997. Isolation and characterization of a cDNA encoding a sulfate transporter from *Arabidopsis thaliana. Plant Physiol.* 113: 1463

170. Yokota A, Shigeoka S, Onishi T, Kitaoka S. 1988. Selenium as inducer of glutathione peroxidase in low carbon dioxide grown *Chlamydomonas reinhardtii. Plant Physiol.* 86:649–51

171. Zayed A, Lytle CM, Terry N. 1998. Accumulation and volatilization of different chemical species of selenium by plants. *Planta* 206:284–92

172. Zayed AM, Pilon-Smits EAH, de Souza MP, Lin ZQ, Terry N. 1999. Remediation of selenium-polluted soils and waters by phytovolatilization. In *Phytoremediation of Metal-Contaminated Water and Soils*, ed. N Terry, G Bañuelos, pp. 61–83. Boca Raton, FL: CRC Press

173. Zayed AM, Terry N. 1994. Selenium volatilization in roots and shoots: effects of shoot removal and sulfate level. *J. Plant Physiol.* 143:8–14

174. Zieve R, Peterson PJ. 1984. The accumulation and assimilation of dimethyl selenide by four plant species. *Planta* 160:180–84

175. Zijian W, Lihua Z, Li Z, Jingfang S, An P. 1991. Effect of the chemical forms of selenium on its volatilization from soils in Chinese low-selenium-belt. *J. Environ. Sci.* 3:113–19

Annu. Rev. Plant Physiol. Plant Mol. Biol. 2000. 51:433–62

DIVERSITY AND REGULATION OF PLANT Ca^{2+} PUMPS: Insights from Expression in Yeast

Heven Sze, Feng Liang[1], and Ildoo Hwang[2]
Department of Cell Biology and Molecular Genetics, and Maryland Agricultural Experiment Station, University of Maryland, College Park, Maryland 20742; e-mail: hs29@umail.umd.edu

Amy C. Curran and Jeffrey F. Harper
Department of Cell Biology, The Scripps Research Institute, La Jolla, California 92037

Key Words P-type ATPase, calcium, calmodulin, ER, phosphorylation

■ **Abstract** The spatial and temporal regulation of calcium concentration in plant cells depends on the coordinate activities of channels and active transporters located on different organelles and membranes. Several Ca^{2+} pumps have been identified and characterized by functional expression of plant genes in a yeast mutant (K616). This expression system has opened the way to a genetic and biochemical characterization of the regulatory and catalytic features of diverse Ca^{2+} pumps. Plant Ca^{2+}-ATPases fall into two major types: AtECA1 represents one of four or more members of the type IIA (ER-type) Ca^{2+}-ATPases in Arabidopsis, and AtACA2 is one of seven or more members of the type IIB (PM-type) Ca^{2+}-ATPases that are regulated by a novel amino terminal domain. Type IIB pumps are widely distributed on membranes, including the PM (plasma membrane), vacuole, and ER (endoplasmic reticulum). The regulatory domain serves multiple functions, including autoinhibition, calmodulin binding, and sites for modification by phosphorylation. This domain, however, is considerably diverse among several type IIB ATPases, suggesting that the pumps are differentially regulated. Understanding of Ca^{2+} transporters at the molecular level is providing insights into their roles in signaling networks and in regulating fundamental processes of cell biology.

CONTENTS

[1]Current address: Department of Bioinformatics, The Institute for Genomic Research, 9712 Medical Center Drive, Rockville, Maryland 20850.

[2]Current address: Department of Molecular Biology, Massachusetts General Hospital, and Department of Genetics, Harvard Medical School, Boston, Massachusetts 02114.

040-2519/00/0601-0433$14.00

INTRODUCTION

Regulating Ca^{2+} Dynamics

Calcium is essential for many aspects of plant growth, development, signaling, and survival (70). Regardless of the specific role of Ca^{2+}, the critical issue for a plant and for each cell is to get Ca^{2+} in the right concentration at the right place and at the right time. Ca^{2+} is taken up by plants via the apoplastic pathway; however, unlike most nutrients, it has low physiological mobility, as reflected by little or no movement between cells and through the phloem (70). Thus each growing cell needs to take up any available Ca^{2+} from the transpiration stream. One of the most prominent roles of Ca^{2+} is as a signal transduction element, and the concentration of cytosolic free $[Ca^{2+}]_{cyt}$ is critically important to control many cellular responses. One likely reason for the unique role of Ca^{2+} is that it must be maintained at submicromolar levels (0.1–0.6 μM) in the cytosol, since it precipitates phosphate the energy currency of cells (21). Elevations in $[Ca^{2+}]_{cyt}$ arise from Ca^{2+} entry via Ca^{2+} channels in the plasma membrane, or Ca^{2+} discharge from internal stores, or both (Figure 1). When signals induce the opening of Ca^{2+} channels, the rapid increase of $[Ca^{2+}]_{cyt}$ by tenfold or so would be easily sensed and decoded to give the appropriate response. An astonishing variety of physiological stimuli elevate cytosolic $[Ca^{2+}]$ in plant cells, including cold, red light, drought, Nod

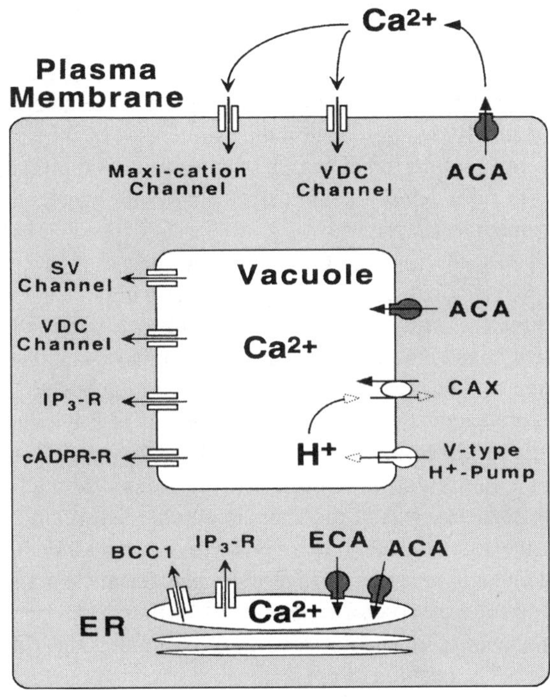

Figure 1 Calcium transport pathways in plant cells. Established calcium transport pathways are shown for energy-coupled transporters (filled arrows) and channels (barbed arrows). Active transporters include H^+/Ca^{2+} antiport (CAX), and two major types of ATP-driven Ca^{2+} pumps. ECA1, an ER-type Ca^{2+}-ATPase (type IIA) has been localized to the ER. ACA, autoinhibited Ca^{2+}-ATPases (type IIB), are found on the PM (ACA8), ER (ACA2), and the tonoplast (BCA1, ACA4) surrounding the vacuole (see text). A proton-calcium exchanger (CAX) is thought to be present in the tonoplast and is energized by a proton gradient generated by a V-type H^+-pumping ATPase. Open arrows indicate the flow of protons. Genes encoding all of the above active transport systems have been identified. Although cloned genes encoding calcium-selective channels have not been verified, many channels have been identified by electrophysiological and biochemical properties in the PM, tonoplast, and ER. These include Maxi-cation channel, nonselective cation channel; VDC channel, voltage-dependent calcium channel; SV channel, slowly activating vacuolar channel, IP_3-R, putative $InsP_3$-activated receptor; and cADPR-R, putative cyclic ADP-ribose activated receptor, BCC1, *Brionica* Ca^{2+} channel. See Reference 88 for further references to calcium channels. Additional transporters and membrane compartments are expected to be involved in calcium signaling and homeostasis but have not yet been well established in plant cells.

factors, gibberellins, and abscisic acid (88). The elevation in Ca^{2+}causes changes in Ca^{2+}-modulated proteins and their targets that act to elicit downstream events in signaling pathways. How each stimulus is distinguished from another has long eluded biologists.

Recent findings in animals demonstrate that Ca^{2+} signals vary in subcellular location, duration, amplitude, and frequency. Ca^{2+} ions entering the cytosol diffuse extremely slowly owing to Ca-binding proteins. Thus the picture of the cytosol as a uniform volume for diffusion is grossly oversimplified, as Ca^{2+} buffering proteins are most likely distributed in a nonuniform manner (21). In neurons, Ca^{2+} can mediate disparate biological effects depending on the location of the Ca^{2+} channels, the route of Ca^{2+} entry, and the cellular context (41). Because various types of Ca^{2+} channels have distinct gating characteristics and kinetics of activation, the influx of Ca^{2+} through different routes is likely to be sensitive to the strength and duration of the stimulation. Downstream effectors can decode information contained in the amplitude and duration of Ca^{2+} signals, as shown by the differential activation of transcription factors in B lymphocytes (28, 29).

Spatially localized and temporally regulated Ca^{2+} signals have been detected in plants [reviewed in (88)]. For instance, Nod factors elevated intracellular Ca^{2+} in the region of the nucleus of root hairs in alfalfa (32). Yet the apex of the root hair in bean is the dominant location for changes in cytoplasmic [Ca^{2+}] within minutes after Nod factor application, with changes near the nucleus appearing later (17). In tip-growing cells, the control of polarized growth is dependent on the establishment and maintenance of a steep tip-focussed Ca^{2+} gradient that oscillates in phase with growth (48). The position of the gradient determines the direction of growth (67). These and other studies support the idea that the kinetics and magnitude of the Ca^{2+} signal, or calcium signature, differ between different stimuli and could contribute to the specificity of the end response (22, 59, 72).

The regulation of any Ca^{2+} transient is determined mainly by two opposing fluxes: Ca^{2+} influx via channels and Ca^{2+} efflux via active transporters. Although the role of active Ca transporters has received less attention than that of channels, the diversity of Ca^{2+} pumps and H^+-coupled Ca^{2+} cotransporters would suggest that these transporters could conceivably participate in determining the overall amplitude, duration, and frequency of Ca^{2+} signals. For example, the frequency of repetitive Ca^{2+} waves induced by $InsP_3$ was increased when SERCA (sarco/endoplasmic reticulum Ca^{2+}-ATPase) pump was overexpressed in frog oocytes (15), indicating that the activity or abundance of intracellular Ca^{2+} pumps could be an important factor in controlling Ca^{2+} oscillations in eukaryotic cells. A current concept is that a change in [Ca^{2+}] due to a stimulus is spatially regulated and occurs in microdomains within the cell. Furthermore, the timing and frequency of a [Ca^{2+}] transient increase results in characteristic forms and patterns. Such localized increases in intracellular Ca^{2+} with distinct forms of spikes, waves, or oscillations could then allow for a large number of cellular responses to be mediated by the activation of specific Ca^{2+}-sensitive biochemical pathways.

In spite of the central issue of transport in regulating intracellular Ca^{2+}, the identification of plant Ca^{2+} transporters at the molecular level has only just begun. Ca^{2+} channels on the plasma membrane, vacuolar membrane, and endomembranes are being identified by electrophysiological and biochemical studies (Figure 1); however, Ca^{2+}-selective channels have yet to be identified at the molecular level (105). In contrast, several Ca^{2+} pumps and one H^+-coupled Ca^{2+} cotransporter have been identified. Here we focus on recent approaches to identify energy-dependent Ca^{2+} transporters functionally, and to determine the regulatory and kinetic properties as a step toward understanding their specific roles in plant growth and signaling. Additional reviews on cellular and biochemical studies (12, 35) and a molecular perspective (40) of active Ca^{2+} transporters are available.

Active Ca^{2+} Transporters: H^+/Ca^{2+} Antiporter and Ca^{2+} Pumps

Active Ca^{2+} transporters are thought to serve several basic functions: (*a*) to replenish Ca^{2+} stores in the cell for release via channels during signaling events; (*b*) to restore cytosolic Ca^{2+} concentration to unstimulated levels, thereby shaping and terminating a Ca^{2+} signal; (*c*) to provide adequate Ca^{2+} and other divalent cations in diverse endolumenal compartments to support specific biochemical reactions, and (*d*) to provide Ca^{2+} required for dynamic membrane interactions, including vesicle trafficking and fusion.

Two types of active transporters drive Ca^{2+} out of the cytosol against the steep Ca^{2+} electrochemical gradient at the PM (plasma membrane) and across the endomembranes: (*a*) H^+-coupled Ca^{2+} antiporter that is driven by a proton electrochemical gradient, and (*b*) Ca^{2+} pumping directly energized by ATP hydrolysis (Figure 1). Both activities were initially detected as ATP-driven $^{45}Ca^{2+}$ uptake into isolated/purified membrane vesicles, though only the H^+-coupled Ca^{2+} antiporter is dependent on a ΔpH. The two types of transporters differ significantly in their kinetic properties. The H^+-coupled Ca^{2+} antiport is a low-affinity ($K_{mCa} = 10$–15 μM) (89), and a high-capacity transporter, whereas Ca^{2+} pumps generally have a high-affinity for Ca^{2+} ($K_m = 0.1$–2 μM) and low capacity. This difference would suggest that H^+-coupled Ca^{2+} antiporters are particularly important for removing cytosolic Ca^{2+} when concentrations are high.

An H^+/Ca^{2+} antiport activity is found most commonly in the vacuolar membrane, though there is evidence that it is also found in the plasma membrane (57). The stoichiometry of $3H^+/1Ca^{2+}$ was estimated for the vacuolar antiporter, implying that the exchanger is competent to drive Ca^{2+} accumulation within the vacuole (6). In the vacuolar membrane, the proton gradient can be established by either the H^+-pumping vacuolar-type ATPase or the H^+-PPase, or both. Thus the H^+/Ca^{2+} antiport activity can be easily distinguished in native membrane vesicles by its sensitivity to bafilomycin, a specific blocker of the V-ATPase, or by proton ionophores that dissipate the ΔpH gradient, and by insensitivity to vanadate, a P-type ATPase inhibitor. The properties of the H^+/Ca^{2+} antiporter can be studied directly by driving Ca^{2+} uptake with an artificially generated pH gradient (acid inside) in isolated vesicles (89). Using this approach, DCCD, La^{3+}, and ruthenium red were found to block the exchange activity from oat roots. However, unlike pumps, the antiporter lacked an associated enzyme activity, which makes the protein difficult to identify after membrane solubilization (90).

The first plant H^+/Ca^{2+} antiporter (AtCAX1, accession number U57411) was cloned by its capability to suppress the Ca^{2+} hypersensitivity of the yeast mutant K665 (*vcx1pmc1*) (47). This yeast strain is defective in vacuolar Ca^{2+} accumulation due to the disruption of both the Vcx1 H^+/Ca^{2+} antiporter (25) and the Pmc1 Ca-pumping ATPase (24), and is intolerant of high levels of exogenous Ca^{2+} (200 mM) (25). AtCAX1 encodes a polypeptide of 459 amino acids belonging to the same family of membrane proteins as microbial H^+/Ca^{2+} antiporters. These proteins contain a central hydrophilic, acidic motif that bisects the polypeptide into two groups of five or six transmembrane spans. Experiments on vacuolar membrane vesicles isolated from yeast expressing AtCAX1 demonstrated that the gene encodes a H^+/Ca^{2+} antiporter with a V_{max} of 12 nmol/mg protein min, and a K_{mCa} of 13 μM, which is similar to the affinity determined with the native plant membranes. It is also not known whether the K_m or V_{max} of this transporter can be shifted by modification of regulatory domains or interaction with other proteins, as the H^+/Na^+ antiporter from animals (104).

A critical role of the antiporter in normal plant growth and adaptation to stress was demonstrated recently. Transgenic plants overexpressing AtCAX1 showed increased sensitivity to cold stress and symptoms of Ca^{2+} deficiency, including leaf necrosis and reduced root mass (46). Exogenous Ca^{2+} (2 mM) enhanced AtCAX1 RNA in Arabidopsis, indicating that the vacuolar-localized antiporter can be regulated transcriptionally. These results also show that antiport activity influenced Ca^{2+} distribution in the plant and perhaps within cells.

FUNCTIONAL AND MOLECULAR CLASSIFICATION OF PLANT Ca^{2+} PUMPS

Two Types of Ca^{2+} Pumping Activity: IIA (ER-Type) and IIB (PM-Type)

Biochemical studies starting from the late 1970s through the late 1990s illustrate the multiplicity of Ca^{2+} pumps in plants. A working hypothesis prevailing in the 1980s was that one type of Ca^{2+} pump was associated with one specific cell membrane. Thus a common approach was to purify a specific membrane, and study the associated Ca^{2+}-ATPase activity or pump activity [see reviews by (12, 35)]. ATP-dependent $^{45}Ca^{2+}$ uptake into native vesicles is a specific and sensitive measure of pump activity. This simple assay is preferred over monitoring Ca^{2+}-dependent ATP hydrolysis, because Ca^{2+}-ATPases are estimated to represent only <0.1% of the membrane protein (19), and thus are 30–100-fold-less abundant than H^+-ATPases in the PM (3%) and the endomembranes (5–10%) (98). The approach has the advantage of working with a defined membrane, and led to the identification of a plasma membrane-bound Ca^{2+} pump (86). However, all Ca^{2+} pumps are inhibited by orthovanadate, indicating they form a phosphorylated intermediate and

TABLE 1 Multiple Ca^{2+}-ATPases detected in isolated membranes of selected plants*

	Ca^{2+} Transport activity or mol. mass of Ca-ATPase				
	PM-type (IIB)			ER (IIA)	
Plant material	**PM**	**E/ER**	**Vacuole**	**ER**	**Reference**
A. *CaM-stimulated Ca^{2+} transport activity*					
Bryonia tendrils	No	Yes	Yes	No	65
Wheat aleurone	Yes	Yes	Yes	No	14
Corn coleoptile/roots	—	Yes	Yes	—	8, 9, 39, 80
Barley aleurone/roots	—	Yes	Yes	—	30, 43
Carrot suspension cells	Yes	Yes	Yes	No	50, 54
Red beet root	Yes	—	—	No	42
B. *Mol. mass of Ca-ATPase in kDa*					
Carrot suspension cells	127	120	—	~120(PE)	50, 54
Cauliflower florets	116	—	111	—	3
Radish seedling roots	124-133	—	—	—	86
Red beet	124 (PE)	—	—	119 (PE)	100
Barley leaf/roots	130	116	—	116 (PE)	26
Maize coleoptile	—	140 (PE)	—	—	9, 10, 99

*Calmodulin-stimulated (ΔCaM) Ca^{2+} transport detected in membranes of the vacuole (Vac), ER and/or endomembrane (ER/E), and the plasma membrane (PM). PM-type (IIB) Ca-ATPases were identified by purification, calmodulin-binding, or reactivity with antibodies against Ca^{2+} pumps. ER-type (IIA) activity was determined using ER vesicles or using biochemical traits. Both types of pumps form a phosphoenzyme (PE). Ca^{2+} transport is either stimulated by CaM (yes) or not by CaM (no); –not determined.

are P-type or E_1E_2 transport ATPases. Because of their similarities to one another, the properties and the relationships among the pumps are often difficult to discern. An alternative approach was to distinguish the type of Ca^{2+} pumps according to functional/ biochemical activities, and then determine the membrane location (54). The latter approach allowed one to detect one or more pump types localized on any membrane, and provided the initial clue that two distinct pumps may reside on one organelle (Table 1).

Two types of Ca^{2+} transport activities emerged from biochemical studies. They are referred to as PM-type and ER-type according to the model of animal Ca^{2+} pumps (16, 66). In the absence of molecular information, these terms are useful in distinguishing pump types based on the biochemical and regulatory properties of the Ca^{2+} transport or Ca^{2+}-ATPase only, regardless of the actual membrane location. The primary distinction is that PM-type is stimulated by calmodulin analogous to animal PM Ca^{2+}-ATPases; however, in plants these pumps are localized on the PM as well as other membranes. In contrast, ER-type Ca^{2+} pumps are insensitive to calmodulin, and sensitive to cyclopiazonic acid (CPA) (50, 54), a diagnostic blocker of animal S/ER Ca^{2+} pumps (82, 92). The CPA concentration (100 nmol/mg membrane protein) required to completely inhibit the CPA-sensitive activity in plant membranes is similar to that needed to block 90% of SERCA activity. However, sensitivity to thapsigargin at concentrations that block animal SERCAs has not been observed to date in plant membranes (19, 100).

ER-Type Ca^{2+} Pumps

The total Ca^{2+} content in the ER lumen of eukaryotic cells is relatively high (5–10 mM), though most of the Ca^{2+} is buffered by lumenal binding proteins, including calreticulin (23). The free [Ca^{2+}] is estimated to range from 0.1 to 1 mM in animal ER/SR (endoplasmic reticulum/sarcoplasmic reticulum) (73), indicating that Ca^{2+} is actively transported there. However, an ER-type Ca^{2+}-ATPase has not been purified from any plant, so this type of pump was poorly understood before molecular studies began. Several important traits were determined by biochemical studies of $^{45}Ca^{2+}$ uptake into purified ER vesicles with presumably right-side-out orientation. In several studies, Ca^{2+} pumping was insensitive to calmodulin (11, 14, 42) and was enhanced several fold (4 to 10 times) by oxalate (14, 54). In one case, Ca^{2+} transport was partially (25%) blocked by cyclopiazonic acid at concentrations that inhibit animal SERCA (50, 54). Because ATP is preferentially hydrolyzed relative to GTP, Ca^{2+} transport from ER-type pumps can also be distinguished from PM-type pumps by assaying for ATP-specific activity (54). However, in several studies, CaM-stimulated Ca^{2+} pump activities were detected in ER fractions (see Table 1). These results suggested that both pump types could be localized on intracellular membranes, like the ER.

A molecular mass of 116–120 kDa was estimated from the acyl [γ^{32}P] phosphoenzyme formed by ER-type Ca^{2+}-ATPases associated with native membranes (Table 1). Ca^{2+}-pumping ATPases belong to the P (phosphorylated)-type or E_1E_2-type transport ATPases. During the reaction cycle, there are two distinct functional states of the enzyme (E_1 and E_2) according to the model based on SERCA pumps (27, 75). In the E_1 state, two high-affinity Ca-binding sites (K_m ~0.2–2 μM) face the cytoplasmic side. E_1 is able to capture Ca^{2+} at submicromolar concentrations in the cytosol, and activates the ATP binding site on E_1. ATP binds and is hydrolyzed, and the terminal P is transferred to an Asp residue to form an acyl-P bond (E_1-P). As a result of the formation of an acyl phosphate bond, the protein undergoes a conformational change to E_2-P. In this conformation, the two Ca^{2+} ions are occluded and can no longer exchange with cytosolic Ca^{2+}. The transition to E_2-P is accompanied by a drop in Ca^{2+} affinity by three orders of magnitude (K_m ~ 1–3 mM), so Ca^{2+} ions are weakly bound to sites facing the exoplasmic side of the pump. The Ca^{2+} ions dissociate, followed by the hydrolysis of the Asp-phosphate bond, causing E_2 to revert to E_1. Thus the formation of Ca^{2+}-dependent phosphoenzyme reflects activity of a Ca^{2+}-pumping ATPase from either type of Ca^{2+} pump.

Broad Distribution of Calmodulin-Stimulated Ca^{2+} Pumps

The roles and distribution of calmodulin-stimulated Ca^{2+} pumps in plants have long perplexed plant biologists. Instead of being exclusively localized to the plasma membrane as in animal cells, these pumps appear in several endomembranes as well. Biochemical studies suggested that plants had several IIB Ca^{2+} pumps, and subsequent molecular studies confirm and extend this idea. Table 1 shows examples of calmodulin-stimulated Ca^{2+} transport activity in the vacuole, the ER, and the PM of a single plant material. Association with the ER was usually demonstrated by cofractionation on a sucrose gradient with an ER-marker, such as NADH-cyt c reductase, as well as a characteristic density shift after stripping ribosomes off the RER using EDTA. Ca^{2+} pumping in vesicles of low density, or in isolated vacuoles, indicated a vacuolar association (4). However, the broad distribution of activity observed on sucrose gradients (54) suggested the possibility of calmodulin-regulated Ca^{2+} pumps on other unidentified membranes, including the prevacuolar compartment, ER-derived compartments, and transport vesicles or compartments.

Calmodulin-stimulated Ca^{2+} pumping is frequently detected in plant materials with young proliferating cells, such as seedlings and suspension-cultured cells (Table 1). This observation agrees with the higher levels of calmodulin found in shoot and root apices compared with more mature regions (108). Several calmodulin-stimulated Ca^{2+}-ATPases have been identified and characterized after partial purification by calmodulin-affinity chromatography. A few examples are given below.

Vacuolar A vacuolar Ca^{2+}-ATPase of 111 kD was first identified from cauliflower florets by phosphoenzyme formation after partial purification and reconstitution (5). The cauliflower Ca^{2+} ATPase showed relatively low affinity for calmodulin, as activity was 50% stimulated by calmodulin at 0.1–0.2 μM (5). Interestingly, there is no evidence for a related Ca^{2+}-ATPase in purified ER of cauliflower inflorescence based on immunostaining with antibody against the vacuolar Ca^{2+} pump (4).

Endomembrane Another calmodulin-stimulated Ca^{2+}-ATPase was partially purified and identified as a 120-kD phosphoenzyme from the endomembranes/ER of carrot suspension cells (54) (see Table 1 for other examples). The vacuolar and the ER pumps showed similar affinity for CaM ($A_{50} = 0.2\ \mu$M), and appeared to be immunologically related inasmuch as the antibody against the cauliflower Ca^{2+}-ATPase recognized the purified carrot protein of 120 kD. Recent molecular studies support the idea for related, but distinct, Ca^{2+} pumps at the ER and at the vacuole (49, 69; see below).

PM PM-bound Ca^{2+}-ATPases are usually 5–14 kD larger (116–133 kD) than those found on intracellular membranes (111–120 kD) from the same plant material (see Table 2). In these studies, the PM was purified either by two-phase partitioning or by sucrose density gradient. The enzyme purified from radish and barley plasma membrane by calmodulin-affinity chromatography are 124–133 kDa and 130 kD, respectively (7, 26). In some cases, Ca^{2+}-ATPase was detected in the PM without enzyme purification by several methods, including calmodulin-binding,

TABLE 2 Multiple Ca^{2+} pumps in plants fall into two major groups*

Source/Gene	Protein size # aa(kDa)	Location	Accession numbers cDNA (protein)	Accession numbers Genome (protein)	Reference
IIA (ER-TYPE)					
Arabidopsis thaliana					
AtECA1/AtAC3	1061 (116)	ER, Golgi	U96455 (aac68819)	I, AC007583	61, 62
			U93845 (aab52420)		
AtECA2/AtACA5	1054(116)	?	AJ132387 (caa10659)	IV, AF013294 (aab62850)	81
AtECA3/AtACA6	998 (109)	?	AF117296 (aad29961)	I, AC004122 (aac34328)	61, 62
			AJ132388 (caa10660)		81
AtECA4	1061	?	AF117125 (aad29957)	I, AC007583	61, 62
Other Plants					
LCA1 (tomato)	1048 (116)	Vac, PM	M96324 (aaa34138)		106
	(120)		AF050495 (aad11617)		76
OsCA1 (rice)	1048 (115)	?	U82966 (aab58910)		20
DbCA1 (Dunaliella)	1037 (114)	?	X93592 (caa63790)		83
IIB (PM-TYPE)					
Arabidopsis thaliana					
AtACA1/PEA1	1020 (111)	Plastid envelope	L08468 (aad10211)	I, L08469 (aad10212)	51
AtACA2	1014 (110)	ER	AF025842 (aac26997)	IV, AL035605 (cab3803)	44
AtACA4	1030 (113)	Vacuole	AF200739	II, AC002510 (aab84338)	M Geisler, personal communication
AtACA7	1015 (111)	?		II, AC004786 (aac32442)	
AtACA8	1074 (116)	PM	AJ249352	V, AB023042 (?)	MI De Michelis, personal communication
AtACA9	≥119	?		III, AB023045 (?)	
AtACA10	1093 (117)	?	T43417 (est)	IV, AL050352 (cab43665)	
Other Plants					
BCA1 (cauliØower)	1025 (111)	Vac	X99972 (caa68234)		69

*Type IIA refers to predicted sequences that share higher identity with SERCA (50±56%) than to PMCA (30%). Type IIB refers to sequences that are more related to mammalian PMCA (54%) than to SERCA (31% identity). ER, endoplasmic reticulum; Vac, vacuolar membrane; PM, plasma membrane; ?, not yet determined. Roman numeral preceding genomic accession number refers to [illegible] immunoreactivity with antibodies against plant PM-type Ca^{2+} pumps, formation of a Ca^{2+}-dependent phosphoenzyme, or labeling with fluorescein isothiocyanate (FITC). FITC inhibits several plant PM Ca^{2+}-ATPase at low concentration (86), most probably by binding with a Lys residue near the ATP-binding site of P-type ATPases (16).

The PM-localized Ca^{2+} pump may differ in stoichiometry from related pumps on the endomembrane. Because plant cells have highly negative membrane potential (-120 to -160 mV inside) across the PM compared to that ($\geq +30$ mV inside) across the vacuolar membrane, thermodynamic constraints would suggest a minimum stoichiometry of $2H^+$ entering the cell for each Ca^{2+} pumped out by the PM Ca^{2+}-ATPase of plants, as in fungi (74). Indeed, Ca^{2+} pumping into presumably inside-out PM vesicles stimulated decay in the ΔpH (acid inside) when the H^+-pumping ATPase was blocked. The results provide support that the PM-bound pump mediates a nH^+/Ca^{2+} exchange (87).

Molecular Evidence for Two Pump Types: IIA and IIB

The molecular identification of Ca^{2+} pumps became imperative given their multiplicity. Genes thought to encode Ca^{2+} pump homologues have been isolated or identified from a variety of plants since 1992 (106), though their functional activities were not determined until 1997. The gene products can be classified as IIB (PM-type) or IIA (ER-type) based on their protein sequence similarity to animal PMCA (16) or SERCA (40, 66), respectively (Table 2). Plant ER-type IIA homologs share 50–56% identity with animal SERCA, but only 28-33% with PMCA (40, 64, 81). In contrast, plant type IIB proteins share 54% identity with hPMCA4, and only 31% with SERCA. Unlike animal PMCA, plant type IIB homologues lack a long carboxyl tail; rather they have an unusually long amino terminal region (Figure 2). *Arabidopsis thaliana* has at least four and seven genes encoding homologues corresponding to IIA- and IIB-type ATPases, respectively (Table 2). To understand the role of each, it is necessary to determine the specificity of the cation translocated, the affinity for the cation, and the mode of regulation. The presence of multiple genes indicates that expression of each pump could be developmentally regulated in a tissue-, cell-, and membrane-specific manner. Apart from a few Ca^{2+}-ATPases (e.g. AtECA1, AtACA2, AtACA4, and BCA1), many of the gene products listed in Table 2 have yet to be characterized at the biochemical or cellular level.

EXPRESSION OF PLANT Ca^{2+} PUMPS IN YEAST

Two Ca^{2+} Pumps in Wild-Type Yeast

Interestingly, wild-type *Saccharomyces cerevisiae* has only two Ca^{2+} pumps, PMR1 at the Golgi and PMC1 on the vacuolar membrane (Figure 3), according to predictions based on the complete inventory of P-type ATPases in the genome (18).

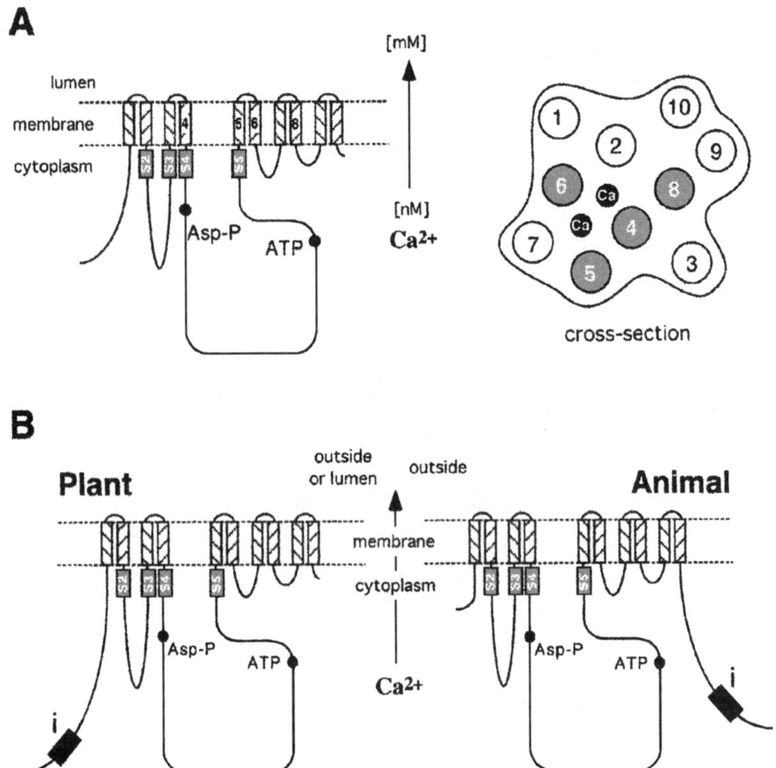

Figure 2 Topology of Ca^{2+}-pumps. *A*. The predicted topology of a Type IIA pump. The diagram to the left shows a ten-transmembrane topology that is supported by structural studies on a sarcoplasmic reticulum calcium pump (107). One possible arrangement of these transmembranes is shown in a cross-section to the right. Genetic evidence has implicated charged residues in transmembranes 4, 5, 6, and 8 as essential for translocation of calcium through the membrane. An 8-Å structure supports the hypothesis that a stalk is formed by the extension of alpha helices from transmembranes 2–5 (shown as gray boxes marked S2–S5). The large central cytoplasmic loop contains an ATP binding site and the aspartate that becomes phosphorylated during the reaction cycle (marked by filled circles). *B*. The predicted topologies of Type IIB pumps identified in plants and animals. The general features of Type IIB pumps are expected to be similar to a Type IIA shown above. The black rectangles marked "i" indicate the two alternate positions of autoinhibitor sequences identified in plant and animal pumps.

Pmr1 is implicated in intracellular transport of Ca^{2+} and Mn^{2+} into the Golgi and perhaps ER. A mutant in which the *PMR1* gene is displaced grows poorly on Ca^{2+}-deficient medium and high (3 mM) Mn^{2+}-containing medium (60). Wild-type yeast is able to grow on media containing $\leq 1\ \mu M\ Ca^{2+}$ by sequestering Ca^{2+} into endomembrane compartments for various functions of the secretory pathway (2, 31).

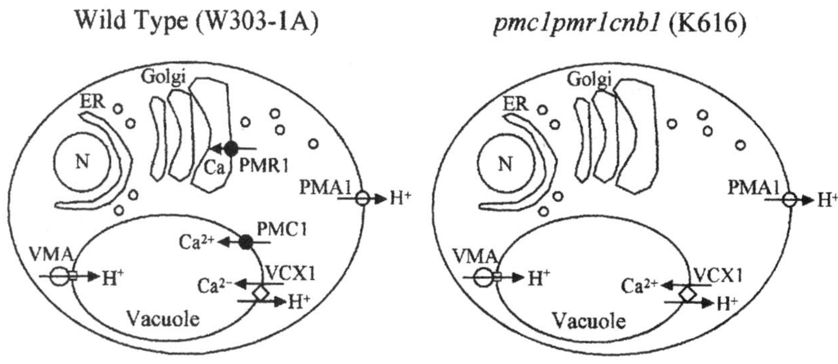

Strain	Relative Growth on Various Media		
	1 mM Ca	10 mM EGTA	3 mM Mn
Wild type	+++++	+++++	+++++
pmr1 AA542	+++	+	+
K616	+++	+	+
K616+AtECA1	++++	++++	+++
K616+AtACA2	++++	+	+
K616+AtACA2ΔN	++++	++++	+

Figure 3 Ca^{2+} transporters of wild-type yeast and in K616 mutant wild-type and mutant yeast grow on SC medium containing 1 mM Ca^{2+}; however, *pmr1* mutant and K616 mutant grow very poorly on SC medium in Ca^{2+}-depleted medium (supplemented with 10 mM EGTA) or in Mn^{2+}-containing medium. +AtECA1 or + AtACA2, indicates K616 expressing Arabidopsis ECA1 or ACA2. PMR1, Golgi Ca^{2+} pump; PMC1, vacuolar Ca^{2+} pump; VCX1, H^+/Ca^{2+} antiport; VMA, vacuolar H^+-ATPase; PMA1, plasma membrane H^+ pump; N, nucleus; *cnb1,* calcineurin subunit B gene mutated.

Pmr1 shows little identity (~35%) with animal SERCA or AtECA1. Moreover, Pmr1 Ca^{2+} pump is insensitive to both thapsigargin and the mycotoxin, cyclopiazonic acid (95). PMC1 encodes a protein most similar to a mammalian PMCA, although a calmodulin-binding domain is absent at the C-terminal domain. Genetic and transport studies suggest that Pmc1 sequesters Ca^{2+} into the vacuole and plays a role in tolerating high extracellular Ca^{2+}. Pmc1 null mutants grow as well as wild-type cells on media containing low levels of calcium (<1 μM Ca^{2+}); but the mutants fail to grow in medium containing high levels of calcium (200 mM Ca^{2+}) (24).

Although *pmr1 pmc1* mutants, in theory, would be ideal for expression of heterologous genes encoding Ca^{2+} pumps, the double mutants are not viable unless calcineurin is inactivated. *CNB1* encodes a regulatory subunit B of calcineurin. The triple mutant (*pmr1 pmc1 cnb1*) can grow on medium containing high (≥1 mM) Ca^{2+}, and ΔpH-dependent Ca^{2+} uptake is more active in vesicles isolated from this K616 strain than from that of wild type. It is still not clear how calcineurin regulates the H^+/Ca^{2+} antiporter, Vcx1, since the transcriptional factor, activated by calcineurin through direct dephosphorylation (97), enhances the expression of Ca^{2+} pumps or Na^+ ATPases (PMC1, PMR1, and PMR2), but not the expression of Vcx1 antiporter (71, 96). A mutant lacking all three active Ca^{2+} transporters (*pmr1 pmc1 vcx1*) is not viable under any conditions tested so far.

Expression in Yeast Mutant K616 The first functional expression in yeast mutants of a plant Ca^{2+} pump, AtECA1 (62), opened the way to a genetic and biochemical characterization of other putative Ca^{2+} pumps. Expression of *AtECA1* in either *pmr1* mutant (strain AA542) or K616 restored mutant growth on media with only 1 μM Ca^{2+} (62) (Figure 3), providing the first genetic evidence that *AtECA1* encodes a Ca^{2+} pump. The triple mutant strain K616 has several advantages and is now widely used to express plant genes. First, since the mutant lacks any calcium pumps, heterologous pumps can be expressed and characterized without the background activity from endogenous pumps (53, 62, 63). Second, the mutant provides a powerful genetic complementation system for studying the functional and regulatory domains of heterologous Ca^{2+}-ATPase, if the expressed pump targets to the ER/Golgi system.

In spite of these advantages, Ca^{2+} transport from the expressed AtECA1 pump is often masked by the upregulated activity of the yeast Vcx1 H^+/Ca^{2+} antiport due to loss of calcineurin function in K616 mutant (62, 63). In other cases, overexpression of ACA2 in K616 can lead to a compensatory decrease in antiport activity (53). Several strategies were used to separate pump activity from the H^+-coupled Ca^{2+} antiport using native yeast vesicles. First, the overexpressed pump is associated mainly with intracellular membranes, so vesicles were collected with a simple step gradient at the 26/45% (w/w) sucrose interface to minimize vacuolar membranes. Second, pump activity was specifically measured after inhibition of antiport activity with Bafilomycin A_1 and gramicidin to block the vacuolar H^+-ATPase, and dissipate any residual pH gradient, respectively. Third, the free Ca^{2+} concentration was buffered to 0.1 μM to reduce activity from the relatively low-affinity H^+/Ca^{2+} antiport (K_m = 0.4 to 10 μM). To specifically measure calmodulin-stimulated Ca^{2+} transport by type IIB ATPases, it is important to measure pump activity as that inhibited by orthovanadate (200 μM) to eliminate a significant level of calmodulin-dependent Ca^{2+} binding to membranes (53).

AtECA1 Encodes an ER-Type Ca^{2+} Pump

Transport Activity of AtECA1 Biochemical characterization of AtECA1 expressed in yeast membranes provided direct evidence that it was a Ca^{2+} pump: (*a*) The high-affinity Ca^{2+} pump is blocked 50% by cyclopiazonic acid at 3 nmol/mg protein (~0.1 μM); (*b*) transport is insensitive to calmodulin; and (*c*) AtECA1 protein is localized to the ER, and perhaps Golgi, of Arabidopsis plants, as shown by Western blotting of membranes fractionated with a sucrose gradient (62, 63) (Table 3). Thus the properties of AtECA1 pump are consistent with those described for ER-type Ca^{2+} transport in native plant membranes (50, 54).

Moreover, two lines of evidence suggested that AtECA1 transports Mn^{2+} into intracellular compartments. First, AtECA1 expression restored growth of *pmr1* mutants on Mn^{2+}-containing medium. Second, Mn^{2+} as well as Ca^{2+} stimulated the formation of an acyl phosphoenzyme in yeast membranes expressing AtECA1 (62). In contrast, another Ca^{2+} pump, AtACA2, is unable to restore growth of mutant K616 or *pmr1* on media supplemented with 3 mM Mn^{2+} (Figure 3). A proposed function of AtECA1 is to supply Mn^{2+} for Mn^{2+}-activated enzymes involved in protein processing in the Golgi (58).

Homologues of AtECA1 At least three homologues of AtECA1 have been identified in Arabidopsis (64, 81) (Table 2). AtECA3 shows the highest similarity (56%) to mammalian SERCA. AtECA1 is highly expressed in roots, shoots, and flowers, as detected by RNA gel blot (61; B Hong & JF Harper, unpublished data). Transcript levels of AtECA2 and AtECA3 appear to be relatively low in whole plants, as detected by RT-PCR (81). AtECA2 and LCA1 share the highest identity (78%)

TABLE 3 Two distinct Ca pumps on the ER of Arabidopsis: AtECA1 and AtACA2*

Properties	AtACA2 (IIB)	AtECA1 (IIA)
Molecular Mass	110 kDa	116 kDa
Ca^{2+} affinity ($K_{1/2}$)	0.6 μM (full-length)	0.03 μM
Deregulated	0.2 μM (ΔN2-80)	
Activator: Calmodulin	Act_{50} = 35 nM (full-length)	No
$K_{m\ ATP}$: High-affinity	0.1 mM	<20 μM
Low-affinity	1.3 mM	260 μM
Phosphoenzyme (PE)	Weak PE with La	Strong PE $\pm$ La (Asp-P)
Inhibitor		
Orthovanadate	I_{50} (40 μM)	I_{50} (1.5 μM)
Erythrosin B	I_{50} (0.4 μM)	I_{50} (0.5 μM)
Cyclopiazonic acid	No effect (3 μM)	I_{50} (3 nmol/mg or 0.1 μM)
Thapsigargin	No effect (0.1 μM)	No effect (1 μM)

Reference; AtECA1, (62), (61); AtACA2, (44), (53).

*Activities were determined using membrane vesicles isolated from yeast K616 cells transformed with AtECA1 or AtACA2. Transport by AtACA2 was determined with the N-terminal truncated (ΔN2-80) or deregulated pump unless otherwise indicated. Act_{50} refers to the concentration of the calmodulin required to stimulate activity 50% of the full-length pump (ACA2-1). I_{50} refers to the concentration of the inhibitor required to activate or block 50% of the control activity. Cyclopiazonic acid concentration is expressed as nmole per mg membrane protein.

among the type IIA pumps from plants (Table 2), and may therefore be functional orthologs. However, localization of LCA1 to the tonoplast and the plasma membrane by sucrose gradient fractionation and immunostaining (38) is surprising, and needs to be confirmed by other approaches.

Biochemical Properties of AtECA1 (IIA) and AtACA2 (IIB)

A type IIB pump, AtACA2, was similarly characterized after expression in K616 strain (Table 3). In contrast to the ER-type AtECA1, the full-length ACA2 pump is stimulated by calmodulin, and is insensitive to the SERCA inhibitor, cyclopiazonic acid (53). Furthermore, it is extremely difficult to detect the steady-state phosphoenzyme (PE) level formed by AtACA2 even with La^{3+} (I Hwang, unpublished data), whereas the PE level of AtECA1 is prominent without La^{3+} (62). La^{3+} also stimulates PE formed by the animal PMCA by blocking the conversion of E_1P to E_2P, thus preventing dephosphorylation (16). The differential effect of La^{3+} on the steady-state PE of plant ER-type and PM-type Ca-ATPases is not understood, even though the velocity of the AtECA1 and AtACA2 pumps are comparable.

REGULATION OF TYPE IIA PUMPS

Transcriptional Regulation

Multiple ECA genes (at least four) in *Arabidopsis thaliana* would suggest that expression of distinct pumps is spatially or temporally regulated. In tomato, *LCA1* is unlikely to be a single gene (38) as several homologues of ECA are found in the tomato EST database (www.tigr.org/tdb/tgi.html). Yet *LCA1* gives rise to one transcript in leaves (106) and two LCA1 transcripts in phosphate-starved roots (76). Two transcripts of 4.7 and 3.6 kb correspond to mRNA that originated from transcription initiation sites −1392 and −72. The corresponding cDNAs, LCA1A and LCA1B, showed 100% sequence identity except for the differential 5′-UTR. The significance of two mRNAs derived by differential transcription initiation is not clear.

Modulation of Activity

We know almost nothing about how plant type IIA pumps are regulated. It is tempting to speculate that plant AtECA pumps are modulated by an unidentified regulatory protein(s) for two reasons. First, AtECA1 is a high-affinity Ca^{2+} pump, and second, its activity is blocked by a synthetic peptide (7 μM) corresponding to an autoinhibitory domain of AtACA2 (53). A working model is that AtECA1 activity (K_m or V_{max}) is modulated by a reversible interaction with a regulatory protein(s). Cardiac SERCA1 pump activity is modulated by a small transmembrane protein of 52 amino acids, phospholamban (PLB). Dephosphorylated PLB monomers inhibit SERCA2a by binding to cytoplasmic and membrane domains of the pump-stabilizing the enzyme in the E_2 conformation, which results in enzyme inactivation by lowering Ca^{2+} affinity (94). Phosphorylation of PLB at a Ser or Thr residue in the cytoplasmic region reverses pump inhibition favoring association of PLB monomers to pentamers. Proteins possessing an amphipathic α-helix, like phospholamban or sarcolipin (77), are attractive candidates for regulating plant ECAs, as a domain of SERCA2 ($KDDK^{400}$) interacting with phospholamban (55, 101) appears to be partially conserved in AtECA1.

REGULATION OF PM-TYPE (IIB) Ca^{2+} PUMPS

Removal of an Autoinhibitory Region Activates Ca^{2+} Pumps at the PM and the Vacuole

Biochemical studies first showed that a PM-localized Ca^{2+} pump from plants is regulated by an autoinhibitory domain. The activity of a 133-kD PM ATPase from radish is stimulated twofold by calmodulin. The PM ATPase is also activated by proteolysis in the absence of calmodulin. Trypsin cleavage yielded a protein of 118 kD that no longer bound calmodulin (84–86). Similarly, a vacuolar Ca^{2+}-ATPase from cauliflower was activated by trypsin in the absence of calmodulin (3). The velocity of the enzyme increased threefold, and the affinity for Ca^{2+} increased from a K_m of 2 μM to 0.5 μM. Proteolysis cleaved the protein from 111 to ~100 kDa, and removed calmodulin binding activity. These studies support a model in which both the PM- and the vacuolar Ca^{2+}-ATPase are autoinhibited by an intramolecular interaction. Calmodulin binding releases the inhibition, similar to the removal of the autoinhibitory region by trypsin. However, these biochemical studies did not resolve the question of whether the autoinhibitory sequence and the calmodulin binding region in the plant pumps resided at the carboxyl terminal domain, as they do in the animal PMCAs (78).

N-Terminal Autoinhibitory and Calmodulin-Binding Domains

In contrast to the known examples of the animal type IIB pumps, two independent studies provided compelling evidence that IIB pumps from plants are regulated by an N-terminal autoinhibitory domain. First, *BCA1* encoded a protein with homology to animal PMCA, although it had an unusually long amino terminal domain. *BCA1* was cloned using peptide sequences obtained from a purified Ca^{2+} pump from cauliflower vacuoles (5, 69). A synthetic peptide corresponding to Ala^{19}-Leu^{43} bound calmodulin providing a provocative suggestion for a regulatory domain at the N-terminal region (69). While the protein encoded by *BCA1* has not been functionally expressed, the correspondence of its predicted protein sequence to actual peptide sequences obtained from a purified and well-characterized enzyme provide strong evidence that BCA1 does encode a calmodulin-regulated pump.

Second, the functional expression of *AtACA2* in yeast K616 revealed several interesting regulatory domains that are being studied by molecular genetic and biochemical methods. *AtACA2* was cloned by homology to P-type Ca^{2+} pumps. The protein had a hydrophilic amino domain that was 76 residues longer than the human PMCA4 (Accession number M25874). Experiments have revealed that the N-terminal domain includes (*a*) an autoinhibitory region, (*b*) a calmodulin-binding sequence, and (*c*) a Ser^{45} residue that when phosphorylated can block calmodulin activation.

Autoinhibitory Domain Expression of *AtACA2-1* in yeast K616 failed to restore growth on Ca^{2+}-depleted medium ($\leq 1\ \mu M$), giving the first clue that this Ca^{2+}-ATPase was autoinhibited (44). The full-length protein (AtACA2-1) showed low Ca^{2+}-ATPase and Ca^{2+} transport activities in vitro. In contrast, a truncated AtACA2-2 protein lacking residues 2–80 was four- to tenfold more active, and its expression in K616 strain increased the growth rate of mutant to that of wild-type yeast (44, 53). An inhibitory sequence was localized to a region within Val^{20}-Leu^{44} as a peptide corresponding to this sequence lowered the V_{max} twofold and increased the K_m for Ca^{2+} (two- to threefold) of the constitutively active AtACA2-2 to values comparable with the full-length pump (Table 3) (53). Mutational analyses are providing insights about the intramolecular interactions that result in inactivation.

Calmodulin Binding Region The full-length AtACA2 enzyme or pump is activated two- to fourfold by calmodulin; however, the deregulated truncated protein (AtACA2-2) is fully active and insensitive to calmodulin. A calmodulin-binding site was identified between Val^{20}-Arg^{36}, as shown by the calcium-dependent binding of calmodulin to a series of fusion proteins from the N-terminal domain (44). Thus calmodulin activation of AtACA2 involves an interaction with an autoinhibitory domain (within Val^{20}-Leu^{44}) (53).

Phosphorylation at the N-Terminal Domain Blocks Calmodulin Activation of AtACA2

Both BCA1 and AtACA2 contain potential phosphorylation sites at their N-terminal domains. However, because regulatory regions are not well conserved, type IIB pumps may be differentially regulated by protein kinases and phosphatases. When AtACA2 was phosphorylated by a Ca^{2+}-dependent protein kinase (CDPK) in vitro, the phosphorylated pump showed a 70% decrease in calmodulin stimulation (I Hwang, H Sze & JF Harper, submitted). A phosphorylation site was mapped to Ser^{45} near a calmodulin-binding site. In a full-length pump, an Ala substitution of Ser^{45} (S45/A) completely blocked the observed CDPK-inhibition of both basal and calmodulin-stimulated activities. An Asp substitution (S45/D) mimicked phospho-inhibition, indicating that a negative charge at this position is sufficient to account for phospho-inhibition. Interestingly, prior binding of calmodulin blocked phosphorylation. This suggests that once ACA2 binds calmodulin, its activation state becomes resistant to phospho-inhibition. How a phospho-Ser blocks calmodulin activation is unclear. One explanation is that a phosphorylation at position Ser^{45} somehow stabilizes an autoinhibitory conformation, perhaps by decreasing the flexibility of a hinge structure located in this region.

Molecular Genetic Analyses of Functional and Regulatory Residues of AtACA2

The observation that a deregulated, but not wild-type, AtACA2 allowed the yeast host K616 to grow on calcium-depleted media (Figure 3) provided the basis for a powerful genetic screen to identify mutations that "deregulate" a calmodulin-activated pump. With this screen, more than 14 different mutations of AtACA2 have been identified that confer growth to K616 when grown on calcium-depleted media. These mutations fall into two groups. In the first group, members are clustered in the N-terminal domain, at locations adjacent or overlapping with a calmodulin-binding site between residues 23—42 (Figure 4). This entire region was deleted and resulted in a constitutively active pump (or deregulated pump), with high levels of calmodulin-independent activity equal to or greater than a calmodulin-activated wild-type enzyme (44, 53). The second group is composed of mutations located in the stalk (S2 and S3) domain (25a) (Figure 4). Three mutant enzymes harboring mutations from this second group have been characterized at the biochemical level and, like the mutant pump with a deletion of its N-terminal domain, also show high levels of calmodulin-independent activity (i.e. deregulated activity). This indicates that a structure associated with the stalk region is involved in regulation. Interestingly, no mutations have thus far been found in regions originally predicted to interact with the autoinhibitor, based on studies with a human calmodulin-regulated pump in which a peptide sequence derived from the autoinhibitor was found to cross-link to two regions in the catalytic loops (36, 37), corresponding to residues 254–318 and 526–533 in Figure 4 of AtACA2. This negative result appears worthy of attention since no mutations in this region were found in a collection of more than 2000 independent EMS-generated mutations (all of which allowed the AtACA2 to confer growth to K616 on calcium-depleted media).In summary, genetic studies with AtACA2 have provided two important new insights into calmodulin-regulated pumps. First, they produced the first evidence for a regulatory structure associated with the stalk region of a calcium pump. Second, they have so far failed to identify mutations that directly support pre-existing models of how calmodulin-regulated pumps are turned off. This result emphasizes a need for further research to determine if the regulation of ACA2-like pumps is fundamentally different from other calmodulin-dependent pumps, or if a new general model of regulation is needed.

Model of Calmodulin-Stimulated Ca^{2+} Pumps from Plants

Model Based on AtACA2 These results support a proposed model for regulation of type IIB pumps. At low cytosolic $[Ca^{2+}]$ of $\leq 0.1\ \mu M$, ACA2 is kept in a state of low basal activity by an intramolecular interaction between an autoinhibitory sequence located between residues 20–44, and a site in the Ca^{2+} pump core

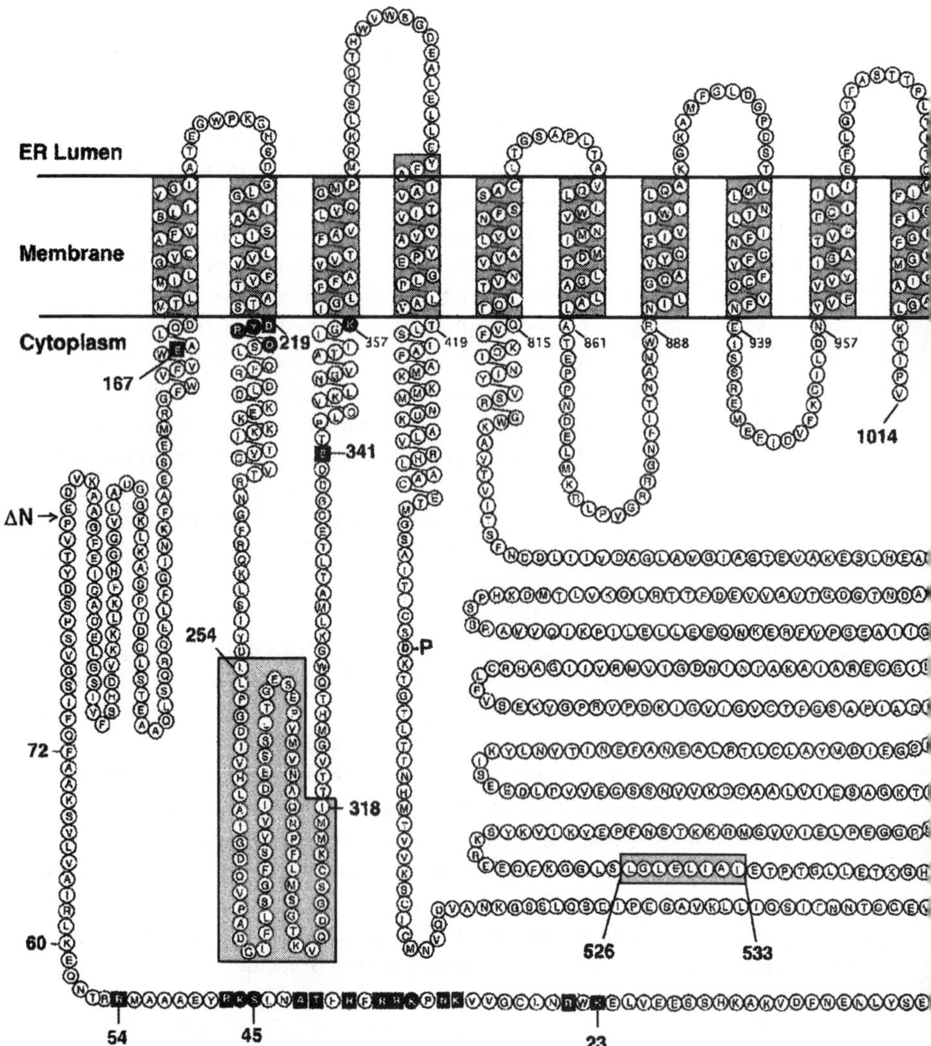

Figure 4 Genetic identification of residues involved in autoinhibition of AtACA2. AtACA2 is diagramed with predicted α-helices shown as stacked sequences of three residues. Residues that have been mutated are shown in black circles (no phenotype) or boxes (deregulation phenotype). Two areas corresponding to regions implicated in autoinhibition of animal pumps are shaded by gray (254–318, and 526–533).

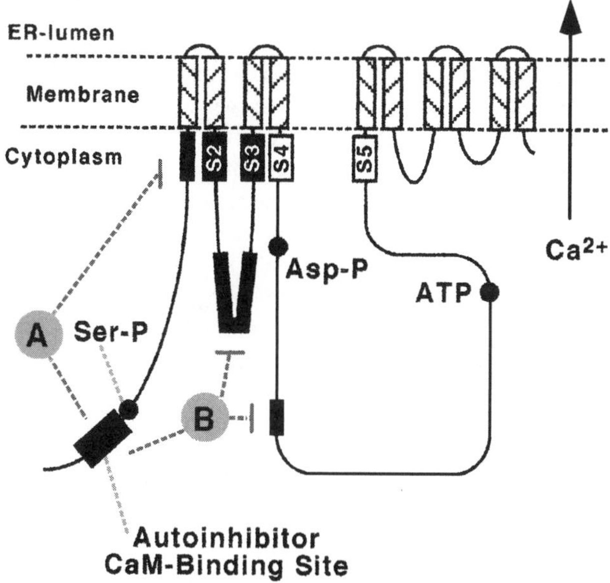

Figure 5 Regulation of type IIB Ca^{2+} pumps: alternative models based on AtACA2. At low cytosolic [Ca^{2+}], the full-length AtACA2 is kept in a state of low basal activity by an intramolecular interaction between an autoinhibitory sequence (including residues Lys^{23} to Arg^{54}) with a site(s) in the Ca^{2+} pump. Potential sites of interactions are marked by thick black lines. Model "B" indicates predicted interactions based on biochemical studies of an animal pump. Model "A" indicates a potential interaction based on genetic studies of AtACA2. The two models are not mutually exclusive. Regardless of the sites of autoinhibitory interactions, the pump is activated as a result of Ca^{2+}-induced binding of calmodulin to a site overlapping or adjacent to the autoinhibitory sequence (marked Autoinhibitor/CaM-Binding Site). Phosphorylation at Ser^{45} (marked Ser-P) in AtACA2 by CDPK blocks calmodulin activation.

(Figure 5). When [Ca^{2+}] levels increase, the pump is then activated as a result of Ca^{2+}-induced binding of calmodulin to a site overlapping or immediately adjacent to the autoinhibitory sequence. This binding confers a conformational change to the N-terminal domain that disrupts the association of an inhibitory sequence with the stalk region or hydrophilic loops 1 and 2 leading to an activated state.

Biochemical and molecular studies of other plant type IIB pumps characterized to date support this general model. For example, a vacuolar Ca^{2+} pump from cauliflower is activated by calmodulin that binds with high affinity to a peptide corresponding to Ala^{19}-Leu^{43} at the N-terminal domain of BCA1 (69). The peptide also inhibits a vacuolar Ca^{2+} pump activated by trypsin (68), supporting the idea that calmodulin activates BCA1 pump by interacting with an autoinhibitory region at the N terminus. Two other Ca^{2+}-ATPases, AtACA4 and AtACA8, which are associated with small vacuoles and plasma membrane, respectively, also bind calmodulin at the N-terminal region (M Geisler, N Frangne, S Malstrom, E Gomes, AC Smith, E Martinoia, MG Palmgren, personal communication; MC Bonza, P Morandini, L Luoni, M Geisler, MG Palmgren & MI De Michelis, personal communication). Furthermore, a plasma membrane Ca^{2+} pump from soybean (SCA1 Acc. no AF195028) appears to have two calmodulin-binding domains (WS Chung et al, submitted). In spite of the similarities (Table 2), N-terminal regulatory domains show considerable diversity, which suggests that the pumps may be differentially regulated by distinct modifications and protein interactions depending on the cellular context. For example, Ser^{45} that is phosphorylated in AtACA2 is conserved in only AtACA1 and AtACA7 adjacent to the calmodulin-binding domain. In contrast, BCA1 and AtACA4 possess a Ser^{28} that lies within the calmodulin-binding domain (68; M Geisler, personal communication).

The observation that calmodulin activation of AtACA2 can be blocked by phosphorylation is important because it illustrates how different signaling pathways can have opposing effects on the activity of a pump, and therefore potentially modify the duration and magnitude of a calcium signal. In principle, any increase in the activity of a Ca^{2+} pump, such as occurs by calmodulin stimulation, should cause a more rapid decrease in the cytosolic free [Ca^{2+}]. Conversely, the effect of blocking this calmodulin activation would result in calcium signals that have a longer duration and possibly an increased magnitude (Figure 6). Thus, the activity of CDPKs, or other protein kinases and protein phosphatases, may be able to change the information content of a calcium signal by changing the activity of ACA2 and its homologs. However, further research is needed to determine if phosphoregulation actually occurs in vivo, and to what extent this regulation changes the nature of calcium signaling dynamics.

The proposed model for the regulation of plant IIB pumps shows striking similarities with animal PM Ca^{2+}-pumping ATPases, which are regulated by a carboxyl autoinhibitor domain (16, 56). First, in a PMCA, calmodulin binding to the C tail enhances the velocity and the affinity for Ca^{2+} (78). Second, a truncation of the C-terminal domain results in an activated pump that is insensitive to calmodulin (33). Third, like Val^{20}-Leu^{44} of Arabidopsis, a sequence overlapping with the calmodulin-binding domain of animal PM Ca^{2+} pumps can also function as an inhibitor (34). In isoform PMCA4a of human, the calmodulin-binding region overlaps to a large degree with the autoinhibitory region (103), where a two-part calmodulin-binding domain is interrupted by a short non-binding region. In PMCA4b isoform, there are two inhibitory domains in which one overlaps with the calmodulin-binding domain. Fourth, PMCA isoforms are differentially regulated by phosphorylation at the carboxyl terminal domain (78). Fifth, the autoinhibitory domain of a PMCA also inhibits SERCA activity (33). In spite of many similarities, recent genetic studies are providing new insights regarding critical regulatory structures in Arabidopsis ACA2 (25a) not previously reported in animal PMCAs.

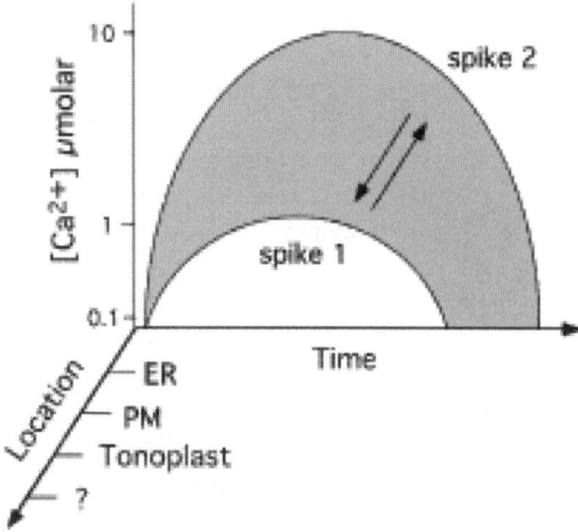

Figure 6 Calcium spikes can potentially be altered by the activity of calcium pumps. The diagram shows how the magnitude and duration of a calcium spike in the cytosol can theoretically be altered by changing the rate of efflux. The influx of Ca^{2+} through different channels generates an increase in cytosolic Ca^{2+}. Efflux is determined by the activity of pumps and antiporters. When a pump, like AtACA2, is stimulated by Ca^{2+}/calmodulin, the resulting Ca^{2+} signal has a relatively low amplitude and short duration as in Spike 1. If the calmodulin-stimulated AtACA2 pump is decreased by phosphorylation through a CDPK pathway, the amplitude and duration of the signal is increased as in Spike 2.

CONCLUDING REMARKS

Although Ca^{2+} often takes center stage as a signaling ion, the roles of Ca^{2+} pumps and channels in regulating basic cellular processes warrant more attention in the future. Early and more recent studies provide evidence for essential roles of Ca^{2+} in the proper functioning of components that govern protein transit and processing in the secretory pathway in eukaryotes. Some important questions for the future include the following:

- What are the specific roles of multiple active Ca^{2+} transporters, such as the 11 or more Ca^{2+} pumps and H^+/Ca^{2+} antiporters in Arabidopsis?
- What regulates the activity of the pumps and antiporter(s)?
- How are the activities integrated in the signaling networks that govern growth, development, and adaptation?
- Are microdomains of cytosolic $[Ca^{2+}]$ signaling formed by colocalization of certain Ca^{2+} channels with Ca^{2+} pumps in membrane patches?

Filling intracellular stores with Ca^{2+} in young cells may be a prerequisite for growth. Given the immobile nature of Ca^{2+}, meristem and elongating cells would be at most risk for Ca^{2+} deficiency. The root and shoot apices receive little Ca^{2+}

from what remains in the transpiration stream. Furthermore, intracellular Ca^{2+} is continuously diluted by cell division and cell elongation. Thus one hypothesis is that young cells must first fill their intracellular stores, including the ER, Golgi, vacuole, and transport vesicles, in order to continue growing. Ca^{2+} within endolumenal compartments serves several roles, including: (*a*) to provide adequate $[Ca^{2+}]_{lumen}$ for the normal synthesis, processing, sorting, and secretion of materials; (*b*) to regulate protein-protein interactions required for vesicle recognition, docking, and perhaps fusion, and (*c*) to modulate $[Ca^{2+}]_{cytosol}$ to regulate cytoskeleton dynamics.

Significantly, both AtECA1 and AtACA2 are highly expressed in roots, and are localized to the ER. AtACA2 was identified as an ER-located pump in root cells, as determined by membrane fractionation, immunocytology, and cytological imaging of a pump tagged with the Green Fluorescent Protein (49). Some cells in the root very likely contain both pumps. As cells in the meristem region, zone of elongation, and root cap are actively synthesizing, processing, and transporting proteins from the ER and polysaccharides from the Golgi to the PM, cell plate, and cell walls, several Ca^{2+} transporters may be involved. Decreasing Ca^{2+} and Mn^{2+} in the Golgi and ER of yeast *pmr1* mutants results in protein degradation in the ER, defects in glycosylation, and protein missorting (31). These results suggest that a certain level of $[Ca^{2+}]_{ER}$ is required for proper protein folding and accurate sorting. GA_3-induction of α-amylase secretion in cereal aleurones would support this idea. Membranes of barley aleurone treated with GA_3 showed increased Ca^{2+} pumping into the ER (12, 13, 43). Moreover, AtECA1 homologues may function in vivo to supply adequate levels of lumenal Mn^{2+} for glycosylation (31, 58) and for cell elongation (70). Interestingly, recent studies suggest roles of Ca^{2+} channels in synaptic vesicle trafficking and exocytosis (91), and of Ca^{2+} pumps and calmodulin in homotypic vacuole docking and fusion in yeast (79, 102).

As additional Ca^{2+} transporters at the nuclear, vacuolar, plasma membrane, and other membranes are identified and characterized at the molecular level, more tools will be available to test their proposed functions in signaling networks and in basic cell biological processes. Whether pumps influence the shapes or frequencies of Ca^{2+} signals can be tested (1) using specific blockers, like cyclopiazonic acid (93), or using transgenic plants or cells that either overexpress or lack a pump (e.g 45). A future challenge is to understand how activities of pumps and channels are coordinated to control the dynamics of intracellular and extracellular calcium.

ACKNOWLEDGMENTS

We thank several colleagues who provided updates and information before publication. Research in the laboratory of HS has been supported through the years by the Department of Energy (DE-FG02-95ER20200), the National Science Foundation, and by the USDA NRI (no. 97-35304-4523). Research in the laboratory of JFH is supported by the Department of Energy (DE-FG03-94ER20152), and a joint grant from the National Aeronautics and Space Administration and National Science

Foundation (IBN-9416038) for the Plant Sensory Systems Collaborative Research Network.

Visit the Annual Reviews home page at www.AnnualReviews.org

LITERATURE CITED

1. Allen GJ, Kwak JM, Chu SP, Llopis J, Tsien RY, et al. 1999. Cameleon calcium indicator reports cytoplasmic calcium dynamics in Arabidopsis guard cells. *Plant J.* 19:735
2. Antebi A, Fink GR. 1992. The yeast Ca^{2+}-ATPase homologue, PMR1, is required for normal Golgi function and localizes in a novel Golgi-like distribution. *Mol. Biol. Cell* 3:633–54
3. Askerlund P. 1996. Modulation of an intracellular calmodulin-stimulated Ca^{2+}-pumping ATPase in cauliflower by trypsin. The use of calcium green-5N to measure Ca^{2+} transport in membrane vesicles. *Plant Physiol.* 110:913–22
4. Askerlund P. 1997. Calmodulin-stimulated Ca^{2+}-ATPases in the vacuolar and plasma membranes in cauliflower. *Plant Physiol.* 114:999–1007
5. Askerlund P, Evans DE. 1992. Reconstitution and characterization of a calmodulin-stimulated Ca-pumping ATPase purified from *Brassica oleracea* L. *Plant Physiol.* 100:1670–81
6. Blackford S, Rea PA, Sanders D. 1990. Voltage sensitivity of H^+/Ca^{2+} antiport in higher plant tonoplast suggests a role in vacuolar calcium accumulation. *J. Biol. Chem.* 265:9617–20
7. Bonza C, Carnelli A, De Michelis MI, Rasi-Caldogno F. 1998. Purification of the plasma membrane Ca^{2+}-ATPase from radish seedlings by calmodulin-agarose affinity chromatography *Plant Physiol.* 116:845–51
8. Brauer D, Schubert C, Tsu SI. 1990. Characterization of a Ca^{2+}-translocating ATPase from corn root microsomes. *Physiol. Plant.* 78:335–44
9. Briars SA, Evans DE. 1989. The calmodulin-stimulated ATPase of maize coleoptiles forms a phosphorylated intermediate. *Biochem. Biophys. Res. Commun.* 159:185–91
10. Briars SA, Kessler F, Evans DE. 1988. The calmodulin-stimulated ATPase of maize coleoptiles is a 140,000-M polypeptide. *Planta* 176:283–85
11. Buckhout TJ. 1984. Characterization of Ca transport in purified endoplasmic reticulum membrane vesicles from *Lepidium sativum* L. roots. *Plant Physiol.* 76:962–67
12. Bush DR. 1995. Calcium regulation in plant cells and its role in signaling. *Annu. Rev. Plant Physiol. Plant Mol. Biol.* 46:95–122
13. Bush DS, Sticher L, van Huystee R, Wagner D, Jones RL. 1989. The calcium requirement for stability and enzymatic activity of two isoforms of barley aleurone alpha-amylase. *J. Biol. Chem.* 264:19392–98
14. Bush DS, Wang T. 1995. Diversity of calcium efflux transporters in wheat aleurone cells. *Planta* 197:19–30
15. Camacho P, Lechleiter JD. 1993. Increased frequency of calcium waves in *Xenopus laevis* oocytes that express a calcium-ATPase. *Science* 260:226–29
16. Carafoli E. 1991. Calcium pump of the plasma membrane. *Physiol. Rev.* 71:129–53
17. Cardenas L, Feijo JA, Kunkel JG, Sanchez F, Holdaway-Clarke T, et al. 1999. *Rhizobium* nod factors induce increases in intracellular free calcium and extracellular calcium influxes in bean root hairs. *Plant J.* 19:347–52

18. Catty P, de Kerchove d'Exaerde A, Goffeau A. 1997. The complete inventory of the yeast *Saccharomyces cerevisiae* P-type transport ATPases. *FEBS Lett.* 409: 325–32
19. Chen FH, Ratterman DM, Sze H. 1993. A plasma membrane-type Ca^{2+}-ATPase of 120 kilodaltons on the endoplasmic reticulum from carrot (*Daucus carota*) cells: properties of the phosphorylated intermediate. *Plant Physiol.* 102:651–61
20. Chen X, Chang M, Wang B, Wu B. 1997. Cloning of a Ca^{2+}-ATPase gene and the role of cytosolic Ca^{2+} in the gibberellin-dependent signaling pathway in aleurone cells. *Plant J.* 11:363–71
21. Clapham DE. 1995. Calcium signaling. *Cell* 80:259–68
22. Clayton H, Knight MR, Knight H, McAinsh MR, Hetherington AM. 1999. Dissection of the ozone-induced calcium signature. *Plant J.* 17:575–79
23. Crofts AJ, Denecke J. 1998. Calreticulin and calnexin in plants. *Trends Plant Sci.* 3:396–99
24. Cunningham KW, Fink GR. 1994. Calcineurin-dependent growth control in *Saccharomyces cerevisiae* mutants lacking PMC1, a homolog of plasma membrane Ca^{2+} ATPases. *J. Cell Biol.* 124:351–63
25. Cunningham KW, Fink GR. 1996. Calcineurin inhibits VCX1-dependent H^+/Ca^{2+} exchange and induces Ca^{2+} ATPases in *Saccharomyces cerevisiae*. *Mol. Cell. Biol.* 16:2226-37

25a. Curran AC. 1999. *Regulatory mechanisms of plant P-type ATPases*. PhD thesis. Univ. Calif., San Diego and San Diego State Univ.

26. Dainese P, James P, Baldan B, Carafoli E. 1997. Subcellular and tissue distribution, partial purification, and sequencing of calmodulin-stimulated Ca^{2+}-transporting ATPases from barley (*Hordeum vulgare* L.) and tobacco (*Nicotiana tabacum*). *Eur. J. Biochem.* 244:31–38
27. de Meis L, Vianna AL. 1979. Energy interconversion by the Ca^{2+}-dependent ATPase of the sarcoplasmic reticulum. *Annu. Rev. Biochem.* 48:275-92
28. Dolmetsch RE, Lewis RS, Goodnow CC, Healy JI. 1997. Differential activation of transcription factors induced by Ca^{2+} response amplitude and duration. *Nature* 386:855–58
29. Dolmetsch RE, Xu K, Lewis RS. 1998. Calcium oscillations increase the efficiency and specificity of gene expression. *Nature* 392:933–36
30. DuPont FM, Bush DS, Windle JJ, Jones RL. 1990. Calcium and proton transport in membrane vesicles from barley roots. *Plant Physiol.* 94:179–88
31. Durr G, Strayle J, Plemper R, Elbs S, Klee SK, et al. 1998. The medial-Golgi ion pump Pmr1 supplies the yeast secretory pathway with Ca^{2+} and Mn^{2+} required for glycosylation, sorting, and endoplasmic reticulum-associated protein degradation. *Mol. Biol. Cell* 9:1149–62
32. Ehrhardt DW, Wais R, Long SR. 1996. Calcium spiking in plant root hairs responding to *Rhizobium* nodulation signals. *Cell* 85:673–81
33. Enyedi A, Penniston JT. 1993. Autoinhibitory domains of various Ca^{2+} transporters cross-react. *J. Biol. Chem.* 68:17120–25
34. Enyedi A, Vorherr T, James P, McCormick DJ, Filoteo AG, et al. 1989. The calmodulin binding domain of the plasma membrane Ca^{2+} pump interacts both with calmodulin and with another part of the pump. *J. Biol. Chem.* 264:12313–21
35. Evans DE, Williams LE. 1998. P-type calcium ATPases in higher plants—biochemical, molecular and functional properties. *Biochim. Biophys. Acta* 1376: 1–25
36. Falchetto R, Vorherr T, Brunner J, Carafoli E. 1991. The plasma membrane Ca^{2+}

pump contains a site that interacts with its calmodulin-binding domain. *J. Biol. Chem.* 266:2930–36

37. Falchetto R, Vorherr T, Carafoli E. 1992. The calmodulin-binding site of the plasma membrane Ca^{2+} pump interacts with the transduction domain of the enzyme. *Protein Sci.* 1:1613–21
38. Ferrol N, Bennett AB. 1996. A single gene may encode differentially localized Ca^{2+}-ATPases in tomato. *Plant Cell* 8:1159–69
39. Gavin O, Pilet PE, Chanson A. 1993. A tonoplast localization of a calmodulin-stimulated Ca-pump from maize roots *Plant Sci.* 92:143–50
40. Geisler M, Axelsen K, Harper JF, Palmgren MG. 2000. Molecular aspects of higher plant P-type Ca-ATPases. *Biochim. Biophys. Acta.* In press
41. Ghosh A, Greenberg ME. 1995. Calcium signaling in neurons: molecular mechanisms and cellular consequences. *Science* 268:239–47
42. Giannini JL, Reynolds-Niesmann I, Briskin DP. 1987. Calcium transport in sealed vesicles from red beet (*Beta vuglaris* L.) storage tissue. Characterization of a Ca^{2+}-pumping ATPase associated with the endoplasmic reticulum. *Plant Physiol.* 85:1129–36
43. Gilroy S, Bethke PC, Jones RL. 1993. Calcium homeostasis in plants. *J. Cell Sci.* 106:453–61
44. Harper JF, Hong B, Hwang I, Guo HQ, Stoddard R, et al. 1998. A novel calmodulin-regulated Ca^{2+}-ATPase (ACA2) from Arabidopsis with an N-terminal autoinhibitory domain. *J. Biol. Chem.* 273:1099–106
45. Hirsch RE, Lewis BD, Spalding EP, Sussman MR. 1998. A role for the AKT1 potassium channel in plant nutrition. *Science* 280:918–21
46. Hirschi KD. 1999. Expression of *Arabidopsis CAX1* in tobacco: altered calcium homeostasis and increased stress sensitivity. *Plant Cell* 11:2113–22
47. Hirschi KD, Zhen RG, Cunningham KW, Rea PA, Fink GR. 1996. CAX1, an H^+/Ca^{2+} antiporter from Arabidopsis. *Proc. Natl. Acad. Sci. USA* 93:8782–86
48. Holdaway-Clarke TF, Eijo JA, Hackett GR, Kunkel JG, Hepler PK. 1997. Pollen tube growth and the intracellular cytosolic calcium gradient oscillates in phase while extracellular clacium influx is delayed. *Plant Cell* 9:1999–2010
49. Hong B, Ichida A, Wang Y, Gens JS, Pickard BG, Harper JF. 1999. Identification of a calmodulin-regulated Ca^{2+}-ATPase in the endoplasmic reticulum. *Plant Physiol.* 119:1165–76
50. Hsieh W-L, Pierce WS, Sze H. 1991. Calcium-pumping ATPases in vesicles from carrot cells. Stimulation by calmodulin or phosphatidylserine and formation of a 120 kDa phosphoenzyme. *Plant Physiol.* 97:1535–44
51. Huang L, Berkelman T, Franklin AE, Hoffman NE. 1993. Characterization of a gene encoding a Ca^{2+}-ATPase-like protein in the plastid envelope *Proc. Natl. Acad. Sci. USA* 90:10066–70
52. Hwang I. 1999. *Characterization and functional identification of a calmodulin-regulated Ca^{2+} ATPase in plants.* PhD thesis, Univ. Maryland, College Park
53. Hwang I, Liang F, Harper JF, Sze H. 2000. Calmodulin interacts with the autoinhibitory domain at the N terminus of ACA2, a novel Ca pump from Arabidopsis. *Plant Physiol.* 122:157–67
54. Hwang I, Ratterman DM, Sze H. 1997. Distinction between ER-type and PM-type Ca pumps: partial purification of a 120 kD Ca-ATPase from endomembranes *Plant Physiol.* 113:535–48
55. James P, Inui M, Tada M, Chiesi M, Carafoli E. 1989. Nature and site of phospholamban regulation of the Ca^{2+} pump of sarcoplasmic reticulum. *Nature* 342: 90–92
56. James P, Vorherr T, Carafoli E. 1995. Calmodulin-binding domains: just two

faced or multi-faceted? *Trends Biochem. Sci.* 20:38–42

57. Kasai N, Muto S. 1990. Ca^{2+} pump and Ca^{2+}/H^{+} antiporter in plasma membrane vesicles isolated by aqueous two-phase partitioning from corn leaves *J. Membr. Biol.* 114:133–42
58. Kaufman RJ, Swaroop M, Murtha-Riel P. 1994. Depletion of manganese within the secretory pathway inhibits O-linked glycosylation in mammalian cells. *Biochemistry* 33:9813–19
59. Knight H, Trewavas AJ, Knight MR. 1996. Cold calcium signaling in Arabidopsis involves two cellular pools and a change in calcium signature after acclimation. *Plant Cell* 8:489–503
60. Lapinskas PJ, Cunningham KW, Liu XF, Fink GR, Culotta VC. 1995. Mutations in PMR1 suppress oxidative damage in yeast cells lacking superoxide dismutase. *Mol. Cell. Biol.* 15:1382–88
61. Liang F. 1998. *Functional identification and biochemical characterization of ER-type Ca-pumping ATPases from* Arabidopsis thaliana. PhD thesis, Univ. Maryland, College Park
62. Liang F, Cunningham KW, Harper JF, Sze H. 1997. ECA1 complements yeast mutants defective in Ca^{2+} pumps and encodes an endoplasmic reticulum-type Ca^{2+}-ATPase in *Arabidopsis thaliana. Proc. Natl. Acad. Sci. USA* 94:8579–84
63. Liang F, Sze H. 1998. A high-affinity Ca^{2+} pump, ECA1, from the endoplasmic reticulum is inhibited by cyclopiazonic acid but not by thapsigargin. *Plant Physiol.* 118:817–25
64. Liang F, Sze H. 1999. AtECA3 (AF117296) encodes a homolog of endoplasmic reticulum-type Ca-ATPase from Arabidopsis. *Plant Physiol.* 120:634
65. Liss H, Weiler EW. 1994. Ion-translocating ATPases in tendrils of *Bryonia dioica* Jacq. *Planta* 194:169–80
66. MacLennan DH, Rice WJ, Green NM. 1997. The mechanism of Ca^{2+} transport by sarco(endo)plasmic reticulum Ca^{2+}- ATPases. *J. Biol. Chem.* 272:28815–18
67. Malhó R, Trewavas AJ. 1996. Localised apical increases of cytosolic free calcium control pollen tube orientation. *Plant Cell* 8:1935–49
68. Malmstrom S, Akerlund H-E, Askerlund P. 2000. Regulatory role of the N-terminus of the vacuolar Ca-ATPase in cauliflower. *Plant Physiol.* In press
69. Malmstrom S, Askerlund P, Palmgren MG. 1997. A calmodulin-stimulated Ca^{2+}-ATPase from plant vacuolar membranes with a putative regulatory domain at its N-terminus. *FEBS Lett.* 400:324–28
70. Marschner H. 1995. *Mineral Nutrition of Higher Plants*. London/San Diego: Academic. 2nd ed.
71. Matheos DP, Kingsbury TJ, Ahsan US, Cunningham KW. 1997. Tcn1p/Crz1p, a calcineurin-dependent transcription factor that differentially regulates gene expression in *Saccharomyces cerevisiae. Genes Dev.* 11:3445–58
72. McAinsh MR, Hetherington AM. 1998. Encoding specificity in Ca signaling systems. *Trends Plant Sci.* 3:32–36
73. Meldolesi J, Pozzan T. 1998. The endoplasmic reticulum Ca^{2+} store: a view from the lumen. *Trends Biochem. Sci.* 23:10–14
74. Miller AJ, Vogg G, Sanders D. 1990. Cytosolic calcium homeostasis in fungi: roles of plasma membrane transport and intracellular sequestration of calcium. *Proc. Natl. Acad. Sci. USA* 87:9348–52
75. Mintz E, Guillain F. 1997. Ca^{2+} transport by the sarcoplasmic reticulum ATPase. *Biochim. Biophys. Acta* 1318:52–70
76. Navarro-Avino JP, Hentzen AE, Bennett AB. 1999. Alternative transcription initiation sites generate two LCA1 Ca^{2+}-ATPase mRNA transcripts in tomato roots. *Plant Mol. Biol.* 40:133–40
77. Odermatt A, Becker S, Khanna VK, Kurzydlowski K, Leisner E, et al. 1998.

Sarcolipin regulates the activity of SERCA1, the fast-twitch skeletal muscle sarcoplasmic reticulum Ca^{2+}-ATPase. *J. Biol. Chem.* 273:12360–69

78. Penniston JT, Enyedi A. 1998. Modulation of the plasma membrane Ca^{2+} pump. *J. Membr. Biol.* 165:101–9
79. Peters C, Mayer A. 1998. Ca^{2+}/calmodulin signals the completion of docking and triggers a late step of vacuole fusion. *Nature* 396:575–80
80. Pfeiffer W, Hager A. 1993. A Ca-ATPase and a Mg/H-antiporter are present on tonoplast membranes from roots of *Zea mays* L. *Planta* 191:377–85
81. Pittman JK, Mills RF, O'Connor CD, Williams LE. 1999. Two additional type (IIA) Ca^{2+}-ATPases are expressed in *Arabidopsis thaliana*: evidence that type IIA sub-groups exist. *Gene* 236:137–47
82. Plenge-Tellechea F, Soler F, Fernandez-Belda F. 1997. On the inhibition mechanism of sarcoplasmic or endoplasmic reticulum Ca^{2+}-ATPases by cyclopiazonic acid. *J. Biol. Chem.* 272:2794–800
83. Raschke BC, Wolf AH. 1996. Molecular cloning of a P-type Ca-ATPase from the halotolerant alga *Dunaliella bioculata*. *Planta* 200:78–84
84. Rasi-Caldogno F, Carnelli A, De Michelis MI. 1992. Plasma membrane Ca-ATPase of radish seedlings. II. Regulation by calmodulin. *Plant Physiol.* 98:1202–6
85. Rasi-Caldogno F, Carnelli A, De Michelis MI. 1993. Controlled proteolysis activates the plasma membrane Ca pump of higher plants. A comparison with the effect of calmodulin in plasma membrane from radish seedlings. *Plant Physiol.* 103:385–90
86. Rasi-Caldogno F, Carnelli A, De Michelis MI. 1995. Identification of the plasma membrane Ca^{2+}-ATPase and of its autoinhibitory domain. *Plant Physiol.* 108:105–13
87. Rasi-Caldogno F, Pugliarello MC, De Michelis MI. 1987. The Ca-transport ATPase of plant plasma membrane catalyzes an H^+/Ca^{2+} exchange. *Plant Physiol.* 83:994–1000
88. Sanders D, Brownlee C, Harper JF. 1999. Communicating with calcium. *Plant Cell* 11:691–706
89. Schumaker KS, Sze H. 1986. Calcium transport into the vacuole of oat roots. Characterization of H^+/Ca^{2+} exchange activity. *J. Biol. Chem.* 261:12172–78
90. Schumaker KS, Sze H. 1990. Solubization and reconstitution of the oat root vacuolar H^+/Ca^{2+} exchanger. *Plant Physiol.* 92:340–45
91. Seagar M, Leveque C, Charvin N, Marqueze B, Martin-Moutot N, et al. 1999. Interactions between proteins implicated in exocytosis and voltage-gated calcium channels. *Philos. Trans. R. Soc. London Ser. B* 354:289–97
92. Seidler NW, Jona I, Vegh M, Martonosi A. 1989. Cyclopiazonic acid is a specific inhibitor of the Ca^{2+}-ATPase of sarcoplasmic reticulum. *J. Biol. Chem.* 264:17816–23
93. Sievers A, Busch MB. 1992. An inhibitor of the Ca^{2+}-ATPases in the sarcoplasmic and endoplasmic reticula inhibits transduction of the gravity stimulus in cress roots. *Planta* 188:619–22
94. Simmerman HK, Jones LR. 1998. Phospholamban: protein structure, mechanism of action, and role in cardiac function. *Physiol. Rev.* 78:921–47
95. Sorin A, Rosas G, Rao R. 1997. PMR1, a Ca^{2+}-ATPase in yeast Golgi, has properties distinct from sarco/endoplasmic reticulum and plasma membrane calcium pumps. *J. Biol. Chem.* 272:9895–901
96. Stathopoulos AM, Cyert MS. 1997. Calcineurin acts through the CRZ1/TCN1-encoded transcription factor to regulate gene expression in yeast. *Genes Dev.* 11:3432–44

97. Stathopoulos-Gerontides A, Guo JJ, Cyert MS. 1999. Yeast calcineurin regulates nuclear localization of the Crz1p transcription factor through dephosphorylation. *Genes Dev.* 13:798–803
98. Sze H, Li X, Palmgren MG. 1999. Energization of plant cell membranes by H^+-pumping ATPases. Regulation and biosynthesis. *Plant Cell* 11:677–90
99. Theodoulou FL, Dewey FM, Evans DE. 1994. Calmodulin-stimulated ATPase of maize cells: functional reconstitution, monoclonal antibodies and subcellular localization. *J. Exp. Bot.* 45:1553–64
100. Thomson LJ, Xing T, Hall JL, Williams LE. 1993. Investigation of the calcium-transporting ATPases at the endoplasmic reticulum and plasma membrane of red beet (*Beta vulgaris*). *Plant Physiol.* 102:553–64
101. Toyofuku T, Kurzydlowski K, Tada M, MacLennan DH. 1994. Amino acids Lys-Asp-Asp-Lys-Pro-Val402 in the Ca^{2+}-ATPase of cardiac sarcoplasmic reticulum are critical for functional association with phospholamban. *J. Biol. Chem.* 269:22929–32
102. Ungermann C, Wickner W, Xu Z. 1999. Vacuole acidification is required for trans-SNARE pairing, LMA1 release, and homotypic fusion. *Proc. Natl. Acad. Sci. USA* 96:11194–99
103. Verma AK, Enyedi A, Filoteo AG, Strehler EE, Penniston JT. 1996. Plasma membrane calcium pump isoform 4a has a longer calmodulin-binding domain than 4b. *J. Biol. Chem.* 271:3714–18
104. Wakabayashi S, Ikeda T, Iwamoto T, Pouyssegur J, Shigekawa M. 1997. Calmodulin-binding autoinhibitory domain controls "pH-sensing" in the Na^+/H^+ exchanger NHE1 through sequence-specific interaction. *J. Biol. Chem.* 36:12854–61
105. White PJ. 1998. Calcium channels in the plasma membrane of root cells. *Ann. Bot* 81:173–83
106. Wimmers LE, Ewing NN, Bennett AB. 1992. Higher plant Ca^{2+}-ATPase: primary structure and regulation of mRNA abundance by salt. *Proc. Natl. Acad. Sci. USA* 89:9205–9
107. Zhang P, Toyoshima C, Yonekura K, Green NM, Stokes DL. 1998. Structure of the calcium pump from sarcoplasmic reticulum at 8 Å resolution. *Nature* 392:835–39
108. Zielinski RE. 1998. Calmodulin and calmodulin binding proteins in plants. *Annu. Rev. Plant Physiol. Plant Mol. Biol.* 49:697–725

Annu. Rev. Plant Physiol. Plant Mol. Biol. 2000. 51:463–99

Plant Cellular and Molecular Responses to High Salinity

Paul M. Hasegawa and Ray A. Bressan
Center for Plant Environmental Stress Physiology, 1165 Horticulture Building, Purdue University, West Lafayette, Indiana 47907-1165; e-mail: paul.m.hasegawa.1@purdue.edu

Jian-Kang Zhu[1] and Hans J. Bohnert[1,2]
Departments of [1]Plant Sciences and [2]Biochemistry, University of Arizona, Tucson, Arizona 85721; e-mail: bonerth@u.arizona.edu

Key Words salt adaptation mechanisms, functional genetics, stress genomics

■ **Abstract** Plant responses to salinity stress are reviewed with emphasis on molecular mechanisms of signal transduction and on the physiological consequences of altered gene expression that affect biochemical reactions downstream of stress sensing. We make extensive use of comparisons with model organisms, halophytic plants, and yeast, which provide a paradigm for many responses to salinity exhibited by stress-sensitive plants. Among biochemical responses, we emphasize osmolyte biosynthesis and function, water flux control, and membrane transport of ions for maintenance and re-establishment of homeostasis. The advances in understanding the effectiveness of stress responses, and distinctions between pathology and adaptive advantage, are increasingly based on transgenic plant and mutant analyses, in particular the analysis of *Arabidopsis* mutants defective in elements of stress signal transduction pathways. We summarize evidence for plant stress signaling systems, some of which have components analogous to those that regulate osmotic stress responses of yeast. There is evidence also of signaling cascades that are not known to exist in the unicellular eukaryote, some that presumably function in intercellular coordination or regulation of effector genes in a cell-/tissue-specific context required for tolerance of plants. A complex set of stress-responsive transcription factors is emerging. The imminent availability of genomic DNA sequences and global and cell-specific transcript expression data, combined with determinant identification based on gain- and loss-of-function molecular genetics, will provide the infrastructure for functional physiological dissection of salt tolerance determinants in an organismal context. Furthermore, protein interaction analysis and evaluation of allelism, additivity, and epistasis allow determination of ordered relationships between stress signaling components. Finally, genetic activation and suppression screens will lead inevitably to an understanding of the interrelationships of the multiple signaling systems that control stress-adaptive responses in plants.

040-2519/00/0601-0463$14.00

CONTENTS

INTRODUCTION

It has been two decades since salinity stress biology and plant responses to high salinity have been discussed in this series (65, 80) and one decade since salinity tolerance in marine algae has been covered (129). Together with a monograph, still unsurpassed in its comprehensive discussion of the biophysical aspects of plant abiotic stresses (141), these reviews covered organismal, physiological, and the then-known biochemical hallmarks of stress and the bewildering complexity of plant stress responses. Much research information has subsequently been gathered about cellular, metabolic, molecular, and genetic processes associated with the response to salt stress, some of which presumably function to mediate tolerance (28–30, 36, 67, 78, 96, 105, 154, 189, 229, 240, 241, 285, 249). Our knowledge of how plants re-establish osmotic and ionic homeostasis after salt stress imposition, and then maintain physiological and biochemical steady states necessary for growth and completion of the life cycle in the new environment is fundamental to our understanding of tolerance (29, 78, 160, 182, 189, 285).

During the past decade, concerted experimental focus targeted the identification of cell-based mechanisms of ion and osmotic homeostasis as essential determinants of tolerance (37, 96, 255). Plant scientists now recognize that underlying cellular mechanisms of salt tolerance are evolutionarily more deeply rooted than previously perceived. Consequently, research that has exploited the molecular genetic advantages of model unicellular organisms (e.g. bacteria, yeast, and algae) has been highly applicable to, and will continue to provide, greater understanding of higher plant salt tolerance (36, 48, 87, 181, 203, 230). However, it is obvious that plant cells become specialized during ontogeny. Cell differentiation and spatial location within plant tissues affect what adaptive mechanisms are most important to proper functioning of specific cells within the organism. In fact, integrative hierarchical functioning among cells, tissues, and organs, in a developmental context, certainly is an essential requisite for salt tolerance of the organism (29, 78, 96, 173, 230, 285). It is acknowledged that whole plant studies are as necessary as always, but will be more insightful, because the prevailing reductionist approaches of the past have provided mechanistic understanding that can be appreciated as the basis for integration of cellular level tolerance to the whole plant.

Algal genera such as *Dunaliella,* which includes many extreme halophytes (flora of saline environments), and higher plants in the halophyte category, for example, species of *Atriplex* and *Mesembryanthemum crystallinum* (2, 18, 19, 59, 89, 184, 263), have provided substantial insight into the response of halophytes to stress, which by comparison with that of non-halophytes (glycophytes) has led to greater comprehension of adaptive mechanisms. More recently, plant molecular genetic models, in particular *Arabidopsis thaliana*, have provided inroads through the investigative power and causal demonstration of gain- and loss-of-function molecular genetic experimentation to elucidate both cellular and organismal mechanisms of salt tolerance (145, 146, 279, 293, 295). We feel that the preceding 20 years of research, coupled with powerful new genetic tools, have brought the field to the brink of understanding the genetic mechanisms underlying the physiological and ecological complexity of salt tolerance. We summarize in this review salient research advances in the topic area since the previous reports in this series (65, 80, 129). We discuss briefly basic physiological stress responses and then outline how new molecular and genetic tools have begun to link and integrate the stress-responsive signaling cascades with the effectors (identified as physiological, biochemical and eventually genetic components) that mediate adaptive processes, indicating that signaling controls and mediates salt adaptation.

SALINE STRESS AND PLANT RESPONSE

Salt Stress

High salinity causes both hyperionic and hyperosmotic stress effects, and the consequence of these can be plant demise (78, 182, 285). Most commonly, the stress is caused by high Na^+ and Cl^- concentrations in the soil solution. Altered water status most likely brings about initial growth reduction; however, the precise contribution of subsequent processes to inhibition of cell division and expansion and acceleration of cell death has not been well elucidated (174, 285). Membrane disorganization, reactive oxygen species, metabolic toxicity, inhibition of photosynthesis, and attenuated nutrient acquisition are factors that initiate more catastrophic events (65, 80, 173, 285).

Salt Movement through Plants

Movement of salt into roots and to shoots is a product of the transpirational flux required to maintain the water status of the plant (66, 285). Unregulated, transpiration can result in toxic levels of ion accumulation in the aerial parts of the plant. An immediate response to salinity, which mitigates ion flux to the shoot, is stomatal closure. However, because of the water potential difference between the atmosphere and leaf cells, and the need for carbon fixation, this is an untenable long-term strategy of tolerance (173, 285).

To protect actively growing and metabolizing cells, plants regulate ion movement into tissues (66, 174). One mode by which plants control salt flux to the shoot is the entry of ions into the xylem stream. Still debated is the extent to which symplastic ion transport through the epidermal and cortical cells contributes to a reduction in Na^+ that is delivered to the xylem (44, 66). However, at the endodermis, radial movement of solutes must be via a symplastic pathway, as the Casparian strip constitutes a physical barrier to apoplastic transport (44, 66).

The accumulation of large quantities of ions in mature and old leaves, which then dehisce, has often been observed under salt stress (66, 174). In a function as ion sinks, old leaves may restrict ion deposition into meristematic and actively growing and photosynthesizing cells. An alternative possibility is that cellular ion discrimination is a natural consequence of transpirational and expansive growth fluxes, cell morphology, and degree of intercellular connection. Meristematic cells, which are not directly connected to the vasculature, are less exposed to ions delivered through the transpiration stream, and their small vacuolar space is not conducive to ion storage. De facto, the solute content of tissues containing cells with little vacuolation (e.g. meristematic regions) is predominated by organic osmolytes and in tissues with highly vacuolated cells by ions (22, 280).

Lessons from Halophyte and Glycophyte Comparisons

Halophytes require for optimal growth electrolyte (typically Na^+ and Cl^+) concentrations higher or much higher than those found in nonsaline soils. How and within which range of NaCl (roughly defined from 20 to 500 mM NaCl) these plants respond best is complex, and has led to a number of classification attempts (65, 78, 80). Halophytes seem to lack unique metabolic machinery that is insensitive to or activated by high Na^+ and Cl^- (65, 181, 182, 214). Instead, plants ultimately survive and grow in saline environments because of osmotic adjustment through intracellular compartmentation that partitions toxic ions away from the cytoplasm through energy-dependent transport into the vacuole (10, 22, 78, 90, 182, 249, 285). Some halophytes exclude Na^+ and Cl^- through glands and bladders, which are specialized structures that seem to be evolutionarily late inventions by which halophytes gain an edge over glycophytes. Osmotic adjustment of both halophytes and glycophytes is also achieved through the accumulation of organic solutes in the cytosol, and the lumen, matrix, or stroma of organelles (Figures 1, 2; see color plates) (182, 213, 285).

A principal difference between halophytes and glycophytes is the capacity of the former to survive salt shock. This greater capacity allows haplophytes to more readily establish metabolic steady state for growth in a saline environment (33, 40, 49, 97, 184). Responsiveness to salinity and at least some ability to establish an adapted new steady state is not unique, however, to halophytes inasmuch as both glycophyte cells and plants exhibit substantial capacity for salt tolerance provided that stress imposition is gradual (8, 37, 96). The question then remains as to whether particular biochemical mechanisms of halophytes are better activated or are preactivated, allowing a more rapid and successful overall response to saline stress. Because of the diversity of halophyte species, there is no simple answer to this question, but unique adaptive responses to NaCl stress are not apparent in the overwhelming majority of halophytes (65, 78, 285).

Whereas glycophytes restrict ion movement to the shoot by attempting control of ion influx into root xylem, halophytes tend more readily to take up Na^+ such that the roots typically have much lower NaCl concentrations than the rest of the plant (3, 37). It seems that a major advantage that halophytes have over glycophytes is not only more responsive Na^+ partitioning but more effective capacity to coordinate this partitioning with processes controlling growth, and ion flux across the plasma membrane, in both cellular and organismal contexts. This may explain why halophytes can use, and perhaps rely on, Na^+ and Cl^- for osmotic adjustment that then supports cell expansion in growing tissues and turgor in differentiated organs (3, 78, 80, 285). In a saline environment, the ability to take up and confine Na^+ to leaves lowers the osmotic potential of aerial plant parts; this then facilitates water uptake and transport and lowers the metabolic cost for the production of osmolytes. Contrarily, the necessity for efficient vacuolar deposition of Na^+ exacts a higher cost for H^+ pumping, and possibly requires additional mechanisms for the acquisition of ion nutrients (principally K^+). Also, the osmotic benefits of storing Na^+ and Cl^- as abundant, cheap osmolytes are limited by the available vacuolar space. Therefore, continued growth, i.e. the production of new vacuoles, may be a factor limiting tolerance (139).

SALT STRESS TOLERANCE DETERMINANTS: Effectors and Signaling Components

We define determinants of salt stress tolerance as effector molecules (metabolites, proteins, or components of biochemical pathways) that lead to adaptation and as regulatory molecules (signal transduction pathway components) that control the amount and timing of these effector molecules. Stress adaptation effectors are categorized as those that mediate ion homeostasis, osmolyte biosynthesis, toxic radical scavenging, water transport, and transducers of long-distance response coordination (12, 131, 140, 170, 180, 182, 213, 265). Listed in Table 1 are

TABLE 1 Osmotic regulated genes that encode proteins with function in stress tolerance downstream of signaling cascades

Components	Suggested mechanisms and/or metabolic functions	Gene/Protein	References
Proteins	**Protein stability**		
LEA/dehydrins	Function unknown (protein stability); control of desiccation;	HVA1, various	34, 253, 281
ROS scavenging	preventing the generation of or detoxifying ROS;	classes Fe-SOD,	31, 268, 276
	inhibiting OH*-production from Fenton-reaction;	Mn-SOD	83, 217, 252
	increases in ROS scavenging enzymes	GP, PHGPX,	12, 187, 239
		ASX, Catalase	5, 122, 244
		Gst/Gpx	81, 264
Chaperones	Heat-/cold-/salt-shock proteins; protein folding	Hsp, Csp, Ssp	88, 186
		DnaJ	296
Carbohydrates	**Osmolytes and/or compatible solute**		
Polyols	Inhibiting OH*-production from Fenton-reaction	*Mtld, Imt1, Stldh*	257, 258, 272
	osmotic adjustment; redox control		232, 233, 234, 236, 237, 238
Fructan (levansucrase)	Osmoprotection	*SacB*	200
Trehalose	Osmoprotection; (signaling?)	*Tps; Tpp*, trehalase	79, 106
Quaternary N-compounds	**Osmoprotection**		
Glycine betaine	Protein protection, one-carbon sink	*codA*	98, 161
		BADH	194, 209
		CMO	4, 109, 219
Dimethyl sulfonium compounds	Protein protection; pathogen defense		73, 212, 216
			189

Amino acid/derivatives	**Osmotic adjustment (& possibly other functions)**		
Proline	Substrate for mitochondrial respiration; redox control; nitrogen balance/storage, transport	P5CS/P5CR POX, ProT2	125, 220 145, 175, 287
Ectoine	Osmoprotectant, (signaling?)	*EctA,B,C* (operon)	149, 190
Inadvertent Na^+ uptake	**Control over potassium uptake**		
K^+-transporters	High affinity K^+ uptake; possibly significant contribution to sodium uptake	*Hkt1, Hak1*	182, 218 222
K^+-channels	Low affinity or dual affinity K^+ uptake; minimal contribution to sodium uptake	*Akt1, Akt2*	9, 228
	Sodium stimulation of potassium uptake		247
Sodium partitioning	**Tissue/cell-specific deposition of sodium**		
Na^+/H^+ transport or antiport	Vacuolar storage of Na^+ as an osmoticum and/or plasma membrane Na^+ exclusion/export	Na^+/H^+ antiporters	3, 76 10, 72
H^+-ATPase	Establishing proton gradients		148, 184, 274 18, 263
Na^+/*myo*-inositol transport	Long-distance phloem & xylem transport	ITR-family	Chauhan et al, unpublished, 180
Water relations	**Control over water flux into and out of cells**		
Water channel proteins (AQP, MIP)	Membrane cycling controlling presence & amount; posttranslational modifications	γ-TIP MIP-A, -B, -F	116, 158, 169, 282 17, 121, Kirch et al, unpublished
	Controlling transcript amounts; expected functions in tolerance for homologues that transported other small metabolites (glycerol, urea, polyols)	RD28 SITIP, SIMIP Fps1p	159, 265, 283 131, 199 (yeast) 256

stress-regulated genes that encode stress-tolerance effectors. Regulatory molecules are cellular signal pathway components (Figure 3, see color plates), including transcription factors, that regulate salt tolerance effectors (241).

FUNCTIONAL CATEGORIES OF SALT TOLERANCE EFFECTOR DETERMINANTS

Ion Homeostasis

A hypersaline environment, most commonly mediated by high NaCl, results in perturbation of ionic steady state not only for Na^+ and Cl^- but also for K^+ and Ca^{2+} (182). External Na^+ negatively impacts intracellular K^+ influx, attenuating acquisition of this essential nutrient by cells. High NaCl causes cytosolic accumulation of Ca^{2+} and this, apparently, signals stress responses that are either adaptive or pathological. Ion homeostasis in saline environments is dependent on transmembrane transport proteins (Figure 1) that mediate ion fluxes, including H^+ translocating ATPases and pyrophosphatases, Ca^{2+}-ATPases, secondary active transporters, and channels (23, 26, 182, 254). A role for ATP-binding cassette (ABC) transporters (210) in plant salt tolerance has not been elucidated, but ABC transporters regulate cation homeostasis in yeast (166).

The molecular identity of many transport proteins that putatively mediate Na^+, K^+, Ca^{2+}, and Cl^- transport has been determined. Most of these transport proteins were identified from structure/function information or by functional complementation of transport-deficient yeast mutants (55). It is now clear that many of the transport determinants that mediate ion homeostasis in yeast and plants are very similar (229). Furthermore, it is likely that the rudiments of salt-responsive signal regulatory pathways controlling ion homeostasis in both organisms are analogous (36, 229). The evolutionary conservation of essential ion transport function and the complete compilation of plant genome databases that will be available shortly make it likely that the as-yet unidentified transport proteins involved in Na^+, K^+, Ca^{2+}, and Cl^- homeostasis will emerge shortly. Less well understood is the physiological function of these transport systems during stress adaptation and after steady state is re-established, but again comparison with the yeast model provides a reasonable paradigm. The physiological function of some transport proteins has been confirmed recently by experimentation with plant mutants (77, 102, 288).

Plasma Membrane and Ionoplast H^+ Electrochemical Gradients Secondary active transport and electrophoretic flux across the plasma membrane and tonoplast are driven by the H^+-electrochemical potential gradients established by H^+ pumps (150, 193, 254, 291). The thermodynamic steady-state conditions facilitated by the plasma membrane (H^+-ATPase) and tonoplast (H^+-ATPase and H^+-pyrophosphatase) H^+ pumps in nonsaline environments are assumed to be established after plants adapt to salinity and re-establish ion homeostasis in a saline environment (Figure 1) (25, 182). The H^+ electrochemical potential gradients established could facilitate between 10^2- to 10^3-fold lower concentrations of Na^+ and Cl^- in the cytosol relative to the apoplast or the vacuole, assuming transport proteins are present in nonlimiting amounts. However, cytosolic Na^+ and Cl^- concentrations of cells growing in salt levels near that of seawater have been measured to be about 80 to 100 mM (22, 182). Consequently, even in seawater concentrations of Na^+ and Cl^-, energy-dependent transport of these ions into the apoplast and vacuole (so that toxic levels do not accumulate in the cytosol) can be facilitated by the capacity of the H^+ electrochemical potential gradients established across the plasma membrane and tonoplast. In high NaCl environments, K^+ remains at about 80 mM in the cytosol (22).

The plasma membrane H^+-ATPase is encoded by a multigene family and expression of isogenes is regulated specifically in spatial and temporal contexts as well as by chemical and environmental inducers, including salt (183, 184, 254). Furthermore, increased ATPase-mediated H^+ translocation across the plasma membrane is a component of the plant cell response to salt imposition (33, 273). Both an ATPase (V-type) and a pyrophosphatase are responsible for H^+-translocation into the vacuole and generation of H^+ motive force across the tonoplast (150, 254, 291). Salt treatment induces ATPase activity and H^+ transport of V-type pumps (23, 207, 269). This increased activity has been attributed to greater protein abundance (206), changes in kinetic properties (211), differential subunit composition (23), and transcriptional regulation (23, 25, 178). The pyrophosphatase can contribute substantially to H^+ transport into the vacuole. However, the relative contribution of the ATPase and pyrophosphatase to the H^+ electrochemical gradient across the tonoplast during salt stress or growth in saline environments is not clear (291). The pyrophosphatase has physiological roles in maintaining cytosolic pH status and turnover of pyrophosphate (291). Evidence indicates that salt treatment both up-regulates and down-regulates pyrophosphatase activity (23).

Na^+ and Cl^- Transport Across the Plasma Membrane Na^+ and Cl^- transport across the plasma membrane in a hypersaline environment must be considered in two cellular contexts, after salt stress shock and after re-establishment of ionic homeostasis. Immediately after salt stress, the H^+ electrochemical gradient is altered. Influx of Na^+ dissipates the membrane potential, thereby facilitating uptake of Cl^- down the chemical gradient. An anion channel has been implicated in this passive flux (50, 100, 242). However, after steady-state conditions are re-established, including establishment of an inside negative plasma membrane potential of -120 to -200 mV, Cl^- influx likely requires coupling to downhill H^+ translocation, presumably via a Cl^--H^+ symporter of unknown stoichiometry (201).

The precise transport system responsible for Na^+ uptake into the cell is still unknown. Physiological data indicate that Na^+ competes with K^+ for intracellular influx because these cations are transported by common proteins (Figure 1) (7, 26, 182). Whereas K^+ is an essential co-factor for many enzymes, Na^+ is not. The need for Na^+ as a vacuolar osmolyte in saline environments may be the reason why plants have not evolved transport systems that completely exclude Na^+ relative to K^+. K^+ and Na^+ influx can be differentiated physiologically into two principal categories, one with high affinity for K^+ over Na^+ and the other for which there is lower K^+/Na^+ selectivity. Many K^+ transport systems have some affinity for Na^+ (26, 43, 223). These include inward rectifying K^+ channels (9, 288); Na^+-K^+ symporter (224); K^+ transporters (71, 128, 222); voltage-dependent, nonselective, outward-rectifying cation channel that mediates Na^+ influx upon plasma membrane depolarization (26, 225); and voltage-independent cation channels (7, 26, 275). The size of the Na^+ electrochemical potential gradient across the plasma membrane, resulting from the large inside negative membrane potential and the Na^+ chemical potential, implicates electrophoretic flux as the principal mode of intracellular Na^+ influx. Low-affinity K^+ transport has been ascribed to inward K^+ channels (152). However, at least two inward K^+ channels mediate high-affinity uptake of K^+ (9, 102). The Na^+-K^+ transporter and K^+ transporters, with dual high and low affinity, may contribute substantially to Na^+ influx. Regardless of the transport systems involved in K^+ and Na^+ acquisition, it is clear that a single genetic locus (*SOS3*), which encodes a signal transduction intermediate, modulates high- and low-affinity K^+ uptake (145, 146). Ca^{2+} can facilitate higher K^+/Na^+ selectivity, and interestingly, external Ca^{2+} can suppress the K^+ acquisition deficiency of *sos3*. The transport systems involved in K^+ acquisition by roots and loading to the xylem have been summarized recently (152, 153).

Na^+ efflux across the plasma membrane and compartmentalization into vacuoles or pre-vacuoles is mediated presumably by Na^+/H^+ antiporters, regardless of whether or not the membrane potential is inside positive or negative. Although Na^+-ATPase activity has been described for some algae, there is no evidence that such a pump exists in cells of higher plants (26, 182). To date, a plasma membrane antiporter is implicated from physiological data obtained with isolated membrane vesicles (26, 57, 97). However, since so many plant transport systems are analogous to yeast, it is likely that NHA1/SOD2 orthologs will be identified that are responsible for Na^+ efflux across the plasma membrane (115, 204).

Na^+ and Cl^- Vacuolar Compartmentation Compartmentation analyses indicate that in seawater concentrations of NaCl, both Na^+ and Cl^- are sequestered in the vacuole of plant cells and represent the primary solutes affecting osmotic adjustment in this compartment (22, 90, 249). Na^+ compartmentation in the vacuole requires energy-dependent transport, and an immediate effect of NaCl treatment is vacuolar alkalization (27, 82, 156). Na^+/H^+ antiporter activity has been associated with tonoplast vesicles (27, 57), and this is presumed to be at least partially responsible for the alkalization. Recently, plant cDNAs encoding NHE-like proteins were isolated that can functionally complement a yeast mutant deficient for the endomembrane Na^+/H^+ transporter, NHX1 (10, 72, 76; FJ Quintero, MR Blatt & JM Pardo, unpublished). Overexpression of an NHE-like antiporter substantially enhanced salt tolerance of *Arabidopsis,* confirming the function of the transporter in Na^+ compartmentation (10). The tonoplast Cl^- transport determinants are predicted to be a channel or a carrier that couples Cl^- influx to the H^+ gradient. A +50 mV (inside positive) tonoplast membrane potential would be sufficient to facilitate an almost tenfold concentration of Cl^- in the vacuole based on electrophoretic flux through an anion permeable channel (16, 50, 99). Secondary active transport (H^+/anion antiporter) has been proposed also (254).

Ca^{2+} Homeostasis Experimental evidence implicates Ca^{2+} function in salt adaptation. Externally supplied Ca^{2+} reduces the toxic effects of NaCl, presumably by facilitating higher K^+/Na^+ selectivity (46, 137, 145). High salinity also results in increased cytosolic Ca^{2+} that is transported from the apoplast and intracellular compartments (132, 151). The resultant transient Ca^{2+} increase potentiates stress signal transduction and leads to salt adaptation (132, 163, 197, 221, 277). A prolonged elevated Ca^{2+} level may, however, also pose a stress; if so, re-establishment of Ca^{2+} homeostasis is a requisite. Ca^{2+} transport systems have been summarized in recent reviews (192, 218).

Endocytosis and Prevacuolar Trafficking The extent to which endocytotic or prevacuolar compartments contribute to intracellular ion compartmentation in plants is yet to be assessed (157). The prevacuole compartments may be sorted for exocytosis or fusion with the tonoplast, or retained in the cells without fusion to the central vacuole (70, 142). In fact, a dramatic characteristic of NaCl-adapted tobacco cells is the presence of numerous small vacuoles (prevacuoles), and transvacuolar strands (22, 24, 41). This and other cytological changes of adapted cells are characteristic of meristematic cells (60), and indicative that a form of developmental arrest, perhaps as a result of altered cell cycle progression or cell division to expansion transition, occurs during salt adaptation (24, 41). Note also that cytological characteristics of plasmolyzed root cells of the halophyte *Atriplex nummularia* include plasma membrane invaginations, vesiculation, Hechtian strands, and numerous small vacuoles (prevacuoles) (183). Frommer et al (70) have suggested that *AtNHX1* (encodes a tonoplast Na^+/H^+ antiporter) overexpression in transgenic plants may affect salt tolerance (10) by disrupting the trafficking of the Na^+/H^+ antiporter such that it is localized not only to prevacuoles but also to the tonoplast or plasma membrane, thereby increasing the capacity for compartmentation of Na^+ away from the cytosol.

Osmolyte Biosynthesis

One response, which is probably universal, to changes in the external osmotic potential is the accumulation of metabolites that act as "compatible" solutes, i.e. they do not inhibit normal metabolic reactions (38, 68, 284). With accumulation proportional to the change of external osmolarity within species-specific limits, protection of structures and osmotic balance supporting continued water influx (or reduced efflux) are accepted functions of osmolytes. Frequently observed metabolites with an osmolyte function are sugars (mainly sucrose and fructose), sugar alcohols (glycerol, methylated inositols), and complex sugars (trehalose, raffinose, fructans). In addition, ions (K^+) or charged metabolites [glycine betaine, dimethyl sulfonium propionate (DMSP), proline and ectoine (1,4,5,6-tetrahydro-2-methyl-4-carboxyl pyrimidine)] are encountered (Figure 2). The accumulation of these osmolytes is believed to facilitate "osmotic adjustment," by which the internal osmotic potential is lowered and may then contribute to tolerance (52, 149, 160). Compatible solutes are typically hydrophilic, which suggests they could replace water at the surface of proteins, protein complexes, or membranes, thus acting as osmoprotectants and nonenzymatically as low-molecular-weight chaperones.

This biophysical view of the function(s) of such solutes is supported by many studies [for reviews see (30, 181, 285)]. Compatible solutes at high concentrations can reduce inhibitory effects of ions on enzyme activity (39, 245) to increase thermal stability of enzymes (74), and to prevent dissociation of enzyme complexes, for example, the oxygen-evolving complex of photosystem II (194). One argument raised against such studies is the high effective concentration necessary for protection in vitro, which is usually not matched in vivo. Considering the cellular concentration of proteins, protection may be achieved at lower solute concentrations found in vivo. It may not be the concentration in solution that is important. Glycine betaine, for example, protects thylakoids and plasma membranes against freezing damage or heat destabilization even at low concentration (118, 290). This indicates that the local concentration at a surface may be more important than the absolute amount.

Metabolic Pathways Typically, pathways leading to osmolyte synthesis are connected to pathways in basic metabolism that show high flux rates (28, 160, 189). Examples are the biosynthetic pathways leading to proline (52), glycine betaine (160, 208), D-pinitol (108, 272), or ectoine (75). The pathways from which these osmolytes originate are situated in amino acid biosynthesis from glutamic acid (proline) or aspartate (ectoine), choline metabolism (glycine betaine), and *myo*-inositol synthesis (pinitol). The enzymes required for pathway extensions that lead to these osmolytes are often induced following stress. In higher plants this has been documented for glycine betaine (93), D-pinitol (108, 180, 272), proline (177, 287, 289), and the operon, encoding three enzymes, required for ectoine accumulation is also stress induced (149, 190). Ectoine may be restricted to bacteria; it has not yet been detected in plants, but transgenic expression of the three genes from *Halomonas elongata* and *Marinococcus halophilus* in tobacco, leading to ectoine amounts in the micromolar range, provided some protection (176; M Rai, G Zhu, N Jacobsen, A Somogoyi & HJ Bohnert, unpublished). DMSP synthesis, found in Gramineae and Compositae and in many algae, is increased under conditions of salinity (262). The metabolic requirements for the synthesis of DMSP and the related glycine betaine are providing a paradigm to engineer metabolic pathways for osmotic stress tolerance (189).

Osmolyte Functions Irrespective of the seemingly clear-cut importance of osmolytes, what the various accumulators might signify remains a matter of debate.

The terms osmoprotection and osmotic adjustment focus on biophysical and physiological characters. Following this view, the multiplicity of accumulating metabolites may be interpreted to reflect the relative enzymatic capacities of various biochemical pathways in different species. In reviewing physiological studies, Greenway & Munns (80) come down on the side of pathology, or favor a view of osmolytes as sinks of reducing power following metabolic disturbance that might be mobilized as a source of carbon and nitrogen once stress is relieved. More recently, new arguments interpreting biochemical and molecular analyses, based on mutant analysis and transgenic approaches, have been introduced into the discussion (29, 30, 95, 189, 294). Plants engineered to synthesize and moderately accumulate a number of osmolytes showed marginally improved performance under abiotic stress conditions. The effects seen with modest increases in mannitol, fructans, trehalose, ononitol, glycine betaine, or ectoine, and with strong increases in proline amount indicate that the purely osmotic contribution of these metabolites to stress tolerance may not describe their function completely, i.e. that the pathway leading to a particular osmolyte may be more important than accumulation per se (30, 95, 112, 181).

Proline accumulation provides an example. Under stress, the imbalance between photosynthetic light capture and NADPH utilization in carbon fixation may alter the redox state and lead to photoinhibition. Proline synthesis, following transcriptional activation of the NAD P H-dependent P5C-synthetase (P5CS), could provide a protective valve whereby the regeneration of $NADP^+$ could provide the observed protective effect (52, 95, 125, 271, 286). An argument against accumulation per se can be seen from the analysis of the *sos1* mutant of *Arabidopsis,* defective in K^+ uptake in which proline accumulated to levels twofold higher than in wild type under moderate salt stress but the plants were not tolerant (53, 145). The regulated synthesis of both proline synthesis and degradation enzymes indicate that cycling between precursor and product may be important (130, 175, 270). Supportive evidence for a nonosmotic function of at least some osmolytes comes from transgenic overexertion of osmolyte-producing enzymes (233, 234) with moderate accumulation of osmolytes leading to the protection of the carbon reduction cycle by reducing the production of hydroxyl radicals generated by a Fenton reaction. There may be more than one function for a particular osmolyte and, based on results from in vitro experiments (92, 192), different compatible solutes could have different functions. Among these functions a role in prevention of oxygen radical production or in the scavenging of reactive oxygen species (ROS) may be paramount (12, 185, 187).

Transgenic plants bave been generated to probe the effect of ROS scavenging on salinity stress tolerance, based on observations of gene expression changes in stressed plants. A putative phospholipid hydroperoxide glutathione peroxidase, PHGPX, transcript increased during salt stress in *Arabidopsis* and citrus (83, 252) and, also in citrus, transcripts and enzyme activities of Cu/Zn-SOD, glutathione peroxidase, and a cytosolic APX (83, 104). Catalase-deficient (antisense) tobacco showed enhanced sensitivity to oxidative stress under conditions of high light and salinity (276).

ROS scavenging as an important component of abiotic stress responses is documented by mutant analysis. The ascorbic acid-deficient *Arabidopsis* semidominant, *soz1* accumulates only 30% of ascorbate compared with wild type, and plants show significantly higher sensitivity to oxidative stress conditions (45). Further support comes from the study of transgenic models, which have been generated to study antioxidant defenses (5, 47, 187, 191, 244). Overexpression of genes leading to increased amounts and activities of mitochondrial Mn-SOD, Fe-SOD, chloroplastic Cu/Zn-SOD, bacterial catalase, and glutathione S-transferase/glutathione peroxidase can increase the performance of plants under stress (31, 84, 85, 217, 239, 268).

Water Uptake and Transport

A few studies are pertinent to water movement under salinity stress conditions. For maize roots (but not for tobacco) high salinity caused a considerable reduction in water permeability in the cortex (13, 266), reducing the osmotic water permeability by as much as fivefold. Changes in the osmotic water permeability were reflected in changes in root hydraulic conductivity due to the fact that most of the water was flowing around cells (14). Conceivably, such changes may be caused by reducing the probability of opening of water channels or by a change in their number. Species- and stress-specific changes in water permeability may be caused by AQP phosphorylation, as demonstrated by Johansson et al (116, 117). Low water potential reduces phosphorylation of the plasma membrane PM28A in spinach. Phosphorylation is carried out by a Ca^{2+}-dependent membrane-bound protein kinase (116). PM28A, expressed in *Xenopus* oocytes, is phosphorylated at the site phosphorylated in vivo, and decreased phosphorylation reduced water permeability in oocytes expressing PM28A (117), which suggests that PM28A could be involved in water flow through leaf tissue. Phosphorylation has also been reported for a seed-specific TIP (158).

The control over plant AQP amount or activity would be particularly important during stress, and evidence for mobility under stress conditions is mounting (17, 269). The amount of MIP transcripts (282) and proteins (Kirch et al, unpublished) varies during salt stress in *M. crystallinum,* as do protein locations (269). Drought and salt stresses regulate protein amounts for the location of AQPs in the tonoplast, internal vesicles, and plasma membrane differently, indicating the existence of signaling pathways that exert control over water flux. It is intuitively clear that water channels (or water/solute channels, "aquaglyceroporins") should have significance for water relations in stressed plants, but the dataset available on the control of water uptake in salt-stressed roots is small. Given the wealth of information from the study of AQPs in the last few years, more surprises can be expected (226).

Long-Distance Response Coordination

Although intracellular phenomena of the salt stress response are well described, much less is known about the organismal response coordination in different organs during development in a changing environment and even less is known about sensing mechanisms at membranes. Communication along the plant body has been revealed in the induction of flowering in the systemic signaling of pathogen attack (6, 20) or upon the disturbance, due to osmolyte changes, of source-sink relationships (215). In this context, the interaction and crosstalk between carbohydrate status, involving glucose sensing by hexokinases, and ethylene-based signal transduction (135, 292) could be a paradigm for understanding plant abiotic stress responses. Long-distance signaling under stress has been associated with root to shoot ABA transport (140, 259), possibly also in cytokinin/auxin transport, or through ethylene (58, 94, 188, 260), and by the coordinated interplay of different growth regulators (127, 170). However, more studies are needed. Although the signals appear to be predominantly chemical (132), action potentials derived from local ion movements could also be involved (32).

The evidence for salt stress-specific ABA transport stems from physiology (171, 172, 243, 259) and, more recently, from mutant analysis in *Arabidopsis* (64). Auxin transport, cytokinin conjugates to inositols, glucose, or amino acids, have mainly been viewed under aspects of development and tropisms (61, 120). The analysis of mutants with altered ABA responses indicated regulatory roles for the loci ABA1 (ABA biosynthetic pathway), ABI1 (calcium-modulated protein phosphatase 2C), and AXR2 (resistance to auxin and ethylene) in the differential expression of P5CS genes responsible for proline accumulation under stress, suggesting signals arriving from outside the responding cells (251). A similar conclusion was reached in the study of a nonpermeable ABA-BSA conjugate, which elicited intracellular gene expression and activated K^+ currents (114). Likewise, evidence has been presented for the movement of glutathione and its conjugates and other redox carriers or oxidants into the apoplast (21, 56, 69). It seems that within the signaling cascades described by the classical plant growth regulators, a variety of other compounds will have to be included, for example, sugars and maybe other osmolytes and extracellular enzymes such as invertases, oxalate oxidases, glutathione S-transferases, and oxidoreductases. These generate a variety of chemical signals that may converge on the production of radical oxygen species, and transmembrane signal mediators, sensitive to redox states, could transfer excitation to the cell's interior eliciting stress responses.

Recently, evidence for a metabolic connection between leaf photosynthetic capacity and root Na^+ uptake has been obtained. Salt stress-dependent increases in the transport of *myo*-inositol through the phloem from leaves to the root system could be correlated with increased Na^+ transport in the xylem (179, 180). A hypothetical connection between *myo*-inositol as an indicator of photosynthetic activity in leaves and increased Na^+ transport is provided by the detection of Na^+/*myo*-inositol symporter proteins, ITRs. Salt stress-inducible ITR mRNAs are located to the vascular system of the root and stem (S Chauhan, F Quigley, DE Nelson, Y Ran, HJ Bohnert, unpublished). Increased amounts of inositols may, however, have another function. IAA promotes rapid changes in phosphatidylinositol (PI) metabolites (62), and increased concentrations of inositols could enhanced PI metabolism. In yeast, the pathway leading to PI-based secondary messengers is activated by osmotic stress with a crucial role for phosphatidylinositol-3P 5-hydroxyl kinase leading to PI-(3,5)-diphosphate (54). Altered PI signaling is indicated by the ABA- and osmotic stress-mediated induction of a PI-4-phosphate 5-kinase in *Arabidopsis* (165).

REGULATORY MOLECULES

We have conceptually divided genetic determinants of salt tolerance into two classes: encoding effectors responsible for remodeling the plant during adaptation, and regulatory genes that control the expression and activity of the effectors. Rapid technical developments and insightful use of yeast as a model organism have greatly facilitated our discovery of these regulator genes (155, 230, 246). Plant growth regulator/hormone physiological function(s) in plant stress responses is not surveyed in this report, as relevant and recent treatises are available (51, 127, 169, 170). Instead, plant growth regulator function is described in the context of stress regulatory determinants that have been linked to osmotic tolerance. We now summarize and provide a framework for the organization of these plant regulatory genes, comparing them to the yeast model (Figure 3).

Plant Signaling Genes as Determinants of Salt Tolerance Based on Functional Complementation of Yeast Mutants

Determinants of plant stress tolerance have been identified, in recent years, by functional complementation of osmotic, sensitive yeast mutants. A putative MAP (mitogen-activated protein) kinase (MAPK) has been identified from *Pisum sativum* (PsMAPK), with 47% primary sequence identity to Hog1p, which is the MAP kinase in the yeast osmoregulatory pathway that controls glycerol accumulation. PsMAPK functionally suppressed salt-induced cell growth inhibition of *hog1* (202). Combinations of *Arabidopsis* proteins ATMEKK1 (MAPKKK) and MEK1 (MAPKK), or ATMEKK1 and ATMKK2 (MAPKK) suppressed growth defects of *pbs2Δ* (wild-type allele encodes the MAPKK of the HOG pathway), implicating these as functional components of an osmotic stress MAP kinase cascade (107).

AtDBF2 kinase of *Arabidopsis* mediated functional sufficiency of yeast cells for LiCl tolerance (138). Overexpression of this serine/threonine kinase enhanced tolerance of plant cells to numerous stresses, including salt and drought. By analogy with yeast, it is presumed that At-DBF2 is a cell cycle–regulated protein kinase that is a component of a general transcriptional regulatory complex, like CCR4 (144), that modulates expression of genes involved in osmotic adaptation (138). *AtGSK1* was isolated as a suppressor of the NaCl-sensitive phenotype of a calcineurin-deficient mutant (198). Calcineurin is a protein phosphatase type 2B that is a pivotal intermediate in a signal pathway that regulates ion homeostasis and salt tolerance of yeast (162, 163). AtGSK1 has sequence similarity to the glycogen synthase kinase 3 (GSK3) of mammalian cells and SHAGGY of *Drosophila melanogaster*. *AtGSK1* expression induced transcription of yeast *ENA1,* which encodes a P-type ATPase responsible for Na^+/Li^+ efflux across the plasma membrane. Furthermore,

AtGSK1 suppressed the NaCl-sensitive phenotype of a mutant (*mck1*) deficient for a yeast GSK3 ortholog, implicating that functional complementation is based on kinase activity.

Arabidopsis SAL1 was isolated by complementation of a mutant defective for ENA ATPase activity (205). A family of four genes (*ENA1–4*) that are tandemly arranged in the genome, encodes ENA. *SAL1* complemented the Li^+-sensitive phenotype of a *ena1-4Δ* strain. SAL1 is the plant ortholog of yeast HAL2 and encodes a protein with (2′), 5′-bisphosphate nucleotidase and inositol polyphosphate 1-phosphatase activities. The nucleotidase activity converts adenosine 3′-phosphate 5′-phosphosulfate to 3′(2′)-phosphoadenosine 5′-phosphate and this catalysis functions as part of a sulfur cycle to reduce SO_4^{2-} levels in high Na^+ environments (196). The polyphosphate phosphatase activity is presumed to have a significant function in phosphoinositide signaling (205).

Two *Arabidopsis* transcription factor-encoding cDNAs *STO* (salt tolerance) and *STZ* (salt tolerance zinc finger) suppressed the Na^+-/Li^+-sensitive phenotype of a calcineurin-deficient (*cna1-2Δ* or *cnbΔ*) mutant (143). STO is similar to *Arabidopsis* CONSTANS and is predicted to be a member of a multigene family. *STZ*-mediated salt tolerance is at least partially dependent on *ENA1,* implying a function that involves regulation of the Na^+-ATPase, whereas STO function is independent of *ENA1.*

NtSLT1 is a tobacco protein that suppresses the Na^+-/Li^+-sensitive phenotype of a calcineurin-deficient (*cnbΔ*) mutant (TK Matsumoto, JM Pardo, S Takeda, I Amaya, K Cheah, et al, unpublished). Functional complementation of *cnbΔ* by NtSLT1, and by the *Arabidopsis* ortholog AtSLT1, can be mediated only by a peptide that is truncated at the N terminus, indicating the presence of an autoinhibitory domain on the native protein. Complementation involves transcriptional activation of *ENA1* as well as a function(s) independent of the ATPase.

Plant Signaling Factors that Enhance Salt Tolerance of Plants

Numerous signal or signal-like molecules have now been identified and presumed, through some evidence, to function in plants as mediators of osmotic adaptation (Figure 3). Some of these affect osmotic tolerance in genetically modified plants (Figure 3). Transcriptional modulation has always been predicted to play a major role in the control of plant responses to salt stress (240, 241, 294). Transcription factors have been identified based on interaction with promoters of osmotic/salt stress–responsive genes. These factors participate in the activation of stress-inducible genes, and presumably lead to osmotic adaptation (1, 147, 231, 240, 241, 278). Since the promoters that are controlled by these transcription factors are responsive to several environmental signals, it is not clear which transcription factors, if any, function only in salt stress responses, or if salt-specific transcriptional regulation alone is a requisite component of salt tolerance in planta. ABA-deficient and -insensitive mutants have been used to delineate transcription factors as components of osmotic stress signal transduction pathways that either involve or are independent of the growth regulator (241). Promoters of ABA-dependent osmotic stress-responsive genes include regulatory elements that interact with basic leucine-zipper motif (bZIP), MYB, or MYC domains in DNA binding proteins (1, 235, 241). The bZIP transcription factors interact with the ABA-responsive element (ABRE) (241). The promoter of the osmotic/salt stress-responsive *rd22* gene contains signature recognition motifs for MYC and MYB transcription factors. DNA binding proteins, rd22BP1 (MYC) and Atmyb2 (MYB), interact with their corresponding *cis*-elements and *trans*-activate *rd22*. However, the *rd22* promoter does not contain ABRE motifs, which indicates that ABA dependency is mediated through another process. At least some ABA-dependent transcriptional activation appears to involve Ca^{2+}-dependent protein kinases (CDPKs) (231). Transcription factors thought to function in osmotic/salt stress gene induction, independent of ABA, include dehydration response element (DRE) binding proteins (147, 248). Two gene families have been characterized, *DREB1* (which includes the previously identified *CBF1*), and *DREB2* (which *trans*-activates the osmotic/salt-responsive *rd29A*). Both DREB1 and DREB2 family members also have domains that bind ethylene-responsive elements (147, 248).

DREB1A overexpressing transgenic plants exhibited constitutive activation of stress-responsive genes and enhanced freezing, dehydration, and salt tolerance (123, 147). Driving *DREB1A* with the stress-inducible *rd29A* promoter substantially increases salt tolerance, with minimal adverse growth effects in the absence of stress (123). Ectopic overexpression of *CBF1/DREB1B* in transgenic plants induces cold-responsive genes and enhances freezing tolerance (111, 261). Overexpression of the zinc finger transcription factor ALFIN1 activates *MsPRP2* (NaCl-responsive gene) expression and increases salt tolerance of alfalfa (278).

In addition to transcription factors, some other signaling components affect a salt-tolerance phenotype in plants (Figure 3). Notable among these is the *SOS3* gene that encodes a Ca^{2+} binding CNB-like protein. Mutation of *SOS3* produces salt-sensitive plants with altered K^+ transport characteristics (145, 146). Another CNB-like protein AtCLB1 is able to suppress salt sensitivity in CNB yeast mutants, but only in the presence of a mammalian CNA subunit (134). The yeast *CNA/CNB* genes are also able to confer salinity tolerance when overexpressed in transgenic plants (195), further implicating these regulatory molecules in plant stress signaling. It seems clear that a Ca^{2+} binding, CNB-like molecule plays an important role in osmotic stress responses in plants. Mutation of the *ABI1* gene that renders plants insensitive to ABA also causes an osmotic stress-sensitive phenotype (11, 63, 133), confirming the important role of ABA in stress signaling.

Genomic Bioinformatics Will Allow Rapid Increase in Signal Gene Identification

So far, few genes encoding plant signaling proteins controlling responses to osmotic stress have been found strictly by mining gene sequence databases (Figure 3). Divergence between species may sometimes make this approach difficult. For example, considerable effort to locate the plant version of CNB was expended, and although CNB-like proteins were eventually identified, none can interact with yeast CNA1/2 (134), which indicates that their function is apparently not the same as in yeast. The use of sequence comparisons to identify potential signal genes will undoubtedly increase rapidly and dramatically. As gene knockout technology in *Arabidopsis* becomes increasingly facile, a quick and easy test of fidelity will also be available for bioinformatic mining (15, 126, 246). Eventually this will become the dominant approach to gene discovery. Phenotypic changes such as increases in salt tolerance caused by ectopic overexpression of regulatory genes in plants are also important for crop improvement, but in many cases may not be able to address the question of necessity, i.e. whether a regulatory protein is necessary for salt tolerance in plants. In one scenario, a gene of a multigene family may be sufficient to confer a stress-tolerance phenotype when overexpressed in plants but may not be necessary because other family members can substitute for its function owing to functional redundancies. In another scenario, a gene that normally does not function in salt stress responses may confer a stress-tolerance phenotype when ectopically expressed because signaling specificity can be lost during ectopic overexpression. It is therefore important to isolate or construct loss-of-function mutations even for genes that are known to confer tolerance phenotypes when overexpressed in order to understand precisely their interrelationships.

Organization of Plant Signaling Genes into Pathways

Although we have identified many genes that encode signal transduction proteins that are potentially important to osmotic stress adaptation in plants, we still know little of how these gene products function in a signal pathway. It is evident from Figure 3 that, compared to yeast, we are virtually unable to draw any functional relationship directly between these many genes. Why are we able to place the yeast genes in an ordered functional signal pathway but cannot do so for the plant genes? Even though some plant genes have been shown to affect stress tolerance of yeast or even of plants, their relationships to each other remains mostly a mystery. It is known that the MP2C gene product somehow reduces the activity of SAMK (164). The plant transcription factors listed in Figure 3 all interact with particular *cis*-elements and control expression of target genes, some of which are known. For instance, ABI1 negatively controls downstream target genes, whereas AtCDPK1 positively controls the same target promoter (250). The tyrosine, dual-function phosphatase AtDsDTP1 dephosphorylates and inhibits AtMPK4 (86). *VP14* is a stress-induced gene encoding a carotenoid cleavage enzyme that is a rate-limiting step in ABA synthesis (277), thus exerting control over numerous ABA-controlled genes. NaCl downregulates in tobacco the expression of a member of the 14-3-3 protein family that has interactive functions with many signal components (42).

One of the best ways to gather information regarding interactions between plant signal components has been to determine their ability to activate downstream promoters fused to marker genes such as was shown for AtCDPK1 and ABI1 (231).

This kind of information is needed for many other signal components. In a comprehensive genetic screen, Ishitani et al (110) isolated a large number of *Arabidopsis* mutations that either reduce or enhance salt stress induced gene expression. The authors used the salt-, drought-, cold-, and ABA-responsive *rd29A* promoter fused to the firefly luciferase gene to report plant responses to salt treatments. The firefly luciferase reporter allows large-scale, nondestructive screening of mutants by real-time, high throughout low light imaging technology (110). The recessive *los* (*l*ow expression of *o*smotic stress-responsive genes) mutants define positive regulators of salt stress gene expression, while the recessive *hos* (*h*igh expression of *o*smotic stress-responsive genes) and *cos* (*c*onstitutive expression of *o*smotic stress-responsive genes) mutations are presumed to be in genes that negatively regulate stress responses. Some *los*, *hos*, and *cos* mutations also alter plant responses to ABA or low temperature, thus revealing points of crosstalk between salt, cold, and ABA signaling pathways. Molecular cloning of these mutations as well as their interactors, enhancers, and suppressors (Figure 4, see color plates) will eventually allow plant regulatory pathways to be built, like the ones in yeast (Figure 3).

Direct physical interaction information by yeast two-hybrid or by in vitro assays also is lacking for most plant gene products so far. Yeast two-hybrid assays have been used to demonstrate that the AtMEKK1 protein interacts with AtMEKK2/MEK and AtMPK4 (167, 168). Both genetic and biochemical evidence for interactions between many more of these plant genes and their products is now needed. On the genetic side, allelism, additivity and epistasis data must be obtained now for the plant genes where function is evident (see Figure 4). Conducting biochemical interaction studies in both yeast two-hybrid and in vitro assays as well as ability to modify through phosphorylation, dephosphorylation, etc, are logical next steps for these genes. Additional signal pathway participants can also be identified by activation or suppression screens with known mutants. Such information is now needed to establish the exact roles of signal components in stress-signaling to allow ordered signaling pathways to be revealed.

One potential pathway that is increasingly meeting these criteria involves the genes mutated at the *SOS1, 2,* and *3* loci. Genetic experiments determined that these loci are not additive (same pathway) and that the *sos1* phenotype is epistatic to the other two, placing it furthest downstream. Recent cloning and interaction studies of SOS2 and 3 indicate that SOS3 is a CNB-like Ca^{2+} binding protein that physically interacts with and activates the SOS2 protein, which is similar to the SNF1/AMPK family of kinases (91).

As new genes enter the camp of potential mediators of stress-adaptation through bioinformatic or expression studies, evidence for their function in mediating adaptation through overexpression and gene knock-out studies must follow (Figure 4). Armed with this evidence for function, further genetic studies and interaction determinations will finally allow us to draw complete signaling pathways for plant osmotic stress responses.

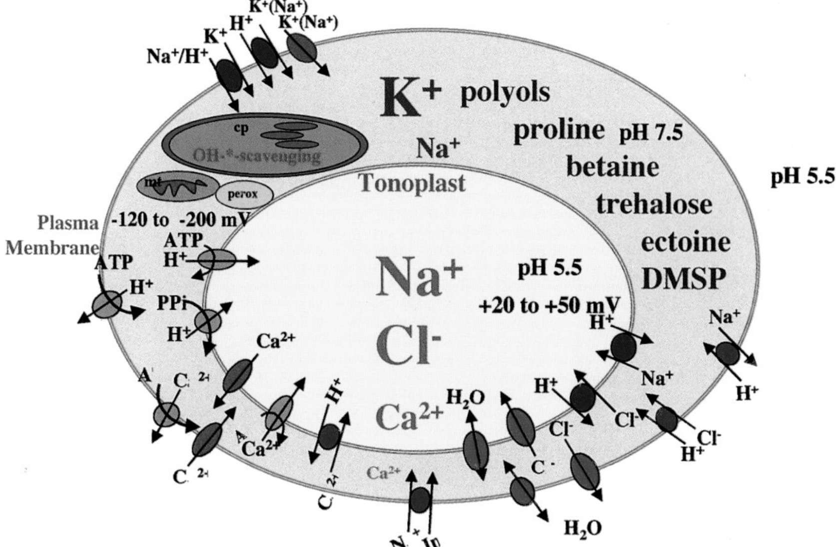

Figure 1 Cellular homeostasis established after salt (NaCl) adaptation. Indicated are the osmolytes and ions compartmentalized in the cytoplasm and vacuole, transport proteins responsible for Na and Cl^- homeostasis, water channels, and electrochemical potentials across the plasma membrane and tonoplast. Included are organelles (chloroplast (cp), mitochondrion (mt), and peroxisome (perox)) for which the importance of ROS-scavenging is implicated.

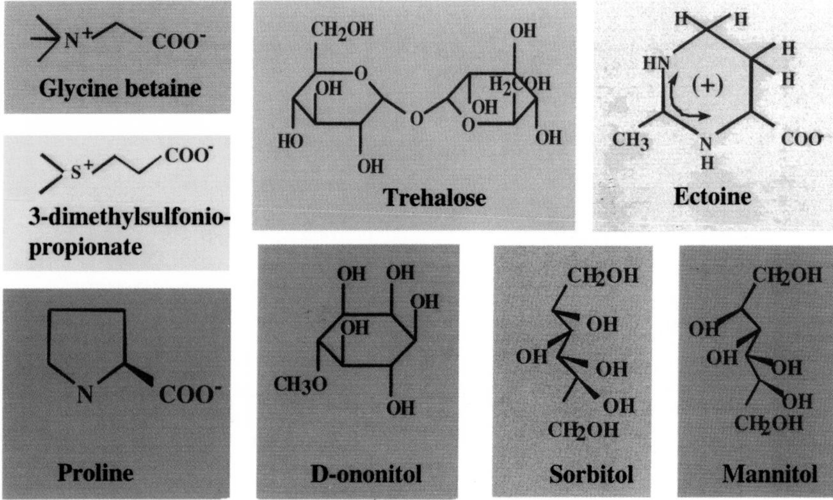

Figure 2 Osmolytes/Osmoprotectants. Listed are common osmolytes involved in either osmotic adjustment or in the protection of structure. In all cases, protection has been shown to be associated with accumulation of these metabolites, either in naturally evolved systems or in transgenic plants (after 28; 189).

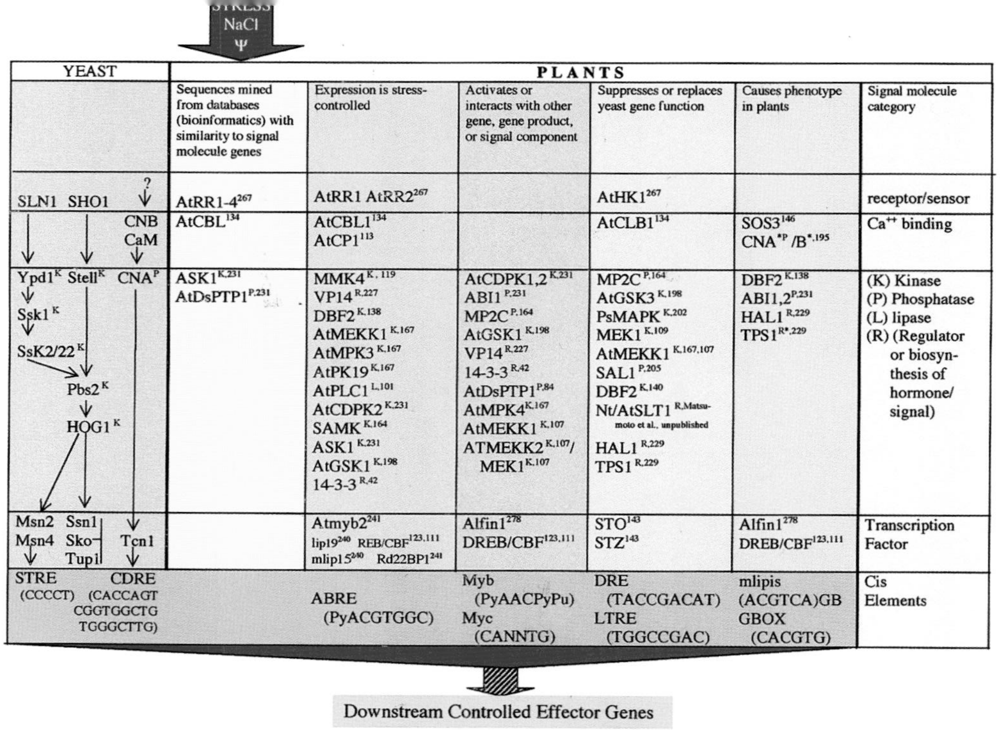

STRESS
NaCl
Ψ

YEAST	PLANTS					
	Sequences mined from databases (bioinformatics) with similarity to signal molecule genes	Expression is stress-controlled	Activates or interacts with other gene, gene product, or signal component	Suppresses or replaces yeast gene function	Causes phenotype in plants	Signal molecule category
SLN1 SHO1 ?	AtRR1-4[267]	AtRR1 AtRR2[267]		AtHK1[267]		receptor/sensor
CNB CaM	AtCBL[134]	AtCBL1[134] AtCP1[113]		AtCLB1[134]	SOS3[146] CNA[*P] /B[*,195]	Ca[++] binding
Ypd1[K] Stell[K] CNA[P] Ssk1[K] SsK2/22[K] Pbs2[K] HOG1[K]	ASK1[K,231] AtDsPTP1[P,231]	MMK4[K, 119] VP14[R,227] DBF2[K,138] AtMEKK1[K,167] AtMPK3[K,167] AtPK19[K,167] AtPLC1[L,101] AtCDPK2[K,231] SAMK[K,164] ASK1[K,231] AtGSK1[K,198] 14-3-3[R,42]	AtCDPK1,2[K,231] ABI1[P,231] MP2C[P,164] AtGSK1[K,198] VP14[R,227] 14-3-3[R,42] AtDsPTP1[P,84] AtMPK4[K,167] AtMEKK1[K,107] ATMEKK2[K,107]/ MEK1[K,107]	MP2C[P,164] AtGSK3[K,198] PsMAPK[K,202] MEK1[K,109] AtMEKK1[K,167,107] SAL1[P,205] DBF2[K,140] Nt/AtSLT1[R,Matsumoto et al., unpublished] HAL1[R,229] TPS1[R,229]	DBF2[K,138] ABI1,2[P,231] HAL1[R,229] TPS1[R*,229]	(K) Kinase (P) Phosphatase (L) lipase (R) (Regulator or biosynthesis of hormone/ signal)
Msn2 Ssn1 Msn4 Sko Tup1 Tcn1		Atmyb2[241] lip19[240] REB/CBF[123,111] mlip15[240] Rd22BP1[241]	Alfin1[278] DREB/CBF[123,111]	STO[143] STZ[143]	Alfin1[278] DREB/CBF[123,111]	Transcription Factor
STRE (CCCCT) CDRE (CACCAGT CGGTGGCTG TGGGCTTG)		ABRE (PyACGTGGC)	Myb (PyAACPyPu) Myc (CANNTG)	DRE (TACCGACAT) LTRE (TGGCCGAC)	mlipis (ACGTCA)GB GBOX (CACGTG)	Cis Elements

Downstream Controlled Effector Genes

Figure 3 Osmotic stress regulatory molecules in plants compared to categories of yeast signal components. Vertical columns indicate type of evidence for participation of gene in plant stress signaling. Horizontal rows indicate category of signal component organized according to the yeast model hierarchy. Superscripts indicate component subtypes and appropriate references. * denotes yeast gene. For all yeast signal genes see Reference (87).

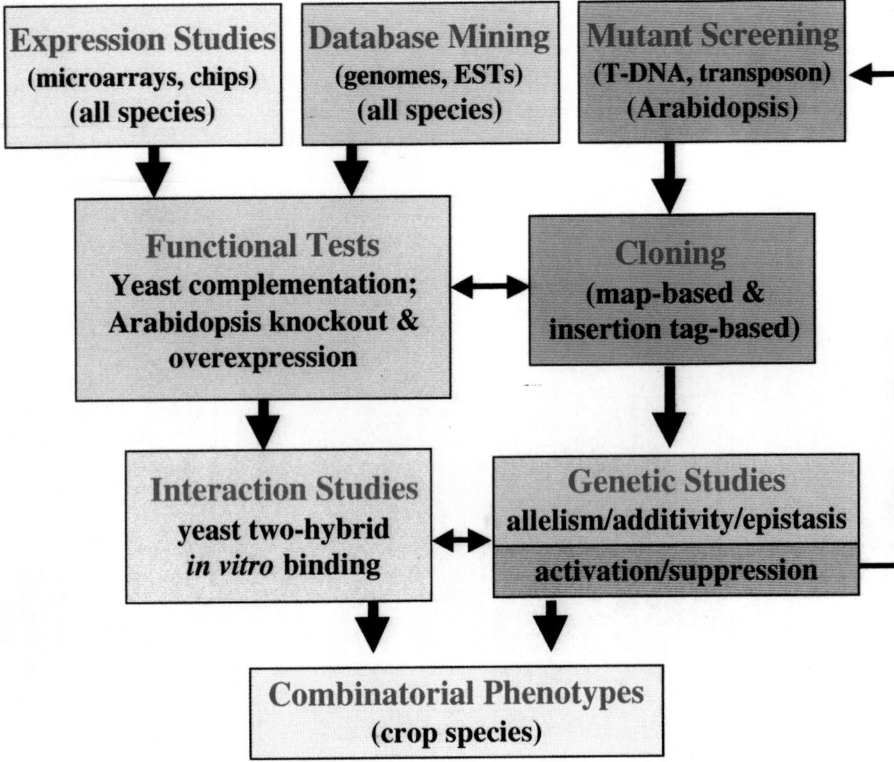

Figure 4 Algorithm for discovering stress tolerance determinants.

KNOWING ALL OSMOTIC STRESS TOLERANCE DETERMINANTS: A Tall Order About to be Filled?

Although genes that participate in signal transduction ought to be constitutively expressed to allow for rapid sensing of a change in the osmotic environment, they are often also induced by osmotic change, as are some of the genes listed in in Figure 3. As with other signal transduction elements and by analogy to yeast responses, this is most likely a mechanism that functions to amplify the signal response (87). It is now obvious that besides genes involved in signaling, a very large number of plant genes are transcriptionally controlled by osmotic stress, and several previous reviews have summarized these genes (35, 294). These transcripts represent "downstream controlled effector genes," as listed in Table 1, and embody the accumulated knowledge from studies of individual genes and mechanisms. With the advent of microarray technology, this period of research is now coming to an end.

Illustrated in Figure 4 is the prevalent strategy for finding the most important genes that control salt adaptation. The focus on individual components of plant stress-tolerance responses is now rapidly being replaced by global analyses. Technological advances, summarized under the "genomics" label, make it possible to monitor the expression of many or even all genes in an organism simultaneously during the entire life cycle of the organism or in response to different stimuli. Genomics-based approaches are, however, more than a collection of novel instruments, techniques, and data production tools. Although at present a learning curve is needed to use these tools effectively, the opportunities provided by genomics are already revolutionizing experimental approaches and are beginning to produce a staggering amount of data and unexpected results. Knowledge of all osmotically induced genes is within reach because of the rapid development of gene microarray technologies (http://stress-genomics.org; http://www.zmdb.iastate.edu; 15, 103, 126). Also, the imminent completion of the *Arabidopsis* genome sequence will provide a roadmap for plant genes at least in type, if not in actual complexity and function. Equally important will be the genome sequence-based isolation of *Arabidopsis* mutants from the rapidly increasing collections of tagged plants (155, 246). By combining microarray analyses with signal transduction mutants, the subarrays of osmotically responsive genes that are under the control of specific signal components can be determined. For example, a mutation of *SOS3,* or activation of CDPK1 must affect the expression of downstream genes and pathways. With microarray technology it is expected that we can determine these and other subsets of activated genes to ascertain which combinations of signal genes control the broadest array of downstream effectors of salt adaptation. This will allow a systematic and logical approach to decide what combinations of gene modifications should be used for engineering signaling and plant metabolism to maximize salinity tolerance through genetic engineering.

ACKNOWLEDGMENTS

Our work has been or is supported by the US Department of Agriculture (NRI) Plant Responses to the Environment Program; by the National Science Foundation Metabolic Biochemistry and International Programs; by the US Department of Energy Biological Energy Research program; Binational Agricultural Research and Development Program, and by Arizona and Purdue Agricultural Experiment Stations. Generous support has recently been provided by the NSF Plant Genome Initiative. Additional support has been provided to colleagues to work in our laboratories by the Deutsche Forschungsgemeinschaft (Germany), the Japanese Society for the Promotion of Science (JSPS), New Energy Development Organization (Japan), and the Rockefeller Foundation (USA). We wish to thank many colleagues for making available unpublished data and for discussions.

LITERATURE CITED

1. Abe H, Yamaguchi-Shinozaki K, Urao T, Iwasaki T, Hosokawa D, Shinozaki K. 1997. Role of *Arabidopsis* MYC and MYB homologs in drought- and abscisic acid-regulated gene expression. *Plant Cell* 9: 1859–58
2. Adams P, Nelson DE, Yamada S, Chmara W, Jensen RG, et al. 1998. Growth and development of *Mesembryanthemum crystallinum* (Aizoaceae). *New Phytol.* 138:170–90
3. Adams P, Thomas JC, Vernon DM, Bohnert HJ, Jensen RG. 1992. Distinct cellular and organismic responses to salt stress. *Plant Cell Physiol.* 33:1215–23
4. Alia Kondo Y, Sakamoto A, Nonaka H, Yayashi H, et al. 1999. Enhanced tolerance to light stress of transgenic *Arabidopsis* plants that express the *codA* gene for a bacterial proline oxidase. *Plant Mol. Biol.* 40:279–88
5. Allen RD, Webb RP, Schake SA. 1997. Use of transgenic plants to study antioxidant defenses. *Free Radic. Biol. Med.* 23:473–79
6. Alvarez ME, Pennell RI, Meijer PJ, Ishikawa A, Dixon RA, Lamb C. 1998. Reactive oxygen intermediates mediate a systemic signal network in the establishment of plant immunity. *Cell* 92:773–84
7. Amtmann A, Sanders D. 1999. Mechanisms of Na^+ uptake by plant cells. *Adv. Bot. Res.* 29:75–112
8. Amzallag GN, Lerner HR, Poljakoff-Mayber. 1990. Induction of increased salt tolerance in *Sorghum bicolor* by NaCl pretreatment. *J. Exp. Bot.* 41:29–34
9. Anderson JA, Huprikar SS, Kochian LV, Lucas WJ, Gaber RF. 1992. Functional expression of a probable *Arabidopsis thaliana* potassium channel in *Saccharomyces cerevisiae*. *Proc. Natl. Acad. Sci. USA* 89:3736–40
10. Apse MP, Aharon GS, Snedden WA, Blumwald E. 1999. Salt tolerance conferred by overexpression of a vacuolar Na^+/H^+ antiport in *Arabidopsis*. *Science* 285:1256–58
11. Armstrong F, Leung J, Grabov A, Brearley J, Giraudat J, Blatt MR. 1995. Sensitivity to abscisic acid of guard cell K^+ channels is suppressed by *abi1-1*, a mutant *Arabidopsis* gene encoding a putative protein phosphatase. *Proc. Natl. Acad. Sci. USA* 92:9520–24
12. Asada K. 1999. The water-water cycle in chloroplasts: scavenging of active oxygens and dissipation of excess photons.

Annu. Rev. Plant Physiol. Plant Mol. Biol. 50:601–39

13. Azaizeh H, Gunse B, Steudle E. 1992. Effects of NaCl and $CaCl_2$ on water transport across root cells of maize (*Zea mays* L.) seedlings. *Plant Physiol.* 99:886–94
14. Azaizeh H, Steudle E. 1991. Effects of salinity on water transport of excised maize (*Z. mays* L.) roots. *Plant Physiol.* 97:1136–45
15. Baldwin D, Crane V, Rice D. 1999. A comparison of gel-based, nylon filter and microarray techniques to detect differential RNA expression in plants. *Curr. Opin. Plant Biol.* 2:96–103
16. Barkla BJ, Pantoja O. 1996. Physiology of ion transport across the tonoplast of higher plants. *Annu. Rev. Plant Physiol. Plant Mol. Biol.* 47:127–57
17. Barkla BJ, Vera-Estrella R, Kirch H-H, Bohnert HJ, Pantoja O. 1999. Aquaporin localization—how valid are the TIP and PIP labels? *Trends Plant Sci.* 4:86–88
18. Barkla BJ, Vera-Estrella R, Maldonado-Gama M, Pantoja O. 1999. Abscisic acid induction of vacuolar H^+-ATPase activity in *Mesembryanthemum crystallinum* is developmentally regulated. *Plant Physiol.* 120:811–20
19. Bental M, Pick U, Avron M, Degani H. 1990. The role of intracellular orthophosphate in triggering osmoregulation in the alga *Dunaliella salina. Eur. J. Biochem.* 188:117–22
20. Bergey DR, Howe GA, Ryan CA. 1996. Polypeptide signaling for plant defensive genes exhibits analogies to defense signaling in animals. *Proc. Natl. Acad. Sci USA* 93:12053–58
21. Berna A, Bernier F. 1999. Regulation of biotic and abiotic stress of a wheat germin gene encoding oxalate oxidase, a H_2O_2–producing enzyme. *Plant Mol. Biol.* 39:539–49
22. Binzel ML, Hess FD, Bressan RA, Hasegawa PM. 1988. Intracellular compartmentation of ions in salt adapted tobacco cells. *Plant Physiol.* 86:607–14
23. Binzel ML, Ratajczak R. 2000. Function of membrane systems under salinity: tonoplast. See Ref. 136. In press
24. Binzel ML, Hasegawa PM, Handa AK, Bressan RA. 1985. Adaptation of tobacco cells to NaCl. *Plant Physiol.* 79:118–25
25. Binzel ML. 1995. NaCl-induced accumulation of tonoplast and plasma membrane H^+-ATPase message in tomato. *Physiol. Plant.* 94:722–28
26. Blumwald E, Aharaon GS, Apse MP. 2000. Sodium transport in plant cells. *Biochim. Biophys. Acta.* In press
27. Blumwald E, Poole RJ. 1985. Na^+/H^+ antiport in isolated tonoplast vesicles from storage tissue of *Beta vulgaris. Plant Physiol.* 78:163–67
28. Bohnert HJ, Jensen RG. 1996. Strategies for engineering water-stress tolerance in plants. *Trends Biotechnol.* 14:89–97
29. Bohnert HJ, Nelson DE, Jensen RG. 1995. Adaptations to environmental stresses. *Plant Cell* 7:1099–11
30. Bohnert HJ, Shen Bo. 1999. Transformation and compatible solutes. *Scientia Hortic.* 78:237–60
31. Bowler C, Slooten L, Vandenbranden S, De Rycke R, Botterman J, et al. 1991. Manganese superoxide dismutase can reduce cellular damage mediated by oxygen radicals in transgenic plants. *EMBO J.* 10:1723–32
32. Bowles DJ. 1997. The wound response of tomato plants: analysis of local and long-range signalling events. *Essays Biochem.* 32:161–69
33. Braun Y, Hassidim M, Lerner HR, Reinhold L. 1986. Studies on H^+-translocating ATPase in plants of varying resistance to salinity. I. Salinity during growth modulates the proton pump in the halophyte *Atriplex nummularia. Plant Physiol.* 81:1050–56
34. Bray EA. 1993. Molecular responses to water deficit. *Plant Physiol.* 103:1035–40
35. Bray EA. 1997. Plant responses to water deficit. *Trends Plant Sci.* 2:48–54

36. Bressan RA, Hasegawa PM, Pardo JM. 1998. Plants use calcium to resolve salt stress. *Trends Plant Sc.* 3:411–12
37. Bressan RA, Nelson DE, Iraki NM, LaRosa PC, Singh NK, et al. 1990. Reduced cell expansion and changes in cell walls of plants adapted to NaCl. See Ref. 124a, pp. 137–71
38. Brown AD, Simpson JR. 1972. Water relations of sugar-tolerant yeasts: the role of intracellular polyols. *J. Gen. Microbiol.* 72:589–91
39. Brown AD. 1990. *Microbial Water Stress Physiology, Principles and Perspectives.* New York: Wiley & Sons
40. Casas AM, Bressan RA, Hasegawa PM. 1991. Cell growth and water relations of the halophyte *Atriplex nummularia* L., in response to NaCl. *Plant Cell Rep.* 10:81–84
41. Chang P-FL, Damsz B, Kononowicz AK, Reuveni M, Chen Z, et al. 1996. Alterations in cell membrane structure and expression of a membrane-associated protein after adaptation to osmotic stress. *Physiol. Plant.* 98:505–16
42. Chen Z, Fu H, Liu D, Chang P-FL, Narasimhan M, et al. 1994. A NaCl-regulated plant gene encoding a brain protein homolog that activates ADP ribosyltransferase and inhibits protein kinase C. *Plant J.* 6:729–40
43. Chrispeels MJ, Crawford NM, Schroeder JI. 1999. Proteins for transport of water and mineral nutrients across the membranes of plant cells. *Plant Cell* 11:661–75
44. Clarkson DT. 1991. Root structure and sites of ion uptake. In *Plant Roots: The Hidden Half*, ed. Y Waisel, A Eshel, U Kafkafi, pp. 417–53. New York: Marcel Dekker
45. Conklin PL, Williams EH, Last RL. 1996. Environmental stress sensitivity of an ascorbic acid-deficient *Arabidopsis* mutant. *Proc. Natl. Acad. Sci. USA* 93:9970–74
46. Cramer GR, Lynch J, Läuchli A, Epstein E. 1987. Influx of Na^+, K^+, and Ca^{2+} into roots of salt-stressed cotton seedlings. *Plant Physiol.* 83:510–16
47. Creissen G, Broadbent P, Stevens R, Wellburn AR, Mullineaux P. 1996. Manipulation of glutathione metabolism in transgenic plants. *Biochem. Soc. Trans.* 24:465–69
48. Csonka LN, Hanson AD. 1991. Prokaryotic osmoregulation: genetics and physiology. *Annu. Rev. Microbiol.* 45:569–606
49. Cushman JC, DeRocher EJ, Bohnert HJ. 1990. Gene expression during adaptation to salt stress. See Ref. 124a, pp. 173–203
50. Czempinski K, Gaedeke N, Zimmermann S, Müller-Röber B. 1999. Molecular mechanisms and regulation of plant ion channels. *J. Exp. Bot.* 50:955–66
51. Davis WJ, Zhang J. 1991. Root signals and the regulation of growth and development in plants. *Annu. Rev. Plant Physiol. Plant Mol. Biol.* 42:55–76
52. Delauney AJ, Verma DPS. 1993. Proline biosynthesis and osmoregulation in plants. *Plant J.* 4:215–23
53. Ding L, Zhu J-K. 1997. Reduced Na^+ uptake in the NaCl-hypersensitive *sos1* mutant of *Arabidopsis thaliana*. *Plant Physiol.* 113:795–99
54. Dove SK, Cooke FT, Douglas MR, Sayers LG, Parker PJ, Michell RH. 1997. Osmotic stress activates phosphatidylinositol-3, 5-bisphosphate synthesis. *Nature* 390:187–92
55. Dreyer I, Horeu C, Lemaillet G, Zimmermann S, Bush DR, et al. 1999. Identification and characterization of plant transporters using heterologous expression systems. *J. Exp. Bot.* 50:1073–87
56. Droog F. 1997. Plant glutathione S-transferases, a tale of theta and tau. *J. Plant Growth Regul.* 16:95–107
57. DuPont FM. 1992. Salt-induced changes in ion transport: regulation of primary pumps and secondary transporters. In *Transport and Receptor Proteins of Plant Membranes*, ed. DT Cooke, DT Clarkson, pp. 91–100. New York: Plenum

83. Gueta-Dahan Y, Yaniv Z, Zilinskas BA, Ben-Hayyim G. 1997. Salt and oxidative stress: similar and specific responses and their relationship to salt tolerance in citrus. *Planta* 203:460–69
84. Gupta AS, Heinen JL, Holaday AS, Burke JJ, Allen RD. 1993. Increased resistance to oxidative stress in transgenic plants that overexpress chloroplastic Cu/Zn-superoxide dismutase. *Proc. Natl. Acad. Sci. USA* 90:1629–33
85. Gupta AS, Webb RP, Holaday AS, Allen RD. 1993. Overexpression of superoxide dismutase protects plants from oxidative stress: induction of ascorbate peroxidase in superoxide dismutase-overexpressing plants. *Plant Physiol.* 103:1067–73
86. Gupta R, Huang Y, Kieber J, Luan S. 1998. Identification of a dual-specificity protein phosphatase that inactivates a MAP kinase from *Arabidopsis*. *Plant J.* 16:581–89
87. Gustin MC, Albertyn J, Alexander M, Davenport K. 1998. MAP kinase pathways in the yeast*Saccharomyces cerevisiae*. *Microbiol. Mol. Biol. Rev.* 62:1264–300
88. Guy CL. 1990. Cold acclimation and freezing stress tolerance: role of protein metabolism. *Annu. Rev. Plant Physiol. Plant Mol. Biol.* 41:187–223
89. Ha KS, Thompson GA Jr. 1992. Biphasic changes in the level and composition of *Dunaliella salina* plasma membrane diacylglycerols following hypo-osmotic shock. *Biochem.* 31:596–603
90. Hajibagheri MA, Harvey DMR, Flowers TJ. 1987. Quantitative ion distribution within root cells of salt-sensitive and salt-tolerant maize varieties. *New Phytol.* 105:367–79
91. Halfter U, Ishitani M, Zhu J-K. 2000. The *Arabidopsis* SOS2 protein kinase physically interacts with and is activated by the calcium-binding protein SOS3. *Proc. Natl. Acad. Sci. USA.* In press
92. Halliwell B, Gutteridge MC. 1990. Role of free radicals and catalytic metal ions in human disease: an overview. *Methods Enzymol.* 186:1–85
93. Hanson AD, Rathinasabapathi B, Rivoal J, Burnet M, Dillon MO, Gage DA. 1994. Osmoprotective compounds in the Plumbaginaceae: a natural experiment in metabolic engineering of stress tolerance. *Proc. Natl. Acad. Sci. USA* 91:306–10
94. Hare PD, Cress WA, van Staden J. 1997. The involvement of cytokinins in plant responses to environmental stress. *Plant Growth Regul.* 23:79–103
95. Hare PD, Cress WA. 1997. Metabolic implications of stress-induced proline accumulation in plants. *Plant Growth Regul.* 21:79–102
96. Hasegawa PM, Bressan RA, Nelsen DE, Samaras Y, Rhodes D. 1994. Tissue culture in the improvement of salt tolerance in plants. In *Soil Mineral Stresses. Approaches to Crop Improvement. Monogr. Theoret. Appl. Genet.*, vol. 21, ed. AR Yeo, TJ Flowers, pp. 83–125. Berlin: Springer-Verlag
97. Hassidim M, Braun Y, Lerner HR, Reinhold L. 1990. Na^+/H^+ and K^+/H^+ antiport in root membrane vesicles isolated from the halophyte *Atriplex* and the glycophyte cotton. *Plant Physiol.* 94:1795–801
98. Hayashi H, Alia Mustardy L, Deshnium P, Ida M, Murata N. 1997. Transformation of *Arabidopsis thaliana* with the *codA* gene for chloine oxidase; accumulation of glycinebetaine and enhanced tolerance to salt and cold stress. *Plant J.* 12: 133–42
99. Hechenberger M, Schwappach B, Fischer WN, Frommer WB, Jentsch T, Steinmeyer K. 1996. A family of putative chloride channels from *Arabidopsis* and functional complementation of a yeast strain with a *CLC* gene disruption. *J. Biol. Chem.* 271:33632–38
100. Hedrich R. 1994. Voltage-dependent chloride channels in plant cells: identification, characterization, and regulation

of guard cell anion channel. *Curr. Topics Membr.* 42:1–33

101. Hirayama T, Ohto C, Mizoguchi T, Shinozaki K. 1995. A gene encoding a phosphatidylinositol-specific phospholipase C is induced by dehydration and salt stress in *Arabidopsis thaliana*. *Proc. Natl. Acad. Sci. USA* 92:3903–7
102. Hirsch RE, Lewis BD, Spalding EP, Sussman MR. 1998. A role for the AKT1 postassium channel in plant nutrition. *Science* 280:918–21
103. Hoefte H, Desprez T, Amselem J, Chiapello H, Rouze P, et al. 1993. An inventory of 1152 expressed sequence tags obtained by partial sequencing of cDNAs from *Arabidopsis thaliana*. *Plant J.* 4:1051–61
104. Holland D, Ben-Hayyim G, Faltin Z, Camoin L, Strosberg AD, Eshdat Y. 1993. Moelcular characterization of salt-stress-associated protein in citrus: protein and cDNA sequence homology to mammalian glutthione peroxidases. *Plant Mol. Biol.* 21:923–27
105. Holmberg N, Bulow L. 1998. Improving stress tolerance in plants by gene transfer. *Trends Plant Sci.* 3:61–66
106. Holmström KO, Mäntylä E, Welin B, Mandal A, Palva ET, et al. 1996. Drought tolerance in tobacco. *Nature* 379:683–84
107. Ichimura K, Mizoguchi T, Irie K, Morris P, Giraudat J, et al. 1998. Isolation of ATMEKK1 (a MAP kinase kinase Kinase)-Interacting proteins and analysis of a MAP kinase cascade in *Arabidopsis*. *Biochem. Biophys. Res. Commun.* 253:532–43
108. Ishitani M, Majumder AL, Bornhouser A, Michalowski CB, Jensen RG, Bohnert HJ. 1996. Coordinate transcriptional induction of *myo*-inositol metabolism during environmental stress. *Plant J.* 9:537–48
109. Ishitani M, Nakamura T, Han SY, Takebe T. 1995. Expression of the betaine aldehyde dehydrogenase gene in barley in response to osmotic stress and abscisic acid. *Plant Mol. Biol.* 27:307–15
110. Ishitani M, Xionig L, Stevenson B, Zhu J-K. 1997. Genetic analysis of osmotic and cold stress signal transduction in *Arabidopsis*: interactions and convergence of abscisic acid-dependent and abscisic acid-independent pathways. *Plant Cell* 9:1935–49
111. Jaglo-Ottosen KR, Gilmour SJ, Zarka DG, Schabenberger O, Thomashow MF. 1998. *Arabidopsis CBF1* overexpression induces *COR* genes and enhances freezing tolerance. *Science* 280:104–6
112. Jain RK, Selvaraj G. 1997. Molecular genetic improvement of salt tolerance in plants. *Biotech. Annu. Rev.* 3:245–67
113. Jang HJ, Pih KT, Kang SG, Lim JH, Jin JB, et al. 1998. Molecular cloning of a novel Ca^{2+}-binding protein that is induced by NaCl stress. *Plant Mol. Biol.* 37:389–47
114. Jeanette E, Rona JP, Bardat F, Cornel D, Sotta B, Miginiac E. 1999. Induction of *RAB18* gene expression and activation of K^+ outward rectifying channels depend on an extracellular perception of ABA in *Arabidopsis thaliana* suspension cells. *Plant J.* 18:13–22
115. Jia ZP, McMullough N, Martel R, Hemmingsen S, Young PG. 1992. Gene amplification at a locus encoding a putative Na^+/H^+ antiport confer sodium and lithium tolerance in fission yeast. *EMBO J.* 11:1631–40
116. Johansson I, Larsson C, Ek B, Kjellbom P. 1996. The major integral proteins of spinach leaf plasma membranes are putative aquaporins and are phosphorylated in response to Ca^{2+} and apoplastic water potential. *Plant Cell* 8:1181–91
117. Johansson I, Karlsson M, Shukla VK, Chrispeels MJ, Larsson C, Kjellbom P. 1998. Water transport activity of the plasma membrane aquaporin PM28A is regulated by phosphorylation. *Plant Cell* 10:451–59
118. Jolivet Y, Larher F, Hamelin J. 1982. Osmoregulation in halophytic higher plants:

The protective effect of glycine betaine against the heat destabilization of membranes. *Plant Sci. Lett.* 25:193–201

119. Jonak C, Kiegerl S, Lighterink W, Barker PJ, Huskisson NS, Hirt H. 1996. Stress signaling in plants: A mitogen-activated protein kinase pathway is activated by cold and drought. *Proc. Natl. Acad. Sci. USA* 93:11274–79
120. Jones A. 1998. Auxin transport: down and out and up again. *Science* 282:2201–2
121. Kaldenhoff R, Källing A, Meyers J, Karmann U, Ruppel G, Ritcher G. 1995. The blue light-responsive *AthH2* gene of *Arabidopsis thaliana* is primarily expressed in expanding as well as in differentiating cells and encodes a putative channel protein of the plasmalemma. *Plant J.* 7:87–95
122. Karpinski S, Reynolds H, Karpinska B, Wingsle G, Creissen G, Mullineaux P. 1999. Systemic signaling and acclimation in response to excess excitation energy in *Arabidopsis*. *Science* 284:654–57
123. Kasuga M, Liu Q, Miura S, Yamaguchi-Shinozaki K, Shinozaki K. 1999. Improving plant drought, salt, and freezing tolerance by gene transfer of a single stress-inducible transcription factor. *Nat. Biotechnol.* 17:287–91
124. Kattermann FR, ed. 1990. *Environmental Injury to Plants*. San Diego: Academic
125. Kaui Kishor PB, Hang Z, Miao G-H, Hu C-AA, Verma DPS. 1995. Overexpression of Δ'–pyrroline-5-carboxylate synthetase increases proline production and confers osmotolerance in transgenic plants. *Plant Physiol.* 108:1387–94
126. Kehoe DM, Villand P, Somerville S. 1999. DNA microarrays for studies of higher plants and other photosynthetic organisms. *Trends Plant Sci.* 4:38–41
127. Kende H, Zeevaart JAD. 1997. The five "classical" plant hormones. *Plant Cell* 9:1197–210
128. Kim EJ, Kwak JM, Uozumi N, Schroeder JI. 1998. *AtKUP1*: an *Arabidopsis* gene encoding high-affinity potassium transport activity. *Plant Cell* 10:51–62
129. Kirst GO. 1989. Salinity tolerance of eukaryotic marine algae. *Annu. Rev. Plant Physiol. Plant Mol. Biol.* 40:21–53
130. Kiyosue T, Yoshiba Y, Yamaguchi-Shinozaki K, Shinozaki K. 1996. A nuclear gene encoding mitochondrial proline dehydrogenase, an enzyme involved in proline metabolism, is upregulated by proline but downregulated by dehydration in *Arabidopsis*. *Plant Cell* 8:1323–35
131. Kjellbom P, Larrson C, Johannson I, Karlsson M, Johanson U. 1999. Aquaporins and water homeostasis in plants. *Trends Plant Sci.* 4:308–14
132. Knight H, Trewavas AJ, Knight MR. 1997. Calcium signalling in *Arabidopsis thaliana* responding to drought and salinity. *Plant J.* 12:1067–78
133. Koornneef M, Reuling G, Karssen CM 1984. The isolation and characterization of abscisic acid-insensitive mutants of *Arabidopsis thaliana. Physiol. Plant.* 61:377–83
134. Kudla J, Xu Q, Harter K, Gruissem W, Luan S. 1999. Genes for calcineurin B-like proteins in *Arabidopsis* are differentially regulated by stress signals. *Proc. Natl. Acad. Sci. USA* 96:4718–23
135. Lalonde S, Boles E, Hellmann H, Barker L, Patrick JW, et al. 1999. The dual function of sugar carriers: transport and sugar sensing. *Plant Cell* 11:707–26
136. Läuchli A, Lüttge U, eds. 2000. *Salinity: Environments-Plants-Molecules*. Dordrecht, Netherlands: Kluwer. In press
137. Läuchli A, Schubert S. 1989. The role of calcium in the regulation of membrane and cellular growth processes under salt stress. *NATO ASI Ser.* G19:131–37
138. Lee JH, Van Montagu M, Verbruggen N. 1999. A highly conserved kinase is an essential component for stress tolerance in yeast and plant cells. *Proc. Natl. Acad. Sci. USA* 96:5873–77
139. Lerner HR. 1999. Introduction to the

response of plants to environmental stresses. In *Plant Responses to Environmental Stresses*, ed. HR Lerner, pp. 1–26. New York/Basel: Marcel Dekker

140. Leung J, Giraudat J. 1998. Abscisic acid signal transduction. *Annu. Rev. Plant Physiol. Plant Mol. Biol.* 49:199–222
141. Levitt J. 1980. *Responses of Plant to Environmental Stress Chilling, Freezing, and High Temperature Stresses*. New York: Academic. 2nd ed.
142. Li Y, Kane T, Tipper C, Spatrick P, Jenness DD. 1999. Yeast mutants affecting possible quality control of plasma membrane proteins. *Mol. Cell. Biol.* 19:3588–99
143. Lippuner V, Cyert MS, Gasser CS. 1996. Two classes of plant cDNAs differentially complement yeast calcineurin mutants and increases salt tolerance of wild-type yeast. *J. Biol. Chem.* 271:12859–66
144. Liu H, Toyn JH, Chiang Y-C, Draper MP, Johnston LH, Denis CL. 1997. DBF2, a cell cycle-regulated protein kinase, is physically and functionally associated with the CCR4 transcriptional regulatory complex. *EMBO J.* 16:5289–98
145. Liu J, Zhu J-K. 1997. An *Arabidopsis* mutant that requires increased calcium for potassium nutrition and salt tolerance. *Proc. Natl. Acad. Sci. USA* 94:14960–64
146. Liu J, Zhu JK. 1998. A calcium sensor homologue required for plant salt tolerance. *Science* 280:1943–45
147. Liu Q, Kasuga M, Sakuma Y, Abe A, Miura S, et al. 1998. Two transcription factors, DREB1 and DREB2, with an EREBP/AP2 DNA binding domain separate two cellular signal transduction pathways in drought- and low-temperature-responsive gene expression, respectively, in *Arabidopsis*. *Plant Cell* 10:1391–406
148. Loew R, Rockel B, Kirsch M, Ratajczak R, Hortensteiner S, et al. 1996. Early salt stress effects on the differential expression of vacuolar H^+-ATPase genes in roots and leaves of *Mesembryanthemum crystallinum*. *Plant Physiol.* 110:259–65
149. Louis P, Galinski EA. 1997. Characterization of genes for the biosynthesis of the compatible solute ectoine from *Marinococcus halophilus* and osmoregulated expression in E. coli. *Microbiology* 143:1141–49
150. Lüttge U, Ratajczak R. 1997. The physiology, biochemistry, and molecular biology of the plant vacuolar ATPase. *Adv. Bot. Res.* 25:253–96
151. Lynch J, Polito VS, Läuchli A. 1989. Salinity stress increases cytoplasmic Ca^{2+} activity in maize root protoplasts. *Plant Physiol.* 90:1271–74
152. Maathuis FJM, Ichida AM, Sanders D, Schroeder JI. 1997. Roles of higher plant K^+ channels. *Plant Physiol.* 114:1141–49
153. Maathuis FJM, Sanders D. 1999. Plasma membrane transport in context-making sense out of complexity. *Curr. Opin. Plant Biol.* 2:236–43
154. Maggio A, Mastumoto TK, Hasegawa PM, Pardo JM, Bressan RA. 2000. The long and winding road to halotolerance genes. See Ref. 136. In press
155. Martienssen RA 1998. Functional genomics: probing plant gene function and expression with transposons. *Proc. Natl. Acad. Sci. USA* 95:2021–26
156. Martinez V, Läuchli A. 1993. Effects of Ca^{2+} on the salt-stress response of barley roots as observed by in-vivo ^{31}P-nuclear magnetic resonance and in-vitro analysis. *Planta* 190:519–24
157. Marty F. 1999. Plant vacuoles. *Plant Cell* 11:587–99
158. Maurel C, Kado RT, Guern J, Chrispeels MJ. 1995. Phosphorylation regulates the water channel activity of the seed specific aquaporin γ-TIP. *EMBO J.* 14:3028–35
159. Maurel C. 1997. Aquaporins and water permeability of plant membranes. *Annu. Rev. Plant Physiol. Plant Mol. Biol.* 48:399–429
160. McCue KF, Hanson AD. 1990. Drought and salt tolerance: towards understanding and application. *Biotechnology* 8:358–62

161. McCue KF, Hanson AD. 1992. Salt-inducible betaine aldehyde dehydrogenase from sugar beet: cDNA cloning and expression. *Plant Mol. Biol.* 18:1–11
162. Mendoza I, Quintero FJ, Bressan RA, Hasegawa PM, Pardo JM. 1996. Activated calcineurin confers high tolerance to ion stress and alters the bud pattern and cell morphology of yeast cells. *J. Biol. Chem.* 271:23061–67
163. Mendoza I, Rubio F, Rodriguez-Navarro A, Pardo JM. 1994. The protein phosphatase calcineurin is essential for NaCl tolerance of *Saccharomyces cerevisiae*. *J. Biol. Chem.* 269:8792–96
164. Meskiene I, Bögre L, Glaser W, Galog J, Brandstötter M, et al. 1998. MP2C, a plant protein phosphatase 2C, functions as a negative regulator of mitogen-activated protein kinase pathways in yeast and plants. *Proc. Natl. Acad. Sci. USA* 95: 1938–43
165. Mikami K, Katagiri T, Iuchi S, Yamaguchi-Shinozaki K, Shinozaki K. 1998. A gene encoding phosphatidylinositol-4-phosphate 5-kinase is induced by water stress and abscisic acid in *Arabidopsis thaliana*. *Plant J.* 15:563–68
166. Miyahara K, Mizunuma M, Hirata D, Tsuchiya E, Miyakawa T. 1996. The involvement of the *Saccharomyces cerevisiae* multidrug resistances transporters Pdr5p and Snq2p in cation resistance. *FEBS Lett.* 399:317–20
167. Mizoguchi T, Ichimura K, Shinozaki K. 1997. Environmental stress response in plants: the role of mitogen-activated protein kinases. *Trends Biotech.* 15:15–19
168. Mizoguchi T, Irie K, Hirayama T, Hayashida N, Yamaguchi-Shinozaki K, et al. 1996. A gene encoding a mitogen-activated protein kinase kinase kinase is induced simultaneously with genes for a mitogen-activated protein kinase and an S6 ribosomal protein kinase by touch, cold, and water stress in *Arabidopsis thaliana*. *Proc. Natl. Acad. Sci. USA* 93: 765–69
169. Morgan PW. 1990. Effects of abiotic stresses on plant hormone systems. In *Stress Responses in Plants: Adaptation and Acclimation Mechanisms*, ed. RG Alscher, JR Cumming, pp. 113–46. New York: Wiley-Liss
170. Morgan PW, Drew MC. 1997. Ethylene and plant responses to stress. *Physiol. Plant.* 100:620–30
171. Munns R, Cramer GR. 1996. Is coordination of leaf growth mediated by abscisic acid? *Plant Soil* 185:33–49
172. Munns R, Sharp RE. 1993. Involvement of abscisic acid in controlling plant growth in soils of low water potential. *Aust. J. Plant Physiol.* 20:425–37
173. Munns R, Termaat A. 1986. Whole-plant responses to salinity. *Aust. J. Plant Physiol.* 13:143–60
174. Munns R. 1993. Physiological processes limiting plant growth in saline soils: some dogmas and hypotheses. *Plant Cell Environ.* 16:15–24
175. Nakashima K, Satoh R, Kiyosue T, Yamaguchi-Shinozaki K, Shinozaki K. 1998. A gene encoding proline dehydrogenase is not only induced by proline and hypoosmolarity, but is also developmentally regulated in the reproductive organs of *Arabidopsis*. *Plant Physiol.* 118:1233–41
176. Nakayama H, Yoshida K, Ono H, Murooka Y, Shinmyo A. 1999. Production of ectoine confers hyperosmotic stress tolerance in cultured tobacco cells. *Int. Bot. Congr., 16th, Abstr.* 2069
177. Nanjo T, Kobayashi M, Yoshiba Y, Sanada Y, Wada K, et al. 1999. Biological functions of proline in morhogenesis and osmotolerance revealed in antisense transgenic *Arabidopsis thaliana*. *Plant J.* 18:185–93
178. Narasimhan ML, Binzel ML, Perez-Prat E, Chen Z, Nelson DE, et al. 1991. NaCl regulation of tonoplast ATPase

70-kilodalton subunit mRNA in tobacco cells. *Plant Physiol.* 97:562–68

179. Nelson DE, Koukoumanos M, Bohnert HJ. 1999. *Myo*-inositol amounts in roots control sodium uptake in a halophyte. *Plant Physiol.* 119:165–72
180. Nelson DE, Rammesmayer G, Bohnert HJ. 1998. The regulation of cell-specific inositol metabolism and transport in plant salinity tolerance. *Plant Cell* 10:753–64
181. Nelson DE, Shen B, Bohnert HJ. 1998. Salinity tolerance—mechanisms, models, and the metabolic engineering of complex traits. In *Genetic Engineering, Principles and Methods*, ed. JK Setlow, 20:153–76. New York: Plenum
182. Niu X, Bressan RA, Hasegawa PM, Pardo JM. 1995. Ion homeostasis in NaCl stress environments. *Plant Physiol.* 109:735–42
183. Niu X, Damsz B, Kononowicz AK, Bressan RA, Hasegawa PM. 1996. NaCl-induced alterations in both cell structure and tissue-specific plasma membrane H^+-ATPase gene expression. *Plant Physiol.* 111:679–86
184. Niu X, Narasimhan ML, Salzman RA, Bressan RA, Hasegawa PM. 1993. NaCl regulation of plasma membrane H^+-ATPase gene expression in a glycophyte and a halophyte. *Plant Physiol.* 103:713–18
185. Niyogi K. 1999. Photoprotection revisited: genetic and molecular approaches. *Annu. Rev. Plant Physiol. Plant Mol. Biol.* 50:333–59
186. Noat D, Ben-Hayyim G, Eshdat Y, Holland D. 1995. Drought, heat, and salt sterss induce the expression of a citrus homologue of an atypical late-embyrogenesis *Lea5* gene. *Plant Mol. Biol.* 27:619–22
187. Noctor G, Foyer CH. 1998. Ascorbate and glutathione: keeping active oxygen under control. *Annu. Rev. Plant Physiol. Plant Mol. Biol.* 49:249–79
188. Normanly J, Bartel B. 1999. Redundancy as a way of life—IAA metabolism. *Curr. Opin. Plant Biol.* 2:207–13
189. Nuccio ML, Rhodes D, McNeil SD, Hanson AD. 1999. Metabolic engineering of plants for osmotic stress resistance. *Curr. Opin. Plant Biol.* 2:128–34
190. Ono H, Sawada K, Khunajakr N, Tao T, Yamamoto M, et al. 1999. Characterization of biosynthetic enzymes for ectoine as a compatible solute in a moderately halophilic eubacterium, *Halomonas elongata. J. Bacteriol.* 181:91–99
191. Orr WC, Sohal RS. 1992. The effects of catalase gene overexpression on life span and resistance to oxidative stress in transgenic *Drosophila melanogaster. Arch. Biochem. Biophys.* 297:35–41
192. Orthen B, Popp M, Smirnoff N. 1994. Hydroxyl radical scavenging properties of cyclitols. *Proc. R. Soc. Edinburgh* 102B:269–72
193. Palmgren MG, Harper JF. 1999. Pumping with plant P-type ATPases. *J. Exp. Bot.* 50:883–93
194. Papageorgiou G, Murata N. 1995. The unusually strong stabilizing effects of glycine betaine on the structure and function of the oxygen-evolveing photosystem II complex. *Photosynth. Res.* 44:243–52
195. Pardo JM, Reddy MP, Yang S, Maggio A, Huh G-H, et al. 1998. Stress signaling through Ca^{2+}/calmodulin-dependent protein phosphatase calcineurin mediates salt adaptation in plants. *Proc. Natl. Acad. Sci. USA* 95:9681–86
196. Peng Z, Verma DPS. 1995. A rice *HAL2*-like gene encodes a Ca^{2+}-sensitive 3′(2′), 5′-diphosphonucleosidase 3′(2′)- phosphohydrolase and complements yeast *met22* and *Escherichia coli cysQ* mutations. *J. Biol. Chem.* 270:29105–10
197. Perez-Prat E, Narasimhan ML, Binzel ML, Botella MA, Chen Z, et al. 1992. Induction of a putative Ca^{2+}-ATPase mRNA in NaCl-adapted cells. *Plant Physiol.* 100:1471–78

198. Piao HL, Pih KT, Lim JH, Kang SG, Jin JB, et al. 1999. An *Arabidopsis GSK3/shaggy*-like gene that complements yeast salt stress-sensitive mutants is induced by NaCl and abscisic acid. *Plant Physiol.* 119:1527–34

199. Pih KT, Kabilan V, Lim JH, Kang SG, Piao HL, et al. 1999. Characterization of two new channel protein genes in *Arabidopsis. Mol. Cells* 9:84–90

200. Pilon-Smits EAH, Ebskamp MJM, Paul MJ, Jeuken MJW, Weisbeek PJ, Smeekens SCM. 1995. Improved performance of transgenic fructan-accumulating tobacco under drought stress. *Plant Physiol.* 107:125–30

201. Poole RJ. 1988. Plasma membrane and tonoplast. In *Solute Transport in Plant Cells and Tissues*, ed. DA Baker, JL Hall, pp. 83–105. New York: Wiley & Sons

202. Pöpping, B, Gibbons T, Watson MD. 1996. The *Pisum sativum* MAP kinase homologue (PsMAPK) rescues the *Saccharomyces cerevisiae hog1* deletion mutant under conditions of high osmotic stress. *Plant Mol. Biol.* 31:355–63

202a. Posas F, Saito H. 1997. Osmotic activation of the HOG MAPK pathway via Ste11p MAPKKK: scaffold role of Pbs2p MAPKK. *Science* 276:1702–5

203. Prieto R, Pardo JM, Niu X, Bressan RA, Hasegawa PM. 1996. Salt-sensitive mutants of *Chlamydomonas reinhardtii* isolated after insertional tagging. *Plant Physiol.* 112:99–104

204. Prior C, Potier S, Souciet JL, Sychrova H. 1996. Characterization of the *NHA1* gene encoding a Na^+/H^+-antiporter of the yeast *Saccharomyces cerevisiae*. *FEBS* 387:89–93

205. Quintero FJ, Garciadebias B, Rodriguez-Navarro A. 1996. The *SAL1* gene of *Arabidopsis*, encoding an enzyme with 3′(2′), 5′- bisphosphate nucleotidase and inositol polyphosphate 1-phosphatase activities, increases salt tolerance in yeast. *Plant Cell* 8:529–37

206. Ratajczak R, Hille A, Mariaux J-B, Lüttge U. 1995. Quantitative stress responses of the V0V1–ATPases of higher plants detected by immuno-electron microscopy. *Bot. Acta* 108:505–13

207. Ratajczak R, Richter J, Lüttge U. 1994. Adaptation of the tonoplast V-type H^+-ATPase of *Mesembryanthemum crystallinum* to salt stress, C3–CAM transition and plant age. *Plant Cell Environ.* 17:1101–12

208. Rathinasabapathi B, Burnet M, Russell BL, Gage DA, Liao PC, et al. 1997. Choline monooxygenase, an unusual iron-sulfur enzyme catalyzing the first step of glycine betaine synthesis in plants: prosthetic group characterization and cDNA cloning. *Proc. Natl. Acad. Sci. USA* 94:3454–58

209. Rathinasabapathi B, McCue KF, Gage DA, Hanson AD. 1994. Metabolic engineering of glycine betaine biosynthesis: Plant betaine aldehyde dehydrogenases lacking typical transit peptides are targeted to tobacco chloroplasts where they confer betaine aldehyde resistance. *Planta* 193:155–62

210. Rea PA. 1999. MRP subfamily ABC transporters from plants and yeast. *J. Exp. Bot.* 50:895–913

211. Reuveni M, Bennett AB, Bressan RA, Hasegawa PM. 1990. Enhanced H^+ transport capacity and ATP hydrolysis activity of the tonoplast H^+-ATPase after NaCl adaptation. *Plant Physiol.* 94:524–30

212. Rhodes D, Hanson AD. 1993. Quaternary ammonium and tertiary sulphonium compounds in higher plants. *Annu. Rev. Plant Physiol. Mol. Biol.* 44:357–84

213. Rhodes D, Samaras Y. 1994. Genetic control of osmoregulation in plants. In *Cellular and Molecular Physiology of Cell Volume Regulation*, ed. K Strange, pp. 347–61. Boca Raton: CRC Press

214. Rhodes D. 1987. Metabolic responses to stress. In *The Biochemistry of Plants*, ed.

DD Davies, pp. 202–42. New York: Academic

215. Roitsch T. 1999. Source-sink regulation by sugar and stress. *Curr. Opin. Plant Biol.* 2:198–206
216. Roosens NH, Thu TT, Iskandar HM, Jacobs M. 1998. Isolation of the ornithine-delta-aminotransferase cDNA and effect of salt stress on its expression in *Arabidopsis thaliana. Plant Physiol.* 117:263–71
217. Roxas VP, Smith RH Jr, Allen ER, Allen RD. 1997. Overexpression of glutathione S-transferase/glutathione peroxidase enhances the growth of transgenic tobacco seedlings during stress. *Nat. Biotechnol.* 15:988–91
218. Rubio F, Gassman W, Schroeder JI. 1995. Sodium-driven potassium uptake by the plant potassium transporter HKT1 and mutations conferring salt tolerance. *Science* 270:1660–63
219. Russell BL, Rathinasabapathi B, Hanson AD. 1998. Osmotic stress induces expression of choline monooxygenase in sugar beet and amaranth. *Plant Physiol.* 116:859–65
220. Sanada Y, Ueda H, Kuribayashi K, Andoh T, Hayashi F, et al. 1995. Novel light-dark change of proline levels in halophyte (*M. crystallinum* L.) and glycophyte (*H. vulgare* L. and *T. aestivum* L.) leaves and roots under salt stress. *Plant Cell Physiol.* 36:965–70
221. Sanders D, Brownlee C, Harper J. 1999. Communicating with calcium. *Plant Cell* 11:691–706
222. Santa-Maria GE, Rubio F, Dubcovsky J, Rodriguez-Navarro A. 1997. The *HAK1* gene of barley is a member of a large gene family and encodes a high-affinity potassium transporter. *Plant Cell* 9:2281–89
223. Schachtman D, Liu W. 1999. Molcular pieces to the puzzle of the interaction between potassium and sodium uptake in plants. *Trends Plant Sci.* 4:281–87
224. Schachtman DP, Schroeder JI. 1994. Structure and transport mechanism of a high-affinity potassium uptake transporter from higher plants. *Nature* 370: 655–58
225. Schachtman DP, Tyerman SD, Terry BR. 1991. The K^+/Na^+ selectivity of a cation channel in the plasma membrane of root cells does not differ in salt-tolerant and salt-sensitive wheat species. *Plant Physiol.* 97:598–605
226. Schäffner AR. 1998. Aquaporin function, structure, and expression: Are there more surprises to surface in water relations? *Planta* 204:131–39
227. Schwartz SH, Tan BC, Gage DA, Zeevaart JAD, McCarty DR. 1997. Specific oxidative cleavage of carotenoids by VP14 of maize. *Science* 276:1872–74
228. Sentenac H, Bonneaud N, Minet M, Lacroute F, Salmon JM, et al. 1992. Cloning and expression in yeast of a plant potassium ion transport system. *Science* 256:663–65
229. Serrano R, Culiañz-Maciá A, Moreno V. 1999. Genetic engineering of salt and drought tolerance with yeast regulatory genes. *Sci. Hortic.* 78:261–69
230. Serrano R. 1996. Salt tolerance in plants and microorganisms: toxicity targets and defense responses. *Int. Rev. Cytol.* 165:1–52
231. Sheen J. 1996. Ca^{2+}-dependent protein kinases and stress signal transduction in plants. *Science* 274:1900–2
232. Shen B, Hohmann S, Jensen RG, Bohnert HJ. 1999. Roles of sugar alcohols in osmotic stress adaptation: replacement of glycerol by mannitol and sorbitol in yeast. *Plant Physiol.* 121:45–52
233. Shen B, Jensen RG, Bohnert H. 1997. Mannitol protects against oxidation by hydroxyl radicals. *Plant Physiol.* 115:527–32
234. Shen B, Jensen RG, Bohnert HJ. 1997. Increased resistance to oxidative stress in transgenic plants by targeting mannitol

biosynthesis to chloroplasts. *Plant Physiol.* 113:1177–83

235. Shen Q, Ho T-HD. 1995. Functional dissection of an abscisic acid (ABA)-inducible gene reveals two independent ABA-responsive complexes each containing a G-box and a novel cis-acting element. *Plant Cell* 7:295–307
236. Sheveleva E, Chmara W, Bohnert HJ, Jensen RG. 1997. Increased salt and drought tolerance by D-ononitol production in transgenic *Nicotiana tabacum* L. *Plant Physiol.* 115:1211–19
237. Sheveleva E, Jensen RG, Bohnert HJ. 2000. Disturbance in the allocation of carbohydrates to regenerative organs in transgenic *Nicotiana tabacum* L. *J. Exp. Bot.* 51: In press
238. Sheveleva E, Marquez S, Zegeer A. 1998. Sorbitol dehydrogenase expression in transgenic tobacco: high sorbitol accumulation leads to necrotic lesions in immature leaves. *Plant Physiol.* 117:831–39
239. Shikanai T, Takeda T, Yamauchi H, Sano S, Tomizawa KI, et al. 1998. Inhibition of ascorbate peroxidase under oxidative stress in tobacco having bacterial catalase in chloroplasts. *FEBS Lett.* 428:47–51
240. Shinozaki K, Yamaguchi-Shinozaki K. 1996. Molecular responses to drought and cold stress. *Curr. Opin. Biotechnol.* 7:161–67
241. Shinozaki K, Yamaguchi-Shinozaki K. 1997. Gene expression and signal transduction in water-stress response. *Plant Physiol.* 115:327–34
242. Skerrett M, Tyerman SD. 1994. A channel that allows inwardly directed fluxes of anions and protoplasts derived from wheat roots. *Planta* 192:295–305
243. Slovik S, Hartung W. 1992. Compartmental distribution and redistribution of abscisic acid in intact leaves. *Planta* 187:37–47
244. Smirnoff N. 1998. Plant resistance to environmental stress. *Curr. Opin. Biotechnol.* 9:214–19
245. Solomon A, Beer S, Waisel Y, Jones GP, Paleg LG. 1994. Effects of NaCl on the carboxylating activity of Rubisco from *Tamarix jordanis* in the presence and absence of proline-related compatible solutes. *Plant Physiol.* 90:198–204
246. Somerville C, Somerville S. 1999. Plant functional genomics. *Science* 285:380–83
247. Spalding EP, Hirsch RE, Lewis DR, Qi Z, Sussman MR, Lewis BD. 1999. Potassium uptake supporting plant growth in the absence of AKT1 channel activity: Inhibition by ammonium and stimulation by sodium. *J. Gen. Physiol.* 113:909–18
248. Stockinger EJ,Gilmour SJ, Thomashow MF. 1997. *Arabidopsis thaliana CBF1* encodes and AP2 domain-containing transcriptional activator that binds to the C-repeat/DRE, a *cis*-acting DNA regulatory element that stimulates transcription in response to low temperature and water deficit. *Proc. Natl. Acad. Sci. USA* 94:1035–40
249. Storey R, Pitman MG, Stelzer R, Carter C. 1983. X-ray micro-analysis of cells and cell compartments of *Atriplex spongiosa*. *J. Exp. Bot.* 34:778–94
250. Straub PF, Shen Q, Ho TD. 1994. Structure and promoter analysis of an ABA- and stress-regulated barley gene, *HVA1*. *Plant Mol. Biol.* 26:617–30
251. Strizhov N, Abraham E, Okresz L, Blickling S, Zilberstein A, et al. 1997. Differential expression of two P5CS genes controlling proline accumulation during salt-stress requires ABA and is regulated by *ABA1, ABI1* and *AXR2* in *Arabidopsis*. *Plant J.* 12:557–69
252. Sugimoto M, Sakamoto W. 1997. Putative phospholipid hydroperoxide glutathione peroxidase gene from *Arabidopsis thaliana* induced by oxidative stress. *Genes Genet. Syst.* 72:311–16
253. Swire-Clark GA, Marcotte WR Jr. 1999. The wheat LEA protein Em functions as an osmoprotective moleculae in

Saccharomyces cerevisiae. Plant Mol. Biol. 39:117–28

254. Sze H, Li X, Palmgren MG. 1999. Energization of plant cell membranes by H^+-pumping ATPases: regulation and biosynthesis. *Plant Cell* 11:677–89
255. Tal M. 1990. Somaclonal variation for salt resistance. In *Biotechnology in Agriculture and Forestry. Somaclonal Variation in Crop Improvement I*, ed. YPS Bajaj, pp. 236–57. Berlin: Springer-Verlag
256. Tamas MJ, Luyten K, Sutherland FC, Hernandez A, Albertyn J, et al. 1999. Fps1p controls the accumulation and release of the compatible solute glycerol in yeast osmoregulation. *Mol. Microbiol.* 31:1087–104
257. Tarczynski MC, Jensen RG, Bohnert HJ. 1993. Stress protection of transgenic tobacco by production of the osmolyte, mannitol. *Science* 259:508–10
258. Tarczynski MC, Jensen RG, Bohnert HJ. 1992. Expression of a bacterial *mtlD* gene in transgenic tobacco leads to production and accumulation of mannitol. *Proc. Natl. Acad. Sci. USA* 89:2600–4
259. Tardieu F, Zhang J, Davies WJ. 1992. What information is conveyed by an ABA signal from maize roots in drying field soil? *Plant Cell Environ.* 15:185–91
260. Theologis A. 1998. Ethylene signalling: redundant receptors all have their say. *Curr. Biol.* 8:R875–78
261. Thomashow MF. 1999. Plant cold acclimation: freezing tolerance genes and regulatory mechanisms. *Annu. Rev. Plant Physiol. Mol. Biol.* 50:571–99
262. Trossat C, Rathinasabapathi B, Weretylnik EA, Shen TL, Huang ZH, et al. 1998. Salinity promotes accumulation of 3-dimethylsulfoniopropionate and its precursor S-methylmethionine in chloroplasts. *Plant Physiol* 116:165–71
263. Tsiantis MS, Bartholomew DM, Smith JAC. 1996. Salt regulation of transcript levels for the c subunit of a leaf vacuolar H^+-ATPase in the halophyte *Mesembryanthemum crystallinum. Plant J.* 9:729–36
264. Tsugane K, Kobayashi K, Niwa Y, Ohba Y, Wada K, Kobayashi H. 1999. A recessive *Arabidopsis* mutant that grows photoautotrophically under salt stress shows enhanced active oxygen detoxification. *Plant Cell* 11:1195–206
265. Tyerman SD, Bohnert HJ, Maurel C, Steudle E, Smith JAC. 1999. Plant aquaporins: their molecular biology, biophysics and significance for plant water relations. *J. Exp. Bot.* 50:1055–71
266. Tyerman SD, Oats P, Gibbs J, Dracup M, Greenway H. 1989. Turgor-volume regulation and cellular water relations of *Nicotiana tabacum* roots grown in high salinities. *Aust. J. Plant Physiol.* 16:517–31
267. Urao T, Yakubov B, Yamaguchi-Shinozaki K, Shinozaki K. 1998. Stress-responsive expression of genes for two-component response regulator-like proteins in *Arabidopsis thaliana. FEBS Lett.* 427:175–78
268. Van Camp W, Capiau K, Van Montagu M, Inzé D, Slooten L. 1996. Enhancement of oxidative stress tolerance in transgenic tobacco plants overproducing Fe-superoxide dismutase in chloroplasts. *Plant Physiol.* 112:1703–14
269. Vera-Estrella R, Barkla BJ, Bohnert HJ, Pantoja O. 1999. Salt stress in *Mesembryanthemum crystallinum* suspension cells activates adaptive mechanisms similar to those observed in the whole plant. *Planta* 207:426–35
270. Verbruggen N, Hua XJ, May M, Van Montagu M. 1996. Environmental and developmental signals modulate proline homeostasis: evidence for a negative transcriptional regulator. *Proc. Natl. Acad. Sci. USA* 93:8787–91
271. Verma DPS. 1999. Osmotic stress tolerance in plants: Role of proline and sulfur metabolisms. In *Molecular Responses to Cold, Drought, Heat and Salt Stress in Higher Plants*, ed. K Shinozaki,

K Yamaguchi-Shinozaki, pp. 155–68. Austin: Landes

272. Vernon DM, Bohnert HJ. 1992. A novel methyl transferase induced by osmotic stress in the facultative halophyte *Mesembryanthemum crystallinum*. *EMBO J.* 11:2077–85
273. Watad AA, Reuveni M, Bressan RA, Hasegawa PM. 1991. Enhanced net K^+ uptake capacity of NaCl-adapted cells. *Plant Physiol.* 95:1265–69
274. Weiss M, Pick U. 1996. Primary structure and effect of pH on the expression of the plasma membrane H^+-ATPase from *Dunaliella acidophila and Dunaliella salina*. *Plant Physiol.* 112:1693–702
275. White PJ. 1999. The molecular mechanism of sodium influx to root cells. *Trends Plant Sci.* 4:245–46
276. Willekens H, Chamnongopol S, Davey M, Schraudner M, Langebartels C, et al. 1997. Catalase is a sink for H_2O_2 and is indispensable for stress defense in C3 plants. *EMBO J.* 16:4806–16
277. Wimmers LE, Ewing NN, Bennett AB. 1992. Higher plant $Ca2^+$-ATPase: primary structure and regulation of mRNA abundance by salt. *Proc. Natl. Acad. Sci. USA* 89:9205–9
278. Winicov I, Bastola DR. 1999. Transgenic overexpression of the transcription factor *Alfin1* enhances expression of the endogenous *MsPRP2* gene in alfalfa and improves salinity tolerance of the plants. *Plant Physiol.* 120:473–80
279. Wu S, Ding L, Zhu J. 1996. *SOS1*, a genetic locus essential for salt tolerance and potassium acquisition. *Plant Cell* 8:617–27
280. Wyn Jones RG. 1981. Salt tolerance. In *Physiological Processes Limiting Plant Productivity*, ed. CB Johnson, pp. 271–92. London: Butterworth
281. Xu D, Duan X, Wang B, Hong B, Ho TDH, Wu R. 1996. Expression of a late embryogenesis abundant protein gene, HVA1, from barley confers tolerance to water deficit and salt stress in transgenic rice. *Plant Physiol.* 110:249–57
282. Yamada S, M Katsuhara M, W Kelly W, CB Michalowski CB, HJ Bohnert HJ. 1995. A family of transcripts encoding water channel proteins: tissue specific expression in the common ice plant. *Plant Cell* 7:1129–42
283. Yamaguchi-Shinozaki K, Koizumi K, Urao S, Shinozaki K. 1992. Molecular cloning and characterization of 9 cDNAs for genes that are responsive to desiccation in *Arabidopsis thaliana*: sequence analysis of one cDNA clone encodes a putative transmembrane channel protein. *Plant Cell Physiol.* 33:217–24
284. Yancey PH, Clark ME, Hand SC, Bowlus RD, Somero GN. 1982. Living with water stress: evolution of osmolyte system. *Science* 217:1214–22

284a. Yeo AR, Lauchli A, Kramer D. 1977. Ion measurements by x-ray microanalysis in unfixed, frozen, hydrated plant cells of species differing in salt tolerance. *Planta* 134:35–38

285. Yeo AR. 1998. Molecular biology of salt tolerance in the context of whole-plant physiology. *J. Exp. Bot.* 49:915–29
286. Yoshiba Y, Kiyosue T, Katagiri T, Ueda H, Mizoguchi T, et al. 1995. Correlation between the induction of agene for D1–pyrroline-5-carboxylate synthetase and the accumulation of proline in *Arabidopsis thaliana* under osmotic stress. *Plant J.* 7:751–60
287. Yoshiba Y, Kiyosue T, Nakashima K, Yamaguchi-Shinozaki K, Shinozaki K. 1997. Regulation of levels of proline as an osmolyte in plants under water stress. *Plant Cell Physiol.* 38:1095–102
288. Young JC, DeWitt ND, Sussman MR. 1998. A transgene encoding a plasma membrane H^+-ATPase that confers acid resistance in *Arabidopsis thaliana* seedlings. *Genetics* 149:501–7
289. Zhang CS, Lu Q, Verma DPS. 1995. Removal of feedback inhibition of delta

1-pyrroline-5-carboxylate synthetase, a bifunctional enzyme catalyzing the first step of proline biosynthesis in plants. *J. Biol. Chem.* 270:20491–96

290. Zhao Y, Aspinall D, Paleg LG. 1992. Protection of membrane integrity in *Medicago sativa* L. by glycine betaine against effects of freezing. *J. Plant Physiol.* 140:541–43

291. Zhien R-G, Kim E, Rea PA. 1997. The molecular and biochemical basis of pyrophosphate-energized ion translocation at the vacuolar membrane. *Adv. Bot. Res.* 27:297–337

292. Zhou L, Jan J-C, Jones TL, Sheen J. 1998. Glucose and ethylene signal transduction crosstalk revealed by an Arabidopsis glucose-insensitive mutant. *Proc. Natl. Acad. Sci. USA* 95:10294–99

293. Zhu BC, Su J, Chan MC, Verma DPS, Fan Y-L, Wu R. 1998. Overexpression of a $\Delta 1$–pyrroline-5-carboxylate synthetase gene and analysis of tolerance to water-stress and salt-stress in transgenic rice. *Plant Sci.* 139:41–48

294. Zhu J-K, Hasegawa PM, Bressan RA. 1997. Moelcular aspects of osmotic stress in plants. Crit. Rev. *Plant Sci.* 16:253–77

295. Zhu JK, Liu J, Xiong L. 1998. Genetic analysis of salt tolerance in Arabidopsis: evidence for a critical role of potassium nutrition. *Plant Cell* 10:1181–91

296. Zhu JK, Shi J, Bressan RA, Hasegawa PM. 1993. Expression of an *Atriplex nummularia* gene encoding a protein homologous to the bacterial molecular chaperone DnaJ. *Plant Cell* 5:341–49

Annu. Rev. Plant Physiol. Plant Mol. Biol. 2000. 51:501–31

GROWTH RETARDANTS: Effects on Gibberellin Biosynthesis and Other Metabolic Pathways

Wilhelm Rademacher
BASF Agricultural Center, 67114 Limburgerhof, Germany; e-mail: wilhelm.rademacher@basf-ag.de

Key Words mode of action, plant hormones, growth regulation, sterols, flavonoids

■ **Abstract** Plant growth retardants are applied in agronomic and horticultural crops to reduce unwanted longitudinal shoot growth without lowering plant productivity. Most growth retardants act by inhibiting gibberellin (GA) biosynthesis. To date, four different types of such inhibitors are known: (*a*) Onium compounds, such as chlormequat chloride, mepiquat chloride, chlorphonium, and AMO-1618, which block the cyclases copalyl-diphosphate synthase and *ent*-kaurene synthase involved in the early steps of GA metabolism. (*b*) Compounds with an N-containing heterocycle, e.g. ancymidol, flurprimidol, tetcyclacis, paclobutrazol, uniconazole-P, and inabenfide. These retardants block cytochrome P450-dependent monooxygenases, thereby inhibiting oxidation of *ent*-kaurene into *ent*-kaurenoic acid. (*c*) Structural mimics of 2-oxoglutaric acid, which is the co-substrate of dioxygenases that catalyze late steps of GA formation. Acylcyclohexanediones, e.g. prohexadione-Ca and trinexapac-ethyl and daminozide, block particularly 3ß-hydroxylation, thereby inhibiting the formation of highly active GAs from inactive precursors, and (*d*) 16,17-Dihydro-GA_5 and related structures act most likely by mimicking the GA precursor substrate of the same dioxygenases. Enzymes, similar to the ones involved in GA biosynthesis, are also of importance in the formation of abscisic acid, ethylene, sterols, flavonoids, and other plant constituents. Changes in the levels of these compounds found after treatment with growth retardants can mostly be explained by side activities on such enzymes.

CONTENTS

1040-2519/00/0601-0501$14.00

INTRODUCTION

Plant growth retardants are synthetic compounds, which are used to reduce the shoot length of plants in a desired way without changing developmental patterns or being phytotoxic. This is achieved primarily by reducing cell elongation, but also by lowering the rate of cell division. In their effect on the morphological structure of plants, growth retardants are antagonistic to gibberellins (GAs) and auxins, the plant hormones that are primarily responsible for shoot elongation. The first growth retardants, certain nicotinium derivatives, became known in 1949 (121). Many other compounds have subsequently been detected, some of which have been introduced into agronomic or horticultural practice. Plant growth retardants represent the commercially most important group of plant bioregulators (PBRs) or plant growth regulators, although compared to herbicides, insecticides, and fungicides, they play a relatively minor role and represent only a few percent of the worldwide sales of crop-protecting chemicals, totaling approximately US $28 billion in 1999.

In addition to other agronomic tools, PBRs can be used relatively flexibly by the farmer to adjust his crop in a desired way to changes in growing conditions. Plant growth retardants have found a number of practical uses: In intensive small grain cultivation in Europe, they have become an integral part of the production system by reducing the risk of lodging due to intensive rainfall and/or wind; in cotton excessive vegetative growth may be controlled, thereby helping adjust a perennial plant species to an annual cycle of cultivation; fruit trees can be kept more compact, thereby reducing costs for pruning and obtaining a better ratio between vegetative growth and fruit production; the quality of ornamental and bedding plants is generally improved by keeping them compact, which also reduces the space in a greenhouse required for production; costs for trimming hedges and trees and for mowing turf grasses may also be reduced by applying plant growth retardants. For more details on applications of plant growth retardants the reader is referred to (43, 50, 80, 118, 129, 134, 151).

Reduction of shoot growth can also be achieved by compounds other than growth retardants. For instance, compounds with a low herbicidal activity or herbicides applied at lower rates may cause a stunted shoot without bringing about visible symptoms of phytotoxicity. Reductions in plant productivity have to be expected, however. Examples of such plant growth suppressants are mefluidide, amidochlor, maleic hydrazide, or chlorflurenol, which might be used, for example, to reduce shoot growth of turf grasses. In principle, breeding offers an alternative way to achieve desired alterations in plant development. However, a fixed genotype is less flexible towards changing growing conditions and does not allow an active steering of growth.

We can classify the existing growth retardants into two main groups: ethylene-releasing compounds, such as ethephon, and inhibitors of GA biosynthesis. This contribution deals with the biochemical mode of action of typical representatives of the latter group. Previous reviews on this subject have been published (22, 34, 68, 55, 135). In light of the availability of new compounds and substantial progress made in the area, an update appears to be useful.

GIBBERELLIN BIOSYNTHESIS

At present, 125 different GAs are known to occur in higher plants and/or GA-producing fungi. A continuously updated list of structures and their occurrence may be found on the internet (http://www.plant-hormones.bbsrc.ac.uk/educatio.html). Only a few of the GAs possess biological activity per se, whereas the majority are precursors or catabolites. The main hormonal functions of GAs are the promotion of longitudinal growth, the induction of hydrolytic enzymes in germinating seeds, the induction of bolting in long-day plants, and the promotion of fruit setting and development. However, some of the many GAs might have functions that are still unknown.

GAs are diterpenoids and consist of 19 or 20 carbon atoms. Their biosynthesis is relatively well understood. The sequential steps involved in GA metabolism have been studied by using cell-free enzymatic systems prepared, for example, from immature seeds of pumpkin or pea. Radiolabeled substrates were converted into their respective products under distinct conditions in the presence of suitable co-factors. These in vitro biosynthesis systems have, in conjunction with analyzing the GAs present in intact plants, also been used to identify the point of interaction of inhibitors in the biosynthetic sequence leading to GAs. More recently, further details of GA biosynthesis were elucidated by employing distinct plant mutants and by the cloning and characterization of genes coding for the GA-biosynthetic enzymes.

The biosynthesis of GAs can be separated into three stages according to the nature of the enzymes involved and the corresponding localization in the cell: Terpene cyclases acting in proplastids, monooxygenases associated with the endoplasmatic reticulum, and dioxygenases located in the cytosol. Only a brief outline is given here for orientation; for more-detailed information on different aspects of GA metabolism the reader is referred to recent review articles (e.g. 51, 70, 71, 74, 75, 101, 112, 146, 157).

Formation of *ent*-Kaurene

It has been assumed until recently that GAs as well as all other isoprenoids are exclusively formed from the C_5 compound isopentenyl diphosphate (IPP) synthesized from mevalonic acid (MVA) (24). In the cytosol MVA is phosphorylated via two steps into MVA-5-diphosphate, which, after decarboxylation, yields IPP. However, new results indicate that IPP can also be formed via a non-mevalonate pathway in plastids (106, 144). In this pathway, D-glyceraldehyde 3-phosphate plus pyruvate yields 1-deoxy-D-xylulose 5-phosphate, which is converted into IPP. The mevalonate pathway gives rise to sterols, sesquiterpenes, and triterpenoids, whereas the pathway involving 1-deoxy-D-xylulose 5-phosphate yields carotenoids, phytol, plastoquinone-9, mono-, and diterpenoids. Some interchange between the pathways seems to exist. In general, it appears likely that GA precursors also are formed primarily via the 1-deoxy-D-xylulose 5-phosphate pathway in plastids of green tissues although this has not yet been conclusively demonstrated.

IPP is transformed via an isomerase-catalyzed reaction into dimethylallyl-PP. In head-to-tail condensations, three molecules of IPP are sequentially added to this compound to form geranyl diphosphate (GPP), farnesyl diphosphate (FPP), and finally, the C_{20} compound geranylgeranyl diphosphate (GGPP). GGPP is cyclized via copalyl diphosphate (CPP) to *ent*-kaurene. The latter steps are catalyzed by two distinct enzymes, namely CPP synthase (formerly known as *ent*-kaurene synthase A) and *ent*-kaurene synthase (formerly known as *ent*-kaurene synthase B). High activities of CPP synthase and *ent*-kaurene synthase were detected in the stroma of proplastids from pea and wheat shoots and in leucoplasts from pumpkin endosperm (2). In contrast, mature chloroplasts are low in such activities and, therefore, it is proposed that *ent*-kaurene is primarily produced in rapidly dividing cells (1).

Oxidation of *ent*-Kaurene to GA_{12}-Aldehyde

The reactions of stage 2 are catalyzed by monooxygenases located on the endoplasmatic reticulum, which require O_2 and NADPH for activity and involve cytochrome P450. The highly lipophilic *ent*-kaurene is oxidized stepwise at C-19 via *ent*-kaurenol and *ent*-kaurenal to *ent*-kaurenoic acid. *ent*-Kaurenoic acid is then hydroxylated to *ent*-7α-hydroxykaurenoic acid. After an oxidative ring contraction with extrusion of C-7, GA_{12}-aldehyde is formed. GA_{12}-aldehyde can be deemed the first intermediate specific for GAs.

Further Oxidation of GA_{12}-Aldehyde to the Different GAs

The conversions of stage 3 take place primarily in the cytosol. Most reactions are catalyzed by soluble dioxygenases, which require 2-oxoglutarate as a co-substrate and Fe^{II} and ascorbate as co-factors for activity. However, depending on the plant species and the tissue, some initial steps may still be catalyzed by monooxygenases. Hydroxylations at positions 13 and 12α may involve both types of enzyme. GA_{12}-aldehyde is oxidized by either a monooxygenase or a dioxygenase at position 7, thereby converting the aldehyde function into a carboxylic acid group and leading to GA_{12}. In the early 13-hydroxylation pathway, which is common to higher plants, GA_{53} would be the next intermediate after GA_{12}. Thereafter, a stepwise oxidation of C-20 and lactone formation (involving C-19 between C-4 and C-10) with the loss of C-20 as CO_2 is catalyzed by the multifunctional GA 20-oxidase. These reactions lead via GA_{44} and GA_{19} to GA_{20}, a C_{19}-GA. Superimposed on these and further steps, species and organ specific hydroxylation patterns occur, which may lead to typical "GA families." Considerable biological activity can be found only among C_{19}-GAs, which are further hydroxylated at position 3ß (e.g. GA_1 as a product of GA_{20}, or GA_3, GA_4, GA_7 and several other GAs). In contrast, hydroxylation at position 2ß (e.g. conversion of GA_1 to GA_8) drastically reduces biological activity. This step, further oxidative reactions, and conjugation with, for example, glucose obviously have the function of terminating the mission of a GA. Evidence is available that several related or isoenzymes of GA 20-oxidase, 3ß-hydroxylase, and 2ß-hydroxylase exist that are relatively low in substrate specificity and that may have overlapping activities.

Figure 1 represents a highly simplified scheme of GA metabolism concentrating on those reactions that are involved in the formation of GA_1, since this GA seems to be of paramount importance for stem elongation in many plant species (133). The structures of some important intermediates are presented on the left side. Numbering of carbon atoms is exemplified in this figure referring to *ent*-gibberellane.

INHIBITORS OF GA BIOSYNTHESIS

Four groups of GA biosynthesis inhibitors are known: "onium" compounds, compounds with an N-containing heterocycle, structural mimics of 2-oxoglutaric acid, and 16,17-dihydro-GAs. Each of these groups inhibits GA metabolism at distinct steps. An overview on the points of interaction with GA biosynthesis is shown in the right part of Figure 1.

Onium-Type Compounds

Several compounds that possess a positively charged ammonium, phosphonium or sulphonium group block the biosynthesis of GAs directly before *ent*-kaurene. The most prominent representatives of this group are chlormequat chloride (163, 164) and mepiquat chloride (178). These compounds, which have a quaternary ammonium group, are primarily used as anti-lodging agents in cereal production and to reduce excessive vegetative growth in cotton. Piproctanyl bromide, which is used to some extent in the production of ornamental plants and AMO-1618 are further growth retardants with a quaternary ammonium function. Chlorphonium and BTS 44584 (49) should be mentioned here as possessing a phosphonium and sulphonium moiety, respectively. Further examples of onium-type compounds may

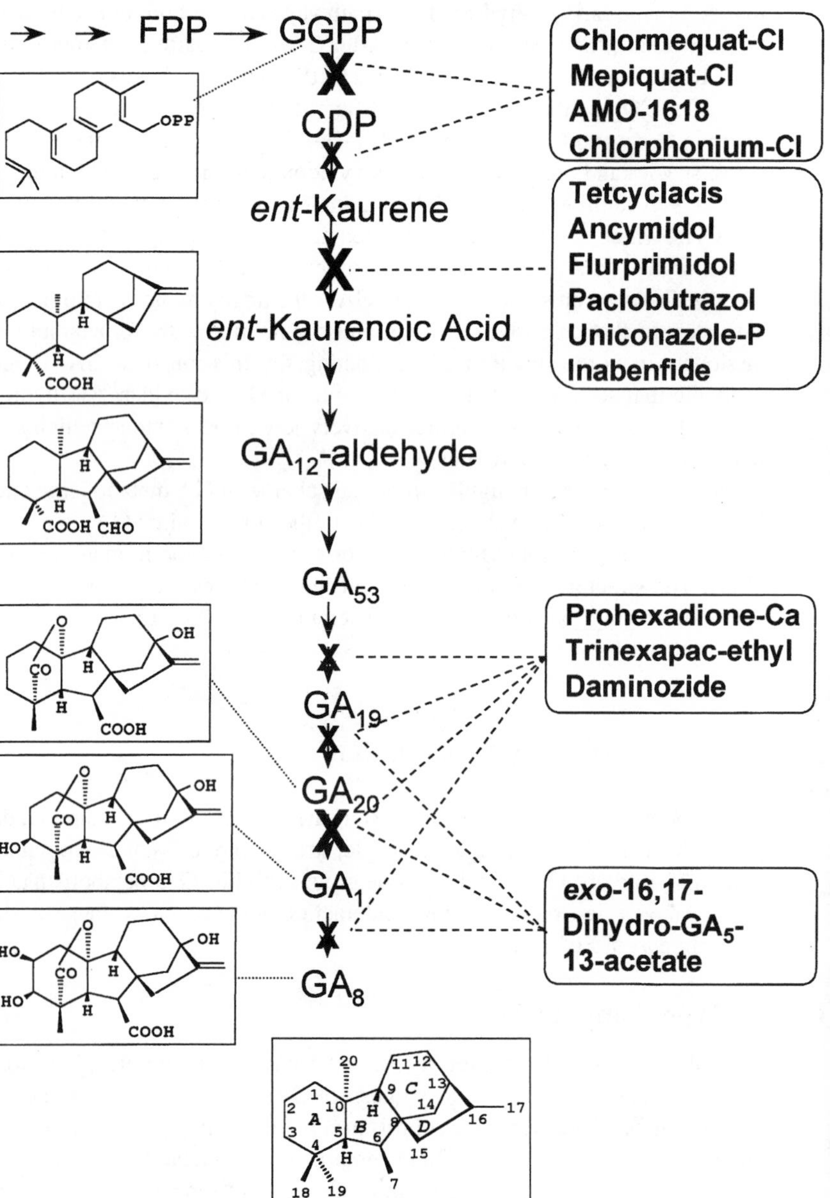

Figure 1 Simplified scheme of biosynthetic steps involved in GA biosynthesis and points of inhibition by plant growth retardants (**X**, **x** = major and minor activity, respectively). (See text for abbreviations.)

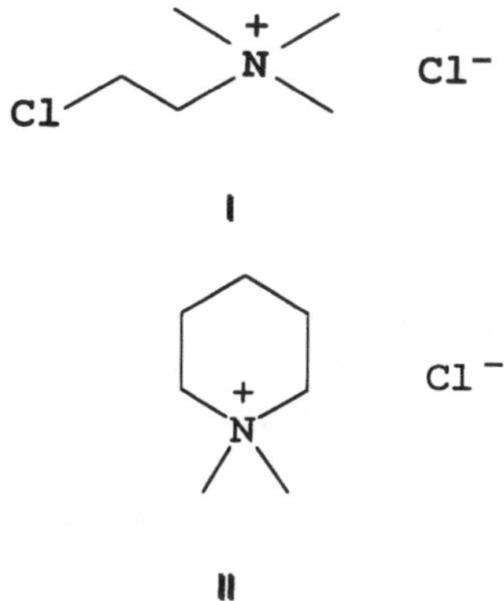

Figure 2 Onium-type plant growth retardants. I. Chlormequat chloride = (2-Chloroethyl)-trimethylammonium chloride {Chlorocholine chloride, ***CCC***}; II. Mepiquat chloride = 1,1-Dimethylpiperidinium chloride {***DPC***}.

be found in the literature (25, 34, 87, 107, 150). The structures of chlormequat chloride and mepiquat chloride are presented in Figure 2. Their official common, systematic chemical and other frequently used names and often used or suggested abbreviations (in bold and italics) are also given in this and the following figures.

Chlormequat chloride, AMO-1618, and chlorphonium inhibit CPP-synthase in both the GA-producing fungus *Gibberella fujikuroi* and cell-free preparations of this fungus and of higher plants. *ent*-Kaurene synthase is also inhibited by these compounds, but mostly at a lower degree of activity (153). The cyclization of GGPP into CPP, catalyzed by CPP-synthase, is analogous to the reaction leading from 2,3-oxidosqualene to lanosterol in mammals and fungi or cycloartenol in higher plants in the respective courses of sterol biosynthesis. Tertiary amine analogs of squalene are efficient inhibitors of oxidosqualene cyclase. It is suggested that such compounds, which are positively charged at physiological pH, mimic carbocationic high-energy intermediates in the cyclization reaction. Such intermediates are expected to bind very tightly to the enzyme, thereby blocking the reaction (8). By analogy, it appears likely that inhibitors of CPP-synthase mimic cationic intermediates in the conversion of GGPP into CPP (68). A similar mechanism would also apply for the succeeding cyclization of CPP into *ent*-kaurene.

To obtain any significant effects in cell-free preparations, relatively high concentrations of chlormequat chloride have to be used and, in some cases, the compound is even inactive (4, 46, 68, 170). The same is true of mepiquat chloride: In an enzyme system derived from pumpkin (*Cucurbita maxima*) endosperm, concentrations as high as 10^{-3} M of this compound, as well as of chlormequat chloride, did not affect the spectrum of GAs and GA precursors (77; L Schwenen & JE Graebe, unpublished data). A possible explanation of this difficulty might be the fact that these compounds are almost inactive in intact pumpkin plants. The same might also be expected for cell-free preparations from pumpkin tissues. Consequently, chlormequat chloride has been tested with enzymes derived from germinating wheat seedlings, where it gave more pronounced effects (53).

More definite results with some of the onium-type growth retardants have also been obtained by studying their effects on GA levels in intact higher plants. Several older investigations exist, in which levels of endogenous GAs had been determined by bioassays. In general, GA levels were found to be decreased by the growth retardants, more or less parallel to reductions in shoot length (cf. 134). With regard to chlormequat chloride, these results could more recently be confirmed employing modern techniques such as combined gas chromatography-mass spectrometry: Chlormequat chloride lowered the levels of GA_1 in both the shoots and grains of *Triticum aestivum* (104). Likewise, it led to a dose-responsive reduction of all GAs (GA_{12}, GA_{53}, GA_{44}, GA_{19}, GA_{20}, GA_1, GA_8) present in two cultivars of *Sorghum bicolor* (103). In *Eucalyptus nitens*, chlormequat chloride caused a reduction of GA_{20} and GA_1 (171).

Compounds with a Nitrogen-Containing Heterocycle

Several growth retardants are known that comprise a nitrogen-containing heterocycle. The pyrimidines ancymidol (165) and flurprimidol are of some commercial relevance, especially in ornamentals. Tetcyclacis, a norbornanodiazetin (88), has been used as a dwarfing agent in the production of rice seedlings for transplanting. Certain triazole-type compounds have attained a relatively high degree of interest. Paclobutrazol (105) and the closely related uniconazole-P (84, 85) are highly active members of this group and have found practical uses in rice, fruit trees and ornamentals. Triapenthenol (110) and BAS 111..W (89) represent further triazole-type growth retardants. A compound being used to lower the risk of lodging in rice is inabenfide, a 4-substituted pyridine (154). Also distinct imidazoles, such as 1-*n*-decylimidazole and 1-geranylimidazole (167) and HOE 074784 (19) possess plant growth-retarding properties. In some instances, plant growth retardation can also be found as a side activity of some triazole-type fungicides such as triadimenol, triadimefon (14), or ipconazole (147). Particularly in oilseed rape, the growth-regulating effect of the fungicides tebuconazole and metconazole is of practical relevance. The structures of typical growth retardants possessing an N-containing heterocycle are shown in Figure 3.

These growth retardants act as inhibitors of monooxygenases catalyzing the oxidative steps from *ent*-kaurene to *ent*-kaurenoic acid (51 and references cited therein, 119). Steps lying after *ent*-kaurenoic acid, which may still involve monooxygenases, do not seem to be affected (51). The structural feature common to all these inhibitors of *ent*-kaurene oxidation is a lone electron pair on the sp^2-hybridized nitrogen of their heterocyclic ring. In each case, this electron pair is located at the periphery of the molecule (140). Most probably, the target monooxygenases contain cytochrome P450 (65, 66) and it appears likely that the lone pairs of electrons of the growth retardants displace oxygen from its binding site at the protoheme iron (30). Evidence for such a type of interaction has been presented

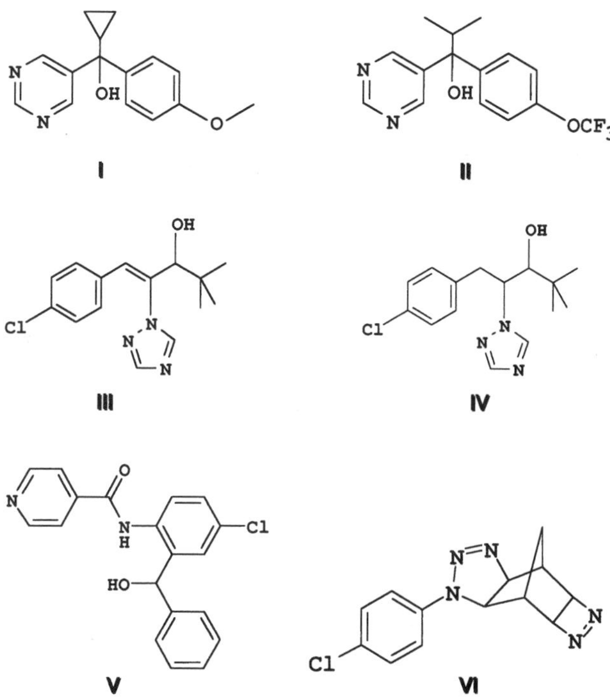

Figure 3 Plant growth retardants with an N-containing heterocycle. I. Ancymidol = α-Cyclopropyl-(p-methoxyphenyl)-5-pyrimidinemethanol {EL-531, ***Anc***}; II. Flurprimidol = α-(1-Methylethyl)-[p-4-(trifluoromethoxy)phenyl]-5-pyrimidinemethanol {EL-500, ***Flp***}; III. Uniconazole-P = (E)-(*RS*)-1-(4-Chlorophenyl)-4,4-dimethyl-2-(1H-1,2,4-triazol-1-yl) pent-1-en-3-ol {S-3307D, XE-1019, ***UCZ***}; IV. Paclobutrazol = (*2RS*, *3RS*)-1-(4-Chlorophenyl)-4,4-dimethyl-2-(1H-1,2,4-triazol-1-yl)pentan-3-ol {PP333, ***PBZ***}; V. Inabenfide = [4-Chloro-2-(a-hydroxybenzyl)]-isonicotinanilide {CGR-811, ***IBF***}; VI. Tetcyclacis = 5-(4-Chlorophenyl)-3,4,5,9,10-pentaazatetracyclo-[5.4.$10^{2,6}$.$0^{8,11}$]-dodeca-3,9-diene {LAB 102 883, BAS 106 W, ***TCY***}.

for ancymidol in microsome preparations of *Marah macrocarpus* (28, 30) and for BAS 111..W, using microsomal membranes isolated from immature pumpkin endosperm (111).

Depending on the presence or absence of a double bond, uniconazole-P and paclobutrazol possess one or two asymmetric carbon atoms, respectively. Since commercial paclobutrazol consists mainly of the (*2RS,3RS*) diastereoisomer (160), this structure allows virtually only two enantiomers, as does uniconazole-P. Detailed experiments carried out with the optical enantiomers of paclobutrazol have shown that the (*2S,3S*)-form exhibits more pronounced plant growth-regulatory activity and blocks GA biosynthesis more specifically, whereas the (*2R,3R*)-enantiomer is more active in inhibiting sterol biosynthesis (15, 73, 160). Fungicidal side activities of paclobutrazol are attributed to its effect on sterol formation (160). It has been demonstrated that the (*2S,3S*)-enantiomer is structurally similar to *ent*-kaurene whereas *the* (*2R,3R*)-form is closely related to lanosterol, the respective intermediates of GA and sterol biosynthesis (160). Similar chiralic specificities have been found for uniconazole-P (84), triapenthenol (109) and inabenfide (119): In all cases, the (*S*)-enantiomer was more inhibitory to *ent*-kaurene oxidation than the respective (*R*)-counterpart. Using computer assisted molecular modeling methods, clear structural similarities could be demonstrated between tetcyclacis and the growth-retarding forms of paclobutrazol and uniconazole-P with *ent*-kaurene and *ent*-kaurenol (94, 123, 160). This indicates that, within certain limits, distinct structural features are required to bind to and thereby block the active site of the enzyme. One may assume that the structures of the other growth retardants possessing an N-containing heterocycle would also fit into this scheme.

Clear evidence is available that reduction of shoot growth by pyridines, 4-pyrimidines, triazoles, imidazoles, and diazetines is caused by a lowered content of biologically active GAs. Reduced levels of GAs have, for instance, been analyzed by modern techniques under the influence of ancymidol in beans (155), tetcyclacis in corn cockle (*Agrostemma githago*) (179), paclobutrazol in barley and wheat (104) and in *Eucalyptus nitens* (171), uniconazole-P in rice (85) and *Sorghum bicolor* (103), BAS 111..W in oilseed rape (72), and inabenfide in rice (119).

Structural Mimics of 2-Oxoglutaric Acid

One part of this group is represented by acylcyclohexanediones such as prohexadione-calcium (prohexadione-Ca) (125), trinexapac-ethyl (3) and the experimental compound LAB 198 999 (134). Virtually all higher plants react with a reduced shoot growth after treatment (cf. 126). Stem stabilization in cereal crops, rice, and oilseed rape, growth control in turf grasses and reduction of vegetative growth in fruit trees are the main applications.

Acylcyclohexanediones interfere with the late steps of GA biosynthesis. Structural similarities between the acylcyclohexanediones and 2-oxoglutaric acid, which is the co-substrate of the involved dioxygenases, are assumed to be responsible for the blocking of GA metabolism (54). Studies with cell-free preparations have revealed that most steps after GA_{12}-aldehyde are inhibited by acylcyclohexanediones (52–54, 69, 93, 124, 126, 143). Enzyme kinetic data indicate that the retardants act largely competitively with respect to 2-oxoglutarate (54, 69). The hydroxylations at position 3ß (e.g. the formation of GA_1 from GA_{20}) and also at position 2ß (e.g. the conversion of GA_1 into GA_8) appear to be the primary targets of acylcyclohexanediones (54, 126). These findings are supported by analytical data, generally showing that growth reduction is accompanied by lowered levels of biologically active GA_1 and its metabolite GA_8 but increased concentrations of GA_{20} and earlier precursors of GA_1 (3, 11, 91, 93, 103, 126, 143, 149, 182)

Growth retardation caused by acylcyclohexanediones can be reversed only by GAs that are active per se and need not be metabolically activated, also indicating that late stages of GA formation are blocked (90, 127, 182). In selected cases, compounds like prohexadione-Ca and trinexapac-ethyl may, paradoxically, even lead to increases in shoot growth, most likely by protecting endogenous active GAs from being metabolically inactivated (78). Likewise, the inactivation of exogenously applied GA_1 by 2ß-hydroxylation can be inhibited by simultaneous treatment with an acylcyclohexanedione, resulting in increased GA-like activity (125, 158).

A number of different acylcyclohexadiones and structurally related compounds have been evaluated for their ability to inhibit GA 2ß- (54) and 3ß-hydroxylases (11, 12, 93) in cell-free systems. When the cyclohexane ring was replaced by benzene, an almost complete loss of activity resulted. In contrast, certain pyridine structures displayed a relatively high degree of activity. In structures related to prohexadione, a free carboxylic acid function resulted in higher activity as compared to the corresponding methyl or ethyl esters, most likely due to a higher degree of similarity to 2-oxoglutaric acid. Longer acyl side-chains lead to increased inhibitory activity as compared to shorter ones. However, when applied to intact plants, too long chains caused phytotoxicity. Therefore, substituents such as ethyl or cyclopropyl appear to be optimal for practical uses. In addition to these findings, one has also to consider that esters are often more easily taken up by leaves after spray application than ionized forms. In the plant cell, the acid might be formed again by saponification. Furthermore, esters may be easier to handle for preparing formulated products. Under practical conditions trinexapac-ethyl (an ester) and prohexadione-Ca (a salt of an acid) display similar degrees of activity when applied in appropriate formulation to graminaceous species, such as small grains or turf grasses. However, in dicots prohexadione-Ca generally outperforms trinexapac-ethyl (W Rademacher, unpublished results). This may indicate that trinexapac-ethyl is easily saponified into its active acidic form in grasses, whereas this process is not as pronounced in dicots.

The growth retardant daminozide has been used for many years to reduce excessive shoot growth. Its growth-retarding activity is, however, restricted to relatively few plant species, such as apple, groundnuts and chrysanthemums. Due to toxicological concerns, the importance of daminozide has declined markedly in recent years, particularly for edible crops. Until a few years ago, the mode of action of daminozide had been unclear. In light of structural similarities between daminozide and 2-oxoglutaric acid and re-evaluating older results from the literature, it has been proposed that daminozide, like acylcyclohexanediones, would block GA formation as an inhibitor of 2-oxoglutarate-dependent dioxygenases (137). This hypothesis has later been proven by working with an enzyme preparation derived from cotyledons of *Phaseolus coccineus* and by analyzing the GAs of treated peanut plants (11).

Figure 4 shows the structures of several compounds mentioned and of 2-oxoglutaric acid for comparison.

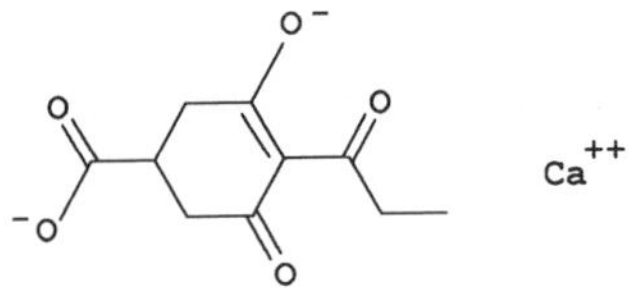

Figure 4 Structural mimics of 2-oxoglutaric acid. I. Prohexadione-calcium = Calcium 3-oxido-4-propionyl-5-oxo-3-cyclohexenecarboxylate {KIM-112, BAS 125..W, ***ProCa***/free acid: KUH 833, BX-112, ***ProH***}; II. Trinexapacethyl (=cimectacarb) = Ethyl-(3-oxido-4-cyclopropionyl-5-oxo) oxo-3-cyclohexenecarboxylate {CGA-163′935, ***TrixE***}; III. Daminozide = Succinic acid 2,2-dimethyl hydrazide {B-995, ***SADH***}; IV. 2-Oxoglutaric acid.

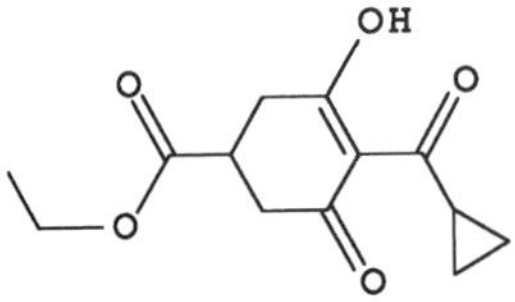

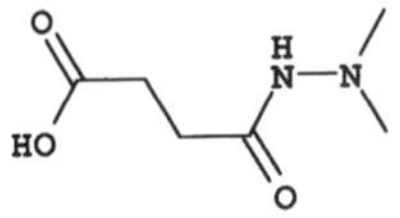

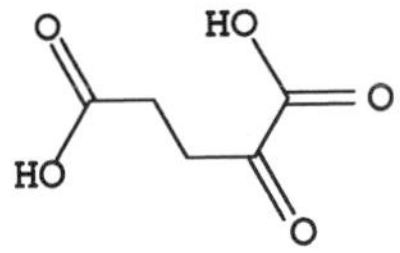

16,17-Dihydro-GAs

16,17-Dihydro-GAs represent the most recent group of growth retardant. A number of different structures of this type, mostly GA_5 derivatives, have been described to reduce shoot elongation in *Lolium temulentum* (41, 114–116) and other grasses (96). Evidence is available that their growth-retarding activity is due to an inhibition of dioxygenases, which catalyze the late stages of GA metabolism, particularly 3ß-hydroxylation (45, 91, 161). Similar to acylcyclohexanediones, such GA derivatives also increase the biological activity of GA_1, when applied simultaneously to seedlings of wheat and barley (W Rademacher, unpublished). Hence, most likely GA_1 2ß-hydroxylation is inhibited as well, although this may be less pronounced in species such as *Lolium temulentum* (91). Treating plants with 16,17-dihydro-GA_5 results in changes of GA levels similar to the ones caused by acylcyclohexanediones: In *Lolium temulentum* (91) and *in Sorghum bicolor* (45) the levels of GA_1 declined, whereas GA_{20} accumulated significantly.

With a view to finding new anti-lodging compounds for small grains, several 16,17-dihydro-GA_5 derivatives have recently been retested. Applying the compounds in conjunction with suitable adjuvants has, in general, significantly raised

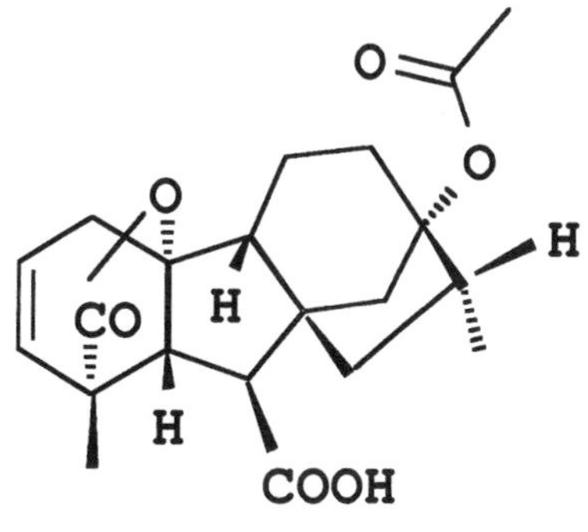

Figure 5 16,17-Dihydro-GAs exo-16,17-Dihydro-GA_5-13-acetate.

their biological activity. Any comparison with older data is almost impossible, however. As a result of these investigations, *exo*-16,17-dihydro-GA_5-13-acetate (Figure 5) represents the most active growth retardant ever known for graminaceous plants. Under greenhouse conditions effects of as little as 500 mg per hectare can be detected in wheat and barley (141). However, in order to reduce the risk of lodging under practical conditions, rates in the range of 20 g per hectare have to be used (141). In contrast to graminaceous plants, *exo*-16,17-dihydro-GA_5-13-acetate and related structures are virtually inactive in reducing shoot growth in any other plant species tested (141). Likewise, 16,17-dihydro derivatives of GA_{19}, GA_{20}, and GA_1 did not cause growth retardation in willow (*Salix pentandra*). As compared to their naturally occurring counterparts, GA-like activity of these compounds was significantly reduced (132). These results demonstrate that 16,17-dihydro derivatives, particularly of GA_5, interact very specifically with GA formation only in graminaceous species. This could be due to distinct peculiarities of GA metabolism or uptake, translocation, and degradation in these species.

It appears logical that *exo*-16,17-dihydro-GA_5-13-acetate and related structures are highly specific in competing in grasses with the natural GA substrates, e.g. GA_{20}, for the respective enzymatic sites (161). The *endo* form of 16,17-dihydro-GA_5-13-acetate is somewhat less active than its *exo* counterpart (141). Similar observations have also been made with slightly different and, in general, less active structures, such as 16,17-dihydro-GA_5 (141) and 17-alkyl derivatives (114). A number of substituents of 16,17-dihydro-GA_5 at C-13, in particular esters and ethers of different chain length, have been assayed in wheat and barley. The 13-acetate function was clearly the most active one. However, a fairly high degree of activity is still observed with groups such as *n*-propionate or O-ethylether (141). According to our current knowledge, it appears that a double bond between C-2 and C-3 is of importance for high growth-retarding activity. Also, the absence of hydroxy groups on these carbon atoms seems to be an essential element for pronounced growth-retarding activity: In sharp contrast to its GA_5 analog, *exo*-16,17-dihydro-GA_1-13-acetate, which displays a single bond between C-2 and C-3 and is 3ß-hydroxylated, is virtually inactive in wheat and barley seedlings

(LG Mander & W Rademacher, unpublished results). Several naturally occurring GAs, in particular GA_5, also reduce the activity of 3ß-hydroxylases obtained from immature *Phaseolus vulgaris* seeds. Unlike several other GAs tested, the inhibiting GAs did not possess any carbonyl functions in the A-ring of the molecule except for the lactone group (100, 156). GA_5 also reduced the conversion of GA_{20} into GA_1, although at a clearly lower degree of activity than 16,17-dihydro-GA_5 (91). Inhibition of GA metabolism has also been reported for other structural GA variants. For instance, deoxygibberellin C, which is an isomer of GA_{20} and displays a keto function at C-16 and a methyl group at C-13, inhibits shoot growth in normal rice and maize (64, 148). Earlier suggestions that deoxygibberellin C would act by inhibiting 3ß-hydroxylation (63) were proven in an enzyme system derived from embryos of immature *Phaseolus vulgaris* seeds (92). In contrast, 3ß-hydroxylase was unaffected in a cell-free system derived from pumpkin endosperm (92), which indicates that species-specific differences may exist. Note also that deoxygibberellin C inhibits shoot growth in rice and maize, but not in cucumber (148) and pea (Y Kamiya, personal communication). This would indicate that, similar to 16,17-dihydro-GA_5 derivatives, graminaceous species, but not dicots, respond with retarded growth. Thus it appears that a double bond between C-2 and C-3 and the absence of hydroxy groups on these carbon atoms, combined with the 16,17-dihydro function, are the main important structural elements of this new class of growth retardant. Derivatization of the hydroxy function at C-13 seems to be of secondary relevance only.

Several 16,17-dihydro GAs occur naturally in higher plants or in GA-producing fungi (GA_2, GA_{10}, GA_{41}, GA_{42}, GA_{82}, GA_{83}). Likewise, some synthetically produced 16,17-dihydro-GAs were dealt with a number of years ago (10, 79). None of these compounds has ever been described as possessing growth-retarding activity. At that time, testing was rather performed to determine GA-like activity and, except for some dwarfing genotypes, graminaceous species were not employed in the assays. As a consequence, the growth-retarding properties of compounds such as 16,17-dihydro-GA_5 (10) did not show up.

EFFECTS OF GROWTH RETARDANTS ON GA METABOLISM IN GA-PRODUCING FUNGI

Many investigations on growth retardants have been conducted with GA-producing fungi, since the analysis of GAs, for instance, from cultures of *Gibberella fujikuroi* and *Sphaceloma manihoticola*, is relatively easy owing to the presence of much higher amounts than in higher plants. In general, the steps of fungal GA metabolism are deemed to be closely related to those in higher plants, although distinct differences must not be overlooked (138).

Both in *G. fujikuroi* and in *S. manihoticola*, onium compounds such as chlormequat chloride, AMO-1618 and mepiquat chloride cause a clear inhibition of GA formation. Only chlorphonium is relatively inactive in these fungi, which is most likely due to rapid disintegration throughout fermentation (61, 95, 130, 136). Fungal GA production is also blocked by a number of growth retardants with a nitrogen-containing heterocycle (29, 136).

Contrasted with the situation in higher plants, GA formation in *G. fujikuroi and S. manihoticola* is not affected by acylcyclohexanediones such as LAB 198 999 (136) or prohexadione-Ca (W Rademacher, unpublished results). This could be explained by a relatively rapid disintegration in fungal cultures, since both compounds are known to be relatively short-lived in biological systems (40). However, daminozide, which has a similar mode of action, did not affect GA synthesis of *G. fujikuroi* even though it remained intact during fermentation (130). *exo*-16,17-dihydro-GA_5-13-acetate does not interfere with GA production in *G. fujikuroi*, nor at the same time is it metabolized (W Rademacher, unpublished results). Altogether, one should not rule out the possibility that the late steps of GA metabolism in fungi are catalyzed by enzymes, which are different from the ones in higher plants. This suggestion is supported by the fact that to date only cyclases and monooxygenases, but not dioxygenases, could be detected in *G. fujikuroi* (166).

SIDE EFFECTS OF GA BIOSYNTHESIS INHIBITORS

The commercially available plant growth retardants have undergone intensive testing in the processes of selection and registration. Therefore, any side effects can be expected to be either neutral or even positive to the growth of treated plants. Early precursors of GA formation, such as IPP, are shared with other terpenoids, and thus there are links, for example, to the biosynthesis of sterols, carotenoids, abscisic acid (ABA), and cytokinins. In addition, related enzymatic reactions may be found in other pathways. Cytochrome P450-dependent monooxygenases would appear to be of particular relevance in this context, since many isoforms exist capable of modifying a variety of substrates (152). Furthermore, the possibility that indirect effects will also influence certain metabolic reactions cannot be ruled out. From the wealth of information available, the following parts of this contribution will concentrate on side effects that are relevant for plant development and plant defense reactions.

Effects on the Levels of Other Phytohormones

Plant growth retardants have often been reported to interfere with the endogenous levels not only of GAs but also of other plant hormones. Here, reference is made only to reports in which reliable methods, as seen from today, have been employed. Thus, many older contributions, most of which involve the long-known onium compounds, have not been considered.

Many investigations have dealt with the effect of growth retardants with a nitrogen-containing heterocycle on levels of hormones other than GAs (see surveys by 43, 55, 134). Typically, these compounds induce increased contents of cytokinins, whereas ethylene levels are lowered. ABA concentrations may be significantly increased under distinct conditions whereas the auxin status is not significantly affected. Resulting primarily from these effects, a delay in senescence and increases in resistance to environmental stresses are often found (43, 55). At present, the observed effects on cytokinin and ethylene levels cannot be explained satisfactorily, since no metabolic links are obvious. It rather appears that non-specific effects are responsible for the hormonal changes observed: Under the influence of growth retardants, assimilates are often shifted into the roots, which are known to be a major site of cytokinin formation. The resulting stimulation of root growth may lead to an increased formation of cytokinins, which are then exported into the shoot (42). Work with the triazole-type retardants BAS 111 W (56) and uniconazole-P (98) and paclobutrazol (120) indicated that ethylene formation might have been reduced by blocking aminocyclopropanecarboxylic acid (ACC) oxidase. Again, this must be an indirect effect since ACC oxidase is a dioxygenase-type enzyme (33, 117) and not a cytochrome P450-dependent monooxygenase, as suggested earlier (98). Inhibited conversion of ACC is also proposed as a reason for increased levels of polyamines (58, 81). The situation is clearer with regard to the mechanisms leading to increased ABA levels: By using detached leaves of *Xanthium strumarium*, it could be shown that tetcyclacis is capable of inhibiting the oxidative metabolism of ABA into phaseic acid (181), which is biologically inactive. As a result, ABA accumulates. Similar observations have been made in embryos of maize (6), primary leaves of barley (168), and in the moss *Riccia fluitans* (76). Since this inactivation involves 8′-hydroxylase, a monooxygenase that is cytochrome P450-dependent (32, 99, 180), the enzyme is likely blocked in a manner similar to *ent*-kaurene oxidase. Blocking ABA 8′-hydroxylase with tetcyclacis may lead to an accumulation of this hormone in a relatively short time (86). Other growth retardants of the group of monooxygenase inhibitors affect ABA metabolism in a similar fashion in other plant species. However, this effect is clearly not achieved by all retardants of this type in all plant species (20, 134, 139). It rather appears that the right compound has to match the right species. This would be in line with other reports, which indicate that many different cytochrome P450-dependent monooxygenases may occur in different species, the substrate specificity of which is not very pronounced (152). Knowing about the existence of monooxygenases that may affect, at the same time, key enzymes of GA and ABA metabolism, it is tempting to suggest that such enzymes could be part of the plant's rapid response mechanism to cope with stressful situations (139). Under favorable growing conditions, these monooxygenases might be "switched on," resulting in low ABA but high GA levels, thereby allowing intensive assimilation and shoot growth. Conversely, under situations such as drought stress, these enzymes would be "switched off," leading to low GA but high ABA levels. As a consequence, shoot growth, photoassimilation, and transpiration would be diminished.

Acylcyclohexanediones, although affecting different types of enzymes in GA metabolism, seem to have similar side effects on other hormones as N-heterocyclic compounds. Prohexadione-Ca, trinexapac-ethyl and LAB 198999 reduce ethylene levels in sunflower cell suspensions and in leaf disks of wheat (55). In shoots of wheat and oilseed rape, prohexadione-Ca leads to increased concentrations of cytokinins and ABA, while no major changes of indole-3-acetic acid contents occur (57). Since no immediate effect of acylcyclohexadiones on the metabolism of cytokinins and ABA are conceivable, indirect effects seem to play a role. In contrast, effects on ethylene levels may, at least partly, be explained by a more direct interaction: Ethylene is generated from aminocyclopropanecarboxylic acid (ACC) in a reaction catalyzed by ACC oxidase. This is a dioxygenase that requires ascorbic acid as a co-substrate. 2-Oxoglutaric acid and similar compounds inhibit its activity (83). It seemed, therefore, appropriate to investigate the effect of prohexadione-Ca, due to its structural relationship to 2-oxoglutaric and, also, ascorbic acid, on this reaction. Employing an enzyme system prepared from ripe pear, it was demonstrated that prohexadione-Ca was inhibitory to ACC oxidase at an I_{50} of approximately 10^{-5} M (142). Daminozide is known to delay the onset of ethylene formation in apple (108). Obviously, this is due to prevention of ACC formation (60) and, unlike the structurally related prohexadione-Ca, daminozide does not affect ACC oxidase.

Effects on Sterol Metabolism

The formation of sterols in fungi and in higher plants involves enzymatic reactions that are similar to certain steps in the biosynthesis of GAs (8, 17, 24, 62, 128). Therefore, it is not surprising that several growth retardants show some side effects on sterol metabolism. In the group of onium compounds, chlormequat chloride, AMO-1618, and chlorphonium, applied at high rates, restricted the biosynthesis of sterols and other terpenoids in tobacco and some further plant species. Growth retardation induced by these compounds could be reversed not only by GAs, but also by emulsions of different phytosterols (35, 36). Most likely these growth retardants inhibit 2,3-oxidosqualene cyclase in the course of plant sterol formation, just as a number of other quaternary ammonium compounds do (8, 18, 68). AMO-1618 applied to tobacco seedlings caused an accumulation of 2,3-oxidosqualene and inhibited the incorporation of radiolabeled mevalonic acid into sterols (37, 38). However, this effect could not be repeated in pea microsomes (39) which may indicate the existence of species-specific differences or the necessity of AMO-1618 to be metabolically activated in vivo. Furthermore, concentrations of the different onium compounds well in excess of 10^{-4} M have been used in most cases, indicating that the growth retardants inhibit sterol formation only at a relatively low degree of specificity.

Fungicides of the pyrimidine-, imidazole- and triazole-type often show a growth-regulatory side activity. These compounds act by blocking the oxidative 14α-demethylation in the course of fungal ergosterol biosynthesis (8, 17). Similarly, such fungicides, as well as tetcyclacis, paclobutrazol, triapenthenol and other triazole-type growth retardants, reduce the formation of 14α-demethylated sterols in higher plants by blocking obtusifoliol 14α-demethylase (13, 14, 16, 162).

58. Ecker JR. 1995. The ethylene signal transduction pathway in plants. *Science* 268: 667–75
59. Einspahr HJ, Peeler TC, Thompson GA Jr. 1988. Rapid changes in polyphosphoinositide metabolism associated with the response of *Dunaliella salina* to osmotic shock. *J. Biol. Chem.* 263:5775–79
60. Esau K. 1977. *Anatomy of Seed Plants.* New York: Wiley & Sons. 550 pp. 2nd ed.
61. Estelle M. 1998. Polar auxin transport: new support for an old model. *Plant Cell* 10:1775–78
62. Ettlinger C, Lehle L. 1988. Auxin induces rapid changes in phosphatidylinositol metabolites. *Nature* 331:176–78
63. Finkelstein RR, Somerville CR. 1990. Three classes of abscisic acid (ABA)-insensitive mutations of *Arabidopsis* define genes that control overlapping subsets of ABA responses. *Plant Physiol.* 94:1172–79
64. Finkelstein RR, Zeevaart JAD. 1994. Gibberellin and abscisic acid biosynthesis and response. In *Arabidopsis*, ed. EM Meyerowitz, CR Sommerville, pp. 1172–79. Cold Spring Harbor, NY: Cold Spring Harbor Press
65. Flowers TJ, Troke PF, Yeo AR. 1977. The mechanism of salt tolerance in halophytes. *Annu. Rev. Plant Physiol.* 28:89–121
66. Flowers TJ, Yeo AR. 1992. *Solute Transport in Plants*. Glasgow, Scotland: Blackie. 176 pp.
67. Flowers TJ, Yeo AR. 1995. Breeding for salinity resistance in crop plants: where next? *Aust. J. Plant Physiol.* 22:875–84
68. Ford CW. 1984. Accumulation of low molecular weight solutes in water stress tropical legumes. *Phytochem.* 23:1007–15
69. Foyer CH, Lopez-Delgado H, Dat JF, Scott IM. 1997. Hydrogen peroxide- and glutathione-associated mechanisms of acclimatory stress tolerance and signalling. *Physiol. Plant.* 100:241–54
70. Frommer WB, Ludewig U, Rentsch D. 1999. Taking transgenic plants with a pinch of salt. *Science* 285:1222–23
71. Fu H-H, Luan S. 1998. AtKUP1: a dual-affinity K^+ transporter from *Arabidopsis. Plant Cell* 10:63–73
72. Fukuda A, Nakamura A, Tanaka Y. 1999. Molecular cloning and expression of the Na^+/H^+ exchanger gene in *Oryza sativa Biochim. Biophys Acta* 1446:149–55
73. Gage DA, Rhodes D, Nolte KD, Hicks WA, Leustek T, et al. 1997. A new route for synthesis of dimethylsulfoniopropionate in marine algae. *Nature* 387:891–94
74. Galinski EA. 1993. Compatible solutes of halophilic eubacteria molecular principles, water-solute interaction, stress protection. *Experientia* 49:487–96
75. Galinski EA. 1995. Osmoadaptation in bacteria. *Adv. Microb. Physiol.* 37:273–328
76. Gaxiola RA, Rao R, Sherman A, Grisafi P, Alper S, Fink GR. 1999. The *Arabidopsis thaliana* proton transporters AtNhx1 and Avp1, can function in cation detoxification in yeast. *Proc. Natl. Acad. Sci. USA* 96:1480–85
77. Gaymard F, Pilot G, Lacombe B, Bouchez D, Bruneau D, et al. 1998. Identification and disruption of a plant shaker-like outward channel involved in K^+ release into the xylem. *Cell* 94:647–55
78. Glenn EP, Brown JJ, Blumwald E. 1999. Salt tolerance and crop potential of halophytes. *Crit. Rev. Plant Sci.* 18:227–55
79. Goddijn JM, van Dun K. 1999. Trehalose metabolism in plants. *Trends Plant Sci.* 4:315–19
80. Greenway H, Munns R. 1980. Mechanisms of salt tolerance in nonhalophytes. *Annu. Rev. Plant Physiol.* 31:149–90
81. Guan L, Scandalios JG. 1998. Two structurally similar maize cytosolic superoxide dismutase genes, *Sod4* and *Sod4A*, respond differentially to abscisic acid and high osmoticum. *Plant Physiol.* 117:217–24
82. Guern J, Mathieu Y, Kurkdjian A. 1989. Regulation of vacuolar pH in plant cells. *Plant Physiol.* 89:27–36

In general, relatively high rates, which induce extreme growth reduction, are required to obtain such changes (140). However, species-specific reactions must be expected: For instance, the triazole fungicide epoxiconazole induces significant growth retardation selectively in cleavers (*Galium aparine*), which is paralleled by a significant accumulation of 14α-methyl sterols (7). Tetcyclacis, applied at moderate rates, totally changes the sterol spectrum of oat, with cholesterol becoming the dominant sterol (19). A similar reaction was induced in roots of fenugreek (*Trigonella foenum-graecum*) (23). This phenomenon cannot be attributed solely to an inhibition of 14α-demethylase, because different sterols had to be expected then. As a possible explanation, the authors suggested that tetcyclacis would inhibit cholesterol 26-hydroxylase, a cytochrome P450-dependent monooxygenase. 26-Hydroxycholesterol, in turn, leads to the synthesis of saponins, which are of major relevance particularly in oats and fenugreek (see 23). Under similar conditions (reduction of shoot length to 50 to 30% of the respective control), tetcyclacis did not induce such changes in the sterol spectrum of maize, pea, bean, and sunflower (140) and of wheat and barley (RS Burden, personal communication).

Using the pure optical enantiomers of paclobutrazol it could be shown that the (*2R,3R*)-enantiomer is more specifically blocking fungal ergosterol biosynthesis while the (*2S,3S*)-form is the more specific inhibitor of GA biosynthesis (160). Likewise, (*2R,3R*)-paclobutrazol reduced plant sterol formation much more intensely than its (*2S,3S*)-analog (16). The (*2R,3S*)-enantiomer, mentioned in this contribution as being even more active as an inhibitor of phytosterol formation, is practically absent in commercial paclobutrazol (see 160). It could be shown that the (*2R,3R*)-form relates closely to lanosterol, while the (*2S,3S*)-enantiomer is structurally similar to *ent*-kaurene (160). Thus it is likely for higher plants also that the different enantiomers compete in distinct cytochrome P45O species with the substrates of sterol or GA biosynthesis, respectively. Similar chiralic specificities have been found for uniconazole-P (48) and triapenthenol (109).

Tetcyclacis, applied at relatively high doses, reduces cell division in cell suspension cultures of maize and simultaneously leads to qualitative and quantitative changes in the sterols present. Adding cholesterol or other plant sterols to the cell cultures can restore normal growth. GA_3, in contrast, has no effect (59). Equivalent results were later reported with cell cultures of celery (*Apium graveolens dulce*) treated with paclobutrazol (67). Conversely, intense growth retardation caused by tetcyclacis and several triazoles in intact wheat and sunflower plants could not even partly be overcome by external application of emulsions of different phytosterols (W Rademacher, unpublished results). This finding may indicate that sterols play only a minor role in longitudinal growth. However, one has to keep in mind the high degree of lipophilicity of such sterols, which inhibits uptake.

In general, one may conclude that influences of plant growth retardants on phytosterol formation will affect longitudinal shoot growth only to a small extent. Cell elongation, primarily occurring in the growth zones outside the meristems, is the more sensitive process as compared to cell division in the meristems itself. The regulation of cell elongation appears to be closely linked to the availability of GAs, which can be affected by relatively low retardant concentrations. Higher rates of retardants will additionally inhibit cell proliferation. Most likely, this is still primarily a result of increased GA deficiency. However, one should not rule out that altered sterol levels and, thus, a change in membrane properties may also be of relevance under such conditions (55, 131).

Effects on Brassinosteroid Formation

Brassinosteroids represent a group of plant steroids known to cause remarkable effects in higher plants (26, 113, 176). Their hormonal status is widely accepted due to results obtained from molecular genetic investigations combined with studies of the biosynthetic pathway. The regular plant sterols, campesterol in particular, function as precursors of brassinosteroids (47, 176). In addition to obtusifoliol 14α-demethylation, which is part of the metabolism of regular plant sterols, many of the reactions committed to brassinosteroid formation are catalyzed by cytochrome P450-type monooxygenases (47). It is therefore not surprising that growth retardants such as uniconazole-P (177) and antimycotic imidazoles such as clotrimazole and ketoconazole (174) inhibit brassinosteroid biosynthesis. However, convincing evidence is not yet available that such interactions are specific for brassinosteroid metabolism. Reducing the formation of sterol precursors by inhibiting obtusifoliol 14α-demethylase would also result in a reduction of brassinosteroid levels. Likewise, it remains to be clearly proven that inhibition of shoot growth in brassinosteroid-deficient mutants is caused by a lack of brassinosteroids or of phytosterols in general. The availability of specific inhibitors of brassinosteroid biosynthesis (5) should certainly help to clarify such questions.

Effects on Flavonoid Metabolism

The biosynthesis of flavonoids and other phenylpropanoids comprises steps that are catalyzed by cytochrome P450-dependent monooxygenases and by dioxygenases requiring 2-oxoglutaric acid as a co-substrate (44, 169). Cinnamate 4-hydroxylase, a cytochrome P450-type monooxygenase, is inhibited by relatively high dosages of tetcyclacis in a cell-free system prepared from soybean cell cultures. A higher degree of activity could be observed employing enzymes prepared from pea apices. Several triazole-type retardants also tested were inactive in the soybean system and gave only weak effects in the pea preparation (97, 140). Results varying between different inhibitors and different plant species were also obtained when ancymidol, tetcyclacis and ketoconazole were tested on flavonoid 3′-hydroxylase and flavone synthase II (159) and chalcone 3-hydroxylase (172). Likewise, anthocyanin formation was inhibited in buckwheat hypocotyls and sorghum coleoptiles by low dosages of tetcyclacis whereas the triazole-type inhibitor BAS 111...W was almost inactive (143). These observations provide further evidence that a great number of isoenzymes exist among cytochrome P450-dependent monooxygenases. Most likely, according to their sterical fit, typical inhibitors of GA metabolism and related compounds may or may not affect such enzymes involved in the metabolism of flavonoids as of sterols or other plant components.

High dosages of prohexadione-Ca and other acylcyclohexanediones inhibit the formation of anthocyanins in flowers and other parts of intact higher plants (143). Inhibition of anthocyanin production could also be observed in carrot cell-suspension cultures to which prohexadione had been added (82). It has been suggested that 2-oxoglutarate-dependent dioxygenases, in particular flavanon 3-hydroxylase (FHT), involved in the biosynthesis of anthocyanidins would represent targets for these growth retardants (143). This hypothesis has meanwhile been supported by the finding that treated young shoots of apple are unable to convert eriodictyol by 3-hydroxylation into flavonoids such as catechin. Instead, eriodictyol accumulates and large amounts of luteoliflavan, which does not normally occur in apple tissue, can be found (142, 145). Apple and pear trees treated with prohexadione-Ca are significantly less affected by fire blight, caused by the bacterium *Erwinia amylovora* (31, 173, 175). Likewise, the incidence of fungal diseases, e.g. scab on apple shoots and gray mold on grape shoots and berries, can also be reduced when plants are pretreated with prohexadione-Ca or trinexapac ethyl (E Ammermann, J Speakman, G Stammler & W Rademacher, unpublished). Morphological and anatomical effects caused by the growth retardants should not be ruled out as being of some relevance for these observations. However, there are several indications of an induction of physiological resistance and it is hypothesized that changes in flavonoids or other phenylpropanoids play a major role (142, 145). Daminozide, recently also identified as an inhibitor of 2-oxoglutarate-dependent dioxygenases involved in GA biosynthesis (11), obviously possesses more parallels to acylcyclohexanediones: It may also block anthocyanin formation, for instance in chrysanthemums (118), and it may also cause a slight induction of resistance against fire blight in apple (KS Yoder, personal communication).

Effects on Other Metabolic Reactions

Further side effects of plant growth retardants have been reported for the inhibitors of cytochrome P450-dependent monooxygenases. Under certain circumstances, compounds of this type may have a relatively high efficiency in blocking the oxidative metabolism of certain herbicides and other xenobiotics (e.g. 21, 27, 102, 122). Tetcyclacis is able to inhibit the formation of jasmonic acid in osmotically challenged barley leaves. It is suggested that allene oxide synthase, which is involved in its biosynthesis, is the target enzyme (168). In a cell-free system derived from the blue-green alga *Aphanocapsa sp.*, tetcyclacis and an experimental triazole-type growth retardant inhibited hydroxylating reactions in the course of xanthophyll formation (9).

Further Important Features of the Different Growth Retardants

Under practical conditions, the growth-retarding effect of a given compound is not necessarily determined by its type of interaction with GA metabolism. Rather, factors such as plant responsiveness, uptake, translocation, persistency, and side effects are of relevance (cf. 43, 134, 151). It is known, for instance, that cotton is relatively sensitive towards mepiquat chloride. A typical dosage would be around 50 g of active ingredient per hectare and season. In contrast, a dosage of approximately 1000 g/ha is required for proper growth regulation in wheat. Even other plant species are virtually insensitive to this retardant. Most growth retardants are applied as a spray, after which they are absorbed via the leaves and translocated to the growing shoot tissues. However, compounds such as paclobutrazol, uniconazole-P, or tetcyclacis are translocated almost entirely acropetally and are absorbed relatively poorly by shoot parts. Hence, in order to obtain appropriate results, they are often applied as a soil drench. Tetcyclacis should even be fed via the roots in hydroponics or similar systems, since it is almost immobile in soil. Extreme differences among different growth retardants are found with regard to their longevity. Whereas the half-life period of paclobutrazol or uniconazole-P in a plant or in a soil is in the range of several months, compounds such as trinexapac-ethyl, prohexadione-Ca, or *exo*-16,17-dihydro-GA_5-13-acetate are much more rapidly degraded. For example, the half-life period of prohexadione-Ca in the soil is in the range of hours rather than days. A long-lasting effect may be desirable, for instance, in order to regulate the growth of perennial ornamentals. In contrast, shorter-lived compounds give more flexibility to the grower, who may have better means for applying a retardant as needed.

CONCLUDING REMARKS

The majority of plant growth retardants are inhibitors of distinct steps of gibberellin biosynthesis. They have been selected from a vast number of other candidate compounds making any negative effects on their target plants highly unlikely. Against this background, it may appear surprising that a number of other metabolic reactions can still be affected by such compounds, albeit only at very high dosages usually. When used in agriculture and horticulture, such side effects often add to the benefit of mere growth retardation. To plant scientists, growth retardants, with their typical and with their additional biochemical effects, are valuable tools to gain better insight into the objects of their work.

Visit the Annual Reviews home page at www.AnnualReviews.org

LITERATURE CITED

1. Aach H, Bode H, Robinson DG, Graebe JE. 1997. *ent*-Kaurene synthase is located in proplastids of meristematic shoot tissues. *Planta* 202:211–19
2. Aach H, Böse G, Graebe JE. 1995. *ent*-Kaurene biosynthesis in a cell-free system from wheat (*Triticum aestivum* L.) seedlings and the localisation of *ent*-kaurene synthetase in plastids of three species. *Planta* 197:333–42
3. Adams R, Kerber E, Pfister K, Weiler EW. 1992. Studies on the action of the

new growth retardant CGA 163′935 (cimectacarb). See Ref. 93a, pp. 818–27
4. Anderson JD, Moore TC. 1967. Biosynthesis of (−)-kaurene in cell-free extracts of immature pea seeds. *Plant Physiol.* 42: 1527–34
5. Asami T, Yoshida S. 1999. Brassinosteroid biosynthesis inhibitors. *Trends Plant Sci.* 4:348–53
6. Belefant H, Fong F. 1991. Abscisic acid biosynthesis in *Zea mays* embryos: influence of tetcyclacis and regulation by osmotic potential. *Plant Sci.* 78:19–25
7. Benton JM, Cobb AH. 1997. Modification of phytosterol profiles and in vitro photosynthetic electron transport of *Galium aparine* L. (cleavers) treated with the fungicide, epoxiconazole. *Plant Growth Regul.* 22:93–100
8. Benveniste P. 1986. Sterol biosynthesis. *Annu. Rev. Plant Physiol.* 37:275–308
9. Böger P. 1989. New plant-specific targets for future herbicides. In *Target Sites of Herbicide Action*, ed. P Böger, G Sandmann, pp. 247–82. Boca Raton, FL: CRC Press
10. Brian PW, Grove JF, Mulholland TPC. 1967. Relationships between structure and growth-promoting activity of the gibberellins and some allied compounds, in four test systems. *Phytochemistry* 6:1475–99
11. Brown RGS, Kawaide H, Yang YY, Rademacher W, Kamiya Y. 1997. Daminozide and prohexadione have similar modes of action as inhibitors of late stages of gibberellin metabolism. *Physiol. Plant.* 101:309–13
12. Brown RGS, Yan L, Beale MH, Hedden P. 1998. Inhibition of gibberellin 3ß-hydroxylase by novel acylcyclohexanedione derivatives. *Phytochemistry* 47:679–87
13. Buchenauer H, Kutzner B, Koths T. 1984. Wirkung verschiedener Triazol-Fungizide auf das Wachstum von Getreidekeimlingen und Tomatenpflanzen sowie auf die Gibberellingehalte und den Lipidstoffwechsel von Gerstenkeimlingen. *J. Plant Dis. Prot.* 91:506–24
14. Buchenauer H, Röhner E. 1981. Effect of triadimefon and triadimenol on growth of various plant species as well as on gibberellin content and sterol metabolism in shoots of barley seedlings. *Pestic. Biochem. Physiol.* 15:58–70
15. Burden RS, Carter GA, Clark T, Cooke DT, Croker SJ, et al. 1987. Comparative activity of the enantiomers of triadimenol and paclobutrazol as inhibitors of fungal growth and plant sterol and gibberellin biosynthesis. *Pestic. Sci.* 21:253–67
16. Burden RS, Clark T, Holloway PJ. 1987. Effects of sterol biosynthesis-inhibiting fungicides and plant growth regulators on the sterol composition of barley plants. *Pestic. Biochem. Physiol.* 27:289–300
17. Burden RS, Cooke DT, Carter GA. 1989. Inhibitors of sterol biosynthesis and growth in plants and fungi. *Phytochemistry* 28:1791–804
18. Burden RS, Cooke DT, White PJ, James CS. 1987. Effects of the growth retardant tetcyclacis on the sterol composition of oat (*Avena sativa*). *Plant Growth Regul.* 5:207–17
19. Bürstell HW, Hacker E, Schmierer R. 1988. HOE 074784 and analogues: a new synthetic group of highly active plant growth retardants in cereals (esp. rice) and rape. In *Proc. 15th Annu. Meet. Plant Growth Regulator Soc. Am.*, ed. AR Cooke, p. 185. Ithaca, NY: Plant Growth Regul. Soc. Am.
20. Buta JG, Spaulding DW. 1991. Effect of paclobutrazol on abscisic acid levels in wheat seedlings. *J. Plant Growth Regul.* 10:59–61
21. Canivenc MC, Cagnac B, Cabanne F, Scalla R. 1989. Induced changes of chlortoluron metabolism in wheat cell suspension cultures. *Plant Physiol. Biochem.* 27:192–201
22. Cathey HM. 1964. Physiology of growth retarding chemicals. *Annu. Rev. Plant Physiol.* 15:271–302

23. Cerdon C, Rahier A, Taton M, Benveniste P. 1995. Effects of tetcyclacis on growth and on sterol and sapogenin content in fenugreek. *J. Plant Growth Regul.* 14:15–22
24. Chappell J. 1995. Biochemistry and molecular biology of the isoprenoid biosynthetic pathway in plants. *Annu. Rev. Plant Physiol. Mol. Biol.* 46:521–47
25. Cho KY, Sakurai A, Kamiya Y, Takahashi N, Tamara S. 1979. Effects of the new plant growth retardants of quaternary ammonium iodides on gibberellin biosynthesis in *Gibberella fujikuroi*. *Plant Cell Physiol.* 20:75–81
25a. Clifford DR, Lenton JR, eds. 1980. *Recent Developments in the Use of Plant Growth Retardants.* Monogr. No. 4. Wantage: Br. Plant Growth Regul. Group
26. Clouse SD, Sasse JM. 1998. Brassinosteroids: essential regulators of plant growth and development. *Annu. Rev. Plant Physiol. Plant Mol. Biol.* 49:427–51
27. Cole DJ, Owen WJ. 1987. Influence of monooxygenase inhibitors on the metabolism of the herbicides chlortoluron and metolachlor in cell suspension cultures. *Plant Sci.* 50:13–20
28. Coolbaugh RC, Hamilton R. 1976. Inhibition of *ent*-kaurene oxidation and growth by α-cyclopropyl-α-(p-methoxyphenyl)-5-pyrimidine methyl alcohol. *Plant Physiol.* 57:245-48
29. Coolbaugh RC, Heil DR, West CA. 1982. Comparative effects of substituted pyrimidines on growth and gibberellin biosynthesis in *Gibberella fujikuroi*. *Plant Physiol.* 69:712–16
30. Coolbaugh RC, Hirano SS, West CA. 1978. Studies on the specificity and site of action of α-cyclopropyl-α-[p-methoxyphenyl]-5-pyrimidine methyl alcohol (ancymidol), a plant growth regulator. *Plant Physiol.* 62:571–76
31. Costa G, Andreotti C, Bucchi F, Sabatini E, Bazzi C, et al. 2000. Prohexadione-Ca (Apogee): growth regulation and reduced fire blight incidence in pear. *HortScience.* In press
32. Cutler AJ, Krochko JE. 1999. Formation and breakdown of ABA. *Trends Plant Sci.* 4:472–78
32a. Davies PJ, ed. 1995. *Plant Hormones.* Dordrecht: Kluwer
33. De Carolis E, De Luca V. 1994. 2-Oxoglutarate-dependent dioxygenase and related enzymes: biochemical characterization. *Phytochemistry* 36:1093–107
34. Dicks JW. 1980. Mode of action of growth retardants. See Ref. 25a, pp. 1–14
35. Douglas TJ, Paleg LG. 1972. Inhibition of sterol biosynthesis by 2-isopropyl-4-dimethylamino-5-methylphenyl-1-piperidine carboxylate methyl chloride in tobacco and rat liver preparations. *Plant Physiol.* 49:417–20
36. Douglas TJ, Paleg LG. 1974. Plant growth retardants as inhibitors of sterol biosynthesis in tobacco seedlings. *Plant Physiol.* 54:238–45
37. Douglas TJ, Paleg LG. 1978. AMO 1618 and sterol biosynthesis in tissues and subcellular fractions of tobacco seedlings. *Phytochemistry* 17:705–12
38. Douglas TJ, Paleg LG. 1981. Inhibition of sterol biosynthesis and stem elongation of tobacco seedlings induced by some hypocholesterolemic agents. *J. Exp. Bot.* 32:59–68
39. Duriatti A, Bouvier-Navé P, Benveniste P, Schuber F, Delprino L, et al. 1985. In vitro inhibition of animal and higher plants 2,3-oxidosqualene-sterol cyclases by 2-aza-2,3-dihydrosqualene and derivatives, and by other ammonium containing molecules. *Biochem. Pharmacol.* 34:2765–77
40. Evans JR, Evans RR, Regusci CL, Rademacher W. 1999. Mode of action, metabolism, and uptake of BAS 125W, prohexadione-calcium. *HortScience* 34:9–10
41. Evans LT, King RW, Mander LN, Pharis RP, Duncan KA. 1994. The differential effects of C-16,17-dihydro gibberellins

and related compounds on stem elongation and flowering in *Lolium temulentum. Planta* 193:107–14

42. Fletcher RA, Arnold V. 1986. Stimulation of cytokinins and chlorophyll synthesis in cucumber cotyledons by triadimefon. *Physiol. Plant.* 66:197–201
43. Fletcher RA, Gilley A, Sankhla N, Davis TD. 1999. Triazoles as plant growth regulators and stress protectants. *Hortic. Rev.* 23:55–138
44. Forkmann G, Heller W. 1999. Biosynthesis of flavonoids. In *Comprehensive Natural Products Chemistry*, ed. D Barton, K Nakanishi, O Meth-Cohn, pp. 713–48. Amsterdam: Elsevier
45. Foster KR, Lee IJ, Pharis RP, Morgan PW. 1997. Effects of ring D-modified gibberellins on gibberellin levels and development in selected *Sorghum bicolor* maturity genotypes. *J. Plant Growth Regul.* 16:79–87
46. Frost RG, West CA. 1977. Properties of kaurene synthetase from *Marah macrocarpus. Plant Physiol.* 59:22–29
47. Fujioka S, Sakurai A. 1997. Biosynthesis and metabolism of brassinosteroids. *Physiol. Plant.* 100:710–15
48. Funaki Y, Yoneyoshi Y, Ishiguri Y, Izumi K. 1984. Optical isomer of triazolylpentenols, and their production and use as fungicide, herbicide and/or plant growth regulant. *Eur. Patent No. 0 121 284*
49. Garrod JF, Hewitt HG, Copping LG, Greenwood D. 1980. A new group of ternary sulphonium growth retardants. See Ref. 25a, pp. 67–74

49a. Gausman HW, ed. 1991. *Plant Biochemical Regulators.* New York: Marcel Dekker

50. Gianfagna T. 1995. Natural and synthetic growth regulators and their use in horticultural and agronomic crops. See Ref. 32a, pp. 751–73
51. Graebe JE. 1987. Gibberellin biosynthesis and control. *Annu. Rev. Plant Physiol.* 38:419–65
52. Graebe JE, Böse G, Grosselindemann E, Hedden P, Aach H, et al. 1992. The biosynthesis of *ent*-kaurene in germinating seeds and the function of 2-oxoglutarate in gibberellin biosynthesis. See Ref. 93a, pp. 545–54
53. Graebe JE, Lange T, Pertsch S, Stöckl D. 1991. The relationship of different gibberellin biosynthetic pathways in *Cucurbita maxima* endosperm and embryos and the purification of a C-20 oxidase from the endosperm. See Ref. 161a, pp. 51–61
54. Griggs DL, Hedden P, Temple-Smith KE, Rademacher W. 1991. Inhibition of gibberellin 2ß-hydroxylases by acylcyclohexanedione derivatives. *Phytochemistry* 30:2513–17
55. Grossmann K. 1992. Plant growth retardants: their mode of action and benefit for physiological research. See Ref. 93a, pp. 788–97
56. Grossmann K, Häuser C, Sauerbrey E, Fritsch H, Schmidt O, et al. 1989. Plant growth retardants as inhibitors of ethylene production. *J. Plant Physiol.* 134:538–43
57. Grossmann K, König-Kranz S, Kwiatkowski J. 1994. Phytohormonal changes in intact shoots of wheat and oilseed rape treated with the acylcyclohexanedione growth retardant prohexadione calcium. *Physiol. Plant.* 90:139–43
58. Grossmann K, Siefert F, Kwiatkowski J, Schraudner M, Langebartels C, et al. 1993. Inhibition of ethylene production in sunflower cell suspensions by the plant growth retardant BAS 111.W: possible relations to changes in polyamine and cytokinin contents. *J. Plant Growth Regul.* 12:5–11
59. Grossmann KE, Weiler W, Jung J. 1985. Effects of different sterols on the inhibition of cell culture growth caused by the growth retardant tetcyclacis. *Planta* 164:370–75
60. Gussman CD, Salas S, Gianfagna TJ. 1993. Daminozide inhibits ethylene production in apple fruit by blocking the conversion of methionine to aminocyclopropane-1-carboxylic acid (ACC). *Plant Growth Regul.* 12:149–54

61. Harada H, Lang A. 1965. Effect of some 2-chloroethyltrimethylammonium chloride analogs and other growth retardants on gibberellin biosynthesis *in Fusarium moniliforme*. *Plant Physiol.* 40:176–83
62. Hartmann MA. 1998. Plant sterols and the membrane environment. *Trends Plant Sci.* 3:170–75
63. Hashimoto T. 1987. Gibberellin structure-dependent interaction between gibberellins and deoxygibberellin C in the growth of dwarf maize seedlings. *Plant Physiol.* 83:910–14
64. Hashimoto T, Tamura S, Mori K, Matsui M. 1967. Inhibitive and promotive effect of deoxygibberellin C and its methyl ester on plant growth. *Plant Physiol.* 42:886–88
65. Hasson EP, West CA. 1976. Properties of the system for the mixed function oxidation of kaurene and kaurene derivatives in microsomes of the immature seed *of Marah macrocarpus*: cofactor requirements. *Plant Physiol.* 58:473–78
66. Hasson EP, West CA. 1976. Properties of the system for the mixed function oxidation of kaurene and kaurene derivatives in microsomes of the immature seed of *Marah macrocarpus*: electron transfer components. *Plant Physiol.* 58:479–84
67. Haughan PA, Lenton JR, Goad LJ. 1988. Sterol requirements and paclobutrazol inhibition of a celery cell culture. *Phytochemistry* 27:2491–500
68. Hedden P. 1990. The action of plant growth retardants at the biochemical level. See Ref. 132a, pp. 322–32
69. Hedden P. 1991. Gibberellin biosynthetic enzymes and the regulation of gibberellin concentrations. See Ref. 161a, pp. 94–105
70. Hedden P. 1997. The oxidases of gibberellin biosynthesis: their function and mechanism. *Physiol. Plant.* 101:709–19
71. Hedden P. 1999. Recent advances in gibberellin biosynthesis. *J. Exp. Bot.* 50:553–63
72. Hedden P, Croker SJ, Rademacher W, Jung J. 1989. Effects of the triazole-type plant growth retardant BAS 111..W on gibberellin levels in oilseed rape, *Brassica napus*. *Physiol. Plant.* 75:445–51
73. Hedden P, Graebe JE. 1985. Inhibition of gibberellin biosynthesis by paclobutrazol in cell-free homogenates of *Cucurbita maxima* endosperm and *Malus pumila* embryos. *J. Plant Growth Regul.* 4:111–22
74. Hedden P, Kamiya Y. 1997. Gibberellin biosynthesis: enzymes, genes and their regulation. *Annu. Rev. Plant Physiol. Plant Mol. Biol.* 48:431–60
75. Hedden P, Proebsting WM. 1999. Genetic analysis of gibberellin biosynthesis. *Plant Physiol.* 119:365–70
76. Hellwege EM, Hartung W. 1997. Synthesis, metabolism and compartmentation of abscisic acid in *Riccia fluitans* L. *J. Plant Physiol.* 150:287–91
77. Hildebrandt E. 1982. *Der Einfluss von ausgewählten Wachstumshemmern auf die Gibberellinbiosynthese in zellfreien Systemen.* Diploma thesis. Univ. Göttingen, Ger.
78. Hisamatsu T, Koshioka M, Kubota S, King RW. 1998. Effect of gibberellin A_4 and GA biosynthesis inhibitors on growth and flowering of stock [*Matthiola incana* (L.) R. Br.] *J. Jpn. Soc. Hortic. Sci.* 67:537–43
79. Hoad GV. 1983. Gibberellin bioassays and structure-activity relationships. In *The Biochemistry and Physiology of Gibberellins*, ed. A Crozier, 2:57–94. New York: Praeger
79a. Hoad GV, Lenton JR, Jackson MB, Atkin RK, eds. 1987. *Hormone Action in Plant Development.* London: Butterworths
80. Hoffmann G. 1992. Use of plant growth regulators in arable crops: survey and outlook. See Ref. 93a, pp. 798–808
81. Hofstra G, Krieg LC, Fletcher RA. 1989. Uniconazole reduces ethylene and 1-aminocyclopropane-1- carboxylic acid and increases spermine levels in mung bean seedlings. *J. Plant Growth Regul.* 8:45–51

82. Ilan A, Dougall DK. 1992. The effect of growth retardants on anthocyanin production in carrot cell suspension cultures. *Plant Cell Rep.* 11:304–9
83. Iturriagagoitia-Bueno T, Gibson EJ, Schofield CJ, John P. 1996. Inhibition of 1-aminocyclopropane-1-carboxylate oxidase by 2-oxoacids. *Phytochemistry* 43: 343–49
84. Izumi K, Kamiya Y, Sakurai A, Oshio H, Takahashi N. 1985. Studies of sites of action of a new plant growth retardant (*E*)-1-(4-chlorophenyl)-4,4-dimethyl-2-(1,2,4-triazol-l-yl)-1-penten-3-ol (S-3307) and comparative effects of its stereoisomers in a cell-free system from *Cucurbita maxima. Plant Cell Physiol.* 26:821–27
85. Izumi K, Yamaguchi I, Wada A, Oshio H, Takahashi N. 1984. Effects of a new plant growth retardant (*E*)-1-(4-chlorophenyl)-4,4-dimethyl-2-(1,2,4-triazol-1-yl)-penten-3-ol (S-3307) on the growth and gibberellin content of rice plants. *Plant Cell Physiol.* 25:611–17
86. Jia WS, Zhang JH. 1999. Stomatal closure is induced rather by prevailing xylem abscisic acid than by accumulated amount of xylem-derived abscisic acid. *Physiol. Plant.* 106:268–75
87. Jung J. 1979. Possibilities for optimization of plant nutrition by new agrochemical substances—especially in cereals. In *Plant Regulation and World Agriculture*, ed. TK Scott, pp. 279–307. New York: Plenum
88. Jung J, Koch H, Rieber N, Würzer B. 1980. Zur wachstumsregulierenden Wirkung von Triazolin- und Aziridinderivaten des Norbornenodiazetins. *J. Agron. Crop Sci.* 149:128–36
89. Jung J, Luib M, Sauter H, Zeeh B, Rademacher W. 1987. Growth regulation in crop plants with new types of triazole compounds. *J. Agron. Crop Sci.* 158:324–32
90. Junttila O, Jensen E, Ernstsen A. 1991. Effects of prohexadione (BX-112) and gibberellins on shoot growth in seedlings of *Salix pentandra. Physiol. Plant.* 83:17–21
91. Junttila O, King RW, Poole A, Kretschmer G, Pharis RP, et al. 1997. Regulation in *Lolium temulentum* of the metabolism of gibberellin A_{20} and gibberellin A_1 by 16,17-dihydro GA_5 and by the growth retardant, LAB 198 999. *Aust. J. Plant Physiol.* 24:359–69
92. Kamiya Y, Kwak SS. 1991. Partial characterization of the gibberellin 3ß-hydroxylase from immature seeds of *Phaseolus vulgaris*. See Ref. 161a, pp. 72–82
93. Kamiya Y, Nakayama I, Kobayashi M. 1992. Useful probes to study the biosynthesis of gibberellins. See Ref. 93a, pp. 555–65
93a. Karssen CM, van Loon LC, Vreugdenhil D, eds. 1992. *Progress in Plant Growth Regulation.* Dordrecht: Kluwer
94. Katagi T, Mikami N, Matsuda T, Miyamoto J. 1987. Structural studies of the plant growth regulator uniconazole (ES pure) and computer-aided analysis of its interaction with cytochrome P-450. *J. Pestic. Sci.* 12:627–33
95. Kende H, Ninnemann H, Lang A. 1963. Inhibition of gibberellic acid biosynthesis in *Fusarium moniliforme* by AMO-1618 and CCC. *Naturwissenschaften* 18:599–600
96. King RW, Blundell C, Evans LT, Mander LN, Wood JT. 1997. Modified gibberellins retard growth of cool-season turfgrasses. *Crop Sci.* 37:1878–83
97. Kochs G, Grisebach H. 1987. Induction and characterization of a NADPH-dependent flavone synthase from cell cultures of soybean. *Z. Naturforsch. Teil C* 42:343–48
98. Kraus TE, Murr DP, Hofstra G, Fletcher RA. 1992. Modulation of ethylene synthesis in acotyledonous soybean and wheat seedlings. *J. Plant Growth Regul.* 11:47–53
99. Krochko JE, Abrams GD, Loewen MK, Abrams SR, Cutler AJ. 1998. (+)-Abscisic acid 8′-hydroxylase is a

cytochrome P450 monooxygenase. *Plant Physiol.* 118:849–60

100. Kwak SS, Kamiya Y, Sakurai A, Takahashi N, Graebe JE. 1988. Partial purification and characterization of gibberellin 3ß-hydroxylase from immature seeds *of Phaseolus vulgaris* L. *Plant Cell Physiol.* 29:935–43
101. Lange T, Graebe JE. 1993. Enzymes of gibberellin synthesis. In *Methods in Plant Biochemistry*, ed. PJ Lea, 9:403–30. London: Academic
102. Leah JM, Worrall TL, Cobb AH. 1989. Metabolism of bentazon in soybean and the influence of tetcyclacis, BAS 110 and BAS 111. In *Brighton Crop Prot. Conf.—Weeds*, pp. 433–40. Croydon: Br. Crop Prot. Counc.
103. Lee IJ, Foster KR, Morgan PW. 1998. Effect of gibberellin biosynthesis inhibitors on native gibberellin content and floral initiation in *Sorghum bicolor. J. Plant Growth Regul.* 17:185–95
104. Lenton JR, Hedden P, Gale MD. 1987. Gibberellin insensitivity and depletion in wheat—consequences for development. See Ref. 79a, pp. 145–60
105. Lever BG, Shearing SJ, Batch JJ. 1982. PP 333—a new broad spectrum growth retardant. In *Brighton Crop Prot. Conf.—Weeds*, 1:3–10. Croydon: Br. Crop Prot. Counc.
106. Lichtenthaler HK. 1999. The 1-deoxy-D-xylulose-5-phosphate pathway of isoprenoid biosynthesis in plants. *Annu. Rev. Plant Physiol. Plant Mol. Biol.* 50:47–65
107. Linser H, Bettner L. 1972. Wachstumsretardantien. *J. Plant Nutr. Soil Sci.* 132:105–43
108. Looney NE. 1968. Inhibition of apple ripening by succinic acid 2,2-dimethylhydrazide. *Plant Physiol.* 43:1133–37
109. Lürssen K. 1987. The use of inhibitors of gibberellin and sterol biosynthesis to probe hormone action. See Ref. 79a, pp. 133–44
110. Lürssen K, Reiser W. 1987. Triapenthenol—a new plant growth regulator. *Pestic. Sci.* 19:153–64
111. Luster DG, Miller PA. 1993. Triazole plant growth regulator binding to native and detergent-solubilized plant microsomal cytochrome P450. *Pestic. Biochem. Physiol.* 46:27–39
112. MacMillan J. 1997. Biosynthesis of the gibberellin plant hormones. *Nat. Prod. Rep.* 1997:221–43
113. Mandava NB. 1988. Plant growth-promoting brassinosteroids. *Annu. Rev. Plant Physiol. Plant Mol. Biol.* 39:23–52
114. Mander LN, Adamson G, Bhaskar VK, Twitchin B, Camp D, et al. 1998. Effects of 17-alkyl-16,17-dihydrogibberellin A_5 derivatives on growth and flowering in *Lolium temulentum. Phytochemistry* 49:1509–15
115. Mander LN, Camp D, Evans LT, King RW, Pharis RP, et al. 1995. Designer gibberellins: the quest for specific activity. *Acta Hortic.* 394:45–55
116. Mander LN, Sherburn M, Camp D, King RW, Evans LT, et al. 1998. Effects of D-ring modified gibberellins on flowering and growth of *Lolium temulentum. Phytochemistry* 49:2195–206
117. McKeon TA, Fernándes-Maculet JC, Yang SF. 1995. Biosynthesis and metabolism of ethylene. See Ref. 32a, pp. 118–39
118. Menhenett R. 1980. Use of retardants on glasshouse crops. See Ref. 25a, pp. 27–39
119. Miki T, Kamiya Y, Fukazawa M, Ichikawa T, Sakurai A. 1990. Sites of inhibition by a plant-growth regulator, 4′-chloro-2′-(α-hydroxybenzyl)-isonicotinanilide (inabenfide), and its related compounds in the biosynthesis of gibberellins. *Plant Cell Physiol.* 31:201–6
120. Min XJ, Bartholomew DP. 1996. Effect of plant growth regulators on ethylene production, 1-aminocyclopropane-1-carboxylic acid oxidase activity, and initiation of inflorescence development of pineapple. *J. Plant Growth Regul.* 15:121–28

121. Mitchell JW, Wirwille JW, Weil L. 1949. Plant growth-regulating properties of some nicotinium compounds. *Science* 110:252–54
122. Moreland DE, Corbin FT, McFarland JE. 1993. Oxidation of multiple substrates by corn shoot microsomes. *Pestic. Biochem. Physiol.* 47:206–14
123. Müller U, Huxley P, Ebert E. 1987. Computer assisted molecular modelling (CAMM)—A tool for structure-activity considerations of inhibitors of the enzymatic oxidation of *ent*-kaurene to *ent*-kaurenoic acid. In *Pesticide Science and Biotechnology. Proc. Int. Congr. Pestic. Chem., 1986, 6th*, ed. R Greenhalgh, TR Roberts, pp. 69–72. Oxford: Blackwell
124. Nakayama I, Kamiya Y, Kobayashi M, Abe H, Sakurai A. 1990. Effect of a plant growth regulator, prohexadione, on the biosynthesis of gibberellins in cell-free systems derived from immature seeds. *Plant Cell Physiol.* 31:1183–90
125. Nakayama I, Miyazawa T, Kobayashi M, Kamiya Y, Abe H, et al. 1990. Effects of a new plant growth regulator prohexadione calcium (BX-112) on shoot elongation caused by exogenously applied gibberellins in rice (*Oryza sativa* L.) seedlings. *Plant Cell Physiol.* 31:195–200
126. Nakayama I, Miyazawa T, Kobayashi M, Kamiya Y, Abe H, et al. 1991. Studies on the action of the plant growth regulators BX-112, DOCHC, and DOCHC-Et. See Ref. 161a, pp. 311–19
127. Nakayama M, Yamane H, Murofushi N, Takahashi N, Mander LN, et al. 1991. Gibberellin biosynthetic pathway and the physiologically active gibberellin in the shoot of *Cucumis sativus* L. *J. Plant Growth Regul.* 10:115–19
128. Nes WD, Venkatramesh M. 1999. Enzymology of phytosterol transformations. *Crit. Rev. Biochem. Mol. Biol.* 34:81–93
129. Nickell LG, ed. 1983. *Plant Growth Regulating Chemicals*, Vols. 1. 2. Boca Raton, FL: CRC Press
130. Ninnemann H, Zeevaart JAD, Kende H, Lang A. 1964. The plant growth retardant CCC as inhibitor of gibberellin biosynthesis in *Fusarium moniliforme. Planta* 61:229–35
131. Nitsche K, Grossmann K, Sauerbrey E, Jung J. 1985. Influence of the growth retardant tetcyclacis on cell division and cell elongation in plants and cell cultures of sunflower, soybean, and maize. *J. Plant Physiol.* 118:209–18
132. Olsen JE, Junttila O. 1997. Growth-promoting activity of gibberellins on shoot elongation in *Salix pentandra* is reduced by 16,17-dihydro derivatisation. *Physiol. Plant.* 99:63–66
132a. Pharis RP, Rood SB, eds. 1990. *Plant Growth Substances 1988.* Berlin: Springer-Verlag
133. Phinney BO. 1984. Gibberellin A_1, dwarfism and the control of shoot elongation in higher plants. In *The Biosynthesis and Metabolism of Plant Hormones*, ed. A Crozier, JR Hillman, pp. 17–41. Cambridge: Cambridge Univ. Press
134. Rademacher W. 1991. Inhibitors of gibberellin biosynthesis: applications in agriculture and horticulture. See Ref. 161a, pp. 296–310
135. Rademacher W. 1991. Biochemical effects of plant growth retardants. See Ref. 49a, pp. 169–200
136. Rademacher W. 1992. Inhibition of gibberellin production in the fungi *Gibberella fujikuroi* and *Sphaceloma manihoticola* by plant growth retardants. *Plant Physiol.* 100:625–29
137. Rademacher W. 1993. On the mode of action of acylcyclohexanediones—a new type of plant growth retardant with possible relationships to daminozide. *Acta Hortic.* 329:31–34
138. Rademacher W. 1997. Gibberellins. In *Fungal Biotechnology*, ed. T. Anke, pp. 193–205. Weinheim: Chapman & Hall

139. Rademacher W. 1997. Bioregulation in crop plants with inhibitors of gibberellin biosynthesis In *Proc. Annu. Meet. Plant Growth Regulation Soc. Am., 24th*, ed. JG Latimer, pp. 27–31. LaGrange: Plant Growth Regul. Soc. Am.
140. Rademacher W, Fritsch H, Graebe JE, Sauter H, Jung J. 1987. Tetcyclacis and triazole-type plant growth retardants: their influence on the biosynthesis of gibberellins and other metabolic processes. *Pestic. Sci.* 21:241–52
141. Rademacher W, Pharis RP, Mander LN. 1999. Agricultural use of new GA derivatives. *Jpn. J. Crop Sci.* 68:362–67
142. Rademacher W, Speakman JB, Evans RR, Evans JR, Römmelt S, et al. 1998. Prohexadione-Ca—a new plant growth regulator for apple with interesting biochemical features. In *Proc. Annu. Meet. Plant Growth Regul. Soc. Am., 25th*, ed. WE Shafer, pp. 113–18. LaGrange: Plant Growth Regul. Soc. Am.
143. Rademacher W., Temple-Smith KE, Griggs DL, Hedden P. 1992. The mode of action of acylcyclohexanediones—a new type of plant growth retardant. See Ref. 93a, pp. 571–77
144. Rohmer M. 1999. The discovery of a mevalonate-independent pathway for isoprenoid biosynthesis in bacteria, algae and higher plants. *Nat. Prod. Rep.* 16:565–74
145. Römmelt S, Treutter D, Speakman JB, Rademacher W. 1999. Effects of prohexadione-Ca on the flavonoid metabolism of apple with respect to plant resistance against fire blight. *Acta Hortic.* 489:359–63
146. Ross JJ. 1994. Recent advances in the study of gibberellin mutants. *Plant Growth Regul.* 15:193–206
147. Saishoji T, Kumazawa S, Chuman H. 1998. Structure-activity relationships of enantiomers of the azole fungizide ipconazole and its related compounds—fungicidal and plant growth inhibitory activities. *J. Pestic. Sci.* 23:129–36
148. Saito T, Kwak SS, Kamiya Y, Yamane H, Sakurai A, et al. 1991. Effects of deoxygibberellin C (DGC) and 16-deoxo-DGC on gibberellin 3ß-hydroxylase and plant growth. *Plant Cell Physiol.* 32:239–45
149. Santes CM, García-Martínez JL. 1995. Effect of the growth retardant 3,5-dioxo-4-butyryl-cyclohexane carboxylic acid ethyl ester, an acylcyclohexanedione compound, on fruit growth and gibberellin content of pollinated and unpollinated ovaries in pea. *Plant Physiol.* 108:517–23
150. Sauter H. 1984. Chemical aspects of some bioregulators. In *Bioregulators: Chemistry and Uses*, ed. RL Ory, FR Rittig, pp. 9–21. Washington, DC: Am. Chem. Soc.
151. Schott PE, Walter H. 1991. Bioregulators: present and future fields of application. See Ref. 49a, pp. 247–321
152. Schuler MA. 1996. Plant cytochrome P450 monooxygenases. *Crit. Rev. Plant Sci.* 15:235–84
153. Shechter I, West CA. 1969. Biosynthesis of gibberellins. IV. Biosynthesis of cyclic diterpenes from *trans*-geranylgeranyl pyrophosphate. *J. Biol. Chem.* 244:3200–9
154. Shirakawa N, Tomioka H, Takeuchi M, Kanzaki M, Fukazawa M, et al. 1986. Effect of a new plant growth retardant N-[4-chloro-2-(α-hydroxybenzyl)phenyl]- isonicotinamide (CGR-811) on the growth of rice plants. In *Plant Growth Regulators in Agriculture*, ed. P Macgregor, pp. 1–17. Taipei: Food Fertil. Technol. Cent. Asian Pac. Reg.
155. Shive JB, Sisler HD. 1976. Effects of ancymidol (a growth retardant) and triarimol (a fungicide) on the growth, sterols, and gibberellins of *Phaseolus vulgaris* (L.). *Plant Physiol.* 57:640–44
156. Smith VA, Albone KS, MacMillan J. 1991. Enzymatic 3ß-hydroxylations of gibberellins A_{20} and A_5. See Ref. 161a, pp. 62–71

157. Sponsel VM. 1995. Gibberellin biosynthesis and metabolism. In *Plant Hormones: Physiology, Biochemistry and Molecular Biology*, ed. PJ Davies, pp. 66–97. Dordrecht: Martinus Nijhoff
158. Sponsel VM, Reid JB. 1992. The effect of the growth retardant LAB 198 999 and its interaction with gibberellins A_1, A_3, and A_{20} in fruit growth of tall and dwarf peas. See Ref. 93a, pp. 578–84
159. Stich K, Ebermann R, Forkmann G. 1988. Einfluß Cytochrom P-450-spezifischer Inhibitoren auf die Aktivität von Flavonoid-3′-hydroxylase und Flavonsynthase II bei verschiedenen Pflanzen. *Phyton (Austria)* 28:237–47
160. Sugavanam B. 1984. Diastereoisomers and enantiomers of paclobutrazol: their preparation and biological activity. *Pestic. Sci.* 15:296–302
161. Takagi M, Pearce DW, Janzen LM, Pharis RP. 1994. Effect of *exo*-16,17-dihydro-gibberellin A_5 on gibberellin A_{20} metabolism in seedlings of dwarf rice (*Oryza sativa* L. cv. Tan-ginbozu). *Plant Growth Regul.* 15:207–13
161a. Takahashi N, Phinney BO, MacMillan J, eds. 1991. *Gibberellins*. New York: Springer-Verlag
162. Taton M, Ullman P, Benveniste P, Rahier A. 1988. Interaction of triazole fungicides and plant growth regulators with microsomal cytochrome P-450-dependent obtusifoliol 14α-methyl demethylase. *Pestic. Biochem. Physiol.* 30:178–89
163. Tolbert NE. 1960. (2-Chloroethyl)-trimethylammoniumchloride and related compounds as plant growth substances. I. Chemical structure and bioassay. *J. Biol. Chem.* 235:475–79
164. Tolbert NE. 1960. (2-Chloroethyl)-trimethylammoniumchloride and related compounds as plant growth substances. II. Effect on growth of wheat. *Plant Physiol.* 35:380–85
165. Tschabold EE, Taylor HM, Davenport JD, Hackler RE, Krumkalns EV, et al. 1970. A new plant growth regulator. *Plant Physiol.* 46:S19
166. Tudzynski B, Hölter K. 1998. Gibberellin biosynthetic pathway in *Gibberella fujikuroi*: evidence for a gene cluster. *Fungal Genet. Biol.* 25:157–70
167. Wada K. 1978. New gibberellin biosynthesis inhibitors, 1-*n*-decyl- and 1-geranylimidazole: inhibitors of (-)-kaurene 19-oxidation. *Agric. Biol. Chem.* 42: 2411–13
168. Wasternack C, Atzorn R, Leopold J, Feussner I, Rademacher W, et al. 1995. Synthesis of jasmonate-induced proteins in barley (*Hordeum vulgare*) is inhibited by the growth retardant tetcyclacis. *Physiol. Plant.* 94:335–41
169. Weisshaar B, Jenkins GI. 1998. Phenylpropanoid biosynthesis and its regulation. *Curr. Opin. Plant Biol.* 1:251–57
170. West CA. 1973. Biosynthesis of gibberellins. In *Biosynthesis and Its Control in Plants*, ed. BV Millborrow, pp. 143–69. New York: Academic
171. Williams DR, Ross JJ, Reid JB, Potts BM. 1999. Response of *Eucalyptus nitens* seedlings to gibberellin biosynthesis inhibitors. *Plant Growth Regul.* 27:125–29
172. Wimmer G, Halbwirth H, Wurst F, Forkmann G, Stich K. 1998. Enzymatic hydroxylation of 6′-deoxychalcones with protein preparations from petals of *Dahlia variabilis*. *Phytochemistry* 47:1013–16
173. Winkler VW. 1997. Reduced risk concept for prohexadione-calcium, a vegetative growth control plant growth regulator in apples. *Acta Hortic.* 451:667–71
174. Winter J, Schneider B, Strack D, Adam G. 1997. Role of cytochrome P450-dependent monooxygenase in the hydroxylation of 24-*epi*-brassinolide. *Phytochemistry* 45:233–37
175. Yoder KS, Miller SS, Byers RE. 1999. Suppression of fireblight in apple shoots by prohexadione-calcium following

experimental and natural inoculation conditions. *HortScience* 34:1202–4

176. Yokota T. 1997. The structure, biosynthesis and function of brassinosteroids. *Trends Plant Sci.* 4:137–43
177. Yokota T, Nakamura Y, Takahashi N, Nonaka M, Sekimoto H, et al. 1991. Inconsistency between growth and endogenous levels of gibberellins, brassinosteroids, and sterols in *Pisum sativum* treated with uniconazole antipodes. See Ref. 161a, pp. 339–49
178. Zeeh B, König KH, Jung J. 1974. Development of a new plant growth regulator with biological activity related to CCC. *Kemia* (Helsinki) 9:621–23
179. Zeevaart JAD. 1985. Inhibition of stem growth and gibberellin production in *Agrostemma githago* L. by the growth retardant tetcyclacis. *Planta* 166:276–79
180. Zeevaart JAD, Creelman RA. 1988. Metabolism and physiology of abscisic acid. *Annu. Rev. Plant Physiol. Plant Mol. Biol.* 39:439–73
181. Zeevaart JAD, Gage DA, Creelman RA. 1990. Recent studies of the metabolism of abscisic acid. See Ref. 132a, pp. 233–40
182. Zeevaart JAD, Gage DA, Talon M. 1993. Gibberellin A_1 is required for stem elongation in spinach. *Proc. Natl. Acad. Sci. USA* 90:7401–5

Subject Index

B

C

D

T

Cumulative Indexes

CONTRIBUTING AUTHORS, VOLUMES 41–51

CHAPTER TITLES, VOLUMES 41–51

Prefatory Chapters

Biochemistry and Biosynthesis

Genetics and Molecular Biology

Tissue, Organ, and Whole Plant Events

Acclimation and Adaptation

Methods